Sustainable Development and Biodiversity

Volume 42

Series Editor

Kishan Gopal Ramawat, Botany Department
Mohanlal Sukhadia University
Udaipur, India

Sustainable Development Goals are best achieved by mechanisms such as research, innovation, and knowledge sharing. This book series aims to help researchers by reporting recent progress and providing complete, comprehensive, and broad subject-based reviews about all aspects of sustainable development and ecological biodiversity.

The series explores linkages of biodiversity with delivery of various ecosystem services and it offers a discourse in understanding the biotic and abiotic interactions, ecosystem dynamics, biological invasion, ecological restoration and remediation, diversity of habitats and conservation strategies. It is a broad scoped collection of volumes, addressing relationship between ecosystem processes and biodiversity. It aims to support the global efforts towards achieving sustainable development goals by enriching the scientific literature. The books in the series brings out the latest reading material for botanists, environmentalists, marine biologists, conservationists, policy makers and NGOs working for environment protection.

We welcome volumes on the themes --

Agroecosystems, Agroforestry, Biodiversity, Biodiversity conservation, Conservation of ecosystem, Ecosystem, Endangered species, Forest conservation, Genetic diversity, Global climate change, Hotspots, Impact assessment, Invasive species, Livelihood of people, Plant biotechnology, Plant resource utilization, Sustainability of the environment, Sustainable management of forests, Sustainable use of terrestrial ecosystems and plants, Traditional methods, Urban horticulture.

Balaji Prasath Barathan • Vijaykumar Velupillai
Santhanam Perumal • Karthikeyan Kannan
Editors

Navigating Climate Change: Impacts on Biodiversity and Ecosystem Resilience

Springer

Editors
Balaji Prasath Barathan
Coastal and Marine Ecology Division
Gujarat Institute of Desert Ecology
Bhuj, Gujarat, India

Vijaykumar Velupillai
Terrestrial Ecology Division and GIS & RS Cell
Gujarat Institute of Desert Ecology
Bhuj, Gujarat, India

Santhanam Perumal
Marine Planktonology and Aquaculture Laboratory
Department of Marine Science
Bharathidasan University
Tiruchirappalli, Tamil Nadu, India

Karthikeyan Kannan
Environmental Monitoring and Assessment Division
Gujarat Institute of Desert Ecology
Bhuj, Gujarat, India

ISSN 2352-474X ISSN 2352-4758 (electronic)
Sustainable Development and Biodiversity
ISBN 978-981-95-0408-4 ISBN 978-981-95-0409-1 (eBook)
https://doi.org/10.1007/978-981-95-0409-1

This Springer imprint is published by the registered company Springer Nature Singapore Pte Ltd.
The registered company address is: 152 Beach Road, #21-01/04 Gateway East, Singapore 189721, Singapore

Foreword

The book *Navigating Climate Change: Impacts on Biodiversity and Ecosystem Resilience* provides an in-depth study of the complex challenges posed by climate change and its cascading effects on biodiversity. In this time where the world faces the most significant issues with the environment, this book serves as a hope and a promise for action. This work covers the complex interconnections of climate change, biodiversity loss, and human health in an analysis that is intricate and needs to be addressed. In my extensive career as an Indian Administrative service officer dealing with governance and policy issues, the holistic approach to global issues in this book deeply resonated with me.

Among these are some of the most difficult ones like biodiversity, climate change, and disruption of ecosystems leading to movement of species. From marine biodiversity to land ecosystems to alien species, the knowledge provided in this book covers how all these systems on Earth are interrelated. Therefore, it seeks to focus attention on the urgent need for systemic interventions, which fundamentally rest on eco-health, sustainable development, and conservation.

GUIDE, with its extensive experience in climate change research and mitigation activities, is bringing out this book for scholars and policymakers who strive to maintain Earth's delicate balances. I commend the authors for profoundly shouldering the role of increasing consciousness and offering answers to these pivotal global issues. Indeed, resilience is one essential component when creating a sustainable future for the next generations. The compliers are to be congratulated at bringing together such an amazing set of chapters and for the accompanying notes on climate change and biodiversity. The end product is a milestone on the impact of climate change in biodiversity worldwide.

Chairman, Gujarat Institute of Desert Ecology
Bhuj, Gujarat, India

V. S. Gadhavi,

Foreword

Preface

The growing imperative of climate change and its consequences to world biodiversity and ecosystem sustainability provided an impetus to this book. *Navigating Climate Change Impacts on Biodiversity and Ecosystems* is intended to help policy-makers, students, and research for scholars advancing the in-depth understanding of the implications of Climate Change.

This book has been designed as an interface between research and practice so as to provide every day professionals with not only an insight into the challenges and opportunities the changing climate brings but also to provide a behind-the-scenes look at how some of the critical work in the area of climate impact and adaptation is being done. The audience for the book is a wide spectrum of professionals, academics, practitioners, and students, from beginners to the experienced who need updated views of the unfolding changes that are taking place.

The book is structured from basics to advanced topics. Each chapter is written by foremost experts in their field, chosen for their ability to explain complex ideas simply. We have taken enormous efforts to variety the content accessible, logical, and up-to-date with current science.

We would like to acknowledge all the authors and the reviewers who have assisted to improve the manuscript chapter. Special thanks to the institutions and people who provided support, resources, and inspiration during the preparation of this book.

We hope this book inspires conversation and action on climate change biodiversity and ecosystem resilience.

Bhuj, Gujarat, India — Balaji Prasath Barathan
Bhuj, Gujarat, India — Vijaykumar Velupillai
Tiruchirappalli, Tamil Nadu, India — Santhanam Perumal
Bhuj, Gujarat, India — Karthikeyan Kannan

General Introduction

The foremost altering drivers of global biodiversity in the twenty-first century include climate change. Activities such as the combustion of fossil fuels tend to bring about transformations in the long term, in temperatures and weather patterns while also increasing the frequency of extreme events. All these changes tend to reshape ecosystems, shift habitats and disrupt equilibrium, which sustains life on earth.

Overview of Climate Change Effects on Biodiversity

Biodiversity is bound to be impacted in several ways due to climate change. The Profound effects include global warming, rise in wildfire frequency causing deaths of remaining living organisms and extensive damage to ecosystems leading to coral bleaching and degradation. Many Species are forced to relocate to cooler areas or higher altitudes with no ability to adapt due to split habitats, rapid extinction due to increases extinction risk. Climate change further disrupts irregular ecological synchrony disruption like unaligned flowering and pollination cycles destroying food webs and interaction, and enhanced ocean acidification from absorbed CO_2 destabilizing marine ecosystems and biodiversity by impairing shell building organisms along with fragmentation which divides populations and decreases resilience toward external stressors dampens the ability to recover from being stressed from shifting. All these effects pave the path toward uncertainty in ecological instability which worsens.

The Importance of Understanding the Impacts of Climate Change on Biological Diversity

Considering the consequence of climate change impacts on biodiversity is significant; subsequently ecosystems supply fundamental services such as nutrition, aquatic, and climate, which are critical for human existence. Ecosystems that are biodiverse tend to be more resilient to environmental fluctuations improving their ability to withstand climate change disruptions and protect them; however, increasing temperatures, severe weather, and habitat loss endanger species survival threatening ecosystems. Strengthening the protected area buffers through enhanced management and the restoration of degraded systems along with establishing climate resilient corridors are a few strategies to protect genetic diversity and biodiversity mobility. Biodiversity loss strengthens climate feedback, such as reduction of carbon sinks due to deforestation and wetland destruction worsening global warming, and loss of plant diversity, which may release up to 146 billion metric tons of stored carbon, intensifying climate change. As exhibited in other parts of the world, there are opportunities for integrated approaches to address these multifaceted challenges. In conclusion, climate change presents diverse concerns for biodiversity, affecting species, ecosystems, and their functions.

Contents

Editors and Contributors

About the Editors

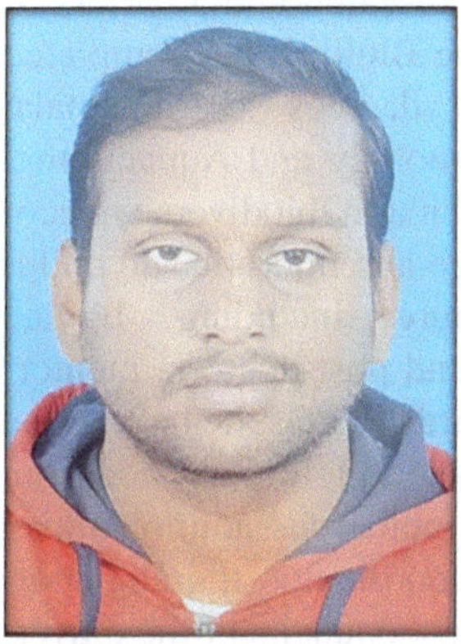

Balaji Prasath Barathan is the Senior Scientist at the Gujarat Institute of Desert Ecology. He obtained his PhD in Marine Science from Bharathidasan University, India. His areas of specialization include Marine Planktonology, Harmful Algal Blooms, Microbial Toxicology, and Wastewater Treatment. He comes with over 15 years of marine scientific research and field-work. He has performed research work in the Indian Ocean and Chinese seas, as well as on-board Taiwan and Hong Kong, using scientific equipment to do sample analysis with Physical Chemical Environmental and Biological Oceanography. He possesses an astounding collection of more than 60 published scholarly works, including research papers, reviews, books, and technical documents spread throughout multiple reputable international journals. He acts as an Academic Editor of many national and international journals and has more than five active memberships in scientific societies.

Vijaykumar Velupillai is serving as Director of Gujarat Institute of Desert Ecology, with 39 years of experience in research, administration, and teaching. He has PhD in Biosciences—Wildlife. He has been part of 251 research/consultancy projects domestically and internationally, and expertise in wildlife research, restoration of pastures, management of watersheds, desertification, and climate change. He has authored 85 published articles in journals, newsletters, book and book chapters, with over 35 presentations at international/national conferences. Having extensively worked with diverse government and nongovernment organizations, his extensive project coordination with leading industrial firms has enabled him to gain proficiency within the corporate sector. In addition, he actively engages in development of infrastructure and collaborating with national and international institutions. Besides the IUCN, he is a member of national and state level committees associated to environment and education. His contributions have also earned him numerous accolades such as the Oryx Award-UK in 1995 and Man of Excellence award 2021.

Santhanam Perumal a Professor in the Department of Marine Science at Bharathidasan University, Tamil Nadu, India, has 22 years of research and 17 years of teaching experience in Marine Biology. He completed his MSc, MPhil, and PhD degrees at Annamalai University in Marine Planktonology and Aquaculture. He specializes in marine plankton biodiversity, ecotoxicology, and climate change and microplastic pollution, sustainable aquaculture including biofloc technology and aquaponics, and post-doctoral mentorship. As the mentor of 19 awarded PhDs, he has guided dozens of scholars and postdoctoral fellows. Aside from these contributions, he holds numerous administrative functions such as placement coordinator, training coordinator, and member of different councils of the university. He moves marine science forward with climate friendly aquaculture innovations and sustainable ecological solutions.

Karthikeyan Kannan is the Assistant Director of Gujarat Institute of Desert Ecology, who holds his PhD in Environmental Sciences and has a working experience of 20 years in the field of Environmental Auditing, Marine Environmental Monitoring, and Microbial biotechnology and 7 years of teaching experience as an Adjunct Faculty in Environmental Science. He is a University First Rank holder in Microbiology and has qualified UGC-NET in Environmental Sciences and State Eligibility Test in Life Sciences. He is a Gujarat Pollution Control Board recognized Environmental Auditor. He has done projects sponsored by national funding agencies such as DST, DBT, MoES, etc. He is a recognized Research Guide for PhD in universities in Gujarat and Tamilnadu. He serves in various Academic and Environmental committees and also in various Conservation Cell and Monitoring committees of Industries.

Contributors

Muhammad Talha Akram Institute of Agronomy, Faculty of Agricultural Sciences and Technology, Bahauddin Zakariya University, Multan, Punjab, Pakistan

Kareem Altaff Aquaculture Research Laboratory, Department of Marine Biotechnology, AMET University, Chennai, India

R. Anburaj Department of Microbiology, Thanthai Hans Roever College (Autonomous), Perambalur, Tamil Nadu, India

M. Arunpandiyan Department of Environmental Science and Management, Bharathidasan University, Trichy, Tamil Nadu, India

Sanjiba Kumar Baliarsingh Indian National Centre for Ocean Information Services, Ministry of Earth Sciences, Government of India, Hyderabad, India

Kartik Baruah Department of Applied Animal Science and Welfare, Swedish University of Agricultural Sciences, Uppsala, Sweden

Ashok Kumar Behera Department of Soil Science, Odisha University of Agriculture & Technology (OUAT), Bhubaneswar, Odisha, India

Sukumar Bhakta Botanical Survey of India, Andaman and Nicobar Regional Centre, Sri Vijaya Puram, Andaman and Nicobar Islands, India

Bedantika Biswas MEME Division, CSIR-Central Salt and Marine Chemicals Research Institute, Bhavnagar, Gujarat, India

Rachna Chandra Terrestrial Ecology Division, Gujarat Institute of Desert Ecology (GUIDE), Bhuj-Kachchh, Gujarat, India

Paromit Chatterjee Zoological Survey of India, Kolkata, West Bengal, India

Habitat Lens Private Limited, Khardah, West Bengal, India

Shruti Chatterjee Institute of Science, Nirma University, Ahmedabad, Gujarat, India

Vivek U. Chauhan Terrestrial Ecology Division, Gujarat Institute of Desert Ecology, Bhuj, Gujarat, India

Tolani College of Arts and Science, Adipur, Gujarat, India

Bhagyashree Dash Indian National Centre for Ocean Information Services, Ministry of Earth Sciences, Government of India, Hyderabad, India

Sudipta Kumar Das Botanical Survey of India, Andaman and Nicobar Regional Centre, Sri Vijaya Puram, Andaman and Nicobar Islands, India

S. Dhanasundaram Marine Planktonology & Aquaculture Laboratory, Department of Marine Science, School of Marine Sciences, Bharathidasan University, Tiruchirappalli, Tamil Nadu, India

Deep D. Dudiya Department of Zoology, R. R. Lalan College, Kachchh, Gujarat, India

Department of Earth and Environmental Science, Krantiguru Shyamji Krishna Verma Kachchh University, Bhuj, Gujarat, India

Prem Kumar Duraivel Aquaculture Research Laboratory, Department of Marine Biotechnology, AMET University, Chennai, India

Santosh Gaikwad Laboratory for Benthic Ecological Trait Analysis (L-BETA), CSIR-National Institute of Oceanography, Regional Centre, Mumbai, India

Nikunj B. Gajera Gujarat Institute of Desert Ecology, Bhuj, Gujarat, India

Deepa Gavali Gujarat Ecology Society, Vadodara, Gujarat, India

Grinson George Indian Council of Agricultural Research-Central Marine Fisheries Research Institute, Kochi, India

Cochin University of Science and Technology, Kochi, India

Shalu George Climate Change and Remediation Research Lab (CCRRL), Department of Environmental Science, Central University of Kerala, Kasaragod, Kerala, India

Achamveetil Gopalakrishnan Indian Council of Agricultural Research-Central Marine Fisheries Research Institute, Kochi, India

Cochin University of Science and Technology, Kochi, India

Vineetha Gopinath Indian Council of Agricultural Research-Central Marine Fisheries Research Institute, Kochi, India

Cochin University of Science and Technology, Kochi, India

Sebaj H. Gori Department of Earth and Environmental Science, Krantiguru Shyamji Krishna Verma Kachchh University, Bhuj, Gujarat, India

Department of Zoology, R. R. Lalan College, Bhuj, Kachchh, Gujarat, India

A. Gowthami Marine Planktonology & Aquaculture Laboratory, Department of Marine Science, School of Marine Sciences, Bharathidasan University, Tiruchirappalli, Tamil Nadu, India

Sara Grace Climate Change and Remediation Research Lab (CCRRL), Department of Environmental Science, Central University of Kerala, Kasaragod, Kerala, India

Soumya Haldar MEME Division, CSIR-Central Salt & Marine Chemicals Research Institute, Bhavnagar, Gujarat, India

Academy of Scientific and Innovative Research (AcSIR), Ghaziabad, India

Azhar Hayat Department of Statistics, Bahauddin Zakariya University, Multan, Punjab, Pakistan

Hafiz Saqib Hayat Department of Agronomy, Faculty of Agriculture and Environmental Sciences, Muhammad Nawaz Shareef University of Agriculture, Multan, Punjab, Pakistan

M. Hema Marine Planktonology & Aquaculture Laboratory, Department of Marine Science, School of Marine Sciences, Bharathidasan University, Tiruchirappalli, Tamil Nadu, India

Kapilkumar Nivrutti Ingle Coastal and Marine Ecology Division, Gujarat Institute of Desert Ecology, Gujarat, India

Department of Ecology, University of Szeged, Szeged, Hungary

Muhammad Umer Iqbal Department of Plant Pathology, Faculty of Agriculture & Environmental Sciences, Muhammad Nawaz Shareef University of Agriculture, Multan, Punjab, Pakistan

Rathinam Arthur James Department of Marine Science, Bharathidasan University, Tiruchirappalli, India

M. Jayaprakashvel Department of Marine Biotechnology, Academy of Maritime Education and Training (AMET), Chennai, Tamil Nadu, India

G. Roseline Jebapriya Marine Planktonology & Aquaculture Laboratory, Department of Marine Science, School of Marine Sciences, Bharathidasan University, Tiruchirappalli, Tamil Nadu, India

Kavya Jeevan Climate Change and Remediation Research Lab (CCRRL), Department of Environmental Science, Central University of Kerala, Kasaragod, Kerala, India

S. Jeyanthi PG and Research Department of Zoology, Ethiraj College for Women (Autonomous), Chennai, Tamil Nadu, India

Sudheer Joseph Indian National Centre for Ocean Information Services, Ministry of Earth Sciences, Government of India, Hyderabad, India

Pallavi Joshi Coastal and Marine Ecology Division, Gujarat Institute of Desert Ecology, Gujarat, India

Agakhan School Mundra, Mundra, Gujarat, India

Pankaj N. Joshi Ecologist, Bhuj Kutch, Gujarat, India

M. Kanmani Presidency Research Innovations in Zoological Eminence (PRIZE) Lab, Department of Zoology, Presidency College (Autonomous), Chennai, Tamil Nadu, India

Vaithianathan Kannan Arulagam, Asokapuram, Coimbatore, Tamil Nadu, India

Indrani Karunasagar FAO Reference Centre for AMR & Aquaculture Biosecurity, DST TEC, Nitte University, Mangalore, India

Thanamegam Kaviarasan National Centre for Coastal Research (NCCR), Ministry of Earth Sciences, Government of India, Chennai, India

Department of Marine Science, Bharathidasan University, Tiruchirappalli, India

Nanda Kishore Indian Council of Agricultural Research-Central Marine Fisheries Research Institute, Kochi, India

Cochin University of Science and Technology, Kochi, India

Jayendra Lakhmapurkar Gujarat Ecology Society, Vadodara, Gujarat, India

Dhanya M. Lal Indian National Centre for Ocean Information Services, Ministry of Earth Sciences, Government of India, Hyderabad, India

B. Manikandan Coral Reef Ecology Lab, Biological Oceanography Division, CSIR-National Institute of Oceanography, Dona Paula, Goa, India

P. Manikandan Division of Ecotechnology, M. S. Swaminathan Research Foundation, Chennai, Tamil Nadu, India

Kannankodantavida Manjusha Climate Change and Remediation Research Lab (CCRRL), Department of Environmental Science, Central University of Kerala, Kasaragod, Kerala, India

S. R. Marigoudar National Centre for Coastal Research (NCCR), Ministry of Earth Sciences, Government of India, Chennai, India

Divya Meril Department of Aquatic Environment Management, St. Devasahayam Institute of Fisheries Science and Technology, Kanniyakumari, Tamil Nadu, India

Dr. J. Jayalalithaa Fisheries University, Kanniyakumari, Tamil Nadu, India

Gourav Kumar Mishra Laboratory for Benthic Ecological Trait Analysis (L-BETA), CSIR-National Institute of Oceanography, Regional Centre, Mumbai, India

Academy of Scientific and Innovative Research (AcSIR), Ghaziabad, India

Soma Mishra Dr. Vishwanath Karad MIT World Peace University, Pune, Maharashtra, India

Krishnendu Mondal Zoological Survey of India, Kolkata, West Bengal, India

Ministry of Environment, Forest and Climate Change, Government of India, New Delhi, India

M. V. Ramana Murthy National Centre for Coastal Research (NCCR), Ministry of Earth Sciences, Government of India, Chennai, India

Muthukumar Muthuchamy Climate Change and Remediation Research Lab (CCRRL), Department of Environmental Science, Central University of Kerala, Kasaragod, Kerala, India

Anbazhagi Muthukumar Climate Change and Remediation Research Lab (CCRRL), Department of Environmental Science, Central University of Kerala, Kasaragod, Kerala, India

T. M. Balakrishnan Nair Indian National Centre for Ocean Information Services, Ministry of Earth Sciences, Government of India, Hyderabad, India

R. Nandakumar Department of Marine Biotechnology, Academy of Maritime Education and Training (AMET), Chennai, Tamil Nadu, India

Tanmoy Nandy Laboratory for Benthic Ecological Trait Analysis (L-BETA), CSIR-National Institute of Oceanography, Regional Centre, Mumbai, India

Muhammad Adnan Naseem Department of Agriculture Business and Marketing, Faculty of Agricultural Sciences and Technology, Bahauddin Zakariya University, Multan, Punjab, Pakistan

Nadirsha P. S. Nawab Climate Change and Remediation Research Lab (CCRRL), Department of Environmental Science, Central University of Kerala, Kasaragod, Kerala, India

Pranav J. Pandya Department of Zoology, R. R. Lalan College, Bhuj, Kachchh, Gujarat, India

Atanu Kumar Panja MEME Division, CSIR-Central Salt & Marine Chemicals Research Institute, Bhavnagar, Gujarat, India

Academy of Scientific and Innovative Research (AcSIR), Ghaziabad, India

Dayesh Parmar Coastal and Marine Ecology Division, Gujarat Institute of Desert Ecology, Gujarat, India

Subhash I. Parmar Department of Earth and Environmental Science, Krantiguru Shyamji Krishna Verma Kachchh University, Bhuj, Gujarat, India

Department of Zoology, R. R. Lalan College, Bhuj, Kachchh, Gujarat, India

Sai Vineeth Perla Coastal and Marine Ecology Division, Gujarat Institute of Desert Ecology, Gujarat, India

Petuna Aquaculture Pty Ltd, Rowella, TAS, Australia

P. Perumal Marine Planktonology & Aquaculture Laboratory, Department of Marine Science, School of Marine Sciences, Bharathidasan University, Tiruchirappalli, Tamil Nadu, India

Santhanam Perumal Marine Planktonology and Aquaculture Laboratory, Department of Marine Science, Bharathidasan University, Tiruchirappalli, Tamil Nadu, India

Raju Piliyan Aquaculture Research Laboratory, Department of Marine Biotechnology, AMET University, Chennai, India

M. Prashanthi Devi Department of Environmental Science and Management, Bharathidasan University, Trichy, Tamil Nadu, India

Kattapuni Suresh Prithvisagar FAO Reference Centre for AMR & Aquaculture Biosecurity, DST TEC, Nitte University, Mangalore, India

P. Ragavan MoEF & CC, Indira Paryavaran Bhawan, New Delhi, India

Aastik D. Rajan Adani Institute of Infrastructure Engineering, Ahmedabad, India

Piliyan Raju Department of Marine Biotechnology, AMET University, Chennai, Tamil Nadu, India

Darwin Ramteke Biological Oceanography Division, CSIR-National Institute of Oceanography, Dona Paula, Goa, India

Academy of Scientific and Innovative Research (AcSIR), Ghaziabad, India

R. Ranjith Jasmine Concrete Export Private Limited, Chennai, Tamil Nadu, India

Raveendhiran Ravinesh Centre for Marine Living Resources and Ecology, Ministry of Earth Sciences, Kochi, India

Coastal and Marine Ecology Division, Gujarat Institute of Desert Ecology, Gujarat, India

Philip Theodore Reginald Aquaculture Research Laboratory, Department of Marine Biotechnology, AMET University, Chennai, India

M. Roshy Ann Department of Environmental Science and Management, Bharathidasan University, Trichy, Tamil Nadu, India

Dilshad Safiullah New Mexico Highlands University, Las Vegas, NM, USA

Ankur Saini Laboratory for Benthic Ecological Trait Analysis (L-BETA), CSIR-National Institute of Oceanography, Regional Centre, Mumbai, India

Academy of Scientific and Innovative Research (AcSIR), Ghaziabad, India

Sheikh Sajan Zoological Survey of India, Kolkata, West Bengal, India

Gujarat Institute of Desert Ecology, Terrestrial Ecology Division, Bhuj, Kachchh, Gujarat, India

Alakes Samanta Indian National Centre for Ocean Information Services, Ministry of Earth Sciences, Government of India, Hyderabad, India

P. Santhanam Marine Planktonology & Aquaculture Laboratory, Department of Marine Science, School of Marine Sciences, Bharathidasan University, Tiruchirappalli, Tamil Nadu, India

R. K. Sarangi Marine Ecosystem Division, BPSG/EPSA, Space Applications Centre (ISRO), Ahmedabad, Gujarat, India

Ranjit Kumar Sarangi AESG/EPSA, Space Applications Centre (ISRO), Ahmedabad, India

M. Saranya Marine Planktonology & Aquaculture Laboratory, Department of Marine Science, School of Marine Sciences, Bharathidasan University, Tiruchirappalli, Tamil Nadu, India

S. Sasidharan Department of Environmental Science and Management, Bharathidasan University, Trichy, Tamil Nadu, India

Sabyasachi Sautya Laboratory for Benthic Ecological Trait Analysis (L-BETA), CSIR-National Institute of Oceanography, Regional Centre, Mumbai, India

Academy of Scientific and Innovative Research (AcSIR), Ghaziabad, India

Muhammad Shaheryar Institute of Agronomy, Faculty of Agricultural Sciences and Technology, Bahauddin Zakariya University, Multan, Punjab, Pakistan

Muhammad Nadeem Shah Institute of Agronomy, Faculty of Agricultural Sciences and Technology, Bahauddin Zakariya University, Multan, Punjab, Pakistan

Department of Agriculture, Government College University, Lahore, Punjab, Pakistan

Venkatesan Shiva Shankar Department of Science, Mahatma Gandhi Government College, Mayabunder, North and Middle Andaman, India

M. Sindhuja Department of Computing Technologies, School of Computing, College of Engineering and Technology, SRM Institute of Science and Technology, Kattankulathur, Tamil Nadu, India

R. Sivasankar Applied Research Center for Environment and Marine Studies, King Fahd University of Petroleum and Minerals, Dhahran, Saudi Arabia

Kalpesh D. Sorathiya Tolani College of Arts and Science, Adipur, Gujarat, India

Ancy C. Stoy Indian Council of Agricultural Research-Central Marine Fisheries Research Institute, Kochi, India

Cochin University of Science and Technology, Kochi, India

Felix Sugantham Dr. J. Jayalalithaa Fisheries University, Kanniyakumari, Tamil Nadu, India

Department of Aquaculture, St. Devasahayam Institute of Fisheries Science and Technology, Kanniyakumari, Tamil Nadu, India

Palanisamy Thanabalan National Centre for Coastal Research (NCCR), Ministry of Earth Sciences, Government of India, Chennai, India

N. Thirunavukkarasu Presidency Research Innovations in Zoological Eminence (PRIZE) Lab, Department of Zoology, Presidency College (Autonomous), Chennai, Tamil Nadu, India

Basudev Tripathy Zoological Survey of India, Western Regional Centre, Pune, Maharashtra, India

Radha Vijayaraj Aquaculture Research Laboratory, Department of Marine Biotechnology, AMET University, Chennai, India

Ketankumar J. Yogi Coastal and Marine Ecology Division, Gujarat Institute of Desert Ecology, Bhuj, Gujarat, India

Terrestrial Ecology Division, Gujarat Institute of Desert Ecology, Bhuj, Gujarat, India

Chapter 1
Blue Carbon Ecosystems of Kachchh Coast, Gujarat: Ecological Treasure of Coastal Biodiversity

Raveendhiran Ravinesh, Dayesh Parmar, Deep D. Dudiya, Sai Vineeth Perla, Pallavi Joshi, and Kapilkumar Nivrutti Ingle

Abstract Coastal ecosystems such as mangroves, tidal marshes, and seagrass belts are generally considered blue carbon ecosystems (BCEs) and are not only nature-based solutions (NBS) for mitigation and adaptation of climate change but also provide vital support to biodiversity. The Kachchh district in Gujarat, India, hosts diverse coastal ecosystems such as mangroves, mudflats, and salt marshes. Coastal ecosystems play a vital role in biodiversity conservation, coastal protection, and support the livelihood of the coastal population. Despite their ecological significance, these ecosystems in Kachchh remain understudied. This study explores the distribution, biodiversity, and a few undiscussed ecosystem services of BCEs in Kachchh, emphasizing their importance as habitats for marine and terrestrial

R. Ravinesh
Centre for Marine Living Resources and Ecology, Ministry of Earth Sciences, Kochi, India

Coastal and Marine Ecology Division, Gujarat Institute of Desert Ecology, Gujarat, India

D. Parmar
Coastal and Marine Ecology Division, Gujarat Institute of Desert Ecology, Gujarat, India

D. D. Dudiya
Department of Zoology, R. R. Lalan College, Kachchh, Gujarat, India

Department of Earth and Environmental Science, Krantiguru Shyamji Krishna Verma Kachchh University, Bhuj, Gujarat, India

S. V. Perla
Coastal and Marine Ecology Division, Gujarat Institute of Desert Ecology, Gujarat, India

Petuna Aquaculture Pty Ltd, Rowella, TAS, Australia

P. Joshi
Coastal and Marine Ecology Division, Gujarat Institute of Desert Ecology, Gujarat, India

Agakhan School Mundra, Mundra, Gujarat, India

K. N. Ingle (✉)
Coastal and Marine Ecology Division, Gujarat Institute of Desert Ecology, Gujarat, India

Department of Ecology, University of Szeged, Szeged, Hungary

B. P. Barathan et al. (eds.), *Navigating Climate Change: Impacts on Biodiversity and Ecosystem Resilience*, Sustainable Development and Biodiversity 42,
https://doi.org/10.1007/978-981-95-0409-1_1

species. Special attention is given to mangrove ecosystems, which dominate the region, as well as the lesser-known contributions of mudflats, salt marshes, and seagrass beds. The study also highlights the interdependence between BCEs and local communities, including fisherfolk and pastoralists, who rely on these ecosystems for their livelihoods. This study also comments on the threats these BCEs are facing these days.

Keywords Coastal ecosystems · Mangroves · Salt marshes · Mudflats · Kachchh biodiversity

1.1 Introduction

Blue carbon (BC) generally encompasses all organic matter captured by marine organisms (Duarte et al. 2013), specifically playing an important role as a producer in an ecosystem. These include mangroves, tidal marshes, seagrass meadows, and recently seaweeds (Choudhary et al. 2024). These vegetated coastal ecosystems possess exceptional abilities to sequester carbon dioxide and are considered the most productive habitats on the planet. Research on these ecosystems, such as mangroves, tidal marshes, and seagrass meadows, has revealed their potential and considered them as blue carbon ecosystems (BCEs). However, their role in compensating global carbon emissions is limited due to their small coverage area, even though they sequester a large amount of carbon (Lovelock and Duarte 2019; Jamilah et al. 2025). BCEs characterize an inimitable interface between marine and terrestrial environments in 0.3% to 1.5% of the biologically productive landmass of the Earth (Macreadie et al. 2021; Bell et al. 2025). Despite a restricted global extent, they can provide multiple benefits by providing crucial resources for various faunal species (Menéndez et al. 2020; Bell et al. 2025). For example, BCEs can provide nesting and denning sites (McConville et al. 2013), nurseries that protect from predators (Du et al. 2020), rich food resources (Jiang et al. 2020), and shelter from various anthropogenic disturbances (Rog et al. 2017; Bell et al. 2025). However, climate change and coastal development lead to a rapid loss of BCEs globally (Lovelock and Reef 2020). On this background, there is a strong need to explore the multiple hidden ecological benefits of BCEs beyond carbon sequestration, specifically from the neglected and remote geographical areas. The marine coastal ecosystems are biologically rich, productive, and extremely valued regions, particularly within the diverse spectrum of ecosystems. For example, the mangrove ecosystem supports uniquely rich and diverse flora and faunal species (Prasanna et al. 2017), throughout the tropical and subtropical intertidal zones (Hookham et al. 2014). They also provide shelter for various coastal species from nutrient-poor soils to highly rich organic fine muds (Nagelkerken et al. 2008; Sousa and Dangremond 2012). The biodiversity of the Gulf of Kachchh, particularly the coastline of Saurashtra, is well discussed for its rich marine and coastal habitats; however, the biodiversity of the coastline of Kachchh district is still neglected.

Kachchh is the largest district in India, which covers almost 45,652 km^2 area of the northwestern part of Gujarat. Kachchh lies between N 22°41′11″ to E 24°41′47″, and covers about 24% of Gujarat's total area. Kachchh shows extreme weather conditions such as a dry and arid-coastal climate with moderate to high winds and limited rainfall, which increases from west to east. These conditions, particularly high temperatures and wind speeds, make the water evaporation rate higher (GUIDE 2002). This region includes a variety of ecological features such as the presence of Rann of Kachchh (desert ecosystem), Banni grassland, a long coastline including mudflats, mangroves, halophytes, etc., and seawater with various commercial fishes and rare marine life. Despite the crucial contributions, studies on the biodiversity of the Kachchh coast remain relatively underexplored. The main focus of the research has been on the Gulf of Kachchh because of its rich marine ecosystems, and adjacent habitats such as corals, seaweeds, and mangroves. However, the roles of BCEs in supporting biodiversity and local livelihoods of the Kachchh district have not yet received adequate attention from the scientific community. This chapter aims to highlight these overlooked aspects of BCEs in the Kachchh district of Gujarat and advocate for a more holistic approach to research efforts.

1.2 Blue Carbon Ecosystems in Kachchh

The unique convergence of diverse habitats contributes meaningfully to Kachchh's rich biological diversity, making it one of the important zones for ecological and environmental studies (Fig. 1.1). Beyond the carbon, the BCEs serve as unique habitats to support the biodiversity by providing essential breeding grounds, nurseries, and feeding areas for various species. They are an integral part of the socioeconomic structure within local communities by providing a variety of resources and economic opportunities. Mangrove is a group of diverse tropical plants that include herbs to trees which are morphologically and physiologically adapted to intertidal zones. These halophytic plants are suited to cope with various adverse conditions such as high salinity levels, strong wind force, tidal fluctuations, and low oxygen availability, and form an ecosystem distinct from coastal regions (Kathiresan 2022). Gujarat has the second largest mangrove cover, with the Gulf of Kachchh showing an extensive area. The region is dominated by the mangrove species *Avicennia marina*, which is found in larger numbers along the coastline. A few other mangrove species, such as *Rhizophora mucronata, Ceriops tagal,* and *Aegicerous corniculatum*, can also be found rarely in very small numbers. Its extreme tolerance to low rainfall, high salinity, evapotranspiration, temperature, etc., makes its mono-species dominance throughout Gujarat. Mangroves are highly significant vegetation when it comes to carbon storage and sequestration, but the ecological importance of this ecosystem extends far beyond it, particularly for maintaining biodiversity.

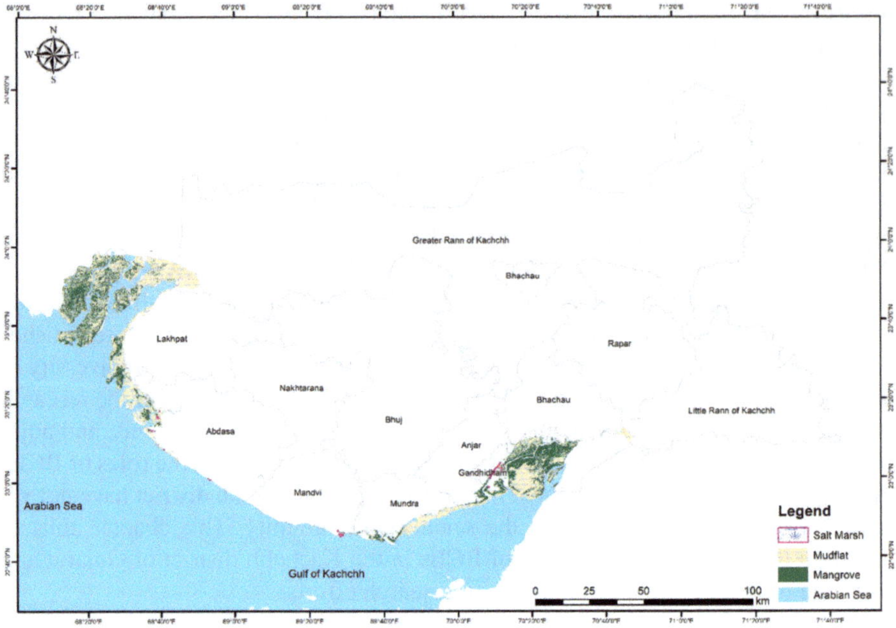

Fig. 1.1 Map of Kachchh showing the locations of BCEs in coastal areas tentatively. The map shows the Arabian Sea and the Gulf of Kachchh in blue, and the areas of salt marsh, mudflat and mangroves marked in purple, yellow and green respectively

The Indian coast shows the presence of widespread mud and tidal flats, particularly along the eastern and northwestern coasts of India. The coast of Kachchh is also located in the northwestern part of India, showing the presence of mud flats in a few areas. However, the major part of these flats is located in the Great and Little Rann of Kachchh, which are vast saline deserts that flood during the monsoon (Baba and Nayak 2002). In the case of the Kachchh coastline, the mudflats are also present in various areas but rarely studied; the mudflats play an important role in serving as a food source for various avifauna species, including the migratory birds visiting during winter in Kachchh. Tidal marshes, also referred to as salt marshes, are saline ecosystems typically found in the upper intertidal zones, which are periodically inundated by tidal flows. These are unique ecosystems vegetated with halophytic plants, such as a diverse array of grasses, rushes, and herbs. Seagrasses are one of the most important BCEs, inhabiting the soft-bottom sediments from the intertidal zone down to depths of 40 meters across the world, including tropical to temperate regions (Gutiérrez et al. 2012). These marine flowering plants offer a diverse range of ecosystem services, such as nutrient cycling, also, serving as a nursery habitat for various fish species, including commercially important fish. They also provide a vital food source for a few marine herbivore animals such as turtles, dugongs, and manatees (Nordlund et al. 2018). Green turtles are primarily herbivores which depend on the seagrass meadows for food. Olive Ridley turtles, which are considered as one of opportunistic feeders, also utilize seagrass habitats, especially as juveniles, benefiting from the shelter as well as an abundance of invertebrates and small fish found in these areas. Both of these turtle species are reported along Kachchh (Bhaskar 1984).

Seaweeds indirectly aid in carbon sequestration by sending their material to the deep sea and sediments as detritus (Hill et al. 2015; Choudhary et al. 2024). They generally act as carbon donors to the accumulators of their organic material from other ecosystems (Krause-Jensen and Duarte 2016). A few seaweed species float on the water surface and can be drifted from one place to another. For example, *Sargassum* sp. can permanently inhabit the ocean's water surface (Godínez-Ortega et al. 2021; van Tussenbroek et al. 2024), and also provide habitat and shelter to various marine organisms, such as invertebrates, including crustaceans and molluscs, to find shelter and protection among the floating mats of *Sargassum* sp. (Coston-Clements et al. 1991). The juveniles of various fish species find habitat in such floating seaweeds (Casazza and Ross 2008). They can serve as a food source to sea turtles (Witherington et al. 2012) and various marine organisms. In terms of Kachchh, there are no studies available on seaweed-associated fauna. There are very limited documents available on the seaweed resources in Kachchh. However, there are only a few pilot-scale attempts done for seaweed cultivation, but this trend can rise in the future. Generally, the seaweed cultivations face the problem of epiphytism (Sahu et al. 2020; Behera et al. 2022), which is a common phenomenon in the marine environment (Ingle et al. 2018) and is certainly responsible for economic loss in seaweed cultivation (Lehahn et al. 2016; Ingle et al. 2020). Seaweeds have a great potential to provide various products (Jiang et al. 2016), which can attract many farmers and industries. But there are various hurdles and challenges for large-scale production of seaweed biomass in Indian seawater (Ingle et al. 2020), including the Kachchh coastline. On the coast of Kachchh, these floating seaweeds can be easily found, particularly in the area of Pingleshwar, Mandvi, etc. These seaweeds come from the Gulf of Kachchh through water waves and enter the Kachchh coast. With this, a few seaweed species, such as *Ulva* sp., *Padina* sp., and *Sargassum* sp., can be easily found in a few areas of the Kachchh coastline particularly in the rocky areas.

1.3 Ecological Importance of BCEs in Kachchh Coastal Areas Beyond the Carbon

The major part of BCEs in Kachchh primarily consist of mangroves with its adjacent mudflats and salt marshes; however, seagrass beds and seaweeds cover less area. They play a crucial role in carbon sequestration, but beyond that they prevent coastal erosion, support fisheries, and maintain ecological balance. They also preserve water quality and also support nutrient recycling. The invertebrates and vertebrates present in BCEs can maintain the ecological balance by serving as a food

Table 1.1 BCEs associated faunal species

Category	Description
1	Herbivores or faunal species depend directly for food
2	Faunal species depend for habitat
3	Faunal species depend for shelter and protection
4	Associated communities depend indirectly for food
5	Top predators that can feed on associated communities

source for larger predators. The detailed list of invertebrate and vertebrate species recorded from the BCEs of Kachchh are given in Table 1.1. The checklist was prepared based on previously published papers, online databases, manuals, reports, and visual and photographic observations. Mangrove ecosystems with their adjacent mudflats are discussed here as they offer immense ecological support to biodiversity and considered invaluable treasures, they also play very important role in various terms by providing a range of ecosystem services.

1.3.1 Support Biodiversity

Coastal ecosystem provides shelter for a diverse range of marine and terrestrial species, which includes various fishes, crustaceans, birds, and other wildlife, fostering rich biodiversity (Fig. 1.2). However, their ecological role is not limited to providing shelter, they also serve as critical habitats for several key species of marine and terrestrial fauna including commercially important fish species. These ecosystems also act as a direct and indirect food source for various organisms. For example, BCEs associated faunal species may be categorized into five distinct groups (Table 1.1), on the basis of their ecological association with mangrove ecosystems which highlights the ecological complexity and significance in supporting to these inverse faunal species.

BCEs served as an important habitat for various faunal species, among these, particularly, the benthic fauna, for example, crabs are one of the dominant groups in mangrove ecosystems (Nagelkerken et al. 2008). These crab species present in the mangrove environment play a crucial role in the biogeochemical cycles, such as nutrient recycles, also help in the formation of detritus and maintain the mangrove food web (Bosire et al. 2004). They influence the overall ecosystem's dynamics through their feeding habits (Bosire et al. 2004). With this, their burrowing activities can alter the topography of the mangrove floor, which further enhances the bioturbation of sediments (Ashton et al. 2003), by improving soil aeration (Warren and Underwood 1986; Ashton et al. 2003). Although the importance of mangrove-associated faunal communities is recognized, the studies on them are limited (Lee 2008), specifically the studies on little-consuming benthic macro-invertebrates

Fig. 1.2 Representative animals associated with various BCEs (**A**–**H**) (**A**) *Diadumene lineata* (Verrill, 1869); (**B**) *Austruca iranica* (Pretzmann, 1971); (**C**) *Macrophthalmus sulcatus* (H. Milne Edwards, 1852); (**D**) *Telescopium telescopium* (Linnaeus, 1758); (**E**) *Bakawan rotundata* (A. Adams, 1850); (**F**) *Echis carinatus* (Schneider, 1801); (**G**) *Dromas ardeola* (Paykull, 1805); and (**H**) *Ardea cinerea* (Linnaeus, 1758)

(Teoh et al. 2018). The various characteristics of the relationship between crabs or other crustaceans and mangroves were unclear and often neglected, which were responsible for the rapid degradation of mangrove ecosystems (Duke et al. 2007).

The majority of benthic animals found in mangrove ecosystems, particularly burrowing decapods such as sesarmidae crabs, and fiddler crabs (Kristensen 2008), are herbivores that play a key role as primary consumers in the mangrove ecosystem. They consume, bury, and break down not only the leaf litter of mangroves present on the floor but also the microalgal mats (Kristensen and Alongi 2006), which further help in nutrient recycling and increase soil fertility by enhancing decomposition processes. In mangrove habitats, gastropods in their juvenile stage act as deposit feeders or detritivores, which consume detritus and other organic matter. In contrast to that, in the adult stage, they feed on leaves, fruits, and propagules present in mangroves (Imamsyah et al. 2020); in this way they get nutrients from mangrove vegetative substrates (Dwijaya et al. 2021). A few species consume the floor leaf litter, propagules, etc., which make detritus in the sediment and a few species feed on the detritus (Penha-Lopes et al. 2009) which are rich in carbon. A few species graze on the algae present in mangrove forests and control them. This also controls the growth of various epiphytes and accelerates the litter decomposition, which further balances the ecosystem (Purnama et al. 2024). Subsequently, gastropods' primary activity involves transporting the organic matter into mangrove ecosystems through enhanced decomposition of litter.

Within the mangrove forests, molluscs show their presence with remarkable diversity and dominance. Gastropods are also found highly abundant in mangrove areas, along with coasts, rocky areas, and intertidal mudflats (Yadav et al. 2019). Their distribution and abundance depend upon several factors, including biotic as well as abiotic factors. Biotic factors are generally patterns of larval distribution and availability of food sources. Abiotic factors: waves, light, substratum, salinity, temperature, etc. (Peng et al. 2017; Imamsyah et al. 2020). Their stages of life cycle, in which they adapt to the environment of BCEs morphologically, psychologically, and behaviorally, also play an important role in their selection of specific habitats (Reis et al. 2021). In mangrove habitats, gastropods in their juvenile stage act as deposit feeders or detritivores, which consume detritus and other organic matter. In contrast to that in the adult stage, they feed on leaves, fruits, and propagules present in mangroves. In the juvenile stage, many species are less mobile and generally attach to the vegetative roots of mangroves, which are substrata for them (Imamsyah et al. 2020). Many, in mature or adult stages are less mobile and stay in mangroves and other BCEs. For example, Sesarmidae and fiddler crabs dig burrows in the sediment, which protect them from predators and harsh environmental conditions. Sometimes these burrows act as storage spaces for food (Dittmann 1996) too for these species.

Mangrove provides shelters to a wide range of organisms, including small to large faunal species, and also microorganisms that are well adapted to intertidal conditions (Neukermans et al. 2008). Mangroves serve as an important habitat for various organisms, particularly many species of fish, molluscs, crustaceans, and birds (Kairo et al. 2008; Nagelkerken et al. 2008), and various species from other habitats also get shelter and habitat in mangrove ecosystems. A few species that are not from mangrove habitat, but adapted to mangrove habitat or depend on the mangrove ecosystem for food, are associated communities that feed on the other organisms present in the mangrove ecosystem. For example, *Echis carinatus* Schneider, 1801, also known as the Indian saw-scaled viper, is a venomous snake commonly found in dry, rocky, and scrubby habitats can also be found adapted to mangrove ecosystems in the coastal areas of Kachchh. In these mangrove habitats, this snake species may find refuge in the roots, and on mangrove vegetation, where it can hide and ambush small prey like mudskippers, mangrove crabs, and small birds. In Kachchh, the top predators depend on mangrove, and overall BCEs are a few organisms such as jackals, and a few species of avifauna, specifically painted stroke and pelicans.

Other BCEs also play an important role in biodiversity support; a total of 314 species have been recorded across various phyla within the blue carbon ecosystems of the Kachchh district, reflecting the region's high biodiversity (Table 1.2). The phylum-wise species variation reveals that Chordata has the highest diversity, with 119 species, encompassing a wide range of vertebrates, including fish, birds, and marine mammals that thrive across multiple ecosystems. Following closely, Mollusca ranks second with 99 species, consisting primarily of gastropods and bivalves, most commonly found in rocky shores and mangroves. Arthropoda comes next, with 75 species, mainly comprising crabs and shrimps that dominate intertidal and mangrove environments. In contrast, Echinodermata accounts for 11 species, such as sea urchins, brittle stars, and sea cucumbers, which are primarily associated with rocky shores. Annelida, represented by eight species, consists mainly of marine worms adapted to soft sediments in mangroves. The Cnidaria phylum is relatively sparse, with only two species of sea anemones recorded across multiple habitats. This distribution highlights that Chordata, Mollusca, and Arthropoda dominate the biodiversity spectrum, while Cnidaria are the least diverse groups in these ecosystems.

1.3.2 Livelihood of Coastal Communities

Kachchh faces various ecological challenges such as sea level rise, leading to soil erosion and habitat loss of marine life that significantly impact the environment and livelihoods of coastal communities in this area that are intensely tangled with these challenges. Fishing activities and pastoralism are the main livelihood of coastal communities in this area (Fig. 1.3).

Table 1.2 Blue carbon ecosystems of associated diversity of Kachchh District, Gujarat

Scientific name	Common name	MG	RS	SG	MF	IUCN status
Phylum: Cnidaria						
Diadumene lineata (Verrill, 1869)		+	+	+	+	DD
Stephensonactis ornata Panikkar, 1936		+	–	–	–	DD
Phylum: Annelida						
Ikeda pirotansis (Menon & Datta Gupta, 1962)		+	–	–	–	NE
Namalycastis indica (Southern, 1921)		+	–	–	–	NE
Nephtys polybranchia Southern, 1921		+	–	–	–	NE
Lumbrineris polydesma Southern, 1921		+	–	–	–	NE
Lumbrineris pseudobifilaris Fauvel, 1932		+	–	–	–	NE
Lumbrineris simplex Southern, 1921		+	–	–	–	NE
Pseudonereis sp.		–	+	–	–	NE
Lepidonotus cristatus (Grube, 1876)		–	+	–	–	NE
Phylum: Arthropoda						
Chaenostoma sinuspersici (Naderloo & Türkay, 2011)		+	–	–	–	NE
Myomenippe hardwickii (Gray, 1831)	Mangrove stone crab	+	–	–	–	NE
Eurycarcinus orientalis A. Milne-Edwards, 1867		+	–	–	–	NE
Pilumnus longicornis Hilgendorf, 1879		–	+	–	–	NE
Pilumnus minutus De Haan, 1835		–	+	–	–	NE
Portunus sanguinolentus (Herbst, 1783)	Three-spot swimming crab, blood-spotted swimming crab or red-spotted swimming crab	+	–	+	–	NE
Portunus segnis (Forskål, 1775)	Swimming blue crab	+		+		NE
Scylla olivacea (Herbst, 1796)	Orange mud crab	+	–	–	+	NE
Scylla serrata (Forskål, 1775)	Mud crab or mangrove crab	+	–	–	+	NE
Thalamita crenata (Latreille, 1829)	Powder blue-clawed swimming crab or crenate swimming crab or spiny rock crab	+	+	+	–	NE

(continued)

Table 1.2 (continued)

Scientific name	Common name	MG	RS	SG	MF	IUCN status
Episesarma mederi (H. Milne Edwards, 1854)		+	–	–	–	NE
Episesarma versicolor (Tweedie, 1940)	The violet vinegar crab	+	–	–	–	NE
Labuanium vitatum Ng & Davie, 2011		+	–	–	–	NE
Nanosesarma jousseaumei (Nobili, 1996)		+	–	–	–	NE
Nanosesarma minutum (De Man, 1887)		+	–	–	–	NE
Nanosesarma sarii Naderloo and Türkay, 2009		–	+	–	–	NE
Neosarmatium smithi (H. Milne Edwards, 1853)		+	–	–	–	NE
Parasesarma persicum Naderloo & Schubart, 2010		+	–	–	–	NE
Parasesarma plicatum (Latreille, 1803)		+	–	–	–	NE
Selatium brockii (De Man, 1887)		+	–	–	–	NE
Metopograpsus cannicci Innocenti, Schubart & Fratini, 2020		+	–	–	–	NE
Metopograpsus latifrons (White, 1847)		+	–	–	–	NE
Metopograpsus messor (Forskål, 1775)		+	–	–	–	NE
Metopograpsus thukuhar (Owen, 1839)	Alamihi crab	+	–	–	–	NE
Metaplax indica H. Milne Edwards, 1852		+	–	–	–	NE
Ptychognathus barbatus (A. Milne-Edwards, 1873)		+	–	–	–	NE
Varuna litterata (Fabricius, 1798)		+	–	–	–	NE
Circulium vitatum (P.K.L. Ng & Davie, 2011)		+	–	–	–	NE
Leptochryseus kuwaitense (Jones & Clayton, 1983)		+	–	–	–	NE
Nasima dotilliformis (Alcock, 1900)		+	–	–	–	NE
Opusia indica (Alcock, 1900)		+	–	–	–	NE
Dotilla blanfordi Alcock 1900.		–	–	–	+	NE

(continued)

Table 1.2 (continued)

Scientific name	Common name	MG	RS	SG	MF	IUCN status
Dotilla myctiroides H. Milne Edwards, 1852		+	–	–	–	NE
Ilyoplax frater (Kemp, 1919)		+	–	–	–	NE
Ilyoplax sayajiraoi Trivedi, Soni, Trivedi & Vachhrajani, 2015		–	–	–	+	NE
Scopimera crabricauda Alcock, 1900		–	–	–	+	NE
Scopimera proxima Kemp, 1919		–	–	–	+	NE
Macrophthalmus brevis (Herbst, 1804 [in Herbst, 1799–1804])		+	–	–	–	NE
Macrophthalmus depressus Rüppell, 1830	Depressed mud crab	+	–	–	–	NE
Macrophthalmus sulcatus H. Milne Edwards, 1852	Sulcated crab	+	–	–	+	NE
Venitus dentipes (Lucas, in Guérin-Méneville, 1838)		+	–	–	–	NE
Austruca albimana (Kossmann, 1877)	Fiddler crab	+	–	–	–	NE
Austruca annulipes (H. Milne Edwards, 1837)	Fiddler crab	+	–	–	–	NE
Austruca iranica Pretzmann, 1971	Fiddler crab	+	–	–	–	NE
Austruca sindensis (Alcock, 1900)	Fiddler crab	+	–	–	–	NE
Gelasimus vocans (Linnaeus, 1758)	Fiddler crab	+	–	–	–	NE
Ocypode brevicornis H. Milne Edwards, 1837	Ghost crab	–	–	–	+	NE
Ocypode ceratophthalmus	Horned ghost crab	–	–	–	+	NE
Ocypode cordimana Latreille, 1818	Smooth-handed ghost crab	–	–	–	+	NE
Ocypode rotundata (Miers, 1882)	Rounded ghost crab	–	–	–	+	NE
Tubuca acuta (Stimpson, 1858)	Acute fiddler crab	+	–	–	–	NE
Tubuca alcocki Shih, Chan & Ng, 2018	Fiddler crab	+	–	–	–	NE
Tubuca dussumieri (H. Milne Edwards, 1852)	Dussumier's fiddler crab	+	–	–	–	NE
Arcotheres placunae (Hornell & Southwell, 1909)	Pea crab	+	–	–	–	NE

(continued)

Table 1.2 (continued)

Scientific name	Common name	MG	RS	SG	MF	IUCN status
Metapenaeus dobsoni (Miers, 1878)	Kadal shrimp	+	–	–	–	LC
Metapenaeus kutchensis P.C. George, M.J. George & Rao, 1963	Ginger shrimp	+	–	+	–	LC
Metapenaeus monoceros (Fabricius, 1798)	Pink shrimp or brown shrimp or speckled shrimp	+	–	+	–	LC
Mierspenaeopsis sculptilis (Heller, 1862)	Rainbow shrimp	+	–	+	–	LC
Penaeus monodon Fabricius, 1798	Tiger shrimp	+	–	+	–	LC
Penaeus indicus H. Milne Edwards, 1837		+	–	+	–	DD
Penaeus semisulcatus De Haan, 1844 [in De Haan, 1833–1850]		+	–	+	–	DD
Alpheus lobidens De Haan, 1849	Pistol shrimp	+	+	+	–	LC
Clibanarius infraspinatus (Hilgendorf, 1869)		+	+	+	–	LC
Clibanarius longitarsus (De Haan, 1849 [in De Haan, 1833–1850])		+	+	+	–	LC
Clibanarius rhabdodactylus Forest, 1953		+	+	+	–	LC
Clibanarius rutilus Rahayu, 1999		+	+	+	–	LC
Clibanarius signatus Heller, 1961		+	+	+	–	LC
Clibanarius. virescens (Krauss, 1843)		+	+	+	–	LC
Diogenes avarus Heller, 1865		+	+	+		LC
Petrolisthes boscii (Audouin, 1826)		–	+	–	–	NE
Petrolisthes lamarckii (Leach, 1820)		–	+	–	–	NE
Amphibalanaus.amphitrite (Darwin, 1854)	Striped barnacle, purple acorn barnacle	+	+	+	+	NE
Megabalanus tintinnabulum (Linnaeus, 1758)	Large barnacle	+	+	+	+	DD
Byblis kachchhensis Thacker, Myers & Trivedi, 2024		+	–	–	–	NE
Nerocila depressa Milne Edwards, 1840		+	–	–	–	NE

(continued)

Table 1.2 (continued)

Scientific name	Common name	MG	RS	SG	MF	IUCN status
Phylum: Mollusca						
Rhyssoplax peregrina (Thiele, 1909)		–	+	–	–	NE
Rhyssoplax affinis (Issel, 1869)		–	+	–	–	NE
Cellana karachiensis (Winckworth, 1930)		–	+	–	–	NE
Clypidina notata (Linnaeus, 1758)		–	+	–	–	NE
Scutus howensis Iredale, 1940		–	+	–	–	NE
Euchelus asper (Gmelin, 1791)		–	+	–	–	NE
Trochus radiatus Gmelin, 1791		–	+	–	–	NE
Trochus stellatus Gmelin, 1791		–	+	–	–	NE
Stomatia phymotis Helbling, 1779		–	+	–	–	NE
Clanculus microdon A. Adams, 1853		–	+	–	–	NE
Umbonium vestiarium (Linnaeus, 1758)	Top shell	+	–	+	–	NE
Lunella coronata (Gmelin, 1791)		–	+	–	–	NE
Turbo bruneus (Röding, 1798)		–	+	–	–	NE
Neripteron violaceum (Gmelin, 1791)	Violet Nerite	+	–	–	–	NE
Nerita albicilla Linnaeus, 1758		–	+	–	–	NE
Nerita balteata Reeve, 1855		+	–	–	–	NE
Nerita polita Linnaeus, 1758		–	+	–	–	NE
Clypeomorus batillariaeformis (Habe & Kosug, 1966)	Necklace or channeled cerith	+	–	+	–	NE
C lypeomorus bifasciata (G.B. Sowerby II, 1855)	Morus cerith	+	–	+	–	NE
Pirenella cingulata (Gmelin, 1791)	Horn snail	+	–	–	–	LC
Telescopium telescopium (Linnaeus, 1758)	Telescope snail	+	–	–	–	LC
Cerithidea obtusa (Lamarck, 1822)	Mud creeper	+	–	–	–	LC
Optediceros breviculum (L. Pfeiffer, 1855)	Red mangrove snail	+	–	–	–	LC
Angustassiminea nitida (Pease, 1865)	Red mangrove snail	+	–	–	–	LC
Paratectonatica tigrina (Röding, 1798)	Moon snail	+	–	+	–	NE
Planaxis sulcatus (Born, 1778)	Tropical periwinkle	–	+	–	–	NE

(continued)

Table 1.2 (continued)

Scientific name	Common name	MG	RS	SG	MF	IUCN status
Ergaea walshi (Reeve, 1859)	White slipper limpet	–	+	–	–	NE
Desmaulus edgarianus (Melvill, 1898)	Cape shell	+	–	+	–	NE
Echinolittorina malaccana (Philippi, 1847)	Knobbly periwinkle snail	–	+	–	–	NE
Echinolittorina vidua (Gould, 1859)	Periwinkle snail	–	+	–	–	NE
Littoraria intermedia (Philippi, 1846)	Intermediate periwinkle snail	–	+	–	–	NE
Rissoina pachystoma Melvill, 1896		–	+	–	–	NE
Erosaria ocellata (Linnaeus, 1758)	Ocellate cowry	–	+	–	–	NE
Erosaria turdus (Lamarck, 1810)	Thrush cowrie	–	+	–	–	NE
Erronea caurica (Linnaeus, 1758)	Thick edged cowry	–	+	–	–	NE
Lyncina vitellus (Linnaeus, 1758)	Calf cowry or deer cowry	–	+	–	–	NE
Mauritia arabica (Linnaeus, 1758)	Arabian cowry	–	+	–	–	NE
Monetaria caputserpentis (Linnaeus, 1758)	Serpent's head cowry or snakehead cowry	–	+	–	–	NE
Monetaria moneta (Linnaeus, 1758)	Money cowrie	–	+	–	–	NE
Palmadusta asellus (Linnaeus, 1758)	Little ass cowry	–	+	–	–	NE
Palmadusta lentiginosa (J.E. Gray, 1825)	Lentiginose cowry	–	+	–	–	NE
Talostolida teres (Gmelin, 1791)	Tapering cowry	–	+	–	–	NE
Dendropoma sp		–	+	–	–	NE
Bursa granularis (Röding, 1798)	Granular frog shell	–	+	–	–	NE
Gyrineum natator (Röding, 1798)	Tuberculara gyre triton	–	+	–	–	NE
Anachis terpsichore (G. B. Sowerby II, 1822)	Striated dove snail	–	+	–	–	NE
Euplica scripta (Lamarck, 1822)	Dotted dove shell	–	+	–	–	NE
Pyrene flava (Bruguiere, 1789)	Yellow dove shell	–	+	–	–	NE
Cantharus spiralis Gray, 1839	Spiral goblet	–	+	–	–	NE
Engina zea Melvill, 1893		–	+	–	–	NE
Pollia undosa (Linnaeus, 1758)	Waved goblet	–	+	–	–	NE

(continued)

Table 1.2 (continued)

Scientific name	Common name	MG	RS	SG	MF	IUCN status
Indothais lacera (Born, 1778)	Rock snail	+	+	–	–	NE
Nassarius deshayesianus (Issel, 1865)	Nassa mud snails	–	+	–	–	NE
Nassarius jacksonianus (Quoy & Gaimard, 1833)	Mud whelk	+	–	+	–	NE
Nassarius olivaceus (Bruguière, 1789)	Olivaceous nassa	+	–	–	–	NE
Vexillum mica (Reeve, 1845)		–	+	–	–	NE
Conus achatinus Gmelin, 1791	Turtle cone or the agate cone	–	+	–	–	LC
Conus biliosus (Röding, 1798)	Bilious cone	–	+	–	–	LC
Conus catus Hwass, 1792	Cat cone	–	+	–	–	LC
Conus nussatella Linnaeus, 1758	Nussatella cone	–	+	–	–	LC
Melampus fasciatus (Deshayes, 1830)		+	–	–	–	DD
Pythia plicata (Férussac, 1821)	Coastal snail	+	–	–	–	LC
Auriculastra saccata (L. Pfeiffer, 1855)		+	+	–	–	DD
Peronia verruculata (Cuvier, 1830)	Air breathing sea slug	+	–	–	–	NE
Bakawan rotundata (A. Adams, 1850)	Mangrove bubble snail	+	–	–	–	NE
Aplysia oculifera A. Adams & Reeve, 1850	Spotted sea hare	–	+	–	–	NE
Dendrodoris fumata (Rüppell & Leuckart, 1830)	Black dorid	–	+	–	–	NE
Dendrodoris nigra (Stimpson, 1855)	Black dorid	–	+	–	–	NE
Doriopsilla miniata (Alder & Hancock, 1864)		–	+	–	–	NE
Jorunna funebris (Kelaart, 1859)	Dotted nudibranch	–	+	–	–	NE
Thecacera pennigera (Montagu, 1813)	Winged thecacera	–	+	–	–	NE
Bornella stellifera (A. Adams & Reeve [in A. Adams], 1848)		–	+	–	–	NE
Phidiana militaris (Alder & Hancock, 1864)		–	+	–	–	NE
Cratena sp.		–	+	–	–	NE
Septifer bilocularis (Linnaeus, 1758)	Box mussel	–	+	–	–	NE
Botula cinnamomea (Gmelin, 1791)		–	+	–	–	NE

(continued)

Table 1.2 (continued)

Scientific name	Common name	MG	RS	SG	MF	IUCN status
Acar plicata (Dillwyn, 1817)		–	+	–	–	NE
Barbatia foliata (Forsskål in Niebuhr, 1775)		–	+	–	–	NE
Barbatia obliquata (Wood, 1828)		–	+	–	–	NE
Lamarcka avellana (Lamarck, 1819)		–	+	–	–	NE
Tegillarca granosa (Linnaeus, 1758)	Ark clam or blood clam	+	–	+	–	NE
Saccostrea cucculata (Born, 1778)	Hooded oyster or Natal rock oyster	+	+	–	–	NE
Hyotissa hyotis (Linnaeus, 1758)		–	+	–	–	NE
Chama asperella Lamarck, 1819	Jewel box	–	+	–	–	NE
Cardita variegata Bruguière, 1792		–	+	–	–	NE
Lepirodes layardi (Deshayes, 1856)		–	+	–	–	NE
Scintilla sp.		–	+	–	–	NE
Glauconome angulata Reeve, 1844		+	–	–	–	DD
Glauconome cerea Reeve, 1844		+	–	–	–	DD
Mactra grandis Gmelin, 1791	Big brown mactra clam	+	–	+	–	NE
Mactra luzonica Reeve, 1854	Luzon trough clam	+	–	+	–	NE
Meretrix casta (Gmelin, 1791)	Asiatic hard clam	+	–	–	–	NE
Gafrarium pectinatum (Linnaeus, 1758)		+	–	+	–	DD
Irus macrophylla (Deshayes, 1853)		–	+	–	–	NE
Petricola lapicida (Gmelin, 1791)		–	+	–	–	NE
Tellina macassariensis Prashad, 1932		+	–	+	–	DD
Solen lamarckii Chenu, 1843	Lamarck razor clam	+	–	+	–	NE
Pholas orientalis Gmelin, 1791	Orient angel's wing	+	–	+	–	NE
Laternula anatina (Linnaeus, 1758)		+	–	+	–	DD
Phylum Echinodermata						
Ophiocomella sexradia (Ducan, 1887)		–	+	–	–	NE

(continued)

Table 1.2 (continued)

Scientific name	Common name	MG	RS	SG	MF	IUCN status
Ophioplocus imbricatus (Müller & Troschel, 1842)		–	+	–	–	NE
Ophionereis dubia dubia (Müller & Troschel, 1842)		–	+	–	–	NE
Amphipholis squamata (Delle Chiaje, 1828)		–	+	–	–	NE
Amphiura ambigua (Koehler, 1905)		–	+	–	–	NE
Ophiactis savignyi (Müller & Troschel, 1842)		–	+	–	–	NE
Ophiothrix savignyi (Müller & Troschel, 1842)		–	+	–	–	NE
Holothuria cinerascens (Brandt, 1835)		–	+	–	–	NE
Holothuria hilla Lesson, 1830		–	+	–	–	NE
Echinothrix calamaris (Pallas, 1774)		–	+	–	–	NE
Aquilonastra lorioli (Koehler, 1909)		–	+	–	–	NE
Phylum: Chordata						
Diplosoma simile (Sluiter, 1909)		–	+	–	–	NE
Phallusia nigra Savigny, 1816		–	+	–	–	NE
Canis aureus Linnaeus, 1758	Asiatic jackal or jackal	+	–	–	–	LC
Chelonia mydas (Linnaeus, 1758)	Green turtle	–	–	+	–	EN
Lepidochelys olivacea (Eschscholtz, 1829)	Olive ridley	–	–	+	–	VU
Varanus bengalensis (Daudin, 1802)	Bengal monitor or Indian monitor	+	–	–	+	NT
Echis carinatus (Schneider, 1801)	Saw-scaled viper, little Indian viper	+	–	–	–	LC
Harpadon nehereus (Hamilton, 1822)	Bombay-duck	+	–	+	–	NT
Protonibea diacanthus (Lacepède, 1802)	Blackspotted croaker	+	–	–	–	NT
Tenualosa toli (Valenciennes, 1847)	Toli shad	+	–	–	–	VU
Oryzias setnai (Kulkarni, 1940)	Malabar ricefish	+	–	–	–	LC
Hyporhamphus limbatus (Valenciennes, 1847)	Congaturi halfbeak	+	–	–	–	LC
Aphanius dispar (Rüppell, 1829)	Arabian pupfish	+	–	–	–	LC

(continued)

Table 1.2 (continued)

Scientific name	Common name	MG	RS	SG	MF	IUCN status
Chanos chanos (Forsskål, 1775)	Milkfish	+	–	–	–	LC
Crenimugil seheli (Forsskål, 1775)	Bluespot mullet	+	–	–	–	LC
Osteomugil cunnesius (Valenciennes, 1836)	Longarm mullet	+	–	–	–	LC
Planiliza macrolepis (Smith, 1846)	Largescale mullet	+	–	–	–	LC
Planiliza planiceps (Valenciennes, 1836)	Tade gray mullet	+	–	–	–	LC
Planiliza melinoptera (Valenciennes, 1836)	Otomebora mullet	+	–	–	–	LC
Planiliza parsia (Hamilton, 1822)	Goldspot mullet	+	–	–	–	LC
Planiliza tade (Fabricius, 1775)	Mullet	+	–	–	–	LC
Chelon planiceps (Valenciennes, 1836)	Tade gray mullet	+	–	–	–	LC
Liza klunzingeri (Day, 1888)	Klunzinger's mullet	+	–	–	–	LC
Mugil cephalus (Linnaeus, 1758)	Flathead grey mullet	+	–	–	–	LC
Rhinomugil corsula (Hamilton, 1822)	Corsula	+	–	–	–	LC
Notopterus notopterus (Pallas, 1769)	Bronze featherback	+	–	–	–	LC
Boleophthalmus dussumieri (Valenciennes, 1837)	Mudskipper	+	–	–	–	LC
Periophthalmus waltoni Koumans, 1941	Mudskipper	+	–	–	+	LC
Taenioides anguillaris (Linnaeus, 1758)	Eel worm goby	+	–	–	+	LC
Lates calcarifer (Bloch, 1790)	Barramundi	+	–	–	–	LC
Lutjanus argentimaculatus (Forsskål, 1775)	Mangrove red snapper	+	–	+	–	LC
Lutjanus johnii (Bloch, 1792)	John's snapper	+	–	+	–	LC
Eleutheronema tetradactylum (Shaw, 1804)	Forefinger threadfin	+	–	+	–	LC
Leptomelanosoma indicum (Shaw, 1804)	Indian threadfin	+	–	+	–	LC
Therapon jarbua (Forsskål, 1775)	Jarbua terapon	+	–	+	–	LC
Arius arius (Hamilton, 1822)	Threadfin sea catfish	+	–	+	–	LC
Arius gagora (Hamilton, 1822)	Gagora catfish	+	–	+	–	LC

(continued)

Table 1.2 (continued)

Scientific name	Common name	MG	RS	SG	MF	IUCN status
Brachirus orientalis (Bloch & Schneider, 1801)	Oriental sole	+	–	+	–	LC
Cynoglossus macrolepidotus (Bleeker, 1851)	Sole	+	–	+	–	DD
Megalops cyprinoides (Broussonet, 1782)	Indo-Pacific tarpon	+	–	–	–	DD
Osteogeneiosus militaris (Linnaeus, 1758)	Soldier catfish	+	–	–	–	NE
Strongylura strongylura (van Hasselt,1823)	Spot tail needlefish	+	–	+	–	NE
Planiliza carinata (Valenciennes, 1836)	Keeled mullet	+	–	–	–	NE
Limosa limosa (Linnaeus, 1758)	Black-tailed Godwit	+	–	–	–	NT
Numenius arquata (Linnaeus, 1758)	Eurasian curlew or common curlew	+	–	–	–	NT
Limosa lapponica (Linnaeus, 1758)	Bar-tailed Godwit	+	–	–	–	NT
Threskiornis melanocephalus (Latham, 1790)	Black-headed Ibis, oriental white ibis, Indian white ibis, and black-necked ibis	+	–	–	+	LC
Mycteria leucocephala (Pennant, 1769)	Painted stork	+	–	–	+	LC
Anhinga melanogaster Pennant, 1769	Oriental darter	+	–	–	+	LC
Phoeniconaias minor (Geoffroy Saint-Hilaire, É, 1798)	Lesser Flamingo	+	–	–	+	NT
Pandion haliaetus (Linnaeus, 1758)	Osprey, sea hawk, river hawk, and fish hawk	+	+		+	LC
Elanus caeruleus (Desfontaines, RL 1789)	Black-winged kite, black-shouldered kite	+	–	–	+	LC
Circus aeruginosus (Linnaeus, 1758)	Western Marsh Harrier	+	–	–	–	LC
Tachyspiza badia (Gmelin, JF, 1788)	Shikra, little banded goshawk	+	–	–	+	LC
Apus affinis (JE Gray, 1830)	Indian House Swift or little swift	+	–	–	–	LC
Actitis hypoleucos (Linnaeus, 1758)	Common sandpiper	+	+	+	+	LC
Tringa stagnatilis (Bechstein, 1803)	Marsh sandpiper	+	+	+	+	LC
Calidris alpina (Linnaeus, 1758)	Dunlin	+	–	–	–	LC
Himantopus himantopus (Linnaeus, 1758)	Black-winged stilt	+	–	–	–	LC

(continued)

Table 1.2 (continued)

Scientific name	Common name	MG	RS	SG	MF	IUCN status
Vanellus malabaricus (Boddaert, 1783)	Yellow-wattled lapwing	+	–	–	–	LC
Vanellus indicus (Boddaert, 1783)	Red-wattled lapwing	+	–	–	–	LC
Calidris minuta (Leisler, 1812)	Little stint	+	–	–	–	LC
Calidris alba (Pallas, 1764)	Sanderling	+	–	–	–	LC
Sterna aurantia Gray, JE, 1831	River Tern	+	–	–	–	VU
Larus fuscus Linnaeus, 1758	Lesser black-backed gull	+	+	+	+	LC
Recurvirostra avosetta Linnaeus, 1758	Pied avocet	+	–	–	–	LC
Charadrius dubius Scopoli, 1786	Little ringed plover	+	+	+	+	LC
Anarhynchus leschenaultii (Lesson, 1826)	Greater sand plover	+	+	+	+	LC
Tringa erythropus (Pallas, 1764)	Spotted redshank	+	–	–	–	LC
Tringa nebularia (Gunnerus, 1767)	Common greenshank	+	–	–	–	LC
Tringa totanus (Linnaeus, 1758)	Common redshank	+	–	–	–	LC
Tringa glareola Linnaeus, 1758	Wood sandpiper	+	+	+	+	LC
Dromas ardeola Paykull, 1805	Crab-plover	+	+	+	+	LC
Chroicocephalus ridibundus (Linnaeus, 1766)	Black-headed gull	+	+	+	+	LC
Sternula albifrons (Pallas, 1764)	Little tern	+	+	+	+	LC
Hydroprogne caspia (Pallas, 1770)	Caspian tern	+	+	+	+	LC
Columba livia Gmelin, JF, 1789	Rock dove or rock pigeon, or common pigeon	+	+	+	+	LC
Spilopelia senegalensis (Linnaeus, 1766)	Laughing dove, palm dove, little brown dove	+	+	+	+	LC
Streptopelia decaocto (Frivaldszky, 1838)	Eurasian collared dove	+	–	–	–	LC
Alcedo atthis (Linnaeus, 1758)	Common kingfisher, Eurasian kingfisher	+	+	+	+	LC
Halcyon smyrnensis (Linnaeus, 1758)	White-throated kingfisher, white-breasted kingfisher	+	+	+	+	LC
Ceryle rudis (Linnaeus, 1758)	Pied kingfisher	+	+	+	+	LC
Merops orientalis Latham, 1801	Asian green bee-eater, little green bee-eater	+	–	–	–	LC
Coracias benghalensis (Linnaeus, 1758)	Indian roller	+	–	–	–	LC

(continued)

Table 1.2 (continued)

Scientific name	Common name	MG	RS	SG	MF	IUCN status
Gallicrex cinerea (Gmelin, JF, 1789)	Watercock		–	–	+	LC
Gallinula chloropus (Linnaeus, 1758)	Common Moorhen or waterhen	+	–	–	+	LC
Fulica atra Linnaeus, 1758	Eurasian coot, common coot, or Australian coot	+	–	–	+	LC
Dicrurus macrocercus Vieillot, 1817	Black Drongo	+	–	–	–	LC
Euodice malabarica (Linnaeus, 1758)	Indian silverbill	+	–	–	–	LC
Ploceus philippinus (Linnaeus, 1766)	Baya weaver	+	–	–	–	LC
Copsychus fulicatus (Linnaeus, 1766)	Indian robin	+	–	–	–	LC
Pastor roseus (Linnaeus, 1758)	Rosy starling	+	–	–	–	LC
Acridotheres tristis (Linnaeus, 1766)	Common myna	+	–	–	–	LC
Hirundo smithii Leach, 1818	Wire-tailed swallow	+	–	–	–	LC
Cecropis daurica (Laxmann, 1769)	Red-rumped swallow	+	–	–	–	LC
Ptyonoprogne concolor (Sykes, 1832)	Dusky Crag Martin	+	–	–	–	LC
Pycnonotus cafer (Linnaeus, 1766)	Red-vented bulbul	+	–	–	+	LC
Pycnonotus leucotis (Gould, 1836)	White-eared bulbul	+	–	–	+	LC
Prinia inornata Sykes, 1832	Plain prinia	+	–	–	–	LC
Galerida cristata (Linnaeus, 1758)	Crested lark	+	–	–	–	LC
Cinnyris asiaticus (Latham, 1790)	Purple sunbird	+	–	–	–	LC
Motacilla flava Linnaeus, 1758	Western yellow wagtail	+	–	–		LC
Motacilla citreola Pallas, 1776	Citrine wagtail	+	–	–	–	LC
Motacilla maderaspatensis Gmelin, JF, 1789	White-browed wagtail	+	–	–	–	LC
Petrochelidon fluvicola (Blyth, 1855)	Streak-throated swallow or Indian cliff swallow	+	–	–	–	LC
Microcarbo niger (Vieillot, 1817)	Little cormorant or Javanese cormorant	+	–	–	+	LC
Phalacrocorax fuscicollis Stephens, 1826	Indian cormorant or Indian shag	+	–	–	+	LC
Ardea cinerea Linnaeus, 1758	Grey heron	+	–	–	+	LC
Ardea alba Linnaeus, 1758	Great egret, or white egret or great white heron	+	–	–	+	LC

(continued)

Table 1.2 (continued)

Scientific name	Common name	MG	RS	SG	MF	IUCN status
Egretta garzetta (Linnaeus, 1766)	Little egret	+	–	–	+	LC
Ardea ibis Linnaeus, 1758	Cattle egret	+	–	–	+	LC
Ardeola grayii (Sykes, 1832)	Indian pond heron	+	–	–	+	LC
Ardea purpurea Linnaeus, 1766	Purple heron	+	–	–	+	LC
Ardea intermedia Wagler, 1829	Intermediate egret, medium egret, median egret, smaller egret	+	–	–	+	LC
Egretta gularis (Bosc, 1792)	Western reef egret	+	–	–	+	LC
Pseudibis papillosa (Temminck, 1824)	Indian black ibis, or black ibis, red-naped ibis	+	–	–	+	LC
Platalea leucorodia Linnaeus, 1758	Eurasian spoonbill or common spoonbill	+	–	–	+	LC
Phalacrocorax carbo (Linnaeus, 1758)	Great cormorant or large cormorant	+	–	–	+	LC
Phoenicopterus roseus Pallas, 1811	Greater flamingo	+	–	–	+	LC

MG mangroves, *MF* mud flats, *RS* rocky shore, *SG* sea grass, *NE* not evaluated, *DD* data deficient, *LC* least concern, *NT* near threatened, *VU* vulnerable, *EN* endangered

1.3.2.1 Pagadia or Pagadiya Fish/Crab Catchers

Pagadiya fishing is traditional and one of the common fishing practices in Kachchh by various fisher folks, particularly along the mangrove, mudflats, and coastal areas. The word "pag" means foot and the term "Pagadiya" refers to crab or fish catcher who wade barefoot into shallow tidal waters, mudflats, mangroves, marshes, creeks, or estuaries to harvest fish, crustaceans, and other marine life, including mudskippers, by hand or using small nets, traps, or traditional tools. In the Gujarat coastline, particularly the Gulf of Kachchh and the Gulf of Khambhat, both areas play a crucial role in the fishing practices approximately 3371 fishermen engaging in pagadiya fishing (Sharma et al. 2017). This fishing type is deeply rooted in the cultural heritage of the region and takes place throughout the year but is more common during the off-season. In Kachchh, the harsh summer and monsoon, which make it unsafe to go into the deep sea with boats, the pagadiya fishing increases. This type of fishing helps them to survive in harsh seasons, and they get a good catch in the off-season. Both men and women who have a profound understanding and knowledge of local tides, fish behavior, and coastal ecosystems can actively work as pagadiya. The catch is generally sold by women in nearby villages and towns. Pagadiyas mostly depend on tidal flows and also the natural movement of fish during low tide (Kumaraguru et al. 2020). In a few areas, pagadiya fishing also involves the use of small boats or rafts made with simple and local equipment. Sometimes, it is largely dependent on family members or small group-based operations.

Fig. 1.3 Fishing and other activities depending on BCEs (**A**) Pagadiya fisherman collecting the mud crab. (**B**) Pagadiya's crab catch. (**C**) Mudskippers. (**D**) Bombay duck fish drying. (**E**) Kharai camel grazing on the mangrove tree leaves

The key fish species targeted by Pagadiyas is the mudskipper or scientifically known as *Boleophthalmus dussumieri* (Valenciennes, 1837), which is common in Kachchh. This fish holds significant nutritional and commercial value, locally known as "levta" or "gola" in Kachchh. Its availability is crucial for coastal communities, providing a steady source of food and income (Kanejiya et al. 2017; Kumaraguru et al. 2020). However, in Kachchh, many pagadiya use this fish as a food-trap for crabs used in crab nets in mangroves and mudflats. The crab genus *Scylla*, which is a very common catch of pagadiya known as mud crabs or swamp crabs, is economically important in mangrove-associated villages. Pagadiya use the crabs, fish, and other catches as a food source for themselves or sell in the market. This fishing provides an important source of livelihood for this coastal community by contributing significantly to the local fish trade. They also support local consumption and commercial fish markets. Unfortunately, pagadiya do not know any other work, also uneducated, which make life challenging for them. They are the most vulnerable to the impacts of climate change and also the victims of various unavoidable developmental activities. Although a few have fish boats, they prefer not to go deep-sea for fishing, instead of that they try to earn daily wages by working as labor near rising saltpans.

1.3.2.2 Local Fisherman Communities

Fisherfolk communities who fish in the seawater of nearby coasts, traditionally, believe that there are more fish available in mangroves than in bare creeks. The creeks adjacent to mangroves show a greater number of fish species in such sites. Mangroves inhibit a rich variety of crabs, bivalves, gastropods, and various species of fish. Fishermen communities get many fish species in the mangrove adjacent creeks and seawater. As the fisheries sector of Kachchh primarily runs within an extensive intertidal zone covering the creeks, mudflats, and estuaries in the region, the majority of the fishermen community is dependent on BCEs. The present BCEs can provide an ideal environment for traditionally operated fishing practices. The major fishing activities consist of the small-scale fisheries, in which diesel motorized boats are involved. The major part of the fish catch can be used for animal feed, particularly for domestic fowls and other chickens. Commercial species of fishery catch are golden anchovies, Bombay duck, and prawns.

1.3.2.3 Camel Herders

In Kachchh, the nomadic pastoralist communities of Rabaris and Jaaths keep a local camel breed called Kharai that feeds on the mangroves. These Kharai camels have special features that enable them to swim across the coastal waters to graze and feed on mangroves. Kharai camel is one of the two camel breeds in the Gujarat state of India, which is genetically different from other Indian camel breeds (Patel et al. 2015). They are well adapted to survive in mangrove habitat, perhaps dependent on mangroves for survival (Sahajeevan 2021). Its adaptation is not limited to food preference but also for extreme climatic conditions in the desert, high salinity environment, and for movement in shallow or deep-sea water. Kharai camel is native to Gujarat, literally meaning "camel of saline environment" (khara means saline in Gujarati), also known as "dariya-taru" meaning "sea swimmers" (dariya means sea and taru means "who can swim"). This is only a breed of camels that shows a special ability to swim in seawater; however, it was found to swim up to three kilometers. This breed of camel can easily swim in search of mangroves and other halophytes. It can easily navigate the wet, salty coastal land through its gently padded hooves. The camel grazer communities claim that their camels have been grazing on *A. marina* for a long time in history. There has been a major decline in the population of this camel in the last few years. The requirement of land for industrial and other developmental activities created a huge threat to the mangroves. Almost all these activities are creating pressure on the mangrove ecosystem, including direct cutting of the mangrove forest, which results in the availability of fodder for the Kharai camels. The number of Kharai camels has declined as their grazing grounds have shrunk. In the harsh summer and dry conditions of Kachchh, mangroves are the only best source of fodder to camels and other cattle. However, a few studies from outside India report that grazing on mangroves by cattle affects the density of mangroves and is also responsible for the changes in the growth characteristics of mangroves in the grazing sites (Minchinton et al. 2019).

1.4 Conclusion

Blue carbon ecosystems in the Kachchh district of Gujarat mainly include mangroves and salt marshes, which habitat halophytic species, a few small patches of seagrasses, and a few seaweed species including the drifted seaweeds. These BCEs can provide essential ecosystem services beyond the coastal resilience, carbon sinks, etc. The most important role of these ecosystems is to support the local biodiversity by supporting a diverse marine life by serving as nurseries for fish and other species. These can enhance the fishery resources that sustain the local economy. These BCEs provide livelihood support, such as economic opportunities through fisheries, sustainable resource use, and also hold cultural and recreational significance. However, there are a few gaps in Kachchh BCE studies.

Acknowledgements DD is thankful to the Government of Gujarat, India, for providing a doctoral research fellowship under the SHODH (Scheme of Developing High-quality Research—Reference Number 202301504) program. RR is grateful to the Government of India, specifically the Secretary of the Ministry of Earth Sciences (MoES), New Delhi, and the Head of the Centre for Marine Living Resources and Ecology (CMLRE), for the support and facilities provided during the study (CMLRE contribution number 215). The authors are thankful to all reviewers for their suggestions to make this work publishable.

Declaration of Conflict of Interest There are no conflicts of interest to declare.

Ethics Approval This work does not contain any experimental studies with live animals.

Declaration Statement The confidentiality and anonymity of the data collected were assured to all participants.

Data Availability Statement The original contributions presented in the study are included in the chapter. Further inquiries can be directed to the corresponding author.

Funding No funding is received for this work.

References

Ashton EC, Hogarth PJ, Macintosh DJ (2003) A comparison of Brachyuran Crab Community structure at four mangrove locations under different management systems along the melaka straits-Andaman Sea Coast of Malaysia and Thailand. Estuaries 26:1461. https://doi.org/10.1007/BF02803654

Baba M, Nayak (2002) Muddy coasts of India. Muddy Coasts World Process Depos Funct

Behera DP, Ingle KN, Mathew DE et al (2022) Epiphytism, diseases and grazing in seaweed aquaculture: a comprehensive review. Rev Aquac 14:14

Bell K, Nuyts S, Bie C et al (2025) Threatened fauna that use blue carbon ecosystems: a review from Australia. Biol Conserv 303:111030. https://doi.org/10.1016/J.BIOCON.2025.111030

Bhaskar S (1984) The distribution and status of sea turtles in India. In: Silas EG (ed) On the workshop on sea turtle conservation. Central Marine Fisheries Institute Special Publication No.18, Kochi, pp 21–35

Bosire JO, Dahdouh-Guebas F, Kairo JG et al (2004) Spatial variations in macrobenthic fauna recolonisation in a tropical mangrove bay. Biodivers Conserv 13:1059. https://doi.org/10.1023/B:BIOC.0000018149.88212.2d

Casazza TL, Ross SW (2008) Fishes associated with pelagic Sargassum and open water lacking Sargassum in the Gulf Stream off North Carolina. Fish Bull 106

Choudhary B, Dhar V, Pawase AS (2024) Blue carbon and the role of mangroves in carbon sequestration: Its mechanisms, estimation, human impacts and conservation strategies for economic incentives. J Sea Res 199:102504. https://doi.org/10.1016/J.SEARES.2024.102504

Coston-Clements L, Center LR, Hoss DE, Cross FA (1991) Utilization of the Sargassum habitat by marine invertebrates and vertebrates: a review. NOAA Tech Memo NMFS-SEFSC-296

Dittmann S (1996) Effects of macrobenthic burrows on infaunal communities in tropical tidal flats. Mar Ecol Prog Ser 134:119. https://doi.org/10.3354/meps134119

Du J, Hu W, Nagelkerken I et al (2020) Seagrass meadows provide multiple benefits to adjacent coral reefs through various microhabitat functions. Ecosyst Heal Sustain 6:1812433

Duarte CM, Losada IJ, Hendriks IE et al (2013) The role of coastal plant communities for climate change mitigation and adaptation. Nat Clim Chang 311(3):961–968. https://doi.org/10.1038/nclimate1970

Duke NC, Meynecke J-O, Dittmann S et al (2007) A world without mangroves? Science (80-) 317. https://doi.org/10.1126/science.317.5834.41b

Dwijaya BA, Krisnanti SA, Ramdhani A (2021) Contribution of the Mangrove ecosystem to the Gastropod community at pantai Bahagia vilage, Muara Gembong, West Java. Genbinesia. J Biol 1. https://doi.org/10.55655/genbinesia.v1i1.2

Godínez-Ortega JL, Cuatlán-Cortés JV, López-Bautista JM, van Tussenbroek BI (2021) A Natural History of Floating Sargassum Species (Sargasso) from Mexico. In: Natural History and Ecology of Mexico and Central America

GUIDE (2002) Kachchh Sub-State Biodiversity Strategy and Action Plan. Prepared under National Biodiversity Strategy and Action Plan, executed by Ministry of Environment and Forests (Government of India), technical implementation by Technical and Policy Core Group (TPC. Bhuj

Gutiérrez JL, Jones CG, Byers JE, et al (2012) Physical ecosystem engineers and the functioning of estuaries and coasts. In: Treatise on Estuarine and Coastal Science

Hill R, Bellgrove A, Macreadie PI et al (2015) Can macroalgae contribute to blue carbon? An Australian perspective. Limnol Oceanogr 60:1689–1706. https://doi.org/10.1002/LNO.10128

Hookham B, Shau-Hwai AT, Dayrat B, Hintz W (2014) A baseline measure of tree and gastropod biodiversity in replanted and natural mangrove stands in Malaysia: Langkawi Island and sungai merbok. Trop Life Sci Res 25:1

Imamsyah A, Arthana IW, Astarini IA (2020) The influence of physicochemical environment on the distribution and abundance of mangrove gastropods in Ngurah Rai Forest Park Bali, Indonesia. Biodiversitas 21. https://doi.org/10.13057/biodiv/d210740

Ingle KN, Polikovsky M, Chemodanov A, Golberg A (2018) Marine integrated pest management (MIPM) approach for sustainable seagriculture. Algal Res 29:223. https://doi.org/10.1016/j.algal.2017.11.010
Ingle KN, Traugott H, Golberg A (2020) Challenges for marine macroalgal biomass production in Indian coastal waters. Bot Mar 63:327
Jamilah M, Becker A, Lee SL et al (2025) Socioeconomic impacts linked to land use and land use changes affecting blue carbon ecosystems in Southeast Asia: a systematic map. Ocean Coast Manag 267:107643. https://doi.org/10.1016/J.OCECOAMAN.2025.107643
Jiang R, Ingle KN, Golberg A (2016) Macroalgae (seaweed) for liquid transportation biofuel production: What is next? Algal Res 14:48
Jiang Z, Huang D, Fang Y et al (2020) Home for marine species: seagrass leaves as vital spawning grounds and food source. Front Mar Sci 7:194
Kairo JG, Lang'at JKS, Dahdouh-Guebas F et al (2008) Structural development and productivity of replanted mangrove plantations in Kenya. For Ecol Manag 255:2670. https://doi.org/10.1016/j.foreco.2008.01.031
Kanejiya J, Solanki D, Beleem I, Gohil B (2017) Mudskipper fishing in the coast of Bhavnagar, Gulf of Khambhat, Gujarat, India. J Fish 5. https://doi.org/10.17017/jfish.v5i1.2017.179
Kathiresan K (2022) Mangrove forests of India: an overview. In: Mangroves: biodiversity, livelihoods and conservation
Krause-Jensen D, Duarte CM (2016) Substantial role of macroalgae in marine carbon sequestration. Nat Geosci 910(9):737–742. https://doi.org/10.1038/ngeo2790
Kristensen E (2008) Mangrove crabs as ecosystem engineers; with emphasis on sediment processes. J Sea Res 59:30. https://doi.org/10.1016/j.seares.2007.05.004
Kristensen E, Alongi DM (2006) Control by fiddler crabs (Uca vocans) and plant roots (Avicennia marina) on carbon, iron, and sulfur biogeochemistry in mangrove sediment. Limnol Oceanogr 51:1557. https://doi.org/10.4319/lo.2006.51.4.1557
Kumaraguru A, Mary RC, Saisaraswathi V (2020) A review about fish walking on land. J Threat Taxa 12:17276. https://doi.org/10.11609/jott.6243.12.17.17276-17286
Lee SY (2008) Mangrove macrobenthos: Assemblages, services, and linkages. J Sea Res 59:16. https://doi.org/10.1016/j.seares.2007.05.002
Lehahn Y, Ingle KN, Golberg A (2016) Global potential of offshore and shallow waters macroalgal biorefineries to provide for food, chemicals and energy: Feasibility and sustainability. Algal Res 17:150. https://doi.org/10.1016/j.algal.2016.03.031
Lovelock CE, Duarte CM (2019) Dimensions of blue carbon and emerging perspectives. Biol Lett 15:20180781
Lovelock CE, Reef R (2020) Variable impacts of climate change on blue carbon. One Earth 3:195–211
Macreadie PI, Costa MDP, Atwood TB et al (2021) Blue carbon as a natural climate solution. Nat Rev Earth Environ 2:826–839
McConville A, Law BS, Mahony MJ (2013) Mangroves as maternity roosts for a colony of the rare east-coast free-tailed bat (Mormopterus norfolkensis) in south-eastern Australia. Wildl Res 40:318–327
Menéndez P, Losada IJ, Torres-Ortega S et al (2020) The global flood protection benefits of mangroves. Sci Rep 10:4404
Minchinton TE, Shuttleworth HT, Lathlean JA et al (2019) Impacts of cattle on the vegetation structure of mangroves. Wetlands 39:1–9
Nagelkerken I, Blaber SJM, Bouillon S et al (2008) The habitat function of mangroves for terrestrial and marine fauna: a review. Aquat Bot 89:155–185. https://doi.org/10.1016/J.AQUABOT.2007.12.007
Neukermans G, Dahdouh-Guebas F, Kairo JG, Koedam N (2008) Mangrove species and stand mapping in gazi bay (Kenya) using quickbird satellite imagery. J Spat Sci 53:75. https://doi.org/10.1080/14498596.2008.9635137
Nordlund LM, Jackson EL, Nakaoka M et al (2018) Seagrass ecosystem services – what's next? Mar Pollut Bull 134:145. https://doi.org/10.1016/j.marpolbul.2017.09.014

Patel AC, Jisha TK, Upadhyay D et al (2015) Molecular characterization of camel breeds of Gujarat using microsatellite markers. Livest Sci 181:85–88

Peng Y, Zhang M, Lee SY (2017) Food availability and predation risk drive the distributional patterns of two pulmonate gastropods in a mangrove-saltmarsh transitional habitat. Mar Environ Res 130:21. https://doi.org/10.1016/j.marenvres.2017.07.005

Penha-Lopes G, Bouillon S, Mangion P et al (2009) Population structure, density and food sources of Terebralia palustris (Potamididae: Gastropoda) in a low intertidal Avicennia marina mangrove stand (Inhaca Island, Mozambique). Estuar Coast Shelf Sci 84:318. https://doi.org/10.1016/j.ecss.2009.04.022

Prasanna J, Seshapriya V, Anand M et al (2017) Biodiversity assessment of crabs, gastropods and bivalves in Chinnapalam creek, of the southeast coast of India. Indian J Geo-Marine Sci 46

Purnama MF, Prayitno SB, Muskananfola MR, Suryanti (2024) Tropical gastropod density and diversity in the mangrove forest of Totobo Village, Southeast Sulawesi, Indonesia. Biodiversitas 25:1663–1675. https://doi.org/10.13057/BIODIV/D250436

Reis A, Alves AT, Dórea A et al (2021) Distribution and movement of the mangrove gastropod Littoraria angulifera. Estuar Coast Shelf Sci 250:107145. https://doi.org/10.1016/j.ecss.2020.107145

Rog SM, Clarke RH, Cook CN (2017) More than marine: revealing the critical importance of mangrove ecosystems for terrestrial vertebrates. Divers Distrib 23:221–230

Sahajeevan (2021) Saving a mangrove forest: camel pastoralists' fight to save the Kachchh mangroves, Bhuj

Sahu SK, Ingle KN, Mantri VA (2020) Epiphytism in seaweed farming: causes, status, and implications

Sharma H, Swain M, Kalamkar SS (2017) Assessment of economic losses due to inadequate post-harvest infrastructure facilities for marine fisheries in Gujarat. Econ Aff 62:1. https://doi.org/10.5958/0976-4666.2017.00020.1

Sousa WP, Dangremond EM (2012) Trophic interactions in coastal and estuarine mangrove forest ecosystems. In: Treatise on estuarine and coastal science

Teoh HW, Sasekumar A, Ismail MH, Chong VC (2018) Trophic discrimination factor and the significance of mangrove litter to benthic detritivorous gastropod, Ellobium aurisjudae (Linnaeus). J Sea Res 131:79. https://doi.org/10.1016/j.seares.2017.11.005

van Tussenbroek BI, Monroy-Velázquez LV, Rodríguez D et al (2024) Monitoring drift and associated biodiversity of nearshore rafts of holopelagic Sargassum spp. in the Mexican Caribbean. Aquat Bot 195:103792. https://doi.org/10.1016/J.AQUABOT.2024.103792

Warren JH, Underwood AJ (1986) Effects of burrowing crabs on the topography of mangrove swamps in New South Wales. J Exp Mar Bio Ecol 102:223. https://doi.org/10.1016/0022-0981(86)90178-4

Witherington B, Hirama S, Hardy R (2012) Young sea turtles of the pelagic Sargassum-dominated drift community: Habitat use, population density, and threats. Mar Ecol Prog Ser 463:1. https://doi.org/10.3354/meps09970

Yadav R, Malla PK, Dash D et al (2019) Diversity of gastropods and bivalves in the mangrove ecosystem of Paradeep, east coast of India: a comparative study with other Indian mangrove ecosystems. Molluscan Res 39:325. https://doi.org/10.1080/13235818.2019.1644701

Chapter 2
Blue Carbon Potential of Indian Mangroves, Seagrasses, and Salt Marshes: A Review of Current Knowledge and Future Directions for Climate Change Mitigation

Thanamegam Kaviarasan, Palanisamy Thanabalan, S. R. Marigoudar, and M. V. Ramana Murthy

Abstract Blue carbon ecosystems, comprising mangroves, seagrasses, and salt marshes play an essential role in combating climate change through carbon sequestration. Blue carbon ecosystems absorb atmospheric carbon dioxide (CO_2) and store it in sediments, and they help mitigate climate change by preventing carbon release. A comprehensive review and understanding of major blue carbon ecosystems, such as mangroves, seagrasses, and salt marshes are crucial for policymaking and the development of mitigation strategies. Hence, this review aims to analyze research publications and reports on the blue carbon ecosystem along the Indian coast to provide valuable inputs for policy and mitigation efforts. Research data from various peer-reviewed journals and the Scopus database have been utilized for this review. The results show that the research along India's coastline has demonstrated the significant carbon storage capacity of these ecosystems. However, blue carbon ecosystems face substantial threats from coastal erosion, climate change, natural disasters, and anthropogenic activities, emphasizing the need for immediate conservation and restoration efforts. To protect these ecosystems, a multi-faceted approach is necessary, involving: raising awareness among local communities and stakeholders; encouraging their active participation; providing economic incentives; and developing and enforcing policies. Ongoing research is crucial to enhance our knowledge of the function of blue carbon ecosystems, and assess the impact of emerging pollutants including plastic litter.

Keywords Blue carbon · Coastal wetlands · Above-ground biomass · Below-ground biomass · Climate change

---

T. Kaviarasan · P. Thanabalan · S. R. Marigoudar (✉) · M. V. R. Murthy
National Centre for Coastal Research (NCCR), Ministry of Earth Sciences, Government of India, NIOT Campus, Chennai, India
e-mail: srmarigoudar@nccr.gov.in

B. P. Barathan et al. (eds.), *Navigating Climate Change: Impacts on Biodiversity and Ecosystem Resilience*, Sustainable Development and Biodiversity 42,
https://doi.org/10.1007/978-981-95-0409-1_2

2.1 Introduction

Mangroves, salt marshes, seaweeds, coral reefs, and seagrass beds are examples of coastal and marine ecosystems that store and sequester carbon, known as "blue carbon." These habitats are essential for reducing the effects of climate change, as they absorb and store atmospheric CO_2 (Choudhary et al. 2024). Blue carbon ecosystems comprise only 0.2% of the ocean surface area, yet provide more than half of the carbon storage, indicating enormous potential and advantages in carbon storage (Duarte et al. 2013). The importance of blue carbon in climate change mitigation cannot be overstated, as blue carbon ecosystems can sequester significant amounts of carbon from the atmosphere, which can help reduce the impacts of climate change, known as carbon sequestration (Sundquist et al. 2008). Soils and sediments store most of the trapped carbon, leading to long-term carbon accumulation (Chmura et al. 2003).

Blue carbon ecosystems provide habitats for a diverse range of plant and animal species (Kathiresan and Bingham 2001; Sandilyan and Kathiresan 2012; Morrison et al. 2014; Silliman 2014; Teagle et al. 2017). Blue carbon ecosystems offer numerous benefits that underpin sustainable livelihoods, including fisheries (Kathiresan and Rajendran 2002; Hutchison et al. 2014), tourism (Spalding and Parrett 2019; Mitra 2020), and shoreline defense (Das and Vincent 2009). Biodiversity plays an essential part in sustaining the stability and resilience of the blue carbon environment. Crabs are the primary bioturbators in mangrove ecosystems, while amphipods, shrimp, lobsters, worms, bivalves, and various fish species aid in sediment particle dispersal. These organisms significantly influence nutrient and carbon cycling in sediments (Sarker et al. 2021). High biodiversity enhances ecosystem resilience and helps overcome various environmental challenges. ShivaShankar et al. (2019) observed that *Rhizophora apiculata*, *R. mucronata*, and *Avicennia* spp. are the most abundant mangroves, with *R. mucronata* present at all eight newly formed sites post-tsunami. Protecting and restoring mangrove biodiversity, particularly by prioritizing site-specific dominant species alongside others with contrasting functional traits, could maximize their capacity for climate change mitigation by increasing carbon storage (Rahman et al. 2021).

The livelihoods of three billion people rely on marine and coastal resources, which contribute US$3 trillion to the global economy annually, accounting for roughly 5% of the global Gross Domestic Product (GDP) (Mejjad et al. 2025). Despite their significance, recent trends or data illustrate the increased threats to blue carbon ecosystems. The extensive devastation of these ecosystems in recent times, including aquaculture, coastal development, and other uses (Duarte 2002; Bhomia et al. 2016; Rastogi et al. 2021), has put at risk their capacity to sequester carbon in the future, in addition to releasing stored carbon. Furthermore, these ecosystems suffer greatly from land-based contamination, particularly plastic and microplastic pollution (Adyel and Macreadie 2022; Kvale 2022; Noman et al. 2024), which reduces their capacity to store carbon. Keeping this in the backdrop, the present review seeks to summarize the current state of blue carbon research, identify key threats and risks, and propose conservation strategies and future directions for blue carbon management in Indian coastal wetlands.

2.2 Methodology

This review examines research publications from various government reports, peer-reviewed journals, and the Scopus database, which were used to collect information on blue carbon ecosystem research.

2.3 Indian Coastal Wetlands

2.3.1 Evolution of Remote Sensing Aspects in India's Coastal Wetlands

In the early stages (1970s–1980s), Landsat satellite imagery was used to map the wetland with large spatial coverage at a global scale. Then, airplanes began to enter existence for medium spatial coverage at the country scale for wetland mapping. After the late 1990s, ISRO launched Indian Remote Sensing (IRS) satellites (IRS-1A, IRS-1B, IRS-P3) with improved spatial and spectral resolution. Initial studies focused on important wetlands on India's east coast, Chilika Lake in Odisha, and Sundarbans in West Bengal. An expansion of the National Wetland Inventory and Assessment (NWIA) was initiated by the Space Applications Centre (SAC-ISRO) to conduct comprehensive wetland mapping using IRS-LISS III optical remote sensing data. The program focused on wetland classification and hydrological modeling, and in the twentieth century, advanced monitoring with high-resolution satellite imagery imaging with two different sensors, such as Hyperspectral and Synthetic Aperture Radar (SAR), a microwave remote sensing datasets are used for wetland vegetation, and hydrology assessments. The hyperspectral data can distinguish continuous different spectral signature to characterize the vegetation associated with wetlands such as mangroves, dunes, and marshland, etc., which are mapped with different spatial and temporal scales. Also, active SAR complemented with day/night observations which are free from cloud cover. These SAR datasets are effectively utilized to understand the root zone level soil moisture estimation. The recently launched NISAR L&S band SAR data having high spatial resolution, enabling accurate soil moisture estimation, wetland mapping with different seasonal observations. It helps to map the wetland area and the standing vegetal growth as well. The European Space Agency also provides free Sentinel satellite datasets for detailed monitoring. In recent times, an Integration of Artificial Intelligence, cloud computing, and deep learning (e.g., Google Earth Engine) has come up with automated wetland change detection. Government initiatives like the

National Wetland Conservation Programme (NWCP) and Wetlands (Conservation and Management) Rules, 2017, strengthen policy-driven wetland protection and monitoring with standard SOPs. In the future, the use of Unmanned Aerial Vehicles (drones) can be taken up for real-time wetland monitoring. The use of drones has twin benefits (cost-effective and high resolution in cm scale) and can give empathized information much more precise for seagrass monitoring for predictive wetland conservation. Hence, India's remote sensing capabilities have significantly improved wetland conservation efforts, ensuring sustainable management and biodiversity protection across diverse wetland ecosystems.

There are several remote sensing indices used to study wetland ecosystems, focused on vegetation fraction, water availability, and soil moisture conditions. Few indices are capable of understanding the wetland through the spectral signature and features.

Normalized Difference Vegetation Index (NDVI).

NDVI = ((NIR – Red) / (NIR + Red)). It helps to differentiate standing vegetation and water bodies and identify the spatial density in wetlands, where NIR is the near-infrared band.

Normalized Difference Water Index (NDWI) and Modified Normalized Difference Water Index (MNDWI).

NDWI = ((NIR – SWIR) / (NIR + SWIR)) & MNDWI = ((Green – SWIR) / (Green + SWIR)). It highlights the water spread area with different time steps, which can be effectively used for watershed mapping and also for tracking wetland water level changes over different times in mixed natural environments, where SWIR is the short-wave-infrared band.

Enhanced Vegetation Index (EVI).

EVI = 2.5 * ((NIR – Red) / (NIR + 6 * Red – 7.5 * Blue) + 1). It helps in analyzing and reducing atmospheric influences and getting clear vegetation fractions in wetland mapping.

Soil-Adjusted Vegetation Index (SAVI).

SAVI = ((NIR – Red) / (NIR + Red + L)) *(1 + L). It helps to address the sparse vegetation areas by adjusting for soil brightness effects, which helps to differentiate current fallow land from barren land in the Land Use Land Cover (LULC) mapping. These indices can distinguish the semi-arid wetland regions with sparse plant cover, where L is a constant factor (0.5), is to minimize the soil brightness on the vegetation Index.

Floating Algae Index (FAI).

FAI = RNIR − R′NIR. where $R'NIR = R_{RED} + (R_{SWIR} - R_{RED}) \times (\lambda_{NIR} - \lambda_{RED}) / (\lambda_{SWIR} - \lambda_{RED})$, where RNIR′ is the important baseline reflectance in the NIR band derived from a linear interpolation between the red and SWIR bands. It is defined to identify floating vegetation like algae blooms and detect algal growth in wetlands.

Also, the Land Surface Temperature (LS), NDVI (N) and Rainfall (R) LSNR model (Thanabalan et al. 2023) uses parameters such as temperature, rainfall, and NDVI to understand the influence of climate parameters and their influence in identifying the wet and dry spells for both pre- and post-monsoon periods. This will be of use in long-term planning and preparedness to reduce vegetation losses in the coastal wetland ecosystems. Hence, the vegetal indices from the optical satellite sensor, surface temperature from the thermal sensor, hyperspectral sensor for mangrove mapping, rainfall from the passive microwave remote sensing sensor, and soil moisture estimate from active microwave remote sensing sensors are being used as indicators for various kinds of applications such as crop assessment mapping, drought monitoring, and coastal vegetation conditions. Overall, each of these remote sensing indices serves a diverse role in wetland monitoring, by detecting water and standing vegetation health conditions and assessing them through hydrological observations and long-term monitoring of wetland conservation.

2.3.2 *Overview and Distribution*

India's coastline stretches 11098.81 kms (www.surveyofindia.gov.in), encompassing nine maritime states and four union territories. The coastline is diverse and, consists of sandy beaches (43%), mudflats (36%), rocky shores (11%), and marshy coasts (10%) (Kumar et al. 2006; Ajai et al. 2012; Kankara et al. 2018). The Indian coastline is home to diverse marine and coastal wetland ecosystems, encompassing a sand/beach, lagoons, creeks, salt marshes, intertidal mudflats, coral reefs, mangroves, aquaculture ponds, and salt pans covering approximately 4,140,116 ha. Intertidal mudflats and mangroves dominate these ecosystems (Indianwetlands.in 2025), playing a vital role in the country's ecosystem services and significantly contributing to global carbon sequestration efforts.

2.3.3 *Mangrove*

Mangrove cover in India has fluctuated over the years. The total area has increased and decreased alternately, indicating a mix of conservation efforts and degradation. From 1957 to 1987, the mangrove cover expanded significantly, peaking at 6819.76 square kilometers (km^2) in 1963 (Sidhu 1963; Blasco 1977; Mathauda 1987; Untawale 1987). In 1987, seven percent of the mangroves occurred in India, the country has about 350 species of mammals, 1200 species of birds, and more than 20,000 species of insects (The State of Forest Report 1987). However, a decline was observed from 1987 to 2005, with the lowest of 2874.58 km^2 in 1980 (Jagtap et al. 1993; Venkataraman and Wafar 2005). Since then, the mangrove cover has gradually increased, reaching 4992 km^2 in 2024 (Kathiresan and Dagar 2024). Notably, the highest mangrove cover area was recorded in 1963, whereas the lowest was recorded in 1980.

Mangroves cover approximately 14.7 million hectares (Mha) globally, comprising 83 true species and over 100 associated species (Tomlinson 2016; Kathiresan and Dagar 2024). India boasts the highest plant diversity, with 43 mangrove species and approximately 100 associated species. India's most impressive mangrove forests are found in West Bengal (Sundarbans), covering 44% of India's total mangrove cover. Gujarat and the Andaman and Nicobar Islands follow, accounting for 23% and 13%, respectively. In terms of density, 31% of India's mangrove forests (1472 km^2) are classified as very dense, while 29% (1391 km^2) are classified as moderately dense (Kandasamy 2017). The forest cover in a mangrove in 2023 was estimated at 4991.68 km^2, and it was found that over a period, *i.e.,* during the decades 2013–2023, the mangrove cover of the country has increased by 296.33 km^2 (Forest Survey of India 2023).

2.3.4 Seagrass

India's coastal states and union territories have substantial seagrass distribution, covering approximately 51,822 ha. This expansive seagrass ecosystem is fragmented into 4162 distinct patches, highlighting the diverse and widespread nature of these critical marine habitats (Ramesh et al. 2024). India is home to a notable array of seagrass species, with 16 varieties found in various locations, including backwaters, lagoons, and estuaries. Despite limited research, these ecosystems support 1250 species, including endangered dugongs and green turtles (Thangaradjou and Bhatt 2018). Notably, the Ramanathapuram coastal district in Tamil Nadu stands out, boasting an impressive 1363 seagrass patches that cover an expansive 22,790 ha. This remarkable concentration accounts for 48% of India's total seagrass distribution area (Ramesh et al. 2024).

2.3.5 Salt Marsh

Salt marshes of India are scattered across various states, including Gujarat, Maharashtra, Tamil Nadu, Puducherry, Andhra Pradesh, Daman and Diu, and the Andaman and Nicobar Islands (Patro et al. 2017), which include 15 species, in 744 numbers covering an area of approximately 1600 km^2 (National Wetland Atlas 2011; Banerjee et al. 2017). But in 2020, India's salt marshes covered an estimated 290 km^2. The overall density of salt marsh vegetation averages around 19 plants per square meter (m^2). However, regional variations are notable, with Lakshadweep recording the highest density at 28 plants/m^2, followed by Tamil Nadu (23 plants/m^2), West Bengal (19 plants/m^2), Maharashtra (18 plants/m^2), Andhra Pradesh (17 plants/m^2), Gujarat (15 plants/m^2), and Puducherry (10 plants/m^2) (Viswanathan et al. 2020), taking into account surveys conducted for status evaluation at designated locations in different seasons, such as November 2014 to July 2018.

2.4 Research on Blue Carbon: A Bibliometric Analysis

2.4.1 *Bibliometric Analysis*

A search of the Scopus database was conducted until December 20, 2024, using the following keywords: "blue carbon," "marine," "mangrove," "salt marsh," "seagrass," and "seaweeds". Search terms were required to appear in the title, abstract, and keywords. Literature selection prioritized studies related to blue carbon, marine, and coastal ecosystems.

The retrieved results were analyzed using VOS Viewer software to perform bibliographic coupling with sources, authors, organizations, and countries. The literary data analysis generated citation, co-occurrence, and coupling networks, providing insights into the research landscape. This was achieved through a quantitative and objective approach, utilizing co-occurrence analysis, co-citation analysis, and assessments of author collaborations, journal sources, and publishing trends.

2.4.2 *Global Bibliometric Analysis on Blue Carbon Research*

The bibliometric analysis revealed that the majority of publications on blue carbon research worldwide are research articles (82.1%), followed by reviews (12.5%) coming in second. About 3.7% are book chapters, with the remaining 0.1 to 0.8% consisting of conference papers, comments, book chapters, brief surveys, data articles, and editorials.

The top journals (Fig. 2.1) publishing blue carbon research are led by Estuarine, Coastal and Shelf Science with 111 papers, followed closely by Science of the Total Environment with 106 papers. Estuaries and Coasts, Marine Ecology Progress Series, and Marine Pollution Bulletin also ranked among the top five, with 70, 51, and 45 papers, respectively. Other notable journals include Frontiers in Marine Science, Global Change Biology, Hydrobiologia, Wetlands, and Marine and Freshwater Research, highlighting the diverse range of scientific outlets that contribute to blue carbon research.

The top affiliations (Fig. 2.2) for blue carbon research are predominantly based in Australia, with Griffith University (89 publications), The University of Queensland (85 publications), Deakin University (61 publications), Southern Cross University (57 publications), James Cook University (53 publications), and Macquarie University (47 publications) ranking among the top institutions. The Ministry of Education of the People's Republic of China (55 publications), Chinese Academy of Sciences (53 publications), and United States Geological Survey (52 publications) also feature prominently, highlighting the global scope of blue carbon research. India has a vast coastline of approximately 11098.81 kms, which is home to diverse marine ecosystems, including mangroves, coral reefs, and salt marshes.

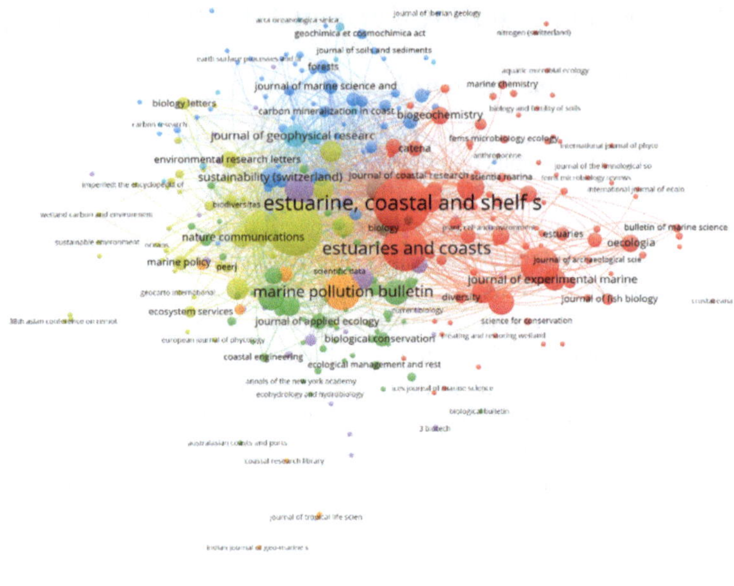

Fig. 2.1 Visualization of top journals publishing blue carbon research

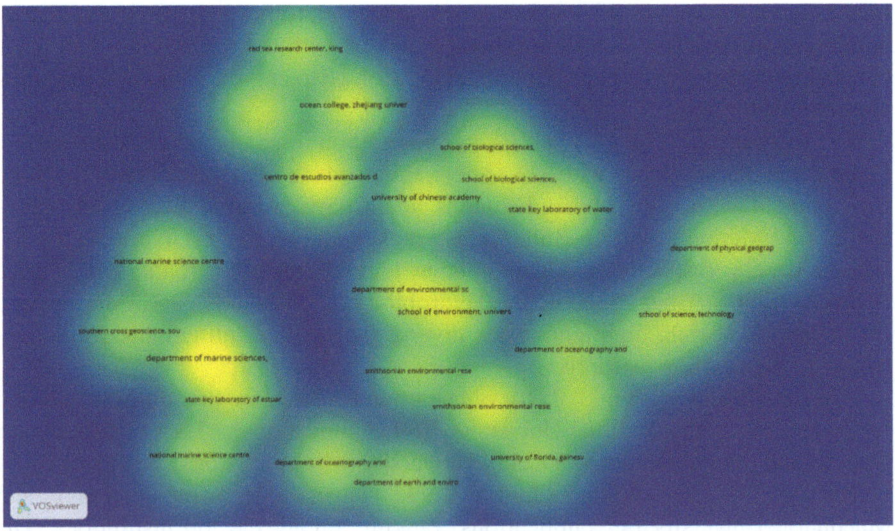

Fig. 2.2 Blue carbon research institutional clusters: a density visualization of leading publishing organizations

Given the significance of these ecosystems in blue carbon sequestration, it is surprising that Indian institutions are not more prominently represented in the top affiliations for blue carbon research. This highlights a potential opportunity for Indian researchers and institutions to contribute to this critical field of study.

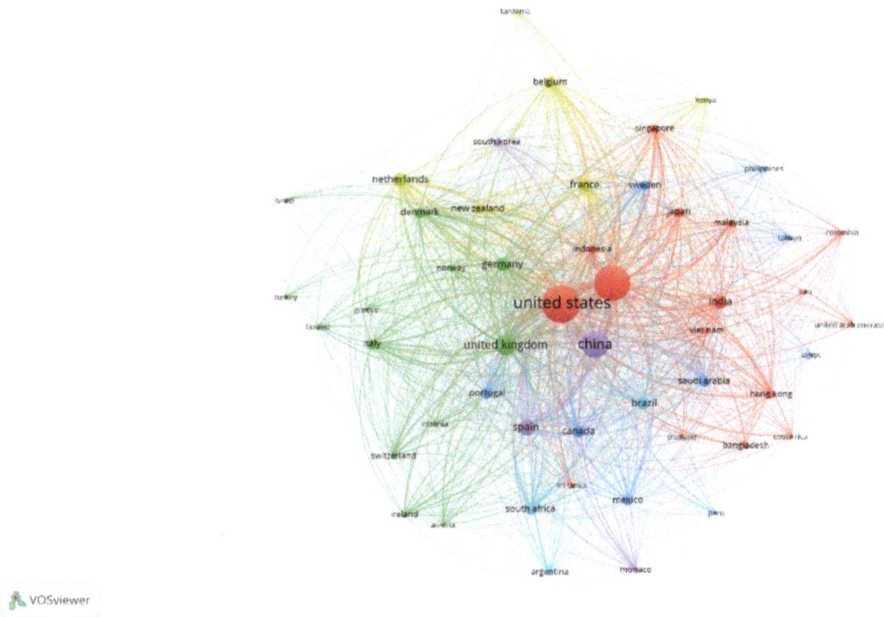

Fig. 2.3 Density visualization of blue carbon research clusters by country of publication

According to a global distribution (Fig. 2.3) of publications on blue carbon research, the United Nations produces the most (525 publications), closely followed by Australia (454 publications). The leading contributors are Spain (108 publications), the United Kingdom (150 publications), and China (265 publications). With 67 articles, India stands eighth, demonstrating a noteworthy but comparatively modest contribution to the worldwide blue carbon research community. And most of the. Keywords which used in publications are mangrove, salt marsh, and wetland seagrass (Fig. 2.4).

2.4.3 India Bibliometric Analysis on Blue Carbon Research

India's blue carbon research publications comprise primarily articles (52.24%), followed by reviews (25.37%) and book chapters (19.40%). Conference papers and data papers account for a smaller proportion, each contributing 1.49% to the total. This breakdown suggests that original research articles and review papers dominate India's blue carbon research landscape. The Indian Institute of Technology Bhubaneswar contributed to India's blue carbon research, accounting for 11.94% of the country's publications. Annamalai University (10.45%) and Jadavpur University (8.96%) closely followed. Other prominent institutions include Anna University (7.46%), Bharathidasan University (7.46%), and the National Institute of Oceanography (5.97%). These groups are spearheading blue carbon research in

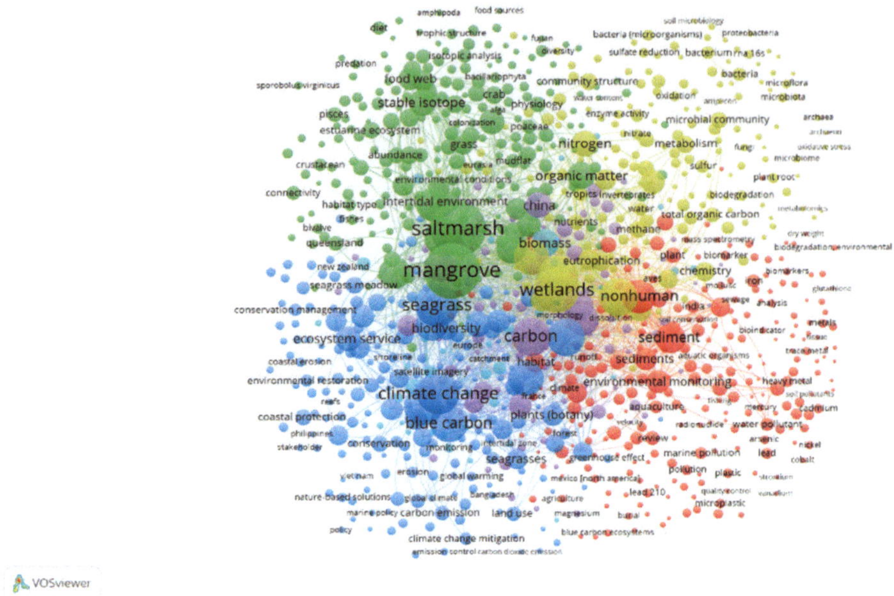

Fig. 2.4 Density visualization of clusters for frequently used keywords in blue carbon research publications

India to tackle the pressing issues of climate change mitigation and sustainable coastal management. Prominent institutions are at the forefront of India's blue carbon research ecosystem, which is distinguished by a heavy emphasis on original research articles and review papers. India's blue carbon research initiatives are a crucial step in tackling the urgent problems of climate change and sustainable coastal management, although there is still an opportunity for development and greater cooperation. India's blue carbon research initiatives will be crucial in guiding policy decisions and promoting sustainable development as it continues to negotiate its place in the global climate governance process.

2.5 Blue Carbon in Indian Coastal Wetlands: Current Knowledge

2.5.1 Mangrove Ecosystem

2.5.1.1 Sundarbans and Other Regions in West Bengal Mangrove Ecosystem

Between 1975 and 2013, the Indian Sundarban lost 107 km^2 of mangrove area, resulting in significant CO_2 emissions (1567.98 gigagram -Gg), which is about a social cost of \$64.29 M (Akhand et al. 2017). Based on the State Forest Report by the Forest Survey of India (2023), the mangrove cover in West Bengal expanded from 2117.75 km^2 in 2021 to 2119.16 km^2 in 2023, reflecting a measurable increase in mangrove cover over this period. Based on data from 2010 to 2011, the Indian Sundarbans, spanning 4264 km^2, sequestered 2.79 Tg C (teragrams of carbon), equivalent to 0.64% of India's annual fossil fuel emissions (Ray and Jana 2017). Chowdhury et al. (2018) found that planting *Avicennia marina* using the "propagule dibbling" method is the most cost-effective approach, yielding significant increases in organic carbon (C_{org}) density, humic acid, and fulvic acid over 5 years in Sundarbans.

A study on mangrove species revealed varying levels of above-ground biomass (AGB) and above-ground carbon (AGC) storage across different species and salinity levels. At high salinity stations, *A. alba* had the highest AGB and AGC, followed by *A. marina* and *A. officinalis*. In contrast, at low salinity stations, *Sonneratia* spp. had the highest AGB and AGC, followed by *A. alba* and *A. marina* found that the impact of aquatic salinity on mangrove diversity, growth, and carbon storage potential (Ahmed et al. 2024).

A study in Sundarban revealed a significant increase in blue carbon pools at a mangrove plantation site, with a four-fold increase in just 4 years (2012–2016) within the top 10 cm of soil depth (Chowdhury et al. 2023a). In contrast, a nearby mangrove grass-dominated community showed merely a one-fold rise throughout the same time frame. The findings suggest that planting iteroparous mangrove species can substantially enhance ecosystem function and services, offering a promising strategy for restoring degraded mudflats (Chowdhury et al. 2023a).

Bichitrapur Mangroves*:* Datta et al. (2022) observed high levels of Soil Organic Carbon (SOC) across various land use/land cover categories in Bichitrapur Mangroves. Notably, dense mangroves, open mangroves, and aquaculture areas exhibited high SOC concentrations, whereas other land use categories recorded significantly lower SOC concentrations.

Medinipur mangroves: An exhaustive assessment of the mangrove ecosystem revealed a total blue carbon (TBC) stock of 246,710.91 megagrams of carbon (Mg C). Upon closer examination, the TBC density exhibited substantial spatial variability, ranging from a low of 0.34 Mg/ha to a high of 881.50 Mg/ha. This notable variation underscores the importance of localized conservation strategies, tailored to address the unique characteristics and challenges of different areas within the mangrove ecosystem (Datta et al. 2023).

2.5.1.2 Bhitarkanika Mangrove Ecosystem (Bhitarkanika National Park—BNP)

Banerjee et al. (2020) conducted a study on five dominant mangrove species in Bhitarkanika and Mahanadi mangrove ecosystems, spanning an area of approximately 142,261 km^2. This study revealed that these ecosystems absorbed a substantial quantity of atmospheric CO_2, with a mean CO_2 of 455.47 ± 110.56 tonnes (t). This translates to a total of approximately 64.80 TgC, equivalent to removing 64,795,617.67 tonnes of CO_2 from the atmosphere. A satellite data study in the

Bhitarkanika mangrove forest estimated the above-ground carbon stocks, revealing a total of approximately 459.82–514.47 kt C, utilizing Enhanced Vegetation Index and Normalized Difference Vegetation Index analyses (Anand et al. 2020). The total carbon (TC) in Bhitarkanika and Mahanadi mangrove ecosystems ranged from 51.35 to 322.47 tons per hectare, with a mean TC of 117.89 t/ha. This corresponds to a mean CO_2 equivalent of 432.64 t (Banerjee et al. 2021). A comprehensive assessment of the Bhitarkanika mangrove revealed a mean stand biomass of 436.89 ± 59.75 t/ha, comprising 318.79 ± 44.42 t/ha of AGB and 118.10 ± 15.39 t/ha of BGB. Extrapolating these values, the total forest biomass of the entire BNP was estimated to be 6.3 million tons (Mt), with a corresponding biomass carbon stock of 2.96 Mt. and CO_2 equivalents of 10.87 Mt. (Panda et al. 2024). A comparative analysis of modeling approaches for estimating AGB revealed notable differences in prediction accuracy. When using individual data values, the XGBoost (XGB) and Random Forest (RF) models yielded Root Mean Square Errors (RMSE) of 149.242 t/ha and 151.149 t/ha, respectively. However, incorporating the average and percentile metrics from a multi-temporal image stack substantially improved the AGB prediction accuracy, with the XGB and RF models achieving RMSE values of 81.882 t/ha and 74.493 t/ha, respectively (Ghosh et al. 2021).

2.5.1.3 Bichitrapur Mangrove Reserve Forest

Panda et al. (2025) found that the higher mean total biomass and SOC content (measured at a 30 cm depth) in older mangrove stands at 586.12 ± 56.74 t C ha^{-1} and 49.68 ± 2.39 t C ha^{-1} than the young mangrove which stands at 165.31 ± 20.89 t C ha^{-1} and 40.20 ± 1.24 t C ha^{-1} in Bichitrapur mangrove reserve forest. *A. marina* had the greatest plant biomass in new (59.72 t ha^{-1}) and old stands (262.28 t ha^{-1}) than the other species.

2.5.1.4 Coringa Mangrove Ecosystem

A study on Coringa mangrove sedimentary carbon stock (SCS) revealed that Coringa mangrove has a mean SCS value of 158 ± 32.60 Mg C/ha (Megagram Carbon per hectare), translating to a total below-ground sedimentary carbon stock of 1.86 Tg C and equivalent CO_2 emission of 579.9 Mg CO_2/ha. Notably, the study found that dense vegetation has the highest carbon stock, followed by moderate vegetation, and the lowest carbon stock in degraded mangroves (Rao et al. 2022).

2.5.1.5 Godavari Estuary Mangrove

The CO_2 flux from the Godavari tidal river zone (0.04 TgC $year^{-1}$) accounts for approximately 0.2% of the global tropical estuary flux (18 TgC $year^{-1}$), it represents around 1.5% of the total carbon (both particulate and dissolved) transported from the Godavari tidal river zone to the Bay of Bengal, falling below the global average contribution of 16% (Sarma 2023). Krishna et al. (2015) observed that the upper estuary has a relatively high concentration of particulate organic carbon (POC), whereas suspended matter and SOC remain low. Conversely, the estuary (lower) is distinguished by a lower abundance of POC, but a greater accumulation of suspended matter and SOC.

2.5.1.6 Pichavaram Mangrove Ecosystem

Kandasamy et al. (2021) found exceptional potential for carbon sequestration in planted sites, with significantly higher carbon storage and sequestration rates than in nonplanted sites. Specifically, mangrove soils showed 22-fold higher carbon storage, whereas combined tree biomass and soil exhibited 56-fold greater carbon storage. Moreover, carbon sequestration rates were 90-fold greater in the soil and 9890-fold higher in plant biomass and sediment. Furthermore, Kandasamy et al. (2021) also found that higher concentrations of moisture, nutrients, silt, and clay resulted in greater carbon sequestration and storage in mangrove soil, and decreased with increasing temperature, salinity, pH, bulk density, and sand content. Chowdhury et al. (2023b) discovered that a 46% reduction in the soil organic blue carbon pool was observed due to a 31% increase in soil salinity in the Indian Sundarbans. Naidu et al. (2022) found that a threefold higher carbon content was observed in mangrove zones than in salt marsh zones in the Pichavaram estuary. The SOC content in *A. marina* on the southeast coast of India recorded at 24.92 grams of carbon per kg (g C kg^{-1}), indicating its potential for carbon sequestration (Sangma et al. 2022).

2.5.1.7 Andaman Nicobar Island Mangrove Ecosystem

Ragavan et al. (2023) found that the mean SOC stock was higher in estuarine mangroves (345.31 ± 97.22 Mg C/ha) compared to marine mangroves (307.02 ± 103.93 Mg C/ha) in the Island ecosystem. Interestingly, SOC stocks displayed a distinct spatial pattern, increasing from the seaward fringe to the interior, and decreasing toward the landward fringe and landward mudflats. A comprehensive assessment of the mangrove forests in the Andaman Islands, India, yielded significant insights into biomass and carbon stocks. The mean AGB was estimated at 469.20 ± 41.25 Mg/ha, while the mean BGB was 166.78 ± 12.79 Mg/ha. Consequently, the total mean biomass of the Andaman mangrove forest was calculated at 635.98 ± 12.79 Mg/ha, with a corresponding vegetation carbon stock of 290.26 ± 24.75 Mg C/ha. Notably, Little Andaman recorded the highest mean biomass per unit area (1081.26 ± 424.67 Mg/ha) and vegetation carbon stock (494.50 ± 194.56 Mg C/ha), followed by Middle Andaman, North Andaman, Havelock, South Andaman, and Baratang (Ragavan et al. 2021).

2.5.1.8 Southwest Coast of India Mangrove Ecosystem

The Kadalundi mangrove wetland stored an estimated 182.15 t C ha^{-1} (equivalent to 668.48 t CO_2 ha-1), with a total carbon sequestration potential of 2409.84 t C (8,844.11 t CO_2) across its 13.23 ha area (Vinod et al. 2018). Harishma et al. (2020) found that mangrove vegetation in Kerala stores a mean biomass of 117.11 t/ha, with *A. marina* having the highest biomass (162.18 t/ha). The mean ecosystem carbon stock is estimated at 139.82 t/ha, corresponding to 513.13 t CO_2 e/ha, with plant and sediment storing 58.56 t C/ha and 81.26 t C/ha, respectively. A study in the Vembanad Lake area found that the mean carbon stock in *Acanthus officinalis* sediments was 136.09 megagrams of carbon per hectare (Mg C/ha). However, this value is significantly lower than the global average of 386 Mg C/ha for SOC stocks in mangrove ecosystems (Arathy et al. 2022). Rani et al. (2023a, b) estimated the Kochi mangroves have a mean total ecosystem carbon stock of 335.33 t C ha^{-1}, comprising AGB (171.68 t C ha^{-1}), BGB (83.30 t C ha^{-1}), litterfall carbon (7.12 t C ha^{-1}), and soil carbon stock (73.22 t C ha^{-1}). The average historical soil carbon sequestration rate is 2.95 t C ha^{-1} $year^{-1}$, equivalent to 10.62 t CO_2 ha^{-1} $year^{-1}$, with a total CO_2 ecosystem carbon stock estimated at 496,311.20 t CO_2. Furthermore, Rani et al. (2023a) discovered that the variability in biomass stocks within a narrow regional area was highly influenced by the main mangrove vegetation type, salinity levels, and the geomorphology of the mangrove habitat. The Edat (Kerala) mangrove ecosystem has a carbon sequestration potential of 286.88 t C/ha, while the Valapattanam mangrove ecosystem has a potential of 212.20 t C/ha. This translates to a CO_2 equivalent of approximately 1051.07 t CO_2 eq/ha for Edat and 778.01 t CO_2 eq/ha for Valapattanam.

2.5.1.9 Goa Mangrove

The organic matter content in sediments (OMs) of Mandovi and Zuari estuaries, Goa, ranges from 2.29 mg/g to 62.61 mg/g, with enrichment attributed to mangrove-derived organic matter, facilitated by the process of outwelling (Seleyi et al. 2024). Mankikar (2024) reported that the total estimated biomass in the mangrove swamps at Chorao, Goa, ranged from 0.37 to 101.78 t/ha, while the carbon stock varied between 0.17 and 48.44 t/ha. Additionally, a comparative analysis of normalized difference vegetation index values from 2001 to 2014 and 2024 was conducted to evaluate changes in vegetation cover, revealing a significant shift over time.

2.5.1.10 Maharashtra Mangrove

Pardeshi et al. (2024) reported that the mangrove carbon pool in Maharashtra, India, varied between 2.78 and 249.64 t C ha^{-1}. In 2005–06, the total carbon pool in the state was estimated at 1.08×10^6 tonnes, which increased to 1.32×10^6 t by 2018–19.

Over this period, Maharashtra's mangroves sequestered approximately 0.24×10^6 t of carbon.

2.5.1.11 Gulf of Kutch Mangrove Ecosystem

The Gulf of Kutch mangroves had a mean above-ground biomass of 96.9 Mg ha^{-1} and below-ground biomass of 19.6 Mg ha^{-1}, totalling 116.5 Mg ha^{-1}. The corresponding carbon stocks were 54.1 Mg C ha^{-1} in vegetation and 25.87 Mg C ha^{-1} in soil. Among the five creeks studied, Navlakhi had the highest total biomass (287.3 Mg ha^{-1}) and vegetative carbon stock (133.9 Mg C ha^{-1}), followed by Tuna, Phan, Kandla, and Jangi creeks (Sigamani et al. 2023). Jadeja and Vachhrajani (2025) reported that the carbon content in coastal sediment samples from the eastern part of the Gulf of Khambhat ranged from 3050 to 8650 mg kg^{-1}, with an average concentration of 4866 mg kg^{-1}.

2.5.2 *Seagrass Ecosystem*

The economic value of seagrass ecosystems in India is estimated to be substantial, with an average annual value of approximately ₹25.94 lakh per hectare (approximately $55,637/ha/yr). The total economic benefits from India's seagrass beds are estimated to be around ₹13,444 crore per year (approximately $2.88 billion/year), highlighting the significant contribution of these ecosystems to the country's economy (Ramesh et al. 2024).

2.5.2.1 Chilika Lagoon

Seagrass meadows are among the most vital coastal and marine ecosystems, and play a crucial role in long-term carbon sequestration and maintaining the health and quality of coastal waters (Banerjee et al. 2019). Ganguly et al. (2018) estimated that the mean ecosystem productivity ranged from 6.31 to 11.9 kg C ha^{-1} d^{-1}, equivalent to sequestering 8.44 to 15.9 Mg CO_2 ha^{-1} $year^{-1}$. Notably, the majority of captured carbon (129 Mg C_{org} ha^{-1}) was stored in the top 1 m of sediment, far exceeding the carbon stored in above-ground (0.20–0.96 Mg C_{org} ha^{-1}) and below-ground biomass (0.30–2.90 Mg C_{org} ha^{-1}). The estimated blue carbon stock in Chilika Lake's seagrass habitats, up to a depth of 30 cm, was found to be 2.018 ± 0.673 Mg C ha^{-1}.

2.5.2.2 Pulicat Lake

Blue carbon stock in Pulicat Lake seagrass habitats (0–30 cm depth) was estimated to be 0.998 ± 0.418 M gC ha^{-1}. The total blue carbon storage in the seagrass meadows ranged from 0.49 ± 0.15 to 9.52 ± 0.15 Mg C/ha (Murugan et al. 2024), Furthermore, the amount of carbon stored in the sediment was found to be significantly higher, ranging from 1.29 ± 0.04 to 11.94 ± 0.15 Mg C/ha.

2.5.2.3 Gulf of Mannar (GoM) and Palk Bay (PB)

Vinithkumar et al. (1999) found higher total organic carbon (TOC) content in the seagrass beds and adjacent areas than in reef sediments. The seagrass meadows of Palk Bay have AGB more than twofold higher than the autotrophic threshold of 41 g dwt m^{-2}, and the total carbon stored as seagrass biomass is 0.94 Mg C_{org} ha^{-1} (94 g m^{-2}) (Ganguly et al. 2017). Both the GoM and PB exhibited higher soil carbon density in subsurface layers. Estimates revealed that the seagrass meadows in the GoM stored approximately 0.001782 Tg of blue carbon, while those in PB stored around 0.043996 Tg. In economic terms, the blue carbon stored in the GoM's seagrass meadows was valued at approximately $17,820, whereas the PB's meadows held a significantly higher value of $4,399,682 (Kaladharan et al. 2020). Seagrass areas with coral reef connectivity exhibited significantly higher SOC levels (90.26 ± 25.68 Mg C_{org}/ha), compared to areas without connectivity (66.96 ± 12.6 Mg C org/ha). Additionally, AGB was maximized in *Syringodium isoetifolium* (35.43 ± 8.50) in connected areas, while *Halophila ovalis* (7.59 ± 0.90) had the lowest AGB in disconnected areas, similar to BGB (Ranith et al. 2024). Overall, the seagrass meadows in the Gulf of Mannar and Palk Bay demonstrated significant carbon sequestration potential, emphasizing the importance of conserving these ecosystems, particularly those with coral reef connectivity, to mitigate climate change.

2.5.2.4 Island Ecosystem

The sediment within the Halophila seagrass bed showed a significant increase in organic carbon content, ranging from 110% to 134% higher than that in the areas outside the seagrass bed (Kaladharan et al. 2021a). Kaladharan et al. (2022) estimated that the mean organic carbon (C_{org}) content varied across locations, with the lowest value recorded at Bangaram (0.1863%) and the highest at Thinnakara (0.3453%) in Lakshadweep Island, and the C_{org} content was higher in the top cores (0–10 cm depth). Notably, the monospecific meadows in the Andaman and Nicobar Islands (ANI) of India recorded ecosystem C_{org} value, with an average of 184.25 ± 23.84 Mg ha^{-1}(Stankovic et al. 2021). The mangrove sites showed a substantial increase in C_{org} content, with levels 1.8 times higher than those found at the mangrove sites in the top 10 cm of sediment. Additionally, the total C_{org} stocks in the mangrove sites' 144 ha of seagrass meadows (988.74 ± 138.77 Mg C) exceeded those in the without mangrove sites' 148 ha by nearly twofold (Mishra et al. 2024). In tropical island settings, seagrass ecosystems situated near mangroves receive a greater influx of organic matter (OM), which in turn supports a rich diversity of seagrass–bivalve interactions and plays a crucial role in maintaining these complex ecosystems (Mishra and Farooq 2023).

2.5.2.5 Southwest Coast of India

A short-term study found that the mean blue carbon stock of Halophila beds in the Kadalundi Estuary was 2.655 ± 0.34 megagrams of carbon per hectare (Mg C/ha) (Kaladharan et al. 2021a).

2.5.3 Salt Marsh Ecosystem

2.5.3.1 Northeast Coast (Sundarbans)

A study in the *Porteresia coarctata* area in the Hooghly estuary revealed that the below-ground carbon in the mudflats ranged from 47.81 to 63.49 grams per square meter, while the above-ground carbon in the estuarine region ranged from 58.33 to 104.30 grams per square meter, highlighting the estuary's significant carbon storage capacity (Jana et al. 2013). The area of *Suaeda maritima* in Frazergaunge and Bali Island reveals variations in ground carbon (64.45 to 71.01 gm/Cm^2) and below-ground carbon (2.46 to 3.02 gm/Cm^2) in biomass and carbon content (Das et al. 2015).

2.5.3.2 Bhitarkanika

A comparative study conducted between 2017–2018 and 2018–2019 found that significant increases in biomass and carbon were seen over the research period in *P. coarctata*, a salt marsh grass located in the Mahanadi Estuary and Bhitarkanika Wildlife Sanctuary. Notably, Bhitarkanika showed increases of 1.03 ± 0.49 Mg ha^{-1} in biomass and 0.48 ± 0.22 Mg ha^{-1} in carbon, while Mahanadi recorded increases of 0.62 ± 0.17 Mg ha^{-1} in biomass and 0.29 ± 0.09 Mg ha^{-1} in carbon (Banerjee et al. 2022).

2.5.3.3 Southeast Coast (Tamil Nadu)

A study along the southeastern coast of India found that the salt marsh species *S. brachiata* has a notable carbon sequestration potential, with a mean soil organic carbon value of 8.04 g C kg^{-1}(Sangma et al. 2022). Four major species of salt marshes *Arthrocnemum indicum, Suaeda maritima, Salicornia brachiate, and Sesuvium portulacastrum* have sediment organic carbon stocks in the Tuticorin coastal area ranging from 8.42 ± 0.64 to 54.46 ± 1.46 (Mg ha^{-1}), with a notable increase of nearly 7 times during the wet season compared to the dry season (Kaviarasan et al. 2019).

2.5.3.4 Mandovi Estuarine

The *P. coarctata* area in the Mandovi estuarine region on India's west coast was found to have significant biomass levels. Specifically, the above-ground biomass ranged from 32.57 to 63.82 grams per square meter, while the below-ground biomass ranged from 26.79 to 54.41 grams per square meter (Jana et al. 2013).

2.5.3.5 Gujarat Coast

The aboveground biomass of *S. brachiate* coastal region of Gujarat was found to range from 2.51 to 6.07 tons/ha (Rathore et al. 2016). The study by Chaudhary et al. (2018) indicates that the summer season, specifically March 2015, had the highest above-ground carbon pools of *S. brachiata* in the Gujarat coastal salt marshes, with a value of 543 kg $ha^{-1.}$

2.6 Threats and Challenges to Blue Carbon in Indian Coastal Wetlands in India

One of the key challenges in implementing blue carbon initiatives is balancing ambitious goals with realistic timeframes for development and implementation (Banerjee et al. 2021).

2.6.1 Aquaculture

The relentless conversion of wetlands to croplands and aquaculture farms poses a significant environmental threat, undermining regional sustainability by substantially reducing carbon sequestration potential (Datta et al. 2022).

2.6.2 Coastal Erosion

Mangrove loss in the Indian Sundarban between 1975 and 2013 was significant, with 107 km^2 of area degraded, erosion accounted for 60% of this loss, while human activities and climate change also played major roles, almost three-fourths of the total mangrove loss occurred in peripheral islands, which are more prone to erosion, highlighting the impact of climate-induced changes and anthropogenic land use changes (Akhand et al. 2017). The Coringa mangroves are experiencing continuous shrinkage, posing a significant threat to the environment. This reduction in mangrove cover could lead to a substantial release of stored CO_2 into the atmosphere. Furthermore, the aggradational characteristics of the Godavari estuary hinder its ability to keep pace with the rising local sea level, exacerbating the vulnerability of the mangrove ecosystem (Rao et al. 2022).

2.6.3 Extreme Events

Blue carbon ecosystems have been found to provide crucial protection during natural disasters such as cyclones (Das and Vincent 2009; Bahinipati and Sahu 2012) and tsunamis (Kathiresan and Rajendran 2005). Although they offer resilience to extreme weather conditions, these ecosystems also face significant challenges during such events. Tropical Cyclone *Yaas* caused widespread destruction to Bhitarkanika National Park, a premier wildlife sanctuary, in 2021, impacting its rich flora and fauna (Mishra et al. 2024). Chowdhury et al. (2023b) noticed the alarming threat posed by cyclones and climate change-induced sea level rise to the delicate mangrove ecosystem. Specifically, these impacts jeopardize two pivotal Sustainable Development Goals: SDG 13 (Climate Action), by diminishing the organic soil blue carbon sink, and SDG 14 (Life Below Water), by driving the local extinction of salinity-sensitive mangrove species. The Andaman Islands are facing escalating threats from natural calamities, including sea-level rise, soil erosion, and intensified tropical cyclones. This vulnerability underscores the urgent need for conservation efforts to protect the island's mangroves, which possess exceptional carbon storage capacity, thereby safeguarding their crucial role in mitigating climate change (Ragavan et al. 2023). The seagrass ecosystem of South Andaman Islands, India, suffered extensive damage due to the intense effects of Cyclone *Lehar*, including high-velocity waves, turbulence, and storm surges (Sachithanandam et al. 2022).

2.6.4 Anthropogenic Activity

Manoj et al. (2024) discovered that in comparison to the mangrove habitats in Valapattanam (Kerala) that are anthropogenically interfering, protected mangrove ecosystems had the maximum potential for sequestering carbon. Sand mining, industrialization, land reclamation, illegal encroachment, and coastal erosion are commonplace in the Valapattanam mangrove habitat (Manoj et al. 2024). Seagrass meadows proved to be an effective carbon sink, offsetting atmospheric CO_2 levels. Nevertheless, human activities can disrupt this balance, potentially transforming them into a modest source of CO_2 emissions (Ganguly et al. 2018). Even a slight depletion in seagrass cover can trigger the release of stored blue carbon back into the atmosphere, undermining its critical role as a carbon sink (Kaladharan et al. 2020).

Table 2.1 highlights the presence of microplastics and macro litter in various blue carbon ecosystems in India, including mangroves and seagrasses. These ecosystems are crucial for sequestering and storing CO_2 and mitigating climate change. However, the presence of plastic pollutants can have devastating effects on these ecosystems, including the following:

(i) Habitat destruction: Macro litter can damage or destroy habitats, reducing the capacity of these environments to sequester carbon.

Table 2.1 Detail study of microplastics and macro-litter in Indian mangrove and seagrass ecosystems

Study area	Type of study	References
Mangrove ecosystem		
Sundarbans	Microplastics	Sengupta and Mitra (2024)
Kota mangrove (KM) in Udupi	Microplastics	Valsan et al. (2024a)
Thoothukudi coast, Gulf of Mannar, southeast coast of India	Macro litter	Ponmani et al. (2024)
Kota mangrove (KM) in Udupi	Microplastics	Valsan et al. (2024b)
Versova Creek, Mumbai	Microplastics	Yousuf Dar et al. (2023)
Central West coast	Macro litter	De et al. (2023)
Mangalavanam, Kerala	Microplastics	Kannankai et al. (2022)
Sundarbans	Microplastics	Kumar et al. (2022)
Kadalundy and Kolavipalam, Kerala	Microplastics	Abhijith et al. (2021)
Mumbai coast	Macro litter	Kesavan et al. (2021)
Ernakulum District, Kerala	Macro litter	Shylaja et al. (2021)
South Juhu creek, Mumbai	Macro litter	Manickavasagam et al. (2021)
South Andaman Island	Macro litter	Kaviarasan and James (2017)
Mumbai coast	Macro litter	Kantharajan et al. (2018)
Seagrass ecosystem		
Tuticorin, the southeast coast of India	Microplastics	Esmeralda and Patterson (2025)
Rameswaram, the southeast coast of India	Microplastics	Jeyasanta et al. (2020)

(ii) Species entanglement and ingestion: Microplastics and macro litter can entangle or be ingested by species, causing harm or even death.
(iii) Reduced carbon sequestration: Plastic pollutants can reduce the ability of blue carbon ecosystems to sequester carbon, thereby exacerbating climate change.
(iv) Disruption of nutrient cycles: Microplastics can disrupt nutrient cycles, affecting the general health and efficiency of these ecosystems.

The presence of plastic pollutants in blue carbon ecosystems can significantly impact their ability to mitigate climate change, emphasizing the necessity of efficient waste management and conservation approaches.

2.7 Conclusion

Overall, the review findings underscore the significant carbon sequestration potential of blue carbon ecosystems. Key factors influencing carbon storage and sequestration rates include soil characteristics such as silt, clay, and moisture content, geomorphology, mangrove species diversity, and density. Notably, increased salinity may cause soil organic blue carbon stores to decrease. Furthermore, planted mangrove sites have been shown to exhibit higher carbon storage and sequestration rates compared to nonplanted sites. The mangrove ecosystems of India, particularly in the Sundarbans, Andaman Islands, and Coringa, face formidable threats from natural hazards, climate change, sea-level rise, soil erosion, and human activities.

The seagrass ecosystems in India provide numerous ecosystem services, including significant carbon sequestration and storage. The economic value of these ecosystems is substantial, with an estimated annual value of approximately ₹25.94 lakh per hectare (~30397.53USD). The total economic benefits from India's seagrass beds are estimated to be around ₹13,444 crore per year (~157542.17USD). Studies from various locations in India, including Chilika Lagoon, Pulicat Lake, the GoM and PB, and the southwest coast of India, have highlighted the importance of seagrass ecosystems in carbon sequestration. These ecosystems have been found to store significant quantities of carbon in both the sediment and biomass. The conservation of seagrass ecosystems is crucial for maintaining their ecosystem services, including carbon sequestration. The outcomes of this review emphasize the need for sustainable management and conservation of seagrass ecosystems in India.

The studies on salt marsh ecosystems along India's coast, including the Sundarban, Bhitarkanika, southeastern coast of Tamil Nadu, and the Gujarat coast, highlight their substantial carbon sequestration potential.

Based on this review, aquaculture, coastal erosion, extreme events, and anthropogenic activity are the major threats that may affect the blue carbon; consequently, an integrated approach is necessary for future climate change mitigation in blue carbon ecosystems.

Conservation and Restoration Blue carbon ecosystems demonstrate significant carbon storage capacities, but the alarming rates of blue carbon ecosystem loss and degradation not only compromise regional sustainability but also jeopardize global climate regulation efforts.

1. *Raising Awareness:* Creating awareness among local communities and tourists is crucial for encouraging the perception of blue carbon ecosystems, the significance of addressing climate change, and the need for sustainable practices. This awareness can inspire collective action to mitigate climate change and preserve these vital ecosystems.
2. *Local communities and private participation*: To protect and revive existing blue carbon ecosystems (seagrasses, mangroves, and salt marshes), by engaging local communities and private sectors in the conservation of present blue carbon ecosystems, and restoration of degraded, destroyed (anthropogenic and extreme events), so they can be benefited by using the livelihoods.
3. *Economic incentives:* Encourage participants with various incentives for the conservation and restoration of blue carbon ecosystems.
4. *Policies development and implementation*: To preserve blue carbon ecosystems and advance sustainable practices, local communities, governments, and the private sector must work together to formulate and enforce policies that conserve blue carbon ecosystems.

Understanding Blue Carbon Ecosystems and Emerging Contaminants

In order to maintain the blue carbon ecosystems' long-term resilience and vitality, the following are essential for reducing climate change and sustaining coastal communities.

1. Monitor ecosystem function: Regularly assess the health and function of blue carbon ecosystems, including their carbon sequestration capacity.
2. Track emerging contaminants: Investigate the impacts of emerging contaminants, particularly plastics, on blue carbon ecosystems.
3. Inform conservation directions: Use the latest research findings to guide conservation efforts and maximize the ecosystem service functions of blue carbon ecosystems.

Acknowledgments The authors are thankful to Secretary, Ministry of Earth Sciences, Government of India for engorgement and support and implementing the Marine Litter and Microplastics Programme/Project at National Centre for Coastal Research (NCCR), Chennai. This book chapter is NCCR contribution no. 2025/MS/437.

Competing Interests The authors declare no competing interests.

Declarations Informed Ethical approval 1) This review has not been previously published elsewhere.

2) This review is not currently being considered for publication elsewhere.

3) This review has been prepared based on the published data, and all sources used are properly disclosed (correct citation).

4) This review properly credits the meaningful contributions of co-authors and co-researchers.

5) All authors have been personally and actively involved in substantial work leading to the paper.

6) The manuscript was checked for plagiarism as well.

Consent to Participate The results from the present review were originally generated from the published data by the first two authors and properly given citations. No data from any other sources/individuals/organization were used in the present research.

Consent for Publication All the authors actively participated in the present review paper. Those who extended the support have been acknowledged.

Data availability All the data and materials are presented as tables and figures.

References

Abhijith R, Kaladharan P, Asokan PK, Vinod K, Shilta MT, Mahesh V (2021) Case study on quantification of plastic litter-A global menace and paving way for unleashing toxins into two mangrove ecosystems of Kerala. Pharma Innov J 10:260–268

Adyel TM, Macreadie PI (2022) Plastics in blue carbon ecosystems: a call for global cooperation on climate change goals. Lancet Planet Health 6:e2–e3. https://doi.org/10.1016/S2542-5196(21)00327-2

Ahmed S, Mitra A, Pramanick P, Bengal W, Zaman S, Bengal W (2024) Chapter 33 spatial variability in carbon storage among dominant mangrove species in the Indian Sundarbans

Ajai NS, Tamilarasan V, Chauhan H, Bahuguna A, Gupta MC, Rajawat A, Chaudhury N, Kumar T, Rao R, Bhattacharya S, Ramakrishnan R, Bhanderi RJ, Mahapatra M (2012) Coastal Zones of India. Space Application Centre-ISRO, Ahmedabad

Akhand A, Mukhopadhyay A, Chanda A, Mukherjee S, Das A, Das S, Hazra S, Mitra D, Choudhury SB, Rao KH (2017) Potential CO2 emission due to loss of above ground biomass from the Indian Sundarban Mangroves during the last four decades. J Indian Soc Remote Sens 45:147–154. https://doi.org/10.1007/s12524-016-0567-4

Anand A, Pandey PC, Petropoulos GP, Pavlides A, Srivastava PK, Sharma JK, Malhi RKM (2020) Use of hyperion for mangrove forest carbon stock assessment in bhitarkanika forest reserve: a contribution towards blue carbon initiative. Remote Sens 12:597. https://doi.org/10.3390/rs12040597

Arathy GS, Prema D, Padua S, Kripa V, Jeyabaskaran R, Varghese E, Lavanya R, Vysakhan P, Syamala MP, John S (2022) Sediment blue carbon stock of Avicennia officinalis in Vembanad Lake ecosystem, Kerala, India. ICAR-Central Marine Fisheries Research Institute, Kochi. https://doi.org/10.6024/jmbai.2022.64.2.2220-01

Bahinipati CS, Sahu N (2012) Mangrove conservation as a sustainable adaptation to cyclonic risk in Kendrapada District of Odisha, India. Asian J Environ Disaster Manag 4:183–202

Banerjee K, Sappal SM, Ramachandran P, Ramesh R (2017) Salt marsh: ecologically important, yet least studied blue carbon ecosystems in India. J Clim Chang 3:59–72. https://doi.org/10.3233/jcc-170014

Banerjee K, Paneerselvam A, Ramachandran P, Ganguly D, Singh G, Ramesh R (2019) Seagrass and macrophyte mediated CO2 and CH4 dynamics in shallow coastal waters. PLoS One 13:1–22. https://doi.org/10.1371/journal.pone.0203922

Banerjee K, Sahoo CK, Bal G, Mallik K, Paul R, Mitra A (2020) High blue carbon stock in mangrove forests of Eastern India. Trop Ecol 61:150–167. https://doi.org/10.1007/s42965-020-00072-y

Banerjee K, Mitra A, Villasante S (2021) Carbon cycling in mangrove ecosystem of western bay of bengal (India). Sustainability 13(12):6740. https://doi.org/10.3390/su13126740

Banerjee K, Mallik K, Sahoo CK, Paul R (2022) Aquatic and edaphic determinants influencing carbon storage in a salt marsh grass, Porteresia coarctata, of the Bhitarkanika wildlife sanctuary and the Mahanadi estuary. Environ Qual Manag 31:11–28. https://doi.org/10.1002/tqem.21735

Bhomia RK, MacKenzie RA, Murdiyarso D, Sasmito SD, Purbopuspito J (2016) Impacts of land use on Indian mangrove forest carbon stocks: implications for conservation and management. Ecol Appl 26:1396–1408. https://doi.org/10.1890/15-2143

Blasco F (1977) Outlines of ecology, botany and forestry of the mangals of the Indian subcontinent. In: Chapman VJ (ed) Ecosystems of the world vol. 1: wet coastal ecosystems. Elsevier/Oxford, Amsterdam/New York, pp 241–260

Chaudhary DR, Rathore AP, Jha B (2018) Aboveground, belowground biomass and nutrients pool in Salicornia brachiata at coastal area of India: interactive effects of soil characteristics. Ecol Res 33:1207–1218. https://doi.org/10.1007/s11284-018-1634-9

Chmura GL, Anisfeld SC, Cahoon DR, Lynch JC (2003) Global carbon sequestration in tidal, saline wetland soils. Global Biogeochem Cycles 17. https://doi.org/10.1029/2002gb001917

Choudhary B, Dhar V, Pawase AS (2024) Blue carbon and the role of mangroves in carbon sequestration: Its mechanisms, estimation, human impacts and conservation strategies for economic incentives. J Sea Res 199:102504. https://doi.org/10.1016/j.seares.2024.102504

Chowdhury A, Naz A, Bhattacharyya S, Sanyal P (2018) Cost–benefit analysis of 'Blue Carbon' sequestration by plantation of few key mangrove species at Sundarban Biosphere Reserve, India. Carbon Manag 9:575–586. https://doi.org/10.1080/17583004.2018.1518105

Chowdhury A, Naz A, Dasgupta R, Maiti SK (2023a) Blue carbon: comparison of chronosequences from Avicennia marina Plantation and Proteresia coarctata dominated Mudflat, at the world's largest Mangrove Wetland. Sustainability 15:368. https://doi.org/10.3390/su15010368

Chowdhury A, Naz A, Sharma SB, Dasgupta R (2023b) Changes in salinity, mangrove community ecology, and organic blue carbon stock in response to cyclones at Indian Sundarbans. Life 13:1539. https://doi.org/10.3390/life13071539

Das S, Vincent JR (2009) Mangroves protected villages and reduced death toll during Indian super cyclone. Proc Natl Acad Sci USA 106:7357–7360. https://doi.org/10.1073/pnas.0810440106

Das S, Zaman S, Pramanick P, Pal N, Mitra A (2015) Suaeda maritima: a potential carbon reservoir of coastal zone. Int Adv Res J Sci Eng Technol 2:61–65

Datta D, Bairagi M, Dey M, Pal AP, Gayen J (2022) Spatially explicit estimation of soil organic carbon stock of an estuarine mangrove wetland of eastern India using elemental analysis and very-fine resolution satellite data. Ecol Process 11. https://doi.org/10.1186/s13717-022-00370-4

Datta D, Dey M, Ghosh PK, Neogy S, Roy AK (2023) Coupling multi-sensory earth observation datasets, in-situ measurements, and machine learning algorithms for total blue C stock estimation of an estuarine mangrove forest. For Ecol Manag 546:121345

De K, Sautya S, Dora GU, Gaikwad S, Katke D, Salvi A (2023) Mangroves in the "Plasticene": High exposure of coastal mangroves to anthropogenic litter pollution along the Central-West coast of India. Sci Total Environ 858:160071. https://doi.org/10.1016/j.scitotenv.2022.160071

Duarte CM (2002) The future of seagrass meadows. Environ Conserv 29:192–206. https://doi.org/10.1017/S0376892902000127

Duarte CM, Losada IJ, Hendriks IE, Mazarrasa I, Marbà N (2013) The role of coastal plant communities for climate change mitigation and adaptation. Nat Clim Chang 3:961–968. https://doi.org/10.1038/nclimate1970

Esmeralda G, Patterson J (2025) Accumulation of microplastics in relation to the sediment grain size distribution in seagrass meadows in the coastal areas of Tuticorin, southeast coast of India. Mar Pollut Bull 212:117557. https://doi.org/10.1016/j.marpolbul.2025.117557

FSI (2023) Forest, forestry, and state of forests. In: India state of forest report 2023 volume I, 18th, 2023rd edn. Forest Survey of India (Ministry of Environment Forest and Climate Change). IPE Dehradun, Uttarakhand, pp 1–368

Ganguly D, Singh G, Ramachandran P, Selvam AP, Banerjee K, Ramachandran R (2017) Seagrass metabolism and carbon dynamics in a tropical coastal embayment. Ambio 46:667–679. https://doi.org/10.1007/s13280-017-0916-8

Ganguly D, Singh G, Purvaja R, Bhatta R, Paneer Selvam A, Banerjee K, Ramesh R (2018) Valuing the carbon sequestration regulation service by seagrass ecosystems of Palk Bay and Chilika, India. Ocean Coast Manag 159:26–33. https://doi.org/10.1016/j.ocecoaman.2017.11.009

Ghosh SM, Behera MD, Jagadish B, Das AK, Mishra DR (2021) A novel approach for estimation of aboveground biomass of a carbon-rich mangrove site in India. J Environ Manag 292:112816

Harishma KM, Sandeep S, Sreekumar VB (2020) Biomass and carbon stocks in mangrove ecosystems of Kerala, southwest coast of India. Ecol Process 9:31. https://doi.org/10.1186/s13717-020-00227-8

Hutchison J, Spalding M, Ermgassen P z (2014) The role of mangroves in fisheries enhancement. The Nature Conservancy and Wetlands International, p 54

Indianwetlands.in (2025). https://indianwetlands.in/wetlands-overview/national-wetlands-statistics/Accessed 16 Dec 2024

Jadeja F, Vachhrajani K (2025) Blue carbon vis-à-vis heavy metal concentration in surface sediments along the coast of Gulf of Khambhat, Gujarat. Soil Sediment Contam 00:1–22. https://doi.org/10.1080/15320383.2025.2467847

Jagtap TG, Chavan VS, Untawale AG (1993) Mangrove ecosystems of India: a need for protection, vol 22. Ambio, pp 252–254

Jana H, Zaman S, Chakraborty S, Pramanick P, Chandra MK, Abhijit M (2013) Spatio-temporal variation of stored carbon in Porteresia coarctata along the east and west coasts of India. Int J Eng Manag Sci 4(3):377–381

Jeyasanta KI, Patterson J, Grimsditch G, Edward JKP (2020) Occurrence and characteristics of microplastics in the coral reef, sea grass and near shore habitats of Rameswaram Island, India. Mar Pollut Bull 160:111674. https://doi.org/10.1016/j.marpolbul.2020.111674

Kaladharan P, Zacharia PU, Thirumalaiselvan S, Anasukoya A, Ratheesh L, Batcha SMS (2020) Blue carbon stock of the Seagrass meadows of gulf of Mannar and Palk bay off Coromandel coast, south India. Indian J Fish 67:149–153. https://doi.org/10.21077/ijf.2020.67.4.98101-18

Kaladharan P, Lavanya R, Akshara C (2021a) Capture and storage of carbon in seagrass beds of a vulnerable species Halophila beccarii (Asch.) at the Kadalundi Estuary, Kerala, India. J Mar Biol Assoc India 63:126–129. https://doi.org/10.6024/jmbai.2021.63.2.2284-19

Kandasamy K (2017) Mangroves in India and climate change: an overview. In: DasGupta R, Shaw R (eds) Participatory mangrove management in a changing climate. Disaster risk reduction. Springer, Tokyo

Kandasamy K, Rajendran N, Balakrishnan B, Thiruganasambandam R, Narayanasamy R (2021) Carbon sequestration and storage in planted mangrove stands of Avicennia marina. Reg Stud Mar Sci 43:101701

Kankara RS, Murthy MVR, Rajeevan M (2018) National assessment of shoreline changes along Indian Coast: status report for 26 years (1990—2016). NCCR

Kannankai MP, Alex RK, Muralidharan VV, Nazeerkhan NP, Radhakrishnan A, Devipriya SP (2022) Urban mangrove ecosystems are under severe threat from microplastic pollution: a case study from Mangalavanam, Kerala, India. Environ Sci Pollut Res 29:80568–80580. https://doi.org/10.1007/s11356-022-21530-1

Kantharajan G, Pandey P, Krishnan P, Bharti V, Deepak Samuel V (2018) Plastics: a menace to the mangrove ecosystems of megacity Mumbai, India. ISME/GLOMIS Electron J 16:1–5

Kathiresan K, Bingham BL (2001) Biology of mangroves and mangrove ecosystems. Adv Mar Biol 40:81–251. https://doi.org/10.1016/S0065-2881(01)40003-4

Kathiresan K, Dagar JC (2024) Mangroves and associated flora: prospects for utilization in coastal agriculture. In: Halophytes vis-à-vis Saline agriculture: perspectives and opportunities for food security. Springer, pp 67–95

Kathiresan K, Rajendran N (2002) Fishery resources and economic gain in three mangrove areas on the. Fish Manag Ecol 9:277–283. https://doi.org/10.1046/j.1365-2400.2002.00306.x

Kathiresan K, Rajendran N (2005) Coastal mangrove forests mitigated tsunami. Estuar Coast Shelf Sci 65:601–606. https://doi.org/10.1016/j.ecss.2005.06.022

Kaviarasan T, Dahms HU, Gokul MS, Henciya S, Muthukumar K, Shankar S, Arthur James R (2019) Seasonal species variation of sediment organic carbon stocks in salt marshes of Tuticorin Area, Southern India. Wetlands 39:483–494. https://doi.org/10.1007/s13157-018-1094-6

Kesavan S, Xavier KAM, Deshmukhe G, Jaiswar AK, Bhusan S, Sukla SP (2021) Anthropogenic pressure on mangrove ecosystems: quantification and source identification of surficial and trapped debris. Sci Total Environ 794:148677. https://doi.org/10.1016/j.scitotenv.2021.148677

Krishna MS, Naidu SA, Subbaiah CV, Gawade L, Sarma VVSS, Reddy NPC (2015) Sources, distribution and preservation of organic matter in a tropical estuary (Godavari, India). Estuar Coasts 38:1032–1047. https://doi.org/10.1007/s12237-014-9859-5

Kumar SV, Pathak KC, Pednekar P, Raju NSN, Gowthaman R (2006) Coastal processes along the Indian coastline. Curr Sci 91:530–536

Kumar R, Sinha R, Refat Jahan Rakib M, Padha S, Ivy N, Bhattacharya S, Dhar A, Sharma P (2022) Microplastics pollution load in Sundarban delta of Bay of Bengal. J Hazard Mater Adv 7:100099. https://doi.org/10.1016/j.hazadv.2022.100099

Kvale K (2022) Implications of plastic pollution on global marine carbon cycling and climate. Emerg Top Life Sci 6:359–369

Kaladharan P, Zacharia PU, Mohammed Koya K, Lavanya R. Anasu Koya A (2022) Organic carbon storage in sediments of seagrass habitats in the lagoons of Lakshadweep Archipelago, west coast of India. JMBAI, 64(1): 61–64

Kaviarasan T, James R A (2017) Assessment of anthropogenic activities and prospect on mangrove and sand dune ecosystem of Carbyn's Cove South Andaman Island, India. J Terr Mar Res 1(4):89–94

Manickavasagam S, Shukla SP, Kumar S, Kumar K, Bhuvaneswari R (2021) Assessment of marine microplastics in floating plastic debris using a fixed sampling device: the example of South Juhu creek, Mumbai coast, India. J Coast Conserv 25. https://doi.org/10.1007/s11852-021-00810-x

Mankikar NN (2024) Satelitte-based carbon stockassessment in mangroves swamps at Chorao, Goa 1–59

Manoj K, Arumugam T, Prakash A (2024) A comparative study on carbon sequestration potential of disturbed and undisturbed mangrove ecosystems in Kannur district, Kerala, South India. Results Eng 21:101716. https://doi.org/10.1016/j.rineng.2023.101716

Mathauda (1987) The mangrove forests of India. Mangrove Symp Calcutta 66–87

Mejjad N, Safhi AEM, Laissaoui A (2025) Insightful Analytical Review of Potential Impacts of Microplastic Pollution on Coastal and Marine Ecosystem Services. J Hazard Mater Adv 17:100578

Mishra AK, Farooq SH (2023) Sediment organic matter content drives bivalve density in tropical oligotrophic seagrass ecosystem. Indian J Geo-Marine Sci 52:91–100. https://doi.org/10.56042/ijms.v52i02.6927

Mishra M, Acharyya T, Halder B, Santos CAG, da Silva RM, Rout NR, Bhattacharyya D (2024) Impact assessment of Cyclone Yaas on the mangrove forest area in the Bhitarkanika National Park (India). J Mar Syst 242:103947

Mitra A (2020) Mangrove forests in India. Springer

Morrison M, Lowe ML, Grant CM, Smith PJ, Carbines GD, Reed J, Bury S, Brown JCS (2014) Seagrass meadows as biodiversity and productivity hotspots. Ministry for Primary Industries, Wellington

Murugan R, Balachandar K, Boopathy NS, Nagarajan R, Ramasubramanian R (2024) Assessment of blue carbon in seagrass meadows in Pulicat Lake, Andhra Pradesh, India. Mar Pollut Bull 198:115868. https://doi.org/10.1016/j.marpolbul.2023.115868

Naidu SA, Kathiresan K, Simonson JH, Blanchard AL, Sanders CJ, Alexander P, Post RM, Subramoniam T, Naidu RA, Narender R (2022) Carbon and Nitrogen Contents Driven by Organic Matter Source within Pichavaram Wetland Sediments

NationalWetlandAtlas (2011) SAC/EPSA/ABHG/NWIA/ATLAS/34/2011. Space Applications Centre (ISRO), Ahmedabad

Noman MA, Adyel TM, Macreadie PI, Trevathan-Tackett SM (2024) Prioritising plastic pollution research in blue carbon ecosystems: A scientometric overview. Sci Total Environ 914:169868. https://doi.org/10.1016/j.scitotenv.2024.169868

Panda M, Lele N, Murthy TVR, Samal RN, Pattnaik AK, Sahu SC (2024) Assessment of mangrove stand biomass in relation to forest structural attributes in Bhitarkanika National Park, India. Vegetos 37:950–960

Panda M, Dash BR, Sahu SC (2025) Ecosystem carbon stock variation along forest stand ages: insight from eastern coast mangrove ecosystem of India. Ecol Process 14. https://doi.org/10.1186/s13717-025-00580-6

Pardeshi SN, Chavan M, Kale M, Khare M, Lele N, Bhattacharya BK (2024) Mangrove Carbon Pool Patterns in Maharashtra, India. J Indian Soc Remote Sens 52:735–746. https://doi.org/10.1007/s12524-024-01823-3

Patro S, Krishnan P, Deepak Samuel V, Purvaja R, Ramesh R (2017) Seagrass and salt marsh ecosystems in South Asia: an overview of diversity, distribution, threats and conservation status. Wetland Science Springer, New Delhi, pp 87–104

Ponmani M, Padmavathy P, Manimekalai D, Shalini R, Ravikumar T, Hariharan G, Manickavasagam S (2024) Vulnerability of mangrove ecosystems to anthropogenic marine litter along the south-east coast of India. Sci Total Environ 956:177224. https://doi.org/10.1016/j.scitotenv.2024.177224

Ragavan P, Kumar S, Kathiresan K, Mohan PM, Jayaraj RSC, Ravichandaran K, Rana TS (2021) Biomass and vegetation carbon stock in mangrove forests of the Andaman Islands, India. Hydrobiologia 848:4673–4693

Ragavan P, Rahman A, Sarkar S, Verma S, Jeeva C, Mohan PM, Kumar S (2023) Variability in soil organic carbon stock and isotopic signature in tropical Island mangrove forests of India. Reg Environ Chang 23:135

Rahman MM, Zimmer M, Ahmed I, Donato D, Kanzaki M, Xu M (2021) Co-benefits of protecting mangroves for biodiversity conservation and carbon storage. Nat Commun 12:1–9. https://doi.org/10.1038/s41467-021-24207-4

Ramesh D, Narayanan D, Karthi N, Infantina A, Parasuram P (2024) Assessment of Seagrass Ecosystems' Goods and Services of India. Asian J Geogr Res 7:149–160

Rani V, Nandan SB, Jayachandran PR, Preethy CM, Sreelekshmi S, Joseph P, Asha CV (2023a) Carbon stock in biomass pool of fragmented mangrove habitats of Kochi, Southern India. Environ Sci Pollut Res 30:96746–96762

Rani V, Schwing PT, Jayachandran PR, Preethy CM, Sreelekshmi S, Joseph P, Bijoy Nandan S (2023b) Carbon stocks and sequestration rate in mangroves and its major influencing factors from highly urbanised port city, southern India. J Environ Manag 335:117542. https://doi.org/10.1016/j.jenvman.2023.117542

Ranith RP, Nandini Menon N, Nobi EP, Raj AA, Sivaraj S (2024) Assessment of coral reef connectivity in improved organic carbon storage of seagrass ecosystems in Palk Bay, India. Mar Pollut Bull 207:116908. https://doi.org/10.1016/j.marpolbul.2024.116908

Rao K, Ramanathan A, Raju NJ (2022) Assessment of blue carbon stock of Coringa mangroves: climate change perspective. J Clim Chang 8:41–58. https://doi.org/10.3233/jcc220013

Rastogi RP, Phulwaria M, Gupta DK (2021) Mangroves: ecology, biodiversity and management. In: Mangroves Ecology Biodiverssity Management. Springer, pp 1–551. https://doi.org/10.1007/978-981-16-2494-0

Rathore AP, Chaudhary DR, Jha B (2016) Biomass production, nutrient cycling, and carbon fixation by Salicornia brachiata Roxb.: a promising halophyte for coastal saline soil rehabilitation

Ray R, Jana TK (2017) Carbon sequestration by mangrove forest: One approach for managing carbon dioxide emission from coal-based power plant. Atmos Environ 171:149–154. https://doi.org/10.1016/j.atmosenv.2017.10.019

Sachithanandam V, Bonthu S, Mageswaran T, Singh KS, Vimala J, Sridhar R, Purvaja R, Ramesh R (2022) Effect of hydrodynamic conditions on seagrass ecosystems during Cyclone Lehar in the South Andaman Islands, India. Ecohydrol Hydrobiol 22:640–659

Sandilyan S, Kathiresan K (2012) Mangrove conservation: A global perspective. Biodivers Conserv 21:3523–3542. https://doi.org/10.1007/s10531-012-0388-x

Sangma PSW, Rani V, Padmavathy P, Zuala V (2022) Differential capacity of carbon sequestration potential: analyzing Avicennia marina and Salicornia brachiata along the South-Eastern Coast of India. J Coast Res 38:377–388

Sarker S, Masud-Ul-Alam M, Hossain MS, Rahman Chowdhury S, Sharifuzzaman SM (2021) A review of bioturbation and sediment organic geochemistry in mangroves. Geol J 56:2439–2450

Sarma VVSS (2023) Role of inter-annual variability in the magnitude of river discharge on CO2 fluxes in the Godavari tidal river zone, India. Estuar Coast Shelf Sci 290:108393. https://doi.org/10.1016/j.ecss.2023.108393

Seleyi SC, Ravindran C, Mohandass C, Patra P, Naik TA (2024) Seasonal dynamics of mangrove outwelling and its impact on estuarine productivity: insights from multi-parameter analysis across tidal variations. Aquat Sci 86:1–18. https://doi.org/10.1007/s00027-024-01092-9

Sengupta T, Mitra A (2024) Abundance of microplastics in the mangrove dominated Indian Sundarbans: a mini-review. In: Plastic pollution challenges and green solutions. Springer, pp 125–136

ShivaShankar V, Narshimulu G, Kaviarasan T, Narayani S, Dharanirajan K, James RA, Singh RP (2019) 2004 Post Tsunami Resilience and Recolonization of Mangroves in South Andaman, India. Wetlands. https://doi.org/10.1007/s13157-019-01211-5

Shylaja G, Kripa V, Ranjith R (2021) Assessment of non-degradable litter and its impact on the benthic community of selected mangrove ecosystems of Kerala, India. J Mar Biol Assoc India 63:129–139. https://doi.org/10.6024/jmbai.2021.63.1.2249-19

Sidhu (1963) Studies on the mangroves of India. In: Ecology and productivity in tropics. Proceedings of the Indian Academy of Sciences, pp 129–136

Sigamani S, Kolandhasamy P, Behera DP, Sarathy PP, Chandra R, Aljarba NH, AL-Hazani TM, Moovendhan M, Dilipan E (2023) Assessment of blue carbon sequestration potential of Avicennia marina in the semi-arid zone of Gulf of Kutch, Gujarat India. Reg Stud Mar Sci 62:102937

Silliman BR (2014) Salt marshes. Curr Biol 24:R348–R350. https://doi.org/10.1016/j.cub.2014.03.001

Spalding M, Parrett CL (2019) Global patterns in mangrove recreation and tourism. Mar Policy 110:103540. https://doi.org/10.1016/j.marpol.2019.103540

Stankovic M, Ambo-Rappe R, Carly F, Dangan-Galon F, Fortes MD, Hossain MS, Kiswara W, Van Luong C, Minh-Thu P, Mishra AK, Noiraksar T, Nurdin N, Panyawai J, Rattanachot E, Rozaimi M, Soe Htun U, Prathep A (2021) Quantification of blue carbon in seagrass ecosystems of Southeast Asia and their potential for climate change mitigation. Sci Total Environ 783:146858. https://doi.org/10.1016/j.scitotenv.2021.146858

Sundquist E, Burruss R, Faulkner S, Gleason R, Harden J, Kharaka Y, Tieszen L, Waldrop M (2008) Carbon sequestration to mitigate climate change

Teagle H, Hawkins SJ, Moore PJ, Smale DA (2017) The role of kelp species as biogenic habitat formers in coastal marine ecosystems. J Exp Mar Bio Ecol 492:81–98

Thangaradjou T, Bhatt JR (2018) Status of seagrass ecosystems in India. Ocean Coast Manag 159:7–15

TheStateofForestReport (1987) Forest Survey of India, Ministry of Environment, Forest and Climate Change. Government of India

Tomlinson PB (2016) The botany of mangroves. Cambridge University Press

Thanabalan P, Vidhya R, Kankara R S, Manonmani R (2023) Time-series analysis of MODIS (LST and NDVI) and TRMM rainfall for drought assessment over India. Appl Geomatics 15:383–405. https://doi.org/10.1007/s12518-023-00505-y

Untawale AG (1987) Exploitation of mangroves in India. AIMS

Valsan G, Paul MM, Warrier AK, Vasudeva M, Unnikrishnan VK, Prabhu S (2024a) Spatiotemporal variations of microplastics in the surface sediments of a tropical mangrove ecosystem in south-western India. Case Stud Chem Environ Eng 9:100632. https://doi.org/10.1016/j.cscee.2024.100632

Valsan G, Warrier AK, Anusree S, Tamrakar A, Rangel-Buitrago N, Chand J (2024b) Seasonal variations in microplastics in a coastal wetland in southwest India as well as their risks to Sillago sihama and Gerres filamentosus. Environ Monit Assess 196:1–21. https://doi.org/10.1007/s10661-024-13400-9

Venkataraman K, Wafar M (2005) Coastal and marine biodiversity of India. Indian J Mar Sci 34:57–75. https://doi.org/10.1016/b978-0-12-801948-1.00019-7

Vinithkumar NV, Kumaresan S, Manjusha M, Balasubramanian T (1999) Organic matter, nutrients and major ions in the sediments of coral reefs and seagrass beds of Gulf of Mannar biosphere reserve, southeast coast of India. Indian J Mar Sci 28:383–393

Vinod K, Koya AA, Kunhi Koya VA, Silpa PG, Asokan PK, Zacharia PU, Joshi KK (2018) Biomass and carbon stocks in mangrove stands of Kadalundi Estuarine Wetland, south-west coast of India. Indian J Fish 65:89–99. https://doi.org/10.21077/ijf.2018.65.2.72473-11

Viswanathan C, Purvaja R, Joyson Joe Jeevamani J, Deepak Samuel V, Sankar R, Abhilash KR, Geevarghese GA, Muruganandam R, Gopi M, Raja S, Das RR, Patro S, Krishnan P, Ramesh R (2020) Salt marsh vegetation in India: Species composition, distribution, zonation pattern and conservation implications. Estuar Coast Shelf Sci 242:106792. https://doi.org/10.1016/j.ecss.2020.106792

Yousuf Dar J, Saha M, Taghizadeh Rahmat Abadi Z, Kantharajan G, Valappil Rajendran K (2023) Microplastics contamination in the gastropod, Telescopium telescopium, from the mangrove area of Versova Creek, Mumbai, India. Int J Aquat Biol 11:417–430. https://doi.org/10.22034/ijab.v11i5.1940

Chapter 3
Climate Change as Drivers of Microbial–Algal Interactions and Their Implications

Grinson George, Ancy C. Stoy, Vineetha Gopinath, Nanda Kishore, Kartik Baruah, Achamveetil Gopalakrishnan, Kattapuni Suresh Prithvisagar, and Indrani Karunasagar

Abstract Climate change is one of the crucial drivers of ecosystem dynamics in both freshwater and marine ecosystems that influences the intricate balance of the algal and microbial populations in the aquatic environment and their interactions thereof. The contribution of the algal and microbial population through their interactions is crucial in maintaining the milieu interior and functioning of aquatic ecosystems since they play a central role in nutrient cycling, primary production, and water quality maintenance. The most prominent and significant impact of climate change is the rise in temperature, especially sea surface temperatures, since the immediate effect of the elevated temperatures is the alteration of the metabolic rates of aquatic organisms due to accelerated biochemical processes such as photosynthesis and nutrient uptake. The increased temperatures also result in a shift of species distribution, creating a new dynamic in microbial–algal interactions, often leading to unpredictable and destructive consequences. Apart from the above, the increased absorption of atmospheric CO_2 by the oceans leads to ocean acidification, a reduction in seawater pH, altering carbonate chemistry, and impacting both algal productivity and microbial processes. Salinity fluctuation, driven by altered precipitation patterns, freshwater inflows, and sea level rise, is another dimension of complexity to microbial–algal dynamics. Extreme weather events, including cyclones, floods, and droughts, intensify nutrient loading in aquatic systems with increased runoff

G. George · A. C. Stoy · V. Gopinath · N. Kishore · A. Gopalakrishnan
Indian Council of Agricultural Research-Central Marine Fisheries Research Institute, Kochi, India

Cochin University of Science and Technology, Kochi, India

K. Baruah
Department of Applied Animal Science and Welfare, Swedish University of Agricultural Sciences, Uppsala, Sweden

K. S. Prithvisagar (✉) · I. Karunasagar (✉)
FAO Reference Centre for AMR & Aquaculture Biosecurity, DST TEC, Nitte University, Mangalore, India
e-mail: prithvi.sagar@nitte.edu.in; indrani.karunasagar@nitte.edu.in

B. P. Barathan et al. (eds.), *Navigating Climate Change: Impacts on Biodiversity and Ecosystem Resilience*, Sustainable Development and Biodiversity 42,
https://doi.org/10.1007/978-981-95-0409-1_3

from agricultural and urban areas, leading to an influx of excessive amounts of nitrogen and phosphorus into water bodies. Additionally, climate changes influence multiple co-stressors within the ecosystem, resulting in a combination of effectors in biological interactions. The alteration of these interactions in the marine environment often occurs in a species-specific way, which could have critical ecological and public health implications.

Keywords Microbial algal interactions · Life below the water · Climate change · Nutrient Cycling · Climate action

3.1 Introduction

Climate change has emerged as a critical driver of ecological transformations in both marine and freshwater ecosystems, significantly influencing the delicate interactions between microbial communities and algal populations. These interactions are essential to the functioning of aquatic ecosystems as they play a central role in nutrient cycling, primary production, and the maintenance of water quality. However, anthropogenic climate change has introduced new stressors, such as rising temperatures, ocean acidification, salinity fluctuations, and extreme weather events, which have disrupted these relationships, triggering far-reaching ecological and economic consequences (Doney et al. 2012).

One of the most pronounced impacts of climate change is the increase in global temperatures. Rising temperatures alter the metabolic rates of aquatic organisms, including algae and microbes, accelerating biochemical processes such as photosynthesis and nutrient uptake (Gomes and Juneau 2017). These changes can create conditions favorable for the proliferation of algal blooms, particularly harmful algal blooms (HABs), which often outcompete other planktonic species under warmer conditions (Griffith and Gobler 2020). Simultaneously, microbial community composition and functionality are reshaped, with some taxa adapting more effectively to elevated temperatures than others. Such shifts can destabilize essential nutrient cycling, including processes like nitrification and denitrification, which are essential for maintaining ecosystem balance (Wallenstein and Hall 2012). Furthermore, temperature-driven changes cause a shift in species distribution, where temperate species migrate poleward and tropical species reach their thermal limits. This creates new dynamics in microbial–algal interactions, often with unpredictable consequences (Jin and Agustí 2018).

Ocean acidification, another significant aspect of climate change, further complicates the interplay between microbes and algae. The increased absorption of atmospheric CO_2 by the oceans has led to a reduction in seawater pH, altering carbonate chemistry and impacting both algal productivity and microbial processes (Riebesell 2004). Algae, particularly calcifying species such as coccolithophores, experience reduced availability of carbonate ions, which hampers calcification and diminishes their ecological role as primary producers. Acidification also influences microbial enzymatic activity and the degradation of organic matter, disrupting nutrient recycling pathways that are critical for algal growth (Zhang et al. 2024). These interconnected processes have the potential to alter the structure and function of aquatic ecosystems, threatening biodiversity and the ecosystem services they provide.

Salinity fluctuations, driven by altered precipitation patterns, freshwater inflows, and sea level rise, add another layer of complexity to microbial–algal dynamics. In estuarine and brackish environments, where salinity gradients are already variable, these changes can have a far-reaching impact on community composition. Elevated salinity can stress freshwater algal and microbial species while favoring halotolerant or halophilic taxa (James et al. 2003). Such shifts can alter primary production rates and microbial activity, destabilizing food webs and reducing the resilience of these ecosystems to other climate-related stressors.

Extreme weather events, including cyclones, floods, and droughts, are becoming more frequent and intense due to climate change. These events intensify nutrient loading in aquatic systems by increasing runoff from agricultural and urban areas, leading to an influx of excessive amounts of nitrogen and phosphorus into water bodies (Bechmann and Stålnacke 2019). This nutrient surplus can drive eutrophication, promoting algal blooms and altering microbial composition. During decomposition, these blooms deplete dissolved oxygen levels, creating hypoxic or anoxic zones that are inhospitable to most aquatic organisms. Moreover, physical disturbances caused by extreme weather events, such as habitat destruction and changes in water quality, further destabilize microbial–algal interactions, with cascading effects on ecosystem functioning. Among the most visible consequences of disrupted microbial–algal interactions are harmful algal blooms (HABs). Fueled by rising temperatures and eutrophication, HABs can produce potent toxins that harm aquatic organisms (Gobler 2020). These toxins, combined with the blooms' ability to reduce oxygen availability through decomposition, pose significant risks to biodiversity and aquaculture. Additionally, the economic impact of HABs is profound, with fisheries and aquaculture operations facing substantial losses due to fish kills, shellfish contamination, and deteriorating water quality. Understanding the implications of climate change on microbial–algal interactions is crucial for predicting and mitigating its broader ecological and socioeconomic impacts. These interactions are not only integral to aquatic ecosystem health but also underpin the sustainability of human activities such as fisheries, aquaculture, and water resource management. This chapter aims to provide insights into the multifaceted effects of climate change on microbial and algal populations, highlighting the need for innovative strategies to enhance ecosystem resilience and support sustainable development in the face of environmental change.

3.2 Algal Bloom Dynamics and Microbial Interactions Due to Rise in Temperature

The increasing prevalence of algal blooms, particularly HABs, in aquatic ecosystems has been closely linked to the ongoing impacts of climate change. One of the key drivers of the increased prevalence of algal blooms is the warming of water bodies, which creates favorable conditions in terms of nutrient levels, temperature, and water chemistry for the rapid proliferation of algae. While algal blooms can occur naturally, anthropogenic climate change has significantly influenced their frequency, intensity, and geographical distribution (Dale et al. 2006). These shifts pose significant threats to both marine and freshwater ecosystems, with far-reaching ecological, economic, and societal implications.

Rising global temperatures alter thermal stratification patterns of water bodies, creating conditions that favor the growth and dominance of specific algal species. Notably, many HAB-forming species, such as *Cyanobacteria* in freshwater systems and *Karenia brevis* in marine systems, flourish in warmer conditions (Gärtner et al. 2021). Furthermore, elevated temperatures can accelerate algal metabolic rates and reproductive cycles, leading to a more rapid onset and longer duration of blooms. The impact of warming is often compounded by changes in precipitation patterns, which affect nutrient loading in aquatic ecosystems. Intense rainfall events, coupled with land-use changes, result in nutrient-rich runoff from agricultural and urban areas, stimulating algal growth (Gao et al. 2024). Conversely, prolonged droughts can concentrate nutrients in stagnant water bodies, further promoting bloom conditions. Climate-driven temperature rise can also expand the ecological niches and growth potential of certain HAB species (Glibert et al. 2014; Gobler et al. 2017). Warmer conditions have been linked to rapid proliferation and extended bloom seasons for several toxin-producing HAB species (Griffith and Gobler 2020). Large-scale metagenomic analyses reveal that temperature-driven biogeographic shifts in microbial–algal assemblages may result in poleward movement of community breakpoints, signaling potential regime shifts under global warming (Martin et al. 2021). However, the relationships between temperature, growth rate, and toxin production vary considerably across species and even strains. For instance, the growth of *Dinophysis* spp. tends to increase up to approximately 25 °C with concurrent elevation in the production of diarrhetic shellfish toxins (DSTs) (Kamiyama et al. 2010). Conversely, some other toxin-producing species showed an inverse relationship between growth rate and toxin production. In *Alexandrium* spp., while growth rates increased with temperature rise until reaching a strain-specific optimum, toxin levels tend to be higher in slower-growing cells maintained at lower temperatures (Hamasaki et al. 2001; Etheridge and Roesler 2005; Griffith and Gobler 2020). Similarly, yessotoxin-producing dinoflagellates within the genus *Protoceratium* exhibited per-cell toxin content in slower-dividing populations (Röder et al. 2012; Griffith and Gobler 2020). For other species, such as *Pseudo-nitzschia* spp., the interactions between temperature, growth, and toxin production remain poorly understood or highly variable. Studies have shown conflicting results, with some reporting variable production of toxins across temperature treatments (Lundholm et al. 1997), and others observing higher toxin content in slower-growing cells (Zhu et al. 2017; Griffith and Gobler 2020). Additionally, temperature-driven shifts in microbial diversity can impact algal bloom composition. Some bacteria, such as those in the genus *Roseobacter*, form mutualistic relationships with algae by providing essential nutrients like vitamins and trace elements. These interactions can enhance algal growth under optimal conditions. Conversely, other microbial taxa produce algicidal compounds that inhibit algal proliferation, potentially curbing bloom size and persistence (Di Costanzo et al. 2023).

3.3 Impact of Climate-Driven Salinity Changes on Algal–Microbial Relationships

Salinity is a key environmental factor that significantly impacts microbial–algal interactions in marine ecosystems, influencing the physiology, growth, and behavior of both microorganisms and algae. The ocean's salinity varies widely depending on location, depth, and environmental factors, such as evaporation rates, precipitation, and freshwater inflows from rivers or melting ice. This variation in salinity creates different ecological niches, where certain microbial and algal species are more competitive or better adapted than others. As a result, salinity plays a fundamental role in shaping the diversity, productivity, and dynamics of microbial–algal communities in marine environments (Röthig et al. 2023). Likewise, microbial responses to warming and hyposalinity in mesocosms highlight that such stressors can alter microbial loop dynamics, nutrient assimilation, and biomass production, with hyposalinity exerting the most pronounced effect on phytoplankton structure (Kaur-Kahlon et al. 2016; Turner et al. 2016).

Microalgae, which form the base of the marine food web, exhibit different tolerances to salinity, and their ability to thrive in specific salinity ranges determines the composition of algal communities. Diatoms and cyanobacteria are typically found in low-salinity regions, such as coastal estuaries, where they dominate the planktonic communities. In contrast, dinoflagellates tend to prefer higher salinities, particularly in open ocean environments or areas with limited freshwater influx (Reynolds 2006). Changes in salinity, whether driven by climate change or other natural processes, can shift the competitive balance between these groups, resulting in alterations to the community structure. Salinity affects not only the growth of algae but also their interactions with microorganisms. The availability of nutrients, such as nitrogen, phosphorus, and trace metals, is often linked to salinity gradients, which can impact the rates of nutrient cycling and the overall productivity of the ecosystem (Lundsør et al. 2022). Some microbial communities, including heterotrophic bacteria, are crucial in nutrient cycling, breaking down organic material, and converting it into forms that algae can utilize. In turn, algae release dissolved organic carbon, which serves as a food source for these bacteria. However, salinity can influence the rate of these processes, with some bacteria being more efficient at nutrient cycling in specific salinity ranges. Salinity also affects the osmotic regulation of microorganisms and algae, as they must maintain a balance between internal and external salt concentrations to survive. Furthermore, salinity can influence the production of HABs, which are often linked to increased nutrient loads, water temperature, and salinity changes. In regions where salinity increases due to decreased freshwater inputs, certain HAB species, such as dinoflagellates, may become more prevalent, leading to toxic blooms that disrupt marine ecosystems and fisheries. These blooms not only impact biodiversity and ecosystem health but also affect the microbial communities associated with them, as bacteria interact with the algal cells, either promoting or inhibiting bloom formation and toxin accumulation (Hallegraeff 2010).

3.4 Ocean Acidification and Algal–Microbial Relationships

Ocean acidification, resulting from increased atmospheric CO_2 absorption, significantly affects microbial–algal interactions. Acidified conditions affect the physiology of many algal species, including their ability to photosynthesize and uptake nutrients, particularly those capable of utilizing dissolved inorganic carbon (Mackey et al. 2015). This advantage can lead to shifts in species composition, favoring harmful algal species over benign ones. These changes influence the entire microbial community, as bacteria that rely on algal exudates must adapt to shifts in algal productivity and composition. Microbial processes such as nitrification and denitrification are also sensitive to changes in pH (Albina et al. 2019). Acidified environments can alter the availability of nitrogen and other nutrients, influencing algal bloom dynamics. Certain nitrogen-fixing bacteria may thrive under lower pH conditions, providing an additional nutrient source that could fuel algal growth (Allison et al. 1937). Ocean acidification can also suppress microbial populations responsible for degrading algal biomass, prolonging the duration of blooms and increasing the risk of hypoxic conditions. Furthermore, pH can influence the metabolic pathways of microbial communities, altering their interactions with algal populations (Joint et al. 2011). The interplay between microbes and algae is further complicated by the increasing prevalence of hypoxic zones in aquatic systems. These low-oxygen areas, often linked to intense algal blooms, can disrupt microbial processes such as nutrient cycling and organic matter decomposition, creating feedback loops that sustain bloom conditions. While ocean deoxygenation is increasingly recognized as a significant climate change stressor (Breitburg et al. 2018; Griffith and Gobler 2020), the interactions between HABs and hypoxia remain poorly understood. It is often assumed that HABs and hypoxia do not occur simultaneously, as the photosynthetic activity of many HABs would typically help maintain oxygen levels in the ecosystem. However, numerous HABs are also capable of heterotrophic behavior, relying on organic matter or prey for growth (Flynn et al. 2018; Glibert and Burkholder 2018), which leads to oxygen depletion through respiration.

This effect is exacerbated in dense blooms, where the high biomass blocks light and reduces photosynthesis (Sunda et al. 2006).

3.5 Nutrient Loading and Eutrophication

Climate change significantly influences nutrient dynamics in aquatic ecosystems, intensifying nutrient loading through increased runoff from extreme precipitation events and rising sea levels. This enhanced nutrient influx, particularly nitrogen and phosphorus, drives eutrophication, promoting conditions conducive to algal blooms (Kausar 2024). These blooms are closely linked to microbial–algal interactions, which play a pivotal role in nutrient cycling and ecosystem stability. Microbial communities, particularly heterotrophic bacteria, exhibit a dual function in these processes—recycling nutrients to sustain algal growth and facilitating nutrient removal through processes, such as denitrification (Song et al. 2021), thereby mitigating nutrient overloading. During algal blooms, interactions between algae and bacteria are especially pronounced. Algae release dissolved organic carbon as a by-product of photosynthesis, providing an energy-rich substrate for bacterial growth (Hall et al. 2022). In turn, bacteria mineralize organic matter, releasing nutrients like ammonium, which algae can readily uptake to fuel further bloom proliferation. This mutualistic exchange creates a feedback loop that sustains the intensity of blooms. However, excessive nutrient cycling driven by these interactions can destabilize ecosystems. Prolonged or intense blooms can cause hypoxic conditions as microbial respiration depletes oxygen, threatening aquatic life (Lan et al. 2024). Additionally, some algal species produce toxins during bloom events, and bacterial activity can exacerbate toxin accumulation by maintaining bloom persistence. These cascading effects emphasize the complexity of microbial–algal interactions in light of climate change.

Eutrophication, influenced by climate change, has led to significant changes in ecosystems worldwide. In particular, the Arabian Gulf and Arabian Sea have witnessed the undeniable expansion of HABs (Goes and Gomes 2016). Urbanization, such as the rapid population growth in Mumbai, has resulted in a sharp increase in nutrient loading, further aggravated by minimal advancements in sewage treatment. Concurrently, regional oceanographic shifts, such as stronger summer monsoonal winds, have enhanced nutrient upwelling, further promoting eutrophication. Since around 2000, the northern Arabian Sea has seen a shift from a winter phytoplankton assemblage dominated by diatoms to one increasingly dominated by *Noctiluca scintillans*, a large dinoflagellate. This mixotrophic species thrives in nutrient-rich waters, relying on symbiotic prasinophyte algae for carbon fixation and preying on other organisms. *Noctiluca* blooms, which were previously unseen in the 1990s, have become over 300% more intense, marking a clear manifestation of eutrophication and its consequences on the marine ecosystem (Goes and Gomes 2016). In polluted waters, algal–microbiome associations within the phycosphere facilitate mid- to long-term adaptation through community restructuring and gene transfer, suggesting a mutualistic "PollutantBiome" that enhances resilience and bioremediation potential (Ashraf et al. 2023).

3.6 Harmful Algal Blooms and Microbial Dynamics

Harmful algal blooms (HABs) pose unique challenges due to the toxins they produce that impact both aquatic organisms and human health. Microbial communities influence HAB dynamics in complex ways. Certain bacteria, such as those in the genus *Pseudoalteromonas*, produce algicidal compounds that suppress HABs (Wang et al. 2022). Conversely, other bacteria may facilitate HABs by degrading organic matter (Lan et al. 2024) and releasing nutrients that sustain algal growth. Climate-induced changes in microbial diversity can also affect the toxicity of HABs. Some bacterial species enhance the production of algal toxins by providing essential cofactors or influencing algal gene expression. These interactions emphasize the crucial roles microbial communities play in determining the ecological and economic impacts of HABs. These interactions are complex, encompassing mutualistic, competitive, and predatory relationships. Some bacteria promote algal growth by supplying essential nutrients like nitrogen and phosphorus (Yao et al. 2019), while in some cases, bacteria form symbiotic associations with algae, exchanging nutrients and metabolites in a mutually beneficial manner (Cole 1982). However, other microbial interactions act as natural checks on bloom development. Specific bacterial taxa produce algicidal compounds that inhibit algal growth or lead to cell lysis (Kuhlisch et al. 2024). Additionally, microbial grazers, including protists and zooplankton, contribute to the top-down control of algal populations, though their effectiveness depends on bloom intensity and environmental conditions.

Microorganisms in the oceans can even trigger positive feedback mechanisms that influence climate. A notable example is the release of dimethyl sulfide (DMS), a compound emitted during the bacterial and eukaryotic breakdown of dimethyl sulfoniopropionate (DMSP), which certain microalgae produce for osmoprotection. Once released into the atmosphere, DMS is converted into sulfate aerosols, which serve as cloud condensation nuclei. These aerosols contribute to an increase in cloud albedo, reflecting more sunlight and subsequently exerting a cooling effect on the surface. This process can negatively impact algae growth (Glöckner et al. 2012), which in turn reduces the production of DMSP and consequently lowers DMS emissions.

3.7 Extreme Weather Events and Algal–Microbial Interactions

Extreme weather events, such as cyclones and floods, introduce additional stressors that influence algal–microbial interactions. These events often lead to abrupt changes in water quality, including increased turbidity, altered salinity, and nutrient loading (da Costa et al. 2022). Such disruptions can create conditions conducive to algal blooms while simultaneously impacting the microbial community structure. Flooding can introduce terrestrial microbes into aquatic systems, altering the balance of microbial–algal interactions (Janauer and Dokulil 2006). In some cases, this can lead to the proliferation of opportunistic pathogens that thrive in nutrient-rich, post-flood environments. Cyclones, on the other hand, can cause physical damage to algal populations, disrupting bloom formation (Phlips et al. 2020). However, the nutrient pulses following such events can lead to secondary blooms, sustained by microbial nutrient recycling.

Waterborne diseases, caused by exposure to contaminated water containing harmful pathogens, such as bacteria, viruses, and parasites (Manetu and Karanja 2021), are a significant public health challenge (Shrestha et al. 2023), particularly in areas with limited access to clean drinking water, sanitation, and hygiene practices. Contaminated water can lead to outbreaks affecting millions worldwide, especially in developing countries (Ashbolt 2004; Pandey et al. 2014). It serves as a breeding ground for pathogens like *Escherichia coli, Vibrio cholerae, Hepatitis* A virus, and parasites such as *Giardia* and *Cryptosporidium*, causing severe illnesses like diarrhea, cholera, dysentery, and hepatitis (Nwabor et al. 2016; Gerba 2009). Bacterial contamination is a major global health issue, particularly in regions with inadequate sanitation and water treatment infrastructure (Omohwovo 2024). It can result in various illnesses transmitted through drinking water, recreational water, or food prepared with contaminated water (Cabral 2010).

Climate change worsens the impact of waterborne diseases (Cissé 2019) through a series of interconnected processes that affect both the environment and human health. Rising water temperatures accelerate the growth and replication of bacteria like *Vibrio cholerae* and *Escherichia coli*, increasing the incidence of waterborne infections (Vezzulli et al. 2013). Increased precipitation and flooding contribute to water contamination by runoff from agricultural, industrial, and urban areas, introducing pathogens into water bodies (Ponting et al. 2021) and overwhelming sewage treatment facilities, which can result in untreated sewage contaminating water sources (Wallender et al. 2014). Extended droughts worsen water quality issues by concentrating contaminants in reduced water volumes and stressing sanitation systems (European Environment Agency 2025; Van Vliet & Zwolsman 2008), which leads to inadequate waste disposal. Increased sea level and coastal inundation can cause seawater intrusion into freshwater sources (Rasmussen et al. 2013), increasing salinity and the concentration of harmful microorganisms. The degradation of ecosystems, such as the loss of wetlands and vegetation due to climate change, also impairs natural filtration systems, resulting in higher pathogen concentrations in water bodies and further compromising water quality. A study by Abdulaziz et al. (2023) examined fecal contamination and acute diarrheal disease in Vembanad Lake, India, over a year from 2018 to 2019. *Escherichia coli*, an indicator of fecal contamination, was detected year-round, with peak levels during the southwest monsoon (June–July), coinciding with increased cases of acute diarrheal disease. Heavy rainfall likely contributed to septic and sewage effluents mixing with surface runoff, elevating *E. coli* levels (Tryland et al. 2011). The study found that rainfall

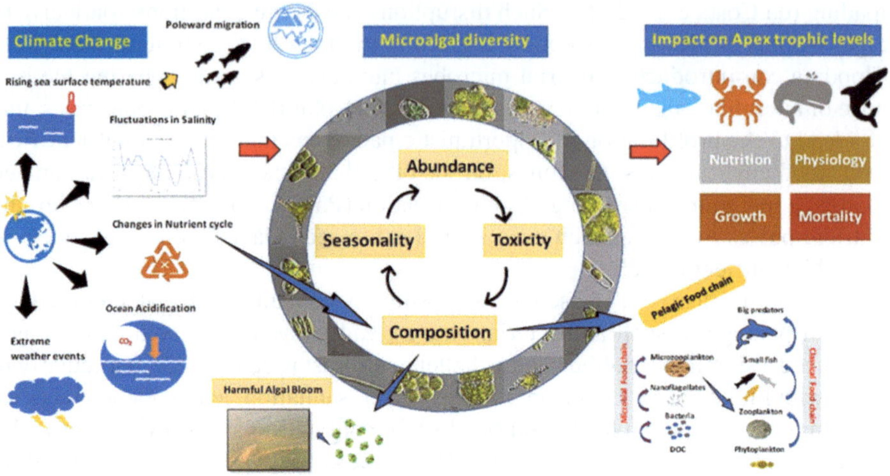

Fig. 3.1 An overview of the combined effect of climate change stressors on algal–microbial interactions and their implications on the marine environments

and flooding can significantly increase fecal contamination, intensified by malfunctioning sewage systems during the monsoon.

3.8 Combined Stressor Effects

The climate change outcomes, such as temperature, salinity, acidification, and nutrient efflux, have rarely been influencing algal–microbial interactions in isolation. These outcomes always exert a combined effect and act in conjunction as a response to climate change (Fig. 3.1). Instead of direct influence on microbial–algal interactions, the climate change influence multiple co-stressors within the ecosystem resulting in a combination of effectors in biological interactions. Evidence suggests rising temperatures as predictors of ocean deoxygenation (Diaz and Rosenberg 2008; Breitburg et al. 2018), eutrophication leading to seasonal hypoxia and acidification (Cai et al. 2011; Wallace et al. 2014; Baumann et al. 2015), habitat destruction (Waycott et al. 2009; Hoegh-Guldberg and Bruno 2010; Friedland et al. 2013), increased disease prevalence (Burge et al. 2014), and persistence of organic pollutants and pesticides (Noyes et al. 2009). The climate change stressors, thus, co-exhibit a complex effect on aquatic life (Griffith and Gobler 2020). The combined effects of these stressors often result in synergistic or antagonistic effects that reshape the microbial–algal community composition, disrupt symbiotic/mutualistic interactions and bring about shift in microbial functions with the consequent change from beneficial microbiomes to opportunistic pathogens (Mensch et al. 2020).

3.9 Understanding the Species-Specific Interactions

Climate change is reshaping microbial–algal interactions in marine ecosystems, often in species-specific ways that influence ecosystem functioning, microbial pathogenicity, and biogeochemical cycles. The alteration of these interactions in the marine environment often occurs in a species-specific way, which could have critical ecological and public health implications. Previous studies from tropical and temperate environments reveal that algal blooms, in particular the diatom-dominated blooms, significantly influence the abundance and distribution of *Vibrio* spp. Several studies have shown that diatom blooms, particularly of *Coscinodiscus* promote the growth of *Vibrio* spp., while other genera like *Chaetoceros* and *Skeletonema* exert a negative or neutral effect, which is a pointer to the species-specific influences within phytoplankton communities and highlights species-specific phytoplankton–bacteria interactions (Peterson et al. 2010; Rehnstam-Holm et al. 2010). The effect of these algae on the population dynamics of *Vibrio* spp. has been observed even under stable environmental conditions. The effect of factors such as temperature and salinity demonstrates that organic substrates from algal exudates can also influence the microbial populations apart from the physical drivers of climate change (Rehnstam-Holm et al. 2010; Asplund et al. 2011). Field observations in tropical oligotrophic coastal waters reveal that total *Vibrio* and *V. parahaemolyticus* abundance peak during diatom-dominated periods, with phosphate and copepod abundance acting as strong environmental correlates, even in the absence of significant temperature or salinity variation (Rehnstam-Holm and Godhe 2012; Rehnstam-Holm et al. 2014). Experimental studies show that diatoms rather than dinoflagellates support higher bacterial growth, with phytoplankton community shifts significantly impacting the abundance and dynamics of *Vibrio* population (Rehnstam-Holm and Godhe 2012; Shahi et al. 2015). Yet another study on *Vibrio* spp. abundance in relation to plankton community conducted during two temporally separated oligotrophic conditions with stable temperature and salinity revealed an increase in the abundance of *Vibrio* spp. population that coincided with a diatom-dominated algal density, indicating that microbial populations are directly influenced by primary producers than microbial predators (Asplund et al. 2011). An interesting study conducted in the Baltic Sea on spatial–temporal coupling between phytoplankton (notably *Skeletonema marinoi*) and bacterial taxa showed less abundant bacterial phyla such as Verrucomicrobia and Planctomycetes to be tightly linked to phytoplankton biomass and the diatom–dinoflagellate ratio, implying that microbial diversity within algal blooms could structure bacterioplankton niches (Bunse et al. 2016). Further, climate change stressors such as warming and hyposalinity can exacerbate or buffer the impacts of pathogenic microbes and toxin-producing algae on bivalve hosts, with complex, often nonadditive effects on host physiology and microbial load. The study observed a significant interaction between climate change and microbes that influenced the metabolic and/or immunological functions, including the accumulation of pathogens/toxins load in bivalve tissues (Turner et al. 2016). Microbial–algal communication is also modulated by quorum sensing. Evidence comes from studies showing *Vibrio harveyi* interacting with *S. marinoi* without disrupting virulence gene regulation, suggesting possible commensalism (Kaur-Kahlon et al. 2021).

3.10 Role of Viruses in Algal–Microbial Networks

Marine viruses are one of the most abundant groups of organisms among the global ecosystems accounting for up to ~10^{30} particles in the world's oceans that is equivalent to ten-fold higher than prokaryotic abundance and play a crucial role in shaping the population composition and dynamics of marine ecosystems (Danovaro et al. 2011). The viral population is greatly influenced by changing physico-chemical parameters of the ocean, such as temperature, salinity, water turbulence and mixing, availability of nutrients, since they are also drivers of population dynamics and metabolism of prokaryotic and eukaryotic hosts (Wommack and Colwell 2000; Weinbauer 2004). Viruses are one of the pivotal players responsible for shaping the algal–microbial networks in response to stressors of climate change, altering microbial diversity, nutrient cycling, and dynamics of water bodies (Wieczynski et al. 2023). Since viruses are recorded pathogens of both microbes and algae, they regulate both algal blooms as well as microbial succession through lysis, resulting in the release of dissolved organic matter (DOM) that fuels the microbial loop. This phenomenon is also known as viral shunt, which redirects carbon and nutrients away from higher trophic consumption toward microbial respiration and recycling (Krishna and Valiya 2023). Viral lysis of blood-forming cyanobacteria like *Nodularia spumigena* can disrupt associated microbiomes, decreasing bacterial fitness and altering nutrient flow through the microbial loop (Šulčius et al. 2025). The climate change stressors exacerbate these algal–microbial–viral interactions through altered temperature, stratification, and physico-chemical composition of the oceans, leading to a shift in host-virus dynamics and acceleration of viral-mediated lysis. As an illustration, warmer temperatures may cause a rise in viral infection rates or enhance host susceptibility, intensifying DOM production and affecting carbon sequestration in marine environments (Locke et al. 2022). In addition, they can mediate horizontal gene transfer, improving microbial adaptability to varying conditions, and contribute to algal bloom termination, influencing both biogeochemical cycles and food web stability (Zhang et al. 2022). Thus, it is important to understand these critical interactions, as viral modulation of algal–microbial communities could amplify or prevent climate change feedback in the marine ecosystem.

3.11 Microbial–Algal Feedback on Climate Processes

The fundamental component of marine biogeochemical cycles are algal-microbial interactions that provides a critical feedback on global climate processes by driving the fate of carbon and other nutrients in marine ecosystems (Wang and Wang 2024).

The phytoplankton play a central role being primary photosynthetic organisms of the marine food web. They convert inorganic CO_2 into organic carbon during photosynthesis by atmospheric CO_2 fixation (Sigman and Hain 2012). Intraspecific variation within bloom-forming algae, such as *Skeletonema marinoi* and *Alexandrium ostenfeldii*, shows that different strains respond variably to increased temperature and pCO_2, affecting not only growth but also toxin production, with implications for HAB dynamics and seafood safety (Kremp et al. 2012). A large proportion of this organic carbon is transferred across the food pyramid to higher trophic levels or exported to deeper ocean through promoting organic particulate matter sinks, thereby resulting in long-term carbon sequestration. This process is strongly mediated by the cohabitant microbial communities, in particular the heterotrophic bacteria that colonize phytoplankton-derived organic matter, degrade, and remineralize them. Thus, a larger proportion of carbon being exported to deeper oceans is consequently prevented, thereby regulating the strength of the carbon sink (Buchan et al. 2014). The algal–microbial feedback mechanisms extend beyond the carbon cycle. The variety of organic matter produced by phytoplanktons, such as DMSP, is metabolized by bacteria (Töpel et al. 2018; O'Brien et al. 2022) to form dimethyl sulfide (DMS), a major source of cloud-condensation nuclei (CCN) that helps in Earth's radiation balance and climate regulation (Charlson et al. 1987). However, when the climate change-induced stressors such as temperature and salinity bring about ocean stratification leading to acidification and altered nutrient composition, it adversely affects the phytoplankton community structure, and results in dominance by smaller and faster-growing picophytoplanktons (Falkowski et al. 2008). This shift in algal population alters the quantity and quality of organic matter produced, which contributes to changes in the selection of microbial consortia affecting the degradation pathways, nutrient cycling, and release of gases to the environment (Azam and Malfatti 2007). The algae-associated microbial community that constitutes symbionts and epibionts influences the host physiology and bloom dynamics by providing essential cofactors (Töpel et al. 2019), scavenging the reactive oxygen species, stimulating the algal growth via signaling molecules, or the inhibition of algal blooms through the production of anti-algal metabolites (Buchan et al. 2014). These algal–microbial intricate interactions may be mutualistic or antagonistic and result in a cascading effect on time, duration, and magnitude of algal blooms. This, in turn, affects the biogeochemical cycles, ultimately leading to climate change (Abate et al. 2024). HABs like those of *Alexandrium fundyense* and *Dinophysis acuminata* host distinct bacterial communities that influence bloom dynamics, sometimes even promoting parasitic feedback loops involving species like *Amoebophrya* (Hattenrath-Lehmann and Gobler 2017). Additionally, bacterial activity such as microbially induced calcite precipitation (MICP) by Bacillaceae illustrates how microbial metabolism can respond to environmental calcium stress, with potential applications in climate-resilient infrastructure development (Seidel et al. 2025). The structure and function of algal-microbial networks are predicted to evolve as climate change continues to alter the physical and chemical characteristics of the ocean, generating intricate feedback that could either lessen or increase the effects of climate change. Therefore, to effectively forecast future ocean-atmosphere feedbacks and biogeochemical fluxes, microbial ecology must be integrated with climate models.

3.12 Conclusion

To conclude, climate change profoundly reshapes the interactions between microbial and algal populations, with broader implications on homeostasis and health of aquatic environments and public health. The increase in temperature, ocean acidification, salinity fluctuations, and extreme weather events are interconnected stressors that influence and disrupt the well-balanced mineralization processes involving nutrient cycling, primary production, nutrient uptake, and overall stability of the ecosystem. The consequences due to these combined stressors, such as harmful algal blooms and imbalance in biodiversity due to loss of habitat or species, underscore the urgency needed for a comprehensive approach and research toward adaptive management and mitigating strategies. With an in-depth understanding of the algal–microbial dynamics, a better ecological response could be predicted to develop a sustainable solution-based algorithm to safeguard aquatic environments, especially in the era of rapid environmental changes happening globally.

Acknowledgments The authors are thankful to the reviewers and editors for constructive feedback, and for helping to improve the quality of the work. The authors from CMFRI would kindly like to acknowledge the project National Innovations in Climate Resilient Agriculture (NICRA) for the institutional facilities provided.

Competing Interests The authors declare no competing interests.

Ethical Approval This review has not been previously published elsewhere.

Consent for Publication All the authors actively participated in the present review chapter. Those who extended the support have been acknowledged.

Data Availability All the data and materials are presented as tables and figures.

References

Abate R, Oon YL, Oon YS, Bi Y, Mi W, Song G, Gao Y (2024) Diverse interactions between bacteria and microalgae: a review for enhancing harmful algal bloom mitigation and biomass processing efficiency. Heliyon 10:e36503

Abdulaziz A, Vikraman HK, Raj D, Menon N, George G, Soman R, Sathyendranath S (2023) Distribution and antibiotic resistance of vibrio population in an urbanized tropical lake—the Vembanad—in the southwest coast of India. Environ Sci Pollut Res 30(54):116066–116077

Albina P, Durban N, Bertron A, Albrecht A, Robinet JC, Erable B (2019) Influence of hydrogen electron donor, alkaline pH, and high nitrate concentrations on microbial denitrification: a review. Int J Mol Sci 20(20):5163

Allison FE, Hoover SR, Morris HJ (1937) Physiological studies with the nitrogen-fixing alga, Nostoc muscorum. Bot Gaz 98(3):433–463

Ashbolt NJ (2004) Microbial contamination of drinking water and disease outcomes in developing regions. Toxicology 198(1–3):229–238

Ashraf N, Ahmad F, Lu Y (2023) Synergy between microalgae and microbiome in polluted waters. Trends Microbiol 31(1):9–21

Asplund ME, Rehnstam-Holm AS, Atnur V, Raghunath P, Saravanan V, Härnström K, Collin B, Karunasagar I, Godhe A (2011 Oct) Water column dynamics of vibrio in relation to phytoplankton community composition and environmental conditions in a tropical coastal area. Environ Microbiol 13(10):2738–2751

Azam F, Malfatti F (2007 Oct) Microbial structuring of marine ecosystems. Nat Rev Microbiol 5(10):782–791

Baumann H, Wallace RB, Tagliaferri T, Gobler CJ (2015 Jan) Large natural pH, CO_2 and O_2 fluctuations in a temperate tidal salt marsh on diel, seasonal, and interannual time scales. Estuar Coasts 38:220–231

Bechmann M, Stålnacke P (2019) Agricultural nitrogen and phosphorus pollution in surface waters. In: Oxford research encyclopedia of environmental science

Breitburg D, Levin LA, Oschlies A, Grégoire M, Chavez FP, Conley DJ, Garçon V, Gilbert D, Gutiérrez D, Isensee K, Jacinto GS (2018) Declining oxygen in the global ocean and coastal waters. Science 359(6371):eaam7240

Buchan A, LeCleir GR, Gulvik CA, González JM (2014 Oct) Master recyclers: features and functions of bacteria associated with phytoplankton blooms. Nat Rev Microbiol 12(10):686–698

Bunse C, Bertos-Fortis M, Sassenhagen I, Sildever S, Sjöqvist C, Godhe A, Gross S, Kremp A, Lips I, Lundholm N, Rengefors K (2016) Spatio-temporal interdependence of bacteria and phytoplankton during a Baltic Sea spring bloom. Front Microbiol 7:517

Burge CA, Mark Eakin C, Friedman CS, Froelich B, Hershberger PK, Hofmann EE, Petes LE, Prager KC, Weil E, Willis BL, Ford SE (2014) Climate change influences on marine infectious diseases: implications for management and society. Annu Rev Mar Sci 6(1):249–277

Cabral JP (2010) Water microbiology. Bacterial pathogens and water. Int J Environ Res Public Health 7(10):3657–3703

Cai WJ, Hu X, Huang WJ, Murrell MC, Lehrter JC, Lohrenz SE, Chou WC, Zhai W, Hollibaugh JT, Wang Y, Zhao P (2011) Acidification of subsurface coastal waters enhanced by eutrophication. Nat Geosci 4:766–770

Charlson RJ, Lovelock JE, Andreae MO, Warren SG (1987) Oceanic phytoplankton, atmospheric Sulphur, cloud albedo and climate. Nature 326(6114):655–661

Cissé G (2019) Food-borne and water-borne diseases under climate change in low-and middle-income countries: further efforts needed for reducing environmental health exposure risks. Acta Trop 194:181–188

Cole JJ (1982) Interactions between bacteria and algae in aquatic ecosystems. Annu Rev Ecol Syst 13:291–314

da Costa ÁKR, Pereira LCC, Jiménez JA, de Oliveira ARG, de Jesus Flores-Montes M, da Costa RM (2022) Effects of extreme climatic events on the hydrological parameters of the estuarine waters of the Amazon coast. Estuar Coasts 45(6):1517–1533

Dale B, Edwards M, Reid PC (2006) Climate change and harmful algal blooms. In: Ecology of harmful algae. Springer Berlin Heidelberg, Berlin/Heidelberg, pp 367–378

Danovaro R, Corinaldesi C, Dell'Anno A, Fuhrman JA, Middelburg JJ, Noble RT, Suttle CA (2011) Marine viruses and global climate change. FEMS Microbiol Rev 35(6):993–1034

Di Costanzo F, Di Dato V, Romano G (2023) Diatom–bacteria interactions in the marine environment: complexity, heterogeneity, and potential for biotechnological applications. Microorganisms 11(12):2967

Diaz RJ, Rosenberg R (2008) Spreading dead zones and consequences for marine ecosystems. Science 321(5891):926–929

Doney SC, Ruckelshaus M, Duffy JE, Barry JP, Chan F, English CA, Galindo HM, Grebmeier JM, Hollowed AB, Knowlton N, Polovina J, Rabalais NN, Sydeman WJ, Talley LD (2012) Climate change impacts on marine ecosystems. Annu Rev Mar Sci 4:11–37

Etheridge SM, Roesler CS (2005) Effects of temperature, irradiance, and salinity on photosynthesis, growth rates, total toxicity, and toxin composition for Alexandrium fundyense isolates from the Gulf of Maine and Bay of Fundy. Deep-Sea Res II Top Stud Oceanogr 52(19–21):2491–2500

European Environment Agency (n.d.) Drought and water scarcity. Climate-ADAPT. Retrieved February 18, 2025, from https://climate-adapt.eea.europa.eu/en/observatory/evidence/health-effects/drought-and-water-scarcity

Falkowski PG, Fenchel T, Delong EF (2008) The microbial engines that drive earth's biogeochemical cycles. Science 320(5879):1034–1039

Flynn KJ, Mitra A, Glibert PM, Burkholder JM (2018) Mixotrophy in harmful algal blooms: by whom, on whom, when, why, and what next. In: Glibert P, Berdalet E, Burford M, Pitcher G, Zhou M (eds) Global ecology and oceanography of harmful algal blooms, vol 232. Springer, Cham, pp 113–132

Friedland KD, Kane J, Hare JA, Lough RG, Fratantoni PS, Fogarty MJ, Nye JA (2013) Thermal habitat constraints on zooplankton species associated with Atlantic cod (Gadus morhua) on the US Northeast Continental Shelf. Prog Oceanogr 116:1–3

Gao G, Xiao R, Li Y (2024) Influences of extreme rainfall events on the nutrient and chlorophyll-a dynamics in coastal regions. In: Current trends in estuarine and coastal dynamics. Elsevier, pp 339–383

Gerba CP (2009) Environmentally transmitted pathogens. In: Environmental microbiology. Academic, pp 445–484

Glibert PM, Burkholder JM (2018) Causes of harmful algal blooms. In: Harmful algal blooms: a compendium desk reference, pp 1–38

Glibert PM, Icarus Allen J, Artioli Y, Beusen A, Bouwman L, Harle J et al (2014) Vulnerability of coastal ecosystems to changes in harmful algal bloom distribution in response to climate change: projections based on model analysis. Glob Chang Biol 20(12):3845–3858

Glöckner FO, Gasol JM, McDonough N, Calewaert JB (2012) Marine microbial diversity and its role in ecosystem functioning and environmental change. European Science Foundation

Gobler CJ (2020) Climate change and harmful algal blooms: insights and perspective. Harmful Algae 91:101731

Gobler CJ, Doherty OM, Hattenrath-Lehmann TK, Griffith AW, Kang Y, Litaker RW (2017) Ocean warming since 1982 has expanded the niche of toxic algal blooms in the North Atlantic and North Pacific oceans. Proc Natl Acad Sci 114(19):4975–4980

Goes JI, Gomes HDR (2016) An ecosystem in transition: the emergence of mixotrophy in the Arabian Sea. In: Aquatic microbial ecology and biogeochemistry: a dual perspective, pp 155–170

Gomes MP, Juneau P (2017) Temperature and light modulation of herbicide toxicity on algal and cyanobacterial physiology. Front Environ Sci 5:50

Griffith AW, Gobler CJ (2020) Harmful algal blooms: a climate change co-stressor in marine and freshwater ecosystems. Harmful Algae 91:101590

Gärtner G, Stoyneva-Gärtner M, Uzunov B (2021) Algal toxic compounds and their aeroterrestrial, airborne and other extremophilic producers with attention to soil and plant contamination: a review. Toxins 13(5):322

Hall JR, Albert G, Twigg IM, Baltar F, Hepburn CD, Martin G (2022) The production of dissolved organic carbon by macroalgae and its consumption by marine bacteria: implications for coastal ecosystems. Front Mar Sci 9:934229

Hallegraeff GM (2010) Ocean climate change, phytoplankton community responses, and harmful algal blooms: a formidable predictive challenge 1. J Phycol 46(2):220–235

Hamasaki K, Horie M, Tokimitsu S, Toda T, Taguchi S (2001) Variability in toxicity of the dinoflagellate Alexandrium tamarense isolated from Hiroshima Bay, western Japan, as a reflection of changing environmental conditions. J Plankton Res 23(3):271–278

Hattenrath-Lehmann TK, Gobler CJ (2017 Sep) Identification of unique microbiomes associated with harmful algal blooms caused by Alexandrium fundyense and Dinophysis acuminata. Harmful Algae 1(68):17–30

Hoegh-Guldberg O, Bruno JF (2010) The impact of climate change on the world's marine ecosystems. Science 328(5985):1523–1528

James KR, Cant B, Ryan T (2003) Responses of freshwater biota to rising salinity levels and implications for saline water management: a review. Aust J Bot 51(6):703–713

Janauer G, Dokulil M (2006) Macrophytes and algae in running waters. In: Biological monitoring of rivers Application and perspectives, pp 89–109

Jin P, Agustí S (2018) Fast adaptation of tropical diatoms to increased warming with trade-offs. Sci Rep 8(1):17771

Joint I, Doney SC, Karl DM (2011) Will ocean acidification affect marine microbes? ISME J 5(1):1–7

Kamiyama T, Nagai S, Suzuki T, Miyamura K (2010) Effect of temperature on production of okadaic acid, dinophysistoxin-1, and pectenotoxin-2 by Dinophysis acuminata in culture experiments. Aquat Microb Ecol 60(2):193–202

Kaur-Kahlon G, Kumar S, Rehnstam-Holm AS, Rai A, Bhavya PS, Edler L, Singh A, Andersson B, Karunasagar I, Ramesh R, Godhe A (2016) Response of a coastal tropical pelagic microbial community to changing salinity and temperature. Aquat Microb Ecol 77(1):37–50

Kaur-Kahlon G, Kumar BK, Ruwandeepika HA, Defoirdt T, Karunasagar I (2021) Quorum sensing regulation of virulence gene expression in Vibrio harveyi during its interaction with marine diatom Skeletonema marinoi. J Pure Appl Microbiol 15(4):2507–2519

Kausar R (2024) Algae at nexus of eutrophication and climate change. In: Algae as a natural solution for challenges in water-food-energy nexus: toward carbon neutrality. Springer, Singapore, pp 851–860

Kremp A, Godhe A, Egardt J, Dupont S, Suikkanen S, Casabianca S, Penna A (2012 Jun) Intraspecific variability in the response of bloom-forming marine microalgae to changed climate conditions. Ecol Evol 2(6):1195–1207

Krishna G, Valiya VM (2023) Marine viruses and their impact on ocean bacterial biomass. In: Handbook of biomass. Springer, Singapore, pp 1–27

Kuhlisch C, Shemi A, Barak-Gavish N, Schatz D, Vardi A (2024) Algal blooms in the ocean: hot spots for chemically mediated microbial interactions. Nat Rev Microbiol 22(3):138–154

Lan J, Liu P, Hu X, Zhu S (2024) Harmful algal blooms in eutrophic marine environments: causes, monitoring, and treatment. Water 16(17):2525

Locke H, Bidle KD, Thamatrakoln K, Johns CT, Bonachela JA, Ferrell BD, Wommack KE (2022) Marine viruses and climate change: Virioplankton, the carbon cycle, and our future ocean. Adv Virus Res 114:67–146

Lundholm N, Skov J, Pocklington R, Moestrup Ø (1997) Studies on the marine planktonic diatom Pseudo-nitzschia. 2. Autecology of P. pseudodelicatissima based on isolates from Danish coastal waters. Phycologia 36(5):381–388

Lundsør E, Eikrem W, Stige LC, Engesmo A, Stadniczeñko SG, Edvardsen B (2022) Changes in phytoplankton community structure over a century in relation to environmental factors. J Plankton Res 44(6):854–871

Mackey KR, Morris JJ, Morel FM, Kranz SA (2015) Response of photosynthesis to ocean acidification. Oceanography 28(2):74–91

Manetu WM, Karanja AM (2021) Waterborne disease risk factors and intervention practices: a review. Open Access Library J 8(5):1–11

Martin K, Schmidt K, Toseland A, Boulton CA, Barry K, Beszteri B, Brussaard CP, Clum A, Daum CG, Eloe-Fadrosh E, Fong A (2021) The biogeographic differentiation of algal microbiomes in the upper ocean from pole to pole. Nat Commun 12(1):5483

Mensch B, Neulinger SC, Künzel S, Wahl M, Schmitz RA (2020 Jun) Warming, but not acidification, restructures epibacterial communities of the baltic macroalga Fucus vesiculosus with seasonal variability. Front Microbiol 26(11):1471

Nwabor OF, Nnamonu EI, Martins PE, Ani OC (2016) Water and waterborne diseases: a review. Int J Trop Dis Health 12(4):1–14

Noyes PD, McElwe MK, Miller HD, Clark BW, Van Tiem LA, Walcott KC Kyle NE, and Edward DL (2009) The toxicology of climate change: Environmental contaminants in a warming world Environment International 35(6):971–986 https://doi.org/10.1016/j.envint.2009.02.006

O'Brien J, McParland EL, Bramucci AR, Siboni N, Ostrowski M, Kahlke T, Levine NM, Brow MV, van de Kamp J, Bodrossy L, Messer LF (2022 Dec) Biogeographical and seasonal dynamics of the marine Roseobacter community and ecological links to DMSP-producing phytoplankton. ISME Commun 2(1):16

Omohwovo EJ (2024) Wastewater management in Africa: challenges and recommendations. Environ Health Insights 18:11786302241289681

Pandey PK, Kass PH, Soupir ML, Biswas S, Singh VP (2014) Contamination of water resources by pathogenic bacteria. AMB Express 4:1–16

Peterson O, Asplund M, Karunasagar I, Godhe A (2010) Phytoplankton community composition and diversity effects on the growth of marine vibrio bacteria. KALLIOPI A PAGOU:147

Phlips EJ, Badylak S, Nelson NG, Havens KE (2020) Hurricanes, El Niño and harmful algal blooms in two sub-tropical Florida estuaries: direct and indirect impacts. Sci Rep 10(1):1910

Ponting J, Kelly TJ, Verhoef A, Watts MJ, Sizmur T (2021) The impact of increased flooding occurrence on the mobility of potentially toxic elements in floodplain soil–a review. Sci Total Environ 754:142040

Rasmussen P, Sonnenborg TO, Goncear G, Hinsby K (2013) Assessing impacts of climate change, sea level rise, and drainage canals on saltwater intrusion to coastal aquifer. Hydrol Earth Syst Sci 17(1):421–443

Rehnstam-Holm AS, Godhe A (2012) Dynamics of vibrio spp. in relation to phytoplankton community composition and environmental conditions. In: Pathogenic vibrio spp. in northern European waters. German Federal Institute of Hydrology, pp 59–63

Rehnstam-Holm AS, Godhe A, Härnström K, Raghunath P, Saravanan V, Collin B, Karunasagar I, Karunasagar I (2010) Association between phytoplankton and vibrio spp. along the southwest coast of India: a mesocosm experiment. Aquat Microb Ecol 58(2):127–139

Rehnstam-Holm AS, Atnur V, Godhe A (2014 Jan) Defining the niche of Vibrio parahaemolyticus during pre-and post-monsoon seasons in the coastal Arabian Sea. Microb Ecol 67:57–65

Reynolds CS (2006) The ecology of phytoplankton. Cambridge University Press

Riebesell U (2004) Effects of CO 2 enrichment on marine phytoplankton. J Oceanogr 60:719–729

Röder K, Hantzsche FM, Gebühr C, Miene C, Helbig T, Krock B et al (2012) Effects of salinity, temperature and nutrients on growth, cellular characteristics and yessotoxin production of Protoceratium reticulatum. Harmful Algae 15:59–70

Röthig T, Trevathan-Tackett SM, Voolstra CR, Ross C, Chaffron S, Durack PJ et al (2023) Human-induced salinity changes impact marine organisms and ecosystems. Glob Chang Biol 29(17):4731–4749

Seidel M, Hamley-Bennett C, Reeksting BJ, Bagga M, Hellmann L, Hoffmann TD, Kraemer C, Ofiţeru ID, Paine K, Gebhard S (2025 Apr) Metabolic insights into Microbially induced calcite formation by Bacillaceae for application in bio-based construction materials. Environ Microbiol 27(4):e70093

Shahi N, Godhe A, Mallik SK, Härnström K, Nayak BB (2015) The relationship between variation of phytoplankton species composition and physico-chemical parameters in northern coastal waters of Mumbai, India

Shrestha A, Bhattarai TN, Acharya G, Timalsina H, Marks SJ, Uprety S, Paudel SR (2023) Water, sanitation, and hygiene of Nepal: status, challenges, and opportunities. ACS ES&T Water 3(6):1429–1453

Sigman DM, Hain MP (2012) The biological productivity of the ocean. Nat Educ Knowl 3(10):21

Song T, Zhang X, Li J, Wu X, Feng H, Dong W (2021) A review of research progress of heterotrophic nitrification and aerobic denitrification microorganisms (HNADMs). Sci Total Environ 801:149319

Šulčius S, Alzbutas G, Lukashevich V (2025 Apr) Cyanophage lysis of the cyanobacterium Nodularia spumigena affects the variability and fitness of the host-associated microbiome. Environ Microbiol 27(4):e70042

Sunda WG, Graneli E, Gobler CJ (2006) Positive feedback and the development and persistence of ecosystem disruptive algal blooms 1. J Phycol 42(5):963–974

Töpel M, Pinder MI, Johansson ON, Kourtchenko O, Godhe A, Clarke AK (2018 Sep) Whole-genome sequence of the novel Antarctobacter heliothermus strain SMS3, found in association with the marine diatom Skeletonema marinoi. J Genomics 4(6):113

Töpel M, Pinder MI, Johansson ON, Kourtchenko O, Godhe A, Clarke AK (2019) Whole genome sequence of Marinobacter salarius strain SMR5, shown to promote growth in its diatom host. J Genomics 7:60

Tryland I, Robertson L, Blankenberg AGB, Lindholm M, Rohrlack T, Liltved H (2011) Impact of rainfall on microbial contamination of surface water. Int J Clim Change Strateg Manag 3(4):361–373

Turner LM, Alsterberg C, Turner AD, Girisha SK, Rai A, Havenhand JN, Venugopal MN, Karunasagar I, Godhe A (2016) Pathogenic marine microbes influence the effects of climate change on a commercially important tropical bivalve. Sci Rep 6(1):32413

Van Vliet MTH, Zwolsman JJG (2008) Impact of summer droughts on the water quality of the Meuse river. J Hydrol 353(1–2):1–17

Vezzulli L, Colwell RR, Pruzzo C (2013) Ocean warming and spread of pathogenic vibrios in the aquatic environment. Microb Ecol 65:817–825

Wallace RB, Baumann H, Grear JS, Aller RC, Gobler CJ (2014) Coastal Ocean acidification: the other eutrophication problem. Estuar Coast Shelf Sci 148:1–3

Wallender EK, Ailes EC, Yoder JS, Roberts VA, Brunkard JM (2014) Contributing factors to disease outbreaks associated with untreated groundwater. Groundwater 52(6):886–897

Wallenstein MD, Hall EK (2012) A trait-based framework for predicting when and where microbial adaptation to climate change will affect ecosystem functioning. Biogeochemistry 109:35–47

Wang L, Wang H (2024) Marine biogeochemical processes and ecosystem evolution: observational and predictive approaches. Int J Mar Sci 14(5):304

Wang J, Yin X, Xu M, Chen Y, Ji N, Gu H et al (2022) Isolation and characterization of a high-efficiency algicidal bacterium Pseudoalteromonas sp. LD-B6 against the harmful dinoflagellate Noctiluca scintillans. Front Microbiol 13:1091561

Waycott M, Duarte CM, Carruthers TJ, Orth RJ, Dennison WC, Olyarnik S, Calladine A, Fourqurean JW, Heck KL Jr, Hughes AR, Kendrick GA (2009) Accelerating loss of seagrasses across the globe threatens coastal ecosystems. Proc Natl Acad Sci 106(30):12377–12381

Weinbauer MG (2004) Ecology of prokaryotic viruses. FEMS Microbiol Rev 28(2):127–181

Wieczynski DJ, Yoshimura KM, Denison ER, Geisen S, DeBruyn JM, Shaw AJ, Weston DJ, Pelletier DA, Wilhelm SW, Gibert JP (2023 Mar) Viral infections likely mediate microbial controls on ecosystem responses to global warming. FEMS Microbiol Ecol 99(3):fiad016

Wommack KE, Colwell RR (2000) Virioplankton: viruses in aquatic ecosystems. Microbiol Mol Biol Rev 64(1):69–114

Yao S, Lyu S, An Y, Lu J, Gjermansen C, Schramm A (2019) Microalgae–bacteria symbiosis in microalgal growth and biofuel production: a review. J Appl Microbiol 126(2):359–368

Zhang QY, Ke F, Gui L, Zhao Z (2022) Recent insights into aquatic viruses: emerging and reemerging pathogens, molecular features, biological effects, and novel investigative approaches. Water Biol Secur 1(4):100062

Zhang J, Zhang F, Dong Z, Zhang W, Sun T, Chen L (2024) Response and acclimation of cyanobacteria to acidification: a comprehensive review. Sci Total Environ 945:173978

Zhu Z, Qu P, Fu F, Tennenbaum N, Tatters AO, Hutchins DA (2017) Understanding the blob bloom: warming increases toxicity and abundance of the harmful bloom diatom Pseudo-nitzschia in California coastal waters. Harmful Algae 67:36–43

Chapter 4
Climate Change Impacts: Insights from Changing Patterns of Sardine Fisheries in Indian Waters

Bhagyashree Dash, Dhanya M. Lal, Sanjiba Kumar Baliarsingh, Alakes Samanta, Sudheer Joseph, and T. M. Balakrishnan Nair

Abstract Small pelagic fishes (SPF) play a crucial role in marine ecosystems by facilitating trophic energy transfer and maintaining food chain linkages. SPF provides food security and supports the livelyhood of coastal communities through fishing (and related) industries such as fish meat, oil, and fishmeal production. However, SPFs, including sardines, are increasingly affected by overfishing and climate change, resulting in population fluctuations, distributional shifts, and often mass mortality events, particularly following extreme climate events, disrupting their ecological roles and functions. India, particularly the southwestern states, has historically contributed significantly to global sardine landings. However, recent observations indicate notable shifts in sardine landings, prompting a decadal analysis of their distribution in Indian waters. An evaluation of sardine landing data from 1985 to 2023 revealed a noticeable relationship between the Oceanic Niño Index (ONI) and oil sardine catches in the primary sardine-landing states: Kerala and Karnataka. Higher oil sardine landings in these states were associated with strong La Niña events (negative ONI), while declines in landings were linked to prolonged El Niño periods (positive ONI). An inverse relationship was also observed between oil sardines and lesser sardines. For instance, during 1989–1999, oil sardine landings declined in both Kerala and Karnataka, whereas lesser sardine landings were comparatively higher. Conversely, between 2000 and 2012, oil sardine landings increased while lesser sardine landings decreased. The decline phase of oil sardines corresponded with prolonged positive ONI periods, while the recovery phase coincided with intense La Niña events. In contrast, high landings of lesser sardines were often associated with strong El Niño periods or sustained positive ONI before their peak landing years. Spatial distribution patterns of oil sardine catches revealed alternating phases of decline and rise between India's eastern and western coasts. Notably, during reduced oil sardine landings along the Kerala and Karnataka coasts,

B. Dash · D. M. Lal · S. K. Baliarsingh (✉) · A. Samanta · S. Joseph · T. M. B. Nair
Indian National Centre for Ocean Information Services, Ministry of Earth Sciences, Government of India, Hyderabad, India
e-mail: baliarsingh.s@incois.gov.in

B. P. Barathan et al. (eds.), *Navigating Climate Change: Impacts on Biodiversity and Ecosystem Resilience*, Sustainable Development and Biodiversity 42,
https://doi.org/10.1007/978-981-95-0409-1_4

a concurrent increase in landings was observed in the eastern states of Tamil Nadu and Andhra Pradesh. These shifts suggest that climate change, including prolonged El Niño conditions, often renders the western coast less favorable for oil sardines while creating more suitable conditions along the eastern coast, facilitating their migration and proliferation in these areas. Therefore, efforts must address the irreversible tipping points in ecosystem functioning and significant fluctuation of oil sardine fishery driven by climate change. This requires programmatically focused and synergetic actions involving scientific research, policy interventions, and active engagement of stakeholders.

Keywords La Niña · El Niño · Sardine · Upwelling · Fishery

4.1 Introduction

Small pelagic fishes (SPF) holds an important ecological role in every global marine ecosystems, and many species are heavily exploited by capture fisheries, including some of the world's largest and most valuable fisheries. Herrings, anchovies, sardines, and other small pelagic clupeoid fishes are essential not only for direct human consumption but also for producing fishmeal and fish oil, which play a vital role in agriculture and aquaculture feeds. This makes SPF crucial for global food security (FAO 2016; Dash et al. 2024). With short generation times and a strong dependence on lower trophic levels, SPF populations exhibit pronounced boom-and-bust cycles that are closely influenced by climate variability.

The western coast of India contributes 60% of the country's total pelagic landings, with the nutrient-rich upwelled waters along the southwest coast covering Goa, Karnataka, and Kerala supporting abundant stocks of SPF (Rohit et al. 2024). Sardines, along with mackerel, are among the most commercially significant pelagic schooling fishes in India. Annually, the sardines formed 2–24% of the total landings, with an average of 13% from 1985 to 2023 (Shyam 2021; Rohit et al. 2024). Among the 13 sardine species, the Indian Oil Sardine (*Sardinella longiceps*) dominates as the largest single species fishery, accounting for approximately 1–13% of the country's whole fish landings, with an average of 7% over the past four decades (Rohit et al. 2024). The remaining 12 species, collectively referred to as "Lesser Sardines," include all sardine species, excluding the Indian Oil Sardine, and the former contributes around 3%–7% of the total marine fish catch (Pillai and Rohit 2003). Lesser Sardines are primarily found in nearshore waters, typically inhabiting depths of 25–30 meters. This group include *Sardinella jussieu*, *S. gibbosa* (Goldstripe sardinella), *S. fimbriata* (Fringe-scale sardine), *S. albella* (Short-bodied sardine), *S. melanura* (Black-tipped sardine), *S. sindensis*, *S. brachysoma*, *Amblygaster sirm* (Spotted sardine), *A. clupeoides* (Bleeker smooth belly sardine), *Escualosa thoracata* (White sardine), *Dussumieria acuta* (Rainbow sardine), and *D. elopsoides* (Bennet et al. 1986; Pillai and Rohit 2003). These 12 species are mostly restricted to Indian waters out of the 15 species of lesser sardines inhabited in the Indo-Pacific region.

In the past decade, prominent climatic phenomena such as El Niño, the Pacific Decadal Oscillation, and Indian Ocean Dipole (IOD)-induced Ocean warming have significantly influenced the distribution and abundance of SPF communities (Hamza et al. 2021; Holmes et al. 2021). These changes have led to alterations in ecosystem structure and function, manifesting as habitat degradation and shifts in species distribution (Vivekanandan et al. 2009; Kripa et al. 2018). Against the aforementioned backdrop, the present study presents a decadal analysis of sardine catch trends in Indian waters, focusing on their spatial and temporal distribution over the past 35 years. The objective is to identify patterns and variations in sardine abundance over time and examine their relationship with long-term environmental phenomena such as oceanographic conditions, climate variability, and ecosystem dynamics. By analyzing these trends, the study aims to provide insights into the factors influencing sardine distribution and assess the potential impacts of long-term environmental and climatic phenomena on their populations.

4.2 Materials and Methods

Annual fish catch data were sourced from the Central Marine Fisheries Research Institute (CMFRI), India, along with reports and online databases that have recorded fish landings along the country's coastline since the early 1950s (Fig. 4.1). Data collection follows a stratified multi-stage random sampling approach, considering variations in boat and gear types (Srinath et al. 2005).

Chlorophyll-*a* concentration, with a 4-km spatial resolution, was retrieved from the Moderate Resolution Imaging Spectroradiometer (MODIS) aboard the Aqua satellite, accessible via the Ocean Color Web (https://oceancolor.gsfc.nasa.gov). Satellite-retrieved Sea Surface Temperature (SST) data at a spatial resolution of 0.04° was obtained from Pathfinder Ver 5.3 (L3C). The Monthly Climate Timeseries of the Oceanic Niño Index (ONI) V2 (available from NOAA Physical Sciences Laboratory) was used as a measure of the El Niño-Southern Oscillation (ENSO) to compare the annual trends in oil sardine catch with respect to ENSO events. The ONI is calculated as the 3-month running mean of SST anomalies (based on ERSST. v5 data) in the Niño 3.4 region (5°N–5°S, 120°–170°W), using the 1971–2000 base period (Shetye et al. 2019). The Monthly Climate Time Series of the Dipole Mode Index (DMI), obtained from the NOAA Physical Sciences Laboratory, was utilized to assess the intensity of the Indian Ocean Dipole (IOD) and its relationship with annual trends in oil sardine catch. The IOD intensity is quantified by the anomalous SST gradient between the western equatorial Indian Ocean (50°E–70°E, 10°S–10°N) and the southeastern equatorial Indian Ocean (90°E–110°E, 10°S–0°N), known as the DMI. A positive DMI indicates a positive IOD phase, while a negative DMI signifies a negative IOD phase. Graphical illustrations were prepared using Matplotlib and Python (version 3.9.12).

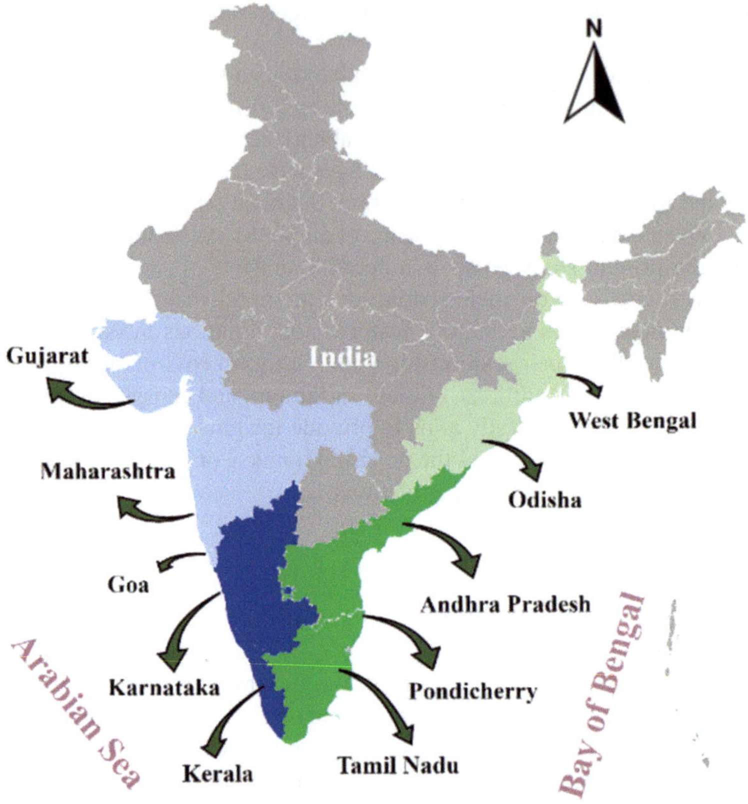

Fig. 4.1 Map showing the study area (coastal states and union territories of mainland India)

4.3 Results and Discussion

4.3.1 Fluctuating Oil Sardine Fisheries: East Versus West Coast of India

Between 1988 and 1994, oil sardine landings along the west coast of India exhibited a notable decline. In contrast, the east coast experienced an increase (Fig. 4.2). This shift coincided with elevated SST in the Arabian Sea relative to the Bay of Bengal, potentially influencing sardine distribution as they prefer relatively cooler waters (Krishnakumar et al. 2008; Kizhakudan et al. 2014).

However, during 1994–1997, an increase in oil sardine catch could be observed along the east coast compared to the west coast. The years 1994 and 1997 were characterized by positive DMI and ONI conditions, along with a weak La Niña, which may have contributed to reduced productivity and unfavorable environmental conditions. Recent studies suggest that coastal upwelling along the southwest coast of India weakens during positive IOD years, with further suppression when El Niño

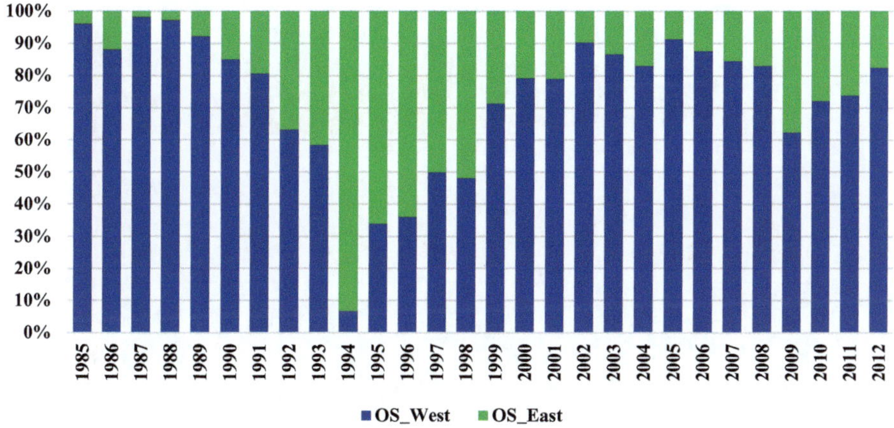

Fig. 4.2 Temporal variation in oil sardine (OS) landing pattern along the east and west coasts of India

and IOD co-occur (Nigam et al. 2018; Nigam and Pant 2024). In contrast, a strong positive IOD has been associated with increased phytoplankton blooms in the southern Bay of Bengal (Girishkumar 2022). This could potentially explain the observed increase in sardine landings along the east coast compared to the western region during this period.

The year 1995 marked a significant reversal in catch trend, as oil sardine populations began to recover along the west coast, while landings on the east coast showed a decline (Fig. 4.2). From 1995 onward, an increase in SST was observed across both the Arabian Sea and the Bay of Bengal, indicating broad-scale oceanographic changes in the region (Fig. 4.3).

Another major transition occurred in 2009 when oil sardine landings on the west coast were comparatively lower than on the east coast, wherein the east coast recorded its highest landings in two decades. This redistribution of sardines was likely driven by environmental variability, including shifts in SST (Lluch-Belda et al. 1989; Kizhakudan et al., 2014; Kripa et al. 2018). Additionally, the years 2009 and 2010 were characterized by anomalously high SST in the Arabian Sea in comparison to the Bay of Bengal (Fig. 4.3), potentially driving sardine populations to migrate toward habitats with more favorable temperature conditions (Vivekanandan et al. 2009; Hamza et al. 2021).

4.3.2 *Oil Sardine Fisheries Trends: Across the Southern Indian States*

A comparative analysis of oil sardine catch trends across the southwestern and southeastern coastal states of India reveals notable shifts in regional dominance over different time periods. Normalized catch trends indicate that oil sardine landings

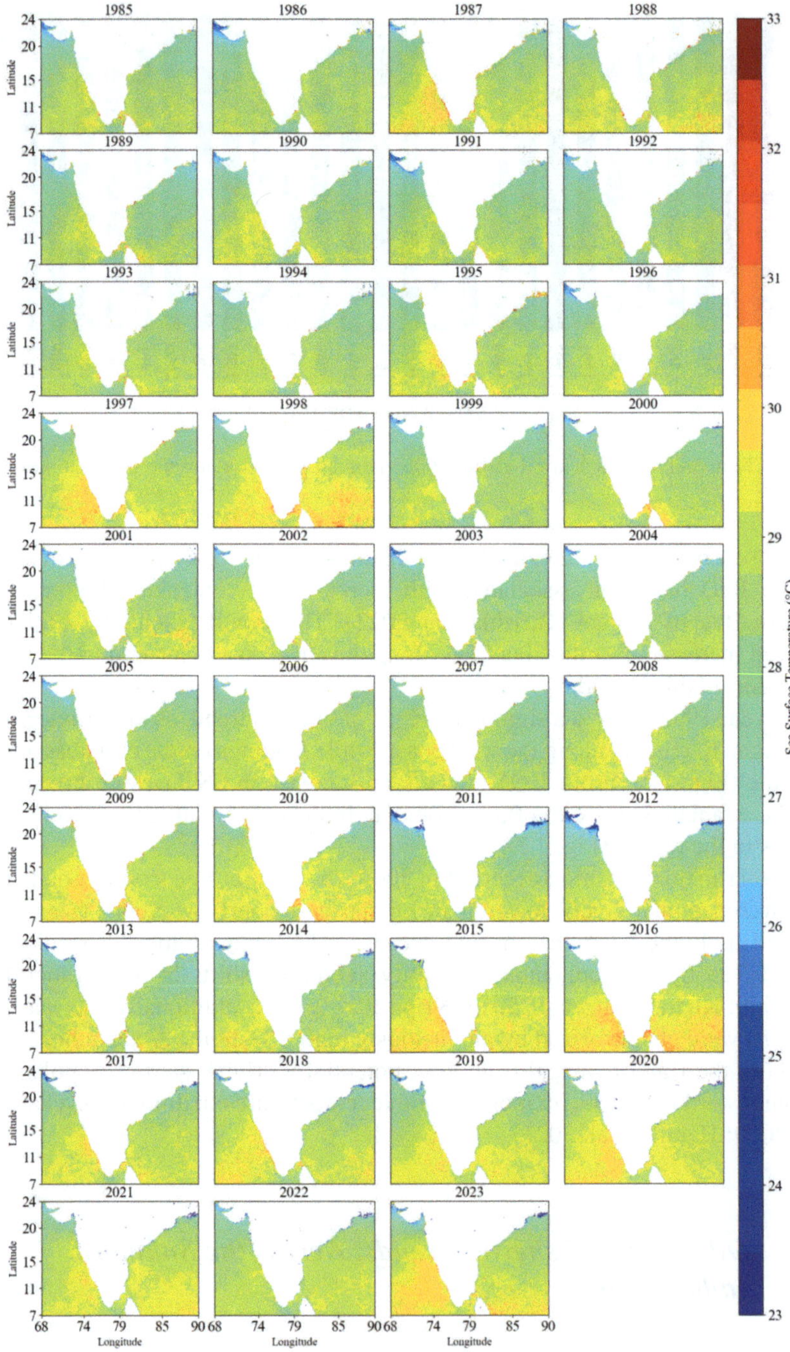

Fig. 4.3 Spatiotemporal variation of sea surface temperature in the Indian waters, 1985–2023

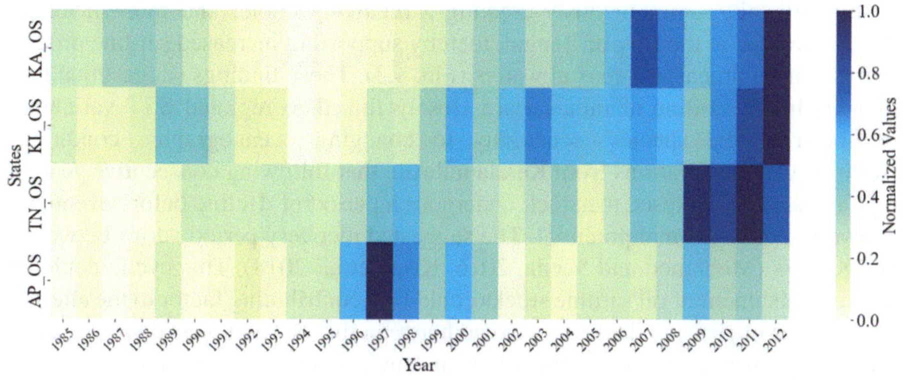

Fig. 4.4 Variation in oil sardine (OS) landing across the southern states. AP Andhra Pradesh, TN Tamil Nadu, KL Kerala; KA Karnataka

along the Andhra Pradesh coast peaked between 1996 and 1998, a period during which the southwestern states, including Kerala and Karnataka, recorded relatively lower catches. This phenomenon could be attributed to the influence of both a positive IOD and El Niño, as 1997 experienced both events. These climatic drivers have been reported to weaken upwelling, thereby reducing primary productivity on the southwestern coast of India (Nigam and Pant 2024). Conversely, a strong positive IOD has been linked to enhanced phytoplankton blooms in the southern Bay of Bengal (Girishkumar 2022). A similar pattern was observed during 2009–2011, when Tamil Nadu reported its highest oil sardine landings, whereas the southwestern states displayed a declining trend (Fig. 4.4).

Between 2009 and 2011, the Arabian Sea experienced a relatively higher degree of warming compared to the Bay of Bengal, attributable to strong El Niño during 2009. Decadal studies on the impact of strong positive DMI events on surface chlorophyll have demonstrated significant algal blooms in the Bay of Bengal following such events, particularly after the strong positive DMI of 2006–2007 (Girishkumar 2022). The increased food availability along the eastern coast, contradictory to the elevated SST and less food availability on the western coast, may have created favorable conditions for oil sardine growth in the Bay of Bengal, contributing to higher catch rates in the region. In contrast, Kerala and Karnataka exhibited peak oil sardine catches in 2012, coinciding with the La Nina period, with their abundance patterns showing a strong proportional relationship (Hamza et al. 2021). However, the southeastern states, particularly Tamil Nadu and Andhra Pradesh, displayed an opposite trend in relation to southwestern states, suggesting an alternating pattern in sardine distribution between the two coastal regions (Fig. 4.4).

A particularly striking contrast is evident from 1991 to 1997 when oil sardine landings were comparatively lower in the southwestern states but significantly higher in the southeastern states. This distributional shift may be attributed to oceanographic changes, specifically the rising SST in the Arabian Sea. The warming of the Arabian Sea likely resulted in suboptimal conditions for oil sardines in the southwest while simultaneously creating a relatively cooler and more favorable thermal regime in the Bay of Bengal, thereby supporting increased sardine productivity in the southeastern coastal waters (Fig. 4.3). These findings suggest that fluctuations in oil sardine abundance are closely linked to regional SST variability, underscoring the species' sensitivity to changing oceanographic conditions. Additionally, observations from Kerala indicate that following consecutive years of high oil sardine harvests, the stock undergoes a period of decline before eventually recovering to its normal potential. The suggested recovery period varies between 6 and 8 years (Mohamed and Veena, 2016; Kripa et al. 2018). This cyclic depletion and replenishment of oil sardine stocks could be a contributing factor to the alternating patterns of rise and decline observed between the east and west coasts of India, potentially driven by a combination of environmental conditions and stock dynamics.

4.3.3 Climate-Induced Changes in Sardine Fisheries Along the Malabar Coast

The Malabar upwelling zone, located off the coasts of Karnataka and Kerala, is among the most significant upwelling zones, where nutrient-rich waters support elevated phytoplankton productivity (Krishnakumar and Bhat 2008). This elevated productivity has historically made the region a key hub for the oil sardine fishery. However, in recent years, sardine abundance has exhibited a cyclical pattern characterized by periods of both appearance and disappearance. Consequently, we aimed to investigate the impact of recent climatic phenomena, particularly the IOD and ENSO, on the sardine landings between 1985 and 2023.

The ONI is a well-established metric for SST anomalies in the east-central Pacific and is commonly used to track the ENSO cycle. Numerous studies worldwide have identified ONI as a key variable influencing SPF landings, establishing its relevance in fisheries research (Ormaza-González et al. 2016; Holmes et al. 2021; Sambah et al. 2021). Figure 4.5 presents the monthly time series of the ONI and DMI indices, along with the annual trends in oil sardine (5a) and lesser sardine (5b) catches from the states of Kerala and Karnataka. Indian oil sardine landings in the Malabar upwelling area have shown a strong linkage with ONI values, with notable fluctuations in catch trends corresponding to different ENSO phases (Fig. 4.5).

The annual trends in ONI and DMI were largely complementary, with positive ONI years during most of the cases corresponding to positive DMI values and vice versa. This pattern was most evident during the periods 1985–2000 and 2015–2023. Between 2001 and 2015, DMI exhibited relatively stable trends with minor fluctuations between weak negative and weak positive values, showing little correlation with ONI in the annual trends.

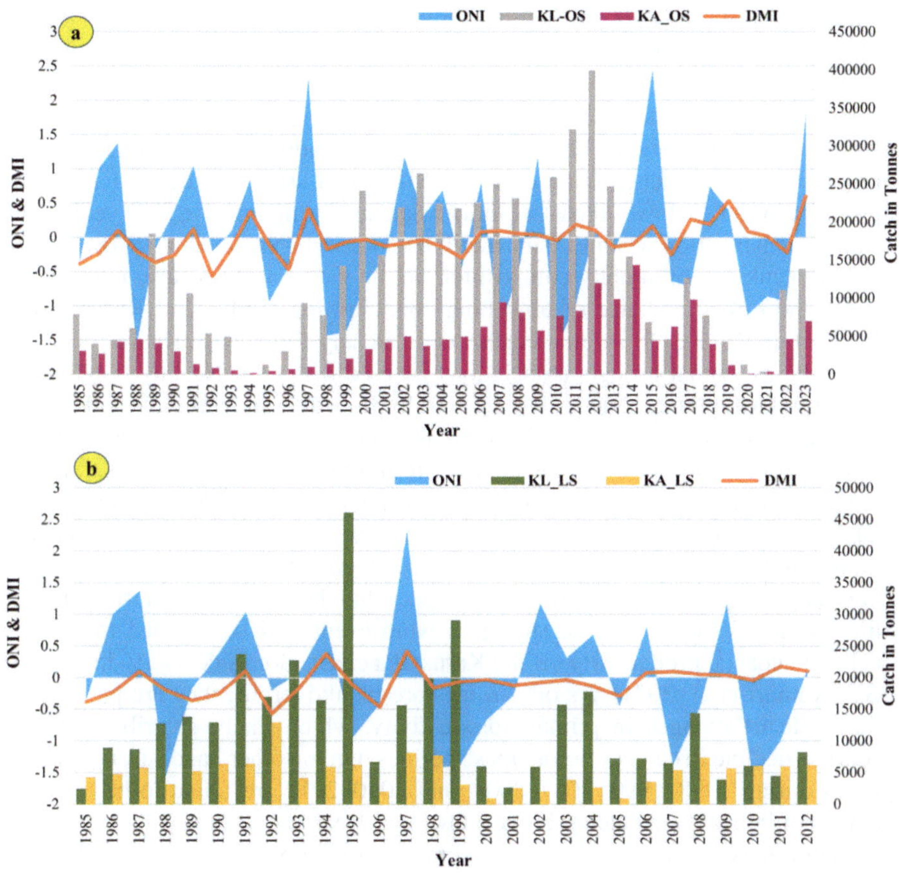

Fig. 4.5 Catch trends in (**a**) oil sardine (OS) and (**b**) lesser sardines (LS) with ONI and DMI variability in Kerala (KL) and Karnataka (KA)

The oil sardine catches in these states exhibit similar trends, with fall and raise often complementing each other. The catch trends also show a strong response to ONI, particularly during El Niño phases (indicated by positive ONI), which appear to reduce catches along both coasts (Kripa et al. 2018). This pattern is especially evident in the years 1985–86, 1993–95, 1997–98, 2008–09, 2015–17, and 2019–21. Conversely, La Niña phases (indicated by negative ONI) preceding El Niño events seem to aid in the recovery of catch trends in the subsequent years. Overall, it can be observed that there has been a relatively lower catch in the first decade over all four states (Fig. 4.4). The strong El Niño events of 1986–1987 and 1990–1991, coupled with a comparatively low fishing fleet strength, may have collectively influenced fish catch trends during this period. Following this phase, the marine fisheries sector has emerged with more degree of mechanization and fleet strength, which also could lead to the increased harvests.

In an attempt to discern any possible shift of upwelling zones due to the effect of ENSO phases, it can be noted that upwelling and resultant algal blooms influence marine primary productivity along India's western coast. However, there is no clear evidence of upwelling shifts between the east and west coasts. For instance, in 2015, weaker upwelling along the Kerala coast may have hindered sardine gonadal development and recruitment (Kripa et al. 2018). Model studies indicate that coastal upwelling at the southwest coast of India weakens during positive IOD years, with further intensification when El Niño and IOD co-occur (Nigam et al. 2018). A shift in phytoplankton community composition on the west coast was also noticed, but there is no direct evidence of upwelling zone shifting (Vidya and Kurien 2018; Shetye et al. 2019). Oil sardine catch trends closely followed variations in ONI–DMI patterns, with a stronger association observed with ONI. Negative ONI values (indicative of La Nina periods) were generally linked to an increase in oil sardine catch trends, whereas positive ONI values (associated with El Niño phases) appeared to correspond with a decline in annual oil sardine catches in both Kerala and Karnataka (Kripa et al. 2018). Specifically, during the strong El Niño events of 1997 and 2015, coupled with positive IOD phases, SST along the Malabar coast was notably higher, and oil sardine catches declined sharply during these years. Unusually low primary productivity was reported in the Arabian Sea during these El Niño events; for example, the strong El Niño event in 2015 has led to a ~24% decline in coastal productivity off the Karnataka coast, alongside a ~1% increase in SST (Kumari et al. 2021). As oil sardines are planktivorous, they are particularly vulnerable to changes in primary productivity, which likely contributed to the observed decline in catches during these periods. On the other hand, strong La Nina phases during 1998–1999, 2007–2008, and 2010–2011 are found to be linked to significant increases in catches in the subsequent years.

Studies have shown that large-scale climatic oscillations like the Pacific Decadal Oscillation (PDO) and the Atlantic Multidecadal Oscillation (AMO) significantly influence Indian oil sardine landings by affecting oceanographic conditions and fish availability (Hamza et al. 2021). Sajna et al. (2019) reported a steep increase in sea surface salinity (SSS) along the Kerala coast during 1994, 1998–2000, 2005–06, and 2013–14, which are observed to be coinciding with La Niña events or transition period to El Niño. The PDO, though operating on a different timescale (compared to ONI), was observed in its negative phase during these high SSS periods, reinforcing the La Niña influence. A study by Sajna et al. (2019) on climatic variation in the oil sardine fishery on the Kerala coast using a GAM approach indicated a positive association of oil sardine CPUE with SSS where CPUE was observed highest at SSS levels between 30.5 and 31.5 PSU, with deviations beyond this range negatively impacting catches. Long-term trend analysis presented in this study also shows increasing sardine catches during these high SSS periods in most cases, except in 1994, when excessively high SSS beyond optimum may have acted as one of the contributing factors for decline despite La Niña conditions. Although the primary focus of sardine fisheries in India is the Indian oil sardine, lesser sardine groups also contribute significantly to fisheries and have notable utilization potential. The dynamics of lesser sardine groups, notably their contrasting abundance patterns with the oil sardine, were worth investigating. The alternating abundance patterns of oil sardine and lesser sardine groups were most prominent along the southwestern coastal states, such as Karnataka and Kerala (Fig. 4.5). For instance, during 1989–1999, oil sardine landings declined in both Kerala and Karnataka, whereas lesser sardine landings were comparatively higher. Conversely, between 2000 and 2012, oil sardine landings increased while lesser sardine landings decreased.

While analyzing the trends in the declining lesser sardine catch, an inverse relationship with the ENSO can be observed (Fig. 4.5). Specifically, during the period from 1990 to 1997, when El Niño events were particularly intense, lesser sardine landings remained consistently high along the coasts of Kerala and Karnataka. This period stands in contrast to the La Niña phases, where low lesser sardine catches were observed in contrast to the high catches of oil sardines (Fig. 4.5).

A general agreement is oil sardines prefer larger sizes of plankton, whereas lesser sardines feed on plankton that are abundant in the environment (Bennet et al. 1986; Pillai and Rohit 2003). A recent analysis of the strong 2015 El Niño phase suggested that a decline in nutrients during the study period has favored the growth of nanophytoplankton and pico-phytoplankton. On the other hand, this low nutrient condition caused a decrease in the micro-phytoplankton population (Vidya and Kurien 2018; Shetye et al. 2019). Oil sardines are reported to feed on microplankton, whereas lesser sardines, being a multispecies group, feed on all available plankton; hence, the effects of ENSO might have minimal impact, and in the absence of oil sardines, it might have given them enough resources to grow abundantly. This relationship highlights the sensitivity of sardine populations to large-scale climatic variability, reinforcing the importance of ENSO-driven oceanographic changes in shaping their abundance and distribution. Despite the general expectation that El Niño, known for disrupting marine ecosystems, which would lead to reduced fish populations, lesser sardine landings remained high during this El Niño-dominant phase. This suggests that other ecological or environmental factors, such as increased resource availability or reduced competition, may have played a significant role in sustaining the lesser sardine population. These observations indicate a possible resilience of the species to the typical impacts of El Niño in this region. This anomaly underpins further investigation of the complex interplay between ENSO events and regional marine productivity.

As reported, chlorophyll-*a* concentrations along the Indian coast also exhibited parallel responses to long-term climatic trends and can be considered one of the direct factors influencing the oscillations in stock decline and recovery along the coast (George et al. 2012; Menon et al. 2019). Between 2003 and 2013, oil sardine catches exhibited a steady increase, coinciding with relatively stable chlorophyll-*a* concentrations along both coasts. However, a significant decline was observed in 2014–15, corresponding with the occurrence of a strong El Niño event and a contraction of the high-chlorophyll zone along the southwest coast (Fig. 4.6). Following this decline, the stock partially recovered but experienced another phase of gradual reduction from 2017 onward. This could be attributed to the 2018–19 co-occurrence of positive IOD as well as El Niño as these scenarios have been reported to be

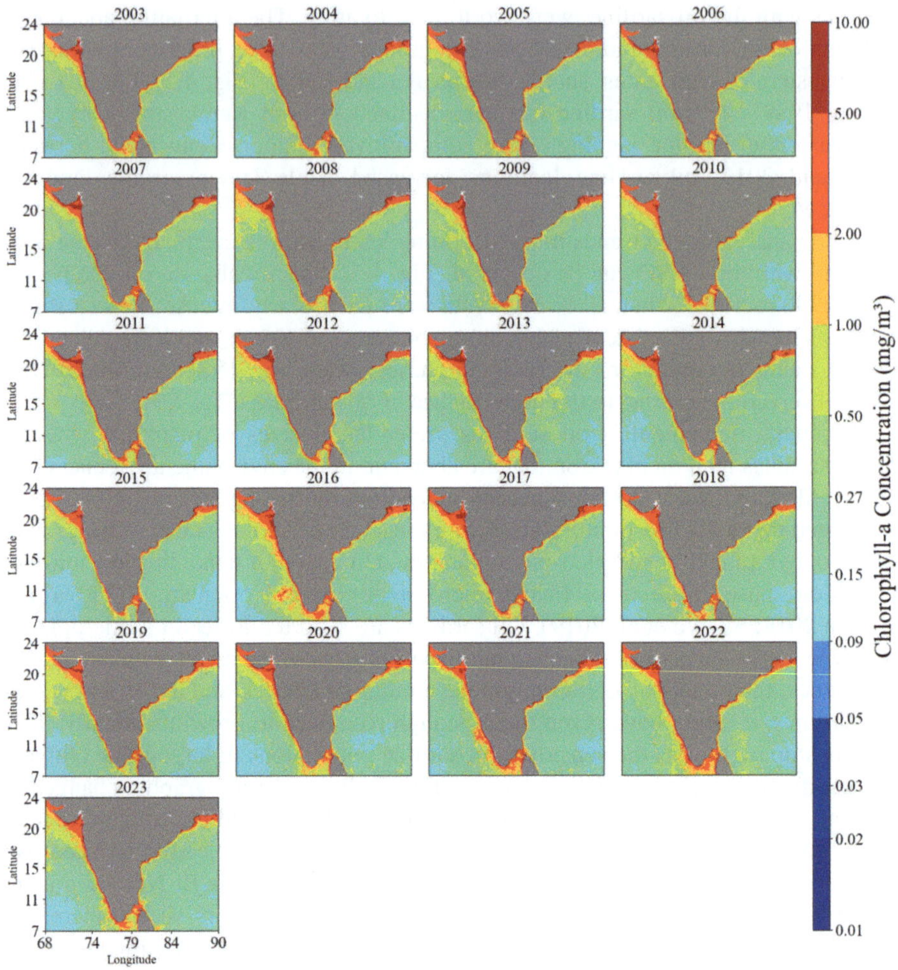

Fig. 4.6 Spatiotemporal variation of chlorophyll-*a* in the Indian waters during 2003–2023

detrimental to upwelling and phytoplankton bloom formation (Shankar et al. 2019; Kumari et al. 2021). Being a planktivorous fish, this might have directly influenced their recruitment and landing (Kripa et al. 2018). A subsequent resurgence was noted in 2020, coinciding with La Niña conditions and an increase in chlorophyll-*a* concentration, highlighting the potential influence of oceanographic variability on sardine stock dynamics. Since oil sardines along the southwestern coast mainly consume micro-phytoplankton, particularly diatoms, and zooplankton like copepod, a decline in these food sources during ENSO years may have played a role in the reduction of sardine populations (Vidya and Kurien 2018; Shetye et al. 2019).

4.4 Conclusion

Investigating the alternating appearance and disappearance of species within the same niche or phylogenetic lineage provides insights into the influence of environmental variation and climatic drivers on species composition and population decline in response to anthropogenic and natural perturbations. The analysis of sardine landings in relation to oceanic and climatic variability, particularly the influence of ENSO, highlights the strong dependence of this fishery on environmental conditions. The observed fluctuations in oil sardine abundance, with higher catches following La Nina phases and significant declines after El Niño years, suggest that large-scale climatic drivers play a crucial role in the population dynamics of these SPF. Additionally, the alternating abundance patterns between oil sardine and lesser sardine groups along the Indian coast further emphasize the need for a broader ecosystem-based approach to fisheries management. Given the increasing frequency and intensity of climate-driven oceanographic changes, it is evident that traditional fishery management approaches may not be sufficient to ensure the sustainability of Indian oil sardine stocks. Proactive strategies that incorporate climate variability into fisheries assessment and management are essential to mitigate the risks associated with unpredictable environmental shifts. By integrating climate-informed decision-making, ecosystem-based approaches, and adaptive fisheries management, the sustainability of sardine fisheries can be better safeguarded against the increasing challenges posed by climate change.

Acknowledgments The authors thank the Director of the Indian National Centre for Ocean Information Services, Hyderabad, for his encouragement in carrying out this work. This is INCOIS contribution no.581.

Declarations Informed Consent Statement All the authors have provided their consent for the publication of this work.

Data Availability Statement Satellite-retrieved Sea Surface Temperature (SST) data at a spatial resolution of 0.04° was obtained from Pathfinder Ver 5.3 (L3C), which is available at https://coastwatch.pfeg.noaa.gov. The inputs for Monthly Climate Timeseries of the Oceanic Niño Index (ONI) V2 and the Monthly Climate Time Series of the Dipole Mode Index (DMI) data are available at NOAA Physical Sciences Laboratory services, https://psl.noaa.gov. MODIS-Aqua retrieved chlorophyll-*a* data, which is available at OceanColorWEB (https://oceancolor.gsfc.nasa.gov). Fish catch data are available at CMFRI reports and database https://www.cmfri.org.in.

Ethics Approval This study does not involve humans or animals requiring ethical approval.

Conflict of Interest The authors declare that they have no known competing financial interests or personal relationships that could have appeared to influence the work reported in this paper.

Funding No external funds, grants, or other support was received to carry out this work.

References

Bennet SP, Lazarus S, Thiagarajan R, Luther G (1986) The present status of our knowledge on the lesser sardines of Indian waters, vol 28. CMFRI Special Publication, pp 1–43

Dash B, Baliarsingh SK, Samanta A, Sahoo S, Joseph S, Nair TB (2024) Evolution and recent trends of Indian oil sardine research: a review. Ocean Coast Manag 258:107396

FAO (2016) The state of world fisheries and aquaculture 2016. In: Contributing to food security and nutrition for all. FAO, Rome, p 200

George G, Meenakumari B, Raman M, Kumar S, Vethamony P, Babu MT, Verlecar X (2012) Remotely sensed chlorophyll: a putative trophic link for explaining variability in Indian oil sardine stocks. J Coast Res 28(1A):105–113

Girishkumar MS (2022) Surface chlorophyll blooms in the Southern Bay of Bengal during the extreme positive Indian Ocean dipole. Clim Dyn 59(5):1505–1519

Hamza F, Valsala V, Anju M, George G (2021) Climate impacts on the landings of Indian oil sardine over the south-eastern Arabian Sea. Fish Fish 21(5):992–1006. https://doi.org/10.1111/faf.12477

Holmes EE, Smitha BR, Nimit K, Maity S, Checkley DM Jr, Wells ML, Trainer VL (2021) Improving landings forecasts using environmental covariates: a case study on the Indian oil sardine (Sardinella longiceps). Fish Oceanogr 30(6):623–642

Kizhakudan SJ, Raja S, Gupta KS, Vivekanandan E, Kizhakudan JK, Sethi SN, Geetha R (2014) Correlation between changes in sea surface temperature and fish catch along Tamil Nadu coast of India-an indication of impact of climate change on fisheries? Indian J Fish 61(3):111–115

Kripa V, Mohamed KS, Koya KS, Jeyabaskaran R, Prema D, Padua S, Kuriakose S, Anilkumar PS, Nair PG, Ambrose TV, Dhanya AM, Abhilash KS, Bose J, Divya ND, Shara AS, Vishnu PG (2018) Overfishing and climate drives changes in biology and recruitment of the Indian oil sardine *Sardinella longiceps* in southeastern Arabian Sea. Front Mar Sci 5:443

Krishnakumar PK, Bhat GS (2008) Seasonal and interannual variations of oceanographic conditions off Mangalore coast (Karnataka, India) in the Malabar upwelling system during 1995–2004 and their influences on the pelagic fishery. Fish Oceanogr 17(1):45–60

Krishnakumar PK, Mohamed KS, Asokan PK, Sathianandan TV, Zacharia PU, Abdurahiman KP, Veena S, Durgekar NR (2008) How environmental parameters influenced fluctuations in oil sardine and mackerel fishery during 1926–2005 along the south-west coast of India? Mar Fish Infor Serv T & E Ser 198:1–5

Kumari PV, Thomas S, Mohanty PC, Jayappa KS, Mahendra RS, Gupta A (2021) Effect of sea surface temperature variation on productivity and fisheries off Karnataka, west coast of India. J Indian Soc Remote Sens 49:3027–3041

Lluch-Belda DRJM, Crawford RJ, Kawasaki T, MacCall AD, Parrish RH, Schwartzlose RA, Smith PE (1989) World-wide fluctuations of sardine and anchovy stocks: the regime problem. S Afr J Mar Sci 8(1):195–205

Menon NN, Sankar S, Smitha A, George G, Shalin S, Sathyendranath S, Platt T (2019) Satellite chlorophyll concentration as an aid to understanding the dynamics of Indian oil sardine in the southeastern Arabian Sea. Mar Ecol Prog Ser 617:137–147

Mohamed KS, Veena S (2016) How long does it take for tropical marine fish stocks to recover after declines? Case studies from the southwest coast of India. Curr Sci 110:584–594. https://doi.org/10.18520/cs/v110/i4/584-594

Nigam T, Pant V (2024) Interannual variability of coastal upwelling features along the eastern and western margins of the Arabian Sea. J Oper Oceanogr 17(2):77–92

Nigam T, Pant V, Prakash KR (2018) Impact of Indian Ocean dipole on the coastal upwelling features off the southwest coast of India. Ocean Dyn 68(6):663–676

Ormaza-González FI, Mora-Cervetto A, Bermúdez-Martínez RM, Hurtado-Domínguez MA, Peralta-Bravo MR, Jurado-Maldonado VM (2016) Can small pelagic fish landings be used as predictors of high-frequency oceanographic fluctuations in the 1–2 El Niño region? Adv Geosci 42:61–72

Pillai NGK, Rohit P (2003) Lesser sardines. In: Status of Exploited Marine Fishery Resources of India, CMFRI, Cochin, Earth Surf Process Landf, 3:25–29

Rohit P, Abdussamad EM, Margaret Muthu Rathinam A, Ganga U, Ghosh S, Rajesh KM, Koya KP, Koya M, Anulekshmi C, Nakhawa AD, Surya S, Roul SK, Azeez PA, Vinoth Kumar R, Manas HM, Jayasankar J, Mini KG, Kuriakose S (2024) Pelagic fisheries of India–an overview. Indian J Fish 71(1):12–28

Sajna VH, Zacharia PU, Liya VB, Rojith G, Somy K, Joseph D, Grinson G (2019) Effect of climatic variability on the fishery of Indian oil sardine along Kerala coast. J Coast Res 86(SI):184–192. https://doi.org/10.2112/SI86-028.1

Sambah AB, Wijaya A, Hidayati N, Iranawati F (2021) Sensitivity and dynamic of Sardinella Lemuru in Bali Strait Indonesia. J Hunan Univ Nat Sci 48(1):97-107

Sankar S, Thondithala Ramachandran A, Franck Eitel KG, Kondrik D, Sen R, Madipally R, Pettersson LH (2019) The influence of tropical Indian Ocean warming and Indian Ocean dipole on the surface chlorophyll concentration in the eastern Arabian Sea. Biogeosci Discuss 2019:1–23

Shetye SS, Kurian S, Gauns M, Vidya PJ (2019) 2015–16 ENSO contributed reduction in oil sardines along the Kerala coast, south-West India. Mar Ecol 40(6):e12568

Shyam SS (2021) Oil sardine landings and revenue realization under a climate change regime in India. J Fish Livest Prod 9(9):1–3

Srinath M, Kuriakose S, Mini KG (2005) Methodology for estimation of marine fish landings in India, vol 86. CMFRI Special Publications, pp 1–57

Vidya PJ, Kurian S (2018) Impact of 2015–2016 ENSO on the winter bloom and associated phytoplankton community shift in the northeastern Arabian Sea. J Mar Syst 186:96–104

Vivekanandan E, Rajagopalan M, Pillai NGK (2009) Recent trends in sea surface temperature and its impact on oil sardine. In: Global climate change and Indian agriculture. Indian Council of Agricultural Research, New Delhi, pp 89–92

Chapter 5
Deep-Sea Harpacticoids in the Indian Ocean: Insights into Biodiversity Amidst Global Climate Challenges

Tanmoy Nandy, Santosh Gaikwad, and Sabyasachi Sautya

Abstract Climate change and global warming are reshaping deep-sea ecosystems, especially impacting harpacticoid copepods that are crucial to deep-sea sediments and dominate meiobenthic communities. Rising ocean temperatures, deoxygenation, and shifts in organic matter deposition are altering their habitats and ecological roles. The Indian Ocean, with its unique monsoon system and expanding oxygen minimum zones (OMZs), is particularly vulnerable. This review examines the distribution and diversity of deep-sea benthic harpacticoids in the Indian Ocean, comparing them with global habitats while considering climate change impacts. We compiled data from the literature on harpacticoid copepods in deep-sea sediments (≥1000 m) from the Indian Ocean and other oceans between 1971 and 2021, focusing on ecology and biodiversity studies. While research on deep-sea harpacticoids varies across habitats, most studies in the Indian Ocean focus on deep-sea basins with polymetallic nodules (PMN). In contrast, areas like trenches, abyssal plains, and seamounts remain largely unexplored in the region, unlike in the Pacific and Atlantic Oceans. The increasing threat of climate change to the Indian Ocean emphasizes the need for further research into its deep-sea biodiversity, as intensified monsoons, declining oxygen levels in OMZs, and warming seas pose challenges for meiofaunal survival. Harpacticoid copepods, due to their sensitivity to environmental changes, are valuable bioindicators of ecosystem health in vulnerable regions. However, significant gaps remain in understanding their adaptability and ecological contributions under climate stress. This review highlights the need for future research to integrate climate change impacts and support the management of anthropogenic activities in sensitive deep-sea regions. Priority should be given to explor-

Supplementary Information The online version contains supplementary material available at https://doi.org/10.1007/978-981-95-0409-1_5.

T. Nandy · S. Gaikwad · S. Sautya (✉)
Laboratory for Benthic Ecological Trait Analysis (L-BETA), CSIR-National Institute of Oceanography, Regional Centre, Mumbai, India
e-mail: sautya.nio@csir.res.in

B. P. Barathan et al. (eds.), *Navigating Climate Change: Impacts on Biodiversity and Ecosystem Resilience*, Sustainable Development and Biodiversity 42,
https://doi.org/10.1007/978-981-95-0409-1_5

ing under-studied habitats such as trenches and seamounts in the Indian Ocean to understand better how climate-induced changes affect harpacticoid copepods and their ecosystem roles. Comprehensive studies on the Indian Ocean's deep-sea ecology will contribute to a broader understanding of biodiversity loss and its long-term consequences for ecosystem function.

Keywords Benthic harpacticoid · Climate-induced changes · Global perspectives · Deep-sea biodiversity

5.1 Introduction

The Earth's climate has exhibited significant variability throughout its history, influenced by natural climate forcings, such as orbital variations, volcanic eruptions, and amplified by feedback processes intrinsic to the climate system (Timmreck 2012). Unlike natural variations, present-day climate change is characterized by an unprecedented rapid pace, driven primarily by anthropogenic emissions of greenhouse gases (GHGs), aerosols, and changes in land use and land cover (LULC) (Han et al. 2024). The resulting global warming has increased the Earth's average near-surface air temperature by approximately 1 °C since pre-industrial times, with severe implications for ecosystems worldwide (Allen et al. 2018).

Human-induced alterations in atmospheric composition have disrupted Earth's energy equilibrium that is essential for climate stability (Laj et al. 2009). This imbalance has triggered various environmental effects, including acidification of oceans, elevated sea levels, and increased global temperatures (Campbell et al. 2016; Chen et al. 2017). The Indian Ocean, the third-largest of the world's oceanic divisions, is bounded by Asia to the north, Africa to the west, Australia to the east, and the Southern Ocean to the south and this Indian Ocean's climatology is characterized by seasonal reversals of winds and currents, driven by the monsoon cycle, which significantly influence regional weather patterns and ocean circulation (Schott et al. 2009). This vast body of water covers approximately 20% of Earth's surface and plays a crucial role in global climate patterns, particularly through its unique monsoon system (Roxy et al. 2020). The deep-sea ecosystems of the Indian Ocean, characterized by varying sediment types, organic matter inputs, and hydrodynamic conditions, host unique habitats that remain largely unexplored. The abyssal plains of the Indian Ocean are divided by meridional ridges (Ninety East Ridges, Mid-Indian Ridge, and Carlsberg Ridge), which form numerous ocean basins (West Australian Basin, Mid-Indian Basin, Somali Basin, Mascarenes Basin, Madagascar Basin, Natal Basin, and Arabian Basin). The Arabian Sea in the northern part of the Indian Ocean is also known as the largest and most intense intermediate-depth oxygen minimum zone (OMZ) (Kamykowski and Zentara 1990; Rogers 2000). The seafloor of the Indian Ocean is differentially covered by calcareous oozes, clay, glacial debris, deep-sea mud, siliceous ooze, shelf, and slope deposits (Berger 1974). The central Indian Ocean Basin is morphologically composed of several abyssal hills and seamounts. The hard substrata of this region consist of basalt rocks and a surface layer composed of ferromanganese concretions. In the south-eastern and southwestern parts of the Indian Ocean (approximately 4000 m depth), there are extensive pavements of polymetallic nodules, and the presence of these nodules at the sediment-water interface is a result of strong bottom currents and benthic biological activity (Berger 1974; Kolla et al. 1980). The deep sea constitutes the largest biome on Earth, encompassing more than 60% of the planet's surface. It harbors diverse ecosystems shaped by unique environmental conditions such as high hydrostatic pressure, low temperatures, limited nutrient availability, and the absence of light. Among myriad organisms inhabiting this extreme environment, benthic harpacticoid copepods are a significant component of the meiofaunal community and play critical roles in ecosystem functioning, nutrient cycling, and energy transfer (Gheerardyn and Veit-Köhler 2009). Despite their ecological importance, the study of harpacticoid copepods in the Indian Ocean remains underexplored, with most research efforts historically concentrated on the Atlantic and Pacific Oceans (Ivanov and Southward 1971; Mascart et al. 2013). The Indian Ocean is renowned for its rich biodiversity, hosting a myriad of marine ecosystems from coral reefs to deep-sea habitats. These ecosystems support a diverse array of life, including numerous endemic species, and provide essential ecosystem services such as food security, climate regulation, and economic opportunities for coastal communities. The region's biodiversity is particularly significant in areas like the Coral Triangle that extends into the eastern Indian Ocean and is recognized as a global biodiversity hotspot (Campbell et al. 2016; Chen et al. 2017). However, the Indian Ocean faces mounting challenges from anthropogenic activities, primarily pollution and climate change. Marine pollution, including plastic waste, chemical contaminants, and nutrient runoff, threatens the health of marine ecosystems and the species they support (Laj et al. 2009). Climate change impacts are particularly pronounced in the Indian Ocean, with rising sea temperatures, ocean acidification, and changing circulation patterns altering marine habitats and species distributions (Roxy et al. 2020).

Globally, studies on harpacticoid copepods have demonstrated their sensitivity to environmental changes, making them valuable indicators of ecosystem's health (Campos et al. 2017; Mascart et al. 2015; and van Damme et al. 1984). The ongoing impacts of climate change, including ocean warming, acidification, and deoxygenation, are already altering deep-sea ecosystems and posing threats to meiofaunal communities (Danovaro et al. 2004). Harpacticoid copepods are a dominant group of meiofauna, accounting for a significant proportion of benthic biomass and abundance in deep-sea sediments (Giere 2009a, 2009b). Their ecological functions include bioturbation, organic matter decomposition, and nutrient recycling (Wang et al. 2010; Zhang et al. 2017), all of which are essential for maintaining ecosystem stability and productivity. Investigations in the North Atlantic and Mediterranean Sea have illuminated the crucial function of harpacticoids in marine food webs.

These organisms act as intermediaries, channeling energy from primary producers to higher trophic levels by serving as food sources for larger benthic and pelagic species (Coull 1999). These findings highlight the urgency to understand how these organisms respond to similar stressors in the Indian Ocean, where climate change effects are particularly pronounced.

Research on deep-sea harpacticoid copepods has predominantly focused on regions such as the North Atlantic, Southern Ocean, abyssal plains of the Pacific Ocean (Menzel et al. 2011; Thistle and Eckman 1990), and shallow depths (Chertoprud et al. 2010). These studies have provided valuable insights into species diversity, biogeographic distribution, and community responses to environmental gradients. The deep sea has revealed high species richness and endemism among harpacticoid copepods, attributed to habitat heterogeneity and isolation (Shimanaga et al. 2004; Thistle et al. 2007). Similarly, studies in the Mediterranean Sea have documented the influence of sediment characteristics and hydrodynamic processes on harpacticoid assemblages (Danovaro et al. 2004).

Despite its ecological significance and vulnerability to anthropogenic impacts, the Indian Ocean remains underrepresented in global meiofaunal research. Existing studies have primarily focused on shallow-water habitats or specific regions, such as the Andaman Sea and the Arabian Sea (Ingole and Singh 2010; Ingole et al. 1998). Comprehensive assessments of harpacticoid copepod diversity and distribution in deep-sea environments are lacking, which limits our understanding of their ecological roles and biogeographic patterns. Furthermore, standardized sampling protocols and taxonomic frameworks are needed to facilitate comparisons with global data and address taxonomic impediments that have historically constrained meiofaunal research (Higgins and Thiel 1988). Given the rapid pace of environmental change and the critical importance of the Indian Ocean's ecosystems, there is an urgent need for comprehensive reviews that synthesize current knowledge and identify research gaps. This review is necessary to provide a holistic understanding of the Indian Ocean's biodiversity, with a particular focus on deep-sea benthic communities, such as harpacticoid copepods that play crucial roles in ecosystem functioning but remain understudied in this region (Gheerardyn and Veit-Köhler 2009).

The importance of this review lies in its potential to inform conservation strategies, guide future research efforts, and contribute to the sustainable management of the Indian Ocean's resources. By examining the diversity, distribution, and ecological roles of deep-sea benthic organisms, particularly harpacticoid copepods, in the context of ongoing environmental changes, the present study aims to: (1) synthesize current knowledge on deep-sea benthic communities in the Indian Ocean, with a focus on harpacticoid copepods, (2) compare findings from the Indian Ocean with global research to contextualize the region's unique characteristics and challenges, and (3) highlight research gaps and priorities for future studies to enhance our understanding of the Indian Ocean's deep-sea ecosystems and their responses to environmental change. By addressing these objectives, this review seeks to contribute to the broader understanding of the Indian Ocean's marine biodiversity and ecosystem functioning, ultimately supporting evidence-based conservation and management strategies in the face of global environmental change.

5.2 Materials and Methodology

We focused on the data collected only from deep-sea sediments (≥1000 m), avoiding terrestrial and shallow water influences. Additionally, it is the largest ecosystem on Earth and is mostly unexplored owing to the financial and technological challenges of working in this area (Giere 2009a, b). Only 1% of the deep seafloor has been systematically studied (Ramirez-Llodra et al. 2010).

The methodology for collecting and analyzing data related to deep-sea harpacticoid copepods encompasses a comprehensive approach spanning several decades and covering vast oceanic regions. The data collection process focused primarily on deep-sea sediments, specifically targeting depths of 1000 meters or greater. This depth criterion was established to ensure a true representation of the deep-sea environment and its unique ecological characteristics. The temporal scope of the study was extensive, covering 50 years from 1971 to 2021. This extended timeframe allowed for the identification of long-term trends and potential shifts in copepod populations or distributions over time. The geographical focus was primarily on the Indian Ocean, but data from other global oceans were also incorporated to provide a broader context and enable comparative analyses. In selecting studies for inclusion, the researchers prioritized those focusing on deep-sea benthos ecology and biodiversity. This specific focus ensured that the data collected were directly relevant to understanding the ecological dynamics of harpacticoid copepods in deep-sea environments. To maintain data quality and consistency, only studies with clear and well-documented sampling methods and precise depth information were included in the analysis. The data screening and analysis process involved a multi-step approach to ensure thoroughness and accuracy. Initially, researchers conducted extensive database searches using carefully selected keywords related to deep-sea copepods, benthic ecology, and relevant oceanographic parameters. This initial search cast a wide net to capture as much of the relevant literature as possible. Following the initial search, a rigorous review process was implemented. Each article was carefully examined to determine if it met the predefined criteria for inclusion in the study. This screening process helped to maintain the focus and relevance of the data set, ensuring that only high-quality, pertinent studies were included in the final analysis. The analysis phase utilized a variety of tools and software to process and interpret the collected data. Microsoft Office suite was employed for basic data organization and preliminary analysis. This allowed for the creation of structured databases and initial data visualization. For more advanced spatial analysis and mapping of copepod distributions, QGIS, a powerful geographic information system, was utilized. This software enabled the researchers to create detailed maps illustrating the spatial patterns of copepod populations across different ocean basins. Statistical analysis and the creation of more complex visualizations were conducted using R, a robust statistical programming language. The methodology placed a strong emphasis on transparency and reproducibility. This was achieved through several means: (1) Clear documentation of all data sources, including the specific databases searched and the keywords used. (2) Detailed explanation of the selection criteria for including studies in the analysis, ensuring that other researchers could understand and potentially replicate the data collection process. (3) Thorough description of the screening process, including how articles were reviewed and the specific criteria used to determine their relevance and quality. (4) Explicit listing of all software and tools used in the analysis, including version numbers where applicable, to facilitate replication of the analytical procedures. The researchers also acknowledged the limitations of their methodology. This might have included potential biases in the available literature, gaps in geographical or temporal coverage, or limitations in the statistical methods employed. By openly addressing these limitations, the researchers demonstrated scientific integrity and provided context for interpreting the results of their analysis. This comprehensive and transparent methodology provides a solid foundation for understanding the distribution, ecology, and biodiversity of deep-sea harpacticoid copepods. It allows for robust analysis of long-term trends and spatial patterns, contributing valuable insights to the field of deep-sea ecology and potentially informing conservation efforts in these unique and often understudied ecosystems.

In addition, the results, figures, and tables were cross-checked, and a detailed examination of the full paper was performed, retaining only those publications that met the screening criteria. Figure 5.1 and Table 5.1, describe a brief description of research done globally, along with its geographical information.

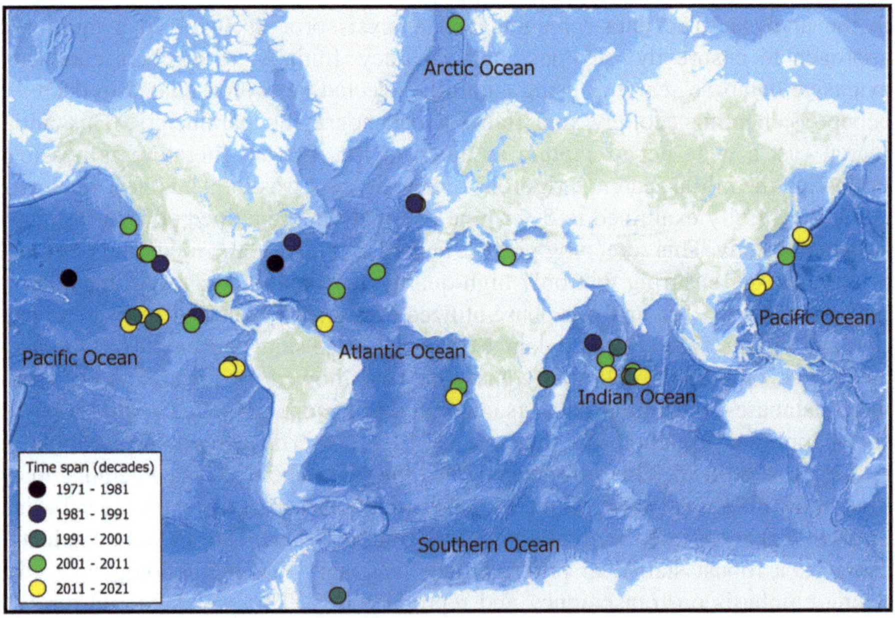

Fig. 5.1 Worldwide work done on deep-sea benthic harpacticoids at different times

5.3 Results and Discussion

5.3.1 Ecological Importance of Deep-Sea Benthic Harpacticoid Copepods

Harpacticoid copepods are aquatic crustaceans belonging to the subclass *Copepoda* and the order *Harpacticoida*. In marine aquatic ecosystems, the majority (90%) of the harpacticoid copepods are benthic and belongs to meiofauna fraction due to their size structure (63–500 μm). Benthic harpacticoid copepods (hereafter harpacticoids) are highly abundant in shallow and deep-sea sediments, with rich species diversity (Thistle et al. 2005, 2007). Deep-sea harpacticoids play a major role in transferring energy and plant materials from producers to higher trophic levels than shallow-water species. Thus, they are considered essential food sources for deep-sea marine ecosystems (Hicks and Coull 1983; Rose et al. 2005). The epibenthic detrivorous nature of harpacticoids makes them a major food item for demersal finfish and shellfish communities (Coull 1990; Service et al. 1992). Studies have shown that early larval stages and juveniles of several demersal fish feed on harpacticoids during their bottom-dwelling life. Their predation on harpacticoids is sometimes species-specific, although multiple harpacticoids exist at high densities. The absence of certain harpacticoids prevents and slows the recruitment process of fish stocks (Schratzberger and Ingels 2018; Schückel et al. 2013). Thus, harpacticoids provide short-term but compelling energetic benefits to these predators (early stages of fish).

The bioturbation activity of harpacticoids enhances nutrient exchange in deep-sea environments. As deposit feeders, they enhance detritus assimilation, which also triggers the remineralization process. Hence, deep-sea harpacticoids indirectly support the ocean biogeochemical cycle through their involvement in the mineralization of carbon and nitrogen (Montagna 1982; Rosli et al. 2018) (Fig. 5.2).

Overall, the benthic harpacticoid copepods serve various ecosystem processes simultaneously by transferring sediment grains and nutrients during bioturbation (Adámek and Maršálek 2013), displacing organic substances and microbial populations within the sediment column during feeding (Ebenhöh et al. 1995), and serving as prey for the juveniles of several demersal fishes (scald fish, dab fish, etc.) (Schückel et al. 2013). Thus, harpacticoids provide energetic benefits and control the recruitment and standing stock of economically important demersal fish populations (Schratzberger and Ingels 2018). Accordingly, benthic harpacticoids can fulfil a predator's nutritional requirements in terms of calories, carbon, nitrogen, amino acids (proteins), micronutrients, and, especially unsaturated fatty acids. Unsaturated fatty acids are crucial for cell membrane and membrane-bound enzyme systems and functions in aquatic organisms (especially, in metazoans), but they cannot synthesize them independently (Tocher 2003). Benthic harpacticoids are known to biosynthesize and accumulate highly unsaturated fatty acids; therefore, harpacticoids have been frequently used as food supplements in mariculture to provide high-quality nutrition to early life stages and juveniles of economically important demersal

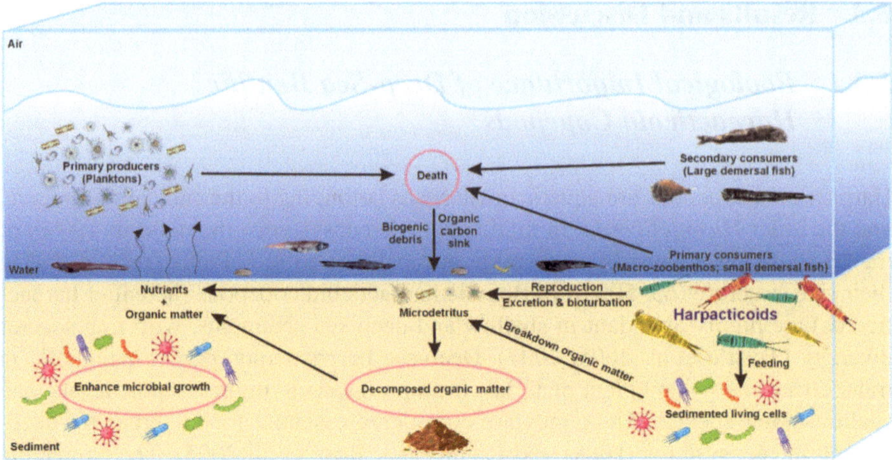

Fig. 5.2 The flow chart shows the role of benthic harpacticoids on ecosystem services. At the top, primary producers like plankton are shown, leading to a cycle of death and organic carbon sinking. Arrows indicate the flow of energy and matter to primary consumers, such as macro-zoobenthos and small demersal fish, and secondary consumers, like large demersal fish. Below, in the sediment, decomposed organic matter enhances microbial growth. Harpacticoids feed on sedimented living cells, contributing to nutrient cycling through reproduction, excretion, and bioturbation. Key terms include primary producers, biogenic debris, organic carbon sink, and microbial growth

fishes (Fleeger 2005). For these ecological values, several studies confirmed the usefulness of harpacticoids as ecological indicators. Their abundance ratio with nematodes is an indication of pollution, anthropogenic disturbances, and climate change (Balsamo et al. 2012; Pusceddu et al. 2014). As the second most dominant meiobenthos, harpacticoids live on much smaller spatiotemporal scales with unique adaptations to maintain their diversity (Bell 1980; Etter and Mullineaux 2001). They are morphologically well adapted to the deep-sea environment and have a proportionally elevated abundance compared to the macrofaunal community. Owing to a lack of planktonic larvae, shorter life spans, and sensitivity to environmental changes, they have limited dispersal capabilities. For this reason, harpacticoids are considered a useful biological tool for deep-sea diversity research over both evolutionary and ecological times (Thistle 2003). Studies have shown that harpacticoids are widespread in marine soft sediments and can be found in various deep-sea environments, including continental slopes, ocean basins, abyssal plains, seamounts, ridges, and trenches. While they represent the most diverse meiofaunal group in deep-sea sediments, their distribution patterns are significantly affected by the spatial variability of three key environmental factors: sediment grain size, biogenic structures, and food availability. (Baguley et al. 2006). Additional research is needed to explore the diversity and ecosystem functions of deep-sea harpacticoids across different deep-sea habitats.

5.3.2 *Distribution of Harpacticoids in Different Deep-Sea Habitats*

The vast deep seafloor encompasses roughly 3 × 108 km^2, accounting for nearly 65% of Earth's solid surface. This expansive region comprises various distinct habitats with unique ecological features, making it unparalleled on our planet (Danovaro et al. 2020). Despite being Earth's largest ecosystem, the deep-sea environment remains largely unexplored and poorly understood compared to other ecosystems. Remote instruments have surveyed only 5% of deep-sea ecosystems, while a mere 1% has undergone detailed sampling and analysis (Gooday et al. 2020). The deep-sea floor is divided into continental margins and abyssal plains (ocean basins), with additional geological formations such as seamounts, ridges, vents, canyons, and trenches interspersed throughout. These diverse deep-sea structures harbor unique faunal communities with high species diversity. Within the benthic system, abyssal plains constitute 76% of the seafloor, followed by continental margins (10%), mid-oceanic ridges (9%), seamounts (3%), and trenches (2%) (Stewart and Jamieson 2018). Other deep-sea habitats, including hydrothermal vents, whale falls, benthic OMZ, and cold-water corals, are comparatively scarce and localized. The deep-sea food web relies entirely on detritus, with limited organic matter influx. Consequently, detritivorous microorganisms play a crucial role in shaping deep-sea ecosystems (Campanyà-Llovet et al. 2017; Danovaro et al. 2014; Maier et al. 2020).

The present review focuses on the study conducted and the habitat-wise distribution of harpacticoids in the deep Indian Ocean and their assessment with other deep-sea habitats of global oceans. Figure 5.3 describes the different studies of harpacticoids across various habitats of global oceans.

5.3.2.1 Continental Margins (Slope and Rise)

The most geologically diverse region of the deep-sea floor is the continental margin, with immense habitat heterogeneity (Levin et al. 2010; Menot et al. 2010). There are very few reports on the global diversity and ecology of deep-sea harpacticoids in these habitats. The first systematic study of deep-sea harpacticoids on continental margins was conducted in the Atlantic Ocean near the northern Carolina coast (Coull 1972). Density and diversity were studied at different depths, and the diversity of meiofauna and harpacticoids increased with depth. Several harpacticoid species show unique deep-sea adaptations with missing eye spots and varied structures (either spade or elongated) of swimming legs. During the Nova Scotian continental rise of the Atlantic Ocean, the effect of benthic storms on the abundance and composition of harpacticoids has been studied in detail by Thistle et al. (1985) and Thistle (1988). The total harpacticoids were observed to be significantly lower during and after the benthic storm, which might be due to their migration into deeper layers and/or habitat shift. Storm-induced erosion of the surface layer was the causative factor for the decrease in harpacticoid abundance. In the Pacific Ocean, the

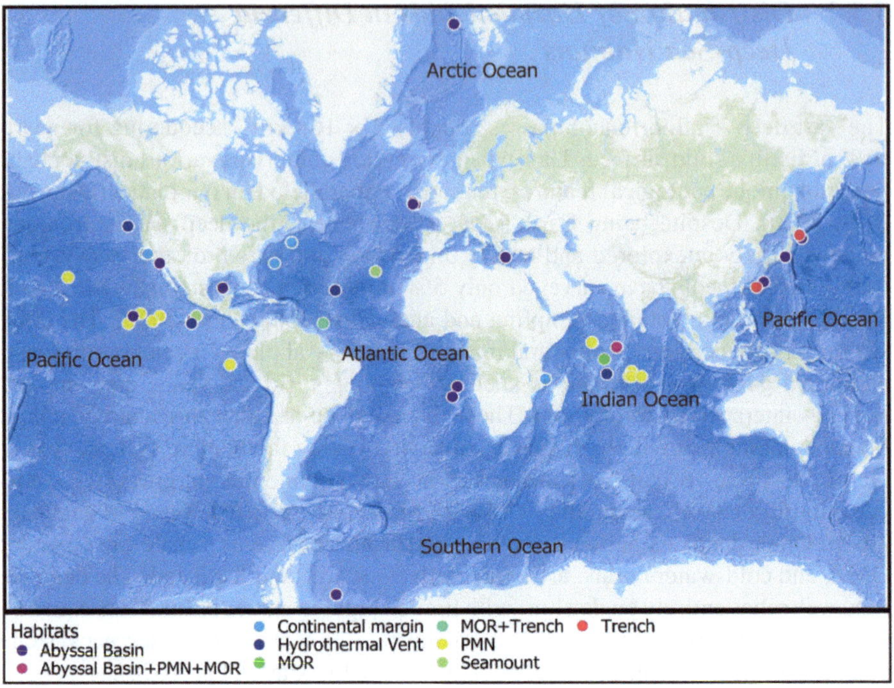

Fig. 5.3 Global distribution of deep-sea studies on harpacticoid copepods across ocean basins and marine habitat types. The figure highlights five major ocean basins (Arctic, Pacific [listed twice for North/South regions], Atlantic, Indian, and Southern Oceans) and ten specialized habitats. Habitat abbreviations: MOR (Mid-Ocean Ridge), PMN (Polymetallic nodules)

only continental margin that has been studied is the continental rise of Central California. Thistle et al. (2006) and Sedlacek (2007) studied the impact of carbon sequestration on harpacticoid species diversity and epibionts association. A total of 67 species of harpacticoids have been reported in these studies. The presence of epibionts was positively associated with their host harpacticoids, which influenced the ability of harpacticoids to deal with unwanted environmental stress (such as CO_2-rich seawater). Thistle et al. (2007) also studied the emergence behavior of harpacticoids in the continental slope of Central California for the first time in deep-sea benthic ecology. A few harpacticoid species have been shown to change their location using near-bottom flow. The impact of carbon sequestration on deep-sea benthic harpacticoids has been studied in Monterey submarine canyons (Sedlacek et al. 2009), and the association between ciliate epibionts and harpacticoids has been described by Sedlacek et al. (2013).

In the Indian Ocean, only one study has been conducted by Duineveld et al. (1997) on the productivity and standing stock of harpacticoids at the Yemen–Somali margin and margin of Kenya. Along with sediment community oxygen consumption, the distribution of benthic biomass and phytodetritus has been studied. Harpacticoids were recorded as the second most dominant meiobenthic taxa with a wide range of density (7–184 ind./10 cm^2). The species diversity, spatial distribution, relationship with environmental factors, impact of benthic storms, and epibiont association on harpacticoids from different continental margins of the deep Indian Ocean remain unexplored.

5.3.2.2 Ocean Basin

Abyssal Plain

Abyssal plains are underwater basins of deep ocean floors (below 3000 m depth) that cover more than 50% of the earth's surface. These regions are fuelled by terrigenous sediments derived from terrigenous rock weathering and biological debris from pelagic organisms. Limited food sources support a diverse group of abyssal fauna with unique adaptations. The top few centimeters of sediments in abyssal plains are colonized by rich macro- and meiofaunal communities with dense populations. Globally, most studies of deep-sea harpacticoids have been conducted in the abyssal region.

The first systematic study was conducted by Thistle (1978) in the San Diego Trough of the Pacific Ocean. The spatial dispersal patterns of 140 harpacticoid species have been reported. Subsequently, several interesting studies have been conducted in the same region, such as the relationship between harpacticoids and other meiobenthic organisms with polychaete functional groups, harpacticoid distribution with mud ball association, and community structure with spatial distribution patterns (Thistle et al. 1993; Thistle and Eckman 1988, 1990). In Sagami Bay, spatio-temporal variation in harpacticoids has been studied at different depths, and their relationship with different environmental factors has been well established (Shimanaga et al. 2004, 2008). Kitahashi et al. (2012) conducted a comparative study of community structure and spatial distribution in the northeastern part of the Pacific Ocean in the Kuril and Ryuku regions (Kitahashi et al. 2012).

In the Atlantic Ocean, a detailed study on species diversity and community structure has been reported in the Angola Basin (George et al. 2014; Rose et al. 2005). More than 700 species have been recorded in the basin, with 600 new species. The distribution and population dynamics of harpacticoids at different depth gradients have been studied by Pfannkuche (1985) and GHeerardyn et al. (2009) from the Porcupine Seabight. Later, a harpacticoid response to phytodetritus was reported in the same region by Gooday et al. (1996). Species diversity and its relationship with associated environmental factors have been reported by Baguley et al. (2006) from the basin of the Gulf of Mexico, and more than 650 species of harpacticoids have been reported. Tselepides and Lampadariou studied the detailed community structure of harpacticoids from the eastern sector of the Atlantic basin near the Mediterranean Sea (Tselepides and Lampadariou 2004). Likewise, different studies on the spatial distribution and community composition of the basins of the Southern and Arctic Oceans have been conducted (Gallucci et al. 2009; Herman and Dahms 1992). However, many multidimensional studies on harpacticoids have already been reported globally; such studies have not been conducted in any abyssal plain of the deep Indian Ocean. Most studies have been centralized based on the polymetallic nodule (PMN area) mining region. Parulekar et al. (1982) studied the overall biomass and abundance of deep-sea benthos (macro- and meio-) from different depths and habitats of the Indian Ocean.

Polymetallic Nodule (PMN) Areas

Polymetallic nodules (PMNs) are rounded accretions of manganese, iron hydroxides, and other metals that cover vast areas of the deep-seafloor (3000–6000 m), mostly abundant on abyssal plains. Different parts of the Pacific and Indian Ocean basins have high concentrations of these deposits. However, very little information is available on the biodiversity associated with nodules, and a wide range of PMN areas are being targeted for future mining. Knowledge about the biodiversity of deep-sea ecosystems and the potential impact of mining on these ecosystems is still limited. Several studies have reported that PMN areas are a hotspot of abundance and diversity for different vulnerable abyssal fauna that inhabit the sediment immediately beneath the nodules or directly attach to the nodule's surface layer.

Hessler and Jumars (1974) conducted the first study on harpacticoids in a PMN area of the central-north Pacific Ocean. The general community structures of macro- and meiobenthic organisms have been recorded. The abundance of harpacticoids was found at 120 ind/0.25 m^2 with 57 species. Another PMN area is located in the south-east near the Peru Basin, where the metazoan meiofaunal community was evaluated by Bussau et al. (1995). High densities of meiofauna (30–100 ind/nodule) were recorded with 10% harpacticoids in this study. In the year 2001, the long-term effect of anthropogenic disturbance on harpacticoids was done by a "disturbance and recolonization experiment" (Ahnert and Schriever 2001) in the Peru Basin. Nineteen harpacticoid families were recorded, contributing >10% of the total meiobenthic community in this study. Artificial disturbances were created during the experimental study to assess the impact of nodule mining in this area. Harpacticoids showed significant differences before, immediately after, and a long time after the disturbance; they also acted as indicator taxa for monitoring benthic disturbance. In addition, Stratmann et al. (2018) investigated the recovery of phytodetritus processing by benthic food web structures through a sediment disturbance experiment and documented high macrofaunal diversity predominated by harpacticoids (33.33%). A crushed nodule particle experiment was performed by Mevenkamp et al. (2019) to evaluate the short-term effects on the meiobenthic community in the Peru Basin. Several studies on benthic impact experiments and benthic community structure have also been carried out in the Clarion-Clipperton Fracture Zone (one of the important PMN areas) in the Pacific Ocean, focusing on the effects of nodule mining, carbon sequestration, and environmental disturbances. Harpacticoids have emerged as the second dominant meiofauna with high species diversity in this region, and a few taxa-specific responses have been reported from various benthic impact experimental assessments (Chuar et al. 2020; De Smet et al. 2017; Radziejewska 2002; Radziejewska et al. 2001; Sweetman et al. 2019).

In the Indian Ocean, most studies on harpacticoids have been conducted in the PMN area for multidimensional purposes. The first survey was conducted by Parulekar et al. (1992) to understand the distribution and density of harpacticoids and their relationship with abiotic factors in a PMN area in the northern central Indian Ocean. The area was reported to be rich in deep-sea meiobenthic diversity, with 27% harpacticoids. The distribution of harpacticoids and their immediate effects due to mining disturbances have been reported by Ingole et al. (2000) in the central Indian Ocean. Significant decreases in abundance and composition were observed before and after benthic disturbance. Recolonization experiments showed that harpacticoids require more time to recolonize and are most sensitive to such physical disturbances. Ingole et al. (2005a, b) conducted an artificial benthic disturbance experiment on the PMN area of the central Indian Ocean, to evaluate the restoration rate of benthic fauna near the PMN area. Physical disturbances due to nodule mining drastically affected the density of harpacticoids and other crustaceans, and they did not reach their previous population density even after 3 years of disturbance. The recovery rate of these fauna is very slow, and only the available food (due to sediment mixing and subsurface deposition) may help them recover and recolonize quickly (Ingole et al. 2005a). Ansari (2000) and Zeng et al. (2018) also described the distribution and density of harpacticoids in a proposed mining area of the central Indian Ocean basin. Pavithran et al. (2007) studied the composition and feeding patterns of deep-sea macrofauna in a PMN area and reported a single species of harpacticoids (*Parameiropsis* sp.) in a considerable amount. As harpacticoids play an important role in indicating environmental changes due to mining and other physical disturbances, they can be used as key indicator taxa for future deep-sea environmental monitoring research. Although different studies have been conducted in the PMN area of the Indian Ocean, studies on indicator taxa, their distribution, and their relationship with abiotic factors are required.

5.3.2.3 Mid-Oceanic Ridges and Seamounts

Mid-oceanic ridges are the most striking features of deep ocean basins, accounting for 33% of the global ocean floor. It is a geological structure of oceanic crust with a tectonically active central rift valley bounded by elevated flanks on both sides sloping down to the abyssal plains (Heezen 1969). Although geological functions are well known in the mid-oceanic ridge system, the biological role of this habitat remains limited. Chemosynthetic life is mostly found around hydrothermal vents that occur along the ridge axis, and the downward export of organic matter from the surface layer supports biological productivity of the ridge system (Gebruk et al. 2000; Van Dover 2000). Seamounts are topographically isolated mountains that rise more than 1000 m above the surrounding level of the seabed (Consalvey et al. 2010; Schlacher et al. 2010). Generally, owing to their unique truncated geological features, ridges and seamounts provide additional habitat for slope-dwelling fauna and enhance their diversity. Moreover, ridges and seamounts act as an ecological barrier to the migration of deep-sea organisms between two halves of the ocean and support the development of endemic species restricted to individual ridge habitats, further enhancing deep-sea biodiversity (Rex and Etter 2010; St Laurent and Thurnherr 2007). Very few studies have been conducted on the ridge and seamount ecosystems.

Schmidt et al. (2018) studied the abundance, composition, and biodiversity of harpacticoids from the Vema Fracture Zone in the Atlantic Ocean mid-oceanic ridge (Schmidt et al. 2018). Spatial diversity was observed in their study, with significant community composition. Qualitative and quantitative investigations of harpacticoids were conducted on the Great Meteor Seamount by George and Schminke (2002). Fifty-six species of harpacticoids species were recorded in their study, of which 54 were new species. Only Levin et al. (1991) conducted studies on harpacticoids in the Pacific Ocean (Levin et al. 1991) from the Eastern Pacific seamount. This study focused on the community structure of harpacticoids with oxygen concentration and organic matter gradients in the seamount.

In the Indian Ocean, only studies of mid-oceanic-ridge harpacticoids have been conducted (Ingole 2003) in the central Indian Ocean. The macrofauna composition, distribution, abundance, and biomass have been examined in ridge ecosystems. A considerable number of harpacticoids have been recorded in their macrofaunal samples, ranging from 20 to 177 ind./m^2. The benthic standing stock showed a positive relationship with the ridge environment. The Indian Ocean Ridge system supports rich benthic fauna and is a very striking environment for benthic biological studies in the future (Ingole 2003).

5.3.2.4 Hydrothermal Vent

Hydrothermal vents are extreme habitats within the natural deep-sea ecosystem associated with tectonically and biologically active zones in the deep-sea. Hydrothermal vents form in volcanically active areas of mid-oceanic ridges, where the earth's tectonic plates are spreading apart from each other. Hydrothermal vents act as natural plumbing systems by transporting heat and chemicals from the Earth's interior to the ocean floor and supporting global ocean chemistry. To date, only few studies have been conducted on harpacticoids in hydrothermal vent ecosystems. First, a systematic study of the biology and distribution of harpacticoids in a hydrothermal vent (Juan De Fuca Ridge) was reported by Tsurumi et al. (2003). Other studies have been conducted by Gollner et al. (2006, 2007) on the East Pacific Rise. Five species of harpacticoids have been reported, and their association with tubeworms and mussels has been well established. The diversity and community structure of harpacticoids in the East Pacific Rise vent along a gradient of vent fluid emissions have been reported (Gollner et al. 2010). A total of 26 species of harpacticoids were documented in their study, and data suggest that less diverse communities are present in more extreme habitats. Zekely et al. (2006) documented the diversity of harpacticoids associated with mytilid mussel aggregations from two important vent ecosystems (East Pacific Rise and Mid-Atlantic Ridge).

In the Indian Ocean, only Kang and Kim (2021) reported the community composition of harpacticoids in a vent area of the central Indian ridge. To date, no other studies have been conducted on harpacticoids in hydrothermal vent ecosystems.

5.3.2.5 Trenches

Trenches are referred to as unique habitats below an abyssal depth of >6000 m with a specific geomorphological feature of the seafloor. The cross-profiles of trenches have "V" shaped characteristics with topographic complexity (Beliaev and Brueggeman 1989). The floor of the trench was mainly composed of fine-grained, noncalcareous sediments. Deep-sea trenches act as giant sedimentation tanks on the sea floor; they accumulate both particles from the surface layers and those transported from the adjoining ocean bottom. Sediments in the trench ecosystem generally contain more organic matter than those in the adjacent abyssal plains. Moreover, trenches are characterized by the presence of high turbidity currents, slides, and collapses, which can have catastrophic consequences on local benthic organisms (Danovaro et al. 2003).

Schmidt et al. (2018) studied the relative abundance and composition of harpacticoids in the Atlantic Ocean (Schmidt et al. 2018) from the Puerto Rico Trench. Owing to high turbidity and benthic disturbances, very few species have been recorded from the trench (four genera) and upper trench slopes (nine genera). Another study reported the trench habitat of the Ryukyu and Kuril regions (NE Pacific) (Kitahashi et al. 2012). Their study investigated community composition and spatial changes in the assemblages of harpacticoids. Harpacticoids were found to be influenced by the quantity of organic matter in the Ryuku region, and sediment textures played a key role in their population in the Kuril region. Although very few studies have been reported globally in trench habitats, no studies have been conducted in the Indian Ocean region to date.

5.3.3 Knowledge Gap and Future Viewpoints Related to Deep Sea-Benthic Harpacticoid Study in the Indian Ocean

Overall, most of the diversity and ecological studies on deep-sea benthic harpacticoid copepods that have been conducted in the Indian Ocean region are restricted to the north-western part of the Ninety-East Ridge and the central part of the Indian Ocean. Table 5.1 shows a comparative statement of the work done on the three major oceans and Fig. 5.4 shows research conducted in the Indian Ocean and the countries involved in these studies.

To summarize the types of studies, habitats, and methodologies applied in deep-sea harpacticoid studies, the categories were compared using an alluvial plot (Fig. 5.5). Most of the studies were conducted in the Pacific Ocean (55.2%),

Table 5.1 Comparative study of deep-sea benthic harpacticoids along the various ecosystems in the three major oceans. "√" and "X" indicate work done and not done at the particular ocean, respectively

Ecosystem settings	Atlantic Ocean	Pacific Ocean	Indian Ocean
Continental margins	√	√	√
Abyssal basin	√	√	X
PMN area	X	√	√
Mid-oceanic ridges	√	√	√
Seamounts	√	√	X
Hydrothermal vents	√	√	√
Trenches	√	√	X

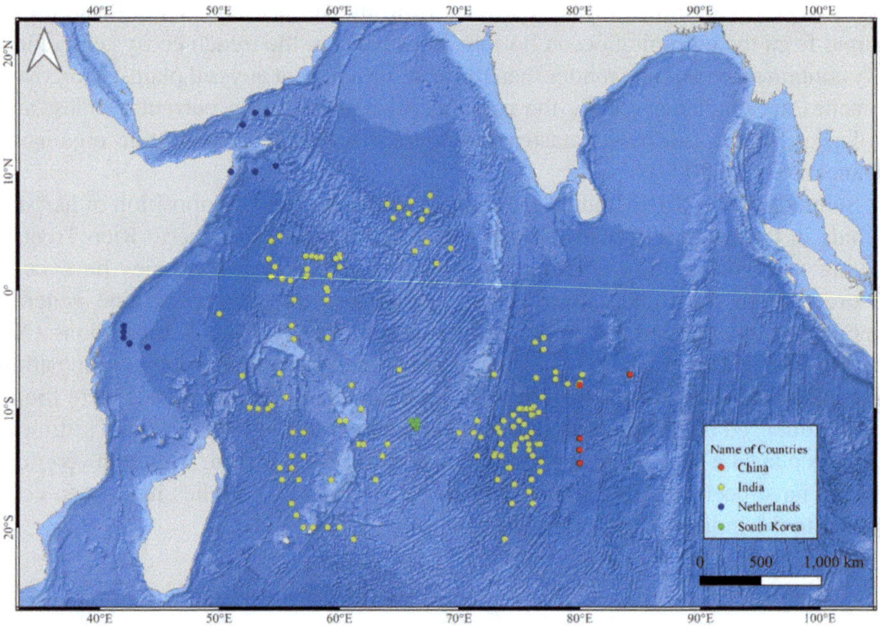

Fig. 5.4 Geographical distribution of deep-sea benthic harpacticoid copepod studies in the Indian Ocean color-coded by contributing research countries. The gridded map highlights sampling effort concentration across deep-sea habitats

followed by the Atlantic Ocean (22.5%), and the Indian Ocean (19%). The primary objectives of most studies were diversity and population ecology (64%), followed by different experimental works such as benthic impact assessment, effects of nodule mining, and carbon sequestration. Globally, more than 55% of studies have mentioned the abundance of harpacticoids, and the rest are reported either as percentages or the presence in their studied area. Further analysis revealed that out of 58 studies, only 19 (not in the Indian Ocean) publications focused on species-level analysis, which might be due to a lack of expertise in deep-sea harpacticoid

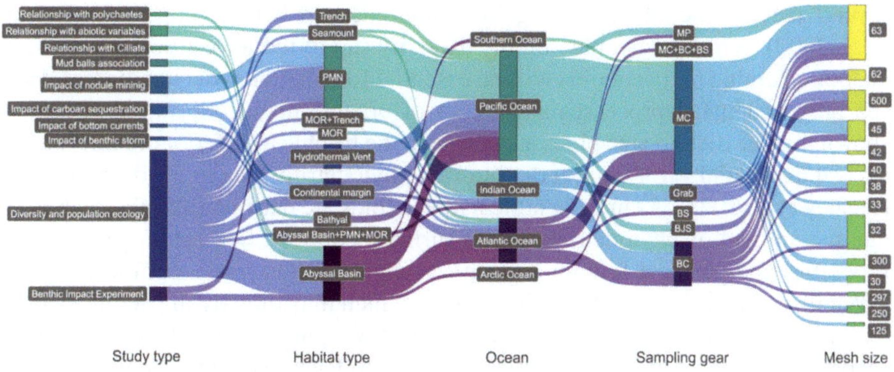

Fig. 5.5 Alluvial plot showing the distribution of research conducted from different ocean habitats based on sampling methodology and objectives of studies. The width of each band denotes the number of studies for that section while the thickness of each alluvium

taxonomy and is not required for the particular study they have done. The unit of abundance also varied in the publications. In Indian Oceans, most of the data were expressed as either ind/m^2 or ind/10 cm^2, as per the density recorded. The abyssal basin and PMN areas have emerged as the most studied areas globally. After the Pacific Ocean, most studies have been conducted from the PMN area in the Indian Ocean (7 out of 11 publications). The category based on the sampling gears revealed that 58.5% of the research was conducted using multiple cores, followed by box cores and grabs. A few studies have used variable gears, such as a benthic sledge, mussel pot, and Bushmaster Jr. sampling devices, as per their study objectives of different benthic impact experiments. Studies in the Indian Ocean, mostly preferred by box core and grab samples, might be due to the focus on the overall benthic community, mostly macrofauna. However, in the Pacific Ocean, 22 out of 32 samplings were done by multiple cores focusing on the meiofaunal community with taxonomic data. The lowest sieves used for harpacticoid analysis also showed variation with % of the studies using a mesh size of 63 µm, followed by 32 µm, 45 µm, and 500 µm. Six out of 11 publications from the Indian Ocean used 500 µm sieves, as their major focus was macrobenthos in a nodule mining area.

In the Indian Ocean, most studies have focused on basins with polymetallic nodule areas in the central Indian Ocean, which might be of both ecological and economic importance or prioritized by funding agencies. Habitats such as trenches, abyssal plains of non-PMN areas, and seamounts from the Indian Ocean have not been explored to date, although other oceans have already explored these habitats, and future studies of deep-sea harpacticoids in those particular areas are vital for global ecological studies. There is no such systematic data available on the species-level diversity of deep-sea benthic harpacticoids from the Indian Ocean, but several studies have been published from the Pacific and Atlantic Oceans focusing on both ecology and taxonomy, as well as various benthic impact experiments. The transport of diatom frustules from surface water to the deep-seabed is dependent on several factors, including grazing by planktonic and benthic copepods (Roman et al. 2000). Accordingly, data on the diversity (up to the lowest possible taxa) of deep-sea benthic harpacticoids will be essential for a better understanding of deep-sea biogeochemical processes and fluxes of organic matter. Recent studies have shown that the diatom size plays a key role in the trophic transfer of organic materials and the spatial variability of diatom-rich phytodetritus flux to the deep-sea floor (Pandey et al. 2023). As most benthic harpacticoids are detritivorous, they can regulate the amount of phytodetritus in the surface sediment. Several biogeochemical properties and physical forcing in deep-sea environments create significant spatial variability in diatom-rich phytodetritus flux, which may directly impact the benthic harpacticoid community (Smith et al. 2008; Sweetman et al. 2017). It has been reported that some benthic harpacticoids prefer to feed on diatom-dominated detritus compared to other organic matter (Nomaki et al. 2021); therefore, the spatial variability of phytodetritus deposition may influence the species diversity of benthic harpacticoids, and it can be expected from the Indian Ocean regime and needs to be addressed in the future. The availability of diatom frustules and/or other phytodetritus in surface sediments is not only dependent on surface phytoplankton and water column biogeochemical processes but is also regulated by the density and diversity of the benthic harpacticoid community. A mismatch between the surface water phytoplankton community and phytodetritus of the upper layer of the sediment bed might be due to the different species and feeding ecology of benthic harpacticoids.

Numerous studies on phytoplankton, zooplankton, and water column chemistry have been conducted in the Indian Ocean; however, the investigation of deep-sea benthic harpacticoids remains limited, potentially representing a significant knowledge gap in the benthic-pelagic coupling of the deep Indian Ocean. Consequently, further comprehensive research on deep-sea benthic harpacticoids is necessary to elucidate organic fluxes, feeding ecology, and associated ecosystem services in the deep Indian Ocean. To address this knowledge gap, there is a pressing need for increased research efforts and international collaboration among countries bordering the Indian Ocean. While India has conducted the majority of studies on benthic harpacticoids in this region, contributions from other countries such as China, the Netherlands, and South Korea have been relatively limited. Engaging surrounding countries like Australia, Africa, Indonesia, Myanmar, Oman, and Yemen in research initiatives and continuous monitoring of their proximal deep-sea environments is crucial for developing a more comprehensive understanding of the density, diversity, and ecological status of deep-sea benthic harpacticoids in the Indian Ocean. By fostering international collaborative deep-sea studies, researchers can work toward filling this knowledge gap and gaining valuable insights into the ecological dynamics of the deep Indian Ocean ecosystem.

5.3.4 Pollution, Climate Change and the Fate of Deep-Sea Harpacticoid Copepods in the Indian Ocean

The deep-sea benthic communities of the Indian Ocean are facing unprecedented challenges due to the combined effects of pollution and climate change. Pollution, primarily in the form of plastic debris, heavy metals, and organic contaminants, has been increasingly detected in the deep-sea environments of the Indian Ocean (Courtene-Jones et al. 2020). Microplastics, in particular, have been found in sediments and organisms at depths exceeding 5000 meters, indicating the pervasive nature of this pollutant (Peng et al. 2018). These contaminants can have direct toxic effects on benthic fauna, altering their physiology, reproduction, and behavior. Moreover, pollutants can bioaccumulate through the food web, potentially affecting entire ecosystems and even human health through the consumption of contaminated seafood (Jamieson et al. 2017).

Climate change is exerting additional pressure on deep-sea benthic communities through various mechanisms. The Indian Ocean is experiencing significant alterations due to climate change, impacting its physical, chemical, and biological environments. The region is warming rapidly with an increase in average sea surface temperature (SST) by 1 °C from 1951 to 2015, with projections indicating further increases of 1.2 to 2.7 °C under various greenhouse gas emission scenarios by the end of the century (Roxy et al. 2020). Deep ocean temperatures are also rising, with significant long-term trends observed in the deeper layers, indicating a comprehensive warming effect throughout the water column (Malavika et al. 2020). Several climate models indicate a noticeable decrease in mixed layer depth (MLD) in the Indian Ocean, with a shoaling of approximately 5–10% under various emission scenarios (Gao et al. 2023) and increased surface buoyancy, which weakens winds, affecting nutrient distribution and marine productivity (Gao et al. 2023). Climate change significantly affects benthic ecosystems in the Indian Ocean, affecting their structure, function, and contribution to global ecological processes. Benthic organisms that play crucial roles in nutrient cycling and carbon uptake are particularly vulnerable to changes in temperature, ocean acidification, and other environmental changes driven by climate change (Dalpadado et al. 2024). These changes can alter the physiology, reproduction, and biodiversity of benthic species, leading to shifts in species distribution and invasive species, and further destabilizing marine ecosystems (Caglar and Mulayim 2024; Sweetman et al. 2017). Climate change is expected to reduce oxygen concentrations and alter nutrient availability in the Indian Ocean, impacting benthic communities and posing risks to marine life and ecosystem services (Roxy et al. 2020). For instance, reducing particulate organic matter flux at the seafloor could decrease food availability for benthic organisms, particularly in abyssal and bathyal zones (Gupta et al. 2008; Sweetman et al. 2017). As climate change poses severe threats to the Indian Ocean, it is essential to consider the compounding effects of both natural and anthropogenic factors, such as overfishing and pollution, which further exacerbate the challenges faced by marine life in the region (Chaturvedi 2022).

The impact of climate change on benthic harpacticoid copepods in the deep sea is a multifaceted issue involving both direct and indirect effects on these organisms and their ecosystems. For example climate change can lead to shifts in bottom water temperature and oxygen concentration, directly affecting the metabolic rates and survival of harpacticoid copepods (Byrne and Przeslawski 2013; Gibson et al. 2011). Additionally, ocean acidification, a result of increased CO_2 levels, can affect the physiological processes of copepods, potentially affecting their reproduction and growth (Danovaro et al. 2017; Hale et al. 2011). Ocean acidification significantly reduces the density of certain harpacticoid species, and the presence of malformations in harpacticoids at low pH suggests sublethal effects that could impact long-term population viability (Sarmento et al. 2017). These climate change-mediated interactions of ocean warming and acidification lead to complex community dynamics, with some species showing resilience and others declining, indicating a potential shift in community structure and biodiversity. The combined stresses of warming, acidification, and deoxygenation threaten the biodiversity and integrity of deep-ocean ecosystems, potentially leading to negative feedback loops that further exacerbate climate change (Levin and Bris 2015). While some harpacticoid species may adapt or acclimatize to changing conditions, the overall trend suggests a risk of biodiversity loss and altered ecosystem functions, emphasizing the need for further research on their long-term resilience in a changing ocean. The direct impacts of climate change on deep-sea harpacticoid copepods are becoming clearer, and the indirect effects of ecosystem change remain complex and less understood. Further research is needed to fully understand how these organisms adapt to the rapidly changing deep-sea environment (Dalpadado et al. 2024; Solan et al. 2020).

5.4 Conclusions

The synergistic effects of pollution and climate change are likely to amplify the impacts on deep-sea benthic communities. For instance, increased temperatures may enhance the toxicity of certain pollutants, while ocean acidification could increase the bioavailability of heavy metals (Schiedek et al. 2007). These combined stressors may lead to reduced biodiversity, altered ecosystem functioning, and diminished resilience of deep-sea communities in the Indian Ocean. Assessing the potential impacts of these stressors on deep-sea benthic communities in the Indian Ocean is challenging due to the logistical difficulties and high costs associated with deep-sea research. However, recent technological advancements, such as autonomous underwater vehicles (AUVs) and environmental DNA (eDNA) sampling, are enabling more comprehensive studies of these remote ecosystems (Danovaro et al. 2020). Long-term observatories, like those established under the Global Ocean Observing System (GOOS), are providing valuable time-series data on environmental parameters and biological responses in the deep sea (Lisa et al. 2018). To effectively assess and mitigate the impacts of pollution and climate change on deep-sea benthic communities in the Indian Ocean, a multifaceted approach is necessary.

This should include: (1) expanding baseline studies to better understand the current state of deep-sea ecosystems and their natural variability, (2) developing and implementing standardized monitoring protocols to track changes in biodiversity, ecosystem functioning, and environmental parameters over time, (3) conducting experimental studies to elucidate the individual and combined effects of various stressors on key benthic species (such as harpacticoid) and communities, (4) improving predictive models to forecast future changes in deep-sea ecosystems under different pollution and climate change scenarios, and (5) strengthening international collaboration and data sharing among Indian Ocean rim countries and beyond to ensure comprehensive coverage of this vast oceanic region. By addressing these research priorities, scientists and policymakers can gain a more comprehensive understanding of the threats facing deep-sea benthic communities in the Indian Ocean. This knowledge is crucial for developing effective conservation strategies and management plans to protect these vulnerable ecosystems in the face of ongoing global changes.

Acknowledgments The authors are thankful to the authorities of CSIR-NIO for providing the facilities to develop this article.

Data Availability Statement The original contributions presented in the study are included in the article/supplementary material. Further inquiries can be directed to the corresponding author.

Declaration Statement The authors declare that the study was conducted following the ethical standards of the institutional and/or national research committees. Informed consent was obtained from all individual participants. All participants were informed about the purpose of the study. The confidentiality and anonymity of the data collected were assured to all participants.

Ethics Approval The research involves a comprehensive analysis of the existing scholarly literature, which does not require ethical approval.

Conflicts of Interest The authors confirm that their study was performed without any commercial or financial connections that could be interpreted as conflicts of interest.

Supplementary Material The supplementary material for this article can be found online at: [Link].

Electronic Supplementary Material Supplementary Table S5.1 Summary of research conducted on deep-sea benthic harpacticoid copepod from different oceans (DOCX 73 kb)

Funding No funding is received for this work.

References

Adámek Z, Maršálek B (2013) Bioturbation of sediments by benthic macroinvertebrates and fish and its implication for pond ecosystems: a review. Aquac Int 21:1–17. https://doi.org/10.1007/S10499-012-9527-3

Ahnert A, Schriever G (2001) Response of abyssal copepoda Harpacticoida (crustacea) and other meiobenthos to an artificial disturbance and its bearing on future mining for polymetallic nodules. Deep. Res Part II Top Stud Oceanogr 48:3779–3794. https://doi.org/10.1016/S0967-0645(01)00067-4

Allen M, Dube OP, Solecki W, Aragón-Durand F, Cramer W, Humphreys S, Kainuma M (2018) Special report: global warming of 1.5 °C. Intergov Panel Clim Chang 27:677. https://doi.org/10.1017/9781009157940.001.Chapter

Ansari ZA (2000) Distribution of Deep-Sea benthos in the proposed mining area of central Indian Basin. Mar Georesour Geotechnol 18:201–207. https://doi.org/10.1080/10641190009353788

Baguley JG, Montagna PA, Lee W, Hyde LJ, Rowe GT (2006) Spatial and bathymetric trends in Harpacticoida (Copepoda) community structure in the Northern Gulf of Mexico deep-sea. J Exp Mar Bio Ecol 330:327–341. https://doi.org/10.1016/j.jembe.2005.12.037

Balsamo M, Semprucci F, Frontalini F, Coccioni R (2012) Meiofauna as a tool for marine ecosystem biomonitoring. Mar Ecosyst 77–104. https://doi.org/10.5772/34423

Beliaev GM, Brueggeman PL (1989) Deep sea ocean trenches and their fauna

Bell SS (1980) Meiofauna-macrofauna interactions in a high salt marsh habitat. Ecol Monogr 50:487–505

Berger WH (1974) Deep-sea sedimentation. In: Burke CA, Drake CD (eds) The geology of continental margins. Springer, New York, pp 213–241

Bussau C, Schriever G, Thiel H, Bussau C, Schriever G, Thiel H (1995) Evaluation of abyssal metazoan Meiofauna from a manganese nodule area of the eastern South Pacific. Vie Milieu 45:39–48

Byrne M, Przeslawski R (2013) Multistressor impacts of warming and acidification of the ocean on marine invertebrates' life histories. Integr Comp Biol 53:582–596

Caglar, S., Mulayim A (2024) The impact of climate change on benthic ecosystems: a perspective. In: Ecological dynamics in the face of climate change: exploring organisms, habitats, and behavioral adaptations. Nobel Tip Kitabevleri, pp. 105–118. https://doi.org/10.69860/nobel.9786053359258.7

Campanyà-Llovet N, Snelgrove PVR, Parrish CC (2017) Rethinking the importance of food quality in marine benthic food webs. Prog Oceanogr 156:240–251. https://doi.org/10.1016/j.pocean.2017.07.006

Campbell LM, Gray NJ, Fairbanks L, Silver JJ, Gruby RL, Dubik BA, Basurto X (2016) Global oceans governance: new and emerging issues. Annu Rev Environ Resour 41:517–543. https://doi.org/10.1146/annurev-environ-102014-021121

Campos CC, Garcia TM, Neumann-Leitão S, Soares MO (2017) Ecological indicators and functional groups of copepod assemblages. Ecol Indic 83:416–426. https://doi.org/10.1016/j.ecolind.2017.08.018

Chaturvedi S (2022) Growing focus on the Western Indian Ocean: evolving equations. J Indian Ocean Reg 18:1

Chen CTA, Lui HK, Hsieh CH, Yanagi T, Kosugi N, Ishii M, Gong GC (2017) Deep oceans may acidify faster than anticipated due to global warming. Nat Clim Chang 7:890–894. https://doi.org/10.1038/s41558-017-0003-y

Chertoprud ES, Garlitska LA, Azovsky AI (2010) Large-scale patterns in marine harpacticoid (Crustacea, Copepoda) diversity and distribution. Mar Biodivers 40:301–315. https://doi.org/10.1007/s12526-010-0054-z

Chuar CH, Tong SJW, Chim CK, Wong HPS, Tan KS (2020) Abyssal macrofaunal community structure in the polymetallic nodule exploration area at the easternmost region of the Clarion-Clipperton Fracture Zone, Pacific Ocean. Deep Res Part I Oceanogr Res Pap 161:103284. https://doi.org/10.1016/j.dsr.2020.103284

Consalvey, M., Clark, M. R., Rowden, A. A., & Stocks, K. I. (2010). Life on seamounts. In: A. D. McIntyre (Ed.), Life in the World's Oceans: Diversity, Distribution, and Abundance, pp. 123–139.

Coull BC (1972) Species diversity and faunal affinities of meiobenthic Copepoda in the deep sea. Mar Biol 14:48–51

Coull BC (1990) Are members of the meiofauna food for higher trophic levels? Trans Am Microsc Soc 109:233–246

Coull BC (1999) Role of meiofauna in estuarine soft-bottom habitats. Aust J Ecol 24:327–343

Courtene-Jones W, Quinn B, Ewins C, Gary SF, Narayanaswamy BE (2020) Microplastic accumulation in deep-sea sediments from the Rockall Trough. Mar Pollut Bull 154:111092. https://doi.org/10.1016/j.marpolbul.2020.111092

Dalpadado P, Roxy MK, Arrigo KR, van Dijken GL, Chierici M, Ostrowski M, Skern-Mauritzen R, Bakke G, Richardson AJ, Sperfeld E (2024) Rapid climate change alters the environment and biological production of the Indian Ocean. Sci Total Environ 906:167342

Danovaro R, Della Croce N, Dell'Anno A, Pusceddu A (2003) A depocenter of organic matter at 7800 m depth in the SE Pacific Ocean. Deep Sea Res Part I Oceanogr Res Pap 50:1411–1420

Danovaro R, Gambi C, Mirto S, Sandulli R, Ceccherelli VU (2004) Meiofauna. In: Gambi MC, Dappiano M (eds) Mediterranean marine benthos: a manual of methods for its sampling and study. Biologia Marina Mediterranea; SIBM, Genova, Italy, pp 55–97

Danovaro R, Snelgrove PVR, Tyler P (2014) Challenging the paradigms of deep-sea ecology. Trends Ecol Evol 29:465–475. https://doi.org/10.1016/j.tree.2014.06.002

Danovaro R, Corinaldesi C, Dell'Anno A, Rastelli E (2017) Potential impact of global climate change on benthic deep-sea microbes. FEMS Microbiol Lett 364. https://doi.org/10.1093/femsle/fnx214

Danovaro R, Fanelli E, Aguzzi J, Billett D, Carugati L, Corinaldesi C, Dell'Anno A, Gjerde K, Jamieson AJ, Kark S, McClain C, Levin L, Levin N, Ramirez-Llodra E, Ruhl H, Smith CR, Snelgrove PVR, Thomsen L, Van Dover CL, Yasuhara M (2020) Ecological variables for developing a global deep-ocean monitoring and conservation strategy. Nat Ecol Evol 4:181–192. https://doi.org/10.1038/s41559-019-1091-z

De Smet B, Pape E, Riehl T, Bonifácio P, Colson L, Vanreusel A (2017) The community structure of deep-sea macrofauna associated with polymetallic nodules in the eastern part of the Clarion-Clipperton Fracture Zone. Front Mar Sci 4:1–14. https://doi.org/10.3389/fmars.2017.00103

Duineveld GCA, De Wilde PAWJ, Berghuis EM, Kok A, Tahey T, Kromkamp J (1997) Benthic respiration and standing stock on two contrasting continental margins in the western Indian Ocean: the Yemen-Somali upwelling region and the margin off Kenya. Deep Res Part II Top Stud Oceanogr 44:1293–1317. https://doi.org/10.1016/S0967-0645(97)00006-4

Ebenhöh W, Kohlmeier C, Radford PJ (1995) The benthic biological submodel in the European regional seas ecosystem model. Netherlands J Sea Res 33:423–452. https://doi.org/10.1016/0077-7579(95)90056-X

Etter RJ, Mullineaux LS (2001) Deep-sea communities. In: Bertness MD, Gaines SD, Hay ME (eds) Marine community ecology. Sinauer Associates, Inc., Sunderland

Fleeger JW (2005) The potential to mass-culture harpacticoid copepods for use as food for larval fish. In: Copepods in aquaculture. Blackwell, Iowa, pp 11–24

Gallucci F, Moens T, Fonseca G (2009) Small-scale spatial patterns of meiobenthos in the Arctic deep sea. Mar Biodivers 39:9–25. https://doi.org/10.1007/s12526-009-0003-x

Gao Z, Long S-M, Shi J-R, Cheng L, Li G, Ying J (2023) Indian Ocean mixed layer depth changes under global warming. Front Clim 5:1112713

Gebruk AV, Chevaldonné P, Shank T, Lutz RA, Vrijenhoek RC (2000) Deep-sea hydrothermal vent communities of the Logatchev area (14 45′ N, Mid-Atlantic Ridge): diverse biotopes and high biomass. J Mar Biol Assoc United Kingdom 80:383–393

George KH, Schminke HK (2002) Harpacticoida (Crustacea, Copepoda) of the great meteor seamount, with first conclusions as to the origin of the plateau fauna. Mar Biol 141:887–895. https://doi.org/10.1007/s00227-002-0878-6

George KH, Veit-Köhler G, Arbizu PM, Seifried S, Rose A, Willen E, Bröhldick K, Corgosinho PH, Drewes J, Menzel L, Moura G, Schminke HK (2014) Community structure and species diversity of Harpacticoida (Crustacea: Copepoda) at two sites in the deep sea of the Angola Basin (Southeast Atlantic). Org Divers Evol 14:57–73. https://doi.org/10.1007/s13127-013-0154-2

Gheerardyn H, Veit-Köhler G (2009) Diversity and large-scale biogeography of Paramesochridae (Copepoda, Harpacticoida) in South Atlantic Abyssal Plains and the deep Southern Ocean. Deep Res Part I Oceanogr Res Pap 56:1804–1815. https://doi.org/10.1016/j.dsr.2009.05.002

Gheerardyn H, De Troch M, Vincx M, Vanreusel A (2009) Harpacticoida (Crustacea: Copepoda) associated with cold-water coral substrates in the Porcupine Seabight (NE Atlantic): species composition, diversity and reflections on the origin of the fauna. Sci Mar 73:747–760

Gibson R, Atkinson R, Gordon J, Smith I, Hughes D (2011) Impact of ocean warming and ocean acidification on marine invertebrate life history stages: vulnerabilities and potential for persistence in a changing ocean. Ocean Mar Biol Annu Rev 49:1–42

Giere O (2009a) Meiobenthology: the microscopic motile fauna of aquatic sediments, Meiobenthology: the microscopic motile Fauna of aquatic sediments. Springer Science & Business Media. https://doi.org/10.1007/978-3-540-68661-3

Giere O (2009b) Meiobenthology. The microscopic motile fauna of aquatic sediments, 2nd edn. Springer, Berlin Heidelberg

Gollner S, Zekely J, Van Dover CL, Govenar B, Le Bris N, Nemeschkal HL, Bright M (2006) Benthic copepod communities associated with tubeworm and mussel aggregations on the East Pacific rise. Cah Biol Mar 47:397–402

Gollner S, Zekely J, Govenar B, Le Bris N, Nemeschkal HL, Fisher CR, Bright M (2007) Tubeworm-associated permanent meiobenthic communities from two chemically different hydrothermal vent sites on the East Pacific rise. Mar Ecol Prog Ser 337:39–49. https://doi.org/10.3354/meps337039

Gollner S, Riemer B, Arbizu PM, le Bris N, Bright M (2010) Diversity of meiofauna from the 9°50′N east pacific rise across a gradient of hydrothermal fluid emissions. PLoS One 5. https://doi.org/10.1371/journal.pone.0012321

Gooday AJ, Pfannkuche O, Lambshead PJD (1996) An apparent lack of response by metazoan meiofauna to phytodetritus deposition in the bathyal North-Eastern Atlantic. J Mar Biol Assoc United Kingdom 76:297–310

Gooday AJ, Schoenle A, Dolan JR, Arndt H (2020) Protist diversity and function in the dark ocean—challenging the paradigms of deep-sea ecology with special emphasis on foraminiferans and naked protists. Eur J Protistol 75:125721. https://doi.org/10.1016/j.ejop.2020.125721

Gupta AK, Das M, Clemens SC, Mukherjee B (2008) Benthic foraminiferal faunal and isotopic changes as recorded in Holocene sediments of the Northwest Indian Ocean. Paleoceanography 23

Hale R, Calosi P, McNeill L, Mieszkowska N, Widdicombe S (2011) Predicted levels of future ocean acidification and temperature rise could alter community structure and biodiversity in marine benthic communities. Oikos 120:661–674

Han H, Zeeshan Z, Talpur BA, Sadiq T, Bhatti UA, Awwad EM, Al-Razgan M, Ghadi YY (2024) Studying long term relationship between carbon emissions, soil, and climate change: insights from a global earth modeling framework. Int J Appl Earth Obs Geoinf 130:103902. https://doi.org/10.1016/j.jag.2024.103902

Heezen BC (1969) The world rift system: an introduction to the symposium. Tectonophysics 8:269–279. https://doi.org/10.1016/0040-1951(69)90037-7

Herman RL, Dahms HU (1992) Meiofauna communities along a depth transect off Halley Bay (Weddell Sea-Antarctica). Polar Biol 12:313–320

Hessler RR, Jumars PA (1974) Abyssal community analysis from replicate □ cores in the central North Pacific. Deep-Sea Res Oceanogr Abstr 21:185–209

Hicks GRF, Coull BC (1983) The ecology of marine meiobenthic harpacticoid copepods. Oceanogr Mar Biol 21:67–175

Higgins R, Thiel H (1988) Introduction to the study of meiofauna. Smithsonian Institution Press. https://doi.org/10.1126/science.243.4896.1376

Ingole B (2003) Macrobenthic abundance in the vicinity of spreading ridge environment in Central Indian Ocean. Curr Sci 85:328–333

Ingole B, Singh R (2010) Biodiversity and community structure of free- living marine nematodes from the Larsemann Ice Shelf, East Antarctica. Curr Sci 99:1413–1419

Ingole BS, Ansari ZA, Parulekar AH (1998) Spatial variation in meiofaunal abundance of some coralline beaches of Mauritius. Trop Ecol 39:103–108

Ingole BS, Ansari ZA, Rathod V, Rodrigues N (2000) Response of meiofauna to immediate benthic disturbance in the Central Indian Ocean Basin. Mar Georesources Geotechnol 18:263–272. https://doi.org/10.1080/10641190009353794

Ingole BS, Goltekar R, Gonsalves S, Ansari ZA (2005a) Recovery of deep-sea meiofauna after artificial disturbance in the Central Indian Basin. Mar Georesources Geotechnol 23:253–266. https://doi.org/10.1080/10641190500446540

Ingole BS, Pavithran S, Ansari ZA (2005b) Restoration of deep-sea macrofauna after simulated benthic disturbance in the Central Indian Basin. Mar Georesources Geotechnol 23:267–288. https://doi.org/10.1080/10641190500446573

Ivanov AV, Southward EC (1971) New Pogonophora from the Atlantic and Pacific Oceans. J Zool 164:271–304. https://doi.org/10.1111/j.1469-7998.1971.tb01320.x

Jamieson AJ, Malkocs T, Piertney SB, Fujii T, Zhang Z (2017) Bioaccumulation of persistent organic pollutants in the deepest ocean fauna. Nat Ecol Evol 1:51. https://doi.org/10.1038/s41559-016-0051

Kamykowski D, Zentara S-J (1990) Hypoxia in the world ocean as recorded in the historical data set. Deep Sea Res Part A Oceanogr Res Pap 37:1861–1874. https://doi.org/10.1016/0198-0149(90)90082-7

Kang T, Kim D (2021) Meiofauna and nematode community composition in a hydrothermal vent and deep-sea sediments in the Central Indian Ridge. Mar Pollut Bull 170:112616. https://doi.org/10.1016/j.marpolbul.2021.112616

Kitahashi T, Kawamura K, Veit-Köhler G, Danovaro R, Tietjen J, Kojima S, Shimanaga M (2012) Assemblages of Harpacticoida (Crustacea: Copepoda) from the Ryukyu and Kuril Trenches, North-West Pacific Ocean. J Mar Biol Assoc United Kingdom 92:275–286. https://doi.org/10.1017/S0025315411001536

Kolla V, Eittreim S, Sullivan L, Kostecki JA, Burckle LH (1980) Current-controlled, abyssal microtopography and sedimentation in Mozambique Basin, Southwest Indian Ocean. Mar Geol 34:171–206. https://doi.org/10.1016/0025-3227(80)90071-7

Laj P, Klausen J, Bilde M, Plaß-Duelmer C, Pappalardo G, Clerbaux C, Baltensperger U, Hjorth J, Simpson D, Reimann S, Coheur PF, Richter A, De Mazière M, Rudich Y, McFiggans G, Torseth K, Wiedensohler A, Morin S, Schulz M, Allan JD, Attié JL, Barnes I, Birmili W, Cammas JP, Dommen J, Dorn HP, Fowler D, Fuzzi S, Glasius M, Granier C, Hermann M, Isaksen ISA, Kinne S, Koren I, Madonna F, Maione M, Massling A, Moehler O, Mona L, Monks PS, Müller D, Müller T, Orphal J, Peuch VH, Stratmann F, Tanré D, Tyndall G, Abo Riziq A, Van Roozendael M, Villani P, Wehner B, Wex H, Zardini AA (2009) Measuring atmospheric composition change. Atmos Environ 43:5351. https://doi.org/10.1016/j.atmosenv.2009.08.020

Levin LA, Bris NL (2015) The deep ocean under climate change. Science (80-.) 350:766–768. https://doi.org/10.1126/science.aad0126

Levin LA, Huggett CL, Wishner KF (1991) Control of deep-sea benthic community structure by oxygen and organic-matter gradients in the eastern Pacific Ocean. J Mar Res 49:763–800

Levin LA, Sibuet M, Gooday AJ, Smith CR, Vanreusel A (2010) The roles of habitat heterogeneity in generating and maintaining biodiversity on continental margins: an introduction, vol 31. Mar Ecol, p 1

Lisa L, Baker M, Anthony T (2018) Deep-ocean climate change impacts on habitat, fish and fisheries. FAO Fisheries and Aquaculture Technical Paper. https://doi.org/10.4060/ca2521en

Maier SR, Kutti T, Bannister RJ, Fang JKH, van Breugel P, van Rijswijk P, van Oevelen D (2020) Recycling pathways in cold-water coral reefs: use of dissolved organic matter and bacteria by key suspension feeding taxa. Sci Rep 10. https://doi.org/10.1038/s41598-020-66463-2

Malavika AR, Pai ML, Saji PK (2020) Analytical evidence for deep ocean warming trend in tropical Indian Ocean. In: IOP conference series: earth and environmental science, vol 491. IOP Publishing, p 12014

Mascart T, Lepoint G, De Troch M (2013) Meiofauna and harpacticoid copepods in different habitats of a Mediterranean seagrass meadow. J Mar Biol Assoc United Kingdom 93:1557–1566. https://doi.org/10.1017/S0025315413000222

Mascart T, Lepoint G, Deschoemaeker S, Binard M, Remy F, De Troch M (2015) Seasonal variability of meiofauna, especially harpacticoid copepods, in Posidonia oceanica macrophytodetritus accumulations. J Sea Res 95:149. https://doi.org/10.1016/j.seares.2014.07.009

Menot L, Sibuet M, Carney RS, Levin LA, Rowe GT, Billett DSM, Poore G, Kitazato H, Vanreusel A, Gal'eron J, Lavrado HP, Sellanes J, Ingole B, Krylova EM (2010) New perceptions of continental margin biodiversity. In: McIntyre AD (ed) Life in the world's oceans: diversity, distribution, and abundance. Wiley-Blackwell, pp 79–103

Menzel L, George KH, Arbizu PM (2011) Submarine ridges do not prevent large-scale dispersal of abyssal fauna: a case study of Mesocletodes (Crustacea, Copepoda, Harpacticoida). Deep Res Part I Oceanogr Res Pap 58:839–864. https://doi.org/10.1016/j.dsr.2011.05.008

Mevenkamp L, Guilini K, Boetius A, De Grave J, Laforce B, Vandenberghe D, Vincze L, Vanreusel A (2019) Responses of an abyssal meiobenthic community to short-term burial with crushed nodule particles in the South-East Pacific. Biogeosciences 16:2329–2341. https://doi.org/10.5194/bg-16-2329-2019

Montagna PA (1982) Morphological adaptation in the deep-sea benthic harpacticoid copepod family Cerviniidae. Crustaceana 42:37–43

Nomaki H, Rastelli E, Alves A, Suga H, Ramos S, Kitahashi T, Tsuchiya M, Ogawa NO, Matsui Y, Seike K (2021) Abyssal fauna, benthic microbes, and organic matter quality across a range of trophic conditions in the western Pacific Ocean. Prog Oceanogr 195:102591

Pandey M, Biswas H, Chowdhury M (2023) Interlinking diatom frustule diversity from the abyss of the central Arabian Sea to surface processes: physical forcing and oxygen minimum zone. Environ Monit Assess 195. https://doi.org/10.1007/s10661-022-10749-7

Parulekar AH, Harkantra SN, Ansari ZA, Matondkar SGP (1982) Abyssal benthos of the Central Indian Ocean. Deep Sea Res Part A Oceanogr Res Pap 29:1531–1537. https://doi.org/10.1016/0198-0149(82)90041-3

Parulekar A.H, Ingole B, Harkantra, SN, Ansari, Z. (1992). Deep sea benthos of the western and central indian ocean. In B. N. Desai (Ed.), Oceanography of the Indian Ocean, pp. 261–267.

Pavithran S, Ingole B, Nanajkar M, Nath NN (2007) Macrofaunal diversity in the Central Indian Ocean Basin. Biodiversity 8:11–16. https://doi.org/10.1080/14888386.2007.9712824

Peng X, Chen M, Chen S, Dasgupta S, Xu H, Ta K, Du M, Li J, Guo Z, Bai S (2018) Microplastics contaminate the deepest part of the world's ocean. Geochemical Perspect Lett 9:1–5. https://doi.org/10.7185/geochemlet.1829

Pfannkuche O (1985) The deep-sea meiofauna of the porcupine seabight and abyssal-plain (NE Atlantic)-population-structure, distribution, standing stocks. Oceanol Acta 8:343–353

Pusceddu A, Bianchelli S, Martin J, Puig P, Palanques A, Masqué P, Danovaro R (2014) Chronic and intensive bottom trawling impairs deep-sea biodiversity and ecosystem functioning. Proc Natl Acad Sci 111:8861–8866

Radziejewska T (2002) Responses of deep-sea meiobenthic communities to sediment disturbance simulating effects of polymetallic nodule mining. Int Rev Hydrobiol 87:457–477. https://doi.org/10.1002/1522-2632(200207)87:4<457::AID-IROH457>3.0.CO;2-3

Radziejewska T, Drzycimski I, Galtsova VV, Kulangieva LV, Stoyanova V (2001) Changes in genus-level diversity of meiobenthic free-living nematodes (Nematoda) and harpacticoids (Copepoda Harpacticoida) at an abyssal site following experimental sediment disturbance. Proc ISOPE Ocean Min Symp:38–43

Ramirez-Llodra, E., Brandt, A., Danovaro, R., De Mol, B., Escobar, E., German, C.R., Levin, L.A., Martinez Arbizu, P., Menot, L., Buhl-Mortensen, P., others, 2010. Deep, diverse and definitely different: unique attributes of the world's largest ecosystem. Biogeosciences 7, 2851–2899

Rex MA, Etter RJ (2010) Deep-sea biodiversity: pattern and scale. Harvard University Press

Rogers AD (2000) The role of the oceanic oxygen minima in generating biodiversity in the Deep Sea. Deep Sea Res Part II Top Stud Oceanogr 47:119–148

Roman M, Smith S, Wishner K, Zhang X, Gowing M (2000) Mesozooplankton production and grazing in the Arabian Sea. Deep Sea Res Part II Top Stud Oceanogr 47:1423–1450

Rose A, Seifried S, Willen E, George KH, Veit-Köhler G, Bröhldick K, Drewes J, Moura G, Arbizu PM, Schminke HK (2005) A method for comparing within-core alpha diversity values from repeated multicorer samplings, shown for abyssal Harpacticoida (Crustacea: Copepoda) from the Angola Basin. Org Divers Evol 5:3–17. https://doi.org/10.1016/j.ode.2004.10.001

Rosli N, Leduc D, Rowden AA, Probert PK (2018) Review of recent trends in ecological studies of deep-sea meiofauna, with focus on patterns and processes at small to regional spatial scales. Mar Biodivers 48:13–34. https://doi.org/10.1007/s12526-017-0801-5

Roxy MK, Gnanaseelan C, Parekh A, Chowdary JS, Singh S, Modi A, Kakatkar R, Mohapatra S, Dhara C, Shenoi SC (2020) Indian ocean warming. Assess Clim Chang over Indian Reg a Rep Minist Earth Sci (MoES). Gov. India, pp 191–206

Sarmento VC, Parreira Santos PJ, Hale R, Ingels J, Widdicombe S (2017) Effects of elevated CO2 and temperature on an intertidal harpacticoid copepod community. ICES J Mar Sci 74:1159–1169

Schiedek D, Sundelin B, Readman JW, Macdonald RW (2007) Interactions between climate change and contaminants. Mar Pollut Bull 54:1845–1856. https://doi.org/10.1016/j.marpolbul.2007.09.020

Schlacher TA, Rowden AA, Dower JF, Consalvey M (2010) Seamount science scales undersea mountains: new research and outlook. Mar Ecol 31:1–13

Schmidt C, Lins L, Brandt A (2018) Harpacticoida (Crustacea, Copepoda) across a longitudinal transect of the Vema Fracture Zone and along a depth gradient in the Puerto Rico trench. Deep Res Part II Top Stud Oceanogr 148:236–250. https://doi.org/10.1016/j.dsr2.2017.12.024

Schott FA, Xie SP, McCreary JP (2009) Indian ocean circulation and climate variability. Rev Geophys 47. https://doi.org/10.1029/2007RG000245

Schratzberger M, Ingels J (2018) Meiofauna matters: the roles of meiofauna in benthic ecosystems. J Exp Mar Bio Ecol 502:12–25

Schückel S, Sell AF, Kihara TC, Koeppen A, Kröncke I, Reiss H (2013) Meiofauna as food source for small-sized demersal fish in the Southern North Sea. Helgol Mar Res 67:203–218

Sedlacek L (2007) Impacts of deep-sea carbon sequestration and of ciliate epibionts on harpacticoid copepods. The Florida State University

Sedlacek L, Thistle D, Carman KR, Fleeger JW, Barry JP (2009) Effects of carbon dioxide on deep-sea harpacticoids revisited. Deep Res Part I Oceanogr Res Pap 56:1018–1025. https://doi.org/10.1016/j.dsr.2009.01.004

Sedlacek L, Thistle D, Fernandez-Leborans G, Carman KR, Barry JP (2013) First report of ciliate (Protozoa) epibionts on deep-sea harpacticoid copepods. Deep Res Part II Top Stud Oceanogr 92:165–171. https://doi.org/10.1016/j.dsr2.2013.01.033

Service SK, Feller RJ, Coull BC, Woods R (1992) Predation effect of three fish species and a shrimp on macrobenthos and meiobenthos in microcosms. Estuar Coast Shelf Sci 34:277–293

Shimanaga M, Kitazato H, Shirayama Y (2004) Temporal patterns in diversity and species composition of deep-sea benthic copepods in bathyal Sagami Bay, Central Japan. Mar Biol 144:1097–1110. https://doi.org/10.1007/s00227-003-1273-7

Shimanaga M, Nomaki H, Iijima K (2008) Spatial changes in the distributions of deep-sea "Cerviniidae" (Harpacticoida, Copepoda) and their associations with environmental factors in the bathyal zone around Sagami Bay, Japan. Mar Biol 153:493–506. https://doi.org/10.1007/s00227-007-0817-7

Smith CR, De Leo FC, Bernardino AF, Sweetman AK, Arbizu PM (2008) Abyssal food limitation, ecosystem structure and climate change. Trends Ecol Evol 23:518–528

Solan M, Bennett EM, Mumby PJ, Leyland J, Godbold JA (2020) Benthic-based contributions to climate change mitigation and adaptation. Philos Trans R Soc B 375:20190107

St Laurent LC, Thurnherr AM (2007) Intense mixing of lower thermocline water on the crest of the Mid-Atlantic Ridge. Nature 448:680–683

Stewart HA, Jamieson AJ (2018) Habitat heterogeneity of hadal trenches: considerations and implications for future studies. Prog Oceanogr 161:47–65. https://doi.org/10.1016/j.pocean.2018.01.007

Stratmann T, Mevenkamp L, Sweetman AK, Vanreusel A, van Oevelen D (2018) Has phytodetritus processing by an abyssal soft-sediment community recovered 26 years after an experimental disturbance? Front Mar Sci 4:1–13. https://doi.org/10.3389/fmars.2018.00059

Sweetman AK, Thurber AR, Smith CR, Levin LA, Mora C, Wei CL, Gooday AJ, Jones DOB, Rex M, Yasuhara M, Ingels J, Ruhl HA, Frieder CA, Danovaro R, Würzberg L, Baco A, Grupe BM, Pasulka A, Meyer KS, Dunlop KM, Henry LA, Roberts JM (2017) Major impacts of climate change on deep-sea benthic ecosystems. Elementa 5:4. https://doi.org/10.1525/elementa.203

Sweetman AK, Smith CR, Shulse CN, Maillot B, Lindh M, Church MJ, Meyer KS, van Oevelen D, Stratmann T, Gooday AJ (2019) Key role of bacteria in the short-term cycling of carbon at the abyssal seafloor in a low particulate organic carbon flux region of the eastern Pacific Ocean. Limnol Oceanogr 64:694–713. https://doi.org/10.1002/lno.11069

Thistle D (1978) Harpacticoid dispersion patterns: implications for deep-sea diversity maintenance. J Mar Res 36:377–397

Thistle D (1988) A temporal difference in harpacticoid-copepod abundance at a deep-sea site: caused by benthic storms? Deep Sea Res Part A Oceanogr Res Pap 35:1015–1020. https://doi.org/10.1016/0198-0149(88)90073-8

Thistle D (2003) On the utility of metazoan meiofauna for studying the soft-bottom deep sea. Vie Milieu/Life Environ. 97–101

Thistle D, Eckman JE (1988) Response of harpacticoid copepods to habitat structure at a deep-sea site. Hydrobiologia 167–168:143–149. https://doi.org/10.1007/BF00026300

Thistle D, Eckman JE (1990) The effect of a biologically produced structure on the benthic copepods of a deep-sea site. Deep Sea Res Part A Oceanogr Res Pap 37:541–554. https://doi.org/10.1016/0198-0149(90)90089-E

Thistle D, Yingst JY, Fauchald K (1985) A Deep-Sea benthic community exposed to strong near-bottom currents on the Scotian Rise (Western Atlantic). Mar Geol 66:91–112

Thistle D, Hilbig B, Eckman JE (1993) Are polychaetes sources of habitat heterogeneity for harpacticoid copepods in the deep sea? Deep Res Part I 40:151–157. https://doi.org/10.1016/0967-0637(93)90058-B

Thistle D, Carman KR, Sedlacek L, Brewer PG, Fleeger JW, Barry JP (2005) Deep-ocean, sediment-dwelling animals are sensitive to sequestered carbon dioxide. Mar Ecol Prog Ser 289:1–4

Thistle D, Sedlacek L, Carman KR, Fleeger JW, Brewer PG, Barry JP (2006) Simulated sequestration of industrial carbon dioxide at a deep-sea site: effects on species of harpacticoid copepods. J Exp Mar Bio Ecol 330:151–158. https://doi.org/10.1016/j.jembe.2005.12.023

Thistle D, Sedlacek L, Carman KR, Fleeger JW, Barry JP (2007) Emergence in the deep sea: evidence from harpacticoid copepods. Deep Res Part I Oceanogr Res Pap 54:1008–1014. https://doi.org/10.1016/j.dsr.2007.03.002

Timmreck C (2012) Modeling the climatic effects of large explosive volcanic eruptions. Wiley Interdiscip Rev Clim Chang. https://doi.org/10.1002/wcc.192

Tocher DR (2003) Metabolism and functions of lipids and fatty acids in teleost fish. Rev Fish Sci 11:107–184. https://doi.org/10.1080/713610925

Tselepides A, Lampadariou N (2004) Deep-sea meiofaunal community structure in the Eastern Mediterranean: are trenches benthic hotspots? Deep Sea Res Part I Oceanogr Res Pap 51:833–847

Tsurumi M, De Graaf RC, Tunnicliffe V (2003) Distributional and biological aspects of copepods at hydrothermal vents on the Juan de Fuca Ridge, North-East Pacific Ocean. J Mar Biol Assoc United Kingdom 83:469–477. https://doi.org/10.1017/S0025315403007367h

van Damme D, Heip C, Willems KA (1984) Influence of pollution on the harpacticoid copepods of two North Sea estuaries. Hydrobiologia 112:143–160. https://doi.org/10.1007/BF00006919

Van Dover CL (2000) The ecology of deep-sea hydrothermal vents. Princeton University Press

Wang JQ, Zhang XD, Jiang LF, Bertness MD, Fang CM, Chen JK, Hara T, Li B (2010) Bioturbation of burrowing crabs promotes sediment turnover and carbon and nitrogen movements in an estuarine salt marsh. Ecosystems 13:586–599. https://doi.org/10.1007/s10021-010-9342-5
Zekely J, Van Dover CL, Nemeschkal HL, Bright M (2006) Hydrothermal vent meiobenthos associated with mytilid mussel aggregations from the Mid-Atlantic Ridge and the East Pacific Rise. Deep Res Part I Oceanogr Res Pap 53:1363–1378. https://doi.org/10.1016/j.dsr.2006.05.010
Zeng Q, Huang D, Lin R, Wang J (2018) Deep-sea metazoan Meiofauna from a Polymetallic nodule area in the Central Indian Ocean Basin. Mar Biodivers 48:395–405. https://doi.org/10.1007/s12526-017-0778-0
Zhang LJ, Qi YA, Buatois LA, Mángano MG, Meng Y, Li D (2017) The impact of deep-Tier burrow systems in sediment mixing and ecosystem engineering in early Cambrian carbonate settings. Sci Rep 7:1–9. https://doi.org/10.1038/srep45773

Chapter 6
Impact of Climate Change on the Sardine Fishery in Indian Subcontinent

Atanu Kumar Panja, Bedantika Biswas, Shruti Chatterjee, and Soumya Haldar

Abstract The Indian oil sardine occupies a vital role in India's marine fisheries, accounting for approximately 15–20% of total marine landings. Recent years have seen a significant decline in stocks due to climate-induced oceanographic changes and human pressures, leading the fishery toward collapse. The southeastern Arabian Sea, especially Kerala, constitutes approximately 45% of India's fish landings; however, variable trends associated with sea surface temperature (SST), salinity, precipitation, and primary productivity have resulted in erratic fishery cycles. Environmental parameters, including sea surface temperatures (27–29 °C), chlorophyll concentration, mixed layer depth, and net primary productivity, are crucial in influencing sardine abundance.

Climate change impacts Net Primary Productivity (NPP), thereby modifying phytoplankton productivity, which directly influences Sardine food availability and recruitment success. Although projections suggest a 7% rise in NPP by the conclusion of the twenty-first century, this beneficial impact is mitigated by increasing SST, which diminishes fish biomass and recruitment efficacy. Extensive climate indices, such as the Pacific Decadal Oscillation (PDO) and Atlantic Multidecadal Oscillation (AMO), have been shown to affect Sardine abundance more profoundly than the previously considered impacts of the El Niño-Southern Oscillation (ENSO). Moreover, overfishing and alterations in recruitment patterns have intensified stock

A. K. Panja · S. Haldar (✉)
MEME Division, CSIR-Central Salt & Marine Chemicals Research Institute, Bhavnagar, Gujarat, India

Academy of Scientific and Innovative Research (AcSIR), Ghaziabad, India
e-mail: shaldar.csmcri@csir.res.in

B. Biswas
MEME Division, CSIR-Central Salt and Marine Chemicals Research Institute, Bhavnagar, Gujarat, India

S. Chatterjee
Institute of Science, Nirma University, Ahmedabad, Gujarat, India

B. P. Barathan et al. (eds.), *Navigating Climate Change: Impacts on Biodiversity and Ecosystem Resilience*, Sustainable Development and Biodiversity 42,
https://doi.org/10.1007/978-981-95-0409-1_6

declines, with the 2012 peak in landings succeeded by an 82% collapse within 3 years due to excessive fishing effort and adverse environmental conditions.

Model projections under climate change scenarios (RCP 4.5 and RCP 6.0) suggest a transient increase in Sardine populations until 2050, succeeded by a significant decline due to elevated sea surface temperatures, irregular precipitation, and habitat degradation. The absence of thorough habitat suitability models under fluctuating oceanographic conditions constitutes a significant research deficiency. Advanced machine learning models, including Generalized Additive Models (GAM), Boosted Regression Trees (BRT), and Random Forest (RF), possess the potential to enhance predictions of Sardine abundance.

The socio-economic consequences of diminishing sardine stocks are significant, influencing coastal livelihoods and food security. Impoverished fishing communities, heavily dependent on sardines for income and sustenance, endure significant economic distress due to stock depletion. This study highlights the critical necessity for adaptive fisheries management, climate-resilient conservation strategies, and interdisciplinary research to guarantee the sustainability of sardine fisheries in Indian waters.

Keywords Sardine · Indian coast · Machine learning · Salinity · Temperature

6.1 Introduction

The atmospheric input affects primary productivity (PP) in the North Indian Ocean (NIO) by altering nutrient availability, influencing oceanic biogeochemistry, and modifying environmental conditions such as temperature and stratification. The key impacts of the atmospheric inputs are dust transport from arid regions (e.g., Thar Desert, Middle East, and northeast Africa) delivers essential nutrients like iron (Fe), phosphorus (P), and silicon (Si) to the ocean. These nutrients stimulate phytoplankton growth, leading to increased Chlorophyll-a (Chl-a) concentrations and enhanced Net Primary Productivity (NPP). Seasonal variations in monsoon winds affect the amount and distribution of dust deposition, impacting regional productivity.

Increased SST (+0.02 °C per year) leads to stronger ocean stratification, preventing the mixing of deeper, nutrient-rich waters with surface waters. This reduces nutrient availability for phytoplankton, causing a decline in NPP. One recent study found a significant decrease in NPP (-0.048 mg m^{-3} day^{-1} yr^{-1}) over the past two decades, mainly due to rising SST (Kuttippurath et al. 2023). A 10–20% shift in air mass transport patterns over the decades (1998–2008 vs. 2009–2019) altered nutrient delivery to the ocean. Reduced dust input in certain regions may have contributed to lower productivity in some areas. Sulfate (SO_4^{2-}) and nitrate (NO_3^-) aerosols from human activities can lead to ocean acidification, affecting phytoplankton growth. Changes in atmospheric composition modify the ocean's pH balance and influence nutrient solubility. Stronger monsoonal winds enhance vertical mixing, bringing deep nutrients to the surface and promoting phytoplankton blooms.

Conversely, excessive wind-induced mixing can suppress phytoplankton growth by dispersing nutrients too widely. Rainfall patterns impact nutrient deposition from both atmospheric and terrestrial sources. Increased precipitation can enhance riverine input of nutrients into the Bay of Bengal, stimulating productivity. Arabian Sea productivity is higher due to frequent dust deposition and upwelling but has shown declining trends in recent decades. While Bay of Bengal has stronger stratification and riverine nutrient inputs make it less responsive to atmospheric inputs compared to Arabian sea.

Rising SST can significantly impact primary productivity. Warmer waters tend to increase stratification (the layering of water based on temperature), which can limit the mixing of surface waters with nutrient-rich deeper waters. This reduces the availability of essential nutrients like nitrogen, phosphorus, and iron for phytoplankton growth. Conversely, in colder waters, or in areas with strong upwelling (such as the coastal regions), nutrients are more readily available, leading to higher productivity. However, sustained warming can inhibit this nutrient upwelling and decrease overall productivity. Primary productivity depends on the availability of nutrients. Nutrient-rich waters, such as those found in upwelling zones or near riverine inputs, generally support higher productivity. However, areas that are nutrient-depleted (like the central gyres of oceans) tend to have lower productivity. Atmospheric deposition of dust that carries nutrients like iron, phosphorus, and nitrogen, can boost productivity, especially in nutrient-limited regions. Changes in dust transport due to shifting wind patterns, climate change, or human activity can alter nutrient availability and, consequently, productivity. Light availability plays a crucial role in photosynthesis. Areas with more sunlight, typically closer to the surface and in more tropical or subtropical regions, support higher productivity. Cloud cover, seasonal changes in sunlight, and water turbidity (due to pollutants, plankton, or sediments) can reduce light penetration and thus affect productivity. In regions with stronger stratification, light may be available to phytoplankton but without the necessary nutrients for growth. In contrast, well-mixed waters allow for better nutrient replenishment, which supports higher productivity even in low-light conditions. Ocean currents and upwelling are vital for bringing nutrient-rich waters from deeper layers to the surface, where phytoplankton can access them. In regions where upwelling is prevalent (e.g., along coasts or in the equatorial Pacific), primary productivity is typically higher. Climate change can impact ocean circulation patterns, altering upwelling areas and nutrient availability. For example, warming seas and changing wind patterns can reduce upwelling, leading to lower productivity in some regions.

Aerosols, especially dust from deserts, contain essential nutrients like iron and phosphorus. In regions where dust deposition is high (such as the Arabian Sea or the Bay of Bengal), this external input can enhance primary productivity by providing essential nutrients. Changes in atmospheric circulation, influenced by both natural and anthropogenic factors, can alter the amount and distribution of aerosols, which in turn affects nutrient availability and ocean productivity. As atmospheric carbon dioxide (CO_2) concentrations rise, a significant portion of this CO_2 is absorbed by the oceans, causing ocean acidification. This can affect marine organisms, including phytoplankton, especially those that rely on calcium carbonate (such as coccolithophores). Acidification may decrease the ability of certain species to thrive and reduce overall productivity, though the full impact is still under study. Excessive nutrient inputs from human activities (like agriculture, wastewater discharge, or fossil fuel combustion) can lead to eutrophication, where nutrient overload stimulates rapid phytoplankton growth. While this can temporarily increase productivity, it often results in harmful algal blooms that deplete oxygen, creating dead zones and harming marine ecosystems. Pollution, including plastics, heavy metals, and chemicals, can also disrupt marine ecosystems and reduce primary productivity by harming marine life and altering food web dynamics.

The Indian Ocean accounted for around 15.7–16.7% of global primary production during the study period. Primary productivity in the Indian Ocean is influenced by various climatic oscillations, including the Indian Ocean Dipole (IOD), which impact nutrient availability and phytoplankton growth. One recent research (Kulk et al. 2020) reports that primary production in the Indian Ocean has experienced variability between 1998 and 2018. Some periods, particularly between 2003 and 2011, saw a decrease in primary production, while other years, especially post-2015, showed slight increases. Variations in primary production were strongly related to climate indices such as the El Niño-Southern Oscillation (ENSO) and the IOD, which affect oceanic conditions like temperature and nutrient mixing. The decrease in primary productivity after 2011, for example was correlated with changes in the ENSO and Pacific Decadal Oscillation (PDO). Primary productivity in the Indian Ocean was also influenced by photosynthetic parameters that were adjusted based on environmental factors like temperature and light. Changes in these parameters led to significant differences in the computed primary productivity for the region. Indian Ocean shows deviations from global trends, with two primary peaks in productivity observed during spring and autumn, and lower productivity in summer.

Significant seasonal variability in NPP has been observed in Indian ocean, with two primary peaks during the southwest monsoon (June–August) and the winter monsoon (December–January), attributed to increased productivity due to summer upwelling and winter mixing. VGPM, Eppley-VGPM, CbPM, CAFE, and SABPM, were used to estimate NPP from MODIS-Aqua satellite data (2003–2021) (Kalita et al. 2024). These models vary in their approach, using parameters like chlorophyll concentration, photosynthetically available radiation (PAR), and sea surface temperature (SST). Spatially, the Arabian Sea and Bay of Bengal (BoB) show different productivity patterns, with the Arabian Sea exhibiting higher NPP due to better nutrient availability from upwelling. In contrast, BoB's productivity is limited by strong stratification and low mixing. The NPP variability was assessed across four provinces in the Northern Indian Ocean (Northwest Arabian Upwelling Province, Indian Monsoon Gyres Province, Eastern Indian Coastal Province region, and Western Indian Coastal Province), with maximum variability seen in the Western Indian Coastal Province region, influenced by physical processes like upwelling. The Eastern Indian Coastal Province region showed high diversity due to riverine input, notably from the Ganges–Brahmaputra delta.

The increase in SST, a direct consequence of global warming, has caused changes in plankton communities. Warmer waters promote thermal stratification, which prevents vertical mixing (Sahu et al. 2018). This disrupts nutrient cycling and can reduce the availability of essential nutrients for phytoplankton growth, negatively impacting primary productivity. Increased carbon dioxide (CO_2) levels in the atmosphere are absorbed by oceans, lowering the pH of seawater. This acidification can affect the ability of calcifying organisms (like certain types of phytoplanktons) to produce calcium carbonate shells, which is crucial for their survival. This can lead to shifts in community composition and impact the broader food web. Climate change has been linked to changes in the species composition of phytoplanktons. Warmer temperatures and altered nutrient conditions can favor certain species over others, leading to shifts in bloom patterns. For example blooms of *Trichodesmium erythraeum* in the Arabian Sea have been attributed to changes in surface water warming, optimal light conditions, and thermal stratification, even when nitrogen nutrients are limited. This chapter highlights a significant increase in the frequency, intensity, and spatial coverage of phytoplankton blooms in the Indian Exclusive Economic Zone (EEZ) since the late 1990s. This increase is primarily attributed to warming SST and the resulting changes in water column stability, which promote bloom conditions. The changes in plankton communities can have cascading effects on marine ecosystems. Phytoplanktons are the bases of the food chains, and disruptions to their population, composition, or timing can affect higher trophic levels, including zooplankton, fish populations, and ultimately human fisheries. As climate change affects temperature and nutrient availability, phytoplankton phenology (the timing of growth and reproduction) has also been affected. This can lead to mismatches in timing between plankton blooms and the availability of food for higher trophic levels, impacting fisheries and marine biodiversity. Uncertainty in climate change projections for ocean NPP is increasing, particularly at regional scales, and remains a major challenge for future ecosystem management, fisheries planning, and climate change adaptation strategies (Tagliabue et al. 2021).

Primary productivity significantly impacts fisheries yield, but factors such as temporal PP variability, fishing pressure, and the trophic level of catches play a crucial role in determining the sustainability of fish stocks and the efficiency of fisheries. Recent findings show that there is a general correlation between primary productivity and fisheries yields, with higher PP often associated with higher fish catches. However, this relationship is not linear or simple (Conti and Scardi 2010). Temporal variability of primary productivity variability plays a significant role in determining fish yield. Ecosystems with higher PP variability (indicating marked seasonal changes in productivity) tend to have lower yields, particularly for demersal species. In contrast, ecosystems with stable PP conditions tend to support higher fisheries yields, especially in pelagic-dominated systems. The average trophic level (TL) of the fish caught is an important factor in the relationship. Low trophic levels (TL <2.75), typically pelagic species like anchovies or sardines, are associated with higher yields in ecosystems with low PP variability. When fishing pressure increases, higher trophic levels (such as predatory fish) are targeted, leading to intermediate or lower yields in ecosystems with higher PP variability. The study finds an inverse relationship between spatial variability and temporal variability in PP. Ecosystems with higher spatial variability (e.g., large seasonal changes) tend to have lower fisheries yields, as seasonal fluctuations in productivity may not support consistent fish availability. Stable ecosystems, with minimal seasonal variability, tend to support higher yields of fish, particularly from low-TL species.

The primary production in marine ecosystems is projected to decline in many tropical and upwelling regions due to the combined effects of warming sea surface temperatures (SSTs) and altered nutrient dynamics. This decline in primary production can have a negative effect on fish biomass and overall productivity, especially in tropical shelf seas and areas influenced by strong upwelling systems like the Humboldt Current and the Canary Current. Conversely, in some high-latitude regions, primary production and fish biomass may increase due to changes in temperature and nutrient availability, with predictions showing potential increases in fish biomass density in areas like the Nordic shelf seas and Guinea Current. Fish production is largely governed by available primary production, and changes in phytoplankton and zooplankton biomass density are often directly mirrored by changes in fish biomass density. Reductions in primary production are expected to result in declines in fish production in regions like the Indo-Pacific, Humboldt, and Canary Currents, potentially by 30–60%. These declines are expected to impact the pelagic predator populations, with cascading effects on fish communities. In contrast, high-latitude ecosystems may see increases in fish biomass, especially for pelagic predators, driven by increased productivity from changing ocean temperatures and nutrient cycling.

6.2 Materials and Methods

6.2.1 Dataset Collection

Sardine fish catching dataset has been collected from different databases, and analyzed for future trends using Python. Net primary productivity and climate datasets (sea surface temperature) have been collected from GLODAP database.

6.2.2 Data Analysis

Sardine fish landing dataset has been analyzed using the linear regression using Python, for the trend analysis. The predicted dataset has been plotted for the future trend of the fish catching.

```
import pandas as pd
import numpy as np
from sklearn.linear_model import LinearRegression
import matplotlib.pyplot as plt
# Data based on the extracted text
data = {
    'Year': [2000, 2001, 2002, 2003, 2004, 2005, 2006, 2007, 2008, 2009,
2010, 2011, 2012],
    'Sardines, Anchovies, etc.': [401.8, 424.8, 344.2, 374.4, 372.8, 359.5,
486, 367.4, 420.1, 412.6, 427.7, 495.6, 518.6]
}
# Create DataFrame
df = pd.DataFrame(data)
# Statistical analysis (mean, median, standard deviation, etc.)
stats = df['Sardines, Anchovies, etc.'].describe()
print(stats)
# Checking the correlation of Sardines catch over the years
correlation = np.corrcoef(df['Year'], df['Sardines, Anchovies, etc.'])
print(f"Correlation: {correlation}")
# Forecasting using a simple linear regression model
# Prepare data for linear regression
X = df['Year'].values.reshape(-1, 1)  # Independent variable (Year)
y = df['Sardines, Anchovies, etc.'].values  # Dependent variable (Sardine
catches)
# Fit the model
model = LinearRegression()
model.fit(X, y)
# Predict future values (let's predict for the next 5 years)
future_years = np.array([2013, 2014, 2015, 2016, 2017]).reshape(-1, 1)
```

```
predictions = model.predict(future_years)
# Combine predictions with years for display
future_predictions = pd.DataFrame({
    'Year': future_years.flatten(),
    'Predicted Sardine Catch': predictions
})
# Display results
print(future_predictions)
# Plotting the sardine catches over the years
plt.figure(figsize=(10, 6))
# Plot the historical sardine catches
plt.plot(df['Year'], df['Sardines, Anchovies, etc.'], label='Actual Sardine
Catches', marker='o', color='b')
# Plot the predicted sardine catches for the next 5 years
plt.plot(future_predictions['Year'], future_predictions['Predicted Sardine
Catch'], label='Predicted Sardine Catches (2013-2017)', marker='o',
color='r', linestyle='--')
# Adding title and labels
plt.title('Sardine Catch Analysis and Predictions (2000-2017)',
fontsize=14)
plt.xlabel('Year', fontsize=12)
plt.ylabel('Sardine Catch (thousands of tonnes)', fontsize=12)
# Set the x-axis labels to show only years (no decimal points)
plt.xticks(np.arange(2002, 2018, 1))
# Adding a legend
plt.legend()
# Saving the plot as a TIFF file
plt.savefig(sardine_catch_analysis_adjusted_years.tiff', format='tiff')
# Show the plot
plt.show()
```

The salinity and temperature datasets have been studied for missing datasets. Thereafter the datasets were processed using predictive mean matching methods for imputation of the missing datasets. The datasets have later undergone trend analysis for the 1995–2020 for Indian coastal area.

These two datasets have been analyzed for future trends and the prediction of the coastal ecosystem impacts on the fish catching. The analysis has been performed using Python, R.

The data imputation codes for missing datasets:

```
library(mice)
# Load your dataset (replace 'your_dataset.csv' with the actual file path)
df <- read.csv("indian_coast_filtered_data.csv")
# Check for missing values in the dataset
sum(is.na(df))  # Total missing values in the dataset
colSums(is.na(df))  # Missing values per column
# Impute missing data using Predictive Mean Matching (PMM) from mice
package
# Method 'pmm' applies Predictive Mean Matching (PMM) for imputation
imputed_data <- mice(df, method = 'pmm', m = 5, seed = 123)
# Check the imputed dataset (summary)
summary(imputed_data)
# Complete the imputed dataset
completed_data <- complete(imputed_data, 1)  # Get the first imputed
dataset
# View the imputed dataset
print(completed_data)
# Save the imputed dataset to a new CSV file (optional)
write.csv(completed_data, "indian_coast_pmm.csv", row.names = FALSE)
```

6.3 Sardine Fishing and Catching

Sardines are one of the most widely caught fish species in India, particularly along the southwestern coast. They are a major part of the country's fishing industry, providing employment to millions of people, including fishermen, processors, and traders. This helps sustain the livelihoods of many coastal communities. Sardines are a popular food item in India, particularly in coastal states like Kerala, where they are an integral part of the local diet. Their affordability and nutritional value make them accessible to a broad segment of the population. India is one of the leading exporters of fish, including sardines. Sardines are exported to various countries, contributing significantly to India's foreign exchange earnings. They are exported in canned, frozen, and dried forms, which are highly valued in international markets. Sardines are a rich source of omega-3 fatty acids, protein, and essential vitamins, making them an important part of the diet, especially in coastal regions. They are an affordable source of nutrition for low-income families, contributing to food security in many regions. Sardines are also a part of sustainable fishing practices, as they are relatively abundant and fast-growing. Their population levels, when managed properly, can be sustainable, ensuring continued economic benefits. Sardines have cultural importance in regions like Kerala, where they are used in traditional dishes like sardine curry, contributing to the tourism and food industry in these areas.

There has been a dramatic increase in the annual fish landings in India from 1925 to 2012. The landings that were as low as 9 tons in 1946 rose sharply to an all-time high of 720,270 tons in 2012. (Rohit et al. 2018). The average annual landing over the period from 1925 to 2015 was 161,960 tons, with a noticeable increase in the later years. The Indian Ocean sector (IOS) has made a significant contribution to the country's overall marine fish landings. It accounted for about 14% of the total marine fish landings (TMFLs) from 2000 to 2015. However, this contribution has fluctuated over time, with a peak of 32% in 1962. By 1994, it had dropped to just 2% of the TMFL, showing a considerable decrease in the proportion of landings from this sector. After the peak in 2012, fish landings from the IOS have declined significantly, from 596,726 tons in 2013 to 544,684 tons in 2014, and further to 265,667 tons in 2015. This suggests a declining trend in the contribution of the IOS to India's marine fish production after 2012. The increase in landings, especially from the late twentieth century to 2012, can be attributed to technological improvements, such as more efficient crafts and fishing gears, along with the use of echosounders, GPS, and mobile phones for locating fish. The availability of PFZ (potential fishing zone) advisories likely played a crucial role in improving fish catch efficiency. The subsequent decline in landings after 2012 could be linked to a variety of factors, such as potential overfishing, depletion of fish stocks, or environmental changes affecting marine productivity in the region.

The decadal trends in landings have fluctuated over time, with significant variability between decades. The landings saw relatively modest figures in the earlier decades (e.g., 1926–35 with 12,834 tons and 1936–45 with 8644 tons). A noticeable surge in landings occurred after 1996–2005, where landings rose sharply to 287,867 tons, almost doubling from the previous decade (1986–95, 132,879 tons). The 2006–2015 decade witnessed an exceptional peak, with the highest recorded landings of 497,607 tons, contributing to 33% of the aggregated landings from the previous nine decades. This represents a significant increase compared to previous decades, suggesting a boost in fish availability or improved fishing practices during this period. The large increase in sardine landings, particularly from 2006 to 2015, can likely be attributed to factors such as improved fishing efficiency, better management practices, advances in technology like echosounders, GPS, and PFZ advisories, as well as potentially favorable environmental conditions for sardine populations.

The overall long-term trend shows a gradual increase in Indian oil sardine landings, with the most dramatic increase occurring during the last decade. However, the earlier periods had far lower landings, indicating either limited fishing effort or lower sardine populations during those times.

The landings from the west coast (represented by the red line) have consistently been higher than those from the east coast (represented by the blue line). This suggests that the west coast has historically been the more productive region for Indian oil sardine (IOS) fisheries. During years when there was a decline in total IOS landings in the country, the southwest coast (particularly states like Kerala, Karnataka, and Goa) showed decreased production. Interestingly, when this happened, the landings from the southeast coast (including Tamil Nadu and Andhra Pradesh) experienced an increase. This could indicate a shift in sardine distribution, possibly as an adaptive response to environmental changes such as increased sea surface temperatures (SSTs), which may have influenced sardine populations to expand northward and eastward. The expansion of IOS distribution to the northern and eastern boundaries of the Indian coast (i.e., along the southeast coast) has been suggested as an adaptation to the effects of climate change, particularly rising SST. This shift in distribution could be a response to warmer waters that are influencing sardine populations to move to new areas. Sardines appear to have a high level of geographical mobility, as they are not restricted by any significant geographical barriers along the Indian coast. This flexibility allows sardines to adapt to changing environmental conditions, such as shifts in water temperature, by moving to more suitable areas for their growth and survival. Sardines are highly sensitive to environmental perturbations, such as changes in temperature and ocean conditions. This sensitivity is reflected in their ability to adjust their distribution in response to climate-related changes, indicating a flexible response to climate change.

6.4 Present Scenario of Sardine Fish Landing

According to the dataset published from CMFRI the average sardine catch over the years is 415.81 thousand tonnes. The catch values vary by approximately 55.16 thousand tonnes from the mean, indicating a moderate level of variation in the sardine catches over the years. The lowest recorded sardine catch was 344.2 thousand tonnes in 2002. The highest recorded sardine catch was 518.6 thousand tonnes in 2012. As the years progress, sardine catches tend to increase, but there are other factors at play that influence this relationship (Fig. 6.1).

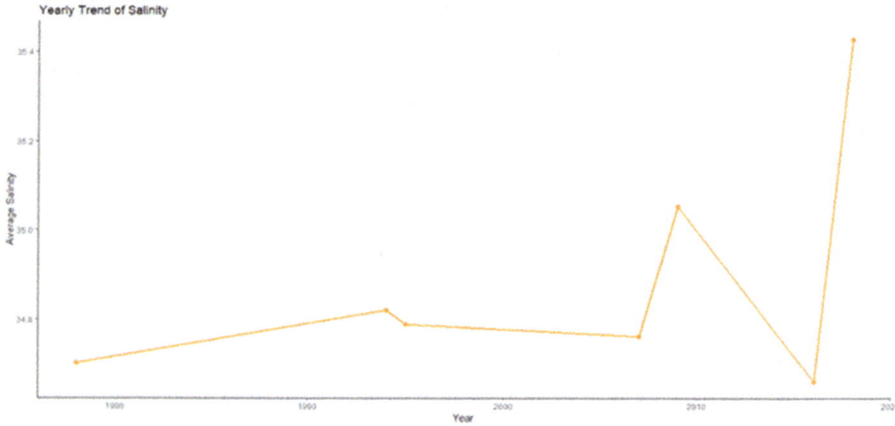

Fig. 6.1 Sardine fish catch prediction in India from 2013 to 2017

6.5 Temperature and Salinity Fluctuation Impact on Sardine Fish Catching

There are noticeable fluctuations with a period of decline in temperature around the early 2000s, followed by a sharp spike after 2010. This may indicate climate variability, possible oceanic influences (such as El Niño and La Niña effects), or anthropogenic factors. The steep rise in temperature post-2010 is concerning. This could be due to factors such as global warming, increased urbanization, sea surface temperature rise, or changing monsoon patterns affecting coastal climate conditions. The temperature remains high in the latest years of the graph, reinforcing concerns about climate change's impact on coastal areas.

The sharp increases and fluctuations in salinity levels over time suggest significant environmental and climate-driven changes in the Indian coastal region. The trend correlates with rising temperatures (as seen in the temperature graph), indicating potential climate change impacts affecting coastal water balance.

The fluctuating salinity trends observed in the provided graph can have significant effects on sardine fish landings in the Indian coastal regions. Sardines, particularly Indian oil sardines (*Sardinella longiceps*), are highly sensitive to environmental changes, including salinity, temperature, and ocean currents. Sardines prefer moderate salinity (~34.5 PSU) (CMFRI 1984). Extreme fluctuations in salinity can disrupt their metabolism, reproductive cycles, and overall survival rates. High salinity (>35 PSU) can cause physiological stress, reducing their spawning success and availability in traditional fishing areas (Agarwal et al. 2024). If salinity becomes too high or too low in coastal regions, sardines may migrate to deeper offshore waters or shift their distribution to areas with more stable conditions (Giannoulaki et al. 2014). This could lead to lower sardine landings in traditional fishing zones along the Indian coast. Sardines spawn in specific environmental conditions, particularly when sea surface temperatures (SSTs) and salinity are optimal. High salinity may reduce the survival of sardine eggs and larvae (Dias et al. 2014), leading to lower recruitment (young fish entering the fishery). Unfavorable conditions during breeding seasons can cause population crashes, which have been observed in previous years (Fig. 6.2). Sardines primarily feed on planktons (Day 2018) that are affected by salinity and temperature shifts. If high salinity disrupts plankton growth, sardines may face food shortages, (Kripa et al. 2018) leading to reduced growth rates and fishery productivity. Sharp fluctuations in salinity (as seen in the graph), combined with rising temperatures, may cause erratic sardine landings (Fig. 6.3).

The sardine catch analysis and predictions (2000–2017) graph shows a fluctuating trend in sardine landings over the years, with periods of both sharp declines and strong recoveries. The actual catch data from 2000 to 2012 (represented by the blue line) exhibit high variability, indicating that sardine populations have been influenced by multiple environmental and climatic factors. A significant dip in the early 2000s is followed by a major increase around 2005, suggesting a period of recovery. However, the overall pattern remains unstable, reflecting the sensitivity of sardine populations to changing oceanic conditions.

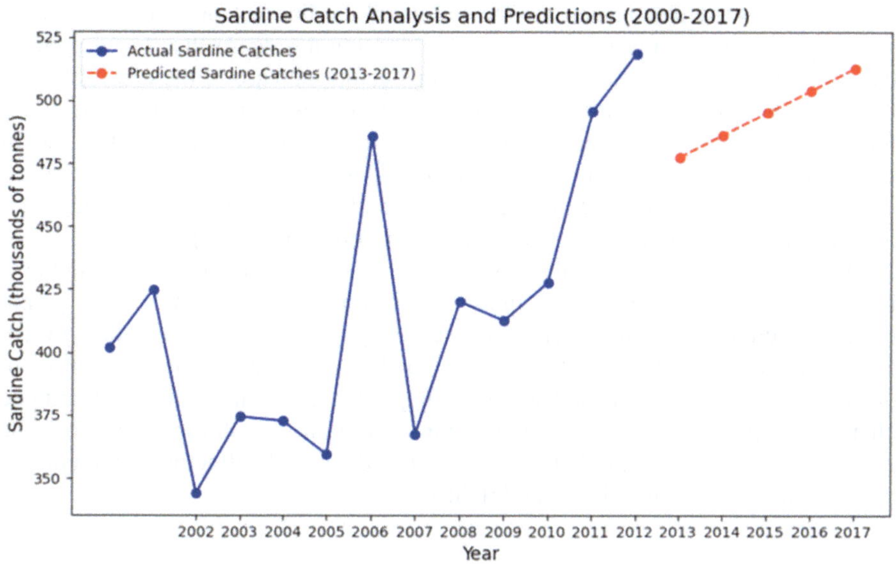

Fig. 6.2 Salinity trend of Indian coastal area

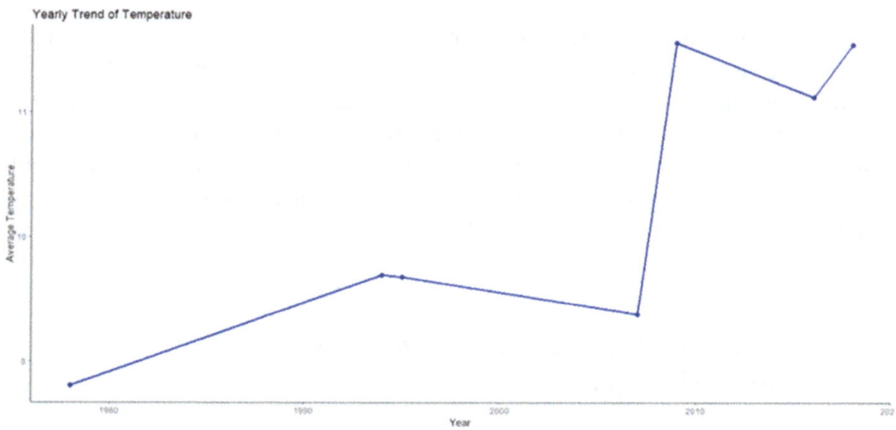

Fig. 6.3 Tempertature trend of Indian coastal area

From 2008 onward, the sardine catch displays a gradual increasing trend, reaching a peak around 2011–2012. This rise could be attributed to favorable environmental conditions, including optimal sea surface temperatures (SSTs) and salinity levels, which play a critical role in sardine spawning and migration. The sharp increase in landings during this period also suggests improved fishery management or natural stock replenishment. However, sardine stocks are known to fluctuate due to variations in plankton availability, ocean currents, and climatic events such as El Niño and La Niña, which directly impact coastal waters (Berger 2020).

The red dotted line represents the predicted sardine catches for 2013–2017, showing a continued upward trajectory. This projection suggests that if environmental conditions remain stable, sardine landings could increase in the coming years. However, the accuracy of these predictions depends on multiple factors, including future climatic changes, overfishing pressure, and habitat conditions. The Indian sardine fishery has historically experienced sudden declines, often linked to rising sea surface temperatures, irregular monsoons, and salinity fluctuations. These factors must be closely monitored to ensure the reliability of future projections.

One of the key concerns in the observed trend is the potential impact of increasing salinity levels on sardine populations. As shown in the salinity trend graph, extreme fluctuations in salinity can stress sardine populations, affecting their growth and reproductive success. Higher salinity levels may force sardines to migrate to deeper or alternative waters, leading to unpredictable fish landings. If climate change continues to alter coastal temperature and salinity conditions, there could be mismatches between the predicted and actual sardine catch, affecting the livelihood of fishermen and the overall seafood industry.

6.6 Future Projections and Challenges

The influence of climate change on sardine fisheries is anticipated to escalate in the forthcoming decades owing to elevated sea surface temperatures (SSTs), ocean acidification, altered monsoon patterns, and heightened variability in primary productivity (Sarkar et al. 2024). These factors are expected to affect sardine populations, migration patterns, and the overall sustainability of fisheries. Current trends and scientific projections reveal several significant challenges and potential future scenarios.

6.6.1 Rising Sea Surface Temperatures (SSTs) and Their Impacts

Sea surface temperature (SST) is anticipated to rise by 0.02–0.03 °C annually in the Indian Ocean, potentially exacerbating ocean stratification and diminishing nutrient mixing from deeper waters (Kuttippurath et al. 2023). Increased sea surface temperatures may modify spawning seasons and reproductive success of sardines, resulting in changes to population dynamics (Ganias 2009). Elevated sea surface temperatures may extend the distribution of sardines northward, compelling conventional fisheries to adjust to novel fishing areas.

6.6.2 Changes in Ocean Currents and Monsoon Variability

Climate models forecast alterations in monsoonal wind patterns, potentially disrupting upwelling and nutrient availability, thereby directly impacting sardine food sources (Jena and Grinson 2017). Alterations in ocean currents may reallocate sardine populations, rendering them less predictable for conventional fisheries (Bolster et al. 2012).

6.6.3 Ocean Acidification and Its Consequences

Enhanced CO_2 absorption is anticipated to decrease ocean pH levels, potentially impacting plankton composition and diminishing sardine food supply (Marshall et al. 2017). Acidification may adversely affect the survival rates of sardine larvae, resulting in diminished recruitment in fisheries (Faleiro et al. 2016).

6.6.4 Salinity Fluctuations and Habitat Shifts

Projected changes in precipitation patterns and freshwater runoff could alter coastal salinity, impacting sardine spawning grounds (Maynou et al. 2020). Extreme salinity variations may push sardines into deeper waters, making them harder to catch using traditional fishing methods (Grant and Bowen 1998).

6.6.5 Decline in Primary Productivity

Certain models indicate that net primary productivity (NPP) in the Indian Ocean may decrease by 5–10% as a result of warming and stratification (Marsac et al. 2024), thereby directly impacting the availability of sardine food sources. A decline in plankton populations may result in the collapse of sardine stocks in specific regions, causing significant economic repercussions for fisheries.

6.6.6 Fisheries Sustainability and Socioeconomic Challenges

Declining sardine populations may result in economic instability for fishery-dependent communities due to variable catches (Giron-Nava et al. 2021). Overfishing due to diminishing sardine stocks may exacerbate population decline, resulting in prolonged fishery collapses. Small-scale fishers dependent on sardines for their livelihoods may find it challenging to compete with industrial fishing enterprises adjusting to shifting sardine distributions.

6.6.7 Potential Adaptation and Resilience Strategies

Monitoring and Modelling: Advanced climate models and real-time monitoring systems can help predict sardine distribution shifts, allowing fishers to adjust their strategies.

Fisheries Management: Implementing quotas, seasonal bans, and habitat protection measures can help sustain sardine stocks.

Alternative Livelihoods: Encouraging aquaculture or alternative fishery options can reduce dependence on sardine catches.

Climate-Resilient Fishing Techniques: Adoption of deep-sea fishing technologies and climate-adaptive policies can mitigate the economic risks of changing sardine populations.

6.7 Conclusions

The influence of climate change on the sardine fishery in the Indian subcontinent constitutes a complex challenge, shaped by increasing sea surface temperatures, variations in salinity, and modifications in oceanic conditions. Environmental changes are influencing primary productivity, which subsequently affects sardine populations and the sustainability of the fishing industry. Sardines exhibit adaptability to specific changes, including temperature variations and salinity alterations; however, the cumulative impacts of climate change, such as ocean acidification and altered monsoon patterns, present substantial risks. Future forecasts suggest possible reductions in sardine catches owing to decreased primary productivity, modified spawning conditions, and the disruption of planktonic food sources. The capacity of sardines to modify their distribution may offer some resilience; however, the escalating unpredictability of environmental factors poses a significant challenge for future fishery management. Implementing sustainable practices, such as enhanced monitoring systems, adaptive management strategies, and diversified livelihood options for fishing communities, is crucial to alleviate the effects of climate change and secure the long-term sustainability of sardine fisheries in the region.

Acknowledgments The authors are thankful to the authorities for providing the facilities to develop this chapter.

Disclosures and declarations All authors of the study consent to publish the book chapter.

Conflict of interests The authors declare that they have no conflict of interests.

Ethics approval Not applicable.

Data availability statement The data used for the study will be available on request.

Funding statement No funders to disclose.

References

Agarwal D, Shanmugam SA, Kathirvelpandian A et al (2024) Unraveling the impact of climate change on fish physiology: a focus on temperature and salinity dynamics. J Appl Ichthyol 2024:5782274. https://doi.org/10.1155/2024/5782274

Berger WH (2020) Chapter 7. Sardines and the California current: on a cold current running South. In: Ocean, pp 185–214. https://doi.org/10.1525/9780520942547-009/HTML

Bolster J, Chavz F, Cournane J, Erlandson J, Field D, Hardt MJ (2012) Shifting baseline: The past and future of ocean fisheries. Island Press

Conti L, Scardi M (2010) Fisheries yield and primary productivity in large marine ecosystems. Mar Ecol Prog Ser 410:233–244. https://doi.org/10.3354/MEPS08630

CMFRI (1984) Annual Report, Ed. Silas

Day Trevor (2018) Sardine. Reaktion

Dias DF, Pezzi LP, Gherardi DFM, Camargo R (2014) Modeling the spawning strategies and larval survival of the Brazilian sardine (Sardinella brasiliensis). Prog Oceanogr 123:38–53. https://doi.org/10.1016/J.POCEAN.2014.03.009

Faleiro F, Pimentel M, Pegado MR et al (2016) Small pelagics in a changing ocean: biological responses of sardine early stages to warming. Conserv Physiol 4. https://doi.org/10.1093/CONPHYS/COW017

Ganias K (2009) Linking sardine spawning dynamics to environmental variability. Estuar Coast Shelf Sci 84:402–408. https://doi.org/10.1016/J.ECSS.2009.07.004

Giannoulaki M, Schismenou E, Pyrounaki MM, Tsagarakis K (2014) Habitat characterization and migrations. In: Biology and ecology of sardines and anchovies, pp 190–241. https://doi.org/10.1201/B16682-7

Giron-Nava A, Ezcurra E, Brias A et al (2021) Environmental variability and fishing effects on the pacific sardine fisheries in the gulf of california. Can J Fish Aquat Sci 78:623–630. https://doi.org/10.1139/CJFAS-2020-0010/SUPPL_FILE/CJFAS-2020-0010SUPPLA.DOCX

Grant WS, Bowen BW (1998) Shallow population histories in deep evolutionary lineages of marine fishes: insights from sardines and anchovies and lessons for conservation. J Hered 89:415–426. https://doi.org/10.1093/JHERED/89.5.415

Jena JK, Grinson G (2017) Agriculture under climate change: threats, strategies and policies. Allied Publishers

Kalita R, Aneesh, Lotliker A, Lotliker AA (2024) Assessment of satellite-based net primary productivity models in different biogeochemical provinces over the northern Indian Ocean. Int J Remote Sens 2024:8807–8826. https://doi.org/10.1080/01431161.2023.2247533

Kripa V, Mohamed KS, Koya KPS et al (2018) Overfishing and climate drives changes in biology and recruitment of the Indian oil sardine Sardinella longiceps in southeastern Arabian Sea. Front Mar Sci 5:418487. https://doi.org/10.3389/FMARS.2018.00443/BIBTEX

Kulk G, Platt T, Dingle J et al (2020) Primary production, an index of climate change in the ocean: satellite-based estimates over two decades. Remote Sens 12:826. https://doi.org/10.3390/RS12050826

Kuttippurath J, Maishal S, Anjaneyan P et al (2023) Recent changes in atmospheric input and primary productivity in the north Indian Ocean. Heliyon 9:e17940. https://doi.org/10.1016/j.heliyon.2023.e17940

Marsac F, Everett B, Shahid U, Strutton PG (2024) Indian Ocean primary productivity and fisheries variability. In: The Indian Ocean and its role in the global climate system, pp 245–264. https://doi.org/10.1016/B978-0-12-822698-8.00019-6

Marshall KN, Kaplan IC, Hodgson EE et al (2017) Risks of ocean acidification in the California current food web and fisheries: ecosystem model projections. Glob Chang Biol 23:1525–1539. https://doi.org/10.1111/GCB.13594

Maynou F, Sabatés A, Raya V (2020) Changes in the spawning habitat of two small pelagic fish in the Northwestern Mediterranean. Fish Oceanogr 29:201–213. https://doi.org/10.1111/FOG.12464

Rohit P, Sivadas M, Abdussamad EM et al (2018) Enigmatic Indian oil sardine: an insight. ICAR-Central Marine Fisheries Research Institute

Sahu BK, Pati P, Panigrahy RC (2018) Impact of climate change on marine plankton with special reference to Indian Seas. IJMS 47:259–268

Sarkar UK, Ajith Kumar TT, Neeraj S et al (2024) Sustainable management of fish genetic resources. Springer, p 295

Tagliabue A, Kwiatkowski L, Bopp L et al (2021) Persistent uncertainties in ocean net primary production climate change projections at regional scales raise challenges for assessing impacts on ecosystem services. Front Clim 3:738224. https://doi.org/10.3389/FCLIM.2021.738224/PDF

Chapter 7
Mussels: A Potent Ecological Indicator to Climate Change

M. Kanmani, S. Jeyanthi, and N. Thirunavukkarasu

Abstract Mussels belonging to phylum Mollusca serve as vital biological indicators for assessing the influence of environmental change on marine ecosystems due to their widespread distribution, sessile nature, and sensitivity to environmental fluctuations. Being filter feeders, mussels accumulate pollutants and reflect changes in water temperature, salinity, and pH, offering valuable insights into the health of aquatic ecosystems. The physiological and ecological responses of mussels to ocean acidification, rising sea temperatures, and altered food availability make them potential indicators of climate-induced stressors. This review highlights how these changes affect mussel growth, reproduction, and survival rates, making them a proxy for broader ecosystem changes. Furthermore, their calcified shells provide historical records of environmental conditions for paleo studies. The integration of mussel-based monitoring with advanced analytical techniques enables the identification of early warning signs of ecosystem degradation, guiding mitigation strategies and marine conservation efforts. By leveraging mussels as bioindicators, this research underscores their critical role in understanding and addressing the multifaceted impacts of climate change on marine biodiversity and ecosystem stability.

Keywords Mussels · Climate change · Bioindicator species · Marine ecosystem · Environmental monitoring

---

M. Kanmani · N. Thirunavukkarasu
Presidency Research Innovations in Zoological Eminence (PRIZE) Lab, Department of Zoology, Presidency College (Autonomous), Chennai, Tamil Nadu, India

S. Jeyanthi (✉)
PG and Research Department of Zoology, Ethiraj College for Women (Autonomous), Chennai, Tamil Nadu, India
e-mail: jeyanthi_s@ethirajcollege.edu.in

B. P. Barathan et al. (eds.), *Navigating Climate Change: Impacts on Biodiversity and Ecosystem Resilience*, Sustainable Development and Biodiversity 42,
https://doi.org/10.1007/978-981-95-0409-1_7

7.1 Introduction

Mollusca is the world's second largest phyla of the invertebrate group with high adaptability in the animal kingdom (Ruppert et al. 2004; Ponder and Lindberg 2008). Bivalves are the wide-ranging classes of species among the living marine molluscans. Shelled molluscs inhabit freshwater, marine, and terrestrial environments, occupying a widespread of biological niches. They have bodies enclosed within a hinged shell like clams, oysters, scallops, and mussels. Bivalves are distributed in a wide range of habitats ranging from intertidal to deep-sea settings and also populate a wide latitudinal gradient in the sea (Tunnicliffe et al. 2009).

One of the most commonly used spatial and temporal trend indicators of contaminants in aquatic environments is bivalve mollusc, which is also a significant part of coastal habitats, closely linked to water quality changes and reflects bioaccumulation near the base of the food web (Melwani et al. 2014). Bivalves are also favorable indicator species; because of their less complex mixed function oxygenase (MFO) capacity, they pile up hydrocarbons rapidly in comparison to the fishes (Hellou et al. 2000; Albayrak et al. 2019). Due to their filter feeding mechanism bivalves have limited ability to metabolize toxic compounds and this makes them an excellent candidate for monitoring emerging contaminants.

Mussels are a category of bivalves, which includes marine mussels of *Mytilidae* family and freshwater mussels of *Unionidae* family. Mussels belonging to the order *Unionoida* form dense assemblages, generally consisting of diversified species, which are critical to heterotrophic production and biogeochemical cycling in freshwater ecosystems (Strayer 2008). Mussels are filter feeding bivalves, frequently employed as a sentinel species in ecotoxicological monitoring. Mussel beds are found throughout the world, living in soft bottoms on rocky substrates along the coast in cold and temperate climates (Seed and Suchanek 1992). Because they outgrow their rivals, they can physically change their environment by linking byssal threads to the substrate (Gutiérrez et al. 2003).

Mytilid species are regarded as predominant organisms of the intertidal zone (Fig. 7.1). Additionally they serve as sustenance for many other creatures, including humans by rapidly filtering pollutants and bacteria that may be present in the pelagic zone. These voracious filter feeders help in the ecosystem recycling (Zippay and

Fig. 7.1 Mussel biology: from the byssus to ecology and physiology, including microplastics ingestion and deep-sea adaptations © 2021 by Koji Inoue et al. is licensed under CC BY 4.0. (**a**) *Mytilus galloprovincialis* (**b**) *Perna viridis* (**c**) *Bathymodiolus septemdierum*

Helmuth 2012). Consequently, many studies have suggested that mussels seemed to be bioindicators of the coastal environments in monitoring-based analysis of the biological effects of pollutants, pesticides, hydrocarbons, metals, and microplastics (Atasaral et al. 2020; Li et al. 2019; Alkan et al. 2020; Ülker et al. 2020). The ubiquitous blue mussel *Mytilus edulis* is a commercially important species that bioaccumulate xenobiotics due to their filter-feeding ability, which may have an impact on a variety of physiological systems and act as a key-stone species for the ecosystem (Fitzer et al. 2015a). In fact, human activities release harmful chemicals that have a complex environmental impact on marine ecosystems. Mussels, sensitive to environmental changes, act as key indicators of climate change helping us understand shifting climate and its implications.

7.2 Effect of Temperature Changes in Mussels

Temperature is a fundamental environmental factor that influences the grouping and behavior of mussel species. In marine and intertidal environments, temperature serves as a key driver of species range, influencing metabolic rates, reproductive cycles, and survival (Somero 2005). Minimum threshold for spawning is around 12–15 °C and an upper threshold is around 28 °C. Mussels grow and metabolize optimally within a specific temperature range but are highly sensitive to fluctuations. Even slight changes in water temperature can significantly affect their metabolic rates, growth, and reproductive cycles. Temperature shifts can lead to higher metabolic rates in mussels, resulting in slower growth rates and spawning cues and also increased demand for oxygen and food. As the planet warms, temperature changes are causing shifts in the geographic distribution of many species including *Mytilus edulis* and *Mytilus californianus*, which are highly sensitive to temperature fluctuations, and their distribution is influenced by temperature tolerances (Jones et al. 2010). When temperatures rise beyond the species' thermal limits, mussels are forced to migrate to cooler regions. This migration often occurs in response to changing sea surface temperatures, as mussels must move toward higher latitudes or deeper waters to avoid thermal stress (Parmesan 2007). As water temperatures rise, mussels are increasingly forced to shift their geographic distribution. This has resulted in some populations moving poleward or to deeper waters to remain within their thermal tolerance limits. However, these migrations are not always seamless and can disrupt local ecosystems.

Mussels play an essential role in habitat formation by providing substrates for various marine organisms and cycling nutrients in the ecosystem. Shifts in their distribution may therefore lead to changes in biodiversity and the functioning of marine ecosystems (Stachowicz et al. 2002), potentially leading to a decline in the populations of species that rely on mussels for shelter or food. For instance, mussels create intertidal habitats that serve as critical feeding grounds for various marine species and disruption in these may trigger cascading effects throughout the food web. While submerged, mussels have low metabolic rates, but during aerial exposure at low tide, their metabolic processes become more complex. As air temperatures rise, mussels can experience body temperatures significantly higher than the surrounding air or substrates. This thermal stress impacts their overall physiology, leading to higher oxygen consumption and increased energy expenditure (Helmuth et al. 2011). Elevated temperatures not only raise metabolic rates but also limit the available energy for growth, reproduction, and immune defense. In particular, temperatures above 20 °C can cause significant metabolic stress in *Mytilus edulis*, reducing the energy available for essential processes. Under these conditions, mussels are more vulnerable to diseases, predation, and other environmental stressors (Beyer et al. 2017).

Temperature has an influencing role in regulating the reproductive cycles of mussels disrupting reproductive success and larval development by inhibiting gamete development (Zippay and Helmuth 2012). The timing of spawning and gamete development is often tightly regulated by environmental conditions, with specific temperature ranges necessary for optimal reproductive success. However, when temperatures deviate from these ranges, mussels may experience delays in gamete development, leading to reduced reproductive success (Rodland et al. 2006). Moreover, the energy required for reproduction competes with the energy needed for other physiological functions, such as immune responses. In warmer conditions, mussels tend to allocate more energy to reproduction, which can weaken their ability to defend against pathogens or other stressors (Petes et al. 2008). This trade-off makes mussels in warmer environments more susceptible to diseases, ultimately affecting their population viability and long-term survival.

Extreme temperature events, particularly heatwaves, have been directly linked to mass mortality events in mussel populations. During the 2003 European heatwave, significant die-offs of *Mytilus galloprovincialis* were reported along the Mediterranean coast due to elevated sea surface temperatures exceeding their thermal tolerance leading to widespread mortality (Garrabou et al. 2009). Ritchie (1927) showed that exposure to aquatic temperatures of 28.9 °C caused death within 14 h. Prolonged exposure to elevated temperatures during heatwaves can lead to physiological failure, as the organisms cannot cope with the extreme stress. Moreover, Temperature fluctuations often coincide with other stressors, such as salinity changes, hypoxia, and increased storm frequency. Certain temperatures can cause mortality of *M. edulis,* which vary with adaption and body size, along with seasonality and brininess.

Verbruggen et al. (2020) proved that higher water temperatures exacerbate the impacts of low oxygen levels in water, leading to more severe physiological stress in mussels. The biochemical responses that nearly uphold the physiological plasticity of bivalves are the heat shock response mediated by heat shock proteins. These proteins are primeval and highly conserved intracellular molecular chaperones constituting multigene super families (Barre et al. 2024). Nomenclature for heat shock protein (HSP) was based on the molecular weight of the protein (HSP70). Temperature coefficient Q_{10} (i.e. the measure of the rate of change as an effect of increasing the temperature by 10 °C in a biological system) is used to quantify responses to thermal stress reflecting the rate of change of physico–chemical and biological processes and it depends on the levels of acclimation (Jansen et al. 2007). Amongst the bivalves, oysters and mussels were frequently used to study the heat shock response and HSP70 control of gene expression as these organisms provide insights into the molecular basis of responses to environmental challenges (Fabbri et al. 2008). Immunological studies give an insight on how mussels adjust heat shock protein (HSP) induction thresholds based on acclimation. In *Mytilus trossulus*, thresholds ranged from 23 °C in winter to 28 °C in summer, with warm-acclimated mussels showing higher HSP70 levels and more protein damage.

7.3 Accumulation of Pollutants in Mussels

Mussels are sedentary organisms living in intertidal areas usually close to estuaries and being extensively filter feeding creatures with a specific mechanism of suspension-feeding that draws in large volumes of water to extract nutrients, inadvertently accumulating pollutants and toxins in their tissues. Consequently, the susceptibility to many contaminants through riverine and sea-based sources are higher. This characteristic makes them excellent bioindicators of marine ecosystems. Assessment on the levels of pollution and the presence of harmful substances in aquatic environments can be verified by analyzing the contaminants found in mussels through field studies and laboratory experiments (Paul-Pont et al. 2016; Van Cauwenberghe et al. 2015). For example mussels have been used to monitor heavy metal concentrations, assess microplastics and organic pollutants, providing crucial information on environmental health and the potential impact of climate changes on pollutant distribution (Von Moos et al. 2012).

In addition to global changes, human interference is significantly altering the aquatic ecosystems by its urban and tourism developments and the unsustainable exploitation of aquatic resources (Häder et al. 2020). Industrial, agricultural, and urban development have inevitably led to water environment pollution. Thousands of chemical substances, including pesticides, artificial radioactive elements, and oil spills make it to the ocean through the atmosphere or continental waters. The impact of these substances on ecosystems is complex, directly and indirectly affecting populations and habitats (Gillis et al. 2017). Metals, both naturally occurring and introduced by human activities, are significant toxic pollutants. Their sources include the following:

- Natural processes: volcanic eruption, vegetation fire, and thermal sources contribute to metal releases in the environment (Zeng et al. 2017).
- Anthropogenic activities: mining, fuel combustion, industrial and urban sewage, and agricultural practices emit large quantities of metals (Duarte et al. 2017).

Mussels are a distinguished group of organisms known to build up broad range of metal concentrations in their soft tissues. As these metals are widely dispersed across coastal ecosystems the detection of pollutant metals in mussels is crucial due to their impact on human health as a result of both natural geological processes and human activity. It has long been recognized that some metals like lead (Pb), chromium (Cr), cadmium (Cd), arsenic (As), and mercury (Hg), are of particular concern due to their toxicity and bioaccumulative nature (Fernández et al. 2010; Sivaperumal et al. 2007).

Combined stressors exacerbate bioaccumulation and reduce species adaptability. These changes could influence aquaculture production and ecosystem stability (Romero-Freire et al. 2020). Metals are deadly at low pH, where they exist as free cations. For example cadmium (Cd) transforms into a free ion at lower pH, increasing its toxicity. At alkaline pH, metals precipitate as insoluble hydroxides, oxides, carbonates, or phosphates, reducing their availability. Salinity is another important variable influencing metal toxicity, especially in estuarine and coastal regions. Metals like Cd are more toxic in low-salinity water, where they become free ions (Gamain et al. 2016). Studies show that *Mytilus edulis* accumulates more Cd in dilute seawater compared to higher salinity conditions, a phenomenon attributed to osmolarity changes in the surrounding medium.

Increased temperatures influence the bioaccumulation of metals. For instance, a rise in temperature decreases copper (Cu) content in mussel foot tissue, affecting byssus gland strength and overall byssus formation (Romero-Freire et al. 2020). Mussels provide essential minerals such as Ca and Fe, and vitamins including niacin, thiamine, and riboflavin necessary for the maintenance of good health. However, they may also pose health risks due to the accumulation of toxic metals such as Pb, Cd, As, and Hg (Fuentes et al. 2009; Özden et al. 2010). Heavy metals accumulate in marine organisms as a nonbiodegradable content posing risks to human health when ingested. They cannot be digested but decompose into innocuous forms that are nonbiodegradable (Stankovic and Jovic 2012). These metals get into the aquatic organisms as a bioaccumulation process and persist in the environment leading to detrimental effects on the health of marine organisms and potentially impacting humans through food chain.

Chronic exposure to these metals can lead to kidney damages, neurological disorders, gastrointestinal issues, and other health problems. Specific metals cause different toxic effects in our body. Cadmium (Cd) highly bioaccumulated in mussels has a long elimination half-life of 10–30 years and accumulates in human kidneys leading to severe toxicity (Amiard et al. 2008). Cd exhibits both acute and chronic levels of toxicity. Kidney damage is the first symptom reflected after the chronic exposure with the metal, which is usually diagnosed by increased excretion of low molecular weight proteins (Widmeyer et al. 2004). Over time, Cd can accelerate an osteoporotic process, since a high calcium dose can inhibit Cd absorption. From the toxicological point of view, excessive consumption of metal-contaminated mussel may result in toxicity to humans (Van Cauwenberghe and Janssen 2014).

Lead (Pb) affects several organs, including the liver and bones. Chronic exposure to lead causes hematopoietic and renal issues (Widmeyer et al. 2004). Mercury is extremely toxic and it damages cellular and tissue functions, crosses blood–brain barriers, and reduces nutrient absorption in the intestine. Mercury exposure limits the mucosal entry of sugars and amino acids to 80–90% of the control levels in the small intestine cells within few minutes and blocks intestinal nutrient transport by interacting directly with the brush-border membrane transport proteins (Crinnion 2000). Arsenic (As) in particular is toxic at relatively low doses. Can cause skin discoloration, respiratory problems, and nerve injuries. Organic forms of arsenic, however, are less toxic (Eisler 1994). Mussels show a wide geographical distribution. Metals like Hg and Cd are transported over long distances, with deposition varying due to climatic factors.

Divergence in global climate leads to uneven deposition of trace metals. Goldberg (1975) and Otchere (2019) concluded that the net gain of Hg, Cd, V, and Mn in Arctic biota has been linked to the relative absence of precipitation, scavenging, and strong atmospheric inversions. Investigations show that contaminants deposited in the tissues of mussels may cause a "stress syndrome" with adaptation to their physiology; in order to quantify this stress, the utilization of "stress indices" has been considered (Viarengo and Canesi 1991). People consuming mussels very frequently with a high accumulation of one or more of these considered toxicants could be exposed to health risks.

7.4 Microplastics Contamination in Mussels

Microplastics pollution is a serious ecological concern in the marine ecosystem. Mussels are benthic feeders that are highly sensitive to microplastics contamination. Microplastics in mussels are a great concern as they inadvertently ingest plastics suspended in the water column by filter feeding mechanism leading to physiological and ecological injuries in different tissue types. Ingested plastics enter the digestive tract, where they can cause blockages, accumulate in tissues, and translocate into the hemolymph or immunocytes further spreading the contamination event. Filter feeding mussels like *Mytilus edulis* draw water through their gills to capture food particles and swallow microplastics (Inoue et al. 2021a, b). Research indicates that microplastics accumulated in *M. edulis* in a way that led to significant alternations in digestive cells, including the formation of granulocytoma, lysosomal destabilization, and enhanced inflammatory responses. These changes increased with increase in time of exposure to microplastics (Avio et al. 2015; Von Moos et al. 2012). It accumulated primarily in the digestive, and specifically, the highest concentration of microplastics was found in the intestine as 9.2 items/g (Li et al. 2018).

In addition to ingestion, adherence plays a significant role in microplastics uptake. Microplastics can adhere to external surfaces such as the foot and mantle, organs not directly involved in digestion. This implies an alternative pathway for contamination, expanding the potential impact of microplastics beyond just ingestion. These physiological disruptions undermine the survival and functioning of mussels in their ecosystem causing intestinal obstruction by preventing breakdown and assimilation of consumed materials, histological alterations inducing lysosomal instability, decreased feeding response affecting reproductive activity, and causing hemocyte mortality and inducing oxidative imbalance (Wright et al. 2013; Canesi et al. 2015; Paul-Pont et al. 2016). Previous research by Von Moos et al. (2012) and

Wegner et al. (2012), demonstrates that the exposure of mussels to polystyrene microbeads (0.1 g/L) resulted in decreased mussel filtering activity, whereas high-density polyethylene particles with 2.5 g/L concentration provoked significant inflammatory reactions.

These polymers act as carriers of toxic chemicals and contaminants including additives in plastic production and pollutants like polycyclic aromatic hydrocarbons (PAHs). When ingested, these chemicals accumulate in mussel tissues, exacerbating cellular and systemic damage. Studies revealed that exposure to microplastics-associated PAHs such as pyrene results in their transfer and accumulation in mussel's tissues, causing biochemical imbalances (Avio et al. 2017; Li et al. 2019). Among ingested plastic polymers, polyethylene is the most frequently detected type across all species, reflecting its widespread global use (Digka et al. 2018). An added complication is the biofouling agents like invertebrates and microorganisms that colonize the surface of microplastics. These biofouled microplastics become vectors for invasive species and pathogens and engage in long-distance ecological risks associated with the dispersion of invasive species and diseases into marine habitats.

The byssus, a thread like protein filament, which helps the mussels to adhere themselves to underwater surfaces (Fig. 7.2), also interacts with microplastics (Waite 2017). Byssal threads and adhesive plaques that allow mussels to attach to materials like rocks, concrete, and plastic can inadvertently trap microplastics (Lin et al. 2007). This interaction highlights another non-ingestive pathway for microplastics contamination. The strength of mussel attachment depends equally on byssus integrity and the mussel's muscular endurance collectively called as "catch phenomenon" and contamination in this structure impacts its ability to form stable aggregations. Investigations validated that microplastics subjected to biofouling act as reservoirs for the spread of pathogens and facilitate the long distance movement of alien species (Andrady 2011).

Fig. 7.2 Green mussels *Perna viridis*, attached to a transport acrylic board, using the byssus. (*Mussel biology: from the byssus to ecology and physiology, including microplastics ingestion and deep-sea adaptations* © 2021 by Koji Inoue et al. is licensed under *CC BY 4.0*)

From an ecological point of view, mussel beds are essential for the ecosystem (Engel et al. 2017). Mussels are considered as a crucial source of top-grade protein with their amino acid score of 107 (compared to other animal meats and eggs which have a score of 100) (Venugopal and Gopakumar 2017). In contrast, microplastics contamination endangers their ecological function by reducing filtering efficiency of mussels and transfer of chemical pollutants and pathogens through food web possessing risks to higher trophic levels through bioaccumulation including humans, as mussels are a significant seafood source (Li et al. 2018).

7.5 Ocean Acidification

In recent years, ocean acidification has been recognized as a significant climate change-related threat. The ocean is a natural sink for carbon dioxide (CO_2) and absorbs nearly 25% of greenhouse gas emissions (Le Quéré et al. 2009). The rise in anthropogenic CO_2 emissions alters ocean chemistry. As seawater absorbs excess CO_2, its pH decreases, making it more acidic. This acidification reduces the availability of calcium carbonate, a crucial component for mussel shell formation (Gazioğlu and Okutan 2016). Hydrogen ion concentration of seawater has decreased by 0.1 units since the industrial revolution and it is predicted to decline by a further 0.4 units conclusively over a 100 year period and such a large change will significantly impact on marine biodiversity (Caldeira and Wickett 2003).

Lowering pH disrupts physiological and metabolic processes in marine organisms, particularly by interfering with intercellular transport mechanisms (Pörtner et al. 2004). It poses a significant threat to ecologically and economically important molluscan species. Acidic conditions can impair shell growth, making mussels more vulnerable to predation and environmental stressors. It also reduces carbonate ion concentrations, essential for calcification (Gazeau et al. 2007). Melzner et al. (2011) suggested that molluscs and other calcifying organisms are more vulnerable in their larval stages, thus making population-level effects worse over time. Further research has shown that mussels exposed to more acidic conditions exhibit thinner and weaker shells, reduced byssal fiber performance causing decreased adhesive strength, and vulnerability to various environmental stress and habitat loss highlighting the impact of ocean acidification on their survival (O'Donnell et al. 2013; Fitzer et al. 2015a, b). Ocean acidification-induced reduction in species numbers and population densities would also lead to habitat reorganization, alterations in trophic levels, depletion of marine resources for human society, and employment losses for people employed in shelled mollusc-related sectors (Newell 2004; Gazeau et al. 2013).

Some species show resilience to fluctuating pH levels, but rapid and unprecedented changes due to anthropogenic CO_2 emissions pose severe challenges. *Mytilidae* mussels are considered bioindicators of ocean change due to their sessile nature, filter-feeding habits, and wide geographic distribution. As examples (Duarte 2014) indicated that *Mytilus chilensis* could tolerate 4 °C of thermal increase for up to 2 months but was impaired in shell growth with pH levels of 0.2–0.3 units. They suggest that whereas the mussel can tolerate a certain level of thermal stress, the combination of ocean acidification and temperature rise provides strong pressure against mussels. *Mytilid* mussels are crucial ecosystem engineers, providing habitat for other species and influencing ecosystem functions (Gutierrez et al. 2003). Their ability to adhere to bare rock makes them competitive dominants on rocky shores worldwide (Thomsen et al. 2017). Ocean acidification can weaken the growth, shell formation and antipredator responses in these organisms making them more vulnerable. Physiological responses to acidification, such as reduced calcification rates and altered energy budgets, further demonstrate their susceptibility and worth as bioindicators of changing marine environments (Beniash et al. 2010).

7.6 Conclusions

Mussels act as foundation species, providing habitat and food for many marine organisms. Changes in mussel populations can be indicative of larger shifts in biodiversity. As climate changes, the distribution and abundance of mussel populations can signal a cascading effect in the ecosystem, including the migration of species and alterations in community structures. Constant monitoring of mussel populations and their response to environmental stressors can form a good starting point to detect and predict trends with regard to climate changes. In the long run it can make it possible to understand gradual and cumulative impacts of climate changes on marine and fresh water habitats. Consequently, it will help to preserve the ecological balance and maintain the health of the ecosystems.

Declaration of conflict of interest There are no conflicts of interest to declare.

Ethics approval This work does not contain any experimental studies with live animals.

Declaration statement The confidentiality and anonymity of the data collected were assured to all participants.

Data availability statement Further inquiries can be directed to the corresponding author.

Funding No funding is received for this work.

References

Albayrak S, Albayrak G, Çelikkol MT, Gazdağlı A, Yörük E, Çağlar S (2019) The effects of polycyclic aromatic hydrocarbon compounds on antioxidant enzyme activities in *Mytilus galloprovincialis* Lamarck, 1819 from Istanbul Strait. Mar Pollut Bull 149:110662

Alkan N, Alkan A, Demirak A, Bahloul M (2020) Metals/metalloid in marine sediments, bioaccumulating in macroalgae and a mussel. Soil Sediment Contam Int J 29(5):569–594

Amiard JC, Amiard-Triquet C, Charbonnier L, Mesnil A, Rainbow PS, Wang WX (2008) Bioaccessibility of essential and non-essential metals in commercial shellfish from Western Europe and Asia. Food Chem Toxicol 46(6):2010–2022

Andrady AL (2011) Microplastics in the marine environment. Mar Pollut Bull 62(8):1596–1605

Atasaral Ş, Khan U, Terzi Y, Seyhan K (2020) Mussel: a potential pollution indicator in the aquatic ecosystem and effect of climate change. Int J Environ Geoinformatics 7(3):300–304

Avio CG, Gorbi S, Milan M, Benedetti M, Fattorini D, d'Errico G, Pauletto M, Bargelloni L, Regoli F (2015) Pollutants bioavailability and toxicological risk from microplastics to marine mussels. Environ Pollut 198:211–222

Avio CG, Gorbi S, Regoli F (2017) Plastics and microplastics in the oceans: from emerging pollutants to emerged threat. Mar Environ Res 128:2–11

Barre A, Briand JF, Vaccher V, Briant N, Briand JM, Dormoy B et al (2024) A comparative biomonitoring study of trace metals and organic compounds bioaccumulation in marine biofilms and caged mussels along the French Mediterranean coast. Environ Pollut 363:125239

Beniash E, Ivanina A, Lieb NS, Kurochkin I, Sokolova IM (2010) Elevated level of carbon dioxide affects metabolism and shell formation in oysters *Crassostrea virginica*. Mar Ecol Prog Ser 419:95–108

Beyer J, Green NW, Brooks S, Allan IJ, Ruus A, Gomes T, Brate ILN, Schoyen M, Schøyen M (2017) Blue mussels (*Mytilus edulis* spp.) as sentinel organisms in coastal pollution monitoring: a review. Mar Environ Res 130:338–365

Caldeira K, Wickett ME (2003) Anthropogenic carbon and ocean pH. Nature 425(6956):365–365

Canesi L, Ciacci C, Bergami E, Monopoli MP, Dawson KA, Papa S, Canonico B, Corsi I (2015) Evidence for immunomodulation and apoptotic processes induced by cationic polystyrene nanoparticles in the hemocytes of the marine bivalve *Mytilus*. Mar Environ Res 111:34–40

Crinnion WJ (2000) Environmental medicine, part three: long-term effects of chronic low-dose mercury exposure. Altern Med Rev 5:209–223

Digka N, Tsangaris C, Torre M, Anastasopoulou A, Zeri C (2018) Microplastics in mussels and fish from the Northern Ionian Sea. Mar Pollut Bull 135:30–40

Duarte CM (2014) Global change and the future ocean: a grand challenge for marine sciences. Front Mar Sci 1:63

Duarte CM, Wu J, Xiao X, Bruhn A, Krause-Jensen D (2017) Can Seaweed Farming Play a Role in Climate Change Mitigation and Adaptation? Front Mar Sci https://doi.org/10.3389/fmars.2017.00100

Eisler R (1994) A review of arsenic hazards to plants and animals with emphasis on fishery and wildlife resources. US National Biological Survey, pp 185–259

Engel FG, Alegria J, Andriana R, Donadi S, Gusmao JB, van Leeuwe MA, Matthiessen B, Eriksson BK (2017) Mussel beds are biological power stations on intertidal fats. Estuar Coast Shelf Sci 191:21–27

Fabbri E, Valbonesi P, Franzellitti S (2008) HSP expression in bivalves. Invertebr Surviv J 5:135–161

Fernández B, Campillo JA, Martínez-Gómez C, Benedicto J (2010) Antioxidant responses in gills of mussel (Mytilus galloprovincialis) as biomarkers of environmental stress along the Spanish Mediterranean coast. Aquat Toxicol 99(2):186–197

Fitzer SC, Vittert L, Bowman A, Kamenos NA, Phoenix VR, Cusack M (2015a) Ocean acidification and temperature increase impact mussel shell shape and thickness: problematic for protection? Ecol Evol 5(21):4875–4884

Fitzer SC, Zhu W, Tanner KE, Phoenix VR, Kamenos NA, Cusack M (2015b) Ocean acidification alters the material properties of *Mytilus edulis* shells. J R Soc Interface 12(103):20141227

Fuentes A, Ferna'ndez-Segovia I, Escriche I, Serra JA (2009) Comparison of physico-chemical parameters and composition of mussels (*Mytilus galloprovincialis* Lmk.) from different Spanish origins. Food Chem 112:295–302

Gamain P, Gonzalez P, Cachot J, Pardon P, Tapie N, Gourves PY, Budzinski H, Morin B (2016) Combined effects of pollutants and salinity on embryo-larval development of the Pacific oyster. Mar Environ Res 113:31–38

Garrabou J et al (2009) Heatwave-induced mortality of *Mytilus galloprovincialis* in the Mediterranean. Glob Chang Biol 15(7):1625–1635

Gazeau F, Quiblier C, Jansen JM, Gattuso JP, Middelburg JJ, Heip CH (2007) Impact of elevated CO_2 on shellfish calcification. Geophys Res Lett 34(7):L07603

Gazeau F, Parker LM, Comeau S, Gattuso JP, O'Connor WA, Martin S, Portner H-O, Ross PM (2013) Impacts of ocean acidification on marine shelled molluscs. Mar Biol 160:2207–2245

Gazioğlu C, Okutan V (2016) Underwater noise pollution at the strait of Istanbul (Bosphorus). Int J Environ Geoinformatics 3(3):26–39

Gillis PL, McInnis R, Salerno J, de Solla SR, Servos MR, Leonard EM (2017) Freshwater mussels in an urban watershed: impacts of anthropogenic inputs and habitat alterations on populations. Sci Total Environ 574:671–679

Goldberg ED (1975) Synthetic organohalides in the sea. Proc R Soc Lond Ser B Biol Sci 189(1096):277–289

Gutiérrez JL, Jones CG, Strayer DL, Iribarne OO (2003) Mollusks as ecosystem engineers: the role of shell production in aquatic habitats. Oikos 101:79–90

Häder PD, Banaszak AT, Villafañe VE, Narvarte MA, González RA, Helbling EW (2020) Anthropogenic pollution of aquatic ecosystems: emerging problems with global implications. Sci Total Environ 713:136586

Hellou J, King T, Willis DE (2000) Seasonal and geographical distribution of PAHs in mussels, *Mytilus edulis*, collected from an urban harbour. Polycycl Aromat Compd 20(1–4):21–38

Helmuth B, Yamane L, Lalwani S, Matzelle A, Tockstein A, Gao N (2011) Hidden signals of climate change in intertidal ecosystems: what (not) to expect when you are expecting. J Exp Mar Biol Ecol 400(1–2):191–199

Inoue K, Onitsuka Y, Koito T (2021a) Mussel biology: from the byssus to ecology and physiology, including microplastic ingestion and deep-sea adaptations. Fish Sci 87(6):761–771

Inoue K, Yoshioka Y, Tanaka H, Kinjo A, Sassa M, Ueda I, Shinzato C, Toyoda A, Itoh T (2021b) Genomics and transcriptomics of the green mussel explain the durability of its byssus. Sci Rep 11:5992

Jansen JM, Pronker AE, Kube S, Sokolowski A, Sola JC, Marquiegui MA, Schiedek D, Bonga SW, Wolowicz M, Hummel H (2007) Geo graphic and seasonal patterns and limits on the adaptive response to temperature of European *Mytilus* spp. and *Macoma balthica* populations. Oecologia 154:23–34

Jones SJ, Lima FP, Wethey DS (2010) Rising environmental temperatures and biogeography: poleward range contraction of the blue mussel, *Mytilus edulis* L., in the western Atlantic. J Biogeogr 37:2243–2259

Le Quéré C, Raupach MR, Canadell JG, Marland G, Bopp L, Ciais P et al (2009) Trends in the sources and sinks of carbon dioxide. Nat Geosci 2(12):831–836

Li J, Green C, Reynolds A, Shi H, Rotchell JM (2018) Microplastics in mussels sampled from coastal waters and supermarkets in the United Kingdom. Environ Pollut 241:35–44

Li J, Lusher AL, Rotchell JM, Deudero S, Turra A, Bråte ILN, Sun C, Hossain MS, Li Q, Kolandhasamy P, Shi H (2019) Using mussel as a global bioindicator of coastal microplastic pollution. Environ Pollut 244:522–533

Lin Q, Gourdon D, Sun C, Holten-Andersen N, Anderson TH, Waite JH, Israelachvili JN (2007) Adhesion mechanisms of the mussel foot proteins mfp-1 and mfp-3. Proc Natl Acad Sci 104(10):3782–3786

Melwani AR, Gregorio D, Jin Y, Stephenson M, Ichikawa G, Siegel E, Crane D, Lauenstein G, Davis JA (2014) Mussel watch update: long-term trends in selected contaminants from coastal California, 1977–2010. Mar Pollut Bull 81(2):291–302

Melzner F, Stange P, Trübenbach K, Thomsen J, Casties I, Panknin U, Gorb SN, Gutowska MA (2011) Food supply and seawater pCO_2 impact calcification and internal shell dissolution in the blue mussel *Mytilus edulis*. PLoS One 6(9):e24223

Newell RI (2004) Ecosystem influences of natural and cultivated populations of suspension-feeding bivalve molluscs: a review. J Shellfish Res 23(1):51–62

O'Donnell MJ, George MN, Carrington E (2013) Mussel byssus attachment weakened by ocean acidification. Nat Clim Chang 3(6):587–590
Otchere FA (2019) A 50-year review on heavy metal pollution in the environment: bivalves as biomonitors. Afr J Environ Sci Technol 13(6):220–227
Özden Ö, Ulusoy Ş, Erkan N (2010) Study on the behaviour of the trace metal and macro minerals in *Mytilus galloprovincialis* as a bioindicator species: the case of Marmara Sea, Turkey. J Verbr Lebensm 5:407–412
Parmesan C (2007) Influences of species, latitudes and methodologies on estimates of phenological response to global warming. Glob Chang Biol 13:1–13
Paul-Pont I, Lacroix C, Fernández CG, Hégaret H, Lambert C, Le Goïc N, Frere L, Cassone AL, Sussarellu R, Fabioux C, Guyomarch J, Albentosa M, Huvet A, Soudant P (2016) Exposure of marine mussels *Mytilus* spp. to polystyrene microplastics: toxicity and influence on fluoranthene bioaccumulation. Environ Pollut 216:724–737
Petes LE, Menge BA, Harris AL (2008) Intertidal mussels exhibit energetic trade-offs between reproduction and stress resistance. Ecol Monogr 78(3):387- 402
Ponder WF, Lindberg DR (2008) Phylogeny and Evolution of the Mollusca. University of California Press
Pörtner HO, Langenbuch M, Reipschläger A (2004) Biological impact of elevated ocean CO_2 concentrations: lessons from animal physiology and earth history. J Oceanogr 60:705–718
Ritchie J (1927) Report on the prevention of the growth of mussels in submarine shafts and tunnels at West bank Electricity Station, Portobello. Trans R Scott Soc Arts 19:1–20
Rodland DL et al (2006) A clockwork mollusc: Ultradian rhythms in bivalve activity revealed by photography. J Exp Mar Biol Ecol 334(2): 316-323
Romero-Freire A, Lassoued J, Silva E, Calvo S, Pérez FF, Bejaoui N, Babarro JMF, Cobelo-García A (2020) Trace metal accumulation in the commercial mussel *M. galloprovincialis* under future climate change scenarios. Mar Chem 224:103840
Ruppert EE, Fox RS, Barnes RD (2004) Invertebrate zoology. Thompson Learning, Inc, Belmont
Seed R, Suchanek TH (1992) Population and community ecology of *Mytilus*. In: Gosling E (ed) The Mussel *Mytilus*: ecology, physiology, genetics and culture. Elsevier Science, Amsterdam, pp 87–169
Sivaperumal P, Sankar TV, Nair PV (2007) Heavy metal concentrations in fish, shellfish and fish products from internal markets of India vis-a-vis international standards. Food Chem 102(3):612–620
Somero GN (2005) Linking biogeography to physiology: evolutionary and acclimatory adjustments of thermal limits. Front Zool 2:1–9
Stachowicz JJ, Fried H, Osman RW, Whitlatch RB (2002) Biodiversity, invasion resistance, and marine ecosystem function: reconciling pattern and process. Ecology 83(9):2575–2590
Stankovic S, Jovic M (2012) Health risks of heavy metals in the mediterranean mussels as seafood. Environ Chem Lett 10:119–130
Strayer DL (2008) Freshwater mussel ecology: a multifactor approach to distribution and abundance, vol 1. University of California Press
Thomsen J, Stapp LS, Haynert K, Schade H, Danelli M, Lannig G, Wegner KM, Melzner F (2017) Naturally acidified habitat selects for ocean acidification–tolerant mussels. Sci Adv 3(4):e1602411
Tunnicliffe V, Davies KT, Butterfield DA, Embley RW, Rose JM, Chadwick WW Jr (2009) Survival of mussels in extremely acidic waters on a submarine volcano. Nat Geosci 2(5):344–348
Ülker D, Bayırhan İ, Burak S (2020) Assessment and comparison of commonly used eutrophication indexes. Turk J Water Sci Manag 4(1):4–30
Van Cauwenberghe L, Janssen CR (2014) Microplastics in bivalves cultured for human consumption. Environ Pollut 193:65–70
Van Cauwenberghe L, Claessens M, Vandegehuchte MB, Janssen CR (2015) Microplastics are taken up by mussels (*Mytilus edulis*) and lugworms (*Arenicola marina*) living in natural habitats. Environ Pollut 199:10–17

Venugopal V, Gopakumar K (2017) Shellfish: nutritive value, health benefits, and consumer safety. Compr Rev Food Sci Food Saf 16:1219–1242

Verbruggen H et al (2020) The interaction of temperature and hypoxia on *Mytilus edulis* physiology. Mar Environ Res 162:105–118

Viarengo A, Canesi L (1991) Mussels as biological indicators of pollution. Aquaculture 94(2–3):225–243

Von Moos N, Burkhardt-Holm P, Köhler A (2012) Uptake and effects of microplastics on cells and tissue of the blue mussel *Mytilus edulis* L. after an experimental exposure. Environ Sci Technol 46(20):11327–11335

Waite JH (2017) Mussel adhesion–essential footwork. J Exp Biol 220(4):517–530

Wegner A, Besseling E, Foekema EM, Kamermans P, Koelmans AA (2012) Effects of nanopolystyrene on the feeding behavior of the blue mussel (Mytilus edulis L.). Environ Toxicol Chem 31(11):2490–2497

Widmeyer JR, Crozier ED, Moore MM, Jurgensen A, Bendell-Young LI (2004) Role of Leptothrix discophora in mediating metal uptake in the filter-feeding bivalve *Mytilus trossulus* (edulis). Environ Sci Technol 38(3):769–774

Wright SL, Thompson RC, Galloway TS (2013) The physical impacts of microplastics on marine organisms: a review. Environ Pollut 178:483–492

Zeng Z, Chen S, Ma Y, Yin X, Wang X, Zhang S, Zhang J, Wu X, Li Y, Dong D, Xiao N (2017) Chemical compositions of mussels and clams from the Tangyin and Yonaguni Knoll IV hydrothermal fields in the southwestern Okinawa Trough. Ore Geol Rev 87:172–191

Zippay ML, Helmuth B (2012) Effects of temperature change on mussel, Mytilus. Integr Zool 7(3):312–327

Chapter 8
Navigating the Shifting Shorelines: Climate Change Effects on Intertidal Fauna and Their Adaptations

Sebaj H. Gori, Deep D. Dudiya, Pranav J. Pandya, Ketankumar J. Yogi, and Subhash I. Parmar

Abstract The present review context investigates how climate change affects intertidal ecosystems, which serve as critical transitional zones between land and sea. Rising CO_2 levels have accelerated ocean warming and acidification, placing extreme stress on this dynamic ecosystem. Key stressors include sea level rise, rising temperatures, changed rainfall patterns, stronger storms, and eutrophication, which are compounded by human activities such as coastal development. Case studies from Gujarat, India, reveal significant coastline retreats of up to 831.4 metres over 40 years. Intertidal ecosystem and fauna therein face habitat loss, physiological stress, shell thinning, and disturbed food webs, although certain species show resistance through zone shifting, physiological adaptations, and altered burrowing behaviours. The present investigation reveals that understanding these complex interactions is crucial for environmental managers and policymakers to develop effective conservation strategies, including the establishment of marine-protected areas and community engagement initiatives.

Keywords Intertidal ecosystem · Climate change impacts · Coastal erosion · Marine biodiversity · Ecological adaptation

S. H. Gori · D. D. Dudiya · S. I. Parmar
Department of Earth and Environmental Science, Krantiguru Shyamji Krishna Verma Kachchh University, Bhuj, Gujarat, India

Department of Zoology, R. R. Lalan College, Bhuj, Kachchh, Gujarat, India

P. J. Pandya (✉)
Department of Zoology, R. R. Lalan College, Bhuj, Kachchh, Gujarat, India

K. J. Yogi
Coastal and Marine Ecology Division, Gujarat Institute of Desert Ecology, Bhuj, Gujarat, India

B. P. Barathan et al. (eds.), *Navigating Climate Change: Impacts on Biodiversity and Ecosystem Resilience*, Sustainable Development and Biodiversity 42,
https://doi.org/10.1007/978-981-95-0409-1_8

8.1 Introduction

In recent decades, aquatic ecosystems have encountered increasing challenges due to environmental and anthropogenic interruptions. One of the key factors of these changes is the excessive release of atmospheric CO_2. This release has contributed significantly to global climatic shifts. As atmospheric greenhouse gas concentrations continue to rise, it is anticipated that by 2100, surface water temperatures will increase by 2–4 °C, while pH levels could decrease by up to 0.4 units (IPCC et al. 2014; IPCC 2019). These changes, collectively known as ocean acidification and ocean warming, serve as key indicators of climate change. Furthermore, the present Anthropocene—an era defined by human activity—has a serious impact on the planet's ecosystems and biota. In this context, flora and fauna in the intertidal zone respond more quickly than higher trophic levels (Mohan and Swathi 2020), and this rapid response can lead to a disruption in the food chain, causing deficiencies (Johnson et al. 2011).

Among the most prominent environmental factors affecting aquatic life is rise in temperature, which directly impacts the physiological energetics of aquatic ectotherms (Huey and Kingsolver 1989; Hochachka and Somero 2002). Overall, rising air and ocean temperatures are anticipated to affect the distribution of coastal species and influence temperature-dependent ecological processes, such as primary production. For instance, corals are highly sensitive to temperature changes, and elevated sea-surface temperatures can trigger coral bleaching events, mass coral mortality, and shifts in the composition of reef communities (Hoegh-Guldberg et al. 2007; Hughes et al. 2007; Mohan and Swathi 2020). Tropical invertebrates are particularly vulnerable to these temperature increases due to their stenothermic nature, which limits their thermal tolerance and places them close to their optimal temperature limits (Deutsch et al. 2008). As temperatures rise, these organisms' ability to cope with environmental changes diminishes, making them more susceptible to additional stressors, including pollutants. Rising temperatures have been shown to alter the sensitivity of animals to toxins, increasing their vulnerability to environmental stress (Coppola et al. 2017; Andrade et al. 2019; Baag et al. 2021). These climate change–contaminant interactions may, in turn, impact the bioaccumulation of two priority contaminant classes: the fat-soluble persistent organic pollutants (POPs), such as polychlorinated biphenyls (PCBs), and the protein-binding methylmercury (MeHg) (Alava et al. 2017). Moreover, global environmental drivers such as CO_2 concentrations, changes in mixing regimes, and shifts in biogeochemical cycles are affecting the bioavailability of pollutants in aquatic ecosystems, which in turn can modify their toxicity potential. These changes have intense implications for both marine and freshwater ecosystems, which provide vital services such as food production, coastal protection, carbon storage, and biodiversity maintenance (Elmqvist et al. 2012). As climate change continues to drive shifts in oceanic physical and chemical properties, marine ecosystems are experiencing significant disruptions that threaten their stability and function (Poloczanska et al. 2016; Hillebrand et al. 2018). These changing shifts, especially in temperature, can negatively impact marine carbon sequestration by reducing the ocean's capacity to absorb CO_2. Also, by decreasing fish stock and accelerating the breakdown of organic matter, ultimately weakening the ocean's role as a carbon sink (Li et al. 2024). Moreover, freshwater ecosystems, often located near human settlements, are also under increasing anthropogenic pressure. These ecosystems are not only vulnerable to temperature changes and altered chemical properties but are also facing direct impacts from pollution and land-use changes (Davies and Vethaak 2012).

Climate change, through ocean warming and acidification, is creating significant challenges for aquatic ecosystems, threatening the essential services they provide and threatening biodiversity. The compounded effects of rising temperatures, altered pollutant dynamics, and environmental stressors highlight the urgency of addressing climate change to protect both marine and freshwater habitats. The continued degradation of these ecosystems, especially marine systems, can have disrupting effects at multiple scales, underscoring the need for more effective global efforts to mitigate anthropogenic impacts on the environment. The pelagic regions are prone to various disturbances caused due to climate change effects, while the effect amplifies in the intertidal areas. Coastal and estuarine ecosystems serve as safeguard transitional zones against natural hazards and environmental disturbances, such as floods, cyclones, tidal surges, and high wave actions.

8.2 Intertidal Zone

The tidal flux makes the intertidal zone an extreme ecosystem because it suffers frequent drastic modifications due to the dynamic, stressful environment and fixed, frequently sessile animals, making them sensitive to quick variations in temperature, sea level, and habitat conditions, leading to probable ecological collapse. Being a transition zone between terrestrial and marine system, the zone receives two states: one at low tide, when it is exposed to the air, and another at high tide, when it is immersed in seawater; therefore, it is referred as an ecotone, a shift from sea to land, where many species spend their lives in such transitional habitats. The zone is totally buried by the tide once or twice daily (Newell 1979; Denny 1988; Walter and Karen 2007). The intertidal regions present a rich diversity of organisms that vary in size and structure. These intertidal forms like crabs, molluscs, polychaetes, and other phyla must adapt to changes in physical and chemical properties. These organisms include sessile and mobile forms occupying this zone based on their salinity and temperature tolerance. These organisms receive their nutrients and energy provided regularly by the tides (Newell 1979; Denny 1988). To avoid desiccation, water is accessible at each high tide; however, the physical properties of the water vary by season. According to Newell (1979), salinity decreases during the monsoon season owing to the influx of freshwater, while increasing during the summer season due to evaporation and salt deposition. To survive in the harsh conditions of the intertidal zone, these species must locate a protected environment that offers refuge and food. These organisms typically create their own communities along the zone's elevation gradient. Some species like ghost crabs *Ocypode* sp. reside higher up on the beach, closer to the high tide line, while other sensitive forms like echinoderms and anthozoans live further down, closer to the low tide line. Anything that lives in the intertidal zone must be able to resist fluctuations in moisture, temperature, and salinity, as well as severe waves. At rocky shorelines, tidal pools can form in holes, cracks, or crevices where seawater accumulates when the tide recedes. These pools can shelter organisms that cannot typically withstand low tide conditions, such as sea stars, shrimp, and fish. Sandy beaches contain sediments in which creatures bury themselves to keep cool and wet during low tides. Where a species lives in the intertidal zone is determined by its tolerance for both underwater and above-water circumstances. The existence of predators and species competing for the same area and food has an influence on where an organism will be located. The rising temperature can alter the tolerant limit of these sensitive organisms and act as a stressor.

The intertidal zone itself is often strongly sub-zoned and further separated into three categories: upper intertidal, middle intertidal, and lower intertidal zone. The upper intertidal zone is only submerged when the tide is high and the whole area is covered by water. The middle intertidal zone is submerged and exposed for equal periods of time. The lower intertidal zone is obtainable during low tide only and contains the most biodiversity of the three zones because it provides better environment for species that cannot relive prolonged air exposure (Newell 1979; Denny 1988). This ecosystem also protects against erosion and stops storm waves from reaching dwellings along the shoreline (Walter and Karen 2007). While the intertidal zone can be harsh, it is home to a diverse range of organisms and serves as a vital feeding ground for both resident and migrating birds. The erosion or loss of these intertidal zones can also impact the feeding ground of waders and mudflat loving shore birds. Numerous niches in the intertidal zone have arisen as a result of variations in marine currents and environmental conditions (Snively 1978), resulting in a vast diversity of organisms in this region. The intertidal zone has a diverse range of habitats, including mangroves, seagrass beds, coral reefs, tidal wetlands, estuaries, barrier islands, peat swamps, and others.

8.2.1 Intertidal Ecosystems and Its Services

Intertidal ecosystems are an important part of marine ecosystems and important transition between land and the sea. This ecosystem's major role is to serve as a buffer, shielding coastal regions from the effects of erosion and storms by absorbing wave energy and stabilizing shorelines (McLean et al. 2001). These natural sources separately and along with entire system perform many ecological functions. These zones are the places of complex interactions encircling environments such as sandy beaches, muddy areas, and rock intertidal areas which may also support vegetation communities like mangrove and saltmarsh halophytes. Intertidal ecosystems provide a wide array of important ecosystem services like coastal protection, biodiversity, carbon sequestration, services via mangrove forest, and critical habitat for a wide range of both marine (Nagelkerken et al. 2008; Yates et al. 2014) and terrestrial biodiversity (Rog and Cook 2017). The vegetated intertidal ecosystems like mangroves and saltmarsh have conglomerate of both marine and terrestrial influence, thus developing physiological adaptations and root modifications (Tomlinson 2016). Intertidal crabs of these mangrove forest utilize these root complexes as their niche to overcome harsh environment.

8.3 The Impact of Global Climate Change on Marine Ecosystems and Biodiversity

Global climate change is characterized by rising temperatures on land and sea surfaces, ocean acidification, the extension of marine dead zones, and more frequent and intense extreme weather experiences (see Sect. 8.3.1.1 for global change in marine zones). These changes are having detrimental effects on marine ecosystems, disrupting their integrity and functions. As a result, the distribution and physiological processes of marine species are altered, leading to profound impacts on marine biodiversity (Ho and Tang 2020).

In recent decades, the accelerating pace of climate change has elevated increasing concerns about its catastrophic effects on biodiversity. Increases in global temperatures resulted in the melting of polar ice caps, leading sea levels to rise, further exacerbating the vulnerability of coastal ecosystems. The warming of the oceans has led to shifts in the distribution of marine species, as they are forced to adapt, migrate, or face the risk of extinction. These changes disrupt the delicate balance of marine food webs (Lindsey and Dahlman 2024).

8.3.1 Intertidal and Shoreline Changes in Climate Change Context

The intertidal zone, in particular, has been significantly affected by these rapid climate shifts. This habitat, which lies between high and low tide marks, experiences dramatic swings in temperature and desiccation as tides rise and fall. Additionally, intertidal organisms are subjected to high wave actions with altering velocities depending on beach morphology. These issues are worsened by global warming and alterations in wind patterns, which produce stronger thermal forcing and larger ocean waves in some regions (Hoegh-Guldberg and Bruno 2010; Denny and Gaylord 2010; Young et al. 2011; Reguero et al. 2019).

Rising temperatures, sea level rise, and extreme weather events have already had an impact on many coastal ecosystems across the world. 'Dead zones' along the US coastline have become a major concern for scientific communities. Nutrient runoff from land and unusual rainfall pattern indirectly depletes oxygen levels in the water, leading to widespread die-offs of marine life, including corals, shellfish, fish, and aquatic plants. As the ocean warms and heavier rains increase nutrient-laden runoff, these dead zones are supposed to expand further (NOAA 2022), many dead zones have been reported in continental seas, such as the Gulf of Mexico, Kattegat, Baltic, East China Sea, and Black Sea, all of which are major fishery areas (Diaz and Rosenberg 2008). The Arabian Sea hosts the largest dead zone in the world, 'The Gulf of Oman'—63,700 Sq. miles and the second largest being 'The Gulf of Mexico'—6000 Sq. miles (Shivakrishna et al. 2020). In India, dead zones are located between 100 and 400 in depth of Bay of Bengal and are continuously expanding, covering an area of around 60,000 square kilometres (Jayaraman 2016). For example, Masselink et al. (2020) found that 28% of the 3700 km of coastline in England and Wales is undergoing erosion of more than 10 cm annually, and that a significant percentage of the UK and Ireland's coastline is now undergoing erosion (17% in the UK and 19.9% in Ireland). There have been cases of shoreline shifting, coastal erosion, and sea water ingression, especially in coastal countries and states.

8.3.1.1 Shifting Shoreline and Intertidal: Coastal Erosion and Sea Water Ingress

At present, coastal erosion is common, it is worldwide. As per the study carried out by Luijendijk et al. (2018), in the State of word beaches, revealed that 24% of the world's sandy beaches are frequently eroding at a pace more than 0.5 m/yr over the study period (between years 1984 and 2016), while 27% are accreting. About 16% (18%) of sandy beaches are undergoing erosion (accretion) rates exceeding 1 m/yr. Sea level rise has also accelerated the process of *transgression* (a large landward movement of coastline caused by relative long-term sea level rise and geological history). On, the contrary, human developmental activities like land reclamation and accretion by countries like China, UAE, and Singapore has added to the total gain of coastal area of 3663 sq. km (Luijendijk et al. 2018). This may favour a chance of inducing an intertidal succession over a long run on these artificial beaches which needs a more elaborative study. The underlying causes and processes for coastal erosion and shoreline shift are shown in Fig. 8.1.

Gujarat is a maritime state of western India having two Gulf systems as well as a huge coastline. Some of its coastal belts and intertidal beaches are showing vulnerability to erosion at a high pace. A study by Patel et al. (2022) reveals significant coastal erosion along parts of the Gujarat coastline. Shoreline analysis over 40 years (1978–2018) shows a substantial landward migration, ranging from 113.9 metres to 831.4 metres. This alarming trend translates to an annual erosion rate of 39.76 metres, posing a serious threat to coastal communities and potentially leading to future disasters. Coastal erosion and vulnerable intertidal areas are presented in Fig. 8.2.

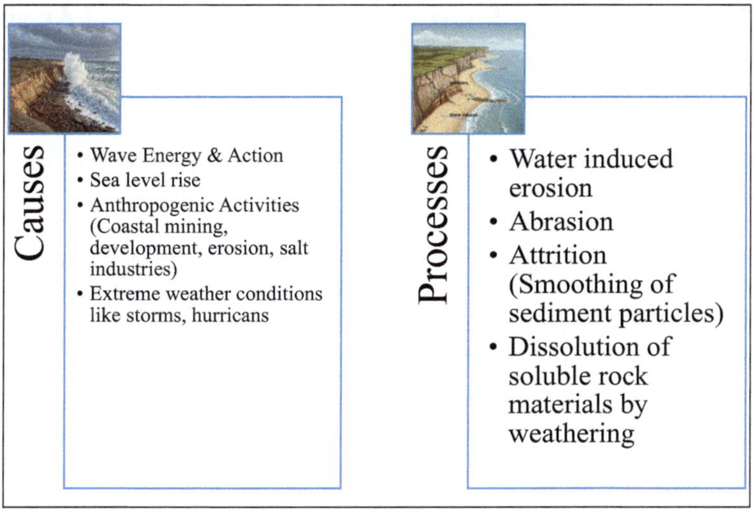

Fig. 8.1 Generalized view of causes and processes responsible for coastal intertidal erosion

Fig. 8.2 Coastal vulnerability assessment using GIS for Gujarat coastline. (Image modified from Kankara et al. 2018)

8.4 Factors Affecting Shoreline Shifts and Intertidal Areas

Climate change is a complex process which is induced by/triggers a cascade of events. These events are interrelated, driving shoreline alterations like rising temperatures raise sea levels, intensify onshore winds and storminess, and alter rainfall patterns, all of which contribute to erosion and floods. Anthropogenic activities alter the natural sediment equilibrium, intensifying these impacts. Ocean acidification damages coral reefs, lowering coastal protection, whilst eutrophication degrades coastal wetlands, further exposing the beach. These variables reinforce one other, hastening coastal changes. Moreover, these factors related to these patterns and directly affecting intertidal fauna and its ecosystem are as under.

8.4.1 Sea Level Rise

Rising sea levels narrow the land–sea interface, increasing erosion, and altering beach dynamics. Sediment loss shrinks beach size and forces the coastline to retreat, leading to habitat destruction and making the environment less stable over time. As a result, species may show more unpredictable orientation and migration patterns. Complex behaviours are replaced by simpler ones, and foraging becomes more opportunistic. These changes increase the risk of local extinctions for supra-littoral invertebrates with limited range and poor dispersal ability.

8.4.2 Temperature Increase

Higher temperature variations between surface and deeper sediments affect moisture levels in the upper littoral zone. Persistent warming of sea surface temperatures changes the properties of sand and water, impacting local hydrodynamics. Species face altered thermal limits and higher energy demands for osmoregulation. Activity rhythms, mating, and recruitment timing may shift, while burrowing behaviour is affected by changes in oxygen and moisture levels. Shallower burrowing increases competition for space. Reduced fitness, disrupted reproduction, and range shifts further strain populations, leading to greater competition and lowered survival rates.

8.4.3 Onshore Winds and Storminess

Warmer air and water result in more frequent and intense storms and stronger onshore winds. These forces expand swash zones, narrow beaches, and speed up erosion. Larger waves can strand intertidal species, causing mass deaths, while supralittoral species lose their habitats. Stress avoidance behaviours like escape and altered migration timing disturb zonation patterns. Increased competition further impacts species. Habitat loss, combined with coastal squeeze, leads to population declines and local extinctions.

8.4.4 Anthropogenic Activities

Rise in human population and its demands have induced pressure on natural resources and their utilizations. Coastlines are the places of human congregations, settlements, recreation, and industrializations due to its easy access and other benefits. These activities have created pressure on intertidal ecosystem and its biota. Macrobenthic infauna are widely recognized as valuable indicators of environmental health in marine ecosystems. Their sedentary nature allows them to reflect long-term environmental conditions, making them sensitive to disturbances like pollution and habitat degradation (Bertocci et al. 2019; Warwick 1993; Wetzel et al. 2015). This large human population and their activities have directly and indirectly affected the biodiversity of coastal ecosystems worldwide, thereby creating anthropogenic stress on the biota. Research has proven lower intertidal diversity in tourism impacted areas in many coastal reaches like western Indian coastline (Pandya et al. 2021), Kimmeridge Bay (Pinn and Rodgers 2005), and Mediterranean islands (Andolina et al. 2021). The strategies driven by scientific insights can allow sustainable use of these coastal zones.

8.4.5 Altered Rainfall Pattern

More freshwater entering the ocean lowers salinity, while rising groundwater levels increase swash distances and worsen beach erosion. Lower salinity challenges species' osmoregulation, affecting vital processes. Habitat loss occurs for species dependent on beaches, and unpredictable environmental cues disrupt tidal and seasonal rhythms. Species distributions can shift along and across the shore. Erosion and floods can harm habitat recovery, and invasive species from freshwater bodies outcompete native ones. This results in failed reproduction, recruitment issues, and changes in community structure, with native species often being replaced by invasive ones.

8.4.6 Ocean Acidification

Lower pH and reduced carbonate levels due to climate change affect the chemistry of surf zones. Species with calcified body parts may face slower calcification and weaker calcium metabolism (Zhao et al. 2020). Acidification impairs burrowing in species with less robust shells, makes them more selective about substrates, disrupted foraging, and reduces survival in sensitive larval stages. These changes increase mortality rates and eventually lead to biodiversity loss. Reproductive success in species with pelagic larvae is affected, and overall fitness declines.

8.4.7 *Eutrophication*

Nutrient overload from climate change and human actions accelerates eutrophication, leading to frequent harmful algal blooms (red tides) and excessive growth of drifting algae (green and golden tides). These blooms alter water and sediment properties, causing mass deaths in suspension feeders and disrupting the food web. Stranded algae reduce intertidal species diversity and impact subsurface deposit feeders. Specialized foraging is impaired, with opportunistic species becoming more dominant. Foraging and habitat selection decline, while chronic metabolic changes affect survival and growth. This contributes to local extinctions and significant biodiversity loss.

8.5 Effects on Marine Life and Food Web Dynamics

Changes in ocean conditions have direct influences on the life history characteristics of marine organisms, including invertebrates and seabirds. For poikilotherms (organisms whose body temperatures vary with the environment), developmental rates increase exponentially with rising temperatures. This has significant ecological consequences, such as altered larval dispersal patterns, population connectivity, local adaptation, and speciation (O'Connor et al. 2007). Studies have reported ecosystem food web fragility and declining ecosystem services providing when specifically addressing the consequences of climate change (Serpetti et al. 2017). For example, acidification has a direct detrimental effect on shell-forming species such as bivalves by lowering $CaCO_3$ saturation state in the sea water. This effect of climate change decreases the maximum bivalve's biomass (Chapman et al. 2020). Reduced developmental times may also cause phenological mismatches, where the timing of larval development does not correspond to the availability of suitable food sources. Such mismatches are similar to those observed in terrestrial ecosystems, where changes in climate lead to disruptions in the synchrony between species' life cycles and food availability (Durant et al. 2007; Walther et al. 2002). These disruptions, when combined with changing patterns of primary productivity and metabolic rates, can substantially modify the dynamics of ocean food webs, from coastal to open-ocean ecosystems.

8.5.1 Ocean Acidification and Its Impact on Intertidal Marine Organisms

Another key effect of climate change on marine ecosystems is ocean acidification. Comparatively higher atmospheric CO_2 concentrations are being absorbed by the oceans, leading to a decrease in ocean pH. This process is expected to adversely affect marine organisms that depend on $CaCO_3$ for their shells and skeletons, such as corals, molluscs, and echinoderms. Studies have concluded that ocean acidification reduces the growth and recruitment of coralline algae, which are vital for the structure of coral reefs, while non-calcareous algae are less affected (Connell and Russell 2010). The effects of ocean acidification vary across different marine species and life stages. Corals, molluscs, and echinoderms have been shown to experience reduced survival rates as ocean pH decreases, whereas crustaceans appear to be less sensitive (Kroeker et al. 2013). In particular, molluscs and sea urchins are exclusively vulnerable during their early life stages. Mollusc larvae, in particular, exhibit a significant decline in the development and metabolism when exposed to acidic conditions, which could lead to a decline in their populations over time (Chambers et al. 2014).

The impacts of climate change on marine ecosystems are diverse and far-reaching. Rising temperatures, ocean acidification, and changing weather patterns disrupt marine biodiversity at multiple levels, from individual species to entire ecosystems. The intertidal zone, in particular, is highly vulnerable to these shifts, while fundamental changes in ocean food webs could have cascading effects throughout marine environments. The survival and adaptation of marine species will depend on their ability to cope with rapidly changing conditions, but many species are already showing signs of stress. Urgent action is needed to mitigate these effects, reduce greenhouse gas emissions, and protect marine biodiversity before these shifts become irreversible.

8.6 Climate Change and Challenges to Intertidal Fauna

Climate change has brought a global perspective to changes in sandy beach ecosystems (Schoeman et al. 2014; McLachlan and Defeo 2018). These ecosystems face growing risks due to the combined effects of various climate change factors (Fig. 8.3). Coralline species, echinoderms, and soft-bodied intertidal fauna like anemones and ascidians are impacted by high erosion and high churning of sediments with wave action. Rising temperatures and sea level increases are expected to have serious ecological impacts, including heightened erosion along sandy shores. Global warming is also linked to more frequent and severe storms, which will accelerate beach erosion, push shorelines back, and damage dunes and vegetation. In some areas, fluctuating rainfall patterns are expected to increase freshwater flow into marine environments, leading to lower salinity. For instance, salinity levels

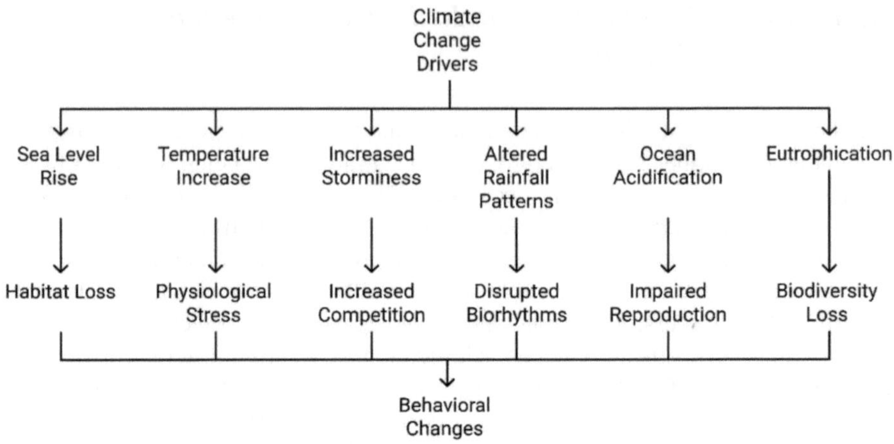

Fig. 8.3 Key climate change drivers and their cascading impacts on sandy beach ecosystems

have dropped in the Baltic Sea due to melting ice caps (Elliott et al. 2015). Warmer air and water have been associated with stronger and more frequent onshore winds (Bakun 1990; Sydeman et al. 2020). These winds generate greater wave energy, shifting beaches toward more erosive conditions (Short 1999). Given these challenges, it is essential to understand how sandy beach ecosystems will adapt to these long-term environmental changes.

Intertidal fauna in general is susceptible to a large number of challenges and environmental factors which directly and indirectly affect intertidal faunal diversity, abundance, and their behavioural responses. Major challenges and their effects on intertidal fauna are presented in Table 8.1.

8.6.1 Vulnerabilities to Intertidal Fauna Due to Climate Change

Over and above the aforesaid mentioned stressors and challenges (as per Table 8.1), intertidal fauna have adapted to these challenges and cope well with the environmental factors. However, this can be up to an optimal level after which the animals might remain under stress, may have to spend their more energy, or even may find difficult to survive. Climate change has actively participated in elevating the challenges to marine ecosystem and especially intertidal zones.

Table 8.1 Effect on intertidal fauna community

Challenge	Effect on intertidal fauna
Wave action	Dislodge animal from its original zone (Target: Echinoderms, gastropods, bivalves), create physical injury to shelled and unshelled animals like gastropods, sabellarid tubes, and sponges
Changing salinities	Can affect certain stenohaline species by disrupting their physiological mechanisms, e.g. Subtidal forms and echinoderms
Long tide-driven exposures	Limited time for feeding and activities for scavengers, grazers, and burrowers. Example: brachyuran crabs, intertidal gastropods, and hermit crabs
Temperature and desiccation stress	Can dry out the animal with long exposure and considerable temperature fluctuation. Impacts soft-bodied animals, especially gastropods, sponges, and coralline species. Dry outs intertidal sandy and muddy areas
Human influence and activities	Burrowing activities and foraging of many intertidal macrofauna are highly impacted. Trampling in intertidal zones can have negative effects on the organisms that live there, including crabs, mussels, and algae

8.6.1.1 Shifting/Decrease of Intertidal Area

Climate-associated sea level rise, water ingression, severe rainfall, etc., are the root causes of shifting shorelines. In intertidal areas, macrofauna shows species-specific zonation preferences (Pandya and Vachhrajani 2010). Intertidal shifts may create negative pressure on intertidal animals and may alter their preferred zones, thereby removing the species from the area. Reduction in intertidal zonations can also decrease the habitat complexities which are root cause of rich biodiversity.

8.6.1.2 Erratic Weather and Long Exposures

The rise global temperature has been a concern not just for humans, but also for global biota. Intertidal animals, in specific, are more prone to high temperatures due to their living between temperature variations during tidal cycles. The high temperatures during the low tides can expose these forms for longer time and may have to spend more energy as a resilience measure (Fig. 8.4).

8.6.1.3 Human Trampling

Changes in climate have increased the human reaches to coastal areas for recreation and reaching favourable weather. Studies from Spain have seen a strong relation between warming of climate and increase in the number of tourists in beach destination (Toubes et al. 2020). Beach tourism can have contrary effects on benthic intertidal organisms by controlling and interfering the species composition, low abundance, diversity and reproductive output, altering their size structure, and more. When impacting ecosystem engineers, they can potentially alter the functioning and structure of ecosystems. For example, disturbing mussels in rocky zones can alter

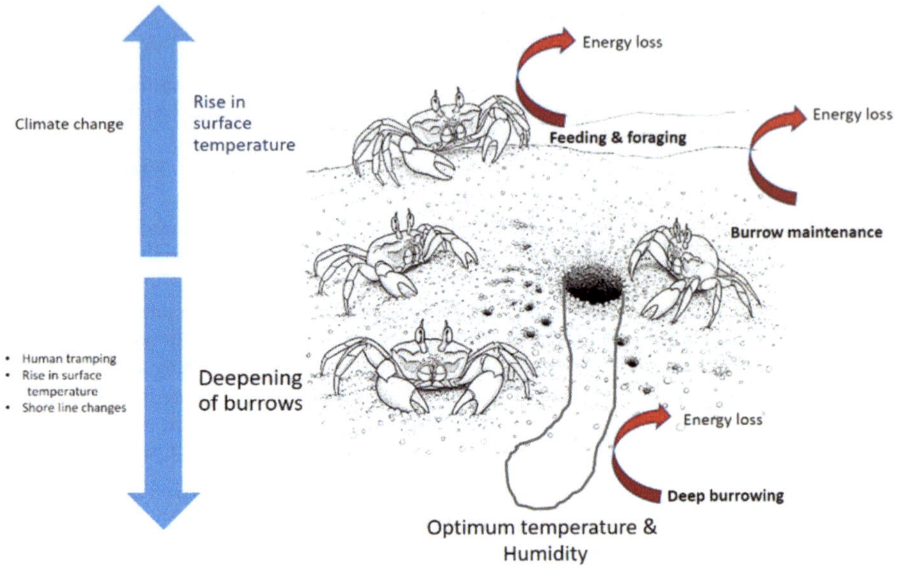

Fig. 8.4 Schematic presentation of the effect of climate change-induced change in intertidal macrofauna (Red arrows shows activities where the species have to spend extra energy due to altered conditions)

the habitat complexity, visits to coral reefs can create accountable damage to reef or other biogenic structures which takes long time to recover.

8.6.1.4 Shell Thinning

As already discussed, ocean acidification has been proved to compromise available calcium carbonate and there by thinning of shelled faunal groups. The same applies to intertidal forms where shell thinning can be more detrimental to shelled animals like gastropods, mussels, and tube-building polychaetes. Owing to the above trampling effect and exposure time, these habitats become more vulnerable to damage.

8.7 The Intertidal Adaptations to Climate Change

Intertidal organisms have developed a range of strategies to cope with the challenges of their environment, such as fluctuating temperatures, exposure to air during low tide, and fluctuations in ocean chemistry. They are vulnerable to climate change and the additional stress that comes with warming climate. As a result, they have exposed some of the profligate responses to global warming. These adaptations include behavioural changes, physical modifications like thicker shells, and physiological adjustments to handle stress. The degree of adaptation varies depending on the organism's position within the intertidal zone, with those living higher up typically better equipped to withstand extreme conditions. However, rising sea levels and ocean acidification poses severe challenges to their survival.

8.7.1 Adjusting or Shifting to Preferred Zones

Many intertidal forms shift to their preferred zones or zones of tolerance to avoid harsh conditions. The Ghost crab *Ocypode* sp. moves to the upper zones and sand dunes for burrowing or may shift to the nearby sandy depositions in the nearby areas. The young ones not much tolerant, prefers the mid-intertidal zones. Many of the species refuge to the creek and enclosed areas to overcome high wave energies and actions. At a geographical scale, biogeographic distributions have also shifted away from warmer low latitude regions toward the cooler poles. This also helps in establishment of resilient invasive species to flourish and establish in favourable areas causing expansion in their geographical distribution, while the same may also cause clearing of more selective species.

8.7.2 Physiological Adaptations

Errant rainfall patterns and altering salinities pressurize stenohaline species and makes more suitable environments for animals preferring a wide range of salinity. The increase in production of Heat shock Proteins (HSPs) as a resilience against increasing climate can be an effective tool for these biota. Many animals, especially polychaetes, can switch to anaerobic metabolism during low oxygen availabilities.

8.7.3 Deep Burrowing

The trampling and surface temperature rise altogether can force the animal to burrow deep in order to find a suitable adobe. This is especially seen in burrowing crabs like *Austruca* sp., *Dotilla* sp., *Scopimera* sp., and many of the burrowing bivalves. However, deep burrowing comes at the expensive cost of high energy demands for digging and maintenance thereafter. This also takes away good time from their limited time window.

8.8 Conclusion

Marine ecosystem and intertidal habitats in specific are vulnerable to climate change and thereby induced human pressure. The range of challenges and stressors can alter the intertidal biodiversity, their abundance, community composition, and behaviour. The control on damage and of shore line deterioration can help the ecosystem to be more robust and firmer towards the climate shifts. Understanding the impact of climate change and alterations on coastal ecosystems is crucial for environmental managers, policy makers, and ecologists to meritoriously and sustainably manage future development. This includes evaluating biodiversity, resistance to disturbance, and the long-term recovery potential of an area. The most essential management measures, according to the experts, are the development of marine-protected zones, community engagement, and public awareness campaigns. This can reduce human influence directly, earmarking faunal-rich area, strict implementation of climate change policies as some of the conservation strategies for sustaining marine ecosystems. Also, understanding scientific basis of these complex faunal patterns help to understand their response to climate change.

Acknowledgements SHG and DDD are thankful to the Government of Gujarat, India, for providing a doctoral research fellowship under SHODH (ScHeme of Developing High-quality Research) (Reference No. 2023015010 and 202301504). SIP is thankful to UGC (University Grants Commission) for providing Junior Research fellowship (Award no. 231610067756) for the doctoral research. Also, the authors are thankful to all reviewers for their suggestions to make this work publishable.

Declaration of Conflict of Interest The authors declare no conflict of interest.

Ethical Approval This work does not contain any experimental study with live animals.

Data Availability Statement Not applicable.

Funding Not applicable.

References

Alava JJ, Cheung WW, Ross PS, Sumaila UR (2017) Climate change–contaminant interactions in marine food webs: toward a conceptual framework. Glob Chang Biol 23(10):3984–4001

Andolina C, Signa G, Tomasello A, Mazzola A (2021) Environmental effects of tourism and its seasonality on Mediterranean islands: the contribution of the Interreg MED BLUEISLANDS project to build up an approach towards sustainable tourism. Environ Dev Sustain 23:8601–8612. https://doi.org/10.1007/s10668-020-00984-8

Andrade M, De Marchi L, Soares AM, Rocha RJ, Figueira E, Freitas R (2019) Are the effects induced by increased temperature enhanced in Mytilus galloprovincialis submitted to air exposure? Sci Total Environ 647:431–440

Baag S, Mahapatra S, Mandal S (2021) An integrated and multibiomarker approach to delineate oxidative stress status of Bellamya bengalensis under the interactions of elevated temperature and chlorpyrifos contamination. Chemosphere 264:128512

Bakun A (1990) Global climate change and intensification of coastal ocean upwelling. Science 247:198–201

Bertocci AD, Anno L, Musco C, Gambi V, Saggiomo M, Cannavacciuolo M, LoMartire A, Passarelli G, Zazoa R, Danovaro. (2019) Multiple human pressures in coastal habitats: variation of meiofaunal assemblages associated with sewage discharge in a post-industrial area. Sci Total Environ 655:1218–1231. https://doi.org/10.1016/j.scitotenv.2018.11.121

Chambers RC, Candelmo AC, Habeck EA, Poach ME, Wieczorek D, Cooper KR, Phelan BA (2014) Effects of elevated CO_2 in the early life stages of summer flounder, Paralichthys dentatus, and potential consequences of ocean acidification. Biogeosciences 11(6):1613–1626

Chapman EJ, Byron CJ, Lasley-Rasher R, Lipsky C, Stevens JR, Peters R (2020) Effects of climate change on coastal ecosystem food webs: implications for aquaculture. Mar Environ Res 162:105103

Connell SD, Russell BD (2010) The direct effects of increasing CO2 and temperature on non-calcifying organisms: increasing the potential for phase shifts in kelp forests. Proc R Soc B Biol Sci 277(1686):1409–1415

Coppola F, Almeida Â, Henriques B, Soares AM, Figueira E, Pereira E, Freitas R (2017) Biochemical impacts of Hg in *Mytilus galloprovincialis* under present and predicted warming scenarios. Sci Total Environ 601:1129–1138

Davies IM, Vethaak D (2012) Integrated marine environmental monitoring of chemicals and their effects. ICES Cooperative Research Reports (CRR)

Denny MW (1988) Biology and the mechanics of the wave-swept environment. Princeton University Press, Paris. 344 pp

Denny MW, Gaylord B (2010) Marine ecomechanics. Annu Rev Mar Sci 2:89–114

Deutsch CA, Tewksbury JJ, Huey RB, Sheldon KS, Ghalambor CK, Haak DC, Martin PR (2008) Impacts of climate warming on terrestrial ectotherms across latitude. Proc Natl Acad Sci USA 105:6668–6672

Diaz RJ, Rosenberg R (2008) Spreading dead zones and consequences for marine ecosystems. Science 321(5891):926–929

Durant JM, Hjermann DØ, Ottersen G, Stenseth NC (2007) Climate and the match or mismatch between predator requirements and resource availability. Clim Res 33(3):271–283

Elliott M, Borja Á, McQuatters-Gollop A, Mazik K, Birchenough S, Andersen JH, Painting S, Peck M (2015) Force majeure: will climate change affect our ability to attain good environmental status for marine biodiversity? Mar Pollut Bull 95:7–27

Elmqvist T, Maltby E, Barker T, Mortimer M, Perrings C, Aronson J, Salles JM (2012) Biodiversity, ecosystems and ecosystem services. In: The economics of ecosystems and biodiversity: ecological and economic foundations. Routledge, pp 41–111

Hillebrand H, Brey T, Gutt J, Hagen W, Metfies K, Meyer B, Lewandowska A (2018) Climate change: warming impacts on marine biodiversity. In: Handbook on marine environment protection: science, impacts and sustainable management. Springer, pp 353–373

Ho K, Tang D (2020) Implications of climate change on marine biodiversity. Glob J Agric Soil Sci 1:1–6

Hochachka PW, Somero GN (2002) Biochemical adaptation: mechanism and process in physiological evolution. Oxford University Press, New York, p 466

Hoegh-Guldberg O, Bruno JF (2010) The impact of climate change on the world's marine ecosystems. Science 328(5985):1523–1528

Hoegh-Guldberg O, Mumby PJ, Hooten AJ, Steneck RS, Greenfield P, Gomez E, Harwell CD, Sale PF, Edwards AJ, Caldeira K, Knowlton N, Eakin CM, Lglesias-Prieto R, Muthiga N, Bradbury RH, Dubi A, Hatziolos ME (2007) Coral reefs under rapid climate change and ocean acidification. Science 318(5857):1737–1742

Huey RB, Kingsolver JG (1989) Evolution of thermal sensitivity of ectotherm performance. Trends Ecol Evol 4(5):131–135

Hughes TP, Rodrigues MJ, Bellwood DR, Ceccarelli D, Hoegh-Guldberg O, McCook L, Moltschaniwskyj N, Pratchett MS, Steneck RS, Willis B (2007) Phase shifts, herbivory, and the resilience of coral reefs to climate change. Curr Biol 17(4):360–365.Epub 2007 Feb 8

IPCC (2019) Summary for policymakers. In: Pörtner HO, Masson-Delmotte DCRV, Zhai P, Tignor M, Poloczanska E, Mintenbeck K, Alegría A, Nicolai M, Okem A, Petzold J, Rama B, Weyer NM (eds) IPCC's special report on the ocean and cryosphere in a changing climate

IPCC, Pachauri RK, Meyer LA (2014) Climate change 2014: synthesis report. In: Core Writing Team (ed) Contribution of working groups I, II and III to the fifth assessment report of the intergovernmental panel on climate change. IPCC, Geneva. 151 pp

Jayaraman KS (2016) Dead zone found in Bay of Bengal. Nature Asia

Johnson CR, Banks SC, Barrett NS, Cazassus F, Dunstan PK, Edgar GJ, Gardner C, Haddon M, Helidoniotis F, Hill KL, Holbrook NJ, Hosie GW, Last PR, Ling SD, Melbourne-Thomas J, Miller K, Pecl GT, Richardson AJ, Ridgway KR, Rintoul SR, Ritz DA, Ross DJ, Sanderson JC, Shepherd SA, Slotwinski A, Swadling KM, Nyan T (2011) Climate change cascades: shifts in oceanography, species' ranges and subtidal marine community dynamics in eastern Tasmania. J Exp Mar Biol Ecol 400:17–32

Kankara RS, Ramana Murthy MV, Rajeevan M (2018) National assessment of shoreline changes along Indian coast. Status report of 26 years (1990–2016). NCCR Publication

Kroeker KJ, Kordas RL, Crim R, Hendriks IE, Ramajo L, Singh GS, Duarte CM, Gattuso JP (2013) Impacts of ocean acidification on marine organisms: quantifying sensitivities and interaction with warming. Glob Chang Biol 19(6):1884–1896. https://doi.org/10.1111/gcb.12179

Li H, Jing L, Tong H, Liu Y (2024) Impact of high temperature heat waves on ocean carbon sinks: based on literature analysis perspective. J Sea Res 198:102487. https://doi.org/10.1016/j.seares.2024.102487.davies

Lindsey R, Dahlman L (2024) Climate change: global temperature. Available at https://www.climate.gov/news-features/understanding-climate/climate-change-global-temperature. Accessed 18 Mar 2025

Luijendijk A, Hagenaars G, Ranasinghe R (2018) The state of the world's beaches. Sci Rep 8:6641. https://doi.org/10.1038/s41598-018-24630-6

Masselink G, Russell P, Rennie A, Brooks S (2020) Impacts of climate change on coastal geomorphology and coastal erosion relevant to the coastal and marine environment around the UK. MCCIP Sci Rev:158–189

McLachlan A, Defeo O (2018) The ecology of sandy shores. Academic Press, London

McLean RF, A. Tsyban V. Burkett JO. Codignott DL. Forbes N. Mimura, RJ, Beamish V. Ittekkot. 2001. Coastal Zones and Marine Ecosystems. In Climate Change 2001: Impacts, Adaptation, and Vulnerability. Contribution of Working Group II to the Third Assessment Report of the Intergovernmental Panel on Climate Change. Edited by James J. McCarthy Osvaldo F. Canziani, Neil A. Leary, David J. Dokken, and Kasey S. White, Cambridge University Press, Cambridge, UK, 343-379.

Mohan PM, Swathi V (2020) Intertidal biodiversity and their response to climatic variables, temperature and pH—what we know. Open J Mar Sci 10:203–217

Nagelkerken I, Blaber SJM, Bouillon S, Green P, Haywood M, Kirton LG, Meynecke J, Pawlik J, Penrose HM, Sasekumar A, Somerfield PJ (2008) The habitat function of mangroves for terrestrial and marine fauna: a review. Aquat Bot 89:155–185

Newell RC (1979) Biology of intertidal animals. Marine Ecological Surveys, Ltd, Kent

NOAA (2022) Climate change likely to worsen U.S. and global dead zones. Retrieved 16 Mar 2022

O'Connor MI, Bruno JF, Gaines SD, Halpern BS, Lester SE, Kinlan BP, Weiss JM (2007) Temperature control of larval dispersal and the implications for marine ecology, evolution, and conservation. Proc Natl Acad Sci 104(4):1266–1271

Pandya PJ, Vachhrajani KD (2010) Spatial distribution and substratum preference of the brachyuran crab, Macrophthalmus Depressus (Decapoda, Ocypodidae) along the lower estuarine

mudflat of Mahi River, Gujarat, India. Crustaceana 83:1055–1067. https://doi.org/10.1163/001121610X521235
Pandya P, Thakkar M, Goswami M (2021) Spatio-temporal comparison of intertidal macrofaunal communities along anthropogenically influenced Mandvi coast, Gulf of Kachchh, India. J Anim Divers 3(2):42–56
Patel K, Jain R, Kalubarme MH, Bhatt T (2022) Coastal erosion monitoring using multi-temporal remote sensing and sea surface temperature data in coastal districts of Gujarat state, India. Geol Ecol Landsc 8(2):194–207. https://doi.org/10.1080/24749508.2022.2127648
Pinn EH, Rodgers M (2005) The influence of visitors on intertidal biodiversity. J Mar Biol Assoc U K 85:263–268. https://doi.org/10.1017/S0025315405011148h
Poloczanska ES, Burrows MT, Brown CJ, García Molinos J, Halpern BS, Hoegh-Guldberg O, Kappel CV, Moore PJ, Rihardson AJ, Schoeman DS, Sydeman WJ (2016) Responses of marine organisms to climate change across oceans. Front Mar Sci 3:62
Reguero BG, Losada IJ, Mendez FJ (2019) A recent increase in global wave power as a consequence of oceanic warming. Nat Commun 10:205
Rog SM, Cook CN (2017) Strengthening governance for intertidal ecosystems requires a consistent definition of boundaries between land and sea. J Environ Manag 197:694–705. https://doi.org/10.1016/j.jenvman.2017.04.052
Schoeman DS, Schlacher TA, Defeo O (2014) Climate-change impacts on sandy beach biota: crossing a line in the sand. Glob Chang Biol 20:2383–2392
Serpetti N, Baudron AR, Burrows MT, Payne BL, Helaouet P, Fernandes PG, Heymans JJ (2017) Impact of ocean warming on sustainable fisheries management informs the ecosystem approach to fisheries. Sci Rep 7(1):13438
Shivakrishna A, Edla AP, Nathaniel TP (2020) Dead zones in the ocean. Pro Aqua Farm Marine Biol 3(2):180032
Short AD (1999) Handbook of beach and shoreface morphodynamics. John Wiley & Sons, Chichester. 379 pp
Snively G (1978) Exploring the seashore in British Columbia, Washington and Oregon. Gordon Soules Book Publishers, Inc
Sydeman WJ, García-Reyes M, Schoeman DS, Rykaczewski RR, Thompson SA, Tang KHD (2020) Implications of climate change on marine biodiversity. Glob J Agric Soil Sci 1(1):1–6
Tomlinson PB (2016) The botany of mangroves. Cambridge University Press
Toubes D, Araújo-Vila N, Fraiz-Brea JA (2020) Influence of weather on the behaviour of tourists in a beach destination. Atmos 11:121. https://doi.org/10.3390/atmos11010121
Walter HA, Karen L (2007) Dynamic aquaria, 3rd edn. Elsevier
Walther G-R, Post E, Convey P, Menzel A, Parmesan C, Trevor JC, Beebee J-MF, Hoegh-Guldberg O, Bairlein F (2002) Ecological responses to recent climate change. Nature 416(6879):389–395
Warwick RM (1993) Environmental impact studies on marine communities: pragmatical considerations. Austral J Ecol 18:63–80
Wetzel MA, Ohe PC, Von Der Manz W, Koop JHE, Wahrendorf D (2015) The ecological quality status of the Elbe estuary. A comparative approach on different benthic biotic indices applied to a highly modified estuary Ecol. Indicator 19:118–129. https://doi.org/10.1016/j.ecolind.2011.08.007
Yates KK, Rogers CS, Herlan JJ, Brooks GR, Smiley NA, Larson RA (2014) Diverse coral communities in mangrove habitats suggest a novel refuge from climate change. Biogeosciences 11:4321–4337. https://doi.org/10.5194/bg-11-4321
Young IR, Zieger S, Babanin AV (2011) Global trends in wind speed and wave height. Science 332(6028):451–455
Zhao X, Han Y, Chen B, Xia B, Qu K, Liu G (2020) CO2-driven ocean acidification weakens mussel shell defense capacity and induces global molecular compensatory responses. Chemosphere 243:125415. https://doi.org/10.1016/j.chemosphere.2019.125415

Chapter 9
Salinity Influence on Sediment Organic Carbon (SOC) Stock in Different Coastal Ecosystems of Southern Andaman Island: An Overlooked Aspect of Blue Carbon Research

Thanamegam Kaviarasan, P. Ragavan, R. Sivasankar, Venkatesan Shiva Shankar, and Rathinam Arthur James

Abstract Surface (30 cm) sediment organic carbon (SOC) stock estimated on the South Andaman Island showed the highest value in mangroves during the three seasons i.e., Pre-monsoon (PM), Southwest monsoon (SWM), Northeast monsoon (NEM) covered, compared to the other three distinct blue carbon ecosystems, namely, seagrass, tidal marsh, and seaweed. The annual mean SOC stock was the highest in mangroves (86.58 ± 26.40 Mg ha^{-1}). The highest SOC content and stock in all ecosystems were observed during the NEM followed by the SWM and pre-monsoon. ANOSIM revealed significant seasonal variation ($P < 0.02$) in SOC stock. The stock declined significantly during the pre-monsoon season, but increased slightly at the beginning of the SWM. Despite the contradictory response of SOC to salinity, the current study indicated that low salinity circumstances linked to NEM rains enhance SOC levels. Pairwise analysis suggested significant variations in SOC between the three seasons (<0.03) indicating that SOC stock in the top 30 cm is

T. Kaviarasan (✉)
Department of Marine Science, Bharathidasan University, Tiruchirappalli, India

Marine Litter and Microplastic Group, National Centre for Coastal Research, Chennai, India

P. Ragavan
MoEF & CC, Indira Paryavaran Bhawan, New Delhi, India

R. Sivasankar
Applied Research Center for Environment and Marine Studies, King Fahd University of Petroleum and Minerals, Dhahran, Saudi Arabia

V. S. Shankar
Department of Science, Mahatma Gandhi Government College, Mayabunder, North and Middle Andaman, India

R. A. James
Department of Marine Science, Bharathidasan University, Tiruchirappalli, India

B. P. Barathan et al. (eds.), *Navigating Climate Change: Impacts on Biodiversity and Ecosystem Resilience*, Sustainable Development and Biodiversity 42,
https://doi.org/10.1007/978-981-95-0409-1_9

inherently unstable on a seasonal scale and hence unlikely to act as a long-term carbon sink. Considering seasonal changes in SOC stocks, a detailed investigation is needed to understand the influence of salinity and other environmental parameters on the carbon storage potential, which would provide new insights into blue carbon dynamics.

Keywords Carbon sequestration · Mangrove · Salt marsh · Seagrass · Seaweeds · Tropical Islands

9.1 Introduction

Blue carbon, a term coined by UNEP in 2009, refers to the carbon captured by marine living organisms in the ecosystems such as mangroves, coral reefs, salt marshes, and seagrass (Choudhary et al. 2024). Notably, a substantial amount of blue carbon is stored in sediments, particularly in coastal wetlands (Chen and Lee 2022; Owers et al. 2022; Wang et al. 2023; Choudhary et al. 2024). Asia is a critical region for the global carbon budget, with a quarter of the world's top 10 largest national emitters of CO_2 (Thompson et al. 2016). Coastal ecosystems not only act as transition zones between land and ocean but also serve as effective carbon sequestration filters; this filtering process occurs through mechanisms such as carbon storage in sediments, carbon sequestration in the water column, and air–sea CO_2 exchange (Watanabe and Kuwae 2015; Wang et al. 2023; Choudhary et al. 2024). Wetlands play a vital role in landscape function, including carbon cycling. They offer significant terrestrial carbon stocks and sinks (Kaviarasan et al. 2019; Li et al. 2024; Nag et al. 2024; Tang et al. 2024). India boasts a rich biodiversity of mangroves (4975 km^2), saltmarsh (290 km^2), and seagrass (517 km^2) (Geevarghese et al. 2018; MoEFCC 2019; Viswanathan et al. 2020).

The foremost global priority in the twenty-first century is to minimize the emission of greenhouse gases including CO_2 and to mitigate climate change implications. In this context, increasing sediment carbon (C) stocks and protecting the C-rich sediments are crucial, as outlined by the Paris climate targets (Rumpel et al. 2018). In view of this, preserving and enhancing the carbon storage potential of blue carbon ecosystems (mangroves, seagrasses, and salt marshes) have become imperative due to their highest level of stock per unit area (Macreadie et al. 2019; Hilmi et al. 2021; Kuwae et al. 2022; Raw et al. 2023). However, the persistent anthropogenic disruptions taking place and the consequent climate change put the blue carbon ecosystems and their carbon storing capability under threat (Kida and Fujitake 2020; Hilmi et al. 2021; Howard et al. 2023; Choudhary et al. 2024; Dye et al. 2024). Sea-level rise associated with an increase in salinity and inundation of vegetated coastal habitats can modify the SOC stock through changes in nitrogen, phosphorus, and carbon mineralization (Stagg et al. 2018; Mazhar et al.2022; Tong et al.

2024). The impact of increasing salinity and inundation on the SOC of blue carbon habitats remains enigmatic due to limited knowledge about the mechanisms governing OC stabilization in these ecosystems (Xue et al. 2019; Kida and Fujitake 2020; Zou et al. 2023), particularly in India. The significance of this study lies in its provision of firsthand data on Soil Organic Carbon (SOC) across various ecosystems, thereby establishing a database that facilitates global comparisons for various stakeholders. This study hypothesizes that seasonal changes SOC occur in various coastal wetlands. Hence, this study aimed to estimate SOC in various coastal ecosystems and investigate its relationships with different environmental and water quality parameters.

9.2 Materials and Methods

Andaman and Nicobar Islands (Fig. 9.1) is located in the Bay of Bengal, within the geographical coordinates of 6–14°N latitude and 92–94°E longitude. This archipelago consists of 572 individual islands, spanning a total area of 8293 km^2 (Kaviarasan et al. 2018). Four coastal ecosystems were selected for this study: Mangrove (MG), Tidal Marsh (TM), Seaweeds (SW), and Seagrass (SG). Within each ecosystem, three sampling sites were identified. Triplicate samples were collected at each site during three different seasons: NEM-December 2016,

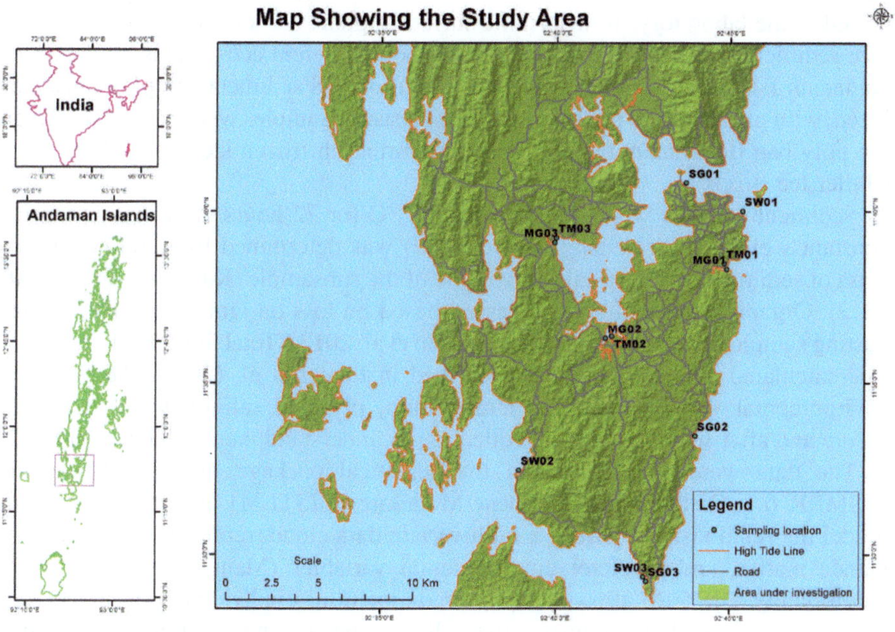

Fig. 9.1 Location of study area in South Andaman Islands, India

pre-monsoon (PM, March 2017), and SWM (September 2017). To analyze the dynamic fluctuations in SOC levels, these specific seasons were chosen as they exemplify distinct hydrological, climatic, and biological conditions. This selection facilitates a thorough investigation of the diverse factors affecting SOC variability in coastal wetland. The sampling stations were designated as follows: CC (Carbyn's Cove), SG (Sippighat), CH (Chouldhari), H (Haddo), BN (Burmanallah), CT (Chidiya Tapu), MP (Marina Park), and W (Wandoor).

The selection of sampling locations was driven by the study's objectives, which aimed to investigate the spatial and temporal variations of SOC in coastal wetlands. Collecting multiple cores (e.g., three or more) within a wetland is essential to accurately characterize soil carbon (C) pools (Bansal et al. 2023), as it allows for the capture of spatial variability and provides a more representative estimate of soil C stocks. So a random sampling approach was employed, covering the entire area and following a split-plot experimental design. Randomizing the sample collection process helps minimize bias, ensures representativeness, and captures the true variability of soil carbon pools within the wetland, leading to more reliable and accurate results (Howard et al. 2014).

During in situ measurements, temperature was recorded using a borosilicate glass mercury thermometer, while, sediment pH levels were measured each plot with a soil pH and moisture tester (Takemura) (Bhatta et al. 2015), and water salinity was determined using an ATAGO refractometer (Rahaman et al. 2013). Sediment cores were collected from each site using a clean PVC corer (4 cm diameter and 30 cm length). Following collection, pore water temperature, salinity, and pH were also measured. Collected sediments were stored in clean polythene bags and transported to the laboratory. To determine surface organic carbon stocks, triplicate surface sediment samples were collected from four different ecosystems on the South Andaman Island during three seasons. A total of 108 sediment samples were analyzed, with nine samples per ecosystem and season. Samples were packed in zipper-top poly bag (local manufacture) and transported on frozen ice packs (Thermocol Chiller Ice Box).

Sediment samples were oven-dried at 60 °C for 72 hours until they reached a constant weight. Bulk density (BD) (g/cm^3) was determined by dividing the dry mass of sediment by the initial wet volume of the subsample (Kauffman and Donato 2012). Organic matter (%) was assessed based on loss on ignition (LOI) by combusting samples overnight at 375 °C (Burden et al. 2013). Total SOC stock ($Mg\ ha^{-1}$) was calculated using the method described in Phang et al. (2015). Additionally, environmental variables such as temperature, pH, and salinity were measured, before and after sediment sample collection to understand their influence on SOC.

The data were analyzed using the statistical package PERMANOVA+ for PRIMER 6.1, Distance-Based Linear Modeling (DISTLM) used to analyze and model the relationship between a multivariate data cloud, represented by a resemblance matrix, and predictor environmental variables (Manokaran et al. 2013). Additionally, DISTLM allowed fitting predictor values individually or in specified sets, providing p-values for testing the null hypothesis of no relationship using permutation methods.

The analysis of similarities (ANOSIM) test was employed to assess differences in SOC between the three seasons across all the ecosystems, using a one-way layout (Manokaran et al. 2013). The test statistic, denoted by R, was calculated as:

$$R = (\mathrm{rB} - \mathrm{rW}) / (M2)$$

where

rB is the average rank similarity between different sites.
rW is the average rank similarity within sites.

$$M = n(n-1)$$

with n representing the total number of samples.

9.3 Results and Discussion

Figure 9.2 displays the results of environmental parameters measured at each site, including temperature, pH, and salinity. Temperature in mangrove environment: The observation shows that there are notable differences in temperature between various metrics across the three seasons. The PM season had the greatest mean temperatures (air, water, soil, and porewater). On the other hand, the NEM season showed comparatively consistent and lower temperatures. The mean temperatures throughout the SWM season ranged moderately. Compared to the NEM season, the SWM season's standard deviations were comparatively higher, suggesting that temperature fluctuation at this season. Overall, the SWT and PWT found to be more than the AT and ST. Similarly, Xu et al. (2025) found that surface soil temperatures along the wetland system are more similar to air temperatures in terms of yearly mean values and ranges. Soil temperature in coastal wetlands is influenced by the interaction between surface heat exchange and internal heat transfer. However, tidal flooding periodically disrupts this balance by altering the heat exchange between wetland sediments, the air, and tidal water. When flooding occurs, temperature differences between the surface water and wetland sediments facilitate heat transfer through conduction (Piccolo 2019; Cho et al. 2005; Kim and Cho 2009; Pivato et al. 2018). The normal temperature for seagrass growth in tropical waters ranges from 24 to 35 °C (Hutomo 1985).

Seawater quality can be inferred from variations in the water's pH, rainfall, land-based impacts, and oxidation processes are some of the elements that affect seawater's pH level. The pH results revealed seasonal fluctuations and significant variances, with PM having the highest sediment pH (7.56 ± 0.40), followed by SWM (6.6 ± 0.2) and NEM (6.733 ± 0.05). PM exhibited the highest surface water pH (7.96 ± 0.50) and pore water pH (7.933 ± 0.37). The pH levels in the seagrass habitat varied significantly, with PM having the greatest sediment pH (7.5 ± 0.36)

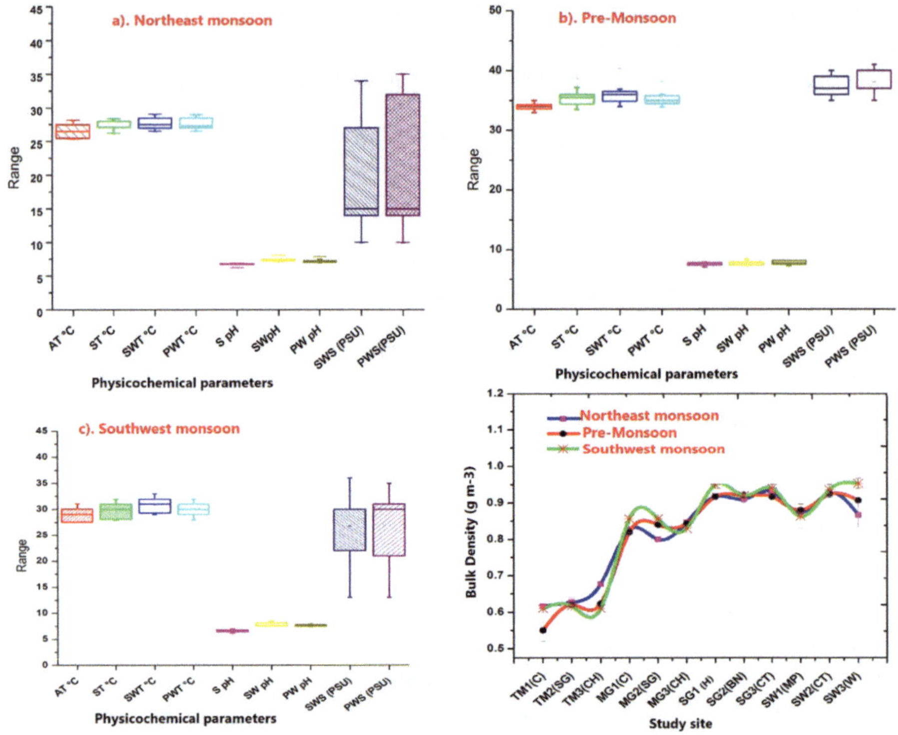

Fig. 9.2 (**a–c**) The results of physicochemical variables at three different seasons, (**d**) displays Bulk density of sediment in the study area

and surface water pH (7.9 ± 0.1). In contrast, SWM has the greatest pore water pH (7.73 ± 0.11). Different pH changes have been observed throughout the seaweed ecosystem. Notably, PM had the greatest sediment and pore water pH, whereas SWM had the highest surface water pH. These results suggest that the studied ecosystem (MG, SG, and SW) exhibit distinct differences in pH, which may be influenced by factors such as tidal patterns, sediment characteristics, and freshwater input. The majority of aquatic species prefer a pH range of 6.5–9.0; however, some can survive in water with a pH outside of this range (Fondriest Environmental Inc 2013). These results align with previous findings by Nobi et al. (2010) and Sreelekshmi et al. (2020), who reported similar pH values for mangrove areas in Sippighat and Carbyn's Cove. However, Dinesh et al. (2004) reported lower pH values (5.72 ± 0.43) in mangrove sediments after the monsoon, indicating a slight increase in pH over the past decade. Salinity levels were highest at PM, with surface water salinity reaching 38 ± 1.73 PSU and pore water salinity reaching 39.33 ± 3.21 PSU. Salinity levels were highest at PM, with both surface water and pore water salinity reaching approximately 38 PSU. Additionally, PM showed the highest salinity levels, with both surface and pore water salinity reaching around 37 PSU. Salinity in coastal wetlands is basically determined by flooding patterns, ocean tides, seawater salt content, shifts in sea level, river discharge, rainfall, and groundwater flow (Barendregt and Swarth 2013; Zedler and Callaway 2001). The effectiveness of carbon storage in soil depends on various physical and chemical factors, as well as the presence or absence of plant life (Chan-Keb et al. 2024; Zou et al. 2023). Rising salinity levels trigger a chain reaction, speeding up microbial decomposition, hindering plant growth, and ultimately decreasing carbon storage (Mitra et al. 2011; Zhao et al. 2018). The pH levels can significantly influence microbial activity, as most microorganisms thrive in neutral pH environments, thereby affecting the breakdown of SOC (Huang et al. 2018).

The BD of sediment samples ranged from 0.55 g cm^{-3} (TM1 during PM) to 0.95 g cm^{-3} (SW2 during SWM) (Fig. 9.2). Higher BD values could be attributed to greater sand content (Lovelock et al. 2014; Phang et al. 2015). Sediment bulk density varied between ecosystems and sampling sites but did not show steady differences between stations within the same ecosystem and season. Notably, SW and SG sediments exhibited higher BD values than MG and TM sediments. This finding is consistent with Lovelock et al. (2014) who reported higher BD values for mangrove sediments compared to salt marsh sediments. The average BD values for TM, MG, SG, and SW were 0.61, 0.82, 0.92, and 0.90 g cm^{-3}, respectively. These values are comparable to those reported by Chen et al. (2017). Comparisons with other studies revealed that the mangrove sediment in this study was higher than those reported for Indonesia, Chek Jawa Singapore, and Micronesia, but lower than those for the Ganges–Brahmaputra delta, Mozambique, and Madagascar (Donato et al. 2011; Jones et al. 2014; Sitoe et al. 2014; Phang et al. 2015). The seagrass BD in this study was lower than those reported for Port Curtis central Queensland and Abu Dhabi seagrass (Phang et al. 2015; Ricart et al. 2015; Campbell et al. 2014) but higher than those for Florida Bay and Shark Bay seagrass (Fourqurean et al. 2012). Additionally, the BD of South Andaman tidal marsh was higher than that reported for tidal marshes on the northeast coast of the United States (Drake et al. 2015). The organic matter percentage ranged from 1.05 ± 0.04 (SW2 in PM) to 10.41 ± 0.12 (MG2 in NEM) (Fig. 9.3). Mangrove sediments exhibited the highest OM percentage (4.4 ± 0.2% to 10.41 ± 0.12%), followed by tidal marsh (2.80 ± 0.11% to 6.84 ± 0.15%), seagrass (1.83 ± 0.01% to 4.02 ± 0.50), and seaweeds (1.05 ± 0.01% to 2.90 ± 0.05%). Seasonally, the highest OM percentages occurred during SWM, followed by NEM and PM. These findings are consistent with Kathiresan et al. (2014) who reported higher sediment carbon stocks during pre- and post-monsoon seasons in mangrove sediments of the Vellar–Coleroon estuary, India.

Similarly, Phang et al. (2015) recorded the highest OM in mangroves compared to other coastal wetlands in Chek Jawa, Singapore. Mangroves are also a source of sediment organic carbon for nearby seagrass ecosystems (Chen et al. 2017). In contrast, Campbell et al. (2014) reported lower organic content in seagrass beds from Abu Dhabi, United Arab Emirates, ranging from less than 0.05–2.44%. The higher OM percentage can be attributed to the sediment composition, with clay content typically having more OM than sand and silt (Kaviarasan et al. 2019). Mangrove sediments are generally dominated by clay rather than sand (Phang et al. 2015).

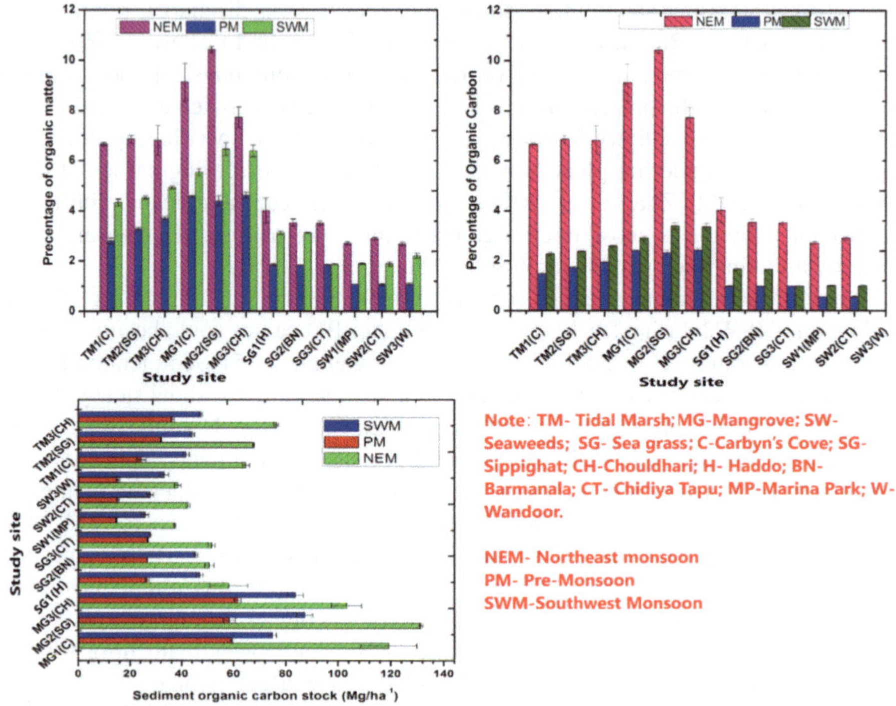

Fig. 9.3 Illustrations of results of percentage of OM, OC and SOCS

The BD and OM exhibit spatial variability across landscapes in various ecosystems, including wetlands, changes in soil texture, age, and depth, as well as variations in plant community composition, all contributing to this variation (Bruland and Richardson 2005). Various environmental factors, ultimately affecting soil BD, including tidal action, sediment delivery from rivers, and the impact of hurricanes and winter storms, can lead to sediment deposition and erosion (Meselhe et al. 2013; Nyman et al. 1990; Turner et al. 2006).

The SOC content and stock varied between ecosystems and seasons. Among the four ecosystems, mangroves had the highest annual mean SOC content and stock (3.47% and 86.58 ± 26.40 Mg ha^{-1}) followed by tidal marsh (2.59% and 48.37 ± 16.96 Mg ha^{-1}), seagrasses (1.45% and 40.12 ± 13.01 Mg ha^{-1}), and seaweed (1.03% and 28.05 ± 10.51 Mg ha^{-1}). The highest SOC content and stock in all ecosystems were observed during the NEM followed by the SWM and pre-monsoon. ANOSIM revealed significant seasonal variation ($P < 0.02$) in SOC stock (Fig. 9.4). Pairwise analysis indicated significant differences between the three seasons ($P < 0.03$), highlighting the seasonal instability of SOC stocks in the top 30 cm. Rainfall-dependent salinity fluctuations play a vital role in seasonal variations in blue carbon storage.

Our findings are consistent with Lima et al. (2020), as statistical analys is revealed that sediment characteristics specifically dry bulk density, grain size, and

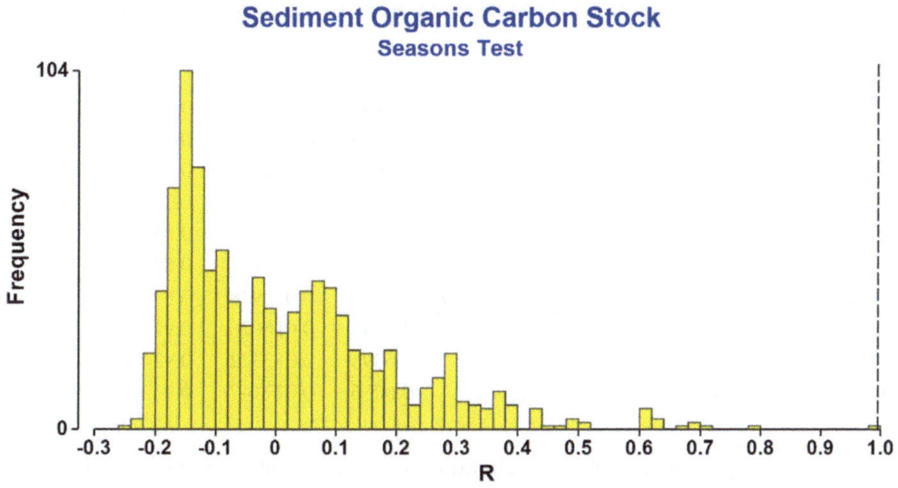

Fig. 9.4 ANOSIM showing significant variations in SOC stock between the seasons

pore water salinity showed the strongest correlations with carbon stocks Fluctuating salinity stimulates various biogeochemical processes. For example, increased salinity raises the concentration of terminal electron acceptors (e.g., iron, manganese, sulfate), which can accelerate microbial mineralization of organic matter and reduce carbon accumulation (Herbert et al. 2015). Moreover, except surface water salinity and pore-water salinity, none of the other parameters varied significantly across ecosystems or seasons in our study. In all four ecosystems, salinity was at the highest during the pre-monsoon season. It decreased during the SWM and reached the lowest during the North-East monsoon. The South-West Monsoon in the Andaman Islands lasts from May to September, whereas the North-East Monsoon occurs from November to December. The study area receives around 85% of its annual mean rainfall (3035 mm/year) between May and November during the SWM and early part of the NEM (Kumar et al. 2012).

Distance-based linear modeling (DISTLM) revealed a significant relationship between SOC content and sediment bulk density ($P < 0.02$) (Table 9.1). The total variability explained in this model was 95.2%. Soil bulk density explained about 41.9% of the total variability in SOC, followed by OM (14.8%) and atmospheric temperature (13.9%). Salinity was shown to be a significant factor in bulk density and sediment organic carbon, since increased salinity results in a change in the oxidation of sediment organic carbon. When organic carbon oxidation takes place, the organic content decreases whereas the bulk density increases, which could be the reason for the correlation with SOC. The results of this study corroborate earlier investigations on tidal wetlands worldwide demonstrating a negative relationship between SOC pools and salinity (Morrissey et al. 2014; Zhao et al. 2018). However, the response of sediment organic matter (SOM) in relation to SOC fluctuations is paradoxical. A thorough investigation on the SOC vis a vis sediment bulk density and organic matter is needed to have a comprehensive idea about the relationship

Table 9.1 Sequential test of DISTLM showing the role of environmental parameters in SOC content

Variable	R^2	SS (trace)	Pseudo-F	*P*	Prop.	Cumul.	Res.df
+BDS	0.41948	18.457	7.2259	0.0191	0.41948	0.41948	10
+AT	0.55859	6.1209	2.8364	0.1069	0.13911	0.55859	9
+OM	0.70655	6.51	4.0334	0.063	0.14795	0.70655	8
+SWpH	0.77333	2.9386	2.0625	0.1657	0.066787	0.77333	7
+OC	0.83668	2.7871	2.327	0.1528	0.063343	0.83668	6
+PWT	0.87253	1.5777	1.4066	0.2753	0.035858	0.87253	5
+PWS	0.89547	1.0093	0.87774	0.416	0.022938	0.89547	4
+PWpH	0.91155	0.70733	0.54522	0.5743	0.016076	0.91155	3
+SpH	0.92487	0.58615	0.35462	0.6396	0.013322	0.92487	2
+SWT	0.95202	1.1945	0.56575	0.6076	0.027147	0.95202	1

between these parameters in the wetland ecosystems (Chambers et al. 2011; Kida et al. 2017).

Long-term sea-level rise and saltwater intrusion typically reduce SOC mineralization rates but may temporarily increase them during early salinity increases (Luo et al. 2019). If this is the case, then a high SOC stock during the dry season and a low SOC stock during the wet season would be envisaged. However, the findings of the present study contradict the aforementioned assumption, i.e., the impact of salinity on SOC mineralization and OC inputs (productivity and allochthonous inputs) could be a possible reason for the high SOC stock. For instance, although raising salinity impairs SOC mineralization on a long-term scale, it enhances the SOC mineralization during early salinity increases and may also diminish the in situ productivity and allochthonous input when freshwater availability is limited. Thus, the overall increase in salinity over the short term results in the loss of SOC owing to the high mineralization rate and low OC input during the dry season. A low saline condition during the rainy season promotes in situ primary production and allochthonous inputs, notwithstanding the dissolution of SOC. Thus, the low saline regime facilitates the accumulation of SOC stock by compensating for OC dissolution through increased OC inputs.

9.4 Conclusion

Since the structure and functions of mangroves are more site-specific and dependent on the local geomorphology and hydrological characteristics, the impact of salinity on the sediment organic carbon pool and quality is not consistent across all wetland settings worldwide. By quantifying the relationships between precipitation, salinity, sediment bulk density, organic matter, plant productivity, and SOM, it has been highlighted that future changes in SOM and plant productivity will be regulated by cascading effects of precipitation on freshwater availability and salinity (Osland et al. 2018). Therefore, a detailed investigation on the impact of seasonal salinity fluctuations on the carbon storage potential of blue carbon ecosystems in a tropical monsoon climate is necessary to gain new insights into the prediction of sea-level rise-induced salinity impacts on blue carbon ecosystems. Furthermore, seasonal impacts on decomposition processes must be considered to accurately assess the carbon storage potential of blue carbon ecosystems under future climate change scenarios.

Acknowledgments T.K. extends heartfelt thanks to the University Grants Commission for providing the UGC Dr. D.S. Kothari Postdoctoral Fellowship (F.4−2/2006 (BSR)/BL/14–15/0226 dated 05.03.2015). Gratitude is also expressed to the authorities at Bharathidasan University for offering the necessary facilities to carry out this work. Additionally, we acknowledge the Head of the Department of Disaster Management, Pondicherry University, Port Blair Campus, for their invaluable assistance during field collections. We dedicate this paper to one of the collaborators who supported statistical analysis, the Late Prof. Seyed Ajmal Khan, Centre of Advanced Study in Marine Biology, Annamalai University, India.

Competing Interests The authors declare no competing interests.

Declarations Informed Ethical approval

1. This material is the authors' own original work, which has not been previously published elsewhere.
2. The paper is not currently being considered for publication elsewhere.
3. The paper reflects the authors' own research and analysis in a truthful and complete manner.
4. The paper properly credits the meaningful contributions of co-authors and co-researchers.
5. The results are appropriately placed in the context of prior and existing research.
6. All sources used are properly disclosed (correct citation).
7. All authors have been personally and actively involved in substantial work leading to the paper.
8. The manuscript was checked for plagiarism as well.

Consent to Participate The results from the present research were originally generated from the fieldwork in the study area by the first authors. No data from any other sources/individuals/organizations were used in the present research.

Consent for Publication All the authors actively participated in the present research activity. Those extended the support are been acknowledged.

Data Availability Data will be made available on reasonable request.

References

Bansal S, Creed IF, Tangen BA, Bridgham SD, Desai AR, Krauss KW, Neubauer SC, Noe GB, Rosenberry DO, Trettin C, Wickland KP, Allen ST, Arias-Ortiz A, Armitage AR, Baldocchi D, Banerjee K, Bastviken D, Berg P, Bogard MJ, Chow AT, Conner WH, Craft C, Creamer C, DelSontro T, Duberstein JA, Eagle M, Fennessy MS, Finkelstein SA, Göckede M, Grunwald S, Halabisky M, Herbert E, Jahangir MMR, Johnson OF, Jones MC, Kelleway JJ, Knox S, Kroeger KD, Kuehn KA, Lobb D, Loder AL, Ma S, Maher DT, McNicol G, Meier J, Middleton BA, Mills C, Mistry P, Mitra A, Mobilian C, Nahlik AM, Newman S, O'Connell JL, Oikawa P, van der Burg MP, Schutte CA, Song C, Stagg CL, Turner J, Vargas R, Waldrop MP, Wallin MB, Wang ZA, Ward EJ, Willard DA, Yarwood S, Zhu X (2023) Practical guide to measuring wetland carbon pools and fluxes. Wetlands. Springer Netherlands 43:105. https://doi.org/10.1007/s13157-023-01722-2

Burden A, Garbutt RA, Evans CD, Jones DL, Cooper DM (2013). Carbon sequestration and biogeochemical cycling in a saltmarsh subject to coastal managed realignment. Estuar Coast Shelf Sci 120: 12–20. https://doi.org/10.1016/j.ecss.2013.01.014

Barendregt A, Swarth CW (2013) Tidal freshwater wetlands: variation and changes. Estuar Coasts 36:445–456

Bhatta KP, Rokaya MB, Muenzbergova Z (2015) Environmental feedbacks of the subalpine ecotone species in the Langtang National Park, central Nepal Himalaya. Pak J Bot 47:2115–2125

Bruland GL, Richardson CJ (2005) Spatial variability of soil properties in created, restored, and paired natural wetlands. Soil Sci Soc Am J 69:273–284

Campbell JE, Lacey EA, Decker RA, Crooks S, Fourqurean JW (2014) Carbon storage in seagrass beds of Abu Dhabi, United Arab Emirates. Estuar Coasts 38:242–251. https://doi.org/10.1007/s12237-014-9802-9

Cho YK, Kim TW, You KW, Park LH, Moon HT, Lee SH, Youn YH (2005) Temporal and spatial variabilities in the sediment temperature on the Baeksu tidal flat Korea. Estuar Coast Shelf Sci 65(1–2):302–308. https://doi.org/10.1016/j.ecss.2005.06.010

Chan-Keb CA, Agraz-Hernández CM, Pérez-Balan RA, Mas-Qui OO, Osti-Sáenz J, Reyes-Castellanos JE (2024) Relationship between carbon sequestration and soil physicochemical parameters in northern Campeche, Mexico. Land 13. https://doi.org/10.3390/land13020139

Choudhary B, Dhar V, Pawase AS (2024) Blue carbon and the role of mangroves in carbon sequestration: Its mechanisms, estimation, human impacts and conservation strategies for economic incentives. Journal of Sea Research, 199: 102504. https://doi.org/10.1016/j.seares.2024.102504

Chen ZL, Lee SY (2022) Tidal Flats as a Significant Carbon Reservoir in Global Coastal Ecosystems. Frontiers in Marine Science 9:900896. https://doi.org/10.3389/fmars.2022.900896

Chambers LG, Reddy KR, Osborne TZ (2011) Short-Term Response of Carbon Cycling to Salinity Pulses in a Freshwater Wetland. Soil Sci Soc Am J 75: 2000–2007. https://doi.org/10.2136/sssaj2011.0026

Chen G, Azkab MH, Chmura GL, Chen S, Sastrosuwondo P, Ma Z, Dharmawan IWE, Yin X, Chen B (2017) Mangroves as a major source of soil carbon storage in adjacent seagrass meadows. Sci Rep 7:42406

Donato DC, Kauffman JB, Murdiyarso D, Kurnianto S, Stidham M, Kanninen M (2011) Mangroves among the most carbon-rich forests in the tropics. Nat Geosci 4:293–297. https://doi.org/10.1038/ngeo1123

Dinesh R, Chaudhuri SG, Ganeshamurthy AN, Pramanik SC (2004) Biochemical properties of soils of undisturbed and disturbed mangrove forests of South Andaman (India). Wetl Ecol Manag 12: 309–320. https://doi.org/10.1007/s11273-004-0777-3

Dye AW, Houtman RM, Gao P et al (2024) Carbon, climate, and natural disturbance: a review of mechanisms, challenges, and tools for understanding forest carbon stability in an uncertain future. Carbon Balance Manag 19:1–25

Drake K, Halifax H, Adamowicz SC, Craft C (2015) Carbon sequestration in tidal salt marshes of the Northeast United States. Environ Manag 56:998–1008. https://doi.org/10.1007/s00267-015-0568-z

Fourqurean JW, Kendrick GA, Collins LS, Chambers RM, Vanderklift MA (2012) Carbon, nitrogen and phosphorus storage in subtropical seagrass meadows: examples from Florida Bay and Shark Bay. Mar Freshw Res 63:967–983. https://doi.org/10.1071/MF12101

Fondriest Environmental Inc (2013) pH of Water. Fundamentals of Environmental Measurements. https://www.fondriest.com/environmental-measurements/parameters/water-quality/ph

Geevarghese GA, Akhil B, Magesh G, Krishnan P, Purvaja R, Ramesh R (2018) A comprehensive geospatial assessment of seagrass distribution in India. Ocean Coast Manag 159:16–25. https://doi.org/10.1016/j.ocecoaman.2017.10.032

Howard J, Hoyt S, Isensee K, Telszewski M, Pidgeon E (2014) Coastal blue carbon: methods for assessing carbon stocks and emissions factors in mangroves, tidal salt marshes, and seagrasses

Huang X, Wang X, Li X, Xin K, Yan Z, Sun Y, Bellerby R (2018) Distribution pattern and influencing factors for soil organic carbon (SOC) in mangrove communities at Dongzhaigang, China. J Coast Res 34:434–442. https://doi.org/10.2112/JCOASTRES-D-16-00207.1

Herbert ER, Boon P, Burgin AJ, Neubauer SC, Franklin RB, Ardón M, Hopfensperger KN, Lamers LPM, Gell P (2015) A global perspective on wetland salinization: ecological consequences of a growing threat to freshwater wetlands. Ecosphere 6(10):1–43

Hilmi N, Chami R, Sutherland MD et al (2021) The role of Blue Carbon in climate change mitigation and carbon stock conservation. Front Clim 102. https://doi.org/10.3389/fclim.2021.710546

Hutomo M (1985) telaah Ekologik Komunitas Ikan pada Padang Lamun (Seagrass, Anthophyta) di Perairan Teluk Banten. Disertasi. Fakultas Pascasarjana, Institut Pertanian Bogor. Bogor.

Howard J et al (2023) Blue carbon pathways for climate mitigation: Known, emerging and unlikely. Mar Pol 156: 105788. https://doi.org/10.1016/j.marpol.2023.105788

Jones TG, Ratsimba HR, Ravaoarinorotsihoarana L, Cripps G, Bey A (2014) Ecological segregation of the late Jurassic stegosaurian and iguanodontian dinosaurs of the Morrison formation in North America: pronounced or subtle? Forests 5:177–205. https://doi.org/10.3390/f5010177

Kathiresan K, Gomathi V, Anburaj R, Saravanakumar K (2014) Impact of mangrove vegetation on seasonal carbon burial and other sediment characteristics in the Vellar-Coleroon estuary. India J For Res 25:787–794. https://doi.org/10.1007/s11676-014-0526-2

Kaviarasan T, Gokul MS, Henciya S, Muthukumar K, Dahms HU, James RA (2018) Trace metal inference on seaweeds in Wandoor area, Southern Andaman Island. Bull Environ Contam Toxicol 100:614–619. https://doi.org/10.1007/s00128-018-2305-9

Kauffman JB, Donato DC (2012) Protocols for the measurement, monitoring and reporting of structure, biomass and carbon stocks in mangrove forests. Center for International Forestry Research (CIFOR) Working Paper 86, Bogor, Indonesia, p 40. https://doi.org/10.17528/cifor/003749

Kumar N, Yadav BP, Tyagi A, Jaswal AK (2012) Trend and spatial distribution of rainfall & rainy days over Andaman & Nicobar Islands. Nat Hazards 63: 575–587. https://doi.org/10.1007/s11069-012-0173-x

Kida M, Tomotsune M, Iimura Y, Kinjo K, Ohtsuka T, Fujitake N (2017) High salinity leads to accumulation of soil organic carbon in mangrove soil. Chemosphere 177:51–55. https://doi.org/10.1016/j.chemosphere.2017.02.074

Kida M, Fujitake N (2020) Organic Carbon Stabilization Mechanisms in Mangrove Soils: A Review. Forests 11:981. https://doi.org/10.3390/f11090981

Kim TW, Cho YK (2009) Heat flux across the surface of a macrotidal flat in southwest Korea. J Geophys Res 114(C7): 1–1. https://doi.org/10.1029/2008JC004966

Kaviarasan T, Dahms HU, Gokul MS, Henciya S, Muthukumar K, Shankar S, Arthur James R (2019) Seasonal species variation of sediment organic carbon stocks in salt marshes of Tuticorin area, Southern India. Wetlands 39:483–494. https://doi.org/10.1007/s13157-018-1094-6

Kuwae T, Watanabe A, Yoshihara S, Suehiro F, Sugimura Y (2022) Implementation of blue carbon offset crediting for seagrass meadows, macroalgal beds, and macroalgae farming in Japan. Mar Pol 138:104996. https://doi.org/10.1016/j.marpol.2022.104996

Li L, Xu H, Zhang Q, Zhan Z, Liang X, Xing J, (2024) Estimation methods of wetland carbon sink and factors influencing wetland carbon cycle: a review. Carbon Res. 3: 50. https://doi.org/10.1007/s44246-024-00135-y

Luo M, Zhu W, Huang J, Liu Y, Duan X, Wu J, Tong C (2019) Anaerobic organic carbon mineralization in tidal wetlands along a low-level salinity gradient of a subtropical estuary: Rates, pathways, and controls. Geoderma 337: 1245–1257. https://doi.org/10.1016/j.geoderma.2018.07.030

Lovelock CE, Adame MF, Bennion V, Hayes M, O'Mara J, Reef R, Santini NS (2014) Contemporary rates of carbon sequestration through vertical accretion of sediments in mangrove forests and saltmarshes of South East Queensland, Australia. Estuar Coasts 37:763–771. https://doi.org/10.1007/s12237-013-9702-4

Lima M do A C, Ward RD, Joyce CB (2020) Environmental drivers of sediment carbon storage in temperate seagrass meadows. Hydrobiologia 847(7): 1773-1792. https://doi.org/10.1007/s10750-019-04153-5

Manokaran S, Khan SA, Lyla S, Raja S, Ansari KGMT (2013) Feeding guild composition of shelf macrobenthic polychaetes of southeast coast of India. Trop Zool 26:120–139. https://doi.org/10.1080/03946975.2013.825425

Meselhe E, McCorquodale JA, Shelden J, Dortch M, Brown TS, Elkan P, Rodrigue MD, Schindler JK, Wang Z (2013) Ecohydrology component of Louisiana's 2012 Coastal Master Plan: mass-balance compartment model. J Coast Res 67:16–28

Morrissey EM, Gillespie JL, Morina JC, Franklin RB (2014). Salinity affects microbial activity and soil organic matter content in tidal wetlands. Glob Chang Biol 20: 1351–1362. https://doi.org/10.1111/gcb.12431

Mazhar S, Pellegrini E, Contin M, Bravo C, De Nobili M (2022) Impacts of salinization caused by sea level rise on the biological processes of coastal soils - A review. Front Environ Sci 10: 909415. https://doi.org/10.3389/fenvs.2022.909415

Mitra A, Sengupta K, Banerjee K (2011) Standing biomass and carbon storage of above-ground structures in dominant mangrove trees in the Sundarbans. For Ecol Manag 261:1325–1335. https://doi.org/10.1016/j.foreco.2011.01.012

Macreadie PI et al (2019) The future of blue carbon science. Nat Commun 10:3998. https://doi.org/10.1038/s41467-019-11693-w

MoEFCC (2019) Press Information Bureau English Releases 1–4

Nyman JAA, DeLaune RD, Patrick WH Jr (1990) Wetland soil formation in the rapidly subsiding Mississippi River deltaic plain: mineral and organic matter relationships. Estuar Coast Shelf Sci 31:57–69

Nag S K, Ghosh B D, Das B K, Sarkar U K (2024) Wetlands function as carbon sink: Evaluation of few floodplains of middle Assam, northeast India in the perspective of climate change. J Environ Manag 373: 123841. https://doi.org/10.1016/j.jenvman.2024.123841

Nobi EP, Dilipan E, Thangaradjou T, Sivakumar K, Kannan L (2010) Geochemical and geostatistical assessment of heavy metal concentration in the sediments of different coastal ecosystems of Andaman Islands, India. Estuar. Coast Shelf Sci 87: 253–264. https://doi.org/10.1016/j.ecss.2009.12.019

Owers CJ, Woodroffe CD, Mazumder D, Rogers K (2022) Carbon storage in coastal wetlands is related to elevation and how it changes over time. Estuar Coast Shelf Sci 267: 107775. https://doi.org/10.1016/j.ecss.2022.107775

Osland MJ (2018) Climate and plant controls on soil organic matter in coastal wetlands. Glob. Chang Biol 24: 5361–5379. https://doi.org/10.1111/gcb.14376

Phang VXH, Chou LM, Friess DA (2015) Ecosystem carbon stocks across a tropical intertidal habitat mosaic of mangrove forest, seagrass meadow, mudflat and sandbar. Earth Surf Process Landf 40:1387–1400. https://doi.org/10.1002/esp.3745

Piccolo MC (2019) Chapter 7 - methods to estimate heat balance in coastal wetlands. In G. M. E. Perillo, E. Wolanski, D. R. Cahoon, & C. S. Hopkinson (Eds.), Coastal wetlands (2nd ed., pp. 263–288). Elsevier.

Pivato M, Carniello L, Gardner J, Silvestri S, Marani M (2018) Water and sediment temperature dynamics in shallow tidal environments: The role of the heat flux at the sediment-water interface. Adv. Water Resour 113: 126–140. https://doi.org/10.1016/j.advwatres.2018.01.009

Rahaman SMB, Sarder L, Rahaman MS, Ghosh AK, Biswas SK, Siraj SMS, Huq KA, Hasanuzzaman AFM, Islam SS (2013) Nutrient dynamics in the Sundarbans mangrove estuarine system of Bangladesh under different weather and tidal cycles. Ecol Process 2:1–13. https://doi.org/10.1186/2192-1709-2-29

Ricart AM, York PH, Rasheed MA, Pérez M, Romero J, Bryant CV, Macreadie PI (2015) Variability of sedimentary organic carbon in patchy seagrass landscapes. Mar Pollut Bull 100:476–482. https://doi.org/10.1016/j.marpolbul.2015.09.032

Raw JL, Van Niekerk L, Chauke O et al (2023) Blue carbon sinks in South Africa and the need for restoration to enhance carbon sequestration. Sci Total Environ 859:160142. https://doi.org/10.1016/j.scitotenv.2022.160142

Rumpel C, Amiraslani F, Koutika LS, Smith P, Whitehead D, Wollenberg E (2018) Put more carbon in soils to meet Paris climate pledges. Nature 564:32–34

Sitoe AA, Mandlate LJC, Guedes BS (2014) Biomass and carbon stocks of Sofala Bay mangrove forests. Forests 5:1967–1981. https://doi.org/10.3390/f5081967

Sreelekshmi S, Nandan SB, Sreejith VK, Harikrishnan M (2020) Floristic Structure, Diversity and Edaphic Attributes of Mangroves of the Andaman Islands, India. Thalassas 36:47–60.https://doi.org/10.1007/s41208-020-00191-2

Stagg CL, Baustian MM, Perry CL, Carruthers TJB, Hall CT (2018) Direct and indirect controls on organic matter decomposition in four coastal wetland communities along a landscape salinity gradient. J Ecol 106: 655–670. https://doi.org/10.1111/1365-2745.12901

Tang Z, Wang D, Tian X, Bi X, Zhou Z, Luo F, Ning R, Li J (2024) Exploring the factors influencing the carbon sink function of coastal wetlands in the Yellow River Delta. Sci Rep 14(1): 1-15. https://doi.org/10.1038/s41598-024-80186-8

Thompson RL, Patra PK, Chevallier F, Maksyutov S, Law RM, Ziehn T, Van Der Laan-Luijkx IT, Peters W, Ganshin A, Zhuravlev R, Maki T, Nakamura T, Shirai T, Ishizawa M, Saeki T, Machida T, Poulter B, Canadell JG, Ciais P (2016) Top-down assessment of the Asian carbon budget since the mid 1990s. Nat Commun 7:1–10. https://doi.org/10.1038/ncomms10724

Tong C, Tan J, Luo M, Huang J, Xiao S, Liu B, Morris JT (2024) Inundation counteracts the promoting effect of nitrogen enrichment on soil organic carbon mineralization in a tidal marsh. Fundam Res. https://doi.org/10.1016/j.fmre.2024.01.013

Turner RE, Baustian JJ, Swenson EM, Spicer JS (2006) Wetland sedimentation from hurricanes Katrina and Rita. Science (80-) 314:449–452

Viswanathan C, Purvaja R, Joyson Joe Jeevamani J, Deepak Samuel V, Sankar R, Abhilash KR, Geevarghese GA, Muruganandam R, Gopi M, Raja S, Das RR, Patro S, Krishnan P, Ramesh R (2020) Salt marsh vegetation in India: species composition, distribution, zonation pattern and conservation implications. Estuar Coast Shelf Sci 242:106792. https://doi.org/10.1016/j.ecss.2020.106792

Watanabe K, Kuwae T (2015) How organic carbon derived from multiple sources contributes to carbon sequestration processes in a shallow coastal system? Glob Chang Biol 21:2612–2623. https://doi.org/10.1111/gcb.12924

Wang F, Liu J, Qin G et al (2023) Coastal blue carbon in China as a nature-based solution towards carbon neutrality. Innovation 4(5):100481. https://doi.org/10.1016/j.xinn.2023.100481

Xu X, Yu X, Wu J, Xiao K, Xin P (2025) Tidal influences on temperature dynamics and heat exchange in coastal wetlands. Water Resources Research. Water Resour Res 61(2): e2024WR038374

Xue L, Jiang J, Li X, Yan Z, Zhang Q, Ge Z, Tian B, Craft C (2019) Salinity Affects Topsoil Organic Carbon Concentrations Through Regulating Vegetation Structure and Productivity. Journal of Geophysical Research: Biogeosciences, 125(1): e2019JG005217. https://doi.org/10.1029/2019JG005217

Zedler JB, Callaway JC (2001) Tidal wetland functioning. J Coast Res:38–64

Zhao Q, Bai J, Zhang G, Jia J, Wang W, Wang X (2018) Effects of water and salinity regulation measures on soil carbon sequestration in coastal wetlands of the Yellow River Delta. Geoderma 319:219–229. https://doi.org/10.1016/j.geoderma.2017.10.058

Zou H, Li X, Li S, Xu Z, Yu Z, Cai H, Chen W, Ni X, Wu E, Zeng G (2023) Soil organic carbon stocks increased across the tide-induced salinity transect in restored mangrove region. Sci Rep 13:1–13. https://doi.org/10.1038/s41598-023-45411-w

Chapter 10
Deep-Sea Benthic Biodiversity and Climate Change: Indian Ocean Perspective

Sabyasachi Sautya, Ankur Saini, and Gourav Kumar Mishra

Abstract The deep sea, encompassing ocean depths below 200 m which covers more than 70% of the Earth's surface, represents one of the largest ecosystem in the planet earth. This expansive and largely uncharted environment is home to an extraordinary diversity of life forms, particularly within benthic communities that inhabit the ocean floor. Despite its vast expanse and ecological significance, the deep-sea benthic ecosystem remains one of the least understood environments on our planet. These benthic organisms are not merely passive inhabitants of their surroundings; they play pivotal roles in deep-sea ecosystems, significantly contributing to critical ecological processes such as nutrient cycling, carbon sequestration, and the sustenance of food webs. Despite their importance, the complexities and dynamics of these communities remain poorly understood, highlighting the urgent need for comprehensive research to elucidate their functions and interactions within the broader context of deep-sea ecology. Indian Ocean, the third largest Ocean, is a significant oceanic basin characterized by diverse deep-sea habitats that harbor an array of benthic diversity. This review aims to provide a comprehensive overview of deep-sea benthic biodiversity (meio, macro, and megafauna) in the Indian Ocean, highlighting the critical need for further research to understand and protect these unique and valuable ecosystems while studying the impact of climate change. Furthermore, we discuss the challenges and opportunities associated with studying deep-sea biodiversity in the Indian Ocean, including the development of innovative technologies and the integration of multidisciplinary approaches.

Keywords Benthos · Biodiversity · Climate change · Conservation · Indian Ocean

S. Sautya (✉) · A. Saini · G. K. Mishra
Laboratory for Benthic Ecological Trait Analysis (L-BETA), CSIR-National Institute of Oceanography, Regional Centre, Mumbai 400053, India

Academy of Scientific and Innovative Research (AcSIR), Ghaziabad 201002, India
e-mail: sautya.nio@csir.res.in

B. P. Barathan et al. (eds.), *Navigating Climate Change: Impacts on Biodiversity and Ecosystem Resilience*, Sustainable Development and Biodiversity 42,
https://doi.org/10.1007/978-981-95-0409-1_10

10.1 Introduction

The deep-sea floor is characterized by extreme conditions, including high pressure, low temperatures, and limited food availability (Canganella 2002; Ingole et al. 2005a, 2005b; Papiol et al. 2014). Despite these challenges, it supports a diverse array of life forms, from microscopic bacteria to large megafauna (De Smet et al. 2021; Pavithran et al. 2007). Benthic organisms play crucial roles in nutrient cycling, carbon sequestration, and maintaining the overall health of marine ecosystems (Griffiths et al. 2017). The distribution of biodiversity in the deep sea is influenced by various factors, majorly governed by depth-related changes in temperature, salinity, oxygen, pressure, and others are substrate type and organic matter input (Bridges et al. 2022; Swanborn et al. 2022). The abyssal plains, which cover vast areas of the ocean floor, are generally considered less diverse than continental margins and seamounts (Laroche et al. 2020). The reason for seamounts and continental margins to be more diverse could be because of higher organic matter presence and varied terrain (from soft sediments to hard rocks) in these habitats, in comparison to abyssal plains. Moreover, the current patterns at seamounts also enhance organic matter accumulation, which could, eventually, support greater diversity of organisms. However, recent studies have revealed unexpected levels of species richness in these seemingly barren environments (O'Hara et al. 2020). One of the most striking features of deep-sea benthic biodiversity is the high proportion of rare species, often referred to as the "rare biosphere." This phenomenon contributes to the difficulty in accurately assessing species richness and distribution patterns in the deep sea. Advanced molecular techniques, such as environmental DNA (eDNA) analysis, are providing new insights into the true extent of deep-sea biodiversity (Laroche et al. 2020).

The deep Indian Ocean, extending from depths of 200 m to the abyssal plains, represents one of the most enigmatic and understudied regions of the global ocean. This vast expanse, covering approximately 20% of the Indian Ocean's total area, plays a crucial role in global climate regulation, biogeochemical cycling, and marine biodiversity. Despite its significance, our understanding of the deep Indian Ocean remains limited with insufficient exploration of the benthic fauna (Figure 1), necessitating further research and exploration (Figure 2). The deep Indian Ocean is characterized by a complex seafloor topography, including mid-ocean ridges, seamounts, and deep-sea trenches. The Southwest Indian Ridge and the Central Indian Ridge are prominent features that influence deep-sea circulation patterns and habitat distribution. These ridges, along with numerous seamounts, create unique ecosystems, such as rift valleys, fracture zones, hydrothermal vents, which support diverse benthic communities. Hydrothermal vents release mineral-rich fluids that support chemosynthetic microbial communities, forming base of food webs for species like tube worms, shrimp, and crabs (Lutz and Kennish 1993). The slopes and submits of seamounts provide niches for benthic filter feeders, like sponge, corals, etc. (Rowden et al. 2010). Moreover, both mid-oceanic ridges and seamounts disrupt water flow, causing upwelling of nutrient-rich deep water (Boehlert and Genin 2013). This

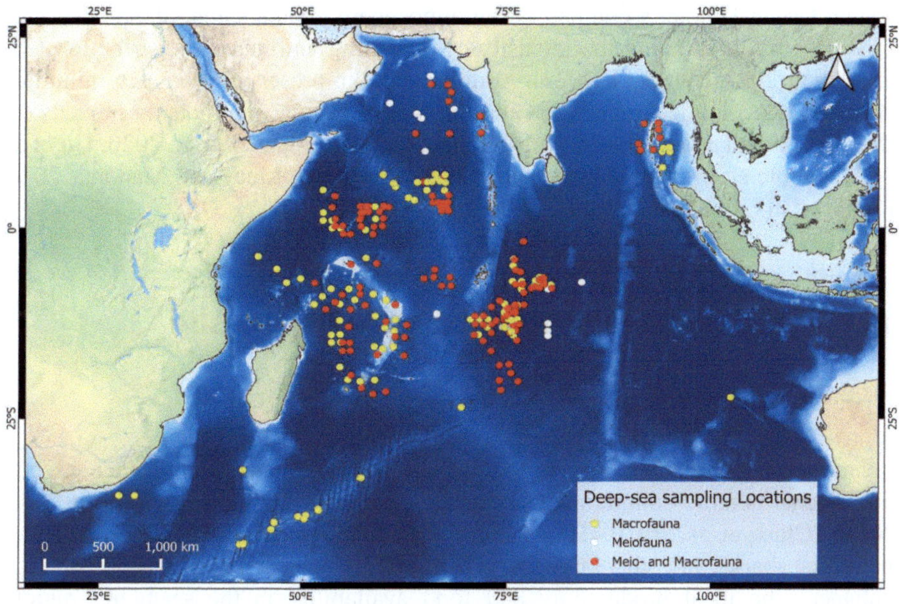

Fig. 10.1 Deep-sea benthos sampling locations in the Indian Ocean

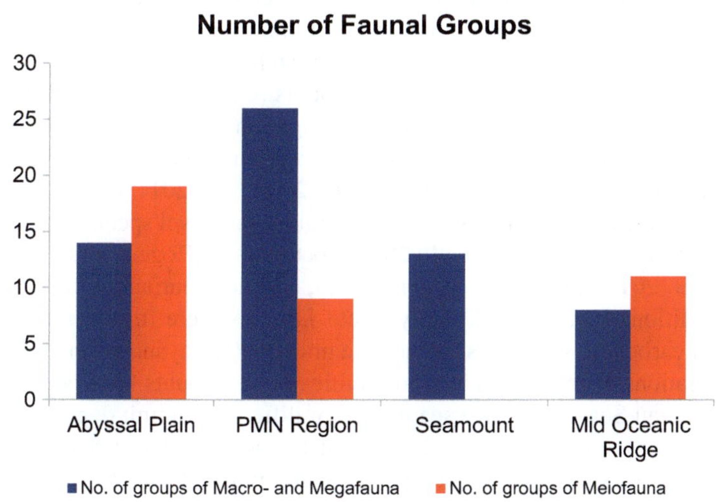

Fig. 10.2 Number of faunal groups reported from various habitats in the Indian Ocean

enhances primary production in the overlying waters, which will lead to more organic matter deposition in these regions, supporting higher biomass and biodiversity (Priede et al. 2013, 2022). The oceanography of the deep Indian Ocean is influenced by several factors, including the monsoon system, thermohaline circulation, and the interaction between surface and deep waters (Phillips et al. 2021). The

Indian Ocean Dipole, a coupled ocean-atmosphere phenomenon, affects deep-sea processes by modulating the intensity of upwelling and nutrient distribution (Shi and Wang 2021; Sun et al. 2022). Additionally, the presence of oxygen minimum zones (OMZs) in the Arabian Sea and the Bay of Bengal significantly impacts the distribution and adaptation of deep-sea organisms (Fernandes et al. 2020; Gu et al. 2022; Ingole et al. 2010; Joydas and Damodaran 2014; Lincy and Manohar 2020; Raman et al. 2015; Rixen et al. 2020).

10.2 Deep-Sea Ecosystem and Benthic Biodiversity

10.2.1 Seamounts

Seamounts are globally recognized as the regions of biological significance, harboring diverse benthic communities along their summits and flanks within the deep ocean (Clark et al. 2010). These underwater features attract mobile megafauna, including sharks, and other large fishes (Rogers 2018), with their topography, which enhances the local productivity and food availability in the seamount regions (Rogers 2012). The seamounts in the Indian Ocean are mostly associated with ridge systems. However, research to understand the dynamics of these ecosystem in the Indian Ocean is scarce, in comparison to other oceans (Clark et al. 2010).

The preponderance of seamounts in the Indian Ocean exhibit summit depths surpassing 1000 m, with a mean of approximately 2400 m (Yesson et al. 2011). Despite their prevalence, these deeper seamounts remain understudied (Clark et al. 2010). In contrast, the limited benthic investigations are from SWIR (Southwest Indian Ridge), which is one of the shallow peaked seamounts (<1000 m depth). Two major scientific expeditions, undertaken in 2009 and 2011 (IUCN and NERC Seamounts Project), led to the discovery of numerous novel species, including carnivorous sponges (Hestetun et al. 2017), octocorals (Rogers and Taylor 2011; Rogers et al. 2017; Taylor and Rogers 2017), and holothurians (O'Loughlin et al. 2013). Additionally, 34 distinct polychaete families were discovered from coral rubble and carbonate substrates on these seamounts (Narayanaswamy et al. 2017).

Investigations of megabenthic communities on seamounts have been conducted in the Andaman Sea, where two seamounts of different summit depths were examined and juxtaposed with the biodiversity of the adjacent basin floor. The highest faunal abundances were recorded on the seamounts relative to the surrounding basin, with the greatest taxonomic diversity reported at their flanks (Sautya et al. 2011a). Distinct faunal assemblages were identified from both the seamounts. The slope of deep-summited (1336 m) seamounts was populated by arthropods and other taxas, whereas the apex was characterized by Porifera and cnidaria (Sautya et al. 2011a). The Andaman seamount surveys yielded six ophiuroid species, all constituting new records from the Andaman Sea, with two novel species (Stöhr et al. 2012).

Benthic ecological studies from the deep-sea mining site from Indian Ocean recently yielded significant biodiversity discoveries related to deep-summited seamounts. This research has facilitated the identification of numerous new species and provided updates on the distribution of species of seamounts from the CIR and SWIR. These findings encompass various taxa, including black corals, squat lobsters, octocorals, carnivorous sponges, bamboo corals, and Spongicoloides shrimp (Periasamy et al. 2022, 2023a, 2023b, 2023c, 2023d; Periasamy and Ingole 2022). To further advance research on deep seamounts and Indian Ocean seamounts in general, future studies should incorporate more extensive examinations of seamount flanks and bases, as these areas appear to harbor unique assemblages (Baco 2007; Sautya et al. 2011a, 2011b) (Figure 3L). An exploration at the base Afanasij Nikitin Seamount (~4700 m) in the Central Indian Ocean Basin resulted in the discovery of seven amphipod species, including two novel abyssal species belonging to the genera Hirondellea and Cyclocaris (Weston et al. 2021). Reported species were *Eurythenes maldoror* (d'Udekem d'Acoz & Havermans, 2015), *Alicella gigantean* (Chevreux, 1899), *Paralicella caperesca* (Shulenberger & Barnard, 1976), *Paralicella tenuipes* (Chevreux, 1908), *Cyclocaris* sp., *Hirondellea* sp. nov., and *Abyssorchomene distinctus* (Birstein & Vinogradov, 1960). At abyssal depths, *P. tenuipes* and *P. caperesca* were numerically dominant, with *P. tenuipes* present at every depth of the study (Weston 2021).

On seamounts, meiofauna serve as important links in the food web, transferring energy from microorganisms to larger benthic and pelagic species. They contribute to the high biodiversity often associated with seamounts, as these underwater mountains can act as "stepping stones" for species dispersal across ocean basins. In the Indian Ocean, seamounts like those along the Southwest Indian Ridge and Andaman Sea host diverse meiofaunal communities that vary with depth, substrate type, and hydrodynamic conditions. The unique topography of seamounts creates a variety of microhabitats, each supporting distinct meiofaunal assemblages. These communities are often characterized by high levels of endemism, with species adapted to the specific conditions of individual seamounts. The role of meiofauna in seamount ecosystems extends beyond their position in the food web. They contribute in biogeochemical cycling, remineralization, biodiversity maintenance, environmental health indicators, etc. (Chertoprud and Novichkova 2023; Schratzberger and Ingels 2018; Wang et al. 2019). They contribute significantly to sediment bioturbation, the process by which sediment particles are reworked and mixed. This activity enhances oxygen penetration into sediments and facilitates the recycling of nutrients, thereby supporting the productivity of the entire seamount ecosystem. Additionally, meiofauna on seamounts play a crucial role in the early stages of organic matter decomposition, breaking down detritus and making nutrients available to other organisms.

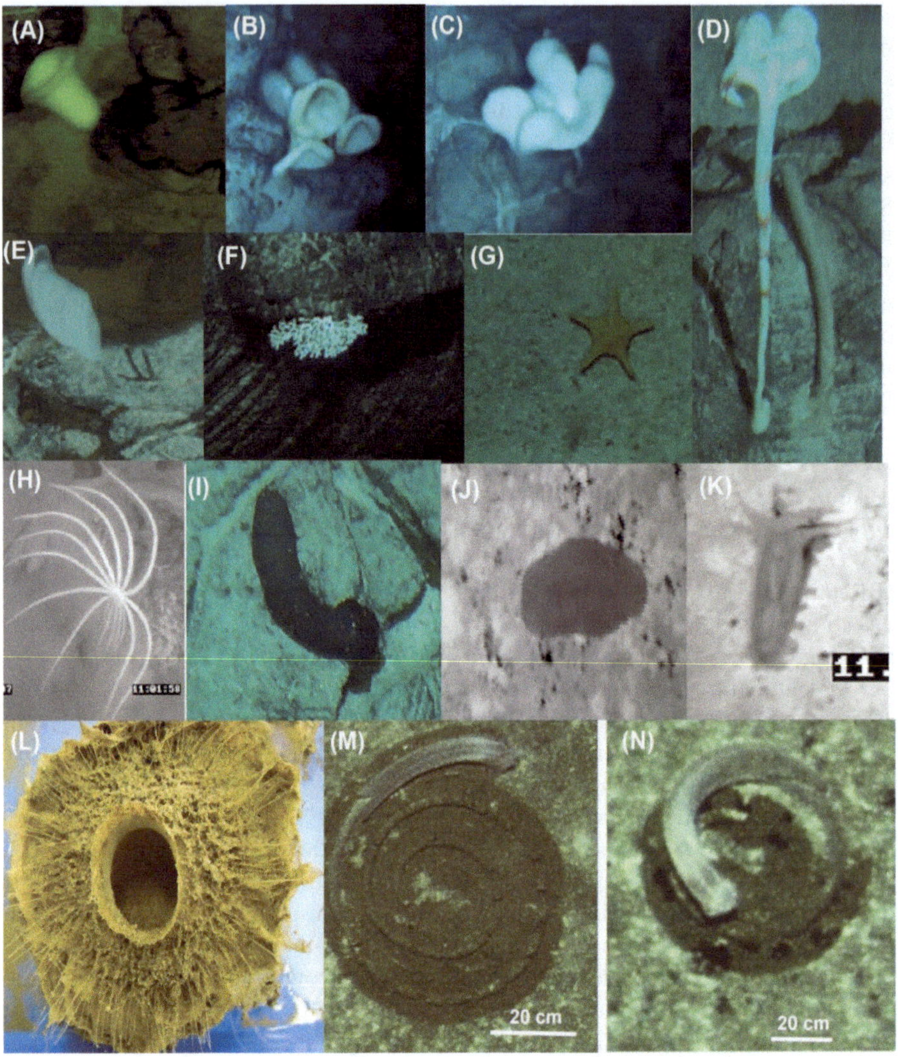

Fig. 10.3 Underwater images from benthic megafauna from the various ecosystem. Images are modified from Sautya et al. 2017, Sautya et al. 2011 and Saini et al. 2025. **Mid-oceanic ridge (Carlsberg ridge)** A. *Hyalonema* sp; B. Rossellidae glass sponge; C-D. *Crateromorpha* sp; E Aulocalycidae glass sponge; F. Octocorallia; G. Asteroid sp; H. Brisingidae sp.; I. *Benthodytes* sp; J. *Paelopatides* sp.; K. *Peniagone* sp; **Seamount (SM2, Andaman Sea)** L. Bird nest sponge *Pheronema* sp.; **Central Indian Ocean Basin (PMN region)** M-N. *Tergivelum* sp.

10.2.2 Mid-Oceanic-Ridge

The International Hydrographic Organization (IHO 2019) characterizes a ridge as an extended elevation with precipitous flanks, varying in intricacy and magnitude. These structures may exhibit seismic activity or remain quiescent. The Indian Ocean hosts four principal active spreading ridges: the Central Indian Ridge (CIR), Carlsberg Ridge (CR), the Southwest Indian Ridge (SWIR), and the Southeast Indian Ridge (SEIR). While each ridge system possesses unique attributes and spreading velocities, their crusts are generally thinner than average due to their gradual expansion rates (Suo et al. 2016).

The SWIR, with an ultraslow spreading rate of approximately 15 mm/year, ranks as the world's second-slowest spreading ridge (Patriat et al. 1997; Van der Most et al. 2023). Investigations of this ridge have yielded discoveries of novel species, such as the sponge-associated polychaete Bathymoorea (Periasamy et al. 2023e). In comparison, the SEIR exhibits a more uniform spreading rate of 69–75 mm/year. The CIR demonstrates a slower to moderate rate of about 48 mm/year (Sempéré and Cochran 1997; Kumagai et al. 2008; Van der Most et al. 2023). The disparities in spreading rates and consequent ridge stability often result in biological variations (Copley et al. 2016). The northern CR expands at 22–32 mm/year gradually (Murton and Rona 2015; Van der Most et al. 2023).

The pioneering quantitative study of benthic megafaunal communities on the CR by Sautya et al. (2017) documented eight distinct substratum types and 90 megafaunal taxa belonging to seven phyla, including chordates, cnidarians, poriferans, and echinoderms (Figure 3A-K). Analogous to the SWIR, CR investigations have led to the identification of new taxa, such as the novel genus and *Indiella ridgenensis*, a glass sponge (Sautya et al. 2011b). Conversely, the SEIR remains among the least explored biological regions in the Indian Ocean, despite its significance as a major active spreading center.

10.2.3 Abyssal Plains

Abyssal plains, acknowledged as the most extensive environment on Earth, encompass approximately 35% of the Indian Ocean seafloor (Ramirez-Llodra et al. 2010). These vast expanses are typically characterized by extensive deposits of fine sediment, ranging from tens to hundreds of meters in thickness, resulting from the accumulation of organic matter from the surface waters (Ebbe et al. 2010). Recent investigations, however, have revealed that the abyssal regions exhibit greater complexity, with the presence of hard substrates such as nodules and rocky formations (Riehl et al. 2020).

Notwithstanding these findings, the biodiversity and biogeography of abyssal plain communities remain poorly understood (Lörz et al. 2023). The limited food supply and nutrient-poor conditions in the abyss generally lead to reduced faunal abundance and biomass across these plains (Romero-Romero et al. 2021; Ruhl et al. 2008; Smith et al. 2008). Paradoxically, abyssal species richness has been documented as remarkably high (Snelgrove and Smith 2002; Ebbe et al. 2010). In contrast, Parulekar et al. (1992) reported abundant fauna and significant biomass in the abyssal benthos of the western and central Indian Ocean. Mean population density ranged from 233,322/m^2 in the 1500–1999 m depth zone to 50,269/m^2 in the

5500–999 m depth zone, with uniformly high biomass ranging from 0.11 to 12.75 g/m^2. Benthic abyssal plain communities comprise two primary components: epifauna and infauna, both crucial for ecosystem functions such as bioturbation, carbon sequestration, and biogeochemical cycling. Epifaunal communities on abyssal plains are generally characterized by a predominance of deposit feeders, including holothurians and acorn worms (Thurston et al. 1994; Iken et al. 2001).

Quantitative analyses of the abyssal benthos (3600–5300 m) in the central Indian Ocean conducted by Parulekar et al. (1982) indicate rich fauna and high standing crops. Substantial densities of 3 meiofaunal and 12 macrofaunal taxa were observed. The macrofauna consists of mainly polychaetes contributing 41.6%, peracarid crustaceans contributing 31.7%, ophiuroids contributing 12.2%, echiuroid-bryozoa contributing 9.7%, molluscs contributing 4.8% of the total population, and agglutinating rhizopod protozoans. Meiofaunal taxa are representing nematodes about 70% of the total population, harpacticoid copepods about 25%, and ostracods contributed 4%. The high benthic biomass and rich fauna are attributed to high organic production in the euphotic zone.

While smaller holothurians are typically associated with abyssal infaunal communities, the abyssal meiofaunal assemblage in the Arabian Sea is predominantly composed of nematodes and copepods with an average density of 234.99 ± 61.60 ind./10 cm^2 (Sautya et al. 2021). In abyssal plains lebensspuren are prevalent, which tells about the organisms responsible for their creation (Jamieson and Onda 2022). Examples include trails, burrows, spiral loops, and fecal casts observed in abyssal environments on the Somali Basin of the Carlsberg Ridge in (Glasby 1973) and spiral-shaped feeding marks of acorn worms from the Central Indian Ocean Basin (Banerjee 2000).

However, there has been limited research on abyssal plains in the Indian Ocean, and the existing studies are unevenly distributed. The most well-known abyssal area in the Indian Ocean is the CIOB (Central Indian Ocean Basin) (Ingole and Koslow 2005; Wafar et al. 2011). Since the 1980s, Indian researchers have conducted seafloor surveys in this area to investigate resource potential and provide insights for EIA studies related to deep-sea mining in the central Indian Ocean nodule field (Ansari 2000; Mukhopadhyay et al. 2002; Wafar et al. 2011). However, these studies have only covered a restricted area within the CIOB and may not accurately represent the faunal patterns for the entire region (Ingole and Koslow 2005).

Parulekar et al. (1992) studied the meiofaunal communities in the bathyal and abyssal regions of the Indian ocean and found an average meiofaunal density of 108 ind./10 cm^2. The recorded a total of 6 meiofaunal communities among which nematodes accounted for about 53% of the total population. Foraminiferans (17%) and harpacticoids (17%) were the other major dominant meiofaunal groups. Meiofaunal studies conducted in the Arabian sea basin showed higher meiofaunal abundance (179 ind./10 cm^2) with much higher dominance of nematodes at about 82–88% of the total meiofaunal population. Other major meiofaunal groups included copepods and ostracods. Other studies in the Central Indian Ocean Basin and Arabian Sea have reported meiofaunal densities from 23 to 230 ind./10 cm^2 (Sautya et al. 2021; Ansari 2000; Ingole et al. 2000, 2005a, 2005b).

10.3 Polymetallic Nodule (PMN) Fields

PMN fields, also referred to as ferro-manganese nodules, are extensively distributed across the abyssal ocean floor, with a notable prevalence in areas characterized by oxygen-rich deep waters and minimal sedimentation rates (Hein et al. 2013). These nodules are formed through the precipitation of metals from seawater, resulting in concentric micro-layers surrounding a nucleus. The resultant deposits, measuring 1–12 cm in size, often contain substantial quantities of manganese, cobalt, nickel, and copper (Kuhn et al. 2017). The presence of these nodules significantly enhances habitat diversity by offering hard substrate for benthic organisms. Notably, regions with abundant nodule coverage exhibit markedly higher densities of both sessile and mobile epifauna compared with the areas devoid of nodules of the abyssal plain (Vanreusel et al. 2016).

In the Indian Ocean, three primary nodule fields have been identified: the central Indian Ocean nodule field (IONF), Cape Leeuwin nodule field, and the Mozambique Basin. Among these, IONF has been the subject of the most extensive research primarily due to its potential for deep-sea mining activities (Hein et al. 2013; Mukhopadhyay et al. 2019). The Indian Ocean nodule field is recognized as the second-largest nodule deposit globally in terms of size, richness, and metal composition (Mukhopadhyay et al. 2002; Mukhopadhyay and Ghosh 2010).

To evaluate the environmental consequences of nodule extraction in the IONF, INDEX (the Indian Deep-sea Environment Experiment) was initiated in 1995 (Desa 1997). This investigation involved the collection of meiofauna and macrofauna communities 1 month prior to a simulated mining disturbance and approximately 3.7 years following the event (Ansari 2000; Ingole et al. 2000, 2001, 2005a, 2005b; Rodrigues et al. 2001). Research conducted by Ansari (2000) revealed that the macrofaunal component was predominantly composed of polychaetes (with 100% prevalence) and peracarid crustaceans, while nematodes and harpacticoid copepods dominated the meiofauna.

Subsequent to the disturbance, benthic communities experienced a significant reduction in both meiofauna and macrofauna densities, with the most substantial physical disruption occurring in the upper 2 cm of sediment (Ingole et al. 2000, 2001). Similarly, benthic megafaunal communities, primarily comprised of xenophyophores, holothurians, and sponges, underwent a 32% decline following the disturbance (Rodrigues et al. 2001). Although nematode and oligochaete populations exhibited signs of partial recovery after 3.7 years, polychaete and crustacean populations failed to return to their pre-disturbance levels (Ingole et al. 2005a, 2005b).

In areas rich in polymetallic nodules, such as the Central Indian Ocean Basin, meiofauna are of particular interest due to their potential vulnerability to deep-sea mining activities. These organisms often colonize the surfaces and crevices of nodules, contributing to the overall biodiversity of these regions (Pape et al. 2021; Singh et al. 2018). The nodules were collected with the help of box-cores and multicores, and later were analyzed in the lab for the presence of associated meiofauna on nodules. Meiofauna in nodule fields may play important roles in biogeochemical processes and serve as indicators of ecosystem health. The relationship between meiofauna and polymetallic nodules is complex and multifaceted. The nodules provide a hard substrate in an otherwise soft-sediment environment, creating unique microhabitats that support distinct meiofaunal communities. These communities often differ significantly from those found in the surrounding sediments, contributing to the overall biodiversity of nodule fields. The potential impacts of deep-sea mining on meiofaunal communities in these regions are a subject of growing concern. Mining activities could lead to the destruction of nodule habitats, sedimentation, and the release of potentially toxic substances, all of which could have severe consequences for meiofaunal populations. Given the important ecological roles played by these organisms, such impacts could have far-reaching effects on the entire deep-sea ecosystem. Mining of polymetallic nodules can significantly impact meiofauna and eventually the deep-sea ecosystem through resuspension and redeposition of the sediment caused during the mining activity (Singh et al. 2018). Moreover, the nodule extraction eliminates both the hard surfaces and crevices habitats on nodules, affecting the nodule-associated fauna. This will lead to a localized loss of meiofaunal abundance and biodiversity, also hampering the ecological functions, such as remineralization, biogeochemical cycling, performed by them (Pape et al. 2021).

The discovery of polymetallic nodules in the Central Indian Ocean Basin have spurred an increase in efforts to understand the biodiversity of the environment. The meiofaunal diversity of the PMN region of the Indian Ocean stands quite low when compared to other abyssal habitats at an average of about 26 ind./10 cm^2 (Ingole et al. 2005b; Singh et al. 2014, 2018; Zeng et al. 2018). Nematodes were the most dominant meiofaunal groups with a contribution of about 88% to the overall meiofaunal population (Zeng et al. 2018).

Gaikwad et al. (2024), analyzing the samples from nine multicore liners and six box corers, revealed a total of 337 macrobenthic individuals. In multicore samples, Nematode dominated, by contributing 52.06%, followed by Harpacticoida (27.72%), Polychaeta (9.74%), Calanoida (3.37%), Isopoda (1.50%), Ostracoda (1.12%), and Tanaidacea (1.12%). From box core samples, Nematode stood out to be the dominant taxa, contributing 23.08%, Harpacticoida (20.51%), Polychaeta (10.26%), Acari (9.40), Isopoda (9.40%), Tanaidacea (5.98%), Ostracoda (5.13%), Bivalvia (3.42%), Sipuncula (3.42%) and Clitellata (3.42%). The recent study evidenced the first record and quantitative assessment of deep-sea acorn worm (Enteropneusta) along the PMN-rich zone in the Indian Ocean (Saini et al. 2025) (Figure 3M-N).

10.4 Climate Change

Climate change is poised to have significant and far-reaching impacts on deep-sea benthic biodiversity in the Indian Ocean. One of the primary ways climate changes will impact deep-sea benthic biodiversity is through ocean warming. As global temperatures rise, the deep sea is also experiencing a gradual increase in temperature

(Yasuhara and Danovaro 2016). Rapid deep water warming, approximately 0.1 °C per decade, has occurred in the western Mediterranean over the last 50 years (Bethoux et al. 1990). The Ladrador sea has experienced dynamic decadal temperature variations for about 60 years, with rates of change reaching up to 0.5 °C per decade (Van Aken et al. 2011). The abyssal depths of the North Atlantic Ocean is roughly 1 °C warmer than the North Pacific Ocean, with even greater temperature differences at bathyal depths. Some marginal seas, like the Mediterranean, Red, and Sulu seas, exhibit extremely high deep-sea temperatures, ranging from approximately 13 °C in the Mediterranean to over 20 °C in the Red Sea at 2000 m depth (Yasuhara and Danovaro 2016). While this warming may seem minimal compared to surface waters, even slight changes can have profound effects on deep-sea ecosystems. Many benthic organisms have evolved to thrive in stable, cold environments, and are highly sensitive to temperature fluctuations. The relationship between temperature and marine diversity is a widespread ecological phenomenon (Tittensor et al. 2010; Yasuhara and Danovaro 2016). As temperatures rise, these species may face physiological stress, reduce reproductive success, and alter growth rates. Deep sea organisms are likely more sensitive to temperature changes compared to coastal taxa because deep sea temperature is relatively stable, particularly on daily and seasonal scales. A temporal analysis in the deep Mediterranean Sea showed that minor temperature shift of 0.1 °C or less can cause significant changes in the biodiversity and community structure of deep-sea nematode assemblages (Danovaro et al. 2004). Some species may be forced to migrate to cooler waters, potentially disrupting established ecological relationships and community structures. Species with a wide range of a wide bathymetric distribution, such as some cold water corals (*Dendrophyllia* sp., and *Desmophyllum* sp.), can show higher thermal tolerance, whose distribution spans 8 to >4000 m water depth. Moreover, smaller organisms such as nematodes, with shorter generation times may react strongly and rapidly to temperature changes, and may adapt to the such changes through generations, but the larger biomass taxa with longer generation time may be more resistant to rapid changes, but less able to adapt quickly (Yasuhara and Danovaro 2016).

Ocean acidification, another consequence of climate change, poses a significant threat to deep-sea benthic biodiversity in the Indian Ocean. The average CO_2 levels has increased to 385 ppm from 280 ppm (pre-industrialization) and is expected to reach 500–1000 ppm by the end of twenty-first century (Meehl et al. 2007; Smith et al. 2009). The estimates indicate that approx. 30% of recently released CO_2 has been absorbed by the ocean thus far (Feely et al. 2009). As atmospheric carbon dioxide levels increase, more CO_2 is absorbed by the ocean, leading to a decrease in pH levels (Guinotte and Fabry 2008). This process is particularly concerning for calcifying organisms such as deep-sea corals and mollusks. These species rely on aragonite and calcite to build their skeletons and shells, a process that becomes increasingly difficult in more acidic conditions. The increased influx of CO_2 to the deep-sea waters is reducing the saturation states of aragonite and calcite, leading to profound implications on the distribution of deep-sea corals and other calcifiers (Hofmann et al. 2010). Measurements of reduced growth is observed in the deep-sea coral (*Lophelia Pertusa*), which supports that the distribution of corals is regulated by aragonite levels (Maier et al. 2009). Ocean acidification also leads to disrupted reproductive processes in organisms such as deep-sea annelid (*Ophryotrocha* sp.), leading to fewer spermatozoa and lower effective fecundity, with slower embryo and larval development (Verkaik et al. 2017). The potential loss or decline of these key species could have cascading effects throughout the entire deep-sea ecosystem, altering food webs and habitat structures.

Changes in ocean circulation patterns, driven by climate change (Sharma et al. 2022; Sun et al. 2022; Toggweiler and Russell 2008), are likely to have profound impacts on deep-sea benthic biodiversity in the Indian Ocean. The Indian Ocean is characterized by complex circulation systems, including the Indian Ocean Dipole (IOD) and the monsoon-driven currents (Chakraborty and Singhai 2021; Schott et al. 2009). Climate change is expected to alter these patterns, potentially affecting the distribution of nutrients, oxygen, and organic matter to the deep sea. Climate models project stronger and more frequent positive IOD events, characterized by warmer western and cooler eastern Indian Ocean temperatures. These events disrupt monsoon-driven upwelling, reducing nutrient supply to eastern regions and increasing stratification, which limits vertical mixing and organic matter flux to deeper waters (Cai et al. 2021; Phillips et al. 2021), eventually affecting the deep-sea ecosystem. Such changes could lead to shifts in productivity patterns, potentially benefiting some species while disadvantaging others. Circulation patterns in the deep sea influence larval dispersal, potentially affecting the connectivity between deep-sea populations and impacting genetic diversity (Galaska et al. 2021; Gary et al. 2020), if altered may affect these processes adversely (Bashevkin et al. 2020).

Deoxygenation is another critical concern for deep-sea benthic biodiversity in the Indian Ocean. As surface waters warm, they become less able to hold dissolved oxygen (Keeling et al. 2010). This, combined with increased stratification of the water column, can lead to reduced oxygen levels in deeper waters. The Indian Ocean already contains several naturally occurring oxygen minimum zones, and climate change is expected to exacerbate these conditions (Fernandes et al. 2020; Gu et al. 2022; Ingole et al. 2010; Joydas and Damodaran 2014; Lincy and Manohar 2020; Raman et al. 2015; Rixen et al. 2020). Expanding low-oxygen areas could lead to habitat compression for many benthic species, potentially resulting in local extinctions or forced migrations (Breitburg et al. 2018). Some opportunistic species adapted to low-oxygen conditions may thrive, potentially leading to shifts in community composition and ecosystem functioning (Noisette et al. 2022).

Climate change is also likely to impact the quantity and quality of organic matter reaching the deep sea, a crucial energy source for benthic communities (Sweetman et al. 2017). Changes in surface productivity, driven by alterations in temperature, nutrient availability, and circulation patterns, will directly affect the food supply to the deep sea (Rogers 2015; Smith et al. 2013). This could lead to changes in the abundance and distribution of benthic organisms, potentially altering food web dynamics and ecosystem services provided by these communities.

The deep Indian Ocean is home to numerous unique and fragile ecosystems, such as hydrothermal vents, cold seeps, and seamounts. These habitats support highly specialized and often endemic species that are particularly vulnerable to environmental changes (Ingole and Koslow 2005; Van der Most et al. 2023). Climate-induced alterations in ocean chemistry, temperature, and circulation could have severe impacts on these ecosystems. For example, changes in the stability of gas hydrates due to warming could affect the energy supply to cold seep communities, while alterations in ocean chemistry could impact the formation and persistence of hydrothermal vent structures. Moreover, climate change may exacerbate the impacts of other anthropogenic stressors on deep-sea benthic biodiversity in the Indian Ocean. As surface waters warm and become less productive, there may be increased pressure to exploit deep-sea resources, including fisheries and mineral deposits. This could lead to direct habitat destruction and further stress on already vulnerable populations. Additionally, the potential for range shifts of surface and mid-water species into deeper waters could introduce new predators or competitors to benthic communities, further disrupting established ecological relationships. The impacts of climate change on deep-sea benthic biodiversity in the Indian Ocean are likely to be complex and multifaceted, with potential for both direct and indirect effects. Some species may benefit from changing conditions, potentially leading to range expansions or increased abundance. However, the overall trend is expected to be negative, with many species facing increased stress, reduced habitat availability, and potential extinction risks.

10.5 Future Directions

Understanding and predicting these impacts is challenging due to the limited baseline data available for many deep-sea ecosystems in the Indian Ocean. To better understand and predict the impacts of climate change on deep-sea benthic biodiversity, several key research areas require attention:

- Long-term monitoring: Establishing and maintaining long-term observatories in diverse deep-sea habitats is crucial for detecting and understanding temporal changes in benthic communities.
- Improved spatial coverage: Expanding research efforts to understudied regions, particularly in the Southern Hemisphere, will provide a more comprehensive picture of global deep-sea biodiversity patterns and their responses to climate change.
- Experimental studies: Conducting in situ and laboratory experiments to assess the physiological tolerances and adaptive capacities of deep-sea organisms under future climate scenarios.
- Integrative approaches: Combining traditional taxonomic methods with advanced molecular techniques and ecological modeling to better characterize deep-sea biodiversity and predict future changes.
- Ecosystem functioning: Investigating the links between biodiversity and ecosystem processes in the deep sea to understand the potential consequences of biodiversity loss on global biogeochemical cycles.

- Cumulative impacts: Assessing the combined effects of climate change and other anthropogenic stressors on deep-sea benthic ecosystems to inform management and conservation strategies.
- Technological advancements: Developing and deploying new technologies for deep-sea exploration and monitoring, including autonomous underwater vehicles, environmental sensors, and high-resolution imaging systems.

10.6 Conservation and Management Implications

The growing recognition of climate change impacts on deep-sea benthic biodiversity has important implications for conservation and management: 1. Marine protected areas: Designing and implementing networks of marine protected areas that account for potential range shifts and changes in habitat suitability under future climate scenarios. 2. Ecosystem-based management: Adopting holistic approaches that consider the interconnectedness of deep-sea ecosystems and their vulnerability to multiple stressors. 3. International cooperation: Enhancing collaboration among nations to address the transboundary nature of deep-sea conservation challenges and to share resources and expertise. 4. Precautionary approach: Implementing precautionary measures in the face of scientific uncertainty, particularly concerning emerging industries such as deep-sea mining. 5. Climate change mitigation: Recognizing the role of deep-sea ecosystems in carbon sequestration and incorporating this knowledge into global climate change mitigation strategies.

10.7 Conclusion

As climate change continues to alter the marine environment, understanding and predicting its impacts on deep-sea benthic biodiversity becomes increasingly important. While significant progress has been made in recent years, many knowledge gaps remain. Climate change poses a significant threat to deep-sea benthic biodiversity in the Indian Ocean. The combined effects of ocean warming, acidification, deoxygenation, and changes in circulation patterns and productivity are likely to result in substantial alterations to these unique and valuable ecosystems. Given the importance of deep-sea benthic communities in global biogeochemical cycles, carbon sequestration, and maintaining overall ocean health, the potential loss of biodiversity in these areas could have far-reaching consequences for the entire planet. Future research efforts should focus on expanding our understanding of deep-sea biodiversity patterns, ecosystem functioning, and the complex interactions between climate change and other anthropogenic stressors. As such, there is an urgent need for increased research, monitoring, and conservation efforts focused on these vulnerable ecosystems to better understand and mitigate the impacts of climate change on deep-sea benthic biodiversity in the Indian Ocean. The conservation and sustainable management of deep-sea ecosystems in the face of climate change will require interdisciplinary collaboration, technological innovation, and adaptive policy frameworks. By advancing our knowledge and taking proactive measures, we can work towards preserving the unique and valuable biodiversity of the deep-sea benthos for future generations.

Acknowledgment The authors are thankful to the authorities of CSIR-NIO for providing the facilities to develop this article.

Declaration Statement The authors declare that the study was conducted following the ethical standards of the institutional and/or national research committee. Informed consent was obtained from all individual participants. All participants were informed about the purpose of the study. The confidentiality and anonymity of the data collected were assured to all participants.

Data Availability Statement The original contributions presented in the study are included in the article/Supplementary Material. Further inquiries can be directed to the corresponding author.

Ethics Approval The research involves a comprehensive analysis of existing scholarly literature, which does not require ethical approval.

Conflict of Interest The authors confirm that their study was performed without any commercial or financial connections that could be interpreted as conflicts of interest.

Funding No funding is received for this work.

References

Ansari ZA (2000) Distribution of deep-sea benthos in the proposed mining area of Central Indian Basin. Mar Georesour Geotechnol 18:201–207. https://doi.org/10.1080/10641190009353788

Baco AR (2007) Exploration for deep-sea corals on North Pacific seamounts and islands. Oceanography 20:108–117

Banerjee R (2000) A documentation on burrows in hard substrates of ferromanganese crusts and associated soft sediments from the Central Indian Basin. Curr Sci:517–521

Bashevkin SM, Dibble CD, Dunn RP, Hollarsmith JA, Ng G, Satterthwaite EV, Morgan SG (2020) Larval dispersal in a changing ocean with an emphasis on upwelling regions. Ecosphere 11(1). https://doi.org/10.1002/ecs2.3015

Bethoux JP, Gentili B, Raunet J, Tailliez D (1990) Warming trend in the western Mediterranean deep water. Nature 347(6294):660–662. https://doi.org/10.1038/347660a0

Boehlert GW, Genin A (2013) A review of the effects of seamounts on biological processes. Seamounts Islands Atolls:319–334. https://doi.org/10.1029/gm043p0319

Breitburg D, Levin LA, Oschlies A, Grégoire M, Chavez FP, Conley DJ, Garçon V, Gilbert D, Gutiérrez D, Isensee K, Jacinto GS (2018) Declining oxygen in the global ocean and coastal waters. Science 359(6371):eaam7240

Bridges AE, Barnes DK, Bell JB, Ross RE, Howell KL (2022) Depth and latitudinal gradients of diversity in seamount benthic communities. J Biogeogr 49(5):904–915. https://doi.org/10.1111/jbi.14355

Cai W, Yang K, Wu L, Huang G, Santoso A, Ng B, Wang G, Yamagata T (2021) Opposite response of strong and moderate positive Indian Ocean Dipole to global warming. Nat Clim Chang 11(1):27–32. https://doi.org/10.1038/s41558-020-00943-1

Canganella F (2002) Deep ocean ecosystems. www.els.net
Chakraborty A, Singhai P (2021) Asymmetric response of the Indian summer monsoon to positive and negative phases of major tropical climate patterns. Sci Rep 11(1):22561. https://doi.org/10.1038/s41598-021-01758-6
Chertoprud E, Novichkova A (2023) Meiofauna: biodiversity, ecology, and role in ecosystems. In: Diversity, vol 15, issue 9. Multidisciplinary Digital Publishing Institute (MDPI). https://doi.org/10.3390/d15090987
Clark MR, Rowden AA, Schlacher T, Williams A, Consalvey M, Stocks KI, Rogers AD, O'Hara TD, White M, Shank TM, Hall-Spencer JM (2010) The ecology of seamounts: structure, function, and human impacts. Annu Rev Mar Sci 2:253–278. https://doi.org/10.1146/annurev-marine-120308-081109
Copley JT, Marsh L, Glover AG, Hühnerbach V, Nye VE, Reid WDK, Sweeting CJ, Wigham BD, Wiklund H (2016) Ecology and biogeography of megafauna and macrofauna at the first known deep-sea hydrothermal vents on the ultraslow-spreading Southwest Indian Ridge. Sci Rep 6:39158. https://doi.org/10.1038/srep39158
Danovaro R, Dell'Anno A, Pusceddu A (2004) Biodiversity response to climate change in a warm deep sea. Ecol Lett 7(9):821–828. https://doi.org/10.1111/j.1461-0248.2004.00634.x
De Smet B, Simon-Ledó E, Mevenkamp L, Pape E, Pasotti F, Jones DOB, Vanreusel A (2021) The megafauna community from an abyssal area of interest for mining of polymetallic nodules. Deep Sea Res Part 1 Oceanogr Res Pap 172:103530. https://doi.org/10.1016/j.dsr.2021.103530
Desa E (1997) Initial results of India's environmental impact assessment of nodule mining. National Institute of Oceanography, India, pp 49–63
Ebbe B, Billett D, Brandt A, Ellingsen K, Glover A, Keller S, Malyutina M, Mart P, Arbizu N, Molodtsova T, Rex M, Smith C, Tselepides A (2010) Diversity of abyssal marine life. In: Life in the world's oceans: diversity, distribution, and abundance, pp 139–160. https://doi.org/10.1002/9781444325508.ch8
Feely R, Doney S, Cooley S (2009) Ocean acidification: present conditions and future changes in a high-CO2 world. Oceanography 22(4):36–47. https://doi.org/10.5670/oceanog.2009.95
Fernandes GL, Shenoy BD, Damare SR (2020) Diversity of bacterial community in the oxygen minimum zones of Arabian Sea and Bay of Bengal as deduced by Illumina sequencing. Front Microbiol 10. https://doi.org/10.3389/fmicb.2019.03153
Gaikwad S, Sautya S, Damare S, Mascarenhas-pereira MBL, Gawas V, Patil J, Nanajkar M, Bukhari S (2024) Macrobenthic communities in the polymetallic nodule field, Indian Ocean, based on multicore and box core analysis. Front Mar Sci. https://doi.org/10.3389/fmars.2024.1395892
Galaska MP, Liu G, West D, Erickson K, Quattrini AM, Bracco A, Herrera S (2021) Seascape genomics reveals metapopulation connectivity network of Paramuricea biscaya in the Northern Gulf of Mexico. Front Mar Sci 8. https://doi.org/10.3389/fmars.2021.790929
Gary SF, Fox AD, Biastoch A, Roberts JM, Cunningham SA (2020) Larval behaviour, dispersal and population connectivity in the deep sea. Sci Rep 10(1):10675. https://doi.org/10.1038/s41598-020-67503-7
Glasby GP (1973) Distribution of manganese nodules and lebensspuren in underwater photographs from the Carlsberg Ridge, Indian Ocean. N Z J Geol Geophys 16:1–17. https://doi.org/10.1080/00288306.1973.10425383
Griffiths JR et al (2017) The importance of benthic–pelagic coupling for marine ecosystem functioning in a changing world. Glob Chang Biol 23(6):2179–2196. https://doi.org/10.1111/gcb.13642
Gu B, Liu J, Cheung S, Ho NHE, Tan Y, Xia X (2022) Insights into prokaryotic community and its potential functions in nitrogen metabolism in the Bay of Bengal, a pronounced oxygen minimum zone. Microbiol Spectr 10(3). https://doi.org/10.1128/spectrum.00892-21
Guinotte JM, Fabry VJ (2008) Ocean acidification and its potential effects on marine ecosystems. In: Annals of the New York Academy of Sciences, vol 1134. Blackwell Publishing Inc, pp 320–342. https://doi.org/10.1196/annals.1439.013

Hein JR, Mizell K, Koschinsky A, Conrad TA (2013) Deep-ocean mineral deposits as a source of critical metals for high- and green-technology applications: comparison with land-based resources. Ore Geol Rev 51:1–14. https://doi.org/10.1016/j.oregeorev.2012.12.001

Hestetun JT, Rapp HT, Xavier J (2017) Carnivorous sponges (Porifera, Cladorhizidae) from the Southwest Indian Ocean Ridge seamounts. In: Deep sea research part II: topical studies in oceanography, advances in deep-sea biology: biodiversity, ecosystem functioning and conservation, vol 137, pp 166–189. https://doi.org/10.1016/j.dsr2.2016.03.004

Hofmann GE, Barry JP, Edmunds PJ, Gates RD, Hutchins DA, Klinger T, Sewell MA (2010) The effect of ocean acidification on calcifying organisms in marine ecosystems: an organism-to-ecosystem perspective. Annu Rev Ecol Evol Syst 41:127–147. https://doi.org/10.1146/annurev.ecolsys.110308.120227

IHO (2019) Standardization of undersea feature names (Edition 4.2.0). International Hydrographic Organization. https://iho.int/uploads/user/pubs/bathy/B-6_e4%202%200_2019_EF_clean_3Oct2019.pdf

Iken K, Brey T, Wand U, Voigt J, Junghans P (2001) Food web structure of the benthic community at the Porcupine Abyssal Plain (NE Atlantic): a stable isotope analysis. In: Progress in oceanography, high resolution temporal and spatial study of the benthic biology and geochemistry of a north-eastern Atlantic abyssal locality (BENGAL), vol 50, pp 383–405. https://doi.org/10.1016/S0079-6611(01)00062-3

Ingole B, Koslow JA (2005) Deep-sea ecosystems of the Indian Ocean. Indian J Mar Sci 34(1):27–34

Ingole BS, Ansari ZA, Rathod V, Rodrigues N (2000) Response of meiofauna to immediate benthic disturbance in the Central Indian Ocean Basin. Mar Georesour Geotechnol 18:263–272. https://doi.org/10.1080/10641190009353794

Ingole BS, Ansari ZA, Rathod V, Rodrigues N (2001) Response of deep-sea macrobenthos to a small-scale environmental disturbance. In: Deep Sea research part II: topical studies in oceanography, Indian Deep-sea Environment Experiment (INDEX), vol 48, pp 3401–3410. https://doi.org/10.1016/S0967-0645(01)00048-0

Ingole BS, Goltekar R, Gonsalves S, Ansari ZA (2005a) Recovery of deep-sea meiofauna after artificial disturbance in the Central Indian Basin. Mar Georesour Geotechnol 23(4):253–266. https://doi.org/10.1080/10641190500446540

Ingole BS, Pavithran S, Ansari ZA (2005b) Restoration of deep-sea macrofauna after simulated benthic disturbance in the Central Indian Basin. Mar Georesour Geotechnol 23(4):267–288. https://doi.org/10.1080/10641190500446573

Ingole BS, Sautya S, Sivadas S, Singh R, Nanajkar M (2010) Macrofaunal community structure in the western Indian continental margin including the oxygen minimum zone. Mar Ecol 31(1):148–166. https://doi.org/10.1111/j.1439-0485.2009.00356.x

Jamieson AJ, Onda DFL (2022) Lebensspuren and müllspuren: drifting plastic bags alter microtopography of seafloor at full ocean depth (10,000 m, Philippine Trench). Cont Shelf Res 250:104867. https://doi.org/10.1016/j.csr.2022.104867

Joydas TV, Damodaran R (2014) Infaunal macrobenthos of the oxygen minimum zone on the Indian western continental shelf. Mar Ecol 35(1):22–35. https://doi.org/10.1111/maec.12052

Keeling RF, Körtzinger A, Gruber N (2010) Ocean deoxygenation in a warming world. Annu Rev Mar Sci 2(1):199–229. https://doi.org/10.1146/annurev.marine.010908.163855

Kuhn T, Wegorzewski A, Rühlemann C, Vink A (2017) Composition, formation, and occurrence of polymetallic nodules. In: Sharma R (ed) Deep-sea mining: resource potential, technical and environmental considerations. Springer International Publishing, Cham, pp 23–63. https://doi.org/10.1007/978-3-319-52557-0_2

Kumagai H, Nakamura K, Toki T, Morishita T, Okino K, Ishibashi J-I, Tsunogai U, Kawagucci S, Gamo T, Shibuya T, Sawaguchi T, Neo N, Joshima M, Sato T, Takai K (2008) Geological background of the Kairei and Edmond hydrothermal fields along the Central Indian Ridge: implications of their vent fluids' distinct chemistry. Geofluids 8:239–251. https://doi.org/10.1111/j.1468-8123.2008.00223.x

Laroche O, Kersten O, Smith CR, Goetze E (2020) Environmental DNA surveys detect distinct metazoan communities across abyssal plains and seamounts in the western Clarion Clipperton Zone. Mol Ecol 29(23):4588–4604. https://doi.org/10.1111/mec.15484

Lincy J, Manohar CS (2020) A comparison of bacterial communities from OMZ sediments in the Arabian Sea and the Bay of Bengal reveals major differences in nitrogen turnover and carbon recycling potential. Mar Biol Res 16(8–9):656–673. https://doi.org/10.1080/17451000.2020.1840593

Lörz A-N, Schwentner M, Bober S, Jażdżewska AM (2023) Multi-ocean distribution of a brooding predator in the abyssal benthos. Sci Rep 13:15867. https://doi.org/10.1038/s41598-023-42942-0

Lutz RA, Kennish MJ (1993) Ecology of deep-sea hydrothermal vent communities: A review. Rev Geophys 31(3):211–242. https://doi.org/10.1029/93RG01280

Maier C, Hegeman J, Weinbauer MG, Gattuso JP (2009) Calcification of the cold-water coral *Lophelia pertusa* under ambient and reduced pH. Biogeosciences 6:1671–1680

Meehl GA, Stocker TF, Collins WD, Friedlingstein P, Gaye AT et al (2007) Global climate projections. In: Solomon S, Qin D, Manning M, Chen Z, Marquis M et al (eds) Climate change 2007: the physical science basis. Contribution Working Group I 4th assessment report of the intergovernmental panel in climate change. Cambridge University Press, Cambridge, UK, pp 747–845

Mukhopadhyay R, Ghosh AK (2010) Dynamics of formation of ferromanganese nodules in the Indian Ocean. J Asian Earth Sci 37:394–398. https://doi.org/10.1016/j.jseaes.2009.09.003

Mukhopadhyay R, Iyer SD, Ghosh AK (2002) The Indian Ocean Nodule Field: petrotectonic evolution and ferromanganese deposits. Earth Sci Rev 60:67–130. https://doi.org/10.1016/S0012-8252(02)00078-8

Mukhopadhyay R, Naik S, De Souza S, Dias O, Iyer SD, Ghosh AK (2019) The economics of mining seabed manganese nodules: a case study of the Indian Ocean nodule field. Mar Georesour Geotechnol 37:845–851. https://doi.org/10.1080/1064119X.2018.1504149

Murton BJ, Rona PA (2015) Carlsberg Ridge and Mid-Atlantic Ridge: comparison of slow spreading centre analogues. In: Deep sea research part II: topical studies in oceanography, exploring new frontiers in deep-sea research: in honor and memory of Peter A. Rona, vol 121, pp 71–84. https://doi.org/10.1016/j.dsr2.2015.04.021

Narayanaswamy BE, Rea T, Serpetti N, Lamont PA (2017) What lies within; annelid polychaetes found in micro-habitats of coral/carbonate material from SW Indian ocean seamounts. In: Deep sea research part II: topical studies in oceanography, advances in deep-sea biology: biodiversity, ecosystem functioning and conservation, vol 137, pp 157–165. https://doi.org/10.1016/j.dsr2.2016.06.018

Noisette F, Pansch C, Wall M, Wahl M, Hurd CL (2022) Role of hydrodynamics in shaping chemical habitats and modulating the responses of coastal benthic systems to ocean global change. In: Global change biology, vol 28, issue 12. Wiley, pp 3812–3829. https://doi.org/10.1111/gcb.16165

O'Hara TD, Williams A, Ahyong ST, Alderslade P, Alvestad T, Bray D et al (2020) The lower bathyal and abyssal seafloor fauna of eastern Australia. Mar Biodivers Rec 13:1–27. https://doi.org/10.1186/s41200-020-00194-1

O'Loughlin PM, Mackenzie M, VandenSpiegel D (2013) New sea cucumber species from the seamounts on the Southwest Indian Ocean Ridge (Echinodermata: Holothuroidea: Aspidochirotida, Elasipodida, Dendrochirotida). Mem Mus Vic 70:37–50. https://doi.org/10.24199/j.mmv.2013.70.04

Pape E, Bezerra TN, Gheerardyn H, Buydens M, Kieswetter A, Vanreusel A (2021) Potential impacts of polymetallic nodule removal on deep-sea meiofauna. Sci Rep 11(1):19996. https://doi.org/10.1038/s41598-021-99441-3

Papiol V, Cartes JE, Fanelli E (2014) The role of food availability in regulating the feeding dynamics and reproductive cycles of bathyal benthopelagic fish in the northwest mediterranean slope. Limnol Oceanogr 59(5):1779–1794. https://doi.org/10.4319/lo.2014.59.5.1779

Parulekar AH, Harkantra SN, Ansari ZA, Matondkar SGP (1982) Abyssal benthos of the central Indian Ocean. Deep Sea Res Part A Oceanogr Res Pap 29:1531–1537. https://doi.org/10.1016/0198-0149(82)90041-3

Parulekar AH, Ingole BS, Harkantra SN, Ansari ZA (1992) Deep sea benthos of the western and central Indian Ocean. In: Oceanography of the Indian Ocean. Oxford and IBH Publication Company, New Delhi, pp 261-267.

Patriat P, Sauter D, Munschy M, Parson L (1997) A survey of the Southwest Indian ridge axis between Atlantis II fracture zone and the Indian Ocean triple junction: regional setting and large scale segmentation. Mar Geophys Res 19:457–480. https://doi.org/10.1023/A:1004312623534

Pavithran S, Ingole B, Nanajkar M, Nath NN (2007) Macrofaunal diversity in the central Indian ocean basin. Biodiversity 8(3):11–16. https://doi.org/10.1080/14888386.2007.9712824

Periasamy R, Ingole B (2022) Three new carnivorous sponge species (Demospongiae: Cladorhizidae) from the seamounts of the Central Indian Ridge. Zootaxa 5162:451–486. https://doi.org/10.11646/ZOOTAXA.5162.5.1

Periasamy R, Kurianurian P, Ingole B (2022) A new report on the deep-sea sponge associated Spongicoloides weijiaensis Xu, Zhou & Wang, 2017 (Decapoda: Spongicolidae) from the Southwest Indian Ocean Ridge. Zootaxa 5195:5. https://doi.org/10.11646/ZOOTAXA.5195.3.5

Periasamy R, Kurian J, Ingole B (2023a) Two black corals (Anthozoa: Antipatharia) from the seamount of Central Indian Ridge (1917 m) and Deeper Southwest Indian Ridge (4007 m). Thalassas. https://doi.org/10.1007/s41208-023-00597-8

Periasamy R, Kurian P, Ingole B (2023b) A new deep-water coral species Telestula ridgensis sp. nov (Scleralcyonacea: Sarcodictyonidae) from the seamount of the Central Indian Ridge. Zootaxa 5254. https://doi.org/10.11646/zootaxa.5254.2.4

Periasamy R, Kurian PJ, Ingole B (2023c) Two new deep-water species of squat lobsters (Crustacea: Anomura: Galatheoidea) from the Central and Southwest Indian Ridge. Zootaxa 5231:3. https://doi.org/10.11646/ZOOTAXA.5231.2.3

Periasamy R, Kurian PJ, Ingole B (2023d) Two new bamboo corals species (Octocorallia: Keratoisididae) from the seamounts of slow-spreading Central Indian Ridge. Deep-Sea Res I Oceanogr Res Pap 201:104158. https://doi.org/10.1016/j.dsr.2023.104158

Periasamy R, Palayil JK, Ingole B (2023e) First record of sponge-associated deep-sea polychaete (Polynoidae: Bathymoorea) on the ultraslow-spreading Southwest Indian Ridge. Deep-Sea Res I Oceanogr Res Pap 191:103923. https://doi.org/10.1016/j.dsr.2022.103923

Phillips HE, Tandon A, Furue R, Hood R, Ummenhofer CC, Benthuysen JA, Menezes V, Hu S, Webber B, Sanchez-Franks A, Cherian D, Shroyer E, Feng M, Wijesekera H, Chatterjee A, Yu L, Hermes J, Murtugudde R, Tozuka T et al (2021) Progress in understanding of Indian Ocean circulation, variability, air–sea exchange, and impacts on biogeochemistry. Ocean Sci 17(6):1677–1751. https://doi.org/10.5194/os-17-1677-2021

Priede IG, Bergstad OA, Miller PI, Vecchione M, Gebruk A, Falkenhaug T, Billett DSM, Craig J, Dale AC, Shields MA, Tilstone GH, Sutton TT, Gooday AJ, Inall ME, Jones DOB, Martinez-Vicente V, Menezes GM, Niedzielski T, Sigurdsson T et al (2013) Does presence of a mid-ocean ridge enhance biomass and biodiversity? PLoS One 8(5). https://doi.org/10.1371/journal.pone.0061550

Priede IG, Muller-Karger FE, Niedzielski T, Gebruk AV, Jones DOB, Colaço A (2022) Drivers of biomass and biodiversity of non-chemosynthetic benthic fauna of the Mid-Atlantic Ridge in the North Atlantic. Front Mar Sci 9. https://doi.org/10.3389/fmars.2022.866654

Raman AV, Damodaran R, Levin LA, Ganesh T, Rao YKV, Nanduri S, Madhusoodhanan R (2015) Macrobenthos relative to the oxygen minimum zone on the East Indian margin, Bay of Bengal. Mar Ecol 36(3):679–700. https://doi.org/10.1111/maec.12176

Ramirez-Llodra E, Brandt A, Danovaro R, De Mol B, Escobar E, German CR, Levin LA, Martinez Arbizu P, Menot L, Buhl-Mortensen P, Narayanaswamy BE, Smith CR, Tittensor DP, Tyler PA, Vanreusel A, Vecchione M (2010) Deep, diverse and definitely different: unique attributes of the world's largest ecosystem. Biogeosciences 7:2851–2899. https://doi.org/10.5194/bg-7-2851-2010

Riehl T, Wölfl A-C, Augustin N, Devey CW, Brandt A (2020) Discovery of widely available abyssal rock patches reveals overlooked habitat type and prompts rethinking deep-sea biodiversity. Proc Natl Acad Sci 117:15450–15459. https://doi.org/10.1073/pnas.1920706117

Rixen T, Cowie G, Gaye B, Goes J, Do Rosário Gomes HR, Hood R, Lachkar Z, Schmidt H, Segschneider J, Singh A (2020) Reviews and syntheses: present, past, and future of the oxygen minimum zone in the northern Indian Ocean. In: Biogeosciences, vol 17, issue 23. Copernicus GmbH, pp 6051–6080. https://doi.org/10.5194/bg-17-6051-2020

Rodrigues N, Sharma R, Nagender Nath B (2001) Impact of benthic disturbance on megafauna in Central Indian Basin. In: Deep sea research part II: topical studies in oceanography, Indian Deep-sea Environment Experiment (INDEX), vol 48, pp 3411–3426. https://doi.org/10.1016/S0967-0645(01)00049-2

Rogers A (2012) An ecosystem approach to management of seamounts in the Southern Indian Ocean: volume 1: overview of seamount ecosystems and biodiversity. IUCN

Rogers AD (2015) Environmental change in the deep ocean. In: Annual review of environment and resources, vol 40. Annual Reviews Inc, pp 1–38. https://doi.org/10.1146/annurev-environ-102014-021415

Rogers AD (2018) Chapter four – The biology of seamounts: 25 years on. In: Sheppard C (ed) Advances in marine biology. Academic Press, pp 137–224. https://doi.org/10.1016/bs.amb.2018.06.001

Rogers AD, Taylor ML (2011) Benthic biodiversity of seamounts in the Southwest Indian Ocean. Cruise Report – R/V James Cook 066

Rogers AD, Alvheim O, Bemanaja E, Benivary D, Boersch-Supan P, Bornman TG, Cedras R, Du Plessis N, Gotheil S, Hoines A, Kemp K, Kristiansen J, Letessier T, Mangar V, Mazungula N, Mork T, Pinet P, Pollard R, Read J, Sonnekus T (2017) Pelagic communities of the South West Indian Ocean seamounts: R/V Dr Fridtjof Nansen Cruise 2009-410. In: Deep sea research part II: topical studies in oceanography, the pelagic ecology of seamounts of the South West Indian Ocean, vol 136, pp 5–35. https://doi.org/10.1016/j.dsr2.2016.12.010

Romero-Romero S, Miller EC, Black JA, Popp BN, Drazen JC (2021) Abyssal deposit feeders are secondary consumers of detritus and rely on nutrition derived from microbial communities in their guts. Sci Rep 11(1):12594. https://doi.org/10.1038/s41598-021-91927-4

Rowden AA, Schlacher TA, Williams A, Clark MR, Stewart R, Althaus F, Bowden DA, Consalvey M, Robinson W, Dowdney J (2010) A test of the seamount oasis hypothesis: seamounts support higher epibenthic megafaunal biomass than adjacent slopes. Mar Ecol 31(SUPPL. 1):95–106. https://doi.org/10.1111/j.1439-0485.2010.00369.x

Ruhl HA, Ellena JA, Smith KL, Karl DM (2008) Connections between climate, food limitation, and carbon cycling in abyssal sediment communities. www.pnas.org/cgi/content/full/

Saini A, Gaikwad S, Sautya S, Sarun Lal MA, Subin Raj VB, Ramesh NR, Ramesh R, Ramesh S, Ramadass GA (2025) First quantitative assessment of deep-sea acorn worm (Enteropneusta) in PMN-rich zones of the Central Indian Ocean Basin Deep Sea Research Part I: Oceanographic Research Papers 224:104560. https://doi.org/10.1016/j.dsr.2025.104560

Sautya S, Ingole B, Ray D, Stöhr S, Samudrala K, Raju KAK, Mudholkar A (2011a) Megafaunal community structure of Andaman seamounts including the Back-arc Basin – a quantitative exploration from the Indian Ocean. PLoS One 6:e16162. https://doi.org/10.1371/journal.pone.0016162

Sautya S, Tabachnick KR, Ingole B (2011b) A new genus and species of deep-sea glass sponge (Porifera, Hexactinellida, Aulocalycidae) from the Indian Ocean. Zookeys:13–21. https://doi.org/10.3897/zookeys.136.1626

Sautya S, Ingole B, Jones DOB, Ray D, Kameshraju KA (2017) First quantitative exploration of benthic megafaunal assemblages on the mid-oceanic ridge system of the Carlsberg Ridge, Indian Ocean. J Mar Biol Assoc UK 97:409–417. https://doi.org/10.1017/S0025315416000515

Sautya S, Gaikwad S, Khokher S, Pradhan UK, Chatterjee S, Choudhury A, Sahu B, Attri S (2021) Distribution pattern of the benthic meiofaunal community along the depth gradient of the Western Indian continental margin, including the OMZ and abyssal plain. Front Mar Sci 8:671444

Schott FA, Xie SP, McCreary JP (2009) Indian ocean circulation and climate variability. Rev Geophys 47(1). https://doi.org/10.1029/2007RG000245

Schratzberger M, Ingels J (2018) Meiofauna matters: the roles of meiofauna in benthic ecosystems. J Exp Mar Biol Ecol 502:12–25. https://doi.org/10.1016/j.jembe.2017.01.007

Sempéré J-C, Cochran JR (1997) The Southeast Indian Ridge between 88°E and 118°E: variations in crustal accretion at constant spreading rate. J Geophys Res Solid Earth 102:15489–15505. https://doi.org/10.1029/97JB00171

Sharma S, Ha KJ, Cai W, Chung ES, Bódai T (2022) Local meridional circulation changes contribute to a projected slowdown of the Indian Ocean Walker circulation. npj Clim Atmos Sci 5(1). https://doi.org/10.1038/s41612-022-00242-w

Shi W, Wang M (2021) A biological Indian Ocean Dipole event in 2019. Sci Rep 11(1). https://doi.org/10.1038/s41598-021-81410-5

Singh R, Miljutin DM, Miljutina MA, Martinez Arbizu P, Ingole BS (2014) Deep-sea nematode assemblages from a commercially important polymetallic nodule area in the Central Indian Ocean Basin. Mar Biol Res 10(9):906–916. https://doi.org/10.1080/17451000.2013.866251

Singh R, Sautya S, Ingole BS (2018) The community structure of the deep sea nematode community associated with polymetallic nodules in the Central Indian Ocean Basin. Deep-Sea Res II Top Stud Oceanogr 161(16):28. https://doi.org/10.1016/j.dsr2.2018.07.009

Smith CR, De Leo FC, Bernardino AF, Sweetman AK, Arbizu PM (2008) Abyssal food limitation, ecosystem structure and climate change. Trends Ecol Evol 23(9):518–528. https://doi.org/10.1016/j.tree.2008.05.002

Smith JB, Schneider SH, Oppenheimer M, Yohe GW, Hare W et al (2009) Assessing dangerous climate change through an update of the Intergovernmental Panel on Climate Change (IPCC) "reasons for concern". Proc Natl Acad Sci USA 106:4133–4137

Smith KL, Ruhl HA, Kahru M, Huffard CL, Sherman AD (2013) Deep ocean communities impacted by changing climate over 24 y in the abyssal northeast Pacific Ocean. Proc Natl Acad Sci USA 110(49):19838–19841. https://doi.org/10.1073/pnas.1315447110

Snelgrove P, Smith C (2002) A riot of species in an environmental calm: the paradox of the species-rich deep-sea floor. Oceanogr Mar Biol 40:311–342

Stöhr S, Sautya S, Ingole B (2012) Brittle stars (Echinodermata: Ophiuroidea) from seamounts in the Andaman Sea (Indian Ocean): first account, with descriptions of new species. J Mar Biol Assoc U K 92:1195–1208. https://doi.org/10.1017/S0025315412000240

Sun S, Fang Y, Zu Y, Liu L, Li K (2022) Increased occurrences of early Indian Ocean Dipole under global warming. Sci Adv 8:eadd6025

Suo Y, Li S, Li X, Guo L, Wang Y (2016) Crustal thickness anomalies in the Indian Ocean inferred from gravity analysis. Geol J 51:634–643. https://doi.org/10.1002/gj.2786

Swanborn DJ, Huvenne VA, Pittman SJ, Woodall LC (2022) Bringing seascape ecology to the deep seabed: A review and framework for its application. Limnol Oceanogr 67(1):66–88. https://doi.org/10.1002/lno.11976

Sweetman AK, Thurber AR, Smith CR, Levin LA, Mora C, Wei CL, Gooday AJ, Jones DOB, Rex M, Yasuhara M, Ingels J, Ruhl HA, Frieder CA, Danovaro R, Würzberg L, Baco A, Grupe BM, Pasulka A, Meyer KS et al (2017) Major impacts of climate change on deep-sea benthic ecosystems. In: Elementa, vol 5. University of California Press. https://doi.org/10.1525/elementa.203

Taylor ML, Rogers AD (2017) Primnoidae (Cnidaria: Octocorallia) of the SW Indian Ocean: new species, genus revisions and systematics. Zool J Linnean Soc 181:70–97

Thurston MH, Bett BJ, Rice AL, Jackson PAB (1994) Variations in the invertebrate abyssal megafauna in the North Atlantic Ocean. Deep-Sea Res I Oceanogr Res Pap 41:1321–1348. https://doi.org/10.1016/0967-0637(94)90100-7

Tittensor DP, Mora C, Jetz W, Lotze HK, Ricard D, Berghe EV, Worm B (2010) Global patterns and predictors of marine biodiversity across taxa. Nature 466(7310):1098–1101. https://doi.org/10.1038/nature09329

Toggweiler JR, Russell J (2008) Ocean circulation in a warming climate. In: Nature, vol 451, issue 7176. Nature Publishing Group, pp 286–288. https://doi.org/10.1038/nature06590

Van Aken HM, Femke de Jong M, Yashayaev I (2011) Decadal and multi-decadal variability of Labrador Sea Water in the north-western North Atlantic Ocean derived from tracer distributions: heat budget, ventilation, and advection. Deep Sea Res Part 1 Oceanogr Res Pap 58(5):505–523. https://doi.org/10.1016/j.dsr.2011.02.008

Van der Most N, Qian PY, Gao Y, Gollner S (2023) Active hydrothermal vent ecosystems in the Indian Ocean are in need of protection. In: Frontiers in marine science, vol 9. Frontiers Media S.A. https://doi.org/10.3389/fmars.2022.1067912

Vanreusel A, Hilario A, Ribeiro PA, Menot L, Arbizu PM (2016) Threatened by mining, polymetallic nodules are required to preserve abyssal epifauna. Sci Rep 6:26808. https://doi.org/10.1038/srep26808

Verkaik K, Hamel JF, Mercier A (2017) Impact of ocean acidification on reproductive output in the deep-sea annelid Ophryotrocha sp. (Polychaeta: Dorvilleidae). Deep-Sea Res II Top Stud Oceanogr 137:368–376. https://doi.org/10.1016/j.dsr2.2016.05.022

Wafar M, Venkataraman K, Ingole B, Khan SA, LokaBharathi P (2011) State of knowledge of coastal and marine biodiversity of Indian Ocean countries. PLoS One 6:e14613. https://doi.org/10.1371/journal.pone.0014613

Wang X, Liu X, Xu J (2019) Distribution patterns of meiofauna assemblages and their relationship with environmental factors of Deep Sea adjacent to the yap trench, Western Pacific Ocean. Front Mar Sci 6. https://doi.org/10.3389/fmars.2019.00735

Weston J (2021) Diversity and biogeography of Hadal Amphipoda. Doctoral dissertation. Newcastle University

Weston JNJ, Peart RA, Stewart HA, Ritchie H, Piertney SB, Linley TD, Jamieson AJ (2021) Scavenging amphipods from the Wallaby-Zenith Fracture Zone: extending the hadal paradigm beyond subduction trenches. Mar Biol 168:1. https://doi.org/10.1007/s00227-020-03798-4

Yasuhara M, Danovaro R (2016) Temperature impacts on deep-sea biodiversity. Biol Rev 91(2):275–287. https://doi.org/10.1111/brv.12169

Yesson C, Clark MR, Taylor ML, Rogers AD (2011) The global distribution of seamounts based on 30 arc seconds bathymetry data. Deep-Sea Res I Oceanogr Res Pap 58:442–453. https://doi.org/10.1016/j.dsr.2011.02.004

Zeng Q, Huang D, Lin R, Wang J (2018) Deep-sea metazoan meiofauna from a polymetallic nodule area in the Central Indian Ocean Basin. Mar Biodivers 48(1):395–405. https://doi.org/10.1007/s12526-017-0778-0

Chapter 11
The Contribution of the Blue Carbon Ecosystem in Reducing Climate Change and Conserving Carbon Stocks: An Indian Perspective

Darwin Ramteke

Abstract According to reports, Indian blue carbon ecosystems store about 67.35 Tg (Teragram) of carbon. Mangroves, seagrass, and salt marshes are the three principal Blue Carbon Ecosystems (BCEs) found throughout the 7516.6 km of Indian coastline, which includes the mainland, Andaman and Nicobar Islands, and Lakshadweep Islands. The Sundarbans, the biggest mangrove forest in the world, are situated between Bangladesh and India. The mangrove ecosystem is the largest in terms of area and has been shown to store 67 Tg of carbon. On the other hand, the estimated areas of seagrass and salt marsh are 500 km^2 and 1611 km^2, respectively, and are reported to store 0.0679 Tg of carbon combined. However, the baseline data regarding carbon sequestered by BCEs are highly limited, and more systematic and extensive field research is needed to understand the carbon sequestration potential. The extent of BCEs based on published literature indicates that the Indian coastline has potential BCEs that can help act as a natural solution to climate change by effectively absorbing and storing large amounts of carbon dioxide from the atmosphere.

Keywords Blue carbon · Climate change · Mangrove · Seagrass · Saltmarsh

11.1 Introduction

The phrase "blue carbon" was first used in 2009 to bring attention to the significance of conserving and restoring marine and coastal ecosystems to reduce climate change and the other ecosystem services they provide. Although the term "blue

D. Ramteke (✉)
Biological Oceanography Division, CSIR-National Institute of Oceanography, Dona Paula, Goa, India

Academy of Scientific and Innovative Research (AcSIR), Ghaziabad, India
e-mail: darwin@nio.org

B. P. Barathan et al. (eds.), *Navigating Climate Change: Impacts on Biodiversity and Ecosystem Resilience*, Sustainable Development and Biodiversity 42,
https://doi.org/10.1007/978-981-95-0409-1_11

carbon" has many different meanings, its original definitions included all organic matter that marine organisms collect as well as how marine ecosystems may be managed to lower greenhouse gas emissions and thereby aid in the mitigation and conservation of climate change (Lovelock and Duarte 2019). Mangroves, seagrass beds, and salt marshes are coastal vegetation and globally considered traditional blue carbon ecosystems (BCEs) (Fig. 11.1) since their net autotrophic ratio is higher than 1. This is because, according to the ratio of gross primary production (GPP or P_{GPP}) to the sum of all respiration (R_E), more plant organic carbon is produced in these ecosystems than is lost through respiration. Despite reports of high productivity, the coastal ocean is not regarded as a BCE (blue carbon ecosystem) because its mean P_{GPP}/R_E ratio is just 0.98 (Alongi 2020).

BCE's high carbon sequestration rates are contributed by three primary factors: their potential for organic matter production, carbon absorption, and storage. Their location at the land–sea interface and high capacity for carbon trapping make them extremely prolific ecosystems that produce large quantities of carbon (also known as "autochthonous carbon"), they can also accrete sediment and store carbon from other ecosystems (also known as "allochthonous carbon"), due to various reasons, but mostly because of their anaerobic, waterlogged environments that restrict decomposition, they can collect and preserve organic matter (carbon) in forms that are resistant to microbial degradation. However, following carbon deposition, three things could happen to carbon after it is placed in a blue carbon habitat: either

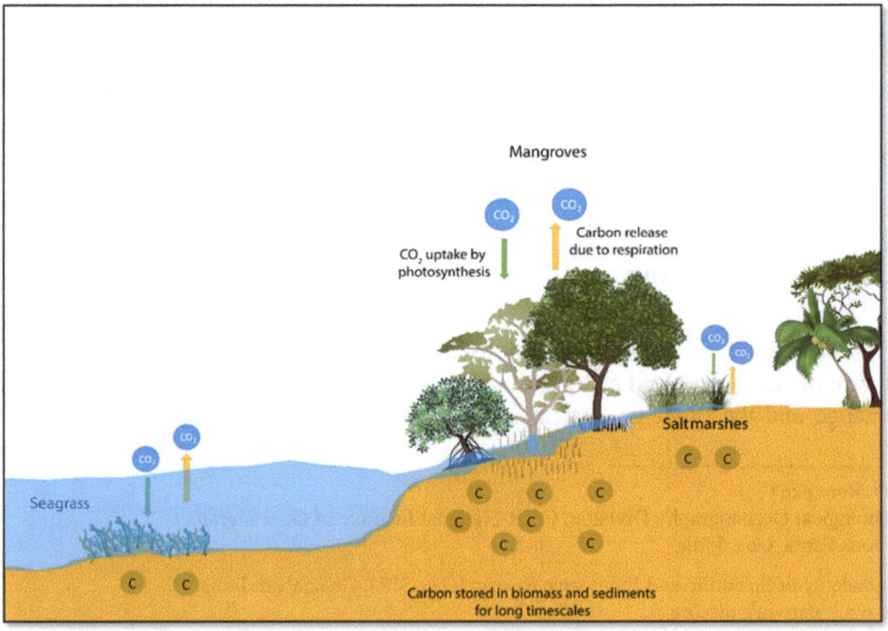

Fig. 11.1 Mangroves, seagrass, and salt marshes are three different kinds of blue carbon ecosystems that sequester carbon and act as natural carbon sinks in bottom sediments

microbes will break it down and absorb it into new biomass, they will either breathe again as CO_2 or the dissolved or particulate carbon in a form that is inaccessible to microorganisms and may contribute to the long-standing carbon store (Trevathan-Tackett et al. 2015). It is estimated that BCEs store more than 30,000 Tg ~185 million hectares worldwide; conserving them could prevent 304 (141–466) Tg of carbon dioxide equivalent (CO_2e) emissions annually. Estimates of potential BCE restoration in tidal marshes range from 0.2 to 3.2 million ha, in seagrass beds 8.3 to 25.4 million ha, and mangroves 9 to 13 million ha. These restorations could absorb an additional 841 (621–1064) Tg CO_2e annually by 2030, representing approximately 3% of global releases (global annual fossil fuel emissions for 2019 and 2020) (Macreadie et al. 2021).

Around 13.7 million hectares of mangrove forests can be found along tropical, subtropical, and warm temperate coasts. Mangrove forests store carbon in live biomass and soil, the range of carbon stocks in their live biomass pool is around 1230 Tg C to 3900 Tg C. Around 70% of carbon in mangrove forests is stored in the soil, with estimations for the top meter of soil varying around 1900 Tg C to 8400 Tg C. In Seagrass, an estimated 1732 to 21,000 Tg C of soil carbon stock is predicted to be stored from the projected global coverage of 16 to 165 million ha of seagrass meadows. In the case of tidal marshes, previous research estimated that the global area is between 2 and 40 million hectares (this estimate is likely conservative and incomplete), and estimated blue carbon stocks in tidal marsh soil are approximately 862 to 1350 Tg C (Macreadie et al. 2021). Macroalgae, compared to salt marshes, sea grasses, and mangroves, have been underestimated in the blue carbon purview because they are less effective at sequestering/storing carbon underground. However, they have been considered to create a big impact on the environment as "carbon donors," as they release carbon to adjoining ecosystems (Choudhary et al. 2024).

In recent decades, many blue carbon ecosystems have been subject to degradation and are at risk due to rising sea levels, loss of tidal habitats due to anthropogenic structures or measures that prevent land migration of coastal ecosystems, and reduced sediment load on coastal systems due to river dams. For example, it was predicted that around 17% of the 973,640 ha of mangroves lost between 1996 and 2016 could not be restored due to urbanization and erosion. Another example is that 1 ha of a BCE if lost corresponds to losing 10–40 hectares of native forest in terms of carbon emissions. The long-standing carbon storage in the sediment compartment of BCEs is primarily responsible for this discrepancy (McLeod et al. 2011).

Countries all across the world have decided that their main goals until 2030 will be the Sustainable Development Goals (SDGs) of the UN. The 17 Sustainable Development Goals include goals directly related to addressing climate change, SDG 13, and protecting and maintaining the use of oceans, coastlines, and aquatic resources, SDG 14. Restoration of mangroves would help achieve Sustainable Development Goal 13—"strengthening resistance and resilient potential of all nations to climate-related threats and catastrophic events" (Connor 2015). The Paris Agreement's 2016 Nationally Determined Contributions (NDCs) encourage countries to choose what steps they will take to reduce their greenhouse gas emissions. There are not many blue carbon pledges in NDCs at the moment. According to recent estimates, 64 nations have mentioned marine and coastal ecosystems in their NDCs when discussing climate adaptation and mitigation. Most of these nations have NDC obligations about mangroves (45), while seagrass has fewer (10) commitments. Although many nations allude to wetlands in their NDCs or pledge to use the Intergovernmental Panel on Climate Change (IPCC) Wetland appendage, which offers guidelines in estimating greenhouse gas (GHG) emissions and removals for certain actions related to seagrass, mangroves, and tidal marshes, are not particularly listed.

11.2 Indian Scenario

11.2.1 Mangroves

Around 3.3% of the world's mangrove cover is found in India, making up 0.15% of the country's total land area. Mangrove forests can be found along the coasts of nine Indian states and four Union Territories. With roughly 57% of its entire area situated alongside the east coast, 31% alongside the west coast, and the remaining 12% in the Andaman and Nicobar Islands, its estimated total area is 4992 km^2 (FSI 2021). The mangrove ecosystem of India can be broadly classified into three types: insular, which is located on the Andaman and Nicobar Islands; deltaic, which is found on the country's eastern coast; and estuarine and backwater, which are found on the country's western coast (Mandal and Naskar 2008). India is home to 46 true mangrove species, including 42 species and 4 natural hybrids, spread throughout 14 groups and 22 genera. The East Coast is home to 40 mangrove species from 14 families and 22 genera. The West Coast is home to 27 species, including 16 genera and 11 families. The Andaman and Nicobar Islands' mangroves are the most diverse in India, with 38 species spread across 13 families and 19 taxa (Ragavan et al. 2016).

Mangroves contribute 67 Tg of the total 67.35 Tg of carbon stored by Indian blue carbon ecosystems. In their meta-analysis, Akhand et al. (2023) offer a valuable summary of research on blue carbon in salt marshes, seagrass beds, and mangroves. In our review study (Ajay et al. 2025, under review), we found that Kerala mangrove forest has the highest values of ecosystem carbon contained in Indian mangroves. However, only 16 of the total 31 studies have assessed carbon from all three compartments—sediment, belowground, and aboveground—so the levels are questionable. Although Kerala mangroves account for the majority of blue carbon reports, there is comparatively little data on the blue carbon values of mangroves in several coastal regions, such as Karnataka, Goa, Maharashtra, Gujarat, and Andhra Pradesh.

11.2.2 Seagrass

Rich seagrass meadows along the Indian coast are found majorly in Palk Bay and the Gulf of Mannar with a total of 14 species, the Andaman and Nicobar Islands with a total 12 species, the Lakshadweep Islands (10 species), Odisha (8 species), and Gujarat (8 species) totaling around 16 seagrass species are found in India. According to studies based on remote sensing, around 225 km^2 of seagrass cover is found in four significant coastal regions (Palk Bay, Gulf of Mannar, Lakshadweep, and Andaman & Nicobar Islands). This does not include seagrass meadows in backwaters and estuaries, where substantial seagrass meadows are found deeper than 5 m. When these regions are taken into account, the country's seagrass extent is estimated to be around 500 km^2 (Thangaradjou and Bhatt 2018).

Akhand et al. (2023) reported a collective carbon stock of 0.0630 Tg C stored in Indian seagrass beds. However, the results of the researchers' studies in various regions revealed values that vary from one another. Ganguly et al. (2018) in their study reported carbon stock stored in the aboveground biomass of seagrass in the range of 0.20–0.96 Mg ha^{-1}, which was lower than the below-ground biomass of 0.31–2.93 Mg ha^{-1}. *Cymodocea* sp. and *Halophila* sp. showed the highest and lowest biomass, respectively, among the five species sampled. In the Gulf of Mannar, studies showed that seagrass sediments hold 196.84 Tg C ha^{-1}, and the potential of 4125 ha of seagrass beds sequester an annual average of 572.6 Mg ha^{-1} (Thangaradjou and Bhatt 2018). The carbon pool in the biomass and sediments of the Chilika lagoon on India's east coast was found to vary between 2.3 and 2.5 Gg C and 4.5 and 5.7 Gg C, respectively (Singh et al. 2015).

According to Kaladharan et al. (2020), a study in Palk Bay and the Gulf of Mannar (GoM), the blue carbon stock of seagrass meadows in Palk Bay was found to be around 0.001782 Tg C, whereas that of GoM was assessed to be 0.043996 Tg C. In Pulicat Lake, blue carbon stock stored in seagrass sediment (30 cm depth) was found to be 0.998 ± 0.418 Mg ha^{-1}, whereas in Chilika Lake, blue carbon stock down to 30 cm was estimated to be 2.018 ± 0.673 Mg ha^{-1} (Kaladharan et al. 2021a). In Kerala, at Kadalundi estuary, seagrass beds were reported to occur along the mangrove area. Kaladharan et al. (2021b) assessed blue carbon stock from seagrass beds and found that *Halophila* beds in the estuary had an average blue carbon stock of 2.655 ± 0.34 Mg ha^{-1}. Murugan et al. (2024) conducted a study on the carbon stock in the Pulicat seagrass beds situated in Andhra Pradesh and discovered that the average biomass of seagrass within the lake ranged from 1.43 to 2.01 Mg ha^{-1}, and the overall amount of blue carbon stored in the seagrass beds ranged from 0.49 to 9.52 Mg ha^{-1} and in the sediment stored carbon varied from 1.29 to 11.94 Mg ha^{-1}.

11.2.3 Salt Marsh

According to ground surveys and satellite sensing data, salt marshes encompass about 1611 km^2 in India. In total, Gujarat covers 1443 km^2, Tamil Nadu covers 61 km^2, the Andaman and Nicobar Islands cover 60 km^2, Andhra Pradesh covers 40 km^2, Maharashtra covers 6 km^2, Puducherry covers 0.7 km^2, and Daman and Diu cover 0.6 km^2 (SAC 2011). According to recently published research by Viswanathan et al. (2020), an estimated 290 km^2 of salt marsh is spread out along the Indian coastline, based on both ground surveys and remote sensing applications. However, data from Karnataka, Kerala, the Andaman and Nicobar Islands, and the Lakshadweep Islands were not included in their study. The salt marsh of India is composed of a total of 14 species belonging to 11 genera under six families. *Sesuvium portulacastrum,* due to its increased resistance to harsh environments, was found to be one of the most prevalent salt marsh species in tropical areas and is extensively scattered throughout India (Viswanathan et al. 2020).

Studies about the blue carbon stock of Indian salt marshes are very limited. Das et al. (2015), in their study about *Suaeda maritima* beds in Sundarbans, found carbon stored in above-ground biomass to be 0.66 to 0.71 Mg ha^{-1}, whereas in below-ground biomass it was 0.02–0.03 Mg ha^{-1}. In soil up to a depth of 30 cm, the carbon stock stored was 9.4–13.4 Mg ha^{-1}. In their 2016 study on *Salicornia brachiata* beds in Gujarat, Rathore et al. discovered that the amount of carbon retained in above-ground biomass ranged from 0.77 to 1.93 Mg ha^{-1}, and the carbon stock contained in soil up to 30 cm was 4.5–8.2 Mg ha^{-1}. Chaudhary et al. (2018) studied *Salicornia brachiata* beds in Gujarat. They found that the carbon stock preserved in above-ground biomass was around 0.05 to 2.1 Mg ha^{-1}, whereas in below-ground biomass, it ranged from 0.009 to 0.335 Mg ha^{-1}. In their investigation, Kaviarasan et al. (2019) determined that the rainy season's sediment organic carbon stored in the Tuticorin salt marsh was at its highest (54.46 Mg ha^{-1}), about seven times more than in the dry season. *A. indicum* had the highest above-ground biomass (14.87 g/cm^2) during the rainy season (January 2016), whereas *S. portulacastrum* had the lowest AGB (6.32 g/cm^2) during the dry season (July 2015). In *S. maritima* and *A. indicum*, the below-ground biomass ranged from 4.49 ± 0.35 to 1.86 ± 0.21 g/cm^2, respectively, as did the shares for the peak and low wet (January 2016) and dry (July 2015) season values.

Carbon sequestration rate can be accelerated, and a nature-based approach can be provided by increasing the density and cover of coastal vegetation and recovering BCEs. Protecting BCE habitats can reduce emissions and avoid their destruction, which can release methane (CH_4) and CO_2 (potent GHGs) into the atmosphere. According to Akhand et al. (2023), 99.5% of Indian blue carbon stock is stored in mangrove ecosystems, which are an important sink of carbon as they store huge amounts of carbon in their ecosystem. As a result, their destruction disrupts the carbon sink and may release a significant quantity of carbon back into the atmosphere, which speeds up climate change. Therefore, restoring and protecting mangrove habitats can mitigate the effects of climate change.

Mangroves that occur as riverine are considered to have the highest potential for carbon sequestration, followed by mangroves at the fringe, whereas the lowest potential of carbon sequestration is basin mangroves (Marchio Jr et al. 2016). Riverine mangroves with the maximum sequestration levels should be prioritized for conservation, preservation, protection, and density enhancement to increase sequestration levels. To reach their full potential or improve their capacities, restoration, and/or afforestation projects must also be implemented for basin and fringe mangroves. This can be done by mitigating factors that contribute to poor sequestration rates or by introducing species with better sequestration levels. Restoration can be adopted into two types: hydrological restoration and ecological restoration.

11.3 Conclusion

Evaluating the carbon stock stored in different blue carbon ecosystems is the first step towards comprehending the significance of any blue carbon ecosystem in reducing climate change. The information will serve as a starting point for understanding the current state of carbon sequestered by BCEs and will support the implementation of critical and vital actions for long-term restoration and improved carbon storage, both of which contribute to climate mitigation. Compared to the global scenario, there is a very small amount of research on blue carbon stock data in India. The overall projected value of carbon stock stored in Indian BCEs as stated in the review literature is underestimated because there are fewer than eight publications each describing blue carbon deposited in seagrass and tidal saltmarsh ecosystems, while only 48.5% of the total area of mangroves is sampled. There may be significant underestimation or overestimation in certain BCEs due to prejudiced sampling and a total lack of data in other areas. Future research should be oriented appropriately to reduce uncertainties and obtain a more confident estimate of the blue carbon stock. Resolving this issue will establish solid baseline data for India's "Blue Carbon Inventory" and assist in implementing precise strategies for utilizing the blue carbon ecosystem to lessen the effects of climate change.

Acknowledgments The author thanks the Director of CSIR-National Institute of Oceanography for providing the necessary facilities.

Funding Statement No external funds, grants, or other support was received to carry out this work.

Data Availability Statement Data are available from the authors on request.

Conflict of Interest The authors declare that they have no known competing financial interests or personal relationships that could have appeared to influence the work reported in this paper.

Ethics Approval This study does not involve humans or animals requiring ethical approval.

References

Akhand A, Chanda A, Jameel Y, Dasgupta R (2023) The present state-of-the-art of blue carbon repository in India: a meta-analysis. Sustain Sci 18(2):1031–1042

Alongi DM (2020) Global significance of mangrove blue carbon in climate change mitigation. Science 2(3):67

Chaudhary DR, Rathore AP, Jha B (2018) Aboveground, belowground biomass and nutrients pool in Salicornia brachiata at coastal area of India: interactive effects of soil characteristics. Ecol Res 33:1207–1218

Choudhary B, Dhar V, Pawase AS (2024) Blue carbon and the role of mangroves in carbon sequestration: its mechanisms, estimation, human impacts and conservation strategies for economic incentives. J Sea Res 199:102504

Connor R (2015) The United Nations world water development report: water for a sustainable world Vo l(1)UNESCO publishing

Das S, Zaman S, Pramanick P, Pal N, Mitra A (2015) Suaeda maritima: a potential carbon reservoir of coastal zone. Int Adv Res J Sci Eng Technol 2(5):61–65

FSI (2021) India state of forest report 2021. Forest Survey of India (Ministry of Environment, Forest & Climate Change), Dehradun. https://fsi.nic.in/isfr-2021/chapter-3.pdf

Ganguly D, Singh G, Purvaja R, Bhatta R, Selvam AP, Banerjee K, Ramesh R (2018) Valuing the carbon sequestration regulation service by seagrass ecosystems of Palk Bay and Chilika, India. Ocean Coast Manag 159:26–33

Kaladharan P, Zacharia PU, Thirumalaiselvan S, Anasu Koya A, Lavanya R, Batcha SM (2020) Blue carbon stock of the seagrass meadows of Gulf of Mannar and Palk Bay off Coromandel Coast, South India. Indian J Fish 67(4):149–153

Kaladharan P, Vijayakumaran K, Edward L, Lavanya R, Dash B (2021a) Blue carbon stock of seagrass meadows of Chilika and Pulicat Lakes along the Eastern coast of India. Fish Technol 58:143–146

Kaladharan P, Lavanya R, Akshara C (2021b) Capture and storage of carbon in seagrass beds of a vulnerable species Halophila beccarii (Asch.) at the Kadalundi Estuary, Kerala, India. J Mar Biol Assoc India 63(2):127–129

Kaviarasan T, Dahms HU, Gokul MS, Henciya S, Muthukumar K, Shankar S, Arthur James R (2019) Seasonal species variation of sediment organic carbon stocks in salt marshes of Tuticorin area, Southern India. Wetlands 39:483–494

Lovelock CE, Duarte CM (2019) Dimensions of blue carbon and emerging perspectives. Biol Lett 15(3):20180781

Macreadie PI, Costa MD, Atwood TB, Friess DA, Kelleway JJ, Kennedy H, Lovelock CE, Serrano O, Duarte CM (2021) Blue carbon as a natural climate solution. Nat Rev Earth Environ 2(12):826–839

Mandal RN, Naskar KR (2008) Diversity and classification of Indian mangroves: a review. Trop Ecol 49(2):131–146

Marchio DA Jr, Savarese M, Bovard B, Mitsch WJ (2016) Carbon sequestration and sedimentation in mangrove swamps influenced by hydrogeomorphic conditions and urbanization in Southwest Florida. Forests 7(6):116

Mcleod E, Chmura GL, Bouillon S, Salm R, Björk M, Duarte CM, Lovelock CE, Schlesinger WH, Silliman BR (2011) A blueprint for blue carbon: toward an improved understanding of the role of vegetated coastal habitats in sequestering CO2. Front Ecol Environ 9(10):552–560

Murugan R, Balachandar K, Boopathy NS, Nagarajan R, Ramasubramanian R (2024) Assessment of blue carbon in seagrass meadows in Pulicat Lake, Andhra Pradesh, India. Mar Pollut Bull 198:115868

Ragavan P, Saxena A, Jayaraj RSC, Mohan PM, Ravichandran K, Saravanan S, Vijayaraghavan A (2016) A review of the mangrove floristics of India. Taiwania 61(3):224–242

Rathore AP, Chaudhary DR, Jha B (2016) Biomass production, nutrient cycling, and carbon fixation by Salicornia brachiata Roxb.: a promising halophyte for coastal saline soil rehabilitation. Int J Phytoremediation 18(8):801–811

Singh G, Ganguly D, Paneer Selvam A, Kakolee K, Purvaja R, Ramesh R (2015) Seagrass ecosystem and climate change: an Indian perspective. J Clim Change 1(1–2):67–74

SAC (2011) Coastal Zones of India.Space Applications Centre, ISRO, Ahmedabad, India p. 597

Thangaradjou T, Bhatt JR (2018) Status of seagrass ecosystems in India. Ocean Coast Manag 159:7–15

Trevathan-Tackett SM, Kelleway J, Macreadie PI, Beardall J, Ralph P, Bellgrove A (2015) Comparison of marine macrophytes for their contributions to blue carbon sequestration. Ecology 96(11):3043–3057

Viswanathan C, Purvaja R, Jeevamani JJJ, Samuel VD, Sankar R, Abhilash KR, Geevarghese GA, Muruganandam R, Gopi M, Raja S, Das RR (2020) Salt marsh vegetation in India: species composition, distribution, zonation pattern and conservation implications. Estuar Coast Shelf Sci 242:106792

Chapter 12
Coral Reefs in Transition: Coral–Algal Interactions and Ecosystem Resilience

B. Manikandan

Abstract Coral Reefs are biodiversity hotspots and are highly productive ecosystems offering numerous benefits to mankind. However, they are among the most vulnerable ecosystems to the ongoing climate change trajectories. Recurrent mass bleaching events driven by anthropogenic climate change and natural phenomena like El-Niño have led to a significant decline in live coral cover across the reefs worldwide. This decline has created an ecological space for macroalgae (seaweeds) and offered a competitive advantage to them to outcompete corals and alter the overall benthic community structure. Globally, macroalgal cover on coral reefs has increased by 20% over the past decade. While macroalgae play important ecological roles contributing to primary production and nitrogen fixation, an unusual increase in their biomass, typically driven by environmental disturbances, can trigger cascading effects that compromise the reef stability, biodiversity, and reef services. Understanding the environmental feedbacks that significantly contribute to macroalgal dominance in reef ecosystems is key to predict and manage reef transitions. This chapter provides an overview about the nature of interaction between the coral and macroalgal communities and explores the mechanisms that underpin their competition and their possible implications on coral reef health and resilience. Also, the chapter reviews the existing management and restoration strategies aimed at mitigating macroalgal dominance and promoting coral recovery and highlights pathways for enhancing the adaptive capacity of reef ecosystems in Anthropocene.

Keywords Corals · Reefs · Climate change · Macroalgae · Phase-shifts · Resilience

B. Manikandan (✉)
Coral Reef Ecology Lab, Biological Oceanography Division, CSIR-National Institute of Oceanography, Dona Paula, Goa, India
e-mail: manikandan.nio@csir.res.in

B. P. Barathan et al. (eds.), *Navigating Climate Change: Impacts on Biodiversity and Ecosystem Resilience*, Sustainable Development and Biodiversity 42,
https://doi.org/10.1007/978-981-95-0409-1_12

12.1 Introduction

12.1.1 Coral Reef Ecosystems

The coral reef ecosystems are the most ancient, dynamic, biologically diverse, and highly productive ecosystems on the planet (Odum and Odum 1955). 'Reefs' refers to the continuous underwater limestone structures built by marine organisms or formed due to geological processes. Coral reefs are unique in being formed entirely by the biological activity of living organisms that are capable of producing and depositing calcium carbonate. Among the thousands of species residing in the reef, only very few species contribute to reef formation by depositing a limestone skeleton. Scleractinian corals are the principal reef-building organisms belonging to the phylum Cnidaria and form the reef framework that serves as a habitat for numerous life-forms (Owen et al. 2012). Coral reefs are among the oldest of marine communities, with a geological history stretching back for more than 500 million years.

Corals are colonies of genetically identical individual animals known as 'polyps' which secrete an external calcium carbonate skeleton by the calcification process (Marshall 1996). Corals exist in symbiosis with dinoflagellates 'Symbiodinium' that perform photosynthesis and cater to the nutritional requirements of the corals by transferring photosynthetically derived organic compounds. In return, the corals host dinoflagellates in their tissues and supplement them with CO_2 and metabolic waste. This mutual symbiotic relationship fuels the productivity and growth of coral reefs in nutrient-poor oligotrophic waters.

Coral reefs are predominantly present along the tropical coast lines between the latitudes 30 °N–30 °S which constitute half of the world's coast lines (Spalding et al. 2001). Corals thrive well in the shallow, clear waters with maximum light penetration since light is essential for photosynthesis by their symbiotic partner, zooxanthellae. Besides light, other environmental factors such as temperature, sediment load, substrate type, depth, and turbulence play a critical role in promoting or preventing the reef growth (Veron 2000). The upper temperature limit for the growth of corals is 30 °C, although certain corals can withstand temperatures up to 35 °C. Cold water corals, which are found globally at depths >70 m from the coastal Antarctica to the Arctic Circle, can thrive at a temperature of 4 °C.

12.1.2 Values of Coral Reefs

Of the 70% of the planet's area occupied by Oceans, coral reefs cover <0.5% (Lough 2008) and serve as home to 25% of the known marine population. Healthy coral reefs are a source of food for millions; protect coastlines from waves, storms, and erosion; provide habitat for other organisms, spawning, and nursery grounds for economically important fish species; provide jobs and income to local people from fishing, recreation, and tourism and source of new medicines; and are hotspots of marine biodiversity (Cesar et al. 2003; Moberg and Folke 1999). Estimates suggest that coral reefs in the Asia-Pacific region alone contributed an average of 25 billion $ annually between 2008 and 2012 through fisheries and tourism. Overall, the average economic productivity of the coral reefs was estimated to be 112,000 US$ per square km, though these values might change between different geographic regions based on their economic utilization of reefs (Bartelet et al. 2024). The export of organic matter and combined nitrogen from coral reefs contributes to the productivity of the Ocean. Coral reefs form Islands facilitating the settlement of human population (Stoddart 1973); promote the growth of seagrass and mangrove ecosystems by dissipating the wave energy and creating a lagoon and sedimentary environment (Birkeland 1985; Ogden 1988).

12.1.3 Coral Reefs: The Falling Kingdom

The coral reefs are under continuous threat due to anthropogenic climate change and direct human activities (Riegl et al. 2012; Souter et al. 2021). As a result, the coral cover on tropical reefs continued to decline substantially, with an estimated loss of 14% of the world's coral reefs since 2009 (Souter et al. 2021; Eddy et al. 2021). Warming seas are the most significant threat leading to more severe and recurrent bleaching events, where corals lose their symbiotic algae, causing them to turn white, starve for energy, and potentially die if the stress is prolonged. Ocean acidification due to high CO_2 emissions reduces the ocean pH, which dissolute the calcium carbonate skeleton of corals, weakening the reef structures. The effects of these global stressors are exaggerated by region-specific local threats, including nutrient run-off, unsustainable exploitation of resources, and increasing reef tourism beyond its carrying capacity. Collectively, these regional stressors contribute to reef destruction and degradation through sedimentation, physical damage, and pollution. The outbreak of stony coral tissue loss disease has devastated the coral reefs in the Caribbean, and it has been observed to rapidly spread across the West Atlantic, affecting the reef services (Alvarez-Filip et al. 2022).

12.1.4 Coral–Algal Dynamics in the Context of Global Reef Decline

Globally, many coral reefs are shifting away from coral dominance to algal dominance with serious implications for the sustainability of reef functions and services (Hughes et al. 2007a, 2007b; Mumby and Steneck 2008). Mirroring the loss in live coral cover, the amount of algae on world coral reefs has increased by 20% since 2010 (Souter et al. 2021). As biodiversity hotspots, the corals and other reef organisms exist in a highly balanced state to maintain the community structure of the reef and its functions. Under optimum environmental conditions, reef building corals thrive, form complex habitats supporting diverse communities, and dominate reef ecosystems. However, disturbances associated with extreme conditions like El-Nino events can cause widespread coral mortality, disrupting this balance. When elevated sea temperatures persist and corals fail to recover, macroalgae often proliferate and gradually establish dominance on reefs (McCook 1999; Nyström et al. 2001; Szmant 2002). Once established, macroalgae can prevent the recovery of reefs by hindering coral recruitment and growth (Kuffner et al. 2006; Hoey et al. 2011). This shift is further exacerbated by overexploitation of herbivorous reef populations (Bejarano et al. 2013) and chronic nutrient input from land-based sources, both of which allow the macroalgae to overgrow and kill the existing coral colonies through physical contact and allelopathic interactions (Manikandan et al. 2017; Rasher et al. 2011.). This reinforces a feedback loop that maintains the reef in a degraded, macroalgal-dominated state. Such persistent shifts contribute to the global decline in coral reef cover.

12.1.5 Objectives and Scope of the Chapter

This chapter seeks to provide a holistic understanding on the dynamic interaction between corals and macroalgae and their possible outcomes on the overall health and resilience of coral reefs. It identifies various environmental drivers of algal dominance and describe the ecological consequences of shift from coral to algal dominance. Also, the chapter evaluates the current management practices and mitigative strategies being followed and identify the gaps and predict the future of the reefs in the Anthropocene.

12.2 Coral–Macroalgal Interactions: A Historical Context

Historically, corals and macroalgae were coexisting in the reef systems and the relative abundance of these communities depends on the reef status and geographical location. In general, the algae existing on reefs were categorized in to three major groups as Crustose coralline algae (CCA), fleshy macroalgae, and turf algae with each having their own functional position in the community. CCA precipitates calcium carbonate similar to corals and contributes to reef framework by cementing them together. CCA also contribute to reef sediment production by erosion and produce chemical cues that induce the settlement of coral larvae (Jorissen et al. 2021). Whereas fleshy macroalgae contributes to the overall biological productivity of the ecosystem alongside the symbiotic algae living within corals. Turf algae also contributes to reef productivity and remains as a food source for a wide range of reef organisms including herbivorous fish.

Steneck (1983) affirmed that large stands of macroalgae were dominating the shallow benthic communities during the Palaeozoic era, although there is a lack of any fossil records to confirm this hypothesis. Existing fossil records revealed that macroalgae were interacting with corals since late Palaeocene (Danian to Late Thanetian) (Moussavian and Vecsei 1995; Scheibner and Speijer 2008) when the modern corals and macroalgae started coexisting in the ecosystem between Palaeocene and mid Cenozoic (Wood 1998). During this period, corals and reef macroalgae have evolved and adapted to the local environment as we see today. The evolution of herbivore fish during this time (Bellwood et al. 2014) and their grazing behaviour might have dramatically changed the reef structure (Steneck 1983). Interaction between macroalgae and herbivores has evolved to reach a stable state in a healthy reef. Although macroalgae compete with corals and other benthic organisms for space, light, and resources, their populations are regulated by herbivory from fish, echinoids, and other invertebrates, as well as by the nutrient-poor oligotrophic conditions. These factors have helped the corals to be dominant on the reefs (Knowlton et al. 1992; Hay 1981; Lewis 1986).

A shift from coral to macroalgal-dominated states was earlier reported since the 1980s leading to the speculations whether the macroalgae were the cause or consequence of the decline in the coral population (McCook 1997). In general, macroalgae vary in their biomass and diversity on temporal scale owing to the seasonality in their growth and reproduction. Also, spatial variation in the grazing intensity by herbivores also influence the biomass of macroalgae. Recent studies have identified several mechanisms such as shading, abrasion, release of allelo-chemicals, and microbial mediation has been described through which the macroalgae can outcompete corals (Jompa and McCook 2003). However, these negative effects are not high enough to fully replace the corals in reef ecosystems, since the macroalgal biomass was regulated by herbivores and, to some extent, by the coral defence mechanisms. Besides, corals also demonstrated their ability to prevent the attachment and survival of algal recruits (Diaz-Pulido and McCook 2004) and inhibit the algal growth (van Steveninck et al. 1988). Mass coral mortality under extreme conditions creates new substrates in the form of dead corals for macroalgae to attach and overgrow. Additionally, reduced herbivory will enable the macroalgae to bloom and opportunistically overgrow the dead corals. From the foregoing account, it is clear that macroalgae are generally a consequence rather than the cause for decline in the coral population and reef degradation. However, accidental release of invasive algae such as *Kappaphycus alvarezii* and their subsequent entrapment in coral reef ecosystems has been reported to cause coral mortality by smothering effect (Chandrasekaran et al. 2008).

12.3 Drivers of Coral–Algal Phase Shifts

The shift from the coral-dominant to algal-dominant state is generally associated with a combination of environmental factors such as nutrient enrichment, reduced herbivory, and recurrent stress events that permanently lock the reef in a degraded state. The relative importance of each of these factors varies across different geographical reefs, depending on the local ecological contexts, historical baselines, and the intensity of anthropogenic pressure. Additionally, recurrent disturbances due to thermal stress, cyclones, and disease outbreaks interact with these drivers to accelerate the pace and permanence of coral–algal phase shifts.

Overexploitation of herbivore reef fish Herbivory by reef fishes is an important reef process that plays a crucial role in maintaining the equilibrium between the corals and macroalgae (Mumby 2006). The extent to which different groups of herbivorous fishes promote reef recovery will depend on their functional role and the algae that they graze. Certain herbivore fishes regulate the algae population and other specific herbivores play an important role in reversing the phase-shifts. The herbivore reef fish are classified as scrapers, grazers, and browsers based on their function in the reef (Bellwood 1994). Scrapers remove algae, sediment, and other materials by taking non-excavating bites and create a substrate for the settlement of new coral recruits. Similarly, the grazers feed on epilithic turf algae which in turn limit the growth of macroalgae (Bellwood and Choat 1990). Browsers feed on individual algal components and remove only the algae and its associated epiphytes. They play a critical role in controlling the algal population over corals and also in reversing the coral–algal phase shifts (Green and Bellwood 2009). The absence of herbivore fishes due to overfishing enables the macroalgae to overtake corals, affecting the reef structure and integrity (Done 1992; Hughes et al. 2007a).

Globally, six million fishermen are involved in coral reef fisheries, and coral reef fish serve as the major source of protein (Teh et al. 2013). More than 80% of the world's shallow reefs were severely exploited (Wilkinson 2002) to meet the growing demand for reef fish. Herbivore reef fish exploitation strengthens the algal competition with corals, leading to their mortality, and turns a coral-dominant reef into an algal reef (Hughes et al. 2007b; Diaz-Pulido et al. 2009). While exploitation of functionally important herbivore fish can have a direct impact on corals, exploitation of predatory fish results in trophic cascades and collapses the food web in a reef ecosystem (Wilson et al. 2008). Exploitation of reef fish has resulted in the outbreak of non-reef-building taxa such as sea urchin *Echinometra mathaei* in Kenyan reefs (McClanahan and Shafir 1990) and crown of thorns starfish in Fiji (Dulvy et al. 2004), which can cause reef erosion and coral mortality, respectively.

Coastal development Almost three billion people live within 100 kms of coast and depend on the ocean and their resources for their livelihood. Therefore, infrastructure development along the coast becomes inevitable to support the growing population, enhance economic opportunities, and improve access to essential services. Coastal development near coral reef areas leads to an increase in the sediment load (Nemeth and Nowlis 2001), nutrients, and sediment run-off (Carpenter et al. 1998) which in turn stimulate algal growth and negatively impact the corals (Arias-González et al. 2017; Piazzi and Gennaro 2011).

Excess of nutrients in coral reefs favours the excess production of macroalgae. Nitrogen (N) is the key element for protein and nucleic acid synthesis, Phosphorus plays a crucial role in energy transfer influencing photosynthesis and growth. Carbon provides building blocks for organic molecules essential for energy storage and structural components. Besides these major nutrients, macroalgae also require other nutrients such as potassium, calcium, magnesium, and sulphur, as well as micronutrients like iron, zinc, and copper (Harrison and Hurd 2001; Suthar et al. 2019). Nutrient enrichment in a reef environment is linked to several factors including rainfall and associated riverine run-off, wave-driven circulation, anthropogenic nutrient discharges (sewage), and large watersheds. Collectively, these sources consistently enriched the reef waters with nutrients, especially N_2, which eventually reduced coral cover and increased the algal cover. In certain reefs like Moorea, the algal phase-shifts had occurred despite the island-wide increase in the density and biomass of herbivorous fish (Adam et al. 2021). This indicates that nutrient loading is an important driver of coral–macroalgal phase-shifts.

Beyond the anthropogenic activities, *climate-related stress,* including Ocean warming, acidification, and extreme events like El-Nino and storms, indirectly favour the algal growth by causing coral mortality. Thermal stress due to Ocean warming causes the corals to starve for energy and become bleached by expelling their symbiotic algae (Brown 1997). This provides a competitive advantage for macroalgae to overtake the corals that exist in a compromised health state. Extreme rainfall associated with storms and cyclones will increase the riverine and land-based run-off, enriching the coastal waters with nutrients and promote the algal growth (McCook 1999). Ocean acidification due to increased CO_2 dissolving in the seawater slows down the coral growth by reducing the availability of carbonate ions (CO_3^{2-}), which in turn reduces the ability of corals to compete with macroalgae for space. Alternatively, macroalgae can directly absorb the CO_2 dissolving in seawater and enhance their growth and photosynthetic efficiency (Zhang et al. 2021). In essence, Ocean acidification slows down the coral growth and promotes the algal growth, leading to macroalgal phase-shifts in coral reefs.

12.4 Ecological Consequences of Macroalgal Dominance

Macroalgae were the most dominant competitors for space and light on coral reefs, often outcompeting corals through several key mechanisms, including physical overgrowth, shading, and release of toxic metabolites. Collectively, these effects can lead to coral tissue damage, increased mortality, long-term degradation of coral habitats, and a significant loss of reef biodiversity.

Macroalgae reduce coral growth and recovery Macroalgae directly compete with corals for essential resources including space and light. Several field trials that experimentally manipulated the direct contact between corals and macroalgae have shown that macroalgae in high density is capable of reducing the coral growth and affecting their long-term survivorship (Clements et al. 2018; Lirman et al. 2014). However, these effects differ with different species of corals and macroalgae (Jompa and McCook 2003). For example, the green filamentous macroalgae *Chlorodesmis fastigiata* significantly reduced tissue recovery in *Acropora pulchra*, whereas the brown algae *Lobophora variegata* had only minor effects on tissue regeneration and coral recovery process (Bender et al. 2012).

Impairment of reproduction and recruitment In general, corals reproduce by both sexual and asexual methods. The sexual method reproduction involves synchronous spawning of corals where they release the gametes in to the water column (sperm and egg) for fertilization. Laboratory experiments by Foster et al. (2008) have shown that direct contact between the corals and the macroalgae like *Dictyota* spp. and *Lopbophora variegata* reduced the fecundity of corals by reducing the size of the eggs and the overall reproductive output. Different groups of macroalgae have different effects on coral settlement. For example, turf algae and crustose coralline algae had minor effects on coral settlement. Whereas, fleshy macroalgae inhibited the settlement of corals' larvae (Diaz-Pulido et al. 2009).

Inhibition of larval settlement Availability of hard substratum and the successful settlement and growth of coral larvae have been the key for recovery of any degraded reef (Shinn 1976; Gilmour et al. 2013). The dead corals can still contribute to the recovery of a degraded reef by serving as a substrate for the settlement of coral larvae (Wallace et al. 1986; Norstrom et al. 2007). Under favourable conditions, macroalgae can quickly colonize these dead coral substrates and leave no space for the settlement and recruitment of coral larvae.

DDAM model The macroalgae need not be necessarily in direct contact with corals to cause any harm. The DDAM model (DOC, Disease, and algae model) suggests that macroalgae can affect corals at distance by altering their microbiome and it is hypothesized to an important contributor to coral decline (Nelson et al. 2023; Haas et al. 2016). Specifically, the laboratory experiments have shown that macroalgae release dissolved organic carbon (DOC) that fuels microbial growth and lowers the oxygen levels in the water column through microbial respiration which creates a local hypoxia (Haas et al. 2016; Walsh et al. 2017). This effect was neutralized by addition of antibiotics (Smith et al. 2006) suggesting that the damage caused to corals is associated with low oxygen levels which eventually kill the coral tissues by hypoxia (Haas et al. 2011).

Reduction in overall biodiversity Corals form complex three-dimensional structures with numerous holes and crevices supporting a myriad of biodiversity, including invertebrates and fish communities. When corals are replaced by macroalgae after a stress event, the reef loses their structural complexity due to the dense, low-lying mats overgrowing corals, which eventually reduce the vertical relief and spatial heterogeneity of the reef. Loss of corals and structural complexity will displace the fish communities that depend on the corals for food and shelter from predators (Pratchett et al. 2011; Ainsworth and Mumby 2015). Additionally, a large group of invertebrates, including Crustaceans, Molluscs, and echinoderms, which find refuge and sustenance within the intricate framework of coral reefs, will be lost. Overall, the competitive dominance of macroalgae will exclude other organisms like crustose coralline algae that rely on the open reef substrate for settlement and growth.

Altered food web dynamics Loss of biodiversity due to algal dominance will disrupt the food web dynamics by reducing the abundance and diversity of communities that depend on corals. This can shift the species composition, favouring the herbivore communities and reducing the abundance of other trophic groups, including commercially important fish communities. For example, in Raja Ampat reef in Indonesia model, simulations of coral loss and subsequent macroalgal dominance predicted a decrease in the biomass of larger, commercially important fish, and an increase in smaller-bodied fish (Ainsworth and Mumby 2015). Similarly, predatory fish like groupers altered their diets from primary pelagic to bottom-dwelling organisms in Keppel Islands, Australia, following a decline in coral cover and increase in macroalgal cover (Hempson et al. 2018). This indicates a change in the prey availability and shortening of food chains within the reef food-web.

12.5 Way Forward and Future Research

The dynamic interplay between corals and macroalgae is an increasing concern for reef across the globe, as their competitive interaction plays a pivotal role in shaping the reef resilience to climate change. As coral reefs continue to decline due to recurrent mass bleaching events and reduced recovery window, macroalgae often gains a competitive advantage over corals existing in a compromised health state and establish themselves as a dominant benthic form in the coral reefs. Once established, macroalgae can dominate benthic landscape, outcompeting corals through a range of mechanisms including biochemical, microbial, and physiological pathways. Although significant progress has been made in understanding these mechanisms, the potential tipping points that lead to ecosystem phase-shifts is still not fully understood. The advent of various omics technologies offers a promising path forward, offering new insights into the complex processes underlying these transitions.

Long-term studies capturing the spatial and temporal dynamics of coral–algal interaction are essential to understand their influence on reef trajectories. Establishing region-specific monitoring frameworks that integrate remote sensing, photogrammetry, and in situ ecological surveys will enable the detection of subtle changes in benthic composition and the spatial expansion of algal-dominated states. This will also help identify refugia and resilient coral assemblages.

Though macroalgal communities possess natural variation in their prevalence and biomass, the influence of anthropogenic activities such as pollution, nutrient enrichment, coastal development, and overfishing on coral–algal dynamics cannot be ignored. Future efforts incorporating socio-ecological models to understand feedback loops between reef degradation and human livelihoods are essential to develop adaptive management strategies that promote both ecological and social well-being. The existing coral restoration technologies should be upgraded to integrative strategies to manage algal overgrowth, besides establishing the coral population in the degraded reefs. Above all, bridging the gap between scientists, local stakeholders, and policymakers is essential to translate the ecological insights into policies.

Data Availability Statement Data are available from the authors on request.

Declaration of Competing Interest The authors declare that they have no known competing financial interest or personal relationships that could have appeared to influence the work reported in this paper.

Conflict of Interest The author declares no conflict of interest.

Ethics Approval Not applicable.

References

Adam TC, Burkepile DE, Holbrook SJ, Carpenter RC, Claudet J, Loiseau C, Thiault L, Brooks AJ, Washburn L, Schmitt RJ (2021) Landscape-scale patterns of nutrient enrichment in a coral reef ecosystem: implications for coral to algae phase shifts. Ecol Appl 31(1):e2227

Ainsworth CH, Mumby PJ (2015) Coral–algal phase shifts alter fish communities and reduce fisheries production. Glob Chang Biol 21(1):165–172

Alvarez-Filip L, González-Barrios FJ, Pérez-Cervantes E, Molina-Hernández A, Estrada-Saldívar N (2022) Stony coral tissue loss disease decimated Caribbean coral populations and reshaped reef functionality. Commun Biol 5(1):440

Arias-González JE, Fung T, Seymour RM, Garza-Pérez JR, Acosta-González G, Bozec YM, Johnson CR (2017) A coral-algal phase shift in Mesoamerica not driven by changes in herbivorous fish abundance. PLoS One 12(4):e0174855

Bartelet HA, Barnes ML, Cumming GS (2024) Estimating and comparing the direct economic contributions of reef fisheries and tourism in the Asia-Pacific. Mar Policy 159:105939

Bejarano S, Golbuu Y, Sapolu T, Mumby PJ (2013) Ecological risk and the exploitation of herbivorous reef fish across Micronesia. Mar Ecol Prog Ser 482:197–215

Bellwood DR (1994) A phylogenetic study of the parrotfishes family Scaridae (Pisces: Labroidei), with a revision of genera. Rec Aust Mus Suppl 20:1–86

Bellwood DR, Choat JH (1990) A functional analysis of grazing in parrot fishes (family Scaridae): the ecological implications. Environ Biol Fish 28(1–4):189–214

Bellwood DR, Goatley CHR, Brandl SJ, Bellwood O (2014) Fifty million years of herbivory on coral reefs: fossils, fish and functional innovations. Proc R Soc B Biol Sci 281(1781):20133046

Bender D, Diaz-Pulido G, Dove S (2012) Effects of macroalgae on corals recovering from disturbance. J Exp Mar Biol Ecol 429:15–19

Birkeland C (1985) Ecological interactions between mangroves, seagrass beds, and coral reefs. In: Birkeland C, Grosenbaugh D (eds) Ecological interactions between tropical coastal ecosystems. UNEP Regional Seas Reports and Studies No. 73, p 126

Brown BE (1997) Coral bleaching: causes and consequences. Coral Reefs 16:S129–S138

Carpenter SR, Caraco NF, Correll DL, Howarth RW, Sharpley AN, Smith VH (1998) Nonpoint pollution of surface waters with phosphorus and nitrogen. Ecol Appl 8:559–568

Cesar H, Burke L, Pet-Soede L (2003) The economics of worldwide coral reef degradation. Cesar Environmental Economics Consulting, The Netherlands

Chandrasekaran S, Nagendran NA, Pandiaraja D, Krishnankutty N, Kamalakannan B (2008) Bioinvasion of *Kappaphycus alvarezii* on corals in the Gulf of Mannar, India. Curr Sci 10:1167–1172

Clements CS, Rasher DB, Hoey AS, Bonito VE, Hay ME (2018) Spatial and temporal limits of coral-macroalgal competition: the negative impacts of macroalgal density, proximity, and history of contact. Mar Ecol Prog Ser 586:11–20

Diaz-Pulido G, McCook LJ (2004) Effects of live coral, epilithic algal communities and substrate type on algal recruitment. Coral Reefs 23(2):225–233

Diaz-Pulido G, McCook LJ, Dove S, Berkelmans R, Roff G, Kline DI, Weeks S, Evans RD, Williamson DH, Hoegh-Guldberg O (2009) Doom and boom on a resilient reef: climate change, algal overgrowth and coral recovery. PLoS One 4(4):e5239

Done TJ (1992) Phase shifts in coral reef communities and their ecological significance. Hydrobiologia 247:121–132

Dulvy NK, Freckleton RP, Polunin NVC (2004) Coral reef cascades and the indirect effects of predator removal by exploitation. Ecol Lett 7(5):410–416

Eddy TD, Lam VW, Reygondeau G, Cisneros-Montemayor AM, Greer K, Palomares ML, Bruno JF, Ota Y, Cheung WW (2021) Global decline in capacity of coral reefs to provide ecosystem services. One Earth 4(9):1278–1285

Foster NL, Box SJ, Mumby PJ (2008) Competitive effects of macroalgae on the fecundity of the reef-building coral *Montastraea annularis*. Mar Ecol Prog Ser 367:143–152

Gilmour JP, Smith LD, Heyward AJ, Baird AH, Pratchett MS (2013) Recovery of an isolated coral reef system following severedisturbance. Sci 340(6128): 69–71

Green AL, Bellwood DR (eds) (2009) Monitoring functional groups of herbivorous reef fishes as indicators of coral reef resilience: a practical guide for coral reef managers in the Asia Pacific region, vol No. 7. IUCN

Haas AF, Nelson CE, Wegley Kelly L, Carlson CA, Rohwer F, Leichter JJ, Wyatt A, Smith JE (2011) Effects of coral reef benthic primary producers on dissolved organic carbon and microbial activity. PLoS One 6(11):e27973

Haas AF, Fairoz MF, Kelly LW, Nelson CE, Dinsdale EA, Edwards RA, Giles S, Hatay M, Hisakawa N, Knowles B, Lim YW (2016) Global microbialization of coral reefs. Nat Microbiol 1(6):1–7

Harrison PJ, Hurd CL (2001) Nutrient physiology of seaweeds: application of concepts to aquaculture. Cah Biol Mar 42(1–2):71–82

Hay ME (1981) Spatial patterns of grazing intensity on a Caribbean barrier reef: herbivory and algal distribution. Aquat Bot 11:97–109

Hempson TN, Graham NA, MacNeil MA, Hoey AS, Wilson SK (2018) Ecosystem regime shifts disrupt trophic structure. Ecol Appl 28(1):191–200

Hoey AS, Pratchett MS, Cvitanovic C (2011) High macroalgal cover and low coral recruitment undermines the potential resilience of the world's southernmost coral reef assemblages. PLoS One 6(10):e25824

Hughes TP, Bellwood DR, Folke CS, McCook LJ, Pandolfi JM (2007a) No-take areas, herbivory and coral reef resilience. Trends Ecol Evol 22(1):1–3

Hughes TP, Rodrigues MJ, Bellwood DR, Ceccarelli D, Hoegh-Guldberg O, McCook L, Moltschaniwskyj N, Pratchett MS, Steneck RS, Willis B (2007b) Phase shifts, herbivory, and the resilience of coral reefs to climate change. Curr Biol 17(4):360–365

Jompa J, McCook LJ (2003) Coral-algal competition: macroalgae with different properties have different effects on corals. Mar Ecol Prog Ser 258:87–95

Jorissen H, Galand PE, Bonnard I, Meiling S, Raviglione D, Meistertzheim AL, Hédouin L, Banaigs B, Payri CE, Nugues MM (2021) Coral larval settlement preferences linked to crustose coralline algae with distinct chemical and microbial signatures. Sci Rep 11(1):14610

Knowlton N, Weil E, Weigt LA, Guzmán HM (1992) Sibling species in *Montastraea annularis*, coral bleaching, and the coral climate record. Science 255(5042):330–333

Kuffner IB, Walters LJ, Becerro MA, Paul VJ, Ritson Williams R, Beach KS (2006) Inhibition of coral recruitment by macroalgae and cyanobacteria. Mar Ecol Prog Ser 323:107–117

Lewis SM (1986) The role of herbivorous fishes in the organization of a Caribbean reef community. Ecol Monogr 56(3):183–200

Lirman D, Schopmeyer S, Galvan V, Drury C, Baker AC, Baums IB (2014) Growth dynamics of the threatened Caribbean staghorn coral *Acropora cervicornis*: influence of host genotype, symbiont identity, colony size, and environmental setting. PLoS One 9(9):e107253

Lough JM (2008) 10th anniversary review: a changing climate for coral reefs. J Environ Monit 10(1): 21–29

Manikandan B, Ravindran J, Vidya PJ, Shrinivasu S, Manimurali R, Paramasivam K (2017) Resilience potential of an Indian Ocean reef: an assessment through coral recruitment pattern and survivability of juvenile corals to recurrent stress events. Environ Sci Pollut Res 24:13614–13625

Marshall AT (1996) Calcification in hermatypic and ahermatypic corals. Science 271(5249):637–639

McClanahan TR, Shafir SH (1990) Causes and consequences of sea urchin abundance and diversity in Kenyan coral reef lagoons. Oceologia 83(3):362–370

McCook LJ (1997) Effects of herbivory on zonation of Sargassum spp. within fringing reefs of the central Great Barrier Reef. Mar Biol 129:713–722

McCook LJ (1999) Macroalgae, nutrients and phase shifts on coral reefs: scientific issues and management consequences for the Great Barrier Reef. Coral Reefs 18(4):357–367

Moberg F, Folke C (1999) Ecological goods and services of coral reef ecosystems. Ecol Econ 29(2):215–233

Moussavian E, Vecsei A (1995) Plaeocene reef sediments from the maiella carbonate platform, Italy. Facies 32(1):213–221

Mumby PJ (2006) The impact of exploiting grazers (Scaridae) on the dynamics of Caribbean coral reefs. Ecol Appl 16(2):747–769

Mumby PJ, Steneck RS (2008) Coral reef management and conservation in light of rapidly evolving ecological paradigms. Trends Ecol Evol 23(10):555–563

Nelson CE, Wegley Kelly L, Haas AF (2023) Microbial interactions with dissolved organic matter are central to coral reef ecosystem function and resilience. Annu Rev Mar Sci 15(1):431–460

Nemeth RS, Nowlis JS (2001) Monitoring the effects of land development on the near-shore reef environment of St. Thomas, USVI. Bull Mar Sci 69(2):759–775

Norstrom AV, Lokrantz J, Nystrom M, Yap HT (2007) Influence of dead coral substrate morphology on patterns of juvenile coral distribution. Mar Biol 150(6):1145–1152

Nyström M, Folke C, Moberg F (2001) Corals and phase shifts: reply from Nyström, Folke & Moberg. Trends Ecol Evol 16(3):127

Odum HT, Odum EP (1955) Trophic structure and productivity of a windward coral reef community on Eniwetok Atoll. Ecol Monogr 25(3):291–320

Ogden JC (1988) The influence of adjacent systems on the structure and function of coral reefs. Proc 2nd Int Coral Reef Symp 1:123–129

Owen AW, Doyle P, Harper DA, Britt BB (2012) Corals and other reef-builders. Lethaia 45(1): 1–1

Piazzi L, Gennaro P, Balata D (2011) Effects of nutrient enrichment on macroalgal coralligenous assemblages. Mar Pollut Bull 62(8):1830–1835

Pratchett MS, Hoey AS, Wilson SK, Messmer V, Graham NA (2011) Changes in biodiversity and functioning of reef fish assemblages following coral bleaching and coral loss. Diversity 3(3):424–452

Rasher DB, Stout EP, Engel S, Kubanek J, Hay ME (2011) Macroalgal terpenes function as allelopathic agents against reef corals. Proc Natl Acad Sci 108(43):17726–17731

Riegl BM, Sheppard CR, Purkis SJ (2012) Human impact on atolls leads to coral loss and community homogenisation: a modeling study. PLoS One 7(6):e36921

Scheibner C, Speijer RP (2008) Late Paleocene–early Eocene Tethyan carbonate platform evolution—a response to long-and short-term paleoclimatic change. Earth Sci Rev 90(3–4):71–102

Shinn EA (1976) Coral reef recovery in Florida and the Persian Gulf. Environ Geol 1(4): 241–254

Smith JE, Shaw M, Edwards RA, Obura D, Pantos O, Sala E, Sandin SA, Smriga S, Hatay M, Rohwer FL (2006) Indirect effects of algae on coral: algae-mediated, microbe-induced coral mortality. Ecol Lett 9(7):835–845

Souter D, Planes S, Wicquart J, Logan M, Obura D, Staub F (eds) (2021) Status of coral reefs of the world: 2020 report. Global Coral Reef Monitoring Network (GCRMN) and International Coral Reef Initiative (ICRI)

Spalding M, Ravilious C, Green EP (2001) World atlas of coral reefs. Univ of California Press, Berkeley

Steneck RS (1983) Escalating herbivory and resulting adaptive trends in calcareous algal crusts. Paleobiology 9(1):44–61

Stoddart DR (1973) Coral reefs of Indian Ocean. In: Biology and geology of coral reefs, vol 1. Academic Press, New York, pp 51–78

Suthar P, Gajaria TK, Reddy CRK (2019) Production of quality seaweed biomass through nutrient optimization for the sustainable land-based cultivation. Algal Res 42:101583

Szmant AM (2002) Nutrient enrichment on coral reefs: is it a major cause of coral reef decline? Estuaries 25:743–766

Teh LS, Teh LC, Sumaila UR (2013) A global estimate of the number of coral reef fishers. PLoS One 8(6):e65397

van Steveninck EDR, Kamermans P, Breeman AM (1988) Transplant experiments with two morphological growth forms of *Lobophora variegata* (Phaeophyceae). Mar Ecol Prog Ser 49:191–194

Veron JEN (2000) Reef evolution. Science 287(5454):811–811

Wallace CC, Watt A, Bull GD (1986) Recruitment of juvenile corals onto coral tables preyed upon by *Acanthaster planci*. Mar Ecol Prog Ser 32:299–306

Walsh K, Haggerty JM, Doane MP, Hansen JJ, Morris MM, Moreira APB, de Oliveira L, Leomil L, Garcia GD, Thompson F, Dinsdale EA (2017) Aura-biomes are present in the water layer above coral reef benthic macro-organisms. PeerJ 5:e3666

Wilkinson C (2002) Status of coral reefs of the world. Australian Institute of Marine Science, Townsville

Wilson SK, Fisher R, Pratchett MS, Graham NAJ, Dulvy NK, Turner RA, Cakacaka A, Polunin NVC, Rushton SP (2008) Exploitation and habitat degradation as agents of change within coral reef fish communities. Glob Chang Biol 14(12):2796–2809

Wood R (1998) The ecological evolution of reefs. Annu Rev Ecol Syst 29(1):179–206

Zhang X, Xu D, Han W, Wang Y, Fan X, Loladze I, Gao G, Zhang Y, Tong S, Ye N (2021) Elevated CO2 affects kelp nutrient quality: a case study of Saccharina japonica from CO2-enriched coastal mesocosm systems. J Phycol 57(1):379–391

Chapter 13
Sociable Lapwing and Beyond: Climate Change and the Imperative to Protect Biodiversity

Vaithianathan Kannan and Nikunj B. Gajera

Abstract The Anthropocene epoch has ushered in an era of rapid biodiversity loss, driven by synergistic forces including habitat degradation, climate change, invasive species, and anthropogenic exploitation. The Sociable Lapwing (*Vanellus gregarius*), a critically endangered migratory shorebird, exemplifies the vulnerability of steppe-dependent species in the face of these global pressures. Once widespread across Central Asia, this species now persists in fragmented populations that traverse complex migratory routes linking breeding grounds in Kazakhstan with wintering areas in South Asia and north-eastern Africa. This chapter presents an integrative analysis of the species population trends, migratory ecology, and habitat requirements, field surveys, and historical sighting records from India. It highlights the multifaceted threats faced by the species including habitat conversion, illegal hunting, pesticide use, and infrastructure collisions and situates its decline within the broader context of planetary biodiversity loss and climate instability. The Sociable Lapwing serves as both a flagship and umbrella species, offering a powerful narrative for transboundary conservation. The chapter concludes by outlining a forward-looking agenda that integrates demographic research, landscape-scale habitat management, community engagement, and policy reform to safeguard this steppe species in India to restore ecological resilience across their flyways as India being the wintering ground.

Keywords Sociable Lapwing · Migratory birds · Biodiversity loss · Climate change · Habitat degradation · Transboundary conservation · Ecosystem resilience

---

V. Kannan
Arulagam, Asokapuram, Coimbatore, Tamil Nadu, India

N. B. Gajera (✉)
Gujarat Institute of Desert Ecology, Bhuj, Gujarat, India

B. P. Barathan et al. (eds.), *Navigating Climate Change: Impacts on Biodiversity and Ecosystem Resilience*, Sustainable Development and Biodiversity 42,
https://doi.org/10.1007/978-981-95-0409-1_13

13.1 Introduction

The Anthropocene a term used to describe the current geological area dominated by human influence has escorted in a period of profound ecological disturbance. Human activities such as deforestation, urbanization, unsustainable agricultural expansion, and fossil fuel burning have accelerated the rate of biodiversity loss to levels unprecedented in recorded history (Díaz et al. 2019). The Intergovernmental Science-Policy Platform on Biodiversity and Ecosystem Services (IPBES) estimates that approximately one million species are now threatened with extinction, many within decades, unless transformative action is taken (IPBES 2019). The biodiversity crisis is not a standalone issue but one that is deeply interwoven with climate change, economic inequality, and environmental degradation. Current extinction rates are estimated to be 10–1000 times higher than the natural background rate (Ceballos et al. 2015), driven by five primary direct causes: habitat destruction, overexploitation, pollution, invasive alien species, and climate change (IPBES 2019; Pyšek et al. 2020). These drivers are exacerbated by underlying indirect drivers such as population growth, unsustainable production and consumption patterns, and weak environmental governance (Secretariat of the Convention on Biological Diversity 2020).

Major scientific interventions and global initiatives have highlighted the urgency of the situation. The *World Scientists' Warning to Humanity* Albert et al. (2021), and its subsequent reiterations (Ripple et al. 2017, 2021), urged policymakers and society at large to recognize the ecological tipping points being crossed. Global frameworks such as the Convention on Biological Diversity (CBD) and the Aichi Biodiversity Targets (2011–2020) provided structured objectives to halt biodiversity loss, but assessments reveal that most targets were missed (CBD 2020). The new *Kunming–Montreal Global Biodiversity Framework* (2022) sets a more ambitious agenda to restore ecosystems, halt species extinction, and ensure fair and equitable benefit-sharing (CBD 2022). Biodiversity loss is not just a biological concern it has severe implications for ecosystem services, food security, climate regulation, and human health (Díaz et al. 2019). For instance, the degradation of pollinator populations can jeopardize crop yields, while loss of forest biodiversity undermines carbon sequestration potential critical to mitigating climate change (IPCC 2021).

Among the species emblematic of this crisis is the Sociable Lapwing (*Vanellus gregarius*), a migratory steppe species whose population has undergone catastrophic declines over the past few decades. The species, once widespread across Central Asia, now occupies a fragmented and declining range and faces threats along every stage of its life cycle from breeding in Kazakhstan to wintering in South Asia and the Middle East (BirdLife International 2024; Sheldon et al. 2012; Vidal and Sheldon 2016). The Sociable Lapwing offers a compelling lens through which to examine the inter-connected threats such as habitat loss, climate change, and human exploitation. Its dependence on multiple ecosystems across international borders makes it a flagship species for transboundary conservation. Furthermore, its story underscores the challenges faced by migratory birds globally, which rely on intact flyways and synchronized environmental signals for survival (Şekercioğlu et al. 2012).

In this chapter, we attempted to explore based on the existing literature and anecdotal accounts on population trends, ecological needs, and conservation efforts associated with the Sociable Lapwing while situating its decline within the broader context of planetary biodiversity loss in the face of climate change. In addition, we examine threats faced by the species across its breeding, migratory, and wintering ranges, analyse the role of climate change and human land-use change, and propose a forward-looking agenda that integrates the species conservation within the ecosystem resilience and climate adaptation strategies.

13.2 The Case of the Sociable Lapwing

The Sociable Lapwing (*Vanellus gregarius*) is a striking, medium-sized migratory shorebird belonging to the family Charadriidae. It is endemic to the Central Asian steppes, particularly Kazakhstan, and is currently listed as Critically Endangered by the International Union for Conservation of Nature (IUCN 2024). Historically, this species bred in vast numbers across southern Russia and western China, but recent estimates place its global population between 13,700 and 55,560 individuals, with the most reliable figures suggesting a central estimate of around 24,000 (BirdLife International 2024; Donald et al. 2021).

13.2.1 Ecology and Life History

The Sociable Lapwing nests semi-colonially in open, arid steppe habitats dominated by short grass and sparse shrubs, typically composed of species like *Stipa* and *Artemisia* (Sheldon et al. 2012). Its name is derived from its gregarious behaviour, as it often congregates in flocks, especially during migration and on wintering grounds. Breeding usually occurs between April and June in Kazakhstan, with nests being simple ground scrapes exposed to both predators and environmental elements. The diet of the Sociable Lapwing primarily consists of insects and other invertebrates, which are abundant in undisturbed grassland ecosystems. The availability of such prey is critical for the survival of chicks during the short breeding season, making the species highly sensitive to shifts in prey abundance due to pesticide use, droughts, or invasive plant species and in the face of climate change (Şekercioğlu et al. 2012).

13.2.2 Migration Routes and Distribution

The Sociable Lapwing is a long-distance migrant, using two major flyways to connect breeding and wintering sites:

- *Western Flyway*: Birds migrate southwest through southern Russia, the Caucasus, and the Middle East to reach wintering sites in Sudan and north-eastern Africa.
- *Eastern Flyway*: A smaller proportion of birds travel southeast through Turkmenistan, Iran, and Pakistan to reach India, especially Gujarat and Rajasthan (Donald et al. 2021; Kannan 2004).

Satellite telemetry has been pivotal in tracing these routes. A long-term study involving tagged birds revealed strong fidelity to stopover sites, especially in Turkey, Syria, and northern Saudi Arabia, even though individuals showed flexibility in breeding site selection (Donald et al. 2021). This indicates that wintering and stopover locations may be more limiting than breeding grounds a key consideration for conservation planning.

13.2.3 Status in India

- *Recent and Historical Sightings of the Sociable Lapwing (Vanellus gregarius) in Gujarat and Rajasthan, India*

The Sociable Lapwing (*Vanellus gregarius*), listed as Critically Endangered by the IUCN, has been sporadically recorded in western India, particularly across Gujarat and Rajasthan. These records are crucial for understanding its current wintering distribution, given its sharp global population decline Rahmani (2012).

Below is a summary table (Table 13.1) compiling all documented sightings reported in recent decades from various sources across Gujarat and Rajasthan:

In a recent field study conducted by Naria and Rathod (2021) in the Banni Grasslands of Kachchh, Gujarat, multiple sightings were reported between January and February 2021. On 22 January, 18 individuals were seen near Bhirandiyara village, followed by three birds in flight near Shervo wetland on 25 January. Further observations included 24 birds in flight on 2 February and a single foraging individual on 6 February, both from the Bhirandiyara area.

Earlier data compiled by Kasambe (2011) provides a historical view of wintering records from Gujarat. Between 2006 and 2010, notable flocks were reported from Dasada and the Little Rann of Kachchh, with up to 45 individuals sighted in November 2007. In the Great Rann, particularly at Chhari Dhandh, 57 individuals were reported on 16 December 2009, followed by 53 birds on 24 January 2010 and smaller flocks in other instances. These regions have been identified as regular wintering sites for the species. At the 3rd Global Bird Watchers' Conference held in January 2013 in Dhordo, Kachchh, a group of 12 Sociable Lapwings was recorded

Table 13.1 Sighting records of the Sociable Lapwing (*Vanellus gregarius*) across Gujarat and Rajasthan (1991–2021)

S. No.	DD/MM/ YYYY	No. of individuals	Location	Sighted by
1	22/01/2021	18	Bhirandiyara, Banni Grassland, Kachchh	Naria and Rathod (2021)
2	25/01/2021	3	Shervo, Banni Grassland, Kachchh	Naria and Rathod (2021)
3	02/02/2021	24	Bhirandiyara, Banni Grassland, Kachchh	Naria and Rathod (2021)
4	06/02/2021	1	Bhirandiyara, Banni Grassland, Kachchh	Naria and Rathod (2021)
5	19/11/2006	14	Dasada, Little Rann of Kachchh	Roger & Denise Lawrence
6	24/11/2006	27	Dasada, Little Rann of Kachchh	Mujhahid Malik
7	November 2007	42–45	Dasada, Little Rann of Kachchh	Yogendra Shah
8	November 2007	45	Dasada, Little Rann of Kachchh	Arpit Deomurari
9	04/12/2010	24	Dasada, Little Rann of Kachchh (Kathju 2010)	Kathju (2010)
10	11/12/2010	4	Dasada, Little Rann of Kachchh	Aditya Roy
11	17/11/2008	30	Chhari Dhand, Banni, Kachchh (Tiwari 2010)	Tiwari (2010)
12	16/12/2009	57	Chhari Dhand, Banni, Kachchh (Tiwari 2010)	Tiwari (2010)
13	24/01/2010	53	Chhari Dhand, Banni, Kachchh	Dr. Tarique Sani
14	29/01/2013	12	Chhari Dhand, Banni, Kachchh	Kasambe et al. (2013)
15	20/02/2013	21	Banni Grassland, Kachchh	Bala Dasari
16	14/10/1995	5	Tal Chhapar, Rajasthan (Sangha 2000)	Sangha (2000)
17	28/01/1998	15	Tal Chhapar, Rajasthan (Sangha 2000)	Sangha (2000)
18	01/02/1998	11	Tal Chhapar, Rajasthan (Sangha 2000)	Sangha (2000)
19	18/01/1999	2	Jaisalmer, Rajasthan	Ben King
20	18/10/2003	6	Baramsar, near Jaisalmer, Rajasthan (Sangha 2005)	Sangha (2005)
21	22/10/1991	1 (juvenile)	Jor-Bir, Bikaner, Rajasthan	R.G. Soni
22	31/12/1998	1	Revasa, Sikar, Rajasthan (Sangha 2002)	Sangha (2002)
23	29/12/1998	2	Lala Bustard Sanctuary, Kachchh	Justus Joshua
24	14/11/2010	90	Thol Village, near Sanand, Ahmedabad (Kathju 2010)	Kathju (2010)
25	21/01/2005	6	Gadara, Surendranagar, Gujarat	Sharma et al. (2005)
26	27/01/2005	4	Sayla, Surendranagar, Gujarat	Sharma et al. (2005)
27	03/02/2005	10	Khimat, Banaskantha, Gujarat	Sharma et al. (2005)
28	05/02/2005	7	Khimat, Banaskantha, Gujarat	Sharma et al. (2005)
29	29/12/2020	3	Patdi, Saurashtra, Gujarat	Joshi (2021)

near Chhari Dhandh. Later, on 20 February 2013, a flock of 21 birds was documented by Bala Dasari in the Banni Grassland (Kasambe et al. 2013), reinforcing the importance of this landscape as a regular wintering ground.

A focused status survey in 2005 by the Gujarat Ecology Society in Saurashtra and North Gujarat found the species at three new locations. These included 6 birds at Gadara (Surendranagar) on 21 January, 4 at Sayla on 27 January, and 10 at Khimat (Banaskantha) on 3 February. A revisit to Khimat on 5 February revealed three individuals, with four more reported by a local farmer (Sharma et al. 2005). More recently, Joshi (2021) recorded a small flock of three Sociable Lapwings near Patdi, Saurashtra, on 29 December 2020. The birds were seen foraging in open harvested fields, consistent with known habitat preferences of the species during the non-breeding season.

Rajasthan has also been a consistent but less intensively surveyed region for the Sociable Lapwing. Sangha (2005) collated several important records from Tal Chhapar Wildlife Sanctuary and beyond. These include sightings of 5 birds on 14 October 1995, 15 birds on 28 January 1998, and 11 on 1 February 1998. Other sightings occurred near Jaisalmer and Bikaner, including a flock of six birds with one juvenile on 18 October 2003, and a juvenile observed at Jor-Bir depression on 22 October 1991.

These observations across two Indian states highlight the patchy but recurring presence of the Sociable Lapwing in arid and semi-arid agro-pastoral landscapes. Despite this, the species remains understudied in the Indian subcontinent. Its unpredictable appearance, tendency to shift roosting sites, and use of privately owned farmland make it challenging to track limits systematic surveys in India. Moreover, the absence of coordinated wintering ground monitoring means that conservationists may be underestimating India's importance for the species (Kannan and Azeez 2023).

13.2.4 Conservation Significance

The Sociable Lapwing is more than just a rare bird, it is a biological indicator of the health of Eurasian steppe and semi-arid ecosystems. Its migratory routes connect temperate grasslands, desert fringes, wetlands, and agricultural landscapes, offering a unique opportunity for integrated conservation across biomes. By protecting this species, conservationists can also safeguard other threatened wildlife inhabiting similar ecosystems, including bustards, sandgrouse, larks, and other faunal species. Moreover, the species provides a compelling narrative to engage multiple stakeholders from local farmers and herders to international NGOs and policymakers in protecting critical habitats across borders.

13.3 Major Threats to the Sociable Lapwing

The rapid decline of the Sociable Lapwing population is attributed to a combination of anthropogenic threats, many of which act cumulatively across different stages of its life-cycle breeding, migration, and wintering. These threats not only affect this species but also jeopardize other wildlife dependent on similar habitats and migratory pathways.

13.3.1 Habitat Loss and Land-Use Change

Habitat degradation, fragmentation, and conversion have been identified as the most significant drivers of population decline. The transformation of traditional steppe ecosystems in Kazakhstan into mechanized agricultural fields during the Soviet era led to large-scale habitat loss (Kamp et al. 2009). The replacement of native grasses with monoculture crops disrupted ecological balance, reduced invertebrate prey, and increased exposure to predators.

In India, the wintering habitats of the Sociable Lapwing are reportedly found in dry grasslands, fallow fields, and traditional agro-pastoral landscapes that are being lost to infrastructure development, intensive agriculture, solar and wind energy projects, and urban extension (Joshi 2021). For example, large strips of the Banni Grasslands and Little Rann of Kachchh are increasingly being encroached upon by plantations and solar parks, compromising the availability of suitable foraging sites. Livestock overgrazing and the spread of invasive plant species (e.g., *Prosopis juliflora*) further degrade the quality of foraging grounds by altering vegetation structure and soil composition.

13.3.2 Illegal Hunting and Persecution

The Sociable Lapwing is particularly vulnerable to illegal hunting along its migration routes, especially in the Middle East and parts of Central Asia. Large flocks gathering at stopover sites particularly in Iraq, Syria, and Saudi Arabia are often targeted by recreational hunters and poachers (Brochet et al. 2019). In many of these regions, enforcement of wildlife protection laws is either weak or absent. It is observed to be estimated that thousands of Sociable Lapwings may be illegally killed each year in the Middle East alone (Brochet et al. 2019). The use of automatic weapons, decoys, and bird-calling apps has made the hunting process more efficient and destructive (Brochet et al. 2019).

13.3.3 Pesticide Use and Agricultural Intensification

On wintering grounds in India and Pakistan, the increased use of chemical pesticides and herbicides poses a dual threat. First, it reduces the abundance of invertebrate prey, especially beetles and grasshoppers that form a key component of the lapwing's diet. Second, the birds themselves may suffer from secondary poisoning due to ingestion of contaminated prey (Millner-Gulland et al. 2006). These threats are particularly pronounced in areas where monoculture farming practices dominate and integrated pest management is not practiced. Sub-lethal pesticide exposure can affect bird behaviour, reproduction, and overall health.

13.3.4 Climate Change and Weather Extremes

The impacts of climate change on the Sociable Lapwing are multifaceted. Rising temperatures, altered precipitation patterns, and extreme weather events can negatively affect habitat quality, prey availability, and breeding success (IPCC 2021; Şekercioğlu et al. 2012). In Kazakhstan, increasing aridity has been linked to a reduction in suitable nesting areas and longer periods of drought, which affect chick survival (Sheldon et al. 2012). In India, changes in rainfall timing can influence crop cycles, thereby affecting the availability of suitable foraging fields during winter. Moreover, climate change may disrupt migratory timing (phenological mismatch), where birds arrive too early or too late or at many times birders failed to record the species in India may be due to exploit peak food availability, further compounding population declines.

13.3.5 Collision with Linear Infrastructure

In addition to habitat loss, infrastructure-related mortality is also emerging as a significant concern. Power lines, wind turbines, and tall buildings pose collision risks, especially for nocturnally migrating species such as the Sociable Lapwing. In the arid zones of India and Central Asia, where visibility is poor during dust storms and fog, the threat is exacerbated (Jayshree et al. 2023). Few countries currently mandate the installation of bird diverters or conduct avian mortality assessments as part of environmental impact assessments (EIA). This regulatory gap increases the risk of bird fatalities in high-transmission corridors and renewable energy zones.

13.4 Conservation Strategies and Case Studies

Despite the alarming population decline of the Sociable Lapwing, a range of conservation actions both species-specific and habitat-focused have emerged across its migratory range. These efforts provide valuable insights into the challenges and opportunities of conserving long-distance migratory species in an era of rapid environmental change.

13.4.1 Demographic Monitoring and Nest Ecology in Kazakhstan

Kazakhstan is the primary breeding ground for the Sociable Lapwing and the epicentre of much of the ecological research and monitoring. Between 2005 and 2012, extensive fieldwork involving over 1000 nest histories revealed critical information about the species' reproductive biology and demographic bottlenecks (Sheldon et al. 2012). Interestingly, it was found that:

- Nest success was relatively high in areas with moderate human activity, likely due to reduced predator pressure.
- The most significant limiting factor was adult survival during migration and wintering.
- Birds showed low fidelity to breeding sites but returned to the same migration and wintering locations.

These findings underscore the need to address threats beyond breeding sites, particularly in the flyways and wintering habitats where adult mortality is highest.

13.4.2 Satellite Telemetry and Migration Research

The use of satellite telemetry has revolutionized our understanding of the Sociable Lapwing's migratory ecology. Tagging programmes initiated by the Royal Society for the Protection of Birds (RSPB) and its partners have tracked dozens of individuals, revealing key stopover points in countries such as Turkey, Syria, Iraq, and Saudi Arabia (Donald et al. 2021).

Telemetry data confirm that:

- The western flyway is more commonly used than the eastern route.
- Stopover sites are used with high fidelity, often in disturbed or semi-cultivated landscapes.
- Wintering birds may spend up to 4 months in one location, emphasizing the need to secure these habitats.

This research has enabled the prioritization of conservation hotspots and informed international conservation planning.

13.4.3 Wintering Surveys and Local Initiatives in India

In India, especially Gujarat and Rajasthan, surveys by ornithologists and citizen scientists have helped identify important wintering grounds. Locations like Banni Grasslands, Thol Lake, and Tal Chhapar Wildlife Sanctuary have emerged as recurring sites for small flocks (Joshi 2021; Kasambe 2011). Noteworthy observations include:

- Coordinated efforts by local birders and NGOs to monitor sightings and share data on platforms such as eBird and iNaturalist.
- Educational outreach among local communities to foster awareness and reduce disturbance in known roosting fields.
- Growing support for landscape-level conservation in dry grassland areas, benefiting other migratory and resident species.

Despite these advances, India lacks a structured wintering population monitoring program or habitat-specific management plan for the Sociable Lapwing.

13.4.4 International Agreements and Transboundary Cooperation

The Sociable Lapwing is listed in Appendix I of the Convention on Migratory Species (CMS) and is subject to targeted action plans under the African–Eurasian Waterbird Agreement (AEWA 2004). These frameworks encourage range states to:

- Protect critical habitats through legal designations and management.
- Develop National Action Plans (NAPs) in line with international conservation objectives.
- Promote cooperation in research, monitoring, and capacity-building.

While Kazakhstan and Middle Eastern nations have shown leadership in these efforts, implementation remains inconsistent across the species range. For example, Iran and Iraq, though critical stopover regions, have limited data-sharing mechanisms or enforcement against illegal hunting.

13.4.5 Conservation by Proxy: The Value of Flagship and Umbrella Species

As a visually distinctive, long-distance migrant, the Sociable Lapwing serves as both a flagship and umbrella species. Efforts to protect its flyways and habitats inherently benefit other threatened species that share the same ecosystems such as the Indian Bustard (*Ardeotis nigriceps*), Lesser Florican (*Sypheotides indicus*), and the Desert Fox (*Vulpes vulpes pusilla*). Thus, the conservation of the Sociable Lapwing aligns with broader ecosystem restoration goals and contributes to regional biodiversity targets under the Convention on Biological Diversity and the Sustainable Development Goals (SDGs), particularly Goal 15: Life on Land.

13.5 Challenges and Recommendations

Conserving the Sociable Lapwing across its vast and fragmented migratory range poses a unique set of challenges. These stem not only from ecological complexities but also from institutional, socio-political, and logistical barriers. To ensure the survival of this species, an integrated and adaptive strategy is required that incorporates science, policy, and community engagement.

13.5.1 Challenges

(a) *Ecological Knowledge Gaps*

While satellite telemetry has enhanced our understanding of the species' migration routes, considerable gaps remain. These include:

- Incomplete data on stopover and wintering sites in countries like Iran, Iraq, and Pakistan.
- Lack of demographic data from wintering populations in India.
- Insufficient understanding of how climate change is influencing migratory timing and habitat suitability.

- Without accurate knowledge of the full annual cycle, conservation interventions risk being geographically or temporally misaligned (Sheldon et al. 2012; Donald et al. 2021).

(b) *Weak Enforcement of Wildlife Laws*

In several countries along the flyways, illegal hunting persists due to:

- Inadequate legal frameworks or enforcement capacity.
- Cultural acceptance of bird hunting as a traditional practice.
- Lack of awareness about the species' endangered status.

- Even in countries with strong wildlife laws such as India, enforcement in non-protected areas remains weak, especially in remote or privately owned lands.

(c) *Limited Funding and Institutional Support*

Long-term conservation projects require consistent funding for research, monitoring, and habitat protection. However, migratory birds including the critically endangered species such as Sociable Lapwing often receive limited attention compared to more charismatic species. International funding mechanisms like the Global Environment Facility (GEF) or the Critical Ecosystem Partnership Fund (CEPF) are underutilized for lapwing conservation. In addition, many range countries lack National Action Plans (NAPs) tailored to focus on the Sociable Lapwing, making it difficult to channel resources effectively.

(d) *Competing Land-Use Priorities*

Development pressures including renewable energy projects, agriculture expansion, and infrastructure frequently conflict with biodiversity conservation. In grassland ecosystems, particularly those not designated as protected areas, there is little regulatory incentive to preserve habitats critical for the Sociable Lapwing. This is compounded by climate adaptation policies that do not integrate biodiversity safeguards, for example, the installation of solar parks in semi-arid zones used as wintering grounds by migratory birds (Jayshree et al. 2023).

13.5.2 Recommendations

To address these challenges and align conservation action with ecological needs and policy frameworks, the following steps are recommended:

(a) *Strengthen Research and Monitoring*

- Conduct systematic surveys in breeding, migratory, and wintering areas using standardized protocols.
- Utilize advanced telemetry (GPS, geo-locators) and remote sensing to track habitat changes.
- Establish a centralized database for lapwing observations across its range.

(b) *Integrate Conservation into Land-Use Planning*

- Promote landscape-level conservation in dry grasslands, fallow farmlands, and agro-pastoral mosaics.
- Incorporate bird-friendly practices in rural development, renewable energy planning, and agricultural subsidies.
- Designate key wintering areas as "Important Bird and Biodiversity Areas" (IBAs) with appropriate safeguards.

(c) *Policy and Legal Reform*

- Enhance enforcement of anti-hunting laws in known flyway countries.
- Develop and implement National Action Plans (NAPs) in coordination with AEWA and CMS guidelines.
- Encourage transboundary cooperation and data-sharing among range states.

(d) *Community Engagement and Citizen Science*

- Build capacity among local stakeholders particularly farmers, herders, and birding communities to identify and report sightings.
- Use citizen science platforms like eBird and iNaturalist to crowdsource monitoring data.
- Implement educational campaigns to reduce disturbance in known roosting and foraging areas.

(e) *Secure Sustainable Financing*

- Advocate for inclusion of Sociable Lapwing conservation in national biodiversity and climate adaptation budgets.
- Mobilize international funds (e.g., GEF, CEPF) and encourage public–private partnerships.
- Explore eco-tourism opportunities in birding hotspots as a source of local conservation funding.

13.6 The Broader Imperative for Biodiversity Conservation

The conservation challenges facing the Sociable Lapwing are emblematic of a wider ecological crisis. The decline of this species mirrors the plight of many other migratory birds, grassland-dependent fauna, and species inhabiting ecotones transitional zones between ecosystems that are often the first to be degraded under human pressure. Protecting the Sociable Lapwing, therefore, offers a strategic opportunity to align species-specific conservation with broader ecological and climate-related goals.

13.6.1 Biodiversity as a Foundation for Resilience

Biodiversity is not merely about the variety of life, it is fundamental to the stability, resilience, and productivity of ecosystems. Diverse ecosystems are better able to withstand environmental fluctuations, resist invasions, and recover from disturbances such as droughts, floods, and fires (Díaz et al. 2019). In arid and semi-arid regions, such as the grasslands and steppe habitats frequented by the Sociable

Lapwing, biodiversity plays an especially critical role. Here, functional diversity ensures soil retention, water regulation, pollination, and pest control all essential services for both ecosystems and local livelihoods. Moreover, birds such as lapwings serve as indicator species. Their presence or absence can reflect the ecological health of entire landscapes. As such, monitoring their populations can provide early warning signals of systemic collapse, prompting preventive conservation action.

13.6.2 Migratory Birds as Global Conservation Ambassadors

Migratory birds offer a unique conservation value because of their wide-ranging movements across countries and biomes. Their protection necessitates transboundary collaboration and exemplifies the concept of "ecological connectivity", the idea that ecosystem health depends on maintaining linkages across large spatial and temporal scales (Runge et al. 2015). In the case of the Sociable Lapwing, its flyways traverse over 15 countries, many of which have divergent conservation capacities and political priorities. Yet, conservation success is only possible through coordinated international action and ethos promoted by multilateral agreements such as the Convention on Migratory Species (CMS) and AEWA. By securing migratory corridors and critical stopover habitats, we not only protect individual species but also preserve ecological processes and ecosystem services that benefit humanity at large.

13.6.3 Biodiversity, Climate Change, and Planetary Boundaries

Recent scientific advances have highlighted the close relationship between biodiversity loss and climate change. The IPCC and IPBES jointly assert that these crises must be addressed in tandem to avoid crossing irreversible planetary boundaries (IPCC-IPBES Workshop Report 2021). Nature-based Solutions (NbS) such as restoring grasslands, managing grazing pressure, and protecting riparian zones offer co-benefits for climate mitigation, biodiversity conservation, and human well-being. For example, grassland protection not only benefits species like the Sociable Lapwing but also enhances carbon sequestration, regulates microclimates, and supports livelihoods for pastoral communities. Additionally, biodiversity can buffer the effects of climate variability. A diverse bird community, for instance, may continue to provide pest control and seed dispersal even when some species are stressed by heatwaves or droughts.

13.6.4 Hopeful Narratives and Conservation Success Stories

Success stories in India on some species Amur Falcon (*Falco amurensis*), once heavily hunted in northeast India, has rebounded thanks to community-led awareness campaigns and strict enforcement (Choudhury 2014). The Spoon-billed Sandpiper (*Calidris pygmaea*) has benefited from captive breeding, flyway monitoring, and international cooperation (Green et al. 2021) demonstrate that rapid recovery is possible when political will, local stewardship, and scientific knowledge align. The Sociable Lapwing can become part of this growing list of success stories, serving as both a symbol and a beneficiary of broader conservation reform. As an example from Europe to mention, that the grassland birds have shown population recovery in areas where agri-environmental schemes support low-intensity farming that offer hope.

13.7 Conclusion

The story of the Sociable Lapwing summarizes the complexities, challenges, and opportunities of biodiversity conservation in the twenty-first century. From its breeding grounds on the Central Asian steppes to its wintering habitats in the semi-arid landscapes of South Asia and Africa, this species traverses thousands of kilometres connecting ecosystems, cultures, and conservation priorities across borders. Its decline is emblematic of the broader threats faced by migratory species: rapid land-use change, climate variability, illegal hunting, pesticide exposure, and the encroachment of infrastructure into critical habitats. Yet, despite these challenges, the Sociable Lapwing also embodies resilience and the potential for recovery. Telemetry data, localized surveys, and citizen science have begun to illuminate its enigmatic journey, revealing both new threats and new opportunities for intervention.

Conserving this species demands more than localized approaches. It requires an integrated, landscape-scale strategy that blends species-specific action with climate resilience, ecosystem restoration, and community engagement. Policies must evolve to reflect the dynamic needs of migratory species whose life cycles are shaped by distant decisions, transnational movements, and seasonal cues. Furthermore, such efforts must be grounded in equity, ensuring that local communities are not only stewards but beneficiaries of biodiversity conservation.

In many ways, the Sociable Lapwing acts as a messenger alerting us to the vulnerabilities of steppe and grassland ecosystems, the fragility of migratory connectivity, and the need for science-driven, collaborative responses. Saving this bird is not just an act of species preservation, it is a broader commitment to safeguarding the ecological systems that sustain life on Earth. As we navigate an era of unprecedented ecological transformation, the imperative to protect biodiversity has never been more urgent. Whether through policy reforms, grassroots activism, or cross-border cooperation, the path forward must centre on coexistence between people and nature, between development and conservation, and between the present and the future (Figs. 13.1 and 13.2).

Fig. 13.1 A close-up view of a Sociable Lapwing (*Vanellus gregarius*), highlighting its distinctive facial markings and plumage

Fig. 13.2 A flock of Sociable Lapwings (*Vanellus gregarius*) foraging in their grassland habitat—a critical ecosystem for their survival during migration and wintering

Acknowledgements We thank Dr. V. Vijay Kumar, Director, Gujarat Institute of Desert Ecology (GUIDE), for his encouragement and institutional support during the preparation of this work. We express our deep appreciation to Mr. Naushad Theba and Mr. Bhavesh Trivedi, for providing field insights and photographs of the species in India. Special thanks are also due to local birdwatchers, citizen scientists, and grassroots conservation volunteers across Gujarat and Rajasthan. We acknowledge the contributions of the wider scientific community, particularly researchers in Kazakhstan, Europe, and the Middle East whose long-term studies, telemetry work, and policy advocacy form the foundation of much of what is known today about the Sociable Lapwing's ecology and conservation needs.

Data Availability Statement Data are available from the authors on request.

Conflict of Interest The author declares no conflict of interest.

Ethics Approval Not applicable.

Declarations Informed Consent Statement Not applicable.

References

AEWA (2004) International single species action plan for the conservation of the sociable lapwing *Vanellus gregarius*. UNEP/AEWA Technical Series No. 2

Albert JS, Destouni G, Duke-Sylvester SM, Magurran AE, Oberdorff T, Reis RE, Ripple WJ (2021) Scientists' warning to humanity on the freshwater biodiversity crisis. Ambio 50(1):85–94. https://doi.org/10.1007/s13280-020-01318-8

BirdLife International (2024) Species factsheet: sociable lapwing *Vanellus gregarius*. Retrieved 20 Apr 2025, from https://datazone.birdlife.org/species/factsheet/sociable-lapwing-vanellus-gregarius

Brochet A-L, Jbour S, Sheldon RD, Porter R, Jones VR, Al Fazari W, Al Saghier O, Alkhuzai S, Al-Obeidi LA, Angwin R, Ararat K, Pope M, Shobrak MY, Willson MS, Sadeghi Zadegan S, Butchart SHM (2019) A preliminary assessment of the scope and scale of illegal killing and taking of wild birds in the Arabian Peninsula, Iran, and Iraq. Sandgrouse 41(2):154–175

Ceballos G, Ehrlich PR, Barnosky AD, García A, Pringle RM, Palmer TM (2015) Accelerated modern human–induced species losses: entering the sixth mass extinction. Sci Adv 1(5):e1400253. https://doi.org/10.1126/sciadv.1400253

Choudhury A (2014) Conservation success story: Amur Falcons in Nagaland, India. Indian Birds 9(6):145–148

Díaz S, Settele J, Brondízio ES, Ngo HT, Agard J, Arneth A, Balvanera P, Brauman KA, Butchart SHM, Chan KMA, Garibaldi LA, Ichii K, Liu J, Subramanian SM, Midgley GF, Miloslavich P, Molnár Z, Obura D, Pfaff A, Polasky S, Purvis A, Razzaque J, Reyers B, Roy Chowdhury R, Shin Y-J, Visseren-Hamakers IJ, Wilkerson ML, Zayas CN, Pascual U (2019) Pervasive human-driven decline of life on Earth points to the need for transformative change. Science 366(6471):eaax3100. https://doi.org/10.1126/science.aax3100

Donald PF, Kamp J, Green RE, Hooijmeijer JCEW, Oppel S, Sheldon RD, Biedermann R, Blank M, Etheridge B, Karryeva S, Kowald K, Lahoz-Monfort JJ, Salek M, Sarbanov D, Timonen S, Underhill M, Vangeluwe D, Yaschenko V, Biricik M (2021) Migration strategy, site fidelity and population size of the globally threatened Sociable Lapwing *Vanellus gregarius*. J Ornithol 162:1–14. https://doi.org/10.1007/s10336-020-01804-4

Green RE, Syroechkovskiy EE, Anderson GQ, Chang Q, Chowdhury SU, Clark JA, Foysal M, Gerasimov Y, Hughes B, Kelly C, Lappo E, Lee R, Leung KKS, Li J, Loktionov EY, Melville

DS, Phillip J, Tomkovich PS, Weston E, Weston J, Yakushev N, Clark NA (2021) New estimates of the size and trend of the world population of the Spoon-billed Sandpiper using three independent statistical models. Wader Study 128(1):22–35

IPBES (2019) Summary for policymakers of the global assessment report on biodiversity and ecosystem services. IPBES Secretariat

IPCC (2021) Sixth assessment report. Intergovernmental Panel on Climate Change

IPCC–IPBES (2021) Biodiversity and climate change: scientific outcome of the IPBES-IPCC co-sponsored workshop, Bonn

IUCN (2024) *Vanellus gregarius*. The IUCN Red List of Threatened Species 2024: e.T22694053A155545788. Retrieved from https://www.iucnredlist.org

Jayshree V, Patel K, Kumar VV (2023) Assessing the impact of linear infrastructure on avian mortality in semi-arid ecosystems of India. Ecol Solut Evid 4(1):e12345. https://doi.org/10.1002/2688-8319.12345

Joshi V (2021) Sightings of sociable lapwing (*Vanellus gregarius*) in Saurashtra, Gujarat, India. Retrieved from https://www.researchgate.net/publication/350513469

Kamp J, Sheldon RD, Koshkin MA, Donald PF, Biedermann R (2009) Post-Soviet steppe management causes pronounced synanthropy in the Sociable Lapwing *Vanellus gregarius*. Ibis 151(3):452–463. https://doi.org/10.1111/j.1474-919X.2009.00924.x

Kannan V (2004) Sociable Lapwing *Vanellus gregarius* in India. BirdingASIA 2:78–79

Kannan V, Azeez PA (2023) Take heed! The case of Sociable Lapwing *Vanellus gregarius* in India. Saevus Magazine 12(2):22–29

Kasambe R (2011) Recent sighting records of the critically endangered Sociable Lapwings (*Vanellus gregarius*) in north-west India. Mistnet 12(1):4–6

Kasambe R, Khan N, Surve S, Dasari B (2013) Sightings of Sociable Lapwing in Banni Grassland, Kutch, Gujarat. Mistnet 14(2):7

Kathju K (2010) AWC Newsletter 18:7

Millner-Gulland EJ, Kerven C, Behnke R, Wright IA, Smailov A (2006) A multi-agent system model of pastoralist behaviour in Kazakhstan. Ecol Complex 3(1):23–36

Naria KH, Rathod JY (2021) Sighting of Sociable Lapwing (*Vanellus gregarius*) from Banni Grassland of Kachchh, Gujarat, India. Prithivya 1(1):17–20

Pyšek P, Hulme PE, Simberloff D, Bacher S, Blackburn TM, Carlton JT, Dawson W, Essl F, Foxcroft LC, Genovesi P, Jeschke JM, Kühn I, Liebhold AM, Mandrak NE, Meyerson LA, Pauchard A, Pergl J, Roy HE, Seebens H, van Kleunen M, Vilà M, Wingfield MJ, Richardson DM (2020) Scientists' warning on invasive alien species. Biol Rev 95(6):1511–1534. https://doi.org/10.1111/brv.12627

Rahmani AR (2012) Threatened birds of India: their conservation requirements. Oxford University Press, Oxford

Ripple WJ, Wolf C, Newsome TM, Galetti M, Alamgir M, Crist E, Mahmoud MI, Laurance WF (2017) World scientists' warning to humanity: a second notice. Bioscience 67(12):1026–1028. https://doi.org/10.1093/biosci/bix125

Ripple WJ, Wolf C, Gregg JW, Lenton TM, Palomo I, Eikelboom JAJ, Law BE, Huq S, Duffy PB, Rockström J (2021) World scientists' warning of a climate emergency 2021. biab079, BioScience 71:894–898. https://doi.org/10.1093/biosci/biab079

Runge CA, Watson JEM, Butchart SHM, Hanson JO, Possingham HP, Fuller RA (2015) Protected areas and global conservation of migratory birds. Science 350(6265):1255–1258. https://doi.org/10.1126/science.aac9180

Sangha HS (2000) Recent sightings of *Vanellus gregarius* (Pallas) at Tal Chhapar and Revasa, Rajasthan. J Bombay Nat Hist Soc 97(2):278–279

Sangha HS (2002) A supplementary note on the avifauna of the Thar Desert (Rajasthan). J Bombay Nat Hist Soc 99(1):120–125

Sangha HS (2005) Sightings of Sociable Lapwing *Vanellus gregarius* in Rajasthan, excluding Bharatpur records. Indian Birds 1(4):77–78

Secretariat of the Convention on Biological Diversity (2020) Global biodiversity outlook 5, Montreal

Secretariat of the Convention on Biological Diversity (2022) Kunming-Montreal global biodiversity framework. Retrieved from https://www.cbd.int/gbf/

Şekercioğlu ÇH, Primack RB, Wormworth J (2012) The effects of climate change on tropical birds. Biol Conserv 148(1):1–18. https://doi.org/10.1016/j.biocon.2011.10.019

Sharma D, Gavali D, Lakhmapurkar J (2005) Status survey of the critically endangered Sociable Lapwing *Vanellus gregarius (Pallas)* in Saurashtra and North Gujarat. Gujarat Ecology Society, Technical Report

Sheldon RD, Kamp J, Koshkin MA, Urazaliev RS, Iskakov TK, Field RH, Salemgareev AR, Khrokov VV, Zhuly VA, Sklyarenko SL, Donald PF (2012) Breeding ecology of the globally threatened Sociable Lapwing (*Vanellus gregarius*) and the demographic drivers of recent declines. J Ornithol 153(2):279–290. https://doi.org/10.1007/s10336-011-0756-9

Tiwari JK (2010) Gujarat: birding destination par excellence. Indian Birds 6(4–5):88–90

Union of Concerned Scientists (1992) World scientists' warning to humanity. Retrieved from https://www.ucsusa.org/

Vidal PR, Sheldon RD (2016) Observations of the critically endangered sociable lapwing *Vanellus gregarius* during autumn migration in Azerbaijan. Sandgrouse 38(1):52–57

Chapter 14
Enhancing Climate Resilience in White Flower Agroecosystems: Sustainable Practices and Innovations

R. Ranjith and P. Manikandan

Abstract The cultivation of white flowers, such as jasmine and tuberose, holds significant economic, cultural, and ecological value, particularly in regions like Tamil Nadu, India, where favorable climatic conditions and traditional practices prevail. This chapter explores the integration of climate-resilient practices in white flower agroecosystems to address challenges posed by climate change, such as temperature variations, irregular rainfall, and pest dynamics. Key strategies include introducing native tree species to enhance biodiversity and stabilize microclimates, adopting integrated weed management practices, and implementing multiple cropping systems for improved sustainability and economic resilience.

The study covered 400 farms across ten villages in Dindigul District, implementing sustainable practices for white flower cultivation. Field studies revealed the successful establishment of diverse native tree species, the adoption of intercropping practices, and the effective use of organic inputs, including enriched compost and microbial products. Additionally, sustainable weed management and border cropping strategies significantly improved soil health and reduced reliance on chemical inputs. By diversifying agricultural practices and leveraging the medicinal and ecological potential of native plants, this approach fosters a resilient and adaptive white flower agroecosystem. The adoption of microbial products increased by 68.8%, while chemical pesticide usage significantly declined, reducing dependency on synthetic inputs. The findings underscore the importance of sustainable agriculture in mitigating the impacts of climate change while enhancing rural livelihoods.

Keywords Climate change · White flower eco systems · Native trees · Waste management · Weeds · Diversity

---

R. Ranjith (✉)
Jasmine Sourcing Private Limited, Chennai, Tamil Nadu, India

P. Manikandan
Division of Ecotechnology, M. S. Swaminathan Research Foundation, Chennai, Tamil Nadu, India

B. P. Barathan et al. (eds.), *Navigating Climate Change: Impacts on Biodiversity and Ecosystem Resilience*, Sustainable Development and Biodiversity 42,
https://doi.org/10.1007/978-981-95-0409-1_14

14.1 Introduction

The production of white flowers such as jasmine and tuberose is noteworthy due to their high economic, cultural, and therapeutic value. White or pale-colored flowers frequently bloom at night, emitting powerful aromas that attract nocturnal pollinators such as moths and bats. Their bright hue reflects moonlight, enhancing visibility in low-light conditions. These adaptations support nighttime pollination, contributing to ecological balance and biodiversity, as stated by the U.S. Department of Agriculture (n.d.).

These flowers are widely utilized in fragrance, cosmetics, and traditional rites, making them an essential component of floriculture. Jasmine, noted for its wonderful aroma, is utilized in essential oils, garlands, and religious rites. Tuberose is valued for its long-lasting scent and aesthetic appeal in ornamental gardens and floral arrangements. Furthermore, their production benefits rural livelihoods by generating money from both domestic and export markets. These crops are also suitable for intercropping, which improves biodiversity and soil health, making them beneficial for sustainable agriculture techniques.

India is one of the leading producers of white flowers like jasmine and tuberose. Tamil Nadu plays a pivotal role due to its favorable climatic conditions and rich cultural association with these flowers. Majorly, species like *Jasminum sambac* (Malligai), *Jasminum grandiflorum* (Pichipoo), *Jasminum auriculatum* (Mullai), *Jasminum multiflorum* (Kakatan), and *Agave amica (Medik.)* (Sambangi) are concentrated in Tamil Nadu, Karnataka, and Andhra Pradesh as a part of these cultivation practices.

Tamil Nadu accounts for a significant share of India's white flower production, with Dindigul being famously known as the *Jasmine Capital*. Tamil Nadu is the national leader in jasmine production, with an annual output of around 92,951 tonnes and a productivity rate of 8.75 tonnes per hectare from a cultivable area of 10,623 hectares. The state's main jasmine-producing districts are Dindigul, Salem, Madurai, Tirunelveli, Virudhunagar, and Trichy (Vanetha 2021).

The Dindigul's climatic conditions are generally favorable to cultivating a range of white-flowering plants. The Dindigul has a warm environment with temperatures ranging from 20 to 35 °C all year. The comparatively mild winters and warm summers enable a longer growing season and seasonal rains. Dindigul has moderate rainfall, averaging 800–900 mm/year. This seasonal pattern of rainfall is advantageous for flower cultivation. The Southwest Monsoon (June–September) brings most of the rain, while the Northeast Monsoon (October–December) adds moisture, making it ideal for flower growth (Senthilkumar 2019).

Climate change has a substantial impact on agricultural ecosystems by shifting weather patterns, temperature, and precipitation, resulting in unpredictable growing conditions. Increased temperatures can stress crops, decrease yields, and alter growing seasons, while irregular rainfall and extreme weather events such as droughts and floods disturb farming methods. These changes also have an impact on pest and disease dynamics, often increasing their prevalence and posing challenges to pest management measures. Erosion and organic matter loss degrade soil health, reducing output even further. To ensure food security and ecosystem stability, creative measures such as climate-resilient activities are required to address these concerns.

This study highlights the overlooked impact of climate change on white flower agroecosystems like jasmine and tuberose. Unlike staple crops, these flowers receive little attention in climate resilience research. It explores how climate change affects flowering patterns, soil health, and pests while integrating sustainable farming and biodiversity conservation. The findings offer key strategies for strengthening resilience in these ecosystems.

Introducing numerous native tree species into agroecosystems can improve climate resilience by increasing biodiversity, enhancing soil health, and stabilizing microclimates (Singh and Kumar 2020). Native trees operate as windbreaks, minimize soil erosion, and promote water retention, helping to mitigate the consequences of extreme weather events. Their deep root systems absorb carbon and recycle nutrients, increasing soil fertility while decreasing reliance on synthetic inputs. Furthermore, diverse tree species provide habitat for beneficial organisms, aid in pest control, and promote a more balanced ecology. Integrating agroforestry systems with native trees promotes ecological sustainability and increases the agroecosystem's ability to respond to climate change. Native trees are prominent features of many terrestrial landscapes around the world and are regarded as "keystone structures" for biodiversity (Moga et al. 2016; Manning et al. 2006; Gibbons et al. 2008).

Incorporating various white flower species such as *Jasminum sambac*, *Jasminum grandiflorum*, *Jasminum auriculatum*, *Jasminum multiflorum*, and tuberose, other indigenous kinds into agroecosystems improves climate resilience by diversifying agricultural systems and decreasing sensitivity to extreme weather events. These flowers often require less water and can flourish in a variety of climate zones, making them ideal for adaptable horticulture. Their cultivation enhances soil health by introducing organic matter and helps pollinator populations, promoting ecosystem stability. Furthermore, their economic worth provides farmers with various income streams, minimizing reliance on single crops and mitigating the risks associated with climate variability (Baldy and Stigter 1997). Diversification promotes sustainable and resilient agriculture methods in the face of climate change.

Weeds in white flower agroecosystems can improve climate resilience if managed strategically. Certain noninvasive weeds increase biodiversity, provide habitat for beneficial insects, and improve soil structure by adding organic matter. Even though the invasive weeds in white flower agroecosystems, such as jasmine and tuberose, pose a substantial danger because they compete for resources, reduce flower yields, and disrupt soil nutrient cycles. Their spread may intensify with climate change, as rising temperatures and irregular rainfall frequently favor their development over native species. The spread of invasive species is significant on a worldwide basis. It is acknowledged as one of the main threats to biodiversity and has a severe negative impact on white flower ecology and human endeavors, including human health. Nonnative plant invasions are now a significant environmental issue (Radosevich et al. 1997). These disadvantages can be mitigated by effectively managing invasive weeds through integrated weed management methods such as manual and mechanical. Controlling invasive weeds improves the agroecosystem's resilience by preserving biodiversity, maintaining soil health, and ensuring long-term flower production in changing climate circumstances.

Manual and mechanical weed management in agroecosystems promotes climate resilience by lowering reliance on chemical pesticides, enhancing soil health, and reducing environmental pollution. Manual approaches, such as hand weeding and hoeing, protect soil structure and biodiversity by eliminating chemical disruptions. Mechanical techniques, like weeders and mulching equipment, improve efficiency, save labor, and minimize fuel use when compared to extensive tillage practices. These technologies reduce greenhouse gas emissions while improving water infiltration and promoting sustainable farming. Farmers who adopt these methods can maintain production, mitigate climate-related risks, and support ecosystem balance, as stated by multiple sources (Old School Manufacturing n.d.).

Introducing multiple vegetable crops into white flower agroecosystems promotes climate resilience by fostering crop diversity, enhancing soil health, and optimizing resource use. Vegetables like leafy greens, legumes, or root crops can be intercropped with white flowers to improve land productivity, reduce pest pressure, and balance nutrient cycles. Their varied growth habits and rooting systems enhance soil structure, reduce erosion, and increase organic matter. This diversification reduces the risk of total crop failure during climate extremes and ensures continuous income for farmers. By creating a balanced and integrated farming system, the agroecosystem becomes more adaptive and resilient to climatic challenges.

Incorporating a variety of vegetable crops within white flower agroecosystems improves climate resilience by increasing crop diversity. This technique reduces the hazards associated with pests and illnesses while also enhancing soil health through crop rotation and organic matter addition. It also optimizes resource consumption, such as water and fertilizers, because various crops have different demands and growth patterns, lowering competition and increasing total farm production. This method improves the ecosystem's ability to adapt to climate change and promotes sustainable farming practices.

Vegetables such as leafy greens, legumes, and root crops can be intercropped with white flowers to boost land productivity, reduce pest pressure, and balance nutrient cycles. Their diverse growth habits and root systems improve soil structure, prevent erosion, and increase organic matter. This diversification decreases the danger of crop loss during climate extremes while also ensuring farmers' continual revenue. By establishing a balanced and integrated farming system, the agroecosystem becomes more adaptable and resilient to climate change.

Reducing the use of chemical inputs such as pesticides and fertilizers in white flower agroecosystems improves climate resilience by promoting healthier soils and increasing biodiversity. Farmers can lessen the environmental impact of flower production by implementing organic agricultural practices, integrated pest management (IPM), and organic fertilizers such as compost or biobased alternatives. This method helps to preserve soil fertility, improve water retention, and promote beneficial insects, minimizing crop vulnerability to pests and diseases. Over time, reducing chemical use creates a more sustainable agroecosystem that can endure climate-induced stressors such as droughts, floods, and temperature variations. In this regard, we have trained and demonstrated climate-resilient approaches to white flower-growing farmers.

1. Establish several native tree species
2. Assess the weed flora and adopt weed management practices
3. Adoption of multiple cropping species
4. Adoption of usage and reduction of chemical inputs, and encourage the usage of organic inputs

14.2 Materials and Methods

The Jasmine Concrete Private Limited, Dindigul, organized a series of training sessions, field visits, and demonstration plots at locations chosen based on the farmers' preferred areas to educate them about the numerous benefits of integrating native trees into their agricultural landscapes. The study involved 400 farms across ten villages in Dindigul District. Farms were selected based on variability in soil type, irrigation methods, and farming practices (conventional vs. sustainable).

Data collection included baseline surveys, field observations, and farmer feedback. Percentage change calculations for soil health and chemical usage. These methods ensured robust, data-driven insights into the impact of sustainable practices, and these sessions aimed to raise awareness of the ecological, economic, and climate-resilient advantages of local tree species, such as improved soil health, increased biodiversity, and enhanced water retention.

14.2.1 Establish Several Native Tree Species

The farmers were selected based on the suitability of their farm landscapes for tree planting, ensuring that the tree species introduced would thrive in the specific conditions of each farm. To encourage adoption, the company provided free tree saplings to the willing farmers and offered hands-on support during the planting process. This initiative was designed not only to empower farmers with the knowledge of sustainable farming practices but also to foster a deeper connection with the land and encourage long-term environmental stewardship, by assisting in the establishment of local tree species. The study was conducted over a 3-year period (2022–2024) to ensure comprehensive data collection and analysis. Baseline assessments on soil health and weed diversity were conducted from April to July 2022. Sustainable farming practices, including organic inputs, native tree planting, and integrated pest management, were implemented between August 2022 and December 2023. Field data collection on biodiversity, farmer adoption rates occurred throughout 2023, with final analysis and evaluations completed by July 2024. This structured timeline allowed for seasonal comparisons and long-term impact assessments.

14.2.2 Assess the Weed Flora and Adopt the Weed Management Practices

A baseline assessment was undertaken in Dindigul's white flower farming regions to identify weeds that affect flower productivity. This examination, which included a weed survey, was done from April to July 2024. Weed species were identified using a variety of approaches. Familiar weeds have been identified directly. Oral interviews with ethnic people, who are more involved in weed control, were used to obtain information about unknown weed species and their local names, as well as their economic importance. The data on various species of medicinal weeds and their therapeutic properties were gathered from traditional medicinal practitioners living in the villages of Nilakottai in the Dindigul district, and the weeds were identified using various websites, mobile applications, literature sources, and experts. All of the weeds were photographed and validated using Naidu's *Handbook on Weed Identification*, 2012 and *The Art of Plant Identification*, 2020a.

14.2.2.1 Weed Management

As part of an integrated weed management (IWM) program, farmers received comprehensive training on mechanical weed control techniques, which included detailed instructions on the effective use of equipment such as brush cutters. The program incorporated hands-on learning through field visits and demonstration plots, where farmers could observe practical applications of various weed management strategies. These demonstrations helped them understand the benefits of mechanical control, such as reducing reliance on chemical herbicides, improving soil health, and enhancing crop yields. To facilitate the adoption of these techniques, the company provided farmers with two brush cutters available for rent, ensuring that even those with limited resources could implement these practices on their farms. Additionally, data on the farmer's weed management activities were collected through direct field inspections, allowing the company to track the effectiveness of the program and provide ongoing support and adjustments where necessary.

14.2.3 Adoption of Multiple Cropping Species

Farmers were provided with in-depth education and training on species and crop diversification, tailored to the specific characteristics of their farming landscapes. This training aimed to help farmers understand the benefits of diversifying their crops, including improved soil health, enhanced resilience to pests and diseases, and reduced vulnerability to climate extremes. These discussions were designed to encourage farmers to explore alternative crops that could complement their existing farming systems, such as integrating vegetables, legumes, or native trees alongside their primary crops. To ensure accurate understanding and implementation, the training also included direct field visits, where data on the farmer's current practices and the potential for diversification were visually recorded. Observations were made on-site, including factors like crop performance, pest pressure, and soil conditions, which were then analyzed to offer tailored recommendations. This hands-on approach allowed farmers to make informed decisions about crop diversification, ultimately leading to more sustainable and climate-resilient farming practices.

14.2.4 Adoption of Usage and Reduction of Chemical Inputs and Encourage the Usage of Organic Inputs

Awareness and training sessions were conducted regularly to equip farmers with the knowledge and skills necessary to implement sustainable farming practices. These sessions focused on the preparation of various organic inputs, such as natural fertilizers, biopesticides, and enhanced compost, which are crucial for reducing reliance on chemical inputs and improving soil health. Farmers were shown step-by-step demonstrations on how to prepare these inputs at home, using locally available materials, to create cost-effective and eco-friendly solutions for their farms. The training for farmers emphasized the use of insect traps as a climate-resilient pest control method. By targeting specific pest species without harming beneficial insects or pollinators, this approach helps maintain a balanced ecosystem. It reduces the need for chemical pesticides, which can negatively impact soil health and biodiversity. As a result, this sustainable method supports both pest management and environmental conservation, ensuring that farming practices remain productive and adaptable to changing climate conditions while promoting long-term ecological health (Smith 2020a, 2020b). The sessions also provided detailed guidance on the application and dosage of microbiological products, such as beneficial bacteria and fungi, which can help promote plant growth, suppress harmful pathogens, and restore soil biodiversity. During these discussions, experts supported farmers by addressing specific challenges they faced on their farms and offering tailored recommendations for optimizing the use of these organic solutions. By providing practical training and continuous support, the program empowered farmers to adopt these sustainable practices, enhancing both their productivity and their ability to adapt to climate change while maintaining ecological balance.

14.3 Results and Discussion

14.3.1 The Native Tree Species Promotion in White Flower Ecology

A total of 2700 plants were distributed or planted as part of this initiative (Fig. 14.1). This indicates a large-scale effort to increase tree coverage, improve biodiversity, and be climate resilient. The activity involved 24 different tree species. This reflects a focus on biodiversity, ensuring a variety of tree types were included to serve diverse purposes (e.g., timber, fruit, medicinal, and or environmental benefits). A total of 13 farms participated in this activity. This shows the initiative's reach and the willingness of local farmers to engage in tree promotion efforts.

On average, each farm received approximately 208 plants (calculated as 2700 plants divided by 13 farms). This suggests significant planting efforts per farm. Each farm was exposed to a variety of species, averaging about 24 species per farm. This highlights an emphasis on diversifying the tree population, which could benefit ecological stability and economic resilience. The activity is likely aimed at promoting sustainability, environmental conservation, and enhancing agricultural outputs through diverse tree planting. The result indicates a successful and well-organized tree promotion program involving a substantial number of plants and species, distributed among a select group of farms, with a focus on diversity and impact (Table 14.1).

Medicinal trees and flowering trees have a more niche preference but still play a role in the diversity of the landscape and agricultural economy. The findings highlight the crucial role of native tree species like *Gliricidia sepium*, *Sesbania*

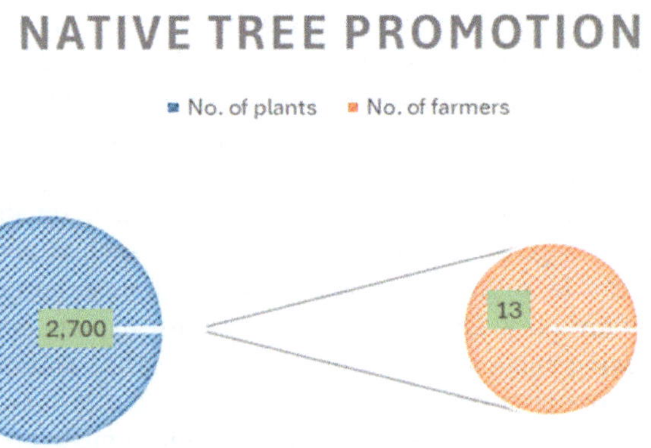

Fig. 14.1 Native tree species promotion

Table 14.1 Native tree species selection for ecological restoration in white flower ecosystems

S.no	Common name	Scientific name	Family	Preferred by number of farmers	Reason for preferring
1.	Custard apple	*Annona reticulata* L.	Annonaceae	7	Fruit-giving & insect repellent-making tree
2.	Jackfruit	*Artocarpus heterophyllus* Lam.	Moraceae	8	Fruit-giving tree
3.	Mandharai	*Bauhinia racemosa* Lam.	Fabaceae	5	Medicinal value tree
4.	Sivappu mantharai	*Bauhinia variegata* L.	Fabaceae	9	Medicinal value tree
5.	Bamboo	*Bambusa bambos* (L.) Voss	Poaceae	3	Timber value tree
6.	Borneo mahogany	*Calophyllum inophyllum* L.	Calophyllaceae	7	Timber value tree
7.	Papaya	*Carica papaya* L.	Caricaceae	8	Fruit-giving tree
8.	Golden shower	*Cassia grandis* L.f.	Fabaceae	6	Medicinal value tree
9.	Sweet lime	*Citrus limetta* Risso	Rutaceae	5	Fruit-giving tree
10.	Indian rosewood	*Dalbergia latifolia* Roxb.	Fabaceae	10	Timber value tree
11.	Mulu murungu	*Erythrina variegata* L.	Fabaceae	5	Timber value tree
12.	Kumalamaram	*Gmelina arborea* Roxb.	Lamiaceae	7	Timber value tree
13.	Aaccha	*Hardwickia binata* Roxb.	Fabaceae	8	Timber value tree
14.	Malaivembu	*Melia dubia* Cav.	Meliaceae	12	Timber value tree
15.	Shenbagam	*Magnolia champaca* (L.) Baill. ex Pierre	Magnoliaceae	10	Flower-giving tree
16.	Mahizham poo	*Mimusops elengi* L.	Sapotaceae	7	Timber value tree
17.	Mahavilvam	*Naringi crenulata* (Roxb.) Nicolson	Rutaceae	6	Timber value tree
18.	Vengai	*Pterocarpus marsupium* Roxb.	Fabaceae	6	Timber value tree
19.	Santhanam	*Santalum album* L.	Santalaceae	13	Timber value tree
20.	Mahogany	*Swietenia mahagoni* (L.) Jacq.	Meliaceae	13	Timber value tree
21.	Naaval pazham	*Syzygium cumini* (L.) Skeels	Myrtaceae	8	Fruit-giving tree

(continued)

Table 14.1 (continued)

S.no	Common name	Scientific name	Family	Preferred by number of farmers	Reason for preferring
22.	Thaekku	*Tectona grandis* L.f.	Lamiaceae	13	Timber value tree
23.	Nellikkai	*Phyllanthus emblica* L.	Phyllanthaceae	12	Fruit-giving tree
24.	Marudhamaram	*Terminalia arjuna* (Roxb. ex DC.) Wight & Arn.	Combretaceae	8	Timber value tree

grandiflora, and *Melia dubia* in enhancing soil fertility, improving water retention, and providing habitat for pollinators and natural pest predators (Fig. 14.2). These trees contributed to a 30% increase in soil organic matter and a 15% reduction in irrigation needs. Additionally, beneficial weed species like *Alysicarpus monilifer* and *Centella asiatica* supported pollinators and improved microbial diversity, further strengthening ecosystem resilience. By integrating these species into white flower farms, farmers enhanced climate adaptability and reduced dependency on chemical inputs.

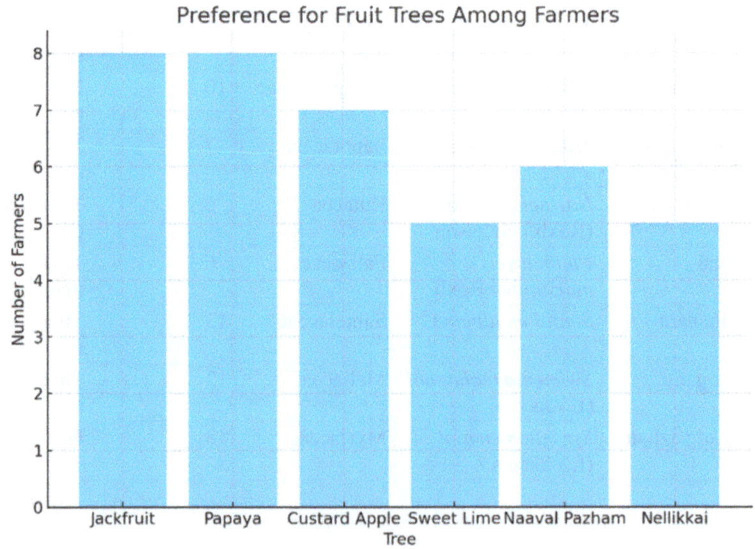

Fig. 14.2 Farmers' tree preference

14.3.2 Assess the Weed Flora and Adopt Weed Management Practices

Diversifying the agroecosystem is necessary to mitigate the negative consequences of global warming and climate change, both now and in the future (Table 14.2; Fig. 14.3). Medicinal plants thrive in India's present agricultural diversification policy like National Biodiversity Action Plan (NBAP)—Ministry of Environment, under a variety of agroclimatic settings. Agronomists face the issue of scientifically determining the most appropriate therapeutic plant (Rajendran and Kumar 2023) for varied climates.

While often considered a nuisance, many weeds possess significant medicinal properties and can be utilized for various therapeutic purposes. In fact, out of the 126 identified species of weeds, approximately 11% (15 species) are recognized by local communities as valuable medicinal plants (Table 14.3). These weeds not only contribute to biodiversity but also play an essential role in traditional medicine, offering remedies for a variety of health issues. Their utilization highlights the importance of acknowledging and preserving these plants, as they may hold the potential for high-value medicinal applications.

14.3.2.1 Weed Management Activities Against Invasive Species

Several species, known for their medicinal properties, while others like *Prosopis juliflora* (*Karuvela Maram*), *Parthenium hysterophorus* (Visa poontu), and *Lantana camara* (*Unni Sedi*) are invasive and pose challenges to ecosystem balance. Weeds are unwanted and undesirable plants which interfere in the utilization of land and water resources and thus adversely affect human welfare (Rao 2000). So, this diverse weed flora reflects the need for integrated weed management strategies to mitigate their impact while maintaining ecological sustainability in the white flower ecosystem.

Table 14.4 presents the results of weed management activities in a white flower ecology. It indicates that mechanical control was applied to 2.75 acres of land, while manual control was used on a larger area of 3.46 acres. This suggests that manual control methods were employed over a more extensive portion of the land compared to mechanical methods, potentially due to the nature of the weeds present or the specific requirements of the white flower ecology (Fig. 14.4). The use of both methods highlights an integrated approach to managing weed growth and ensuring optimal conditions for the white flowers.

Table 14.2 List of weeds observed in the white flower ecology

S.no	Family	Common name	Scientific name	Tamil vernacular name
1.	Malvaceae	White wild musk mallow	*Abelmoschus ficulneus (L.) Wight & Arn.*	Nari vendai
2.	Malvaceae	Musk mallow	*Abelmoschus moschatus (L.) Medikus*	Kanthu kasthuri
3.	Malvaceae	County mallow	*Abutilon indicum (L.) Sweet*	Thuthi
4.	Euphorbiaceae	Indian nettle	*Acalypha indica L.*	Kuppaimeni
5.	Asteraceae	Horn spine	*Acanthospermum hispidum (Willd.) Nees*	Kombumul sedi
6.	Amaranthaceae	Devils horsewhip	*Achyranthes aspera L.*	Naauruvi
7.	Amaranthaceae	Mountain knotgrass	*Aerva lanata (L.) Juss. ex Schult.*	Poolai poo
8.	Fabaceae	Indian jointvetch	*Aeschynomene indica L.*	Nettivagai
9.	Amaranthaceae	Khaki weed	*Alternanthera pungens (Kunth) L.B. Sm.*	Mul ponnangkanni
10.	Amaranthaceae	Sissoo spinach	*Alternanthera sessilis (L.) R.Br. ex DC.*	Ponnangkanni
11.	Fabaceae	Necklace pod alyce clover	*Alysicarpus monilifer (L.) DC.*	Kaasukodi
12.	Fabaceae	Alyce clover	*Alysicarpus rugosus (L.) DC.*	Naama poondu
13.	Amaranthaceae	Minute amaranth	*Amaranthus viridis L.*	–
14.	Amaranthaceae	Amaranthus slender	*Amaranthus minutes (L.)*	Kuppai keerai
15.	Lythraceae	Blistering ammannia	*Ammannia baccifera L.*	Neermel neruppu
16.	Lamiaceae	Indian catmint	*Anisomeles indica (L.) Kuntze*	–
17.	Lamiaceae	Malabar catmint	*Anisomeles malabarica (L.) Kuntze*	Perum tumpai
18.	Poaceae	Mauritian grass	*Apluda mutica L.*	Moongil pul
19.	Poaceae	Common needle grass	*Aristida adscensionis L.*	Seevam pul
20.	Poaceae	Small carpet grass	*Arthraxon hispidus (Thunb.) Hochst.*	–
21.	Acanthaceae	Water leaf	*Asteracantha longifolia (L.) Nees*	Neermulli
22.	Oxalidaceae	Little tree plant	*Biophytum sensitivum (L.) DC.*	Thindanali
23.	Nyctaginaceae	Spreading hogweed	*Boerhavia diffusa L.*	Mookkiratai
24.	Nyctaginaceae	Erect spiderling	*Boerhavia erecta L.*	Seemai mookkiratai
25.	Rubiaceae	Shaggy button weed	*Borreria hispida (L.) G. Don*	Nathai choori

(continued)

Table 14.2 (continued)

S.no	Family	Common name	Scientific name	Tamil vernacular name
26.	Apocynaceae	Milk weed	*Calotropis procera (Aiton) W.T. Aiton*	Erukku
27.	Sapindaceae	Balloon vine	*Cardiospermum halicacabum L.*	Mudakathaan keerai
28.	Leguminosae	Coffee senna	*Cassia occidentalis L.*	Pei avarai
29.	Fabaceae	Sickle pod	*Cassia tora L.*	Oosithagarai
30.	Vitaceae	Bush grap	*Cayratia trifolia (L.) Domin*	Kaattu pirandai
31.	Apiaceae	Indian pennywort	*Centella asiatica (L.) Urb.*	Vallarai
32.	Poaceae	Purple chloris	*Chloris barbata (L.) Sw.*	Mayil kondai
33.	Euphorbiaceae	Giradol	*Chrozophora plicata (L.) A. Juss.*	Purapirakai
34.	Poaceae	Guria grass	*Chrysopogon fulvus (Spreng.) Chiov.*	–
35.	Cucurbitaceae	Colocynth	*Citrullus colocynthis (L.) Schrad.*	Komattikaai
36.	Cleomaceae	Spider plant	*Cleome gynandra L.*	Naaivelai
37.	Cleomaceae	Tick weed	*Cleome viscose L.*	Manja kadugu
38.	Fabaceae	Butterflypea	*Clitoria ternatea L.*	Sangu poo
39.	Cucurbitaceae	Ivy gourd	*Coccinia grandis (L.) Voigt*	Kovai
40.	Commelinaceae	Benghal dayflower	*Commelina benghalensis L.*	Kana vaalai
41.	Malvaceae	East Indian mallow	*Corchorus aestuans L.*	Kaattu thuthi
42.	Malvaceae	Cotton weed	*Corchorus trilocularis L.*	Thalaikaai poondu
43.	Euphorbiaceae	Three-leafed caper	*Croton bonplandianum Baill.*	Aathupoondu
44.	Cucurbitaceae	Wild melon	*Cucumis melo var. agresti (L.) L.*	Sukkangkaai
45.	Commelinaceae	Crested cat ears	*Cyanotis cristata (L.) D. Don*	Kuthirai kulampadi
46.	Amaranthaceae	Pasture weed	*Cyathula prostrate (L.) Blume*	Sivappu naauruvi
47.	Poaceae	Bermuda grass	*Cynodon dactylon (L.) Pers.*	Arugampul
48.	Cyperaceae	Annual sedge	*Cyperus compressus L.*	Kunna korai
49.	Cyperaceae	Variable flatsedge	*Cyperus difformis L.*	Korai
50.	Cyperaceae	Rice flat sedge	*Cyperus iria L.*	Yaanaikitti
51.	Cyperaceae	Purple nut sedge	*Cyperus rotundus L.*	Korai kilangu
52.	Cucurbitaceae	Dactyliandra	*Dactyliandra welwitschii (Benth.) J. Presl*	–
53.	Poaceae	Crowfoot grass	*Dactyloctenium aegyptium (L.) Willd.*	Kakkakaal poondu
54.	Solanaceae	Indian thorn apple	*Datura metel L.*	Oomathai

(continued)

Table 14.2 (continued)

S.no	Family	Common name	Scientific name	Tamil vernacular name
55.	Fabaceae	Tick trefoil desmodium	*Desmodium dichotomum (L.) DC.*	Ottadai sedi
56.	Amaranthaceae	False amaranth	*Digera arvensis (L.) R.Br.*	Thoyil keerai
57.	Poaceae	Southern crabgrass	*Digitaria ciliaris (Retz.) Koeler*	Arisi pul
58.	Poaceae	Jungle rice	*Echinochloa colona (L.) Link*	Nel mirati
59.	Asteraceae	False daisy	*Eclipta prostrata L.*	Karisilang kanni
60.	Acanthaceae	Asian scalystem	*Elytraria acaulis (L.) Vahl*	Pumi katampam
61.	Poaceae	Japanese lovegrass	*Eragrostis tenella (L.) P. Beauv.*	–
62.	Euphorbiaceae	Asthma plant	*Euphorbia hirta L.*	Amman pachcharisi
63.	Euphorbiaceae	Matted sand mat	*Euphorbia serpens L.*	–
64.	Euphorbiaceae	Gulf sand mat	*Euphorbia thymifolia L.*	Amu pachaiarisi
65.	Euphorbiaceae	Wild poinsettia	*Euphorbia geniculata (L.) M.A. Lawson*	Pachaiarisi
66.	Convolvulaceae	Slender dwarf morning glory	*Evolvulus alsinoides (L.) L.*	Vishnu kiranthi
67.	Cyperaceae	Tropical fimbry	*Fimbristylis cymosa (L.) Vahl*	–
68.	Cyperaceae	Forked fimbry	*Fimbristylis dichotoma (L.) Vahl*	–
69.	Cyperaceae	Lesser fimbristylis	*Fimbristylis miliacea (L.) Vahl*	–
70.	Amaranthaceae	Bachelor's button	*Gomphrena celosioides (L.) Mart.*	Neervaadaa malli
71.	Boraginaceae	Indian heliotrope	*Heliotropium indicum L.*	Thelkodukki
72.	Heliotropiaceae	Grey leaf heliotrope	*Heliotropium ovalifolium (L.) Kuntze*	Yaanai kundumani
73.	Apocynaceae	False sarsaparilla	*Hemidesmus indicus (L.) R. Br.*	Naanari
74.	Poaceae.	Spear grass	*Heteropogon contortus (L.) Beauv.*	Oosipul
75.	Rubiaceae	Rough button weed	*Hexasepalum teres (Forssk.) H. Scholz*	–
76.	Malvaceae	Tropical rose mallow	*Hibiscus vitifolius L.*	Siruthuthi
77.	Lamiaceae	Pignut	*Hyptis suaveolens (L.) Poit.*	Naai thumbai
78.	Convolvulaceae	Obscure morning glory	*Ipomoea obscura (L.) Ker Gawl.*	Sirutali
79.	Convolvulaceae	Tiger foot morning glory	*Ipomoea pes-tigridis L.*	Pulisuvadi
80.	Euphorbiaceae	Belly ache bush	*Jatropha gossypiifolia L.*	Kaattu amanakku
81.	Verbenaceae	West Indian lantana	*Lantana camara L.*	Unni sedi

(continued)

Table 14.2 (continued)

S.no	Family	Common name	Scientific name	Tamil vernacular name
82.	Asteraceae	Beach launaea	*Launaea sarmentosa (L.) Hook. f.*	Sentham
83.	Poaceae	Red sprangle top	*Leptochloa chinensis (L.) Nees*	–
84.	Poaceae	Sprangle top	*Leptochloa panicoides (J. Presl) Ohwi*	–
85.	Lamiaceae	Common leucas	*Leucas aspera (Willd.) Link*	Thumbai
86.	Onagraceae	Water prime rose	*Ludwigia parviflora (L.) Greenm.*	Mayilkaathu ilai
87.	Malvaceae	Three-lobed false mallow	*Malvastrum coromandelianum (L.) Garcke*	–
88.	Marsileaceae	Water clover	*Marsilea quadrifolia L.*	Alai keerai
89.	Convolvulaceae	Kidney leaf morning glory	*Merremia emarginata (L.) Hallier f.*	Eli kaathu keerai
90.	Rubiaceae	Tropical girdle pod	*Mitracarpus villosus (L.) DC.*	Kaaya poondu
91.	Molluginaceae	Naked stem carpet weed	*Mollugo nudicaulis L.*	Parpaadagam
92.	Rubiaceae	Indian mulberry	*Morinda tinctoria Roxb.*	Nuna
93.	Cucurbitaceae	Madras pea pumpkin	*Mukia maderaspatana (L.) M. Roem.*	Musumusukai
94.	Lamiaceae	Holy basil	*Ocimum tenuiflorum L.*	Thulasi
95.	Rubiaceae	Two flowermillegraines	*Oldenlandia biflora L.*	–
96.	Rubiaceae	Choy root	*Oldenlandia umbellate (L.) DC.*	Saayaver
97.	Cactaceae	Erect prickly pear	*Opuntia dillenii (Ker Gawl.) Haw.*	Chappathi kalli
98.	Asclepiadaceae	Rosy milk weedvine	*Oxystelma esculentum (L.) Merr.*	Oosipaalai
99.	Asteraceae	Congress grass	*Parthenium hysterophorus L.*	Visa poondu
100.	Poaceae	Yellow watercrown	*Paspalidium flavidum (Retz.) Stapf*	–
101.	Passifloraceae	Passionflower	*Passiflora foetida L.*	Siru poonaikaali
102.	Pedaliaceae	Large caltropis	*Pedalium murex L.*	Yaanai nerunji
103.	Fabaceae	Jungle mat bean	*Phaseolus trilobus L.*	Nari payir
104.	Phyllanthaceae	Madras leaf flower	*Phyllanthus maderaspatensis (L.) Suresh*	Melaa nelli
105.	Phyllanthaceae	Stone breaker	*Phyllanthus niruri L.*	Keelaa nelli
106.	Solanaceae	Native gooseberry	*Physalis minima L.*	Sodakku thakkali
107.	Fabaceae	Mesquite	*Prosopis juliflora (Sw.) DC.*	Karuvela maram

(continued)

Table 14.2 (continued)

S.no	Family	Common name	Scientific name	Tamil vernacular name
108.	Fabaceae	Least snout bean	*Rhynchosia minima (L.) DC.*	Kaliyan thuvarai
109.	Poaceae	Wild sugarcane	*Saccharum spontaneum L.*	Naanal
110.	Malvaceae	Morning mallow	*Sida acuta Burm. f.*	Arivaalmanai poondu
111.	Malvaceae	Heart leaf sida	*Sida cordifolia L.*	Nilathuthi
112.	Solanaceae	Black night shade	*Solanum nigrum L.*	Manathakkali
113.	Solanaceae	Red pea egg plant	*Solanum trilobatum L.*	Thuthuvalai
114.	Solanaceae	Yellow fruit night shade	*Solanum xanthocarpum Schrad.*	Kantang Kathiri
115.	Asteraceae	Smooth sow thistle	*Sonchus oleraceus L.*	Siruvaanai Chuvadi
116.	Plantaginaceae	Sticky blue rod	*Stemodia viscosa (L.) Steud.*	–
117.	Fabaceae	Wild indigo	*Tephrosia purpurea (L.) Pers.*	Kolinji
118.	Euphorbiaceae	Indian stinging nettle	*Tragia involucrata L.*	Senthatti
119.	Aizoaceae	Desert horsepurslane	*Trianthema protulacastrum (L.) L.*	Saaranai
120.	Zygophyllaceae	Puncture vine	*Tribulus terrestris L.*	Nerunji
121.	Asteraceae	Coat button	*Tridax procumbens L.*	Vettukaaya Poondu
122.	Malvaceae	Round leaf burr bush	*Triumfetta rotundifolia L.*	Adayodi
123.	Typhaceae	Southern cattail	*Typha angustata Bory*	Sambu
124.	Asterales	Little iron weed	*Vernonia cinerea (L.) Less.*	Mookkuthi poondu
125.	Asterales	Sonkadi	*Vicoa indica (L.) Less.*	Manja mookkuthi poondu
126.	Asterales	Rough cocklebur	*Xanthium strumarium L.*	Marulumathai

Source: Priyadharshini et al. (2023)

14.3.3 Adoption of Multiple Cropping Species

The result highlights agricultural activities aimed at improving crop diversity and sustainability across various farms. A total of 15 farms participated in cultivating a diverse range of white flower species, including *Jasminum sambac*, *Jasminum grandiflorum*, *Jasminum auriculatum*, and *Jasminum multiflorum*, reflecting a focus on enhancing floriculture and promoting biodiversity in white flower ecology (Table 14.5). Additionally, 27 farms practiced intercropping by planting onions around the basins of jasmine saplings, optimizing land use, and providing an early income source while the jasmine plants mature. Furthermore, 35 farms adopted border cropping with *Sesbania grandiflora*, *Gliricidia*, and *Ricinus communis*, which serve as protective barriers, improve soil fertility, and contribute additional

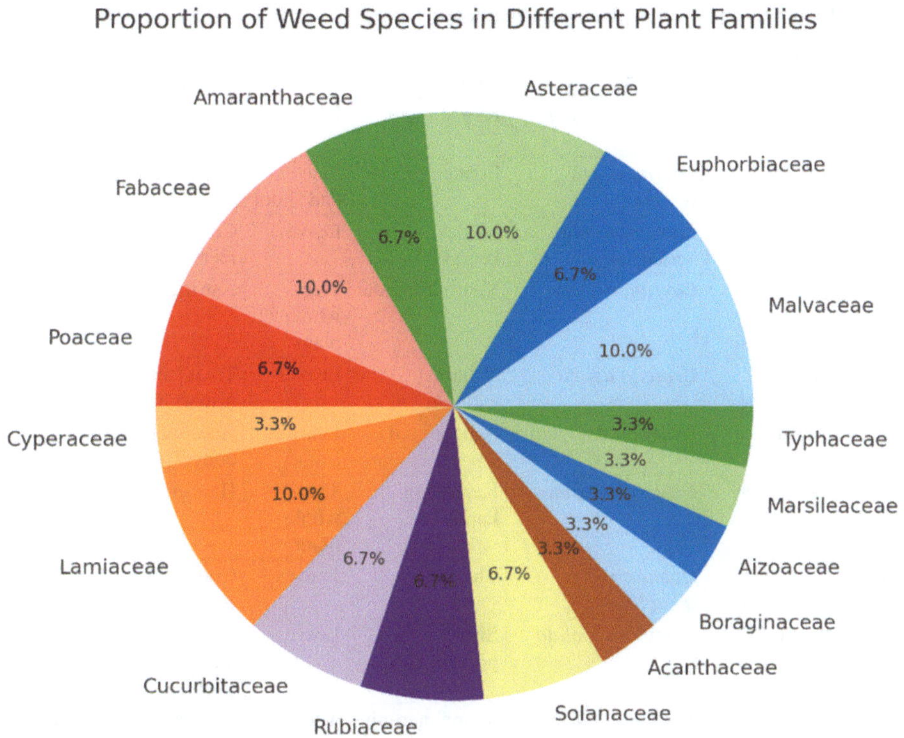

Fig. 14.3 Weed's family diversification

economic benefits. These activities collectively demonstrate an integrated approach to sustainable agriculture, balancing ecological benefits with economic gains while fostering diversification and resource efficiency.

The result outlines the maintenance of white flower species across different farms, focusing on their flowering seasons and yield duration (Table 14.6). Farms cultivating a combination of *J. sambac* and *J. grandiflorum* (two farms) have a flowering season from March to November, providing a yield duration of 9 months. In contrast, farms growing *J. sambac*, *J. auriculatum*, and *J. multiflorum* (six farms) achieve a year-round yield (12 months) by leveraging overlapping flowering seasons: March to October, April to August, and November to March, respectively. Similarly, farms cultivating *J. grandiflorum*, *J. auriculatum*, and *J. multiflorum* (five farms) also secure a 12-month flowering cycle by synchronizing the species' seasons (June to November, April to August, and November to March). A smaller number of farms (two farms) maintain all four species of *J. sambac*, *J. grandiflorum*, *J. auriculatum*, and *J. multiflorum* to maximize diversity and ensure uninterrupted flower yield throughout the year as stated by the National Horticulture Board (n.d.) (Rao and Singh 2012). This data demonstrates how strategic species combinations can optimize yield duration and meet market demands consistently.

Table 14.3 Medicinal value plants in white flower ecology

S.no	Common name	Scientific name	Vernacular name	Part used	Medicinal uses
1.	Indian nettle	*Acalypha indica*	Kuppaimeni	Leaf	Anthelmintic, anti-inflammation
2.	Sissoo spinach	*Alternanthera sessilis*	Ponnang kanni	Leaves and shoot	Hepatitis, tight chest
3.	Balloon vine	*Cardiospermum halicacabum*	Mudakathaan keerai	Leaves	Cough, skin diseases, menstrual cramps
4.	Bush grap	*Cayratia trifolia*	Kaattu Pirandai	The whole plant	Astringent medicine
5.	Indian pennywort	*Centella asiatica*	Vallarai	Leaves	Treat stomach and urinary tract infections
6.	Bermuda grass	*Cynodon dactylon*	Arugam pul	Leaves	Anasarca, cancer, convulsions
7.	False daisy	*Eclipta prostrata*	Karisilang kanni	The whole plant	Hair growth, antidiabetic
8.	Holy basil	*Ocimum tenuiflorum*	Thulasi	Leaves	Cough, asthma, diarrhea, fever
9.	Passion flower	*Passiflora foetida*	Siru poonaikaali	Leaves	Dietary supplement for anxiety and sleep problems
10.	Large caltropis	*Pedalium murex*	Yaanai nerunji	The whole plant	Kidney stone formation, incontinence of urine
11.	Stone breaker	*Phyllanthus niruri*	Keelaa nelli	Leaves	Cure for kidney stones
12.	Native gooseberry	*Physalis minima*	Sodakku thakkali	The whole plant	High cholesterol, "hardening of the arteries"
13.	Black nightshade	*Solanum nigrum*	Manathakkali	Leaves	Anti-insulin resistance and antiobesogenic effects
14.	Red pea egg plant	*Solanum trilobatum*	Thuthuvalai	Leaves and root	Tuberculosis, bronchial asthma, difficulty in breathing
15.	Coat button	*Tridax procumbens*	Vettukaaya poondu	Leaves	Jaundice, bronchial catarrh

Source: Priyadharshini et al. (2023)

Table 14.4 The weed management activities in white flower ecology

Weed management	Area (acre)
Mechanical control	2.75
Manual control	3.46

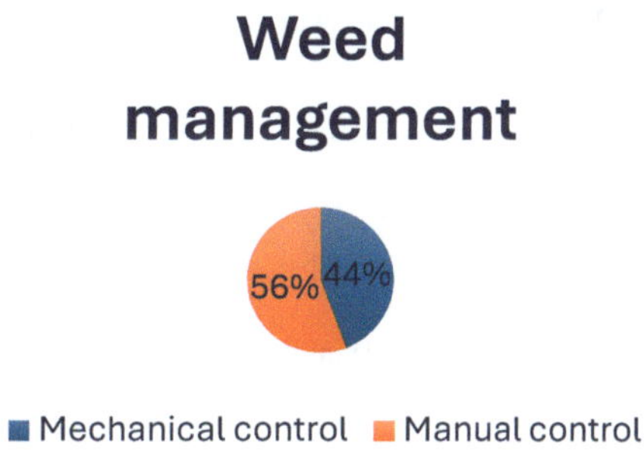

Fig. 14.4 Weed management

Table 14.5 The species and crop diversity in white flower ecology

Activities	Crop pattern	No. of farms	Crops
Cropping diversity	Multiple white flower species	15	*J. sambac*/*J. grandiflorum J. auriculatum*/*J. multiflorum*/tuberose
	Crop adjusting	27	Onion planted around the basin in jasmine saplings.
	Border crop	35	*Sesbania grandiflora, Gliricidia* & *Ricinus communis*

Table 14.6 Diversification of white flower species in white flower ecology

S.no	Maintaining white flower species	Flowering season	No. of farms	Flower yield duration (month)
1	*J. sambac* & *J. grandiflorum*	March to October; June to November	2	9
2	*J. sambac*, *J. auriculatum*, & *J. multiflorum*	March to October; April to August; November to March	6	12
3	*J. grandiflorum*, *J. auriculatum*, & *J. multiflorum*	June to November; April to August; November to March	5	12
4	*J. sambac*, *J. grandiflorum*, *J. auriculatum*, & *J. multiflorum*	March to October; June to November; April to August; November to March	2	12

14.3.4 Adoption of Usage and Reduction of Chemical Inputs and Encourage the Usage of Organic Inputs

The result (Table 14.7) highlights various alternatives to chemical inputs and their adoption by farmers. Enriched compost, with a total usage of 700 kg, is adopted by seven farms as a sustainable soil amendment. Liquid organic inputs, amounting to 1250 L, are the most widely adopted alternative with nine farms utilizing them (Fig. 14.5). Microbial products, weighing 45 kg, are used by five farms, indicating the lowest adoption rate among the alternatives. Lastly, insect traps with 23 units

Table 14.7 Alternate way of chemical inputs

Organic input	No. of farms involved
Enriched compost (700 kg)	7
Liquid organic inputs (1250 L)	9
Microbial products (45 kg)	5
Insect traps (23 No.)	8

Fig. 14.5 Usage organic inputs-2024

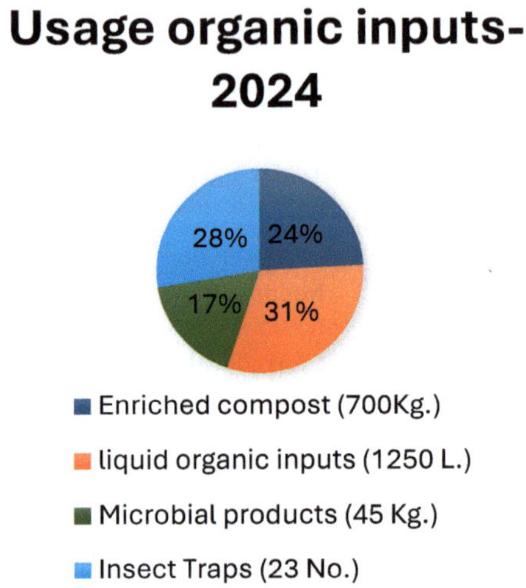

deployed are employed by eight farms as an eco-friendly pest control solution. This data underscores a preference for liquid organic inputs, while microbial products have seen relatively limited use.

The result (Table 14.8) highlights the diverse inputs used by farmers to enhance soil health, manage pests, and improve crop productivity. Enriched compost, derived from organic waste, is primarily utilized for soil health improvement. Liquid organic inputs, including formulations like Panchakavya, Jeevamirtham, Amutha Karaisal, and Dhasapoorani Karaisal, focus on enhancing soil health, while Ainthilai Karaisal and 3G Karaisal serve as pest repellents and The-Mor Karaisal functions as a flower inducer (Rao and Kumar 2023). Microbial products such as *Trichoderma viride* and *Pseudomonas fluorescens* are used for seedling treatment and disease management, while *Lecanicillium lecanii*, *Beauveria bassiana*, and *Metarhizium anisopliae* target specific pests, including sucking pests, caterpillars, and beetles, respectively (Nandini 2019). Additionally, traps like insect light traps, pheromone traps, and sticky traps are employed for pest monitoring purposes. This comprehensive use of organic and biological inputs reflects a sustainable approach to agriculture, addressing key challenges in pest and disease management while promoting soil and plant health.

Statistical analysis confirmed the positive impact of sustainable practices on white flower agroecosystems. Soil organic matter increased by 30%, while chemical pesticide use decreased by 60%, with a 68.8% rise in biocontrol applications.

Table 14.8 Farmer's using an alternate way of chemical inputs

Farmer's using inputs	Particulars	Purpose of usage
Enriched compost	Organic waste	Soil health improvement
Liquid organic inputs	Panchakavya	Soil heath & flower perishability
	Jeevamirtham	Soil health improvement
	Amutha karaisal	Soil health improvement
	Dhasapoorani karaisal	Soil health improvement
	Ainthilai karasal	Pest repellent
	The-mor karaisal	Flower inducer
	3G karaisal	Pest repellent
Microbial products	*Trichoderma viride*	Seedling treatment & disease management
	Pseudomonas fluorescens	Seedling treatment & disease management
	Lecanicillium lecanii	Sucking pest management
	Beauveria bassiana	Caterpillar management
	Metarhizium anisopliae	Beetles' management
Traps	Insect light trap	Pest monitoring purpose
	Pheromone trap	Pest monitoring purpose
	Sticky trap	Pest monitoring purpose

These improvements contributed to an 18% increase in flower yield, demonstrating the effectiveness of biodiversity conservation and climate resilience strategies.

While the study demonstrates the benefits of sustainable practices in white flower agroecosystems, certain limitations must be acknowledged. The findings are region-specific to Dindigul District, where climate, soil conditions, and farming practices may differ from other jasmine-growing regions. Scalability challenges include the availability of organic inputs, farmer adoption rates, and initial costs associated with transitioning to sustainable methods. Additionally, long-term monitoring is needed to assess the continued impact of biodiversity conservation efforts. Future research should explore adaptive strategies for different agroclimatic zones to enhance broader applicability.

14.4 Conclusion

The study highlights the integration of sustainable practices to enhance climate resilience in white flower agroecosystems. It emphasizes the adoption of biodiversity-focused strategies, such as native tree plantations, weed management, multiple cropping systems, and organic inputs, to address the impacts of climate change. Field trials demonstrated improved soil health, reduced reliance on chemicals and enhanced biodiversity, contributing to ecosystem stability and rural livelihoods. These findings underscore the importance of diversifying agricultural practices and leveraging local ecological knowledge to create resilient and sustainable farming systems.

The study's findings strongly support the effectiveness of sustainable practices in white flower agroecosystems. Soil organic matter increased by 30%, while chemical pesticide use decreased by 60%, with a 68.8% rise in biocontrol applications. Additionally, flowers yielded by 18%, and water retention improved by 15% due to native tree planting. These quantitative results confirm that sustainable farming enhances soil health, pest management, and overall climate resilience in white flower cultivation.

Future Research Directions

Future research should focus on developing drought-resistant jasmine varieties, optimizing organic input formulations, and utilizing AI-based climate prediction models to enhance farm resilience. Long-term studies on pollinator diversity and ecosystem services in white flower agroecosystems are also crucial. Policy recommendations include incentivizing farmers to adopt sustainable practices through subsidies for organic inputs, native tree plantations, and integrated pest management training. Strengthening market access for sustainably grown jasmine can further encourage eco-friendly farming. These steps will ensure greater climate resilience, biodiversity conservation, and economic sustainability in white flower cultivation.

Acknowledgments I would really like to thank JCEPL for providing this wonderful opportunity.

Data Availability Statement Data are available from the authors on request.

Declarations Informed Consent Statement Not applicable.

Conflict of Interest The author declares no conflict of interest.

Ethics Approval Not applicable.

References

Baldy C, Stigter CJ (1997) Agrometeorology of multiple cropping in warm climates. INRA, Paris

Gibbons P, Lindenmayer DB, Fischer J, Manning AD, Weinberg A, Seddon J, Barrett G (2008) The future of scattered trees in agricultural landscapes. Conserv Biol 22:1309–1319

Manning AD, Fischer J, Lindenmayer DB (2006) Scattered trees are keystone structures – implications for conservation. Biol Conserv 132:311–321

Moga CI, Samoilă C, Öllerer K, Băncilă RI, Réti K-O, Craioveanu C, Hartel T (2016) Environmental determinants of the old oaks in wood-pastures from a changing traditional social–ecological system of Romania. Ambio 45:480–489

Naidu VSGR (2012) Hand book on weed identification, directorate of weed science research. E Flora of India, p 354

Nandini B (2019) Compatibility studies among different microbial pesticides commonly used for soil application. J Entomol Zool Stud 7(1):244–249

National Horticulture Board (n.d.) Jasmine flower production guidelines. NHB India. Retrieved February 19, 2025, from https://nhb.gov.in/jasmine

Old School Manufacturing (n.d.) Electric weed control for organic farms. The Weed Zapper. Retrieved January 30, 2025, from https://theweedzapper.oldschoolmanufacturing.com/electric-weed-control-organic-farms

Priyadharshini K, Sanjay P, Abinaya A, Jayabharathi K, Kumaravel S, Mathialagan M (2023) Study of weeds in Sethu Bhaskara agricultural college and research foundation, Karaikudi, Tamil Nadu. Pharma Innov J 12(7):1110–1114

Radosevich SR, Holt J, Ghersa CM (1997) Weed ecology, implication for management. Wiley, New York, pp 114–160

Rajendran S, Kumar R (2023) Traditional medicinal uses of *Acalypha indica*. In: Doe J, Smith A (eds) Handbook of medicinal plants in India. Taylor & Francis, pp 102–118

Rao VS (2000) Principles of weed science. Oxford and IBH Publishing Co. Pvt. Ltd, pp 427–436

Rao S, Kumar P (2023) Preparation procedures of botanicals and animal products. In: Sharma V, Gupta A (eds) Organic farming practices in sustainable agriculture. Springer, New Delhi, pp 120–135

Rao AN, Singh V (2012) Influence of climate on jasmine flowering patterns. Hortic Res J 28(3):225–237

Senthilkumar R (2019) A case study of cultivation and marketing of jasmine in Tamil Nadu with reference to Dindugal District: an economic analysis. Int J Res Eng IT Soc Sci 9(6):106–116

Singh R, Kumar A (2020) Role of native trees in ecosystem stability. J Agrofor Res 45(3):215–230

Smith J (2020a) The art of plant identification. Greenleaf Publishing

Smith J (2020b) Sustainable pest management in organic farming. In: Johnson A, Brown B (eds) Advances in agricultural practices. Greenleaf Publishing, New York, pp 50–65

U.S. Department of Agriculture, Forest Service (n.d.) Moths as pollinators. USDA. https://www.fs.usda.gov/wildflowers/pollinators/animals/moths.shtml. Accessed 30 Jan 2025

Vanetha KP (2021) A study on the constraints faced by the jasmine growers in Dindigul District of Tamil Nadu. Int Adv Res J Sci Eng Technol 8(10):190–193. https://doi.org/10.17148/IARJSET.2021.81036

Chapter 15
Satellites to Sanctuaries: An Integrative Study to Assess the Detrimental Impact of Rapid Urbanization on Changes in Land Use and the Avian Fauna of the Selected Wetlands in Tamil Nadu, India

M. Arunpandiyan, S. Sasidharan, M. Roshy Ann, and M. Prashanthi Devi

Abstract Wetlands are vital ecosystems that sustain diverse biodiversity, offering crucial resources like water, food, and habitats for numerous species. They are integral to maintaining the food chain and ecological balance. However, rapid urbanization and human activities are causing significant degradation and loss of wetland areas. This is particularly evident in the decline of bird populations at sanctuaries such as Karaivetti, attributed to construction debris, increased human settlements, and disturbances from vehicular traffic and fishing activities. The study employed Land Use Land Cover (LULC) analysis and satellite imagery to monitor changes in wetlands from 2013 to 2023. Findings indicated substantial reductions in vegetation cover and water quality due to industrial discharges, sewage, and agricultural runoff. These alterations have negatively impacted habitats, resulting in a decrease in migratory bird visits and overall biodiversity. Wetlands provide invaluable ecological services and support rich biodiversity. The degradation of these ecosystems due to anthropogenic activities threatens the species that rely on them. Immediate action is required for effective management and conservation strategies to protect and restore wetland habitats. Recommendations include minimizing fishing activities during migratory seasons, establishing silent zones to reduce disturbances, and enforcing stringent measures against illegal dumping of construction debris. Implementing these measures can help mitigate the adverse impacts on wetlands and promote the sustainability of these essential ecosystems.

Keywords Paper and pulp mill effluent · Halophyte · Phytoremediation · Salinity

---

M. Arunpandiyan · S. Sasidharan · M. Roshy Ann · M. Prashanthi Devi (✉)
Department of Environmental Science and Management, Bharathidasan University, Trichy, Tamil Nadu, India

B. P. Barathan et al. (eds.), *Navigating Climate Change: Impacts on Biodiversity and Ecosystem Resilience*, Sustainable Development and Biodiversity 42,
https://doi.org/10.1007/978-981-95-0409-1_15

15.1 Introduction

Covering over 8% of Earth's land surface, freshwater wetlands play a crucial role in ecosystem functions and provide habitat to diverse plant and animal species (Keddy et al. 2009). Wetlands are considered keystone ecosystems due to their significant influence on landscape processes relative to their size. Estimates suggest that factors like land conversion, climate change, and evolving land use have led to the loss of over half of the wetlands in the contiguous United States (Davidson 2014). Understanding the ecological impacts of wetland alterations requires tools that offer comprehensive, long-term insights. Given resource limitations, biodiversity monitoring commonly relies on indicator species (as reviewed in Caro 2010); however, some indicators may fall short in capturing full ecosystem complexity (Hilty and Merenlender 2000).

Bird migration patterns are shaped by seasonal changes in food availability, with birds traveling back and forth between resting sites and nesting areas. These migrations benefit ecosystems seasonally (Faaborg et al. 2010). However, modern infrastructure and diminishing food sources, like seeds and insects, are causing rapid declines in bird populations. Land use and cover shifts from urban development impact birds' breeding and feeding habitats on both small and large scales (Seress and Liker 2015). Habitat loss pushes migratory birds to seek new destinations, introducing additional risks (Radić and Gavrilovic 2021). At the Karaivetti Bird Sanctuary, a sanctuary notable for its diversity, including ducks, waders, swimmers, divers, migratory, and endangered species, recent land modifications by the Public Works Department (PWD) have contributed to a decline in migratory bird populations. This study aims to identify the drivers of this decline by analyzing migrating bird populations through census data, examining population trends over the past decade, investigating the link between bird populations and land use/land cover changes, and assessing vegetation changes' impacts on migration using remote sensing.

Hence, this study seeks to understand the underlying causes of bird population decline and to necessitate conservation strategies to mitigate the impacts of environmental and infrastructural changes.

15.2 Materials and Methods

15.2.1 Study Area

The Karaivetti Bird Sanctuary is a 4.537 km^2 protected area located at Karaivetti village in Ariyalur District of Tamil Nadu, South India, established on April 5, 1999. It is one of Tamil Nadu's biggest freshwater lakes located in the south. There are at least 188 bird species known to exist in the sanctuary. Karaivetti Wildlife Sanctuary is the name used internationally. Karaivetti, which receives water from the river

Kaveri and rainfall via the Pullambadi-Kattalal canal, is situated around 50 km northeast of Tiruchchirappalli town. It is the biggest water body in the district and attracts hundreds of thousands of birds every year. The Tamil Nadu government designated it as a bird sanctuary in 1999 due to its significance as a refuge for a wide number of avian species. The lake can contain water to an average depth of three meters when it is full. There are 188 bird species known to exist in the sanctuary, 82 of which are aquatic birds. For migrating water birds, the sanctuary is one of the most significant freshwater feeding areas.

15.2.2 Primary Data and Study Period

Landsat 8–9 Operational Land Imager (OLI) and Thermal Infrared satellite images were used to evaluate the land use and land cover pattern for the years 2013, 2016, 2018, 2020, and 2023 (TIRS). The LULC was investigated using remote sensing datasets from Landsat MSS, TM, and OLI (USGS Earth Explorer) (Islam et al. 2021). Google Earth Pro satellite imagery of the chosen wetlands' plant cover from various years was gathered.

15.2.3 Land Use and Land Cover Classification

The land use and land cover (LULC) classification was divided into categories: dry land, sparse vegetation, wetlands, buildings, and dense vegetation (Table 15.1). Throughout the study, the Supervised Maximum Likelihood classifier was applied to detect any changes in LULC. The ERDAS Imagine software was used to process Landsat TM images. Multitemporal Landsat satellite data were employed to assess shifts in LULC over time. Various remote sensing vegetation indices, such as the Normalized Difference Vegetation Index (NDVI), were used to illustrate vegetation characteristics within each region (pixel) (Rouse Jr et al. 1973; Tucker et al. 1985). The NDVI (1) is derived as a single value by dividing the difference in spectral

Table 15.1 The classification scheme used in the study

ID	Land use/land cover	Description
1	Dense vegetation	The areas with natural vegetation cover in the form of herbs, shrubs, grassland, trees, and other plants
2	Wetlands	It comprises all forms of surface water present in drains, canals, ponds, lakes, wetlands, etc.
3	Less	This class consists of cultivated and noncultivated areas
4	Dry land	It contains the land left with no vegetation and harsh terrain with poor moisture
5	Buildings	Areas of constructed land

responses of specific bands by their sum, reflecting vegetation traits in a given pixel. To eliminate errors and achieve precise unsigned integers to the hundredths place, values were scaled from 0 to 20,000. This was done by adding one to the initial NDVI value, multiplying by 10,000, and rounding. The calculations and outputs of the original regression model then utilize these scaled NDVI units, denoted as NDVIT.

$$\text{NDVI} = (\text{Band4} - \text{Band5}) / (\text{Band4} + \text{Band5}) \tag{15.1}$$

The accuracy of the classified images was evaluated using ground truth data. Metrics included User's Accuracy (1), Producer's Accuracy (3), Overall Accuracy (4), and Kappa Coefficient (5). The corrections in the image analysis based on the accuracy were carried as per the illustration (Fig. 15.1).

$$\begin{aligned}\text{Users accuracy} &= (\text{Number of correctly classified pixels in each category}) \\ &/ (\text{Total number of classified pixels in that category (the row total)}) \times 100\end{aligned} \tag{15.2}$$

$$\begin{aligned}\text{Producer accuracy} &= (\text{Number of correctly classified pixels in each category}) \\ &/ (\text{Total number of reference pixels in that category (the column total)}) \times 100\end{aligned} \tag{15.3}$$

$$\begin{aligned}\text{Overall accuracy} &= (\text{Total number of correctly classified pixels (diagonal)}) \\ &/ (\text{Total number of reference pixels}) \times 100\end{aligned} \tag{15.4}$$

$$\begin{aligned}\text{Kappa coefficient}(T) &= ((\text{TS} \times \text{TCS}) - \Sigma[(\text{Column total} \times \text{Row total})]) \\ &/ (\text{TS}^2 - \Sigma[(\text{Column total} \times \text{Row total})]) \times 100\end{aligned} \tag{15.5}$$

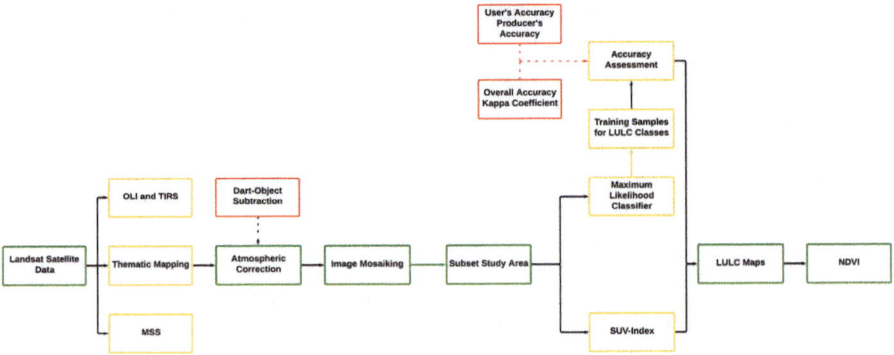

Fig. 15.1 Flow chart of satellite imagery to obtain the NDVI index

15.2.4 Bird Census

Bird surveys were conducted during the Northeast monsoon (October–December) and winter season (January–February) of 2023. Censuses took place in the mornings (6:00 to 10:00 a.m. GMT + 5:30) to minimize the chances of double-counting. Point counts involve observing and recording all birds seen or heard from a fixed location over a set period. Birds were counted at six observation points using binoculars and cameras. Species identification and foraging habitats were documented using the Asian Waterbird Census (AWC) form. Aquatic bird counts were conducted in both morning and evening sessions across six sites, with 15 min spent at each observation point. Data from the Forest Survey of wetland migratory birds was used to aid in interpretation and verification.

15.3 Results and Discussion

In this study, the Land Use/Land Cover (LULC) change has been observed in wetlands from 2013 to 2023. Each wetland region has been categorized into five LULC classes; building, wetland, dense vegetation, less vegetation, and dry land.

15.3.1 Classification Accuracy Assessment

Accuracy assessment is critical in image classification to ensure reliable LULC mapping. In Maximum Likelihood Classification (MLC), some pixels may be misclassified due to uneven data distribution. This study used an error matrix to summarize and quantify accuracy, resulting in four metrics: producer accuracy, user accuracy, overall accuracy, and the kappa coefficient. These measures offer a comprehensive view of classification accuracy. For satellite images of Karaivetti Lake in 2013, 2016, 2018, 2020, 2021, and 2023, overall accuracies were 85%, 85%, 90%, 95.83%, 91.6%, and 87.5%, with corresponding kappa statistics of 75%, 86.36%, 87.5%, 95%, 89.91%, and 85%.

The overall assessment results of each classified image from 2013 to 2023 for the vegetation, water body, density, dry lands, and building at Karaivetti bird sanctuary are shown in Fig. 15.2.

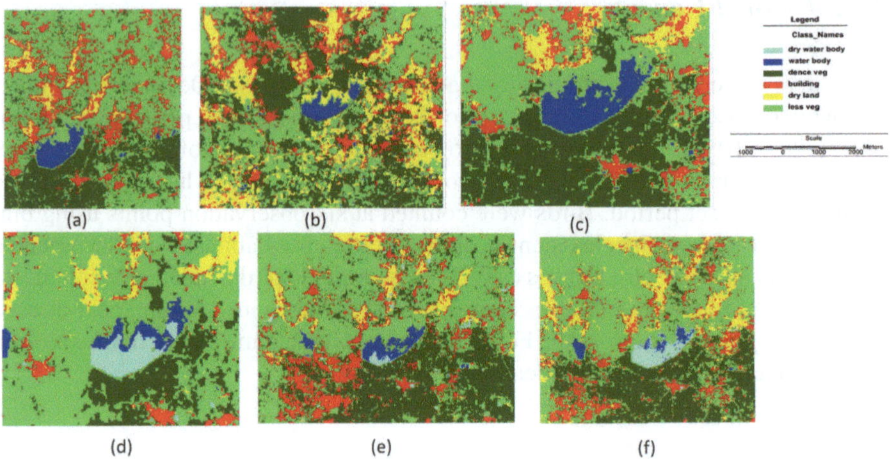

Fig. 15.2 Vegetation, water body, density, dry lands, and building at Karaivetti bird sanctuary and nearby area in 2013 (**a**), 2016 (**b**), 2018 (**c**), 2020 (**d**), 2021 (**e**), and 2023 (**f**)

15.3.2 Normalized Differential Vegetation Index

The NDVI analysis revealed that wetlands show substantial variations in vegetation indices, impacting local habitats and ecosystems. NDVI values indicate differences in water availability across the study areas. Although water availability has increased over time, water quality has degraded due to sewage discharge, construction, and industrial waste, negatively affecting vegetation cover. This valuable wetland faces ongoing deterioration from industrial, agricultural, and residential pollution, threatening both biodiversity and natural beauty. The shrinking water body and vegetation also pose risks to migratory bird habitats. Increasing wastewater discharge and heavy use of fertilizers and pesticides further harm natural vegetation and soil quality.

15.3.3 Avian Fauna Population

The number of migrating birds in the Karaivetti bird sanctuary has reduced in comparison to prior years, and some of the species are not being visited. In 2016, there were 10,943 birds recorded, and in 2023, the number was just 1615. Glossy Ibis' Black-winged Stilt population is drastically declining. Greenshank, Bar-headed Goose, Spoonbill, and a few other birds have not been seen. The total number of migratory birds visiting the bird sanctuary is shown in Fig. 15.3.

The study area is impacted by intensive agricultural practices and high levels of industrial and sewage wastewater discharge. Processed satellite images highlighted the LULC (Land Use and Land Cover) patterns in the region, offering a clearer

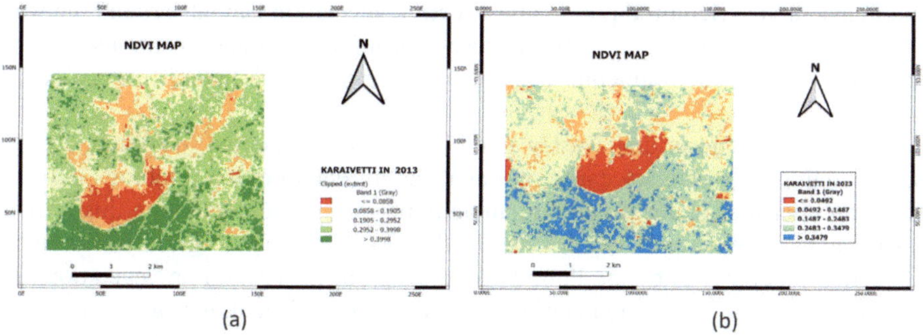

Fig. 15.3 NDVI map of Karaivetti bird sanctuary in 2013 (**a**) and 2023 (**b**)

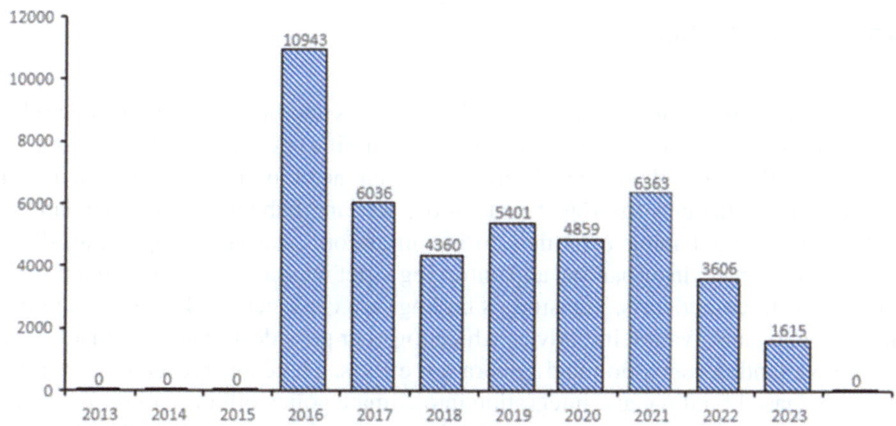

Fig. 15.4 Total number of migratory birds observed in the years 2013–2023

understanding of these classified images. The findings indicate significant environmental issues arising from unchecked and ongoing land use changes. Data show rapid pollution of wetlands and the extensive use of synthetic chemicals in agriculture, both of which pose risks to the environment and public health. Water availability and natural vegetation significantly influence the development of new croplands and the frequency of bird visits. However, reduced water resources and unregulated agricultural practices can detract from the natural landscape, leading to soil erosion, desertification, and a loss of biodiversity, including impacts on bird migration patterns (Fig. 15.4).

Classified images of the study area reveal marked changes, with a gradual but significant decline across key land cover types—wetlands, dense vegetation, sparse vegetation, buildings, and dry land—over the past decade. This trend illustrates the severity of human-induced environmental degradation, which requires urgent attention. NDVI-based analysis from 2013 to 2023 also tracked changes in vegetation and temperature dynamics, showing a marked decrease in vegetation during this period. This decline is largely attributed to the excessive discharge of industrial, municipal, and agricultural pollutants, affecting vegetation habitats around the wetland. Water availability generally increases during the pre- and postmonsoon seasons, but similar environmental challenges are eroding the natural beauty of the wetlands.

The study suggests that wastewater influx from residential, commercial, and saline sources is a primary cause of water quality deterioration. Findings indicate that improper management and human activities are rapidly depleting this critical resource, harming natural habitats, and leading to a decline in migratory bird visits. Construction activities by the Public Works Department (PWD) on vegetation and drylands near the lake's boundaries have also reduced natural vegetation cover, diverting wetlands and diminishing the habitat necessary for bird populations.

15.4 Conclusion

Wetlands are invaluable for their ecological services and biodiversity. The degradation of these ecosystems due to anthropogenic activities poses a significant threat to the species that depend on them. There is an urgent need for better management and conservation strategies to protect and restore wetland habitats. Recommendations include reducing fishing activities during migratory seasons, designating silent zones to minimize disturbances, and enforcing strict measures against illegal dumping of construction debris. The study's findings are consistent with recent scientific insights and conservation initiatives, which together provide a robust foundation for future wetland conservation and restoration efforts. The convergence of historical data, increased funding, and integration into climate policy offers an optimistic outlook for wetland conservation, provided that these efforts are sustained and scaled up globally. By implementing these measures, it is possible to mitigate the negative impacts on wetlands and promote the sustainability of these critical ecosystems.

Acknowledgments The authors thank the Chief Conservator of Forests, Tiruchirappalli, and District Forest officers, Thiruvarur, Ariyalur forest Divisions for extending their support in conducting the research and in providing all resources for performing the studies.

Data Availability Statement Data are available from the authors on request.

Declaration of Competing Interest The authors declare that they have no known competing financial interest or personal relationships that could have appeared to influence the work reported in this paper.

Conflict of Interest The author declares no conflict of interest.

Ethics Approval Not applicable.

References

Caro T (2010) Conservation by proxy: indicator, umbrella, keystone, flagship, and other surrogate species. Island Press

Davidson NC (2014) How much wetland has the world lost? Long-term and recent trends in global wetland area. Mar Freshw Res 65(10):934–941

Faaborg J, Holmes RT, Anders AD, Bildstein KL, Dugger KM, Gauthreaux SA Jr et al (2010) Recent advances in understanding migration systems of New World land birds. Ecol Monogr 80(1):3–48

Hilty J, Merenlender A (2000) Faunal indicator taxa selection for monitoring ecosystem health. Biol Conserv 92(2):185–197

Islam H, Abbasi H, Karam A, Chughtai AH, Ahmed Jiskani M (2021) Geospatial analysis of wetlands based on land use/land cover dynamics using remote sensing and GIS in Sindh, Pakistan. Sci Prog 104(2):00368504211026143

Keddy PA, Fraser LH, Solomeshch AI, Junk WJ, Campbell DR, Arroyo MT, Alho CJ (2009) Wet and wonderful: the world's largest wetlands are conservation priorities. Bioscience 59(1):39–51

Radić B, Gavrilovic S (2021) Natural habitat loss: causes and implications of structural and functional changes. In: Life on land. Springer, Cham, pp 699–712

Rouse JW Jr, Haas RH, Schell JA, Deering DW (1973) Monitoring the vernal advancement and retrogradation (green wave effect) of natural vegetation (No. NASA-CR-132982). NASA

Seress G, Liker A (2015) Habitat urbanization and its effects on birds. Acta Zool Acad Sci Hung 61(4):373–408

Tucker CJ, Townshend JR, Goff TE (1985) African land-cover classification using satellite data. Science 227(4685):369–375

References

[illegible]

Chapter 16
Soil Texture, Salinity, and Grass Composition: A Geospatial and Statistical Investigation

Vivek U. Chauhan, Ketankumar J. Yogi, and Kalpesh D. Sorathiya

Abstract This pilot study examines the spatial distribution of grass species within the ecologically significant grassland of Kachchh, India, with a specific focus on the influence of key soil physicochemical factors: soil texture (specifically, the proportions of sand, silt, and clay) and soil salinity to establish foundational data concerning the interplay between soil physicochemical properties and grasses species. Results indicate that clay-rich soils correlate with higher salinity levels ($r = 0.62$, $p < 0.05$), impacting species composition. *Dicanthium annulatum* exhibits the highest relative density (9.83%), followed by *Dicanthium foveolatum* (9.00%), *Aristida funiculata* (7.64%), and *Cymbopogon iwarancusa* (7.16%).

Keywords Grassland · Phytosociology · Soil analysis · Spatial analysis · Kachchh

16.1 Introduction

Grasslands are areas primarily covered by grasses, with minimal or no tree cover. According to UNESCO, grasslands are defined as "land covered with herbaceous plants and less than 10% tree and shrub cover," while wooded grasslands have 10–40% tree and shrub cover (Hopkins and White 1987). Grasslands are the world's largest ecosystem, covering about 40% of the earth's surface, and 52.5 million km^2 apart from Antarctica and Greenland (White et al. 2000). Grasslands can be divided by different classification systems based on parameters such as human

V. U. Chauhan
Terrestrial Ecology Division, Gujarat Institute of Desert Ecology, Bhuj, Gujarat, India

Tolani College of Arts and Science, Adipur, Gujarat, India

K. J. Yogi
Terrestrial Ecology Division, Gujarat Institute of Desert Ecology, Bhuj, Gujarat, India

K. D. Sorathiya (✉)
Tolani College of Arts and Science, Adipur, Gujarat, India

B. P. Barathan et al. (eds.), *Navigating Climate Change: Impacts on Biodiversity and Ecosystem Resilience*, Sustainable Development and Biodiversity 42,
https://doi.org/10.1007/978-981-95-0409-1_16

interventions, biogeographic zones, ecological factors, and growth patterns. Grasslands can be categorized into three main types based on human interventions: natural, semi-natural, and improved (Bullock et al. 2011; Lemaire et al. 2011). Natural grasslands are primarily shaped by natural forces like climate, fire, and the grazing of wild animals. However, these natural areas are also utilized for livestock grazing. Semi-natural grasslands result from human involvement and rely on continued management, such as livestock grazing or harvesting hay. Without these practices, they are likely to be overtaken by shrubs and trees (Queiroz et al. 2014). People make improved grasslands. They plow the land and plant special types of grass that grow very quickly. These grasses often need extra fertilizer, and people must take good care of them to keep them healthy (Suttie et al. 2005; Pilgrim et al. 2010). Based on ecological factors such as rainfall, temperature, and seasonality, grasslands are classified into tropical and temperate grasslands. Tropical grasslands have distinct dry and wet seasons. This zone remains warm throughout the year with moderate precipitation of around 140 cm. Savannah, Campos, and Llanos are tropical grasslands. On the other hand, Temperate grasslands have extreme climatic conditions (Pathak and Dagar 2016). Pampas, Prairie, Veld, Steppe, and Downs are examples. In India, a different attempt was made by Whyte (1957), by classifying grasslands into eight types based on their floral characteristics. Champion and Seth (1968) also classified grasslands into two types based on climatic conditions. Later on, Dabadghao and Shankarnarayan (1973) classified into five types based on an extensive survey of 500 sites across India, that research marked a significant turning point in the study of grassland ecology. The distribution of these grass covers is largely determined by climatic conditions, particularly latitude. As a result,the Sehima–Dichanthium type occurs in tropical regions; the Dichanthium–Cenchrus–Lasiurus, Phragmites–Saccharum–Imperata, and Themeda–Arundinella types dominate sub-tropical regions; while temperate alpine grass covers are found in temperate regions.(Pathak and Dagar 2016).

Grasslands have played an essential role in the history of humankind. Van Dyne et al. (1978) coined the term since back in the Tertiary period, true grasses have been well developed according to fossils that are being found today. Grasslands and grazing animals have been established as significant life forms. Grasslands are crucial in water management, restocking groundwater reserves, and contributing to a stable and reliable water supply (Oliveira et al. 2005). By anchoring the soil, deep roots significantly improve soil stability and reduce erosion, thereby preventing land degradation (Bardgett et al. 2021). Grasslands are major carbon sinks, storing 34% of the world's terrestrial carbon, making them crucial for climate change mitigation (White et al. 2000). Grasslands offer breeding grounds to many birds, and are preferred by wild animals for movements and migration, eventually playing a significant role as biodiversity hotspots (Nugent et al. 2022). Humans have been consuming cereal crops as part of their diet for ages. Hunting culture existed in or fringe of the grasslands (Pathak and Dagar 2016). The dependency of livestock culture more or less entirely relies on grasslands. Building upon this understanding of grassland classifications and ecological roles, it is necessary to analyze the aspects that substantially affect their health and stability. These natural and anthropogenic factors are vital in shaping the current state of grassland ecosystems. Among the various factors influencing grassland ecosystems, soil conditions play a particularly critical role. The physical and chemical properties of the soil directly impact plant growth, species composition, and overall grassland health. Plant species diversity is influenced by the physicochemical characteristics of the soil. However, ecological circumstances have an impact on the link between plant diversity and soil characteristics, making it crucial to assess the impact of soil physicochemical characteristics on plant species diversity (Grace et al. 2016; Zhu et al. 2020; Han et al. 2022). Soil texture is quite important in determining soil properties, including structure, nutrition retention capacity, and water retention capacity. Thus, soil texture is often known as the fundamental property of soil. It refers to the relative proportions of sand, silt, and clay particles. The ability of sandy soils to hold water and nutrients is low, while clay soils are better at holding both.

Furthermore, soil texture influences soil structure, as sandy soils exhibit a loose structure, whereas clay soils possess a denser structure (Dexter 2004; Dewangan 2023). Soil texture is essential for nutrient availability, as nutrients are generally dissolved in water. Soil Salinity changes the physicochemical features of soils as well as their overall health and disrupts the ecological balance of the area. The effects of salinity include reduced soil fertility, changes in soil structure, and increased soil erosion. Moreover, elevated salinity can affect water retention, nutrient availability, and the diversity of soil organisms, leading to long-term degradation of soil quality (Hu and Schmidhalter 2004). Salinity can create a physiological drought by increasing the osmotic potential. It can be concluded that soil texture and salinity are fundamental properties that exert a cascading effect on other soil parameters.

The long-term average of atmospheric conditions in a specific region constitutes its climate. When these atmospheric elements shift beyond their typical range, it is recognized as climate change. One of the main consequences of climate change is increasing soil salinity, particularly in arid and coastal agricultural lands (Kasim et al. 2013; Ullah et al. 2021). A study by Bannari and Al-Ali (2020) showed that in arid landscapes, a 30-year period of rising temperatures and less rainfall led to higher salt levels in the soil. This investigation came to the conclusion that the cause is less excessive salt leaching in the soil. A 33% increase in salinity was recorded in the soil of coastal agricultural lands over 25 years (Rahman et al. 2018). Salinization of agricultural lands is primarily driven by irrigation with saline water, coupled with low precipitation and high evapotranspiration rates. These factors contribute to a 10% annual increase in soil salinity. Projections indicate that if this trend persists, over 50% of arable land will be affected by salinization by 2050 (Jamil et al. 2011).

Kachchh is a region in the far west of India. It is located between 22° and 24° north of the equator and between 68° and 71° east. Also, the largest district in India, covering about 45,612 km^2. The Arabian Sea borders it to the south. To the north and east, it is surrounded by two substantial salt flats: the Great Rann of Kachchh, which is about 15,127 km^2, and the Little Rann, which is about 4000 km^2 (Kar 2011). The central part of Kachchh district, called the "Kachchh Mainland," is where most people live. Its shape and overall land features look like a tortoise shell.

Fig. 16.1 Landscape view of Godhara grassland

This is probably where the name "Kachchh", Sanskrit for tortoise, comes from (Thakkar 2017). Grasslands are important ecosystems providing valuable services to humans and animals; however, these grasslands face numerous threats, including climate change, conversion to agricultural lands, and other human-induced pressures. Studying climate change has become particularly important because of its effects on a global scale. With a specific purpose, the study area was proposed in the Mandvi taluka of Kachchh.

First and foremost, the Kachch district suffers from soil salinity due to its location. Gujarat alone contributes approximately 75% of saline soil in India (Arora and Sharma 2017; Mandal et al. 2018). The selected grassland is located in the vicinity of Godhara (Figs. 16.1 and 16.2). This region is a reserve forest under the Western Kachchh Division Forest Department, and it is relatively pristine, so there is less scope for error from anthropogenic influence. Finally, this study aims to give initial information needed to fix a gap in earlier research and tries to outline the variety and spread of plants affected by certain soil characteristics, like salinity and texture. This research investigates the effects of soil texture (sand, silt, and clay) and salinity on the spatial distribution of grass species in the Kachchh grasslands. It aims to investigate the main spatial patterns of grass species, the interactions among soil texture, salinity, the degree of species diversity, and distribution in the study area. The results will further be understanding of grassland ecosystems under environmental constraints.

16.2 Methodological Framework

A 500 × 500 m^2 grid system was overlaid on the study area to facilitate systematic and representative sampling, enabling a spatially distributed approach within this gridded framework. A total of 150 random points were generated using GIS software (QGIS 3.34) (Fig. 16.3). At each of these points, a 10 × 10 m^2 quadrat was

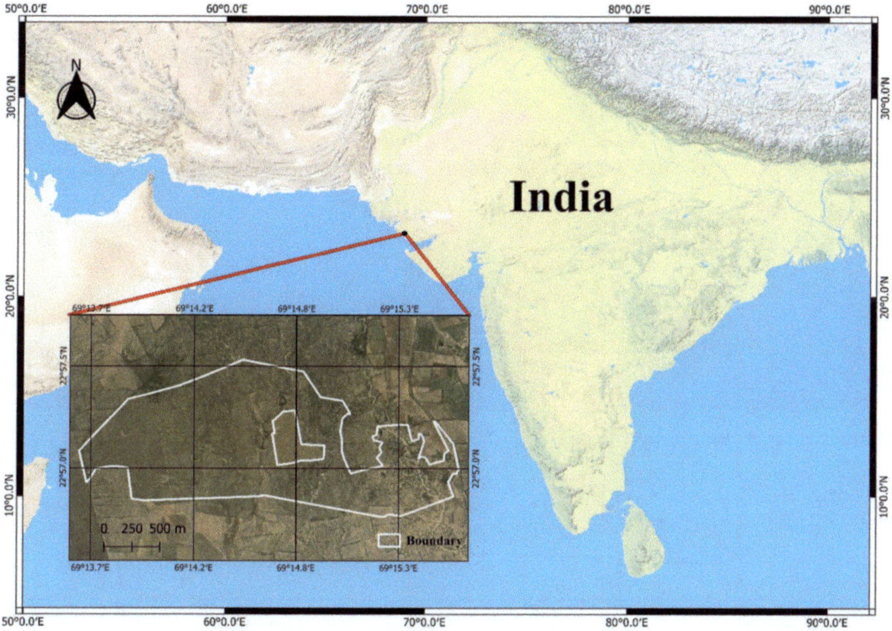

Fig. 16.2 Location of Godhara grassland in Kachchh, Gujarat, India

placed for detailed vegetation and soil data collection, and in each quadrat, four $1 \times 1\ m^2$ quadrats at each end were placed. In order to guarantee that the designated quadrats were dispersed rather than crowded, random points provided structure for point random selection inside a grid and thereby reducing bias and improving diversity of microhabitats captured. Choose the postmonsoon date range of August to October 2024 to provide a robust snapshot of ecological conditions because the vegetation would be at its peak. As stated above, the chosen time range results in vegetation accompanied by the desired peak growth. This combination leads to strong ecological conditions. These two objectives, alongside each other, lead to strong modeling approaches which provide efficient and accurate results. This methodological approach prioritized a spatially balanced, random sampling strategy, allowing for a representative assessment of the vegetation and soil characteristics within the study area despite the limited number of quadrats.

16.2.1 Phytosociological Analysis

After the field data, critical analysis was done on the collected specimens using a stereo microscope (Metlab PST-901) with the regional and national pieces of literature available for identification (Cooke 1908; Shah 1978; Yusufzai and Gandhi 1999; Yadav 2010; POWO 2025). Collected plant specimens were turned into

Fig. 16.3 (**a**) Study area map (**b**) divided into 500 × 500 m² grids (**c**) 150 random points through QGIS software

herbaria following the method outlined by Rao and Sharma (1990). Further collected data were analyzed by adopting phytosociological parameters like relative frequency, relative density, and relative abundance to determine their distribution within the study area (Rao et al. 2015).

$$\text{Frequency} = \frac{\text{Total no.of quadrats in which species occured}}{\text{Total no.of quadrats studied}} \times 100 \tag{16.1}$$

$$\text{Relative frequency} = \frac{\text{Frequency of one species}}{\text{Frequency of all species}} \times 100 \tag{16.2}$$

$$\text{Density} = \frac{\text{Total no.of individuals of a species in all quadrats}}{\text{Total no.of quadrats(sampled unit)studied}} \tag{16.3}$$

$$\text{Relative density} = \frac{\text{Density of one species}}{\text{Density of all species}} \times 100 \tag{16.4}$$

$$\text{Abundance} = \frac{\text{Total no.of individuals of a species}}{\text{Total no.of quadrats in which species occured}} \tag{16.5}$$

$$\text{Relative abundance} = \frac{\text{Abundance of one species}}{\text{Abundance of all species}} \times 100 \tag{16.6}$$

Soil samples were stored at room temperature after being collected at the study location. Sieve Separation Method was used to analyze texture and refractometer (Atago MASTER-S/MillM) for salinity, following the procedure outlined by Gupta (2009). Each sample was analyzed in triplicate and the mean values were considered.

16.2.2 Statistical Analysis

Geospatial and statistical analysis were used to evaluate the characteristics of the soil and their impact on grass diversity with QGIS and R Studio (R Studio 2024.12.1+563). The IDW (Inverse Distance Weighting) method was used in QGIS to spatially map the soil texture (sand, silt, and clay), salinity, and species distribution patterns. The closer known sample points are to the estimated location, the greater their influence is expected to have on the estimation. The soil maps created showed vegetation distribution more clearly as contour maps than the earlier soil characteristic maps, which enhanced the understanding of the soils' ecological impact. Besides geospatial analysis work done in R Studio, a Pearson correlation analysis was also used to investigate the statistical correlations between grass species and soil parameters. The strength and direction of the relationships were evaluated using Pearson's correlation coefficient. In forecasting the degree of relationship between the variables, the correlation coefficient (r) is indicating to lie between (−1) and (+1), which means the relationship would be said to be strongly negatively and positively correlated, respectively, while values toward (0) would suggest weak or no correlation at all.

16.3 Findings and Interpretation

From the current study, a total of 33 species belonging to 27 genera, seven tribes, and two groups were identified (Table 16.1, Plates 16.1, 16.2, and 16.3). In that two tribes, namely, Andropogoneae and Paniceae, were recorded in the Panicoideae group, while five tribes, Arundineae, Aristideae, Chlorideae, Eragrostideae, and Sporoboleae, were recorded in Pooideae tribe. Arundineae tribe was recorded with the highest 9 genus and 11 species, followed by Paniceae (6 genus & 7 species), and Chlorideae (6 genus & 6 species) (Table 16.1). Analysis revealed that *Dicanthium annulatum* recorded the highest relative density value 9.83%, followed by *Dicanthium foveolatum* (9.00%), *Aristida funiculata* (7.64%), and *Cymbopogon iwarancusa* (7.16%). *C. iwarancusa* was likewise recorded as the most widespread among all 33 species with a relative frequency of 9.20% followed by *D. foveolatum*

Table 16.1 List of all recorded grass species along with their phytosociological values and palatability grades (***: Highly palatable, **: Moderate palatable, *) : Less palatable, -: Non palatable

Sr. no.	Sub-group	Tribe	Genus	Scientific name	Common name	R.F. (%)	R.D. (%)	R.A. (%)	P.G.
1.	Panicoideae	Andropogoneae	*Andropogon*	*Andropogon pumilus* Roxb.	Dwarf bluestem grass	4.6	5.86	4.02	***
2.			*Apluda*	*Apluda mutica* L.	Ghaulo	4.48	4.97	3.5	*
3.			*Chrysopogon*	*Chrysopogon fulvus* (Spreng.) Chiov.	Bread grass	2.06	1.36	2.09	***
4.			*Cymbopogon*	*Cymbopogon iwarancusa* (Jones ex Roxb.) Schult.	Rosh ghas, Poyad	9.2	7.16	2.46	*
5.				*Cymbopogon martini* (Roxb.) Will.Watson	Palma rosa	2.18	3.02	4.37	*
6.			*Dichanthium*	*Dichanthium annulatum* (Forssk.) Stapf	Jhijhvo	8.84	9.83	3.51	***
7.				*Dichanthium foveolatum* (Delile) Roberty	Jhijhvo	9.08	9	3.13	**
8.			*Elionurus*	*Elionurus royleanus* Nees ex A.Rich.	–	1.57	1.24	2.49	*
9.			*Heteropogon*	*Heteropogon contortus* (L.) P.Beauv. ex Roem. & Schult.	Dabh shul	0.61	1.12	5.86	*
10.			*Saccharum*	*Saccharum spontaneum* L.	Kaans	0.97	1.36	4.43	–
11.			*Sehima*	*Sehima nervosa* (Rottler) Stapf	Shaniyar	2.3	3.08	4.22	***
12.		Paniceae	*Cenchrus*	*Cenchrus ciliaris* L.	Dhraman	3.51	3.32	2.98	***
13.				*Cenchrus setigerus* Vahl	Dhraman	3.03	2.49	2.59	***
14.			*Digitaria*	*Digitaria ciliaris* (Retz.) Koeler	Tarodiyu	1.33	1.18	2.8	**
15.			*Echinochloa*	*Echinochloa colonum* (L.) Link	Sau	1.09	0.83	2.4	**
16.			*Janochloa*	*Janochloa antidotalis* (Retz.) Zuloaga & Delfini	Gum ghas	0.85	0.83	3.08	***
17.			*Setaria*	*Setaria* sp.	Foxtail	0.73	0.65	2.83	*
18.			*Urochloa*	*Urochloa ramosa* (L.) T.Q.Nguyen	Browntop millet	0.36	0.3	2.57	**

19.	Pooideae	Arundineae	*Phragmites*	*Phragmites karka* (Retz.) Trin. ex Steud.	Devnal	1.09	1.01	2.91	–
20.		Aristideae	*Aristida*	*Aristida adscensionis* L.	Laanp	4.84	6.63	4.32	*
21.				*Aristida funiculata* Trin. & Rupr.	Laanp	6.66	7.64	3.62	*
22.		Chlorideae	*Chloris*	*Chloris barbata* Sw.	Shiyal punch	3.99	3.02	2.38	*
23.			*Cynodon*	*Cynodon dactylon* (L.) Pers.	Chhabar	1.54	1.08	1.8	***
24.			*Melanocenchris*	*Melanocenchris jacquemontii* Jaub. & Spach	Desert black millet	1.45	1.12	2.44	*
25.			*Ochthochloa*	*Ochthochloa compressa* (Forssk.) Hilu	Ganthiro	1.08	0.57	1.85	**
26.			*Schoenefeldia*	*Schoenefeldia gracilis* Kunth	Slender braid grass	4.36	4.97	3.6	*
27.			*Tetrapogon*	*Tetrapogon tenellus* (J.Koenig ex Roxb.) Chiov.	–	0.48	0.47	3.08	**
28.		Eragrosteae	*Dactyloctenium*	*Dactyloctenium aegyptium* (L.) Willd.	Mandhanu	3.99	2.84	2.24	***
29.				*Dactyloctenium scindicum* Boiss.	Nanu mandhanu	2.3	1.42	1.95	**
30.			*Desmostachya*	*Desmostachya bipinnata* (L.) Stapf	Dhrab	2.3	2.55	3.49	–
31.			*Eragrostis*	*Eragrostis ciliaris* (L.) R.Br.	Chichni	5.81	5.98	3.25	**
32.				*Eragrostis tremula* Hochst. ex Steud.	Chichni	1.13	0.90	1.2	
33.		Sporoboleae	*Sporobolus*	*Sporobolus helvolus* (Trin.) T.Durand & Schinz	Khevai	2.18	2.19	2.57	*

Plate 16.1 Recorded grass species from the present study. (1) *Andropogon pumilus*, (2) *Apluda mutica*, (3) *Aristida adscensionis*, (4) *Aristida funiculata*, (5) *Cenchrus ciliaris*, (6) *Cenchrus setigerus*, (7) *Chloris barbata*, (8) *Chrysopogon fulvus*, (9) *Cymbopogon iwarancusa*, (10) *Cymbopogon martini*, (11) *Cynodon dactylon*

Plate 16.2 Recorded grass species from the present study. (12) *Dactyloctenium aegyptium*, (13) *Dactyloctenium scindicum*, (14) *Desmostachya bipinnata*, (15) *Dichanthium annulatum*, (16) *Dichanthium foveolatum*, (17) *Digitaria ciliaris*, (18) *Elionurus royleanus*, (19) *Eragrostis ciliaris*, (20) *Eragrostis tremula*, (21) *Echinochloa colonum*

Plate 16.3 Recorded grass species from the present study. (22) *Heteropogon contortus*, (23) *Janochloa antidotalis*, (24) *Melanocenchris jacquemontii*, (25) *Ochthochloa compressa*, (26) *Phragmites karka*, (27) *Saccharum spontaneum*, (28) *Schoenefeldia gracilis*, (29) *Sehima nervosa*, (30) *Setaria* sp., (31) *Sporobolus helvolus*, (32) *Tetrapogon tenellus*, (33) *Urochloa ramosa*

(7.16%), *D. annulatum* (8.84%), and *A. funiculata* (6.66%). Relative abundance showed up with species like *Heteropogon contortus* (5.86%), followed by *Saccharum spontaneum* (4.43%), *Cymbopogon martini* (4.37%), *Aristida adscensionis* (4.32%), and *Sehima nervosa* (4.22%). The palatability grades of the recorded grasses were examined using secondary literature such as Gandhi et al. (2011), Gorade and Datar (2014), and Odedra et al. (2024). Among them, 9 species were identified as highly palatable, 8 species as moderately palatable, 12 species as less palatable, and 3 species as non-palatable. (Table 16.1 and Fig. 16.4).

After analyzing the soil samples, it was linked to the Soil texture triangle to determine the soil texture class. This soil texture triangle was first developed by the United States Department of Agriculture (USDA) in the early 1900s. Since then, it has been used by soil scientists to understand the texture class of the soil (Dewangan 2023). According to it, half of the study area is classified as having clay texture, and half clay loamy texture (Figs. 16.5 and 16.6).

The IDW interpolation maps reveal significant spatial variability in soil factors, highlighting their interrelationships. Sand content is highest in the northern and central regions, while clay dominates the south-western areas, showing an inverse relationship between the two. Silt is moderately distributed, with higher concentrations in the south-eastern and north-eastern regions (Fig. 16.7).

The spatial correlation suggests that sandier areas have better drainage. Salinity levels vary, with increased concentrations in the south-western and south-eastern parts, likely influenced by clay-rich soils that retain moisture and salts, whereas clayey regions retain more water, contributing to higher salinity. These results highlight the requirement for focused soil management techniques depending on the noted dispersion patterns.

The above analysis of soil texture class and IDW revealed that most of the study area has a clayey texture and less sandy composition. *C. iwarancusa*, *S. nervosa*, and *S. gracilis* were moderately associated with this content (Fig. 16.8). While in

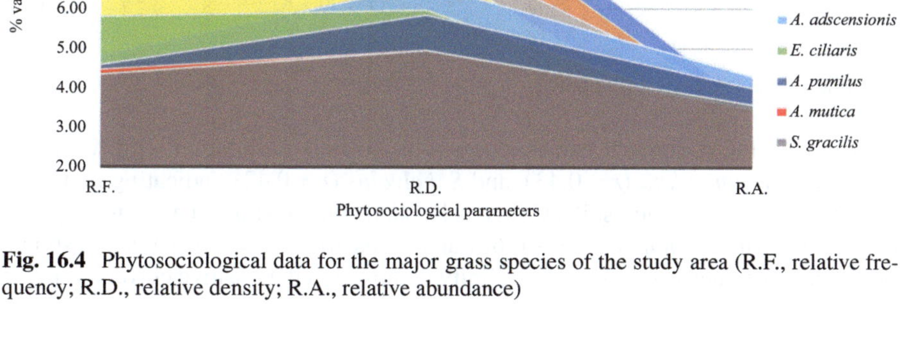

Fig. 16.4 Phytosociological data for the major grass species of the study area (R.F., relative frequency; R.D., relative density; R.A., relative abundance)

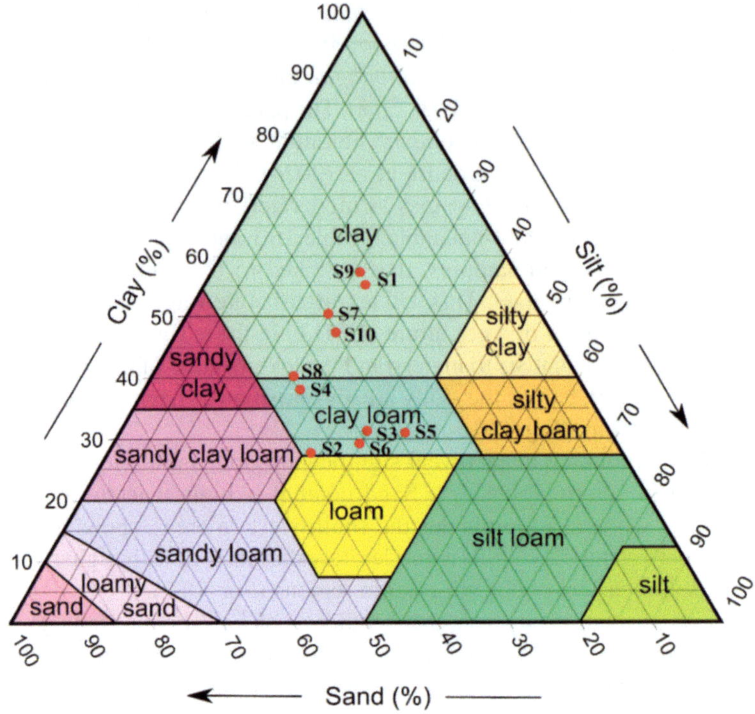

Fig. 16.5 Soil textures of the studied soil samples placed in the soil structure triangle

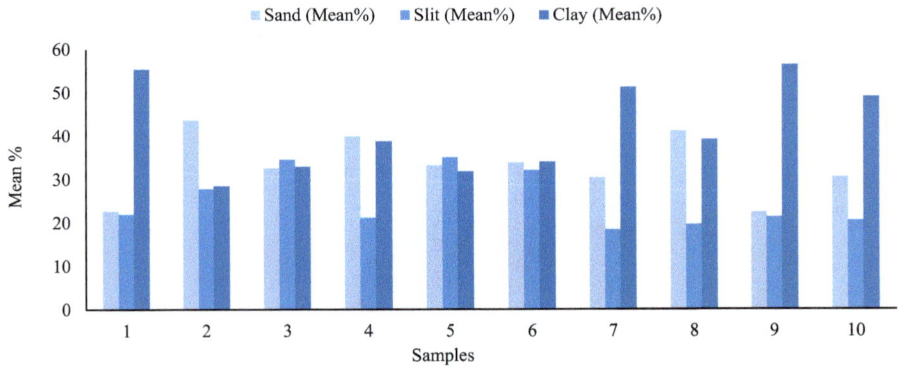

Fig. 16.6 Mean values of soil samples

the case of clay content, it exhibited a strong positive correlation with *A. mutica* ($r = 0.83$), *A. funiculata* ($r = 0.43$), and *S. helvolus* ($r = 0.42$), indicating that these species thrive in clay-rich soils. Conversely, sand content showed a strong negative correlation with *A. mutica* ($r = -0.66$) and *C. martinii* ($r = -0.30$), suggesting a preference for finer-textured soils. Salinity demonstrated a positive correlation with

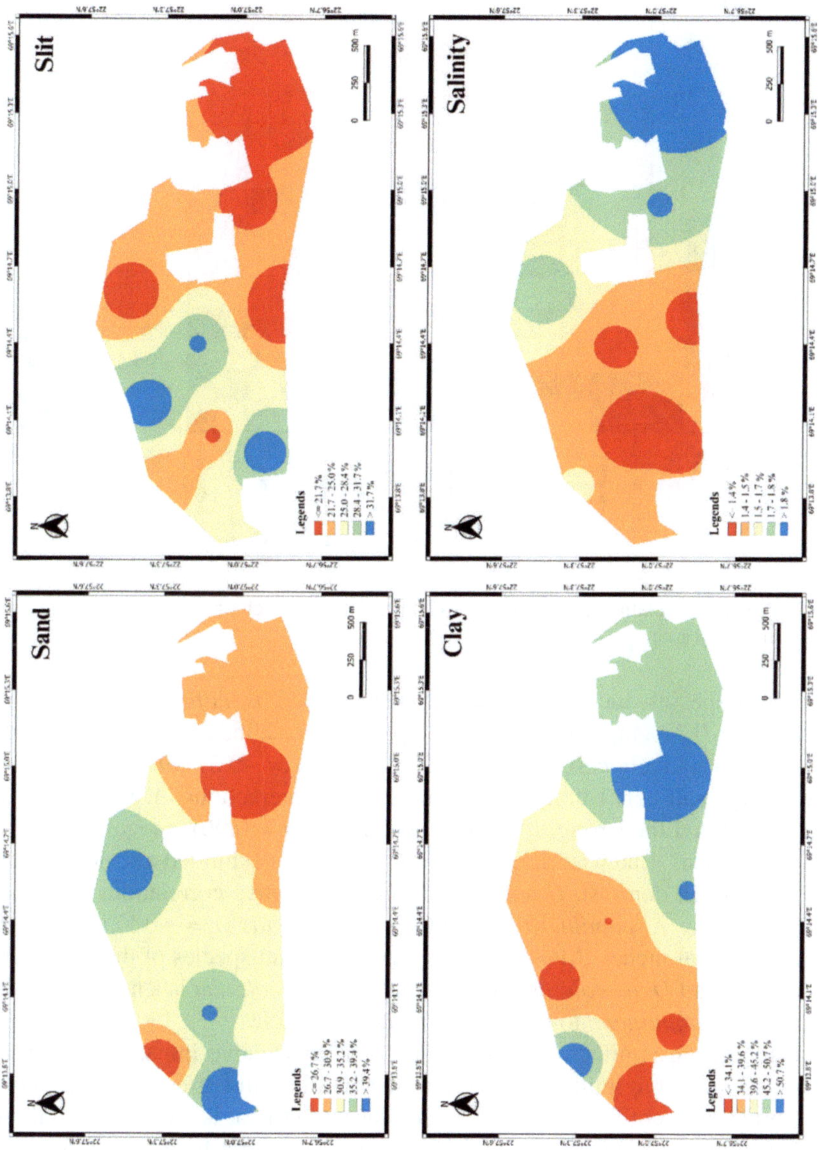

Fig. 16.7 IDW interpolation map revealing the soil texture and salinity across the study area

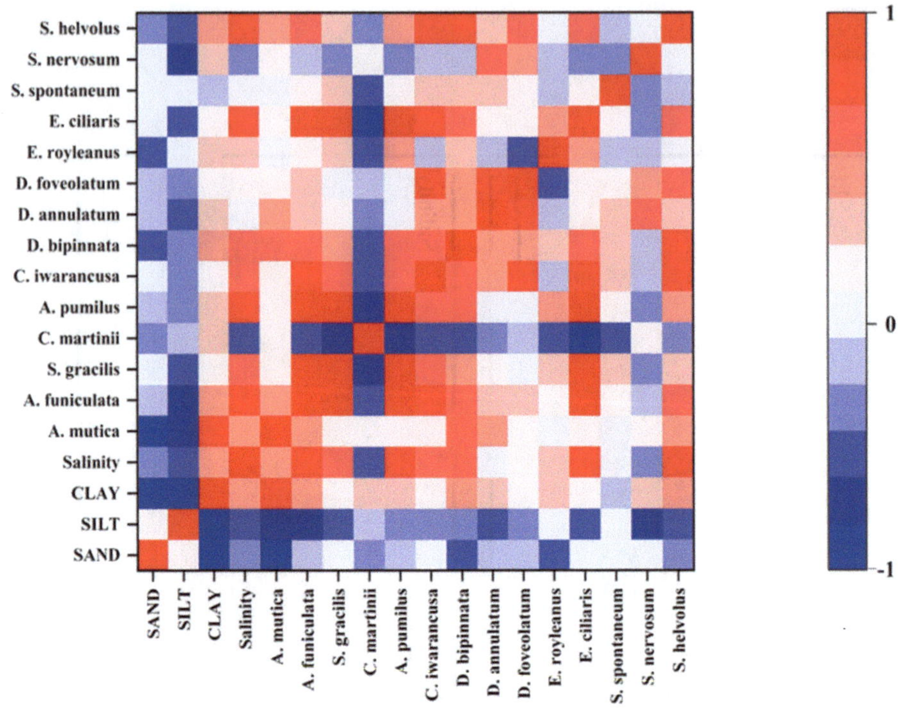

Fig. 16.8 Pearson correlation analysis of the soil physicochemical characters and significant grass species of the study area

several grass species, including *S. helvolus* (r = 0.83), *E. ciliaris* (r = 0.83), and *A. pumilus* (r = 0.77), highlighting their adaptability to saline environments. In contrast, salinity negatively correlated with *C. martinii* (r = −0.37), indicating its sensitivity to high salt concentrations. Among species interactions, *A. funiculata* and *S. gracilis* showed the strongest positive correlation (r = 0.90), suggesting similar habitat preferences. Likewise, *A. pumilus* and *S. gracilis* exhibited a strong association (r = 0.95). In contrast, *C. martinii* displayed negative correlations with most species, including *A. pumilus* (r = −0.60) and *S. gracilis* (r = −0.52), indicating distinct ecological niches. After comparing the dominant species of the study area, *D. annulatum* and *D. foveolatum*, it showed that *D. foveolatum* is a more adaptable species than *D. annulatum* to the salinity and other species.

Climate change is affecting biodiversity, and its negative impact is increasing daily. It is important to imitate actions to retain their harmful effects. To take action, it is important to identify the threats. In this case, the threat is identified as climate change in the form of increasing salinity. By establishing a long-term monitoring program, the progression of salinity-induced changes can be tracked, and implementation of management strategies to mitigate the harmful effects of climate change can be done to safeguard this vital ecosystem. Current findings emphasize that soil texture and salinity can significantly shape species distribution, with some species thriving in high-salinity and clay-rich environments, while others are more sensitive.

This pilot study establishes the baseline data for a long-term monitoring project. An initial overview of the grassland's current state is presented here. Future research will incorporate multiseasonal data and broader physicochemical parameters. Environmental factors determine the precise impacts of climate change on grass distribution and density through salinity and other factors.

16.4 Conclusions and Conservation Measures

Refining approaches to soil property relationships with grass species and conservation strategies for Kachchh's grassland management and restoration need stronger efforts at an ecosystem scale. In order to further enhance the resilience of these grasslands, systematic monitoring of changes in vegetation and soil properties over time within particular study sites should be accompanied by controlled salinity level experiments. Socioeconomic considerations alongside climate change impacts related to grassland degradation and restoration need parallel attention for effective strategy formulation. Using local populations as active participants in researching and managing soils will enhance the scope of soil analyses regarding other macronutrients, and sustainable land use and livelihood improvement efforts in the region. Further, it can be implemented on a grand scale in other grassland ecosystems.

Acknowledgments The authors are grateful to the Respected Deputy Conservator of Forest-Western Kachchh Division, the Respected Range Forest Officer, and the Forest staff of the Mandvi Range, Gujarat Forest Department, for giving necessary permissions and their support throughout the study. We also thank the Director of the Gujarat Institute of Desert Ecology (GUIDE) for their support throughout the study, and the Environmental and Monitoring Division of GUIDE, for allowing and helping in soil analysis.

Funding Statement No funders to disclose.

Data Availability Statement The data used for the study will be available on request.

Disclosures and Declarations All authors of the study consent to publish the book chapter.

Conflict of Interests The authors declare that they have no conflict of interest.

Ethics Approval Not applicable.

References

Arora S, Sharma V (2017) Reclamation and management of salt-affected soils for safeguarding agricultural productivity. J Safe Agric 1:1–10

Bannari A, Al-Ali ZM (2020) Assessing climate change impact on soil salinity dynamics between 1987–2017 in arid landscape using Landsat TM, ETM+ and OLI data. Remote Sens 12(17):2794. https://doi.org/10.3390/RS12172794

Bardgett RD, Bullock JM, Lavorel S et al (2021) Combatting global grassland degradation. Nat Rev Earth Environ 2:720–735

Bullock JM, Jefferson RG, Blackstock TH et al (2011) Semi-natural grasslands. In: UK NEA (ed) The UK National Ecosystem Assessment. UNEP-WCMC, Cambridge

Champion HG, Seth SK (1968) A revised survey of the forest types of India. Government of India. Natraj Publisher, New Delhi

Cooke T (1908) The Flora of the presidency of Bombay, vol 2. Taylor & Francis, London

Dabadghao PM, Shankarnarayan KA (1973) The grass cover of India. Indian Council of Agriculture Research, New Delhi

Dewangan SK (2023) Using the soil texture triangle to evaluate the effect of soil texture on water flow: a review. Int J Res Appl Sci Eng Technol 11:389–393. https://doi.org/10.22214/ijraset.2023.53635

Dexter AR (2004) Soil physical quality: Part I. Theory, effects of soil texture, density, and organic matter, and effects on root growth. Geoderma 120:201–214. https://doi.org/10.1016/J.GEODERMA.2003.09.004

Gandhi DJ, Albert S, Pandya NR, Panchal KR (2011) Socioeconomic study of grasses and legumes in Baria and Godhra Forest Division, Gujarat. Not Sci Biol 3:53–61. https://doi.org/10.15835/nsb336061

Gorade PD, Datar MN (2014) Checklist of palatable grass species from Peninsular India. Not Sci Biol 6:441–447. https://doi.org/10.15835/nsb649376

Grace JB, Anderson TM, Seabloom EW et al (2016) Integrative modeling reveals mechanisms linking productivity and plant species richness. Nature 529:390–393. https://doi.org/10.1038/NATURE16524

Gupta PK (2009) Methods in environmental analysis: water, soil and air. Agrobios Publishing, Jodhpur

Han W, Chen L, Su X et al (2022) Effects of soil physico-chemical properties on plant species diversity along an elevation gradient over alpine grassland on the Qinghai-Tibetan Plateau, China. Front Plant Sci 13:1–13. https://doi.org/10.3389/fpls.2022.822268

Hopkins B, White F (1987) Vegetation map of Africa. The vegetation of Africa: a descriptive memoir to accompany the UNESCO/AETFAT/UNSO vegetation map of Africa. J Ecol 75:1214. https://doi.org/10.2307/2260340

Hu Y, Schmidhalter U (2004) Limitation of salt stress to plant growth. In Hock B, Elstner EF (eds) Plant toxicology(4th Edition). CRC press, Boca Raton. https://doi.org/10.1201/9780203023884

Jamil A, Riaz S, Ashraf M, Foolad MR (2011) Gene expression profiling of plants under salt stress. Crit Rev Plant Sci 30:435–458. https://doi.org/10.1080/07352689.2011.605739

Kar A (2011) Geomorphology of the arid lands of Kachchh and its importance in land resources planning. In: Bandyopadhyay S, Bhattacharji M, Chaudhuri S, Goswami D, Jog SR, Kar A (eds) Landforms processes and environment management. ACB Publications, Kolkata

Kasim WA, Osman ME, Omar MN et al (2013) Control of drought stress in wheat using plant-growth-promoting bacteria. J Plant Growth Regul 32:122–130. https://doi.org/10.1007/S00344-012-9283-7

Lemaire G, Hodgson J, Chabbi A (eds) (2011) Grassland productivity and ecosystem services. CABI, Wallingford. https://doi.org/10.1079/9781845938093.0000

Mandal S, Raju R, Kumar A et al (2018) Current status of research, technology response and policy needs of salt-affected soils in India: a review. J Indian Soc Coast Agric Res 36:40–53

Nugent DT, Baker-Gabb DJ, Green P et al (2022) Multi-scale habitat selection by a cryptic, critically endangered grassland bird—the Plains-wanderer (Pedionomus torquatus): implications for habitat management and conservation. Austral Ecol 47:698–712. https://doi.org/10.1111/AEC.13157

Odedra KN, Odedra NK, Jadeja BA (2024) Grass flora of Porbandar District, Gujarat, India. Proc Natl Acad Sci India Sect B Biol Sci 95:93–103. https://doi.org/10.1007/s40011-024-01629-w

Oliveira RS, Bezerra L, Davidson EA et al (2005) Deep root function in soil water dynamics in cerrado savannas of central Brazil. Funct Ecol 19:574–581. https://doi.org/10.1111/J.1365-2435.2005.01003.X
Pathak PS, Dagar JC (2016) Indian grasslands and their management. In: Grassland: a global resource perspective. ICAR, New Delhi
Pilgrim ES, Macleod CJA, Blackwell MSA et al (2010) Interactions among agricultural production and other ecosystem services delivered from European temperate grassland systems. Adv Agron 109:117–154. https://doi.org/10.1016/B978-0-12-385040-9.00004-9
POWO (2025) Plants of the World Online. Facilitated by the Royal Botanic Gardens, Kew. Available at: http://www.plantsoftheworldonline.org/. Accessed on 11 Mar 2025
Queiroz C, Beilin R, Folke C, Lindborg R (2014) Farmland abandonment: threat or opportunity for biodiversity conservation? A global review. Front Ecol Environ 12:288–296. https://doi.org/10.1890/120348
Rahman AKMM, Ahmed KM, Butler AP, Hoque MA (2018) Influence of surface geology and micro-scale land use on the shallow subsurface salinity in deltaic coastal areas: a case from southwest Bangladesh. Environ Earth Sci 77:1–8. https://doi.org/10.1007/S12665-018-7594-0/FIGURES/6
Rao RR, Sharma BD (1990) A manual for herbarium collections. Botanical Survey of India, Kolkata
Rao DS, Murthy PP, Kumar OA (2015) Plant biodiversity and phytosociological studies on tree species diversity of Khammam District, Telangana State, India. J Pharm Sci Res 7(8):518–522
Shah GL (1978) Flora of Gujarat State. Sardar Patel University, Vallabh Vidyanagar
Suttie JM, Reynolds SG, Batello C (2005) Grassland perspectives. In: Grasslands of the world. Food and Agriculture Organization, Rome
Thakkar MG (2017) Geomorphological field guide book on Kachchh Peninsula. Indian Institute of Geomorphologists, Allahabad
Ullah A, Bano A, Khan N (2021) Climate change and salinity effects on crops and chemical communication between plants and plant growth-promoting microorganisms under stress. Front Sustain Food Syst 5:1–16. https://doi.org/10.3389/fsufs.2021.618092
Van Dyne GM, Smith FM, Czaplewski RL, Woodmansee RG (1978) Analysis and syntheses of grassland ecosystem dynamics. In: Singh JS, Gopal B (eds) Glimpses of ecology. International Scientific Publications, Jaipur
White RP, Murray S, Rohweder M (2000) Pilot analysis of global ecosystems. Grassland ecosystems. World Resources Institute, Washington, DC
Whyte RO (1957) The grassland and fodder resources of India. Indian Council of Agricultural Research, Jhansi
Yadav SR (2010) Know your grass genera through hand lens. Shivaji University, Kolhapur
Yusufzai AS, Gandhi AP (1999) Grasses of Gujarat. Gujarat Agricultural University, Ahmedabad
Zhu Y, Delgado-Baquerizo M, Shan D et al (2020) Diversity-productivity relationships vary in response to increasing land-use intensity. Plant Soil 450:511–520. https://doi.org/10.1007/S11104-020-04516-1/METRICS

Chapter 17
Terrestrial Diatoms as Potential Indicators of the Impacts of Climate Change

Sukumar Bhakta and Sudipta Kumar Das

Abstract Diatoms are ubiquitous across vast habitats and highly sensitive to environmental changes. Thus, these photosynthetic microbial eukaryotes are often considered as indicators of environmental health. In the prevailing climate change and its consequential impacts, periodic monitoring of the environmental changes and their impacts on terrestrial ecosystems is inevitable. Ecological models and indices based on diatoms were used by environmentalists all across the globe to monitor aquatic health and trophic status of both marine and freshwater. On the contrary, terrestrial diatoms and their autecology were not explored much. This chapter discusses the potential of terrestrial diatoms as biological indicators, specifically to assess the impacts of climate change.

Keywords Diatoms · Terrestrial · Indicators · Climate change

17.1 Introduction

Climate change denotes substantial and enduring alterations in the Earth's climate system, predominantly caused by human actions, including the combustion of fossil fuels, deforestation, and various industrial activities. These practices emit greenhouse gases (GHGs) such as carbon dioxide (CO_2), methane (CH_4), and nitrous oxide (N_2O) into the atmosphere, resulting in global warming and associated climate alterations (Wardle 2002; Bardgett 2005). According to the Intergovernmental Panel on Climate Change (IPCC), the impacts of climate change include rising temperatures, sea-level rise, altered precipitation patterns, and increased frequency of extreme weather events (IPCC 2021). Any climate change poses unprecedented challenges to ecosystems and human societies (Malhi et al. 2020). Human-induced

S. Bhakta · S. K. Das (✉)
Botanical Survey of India, Andaman and Nicobar Regional Centre,
Sri Vijaya Puram, Andaman and Nicobar Islands, India

B. P. Barathan et al. (eds.), *Navigating Climate Change: Impacts on Biodiversity and Ecosystem Resilience*, Sustainable Development and Biodiversity 42,
https://doi.org/10.1007/978-981-95-0409-1_17

factors, including greenhouse gas emissions, lead to species extinctions and modify their distributions. Recent findings indicate that these factors have widespread effects on both antagonistic and mutualistic interactions among different species. Climate change affects terrestrial biota, altering ecosystems, species distributions, and biological interactions (Tylianakis et al. 2008). The most evident biological responses to climate change, where rising temperatures have led to changes in phenology, are earlier flowering, leaf-out, and breeding in various species (Parmesan et al. 2000; Parmesan and Yohe 2003; Both et al. 2006; Menzel et al. 2006). Biological indicators are indispensable for understanding these impacts, providing early warnings, and shaping effective mitigation and adaptation strategies. It is evident from the response to various stressors that diatoms are quite responsive to the trophic status of aquatic habitats (Hall and Smol 1999), soil environmental conditions, anthropogenic disturbances, and agricultural practices (Naz et al. 2023).

17.1.1 What Are Terrestrial Diatoms?

Diatoms are microscopic algae that colonize different environments, including freshwater, marine, and terrestrial ecosystems, which possess a unique siliceous cell wall, also known as a frustule (Round et al. 1990). According to the extensive fossil record of diatoms, the origin of diatoms can be traced back to the first Jurassic Period, roughly about 190 million years ago. However, according to molecular footprints, diatoms originated during the Triassic period, more than 240 million years ago (Kooistra and Medlin 1996; Sims et al. 2006; Sorhannus 2007). Because of the 60-million-year gap between the oldest fossils and the molecular clock data, the first diatoms are unlikely to be silicified (Raven and Waite 2004). These highly intricate and species-specific cell walls make them valuable for taxonomic studies. Diatoms perform photosynthesis and are essential components of the primary producers in aquatic and terrestrial ecosystems, forming the base of food webs. Diatoms are abundant phototrophic organisms colonizing marine to fresh environments, extending from the tropics to the poles. They can also be found in soils and aerosols (Johansen 2004). Diatoms are classified as planktonic (suspended in the water column), benthic (growing in the bottom of aquatic habitats), or subaerial (attached to the moist aerial substrates). Epilithic (attached to rocks), epiphytic (attached to aquatic plants and other algae), epipsammic (attached to sand grains), and epizoic (attached to animal bodies) are the different types of benthic communities (Round et al. 1990; Lowe 2011). They have evolved various adaptations to survive in various environments throughout evolution. For example, diatoms can float in water due to their ability of oil and the presence of silica frustules as a protector from predators. Some planktonic diatoms can form long chains that are linked by silica spines. This one is a frequent morphological adaptation to avoid sinking evident in some radial centric diatoms like *Paralia*, *Aulacosiera, Stephanopyxis*, and *Skeletonema*, pennates like *Staurosira, Fragilaria,* and *Diadesmis* (Kooistra et al. 2007). Epilithic diatoms, such as *Licmophora* and *Tabularia*, grow attached to rock surfaces.

Epiphytic diatoms are attached to aquatic plants and also to mosses in terrestrial habitats. Rapid diatoms have the power to move over benthic surfaces, such as fine sand grains and mud surfaces, in a tidal zone. A few diatoms also have mucilage tubes and can move up and down inside the tubes. While diatoms are predominantly associated with aquatic environments, several species have adapted to terrestrial ecosystems, where they occupy various niches, including soils, mosses, and lichens (Krammer and Lange-Bertalot 2000; Serieyssol 2012).

Lund (1945) referred to the term *terrestrial diatoms* as the population of diatoms growing on any substrates where atmospheric humidity is the exclusive source of water. The substrates range from soil to wet walls, rocks, tree barks, moss and lichen patches, etc. The diatoms are omnipresent heterokonts and constitute one of the most diverse groups of organisms utilized as robust ecological indicators for aquatic ecosystems (Naz et al. 2023). While less studied than their aquatic counterparts, they have been shown to inhabit various habitats such as soils, mosses, and peat bogs. Their unique properties, including their sensitivity to environmental conditions, make them ideal candidates for monitoring climate-induced changes. However, a precise sampling method enhances better insight into terrestrial diatoms as a potential bioindicator of the soil ecological quality (Barragán et al. 2018). Their distribution is influenced by environmental conditions such as temperature, moisture, and light availability. Unlike aquatic diatoms, terrestrial species are typically exposed to more extreme environmental fluctuations, making them sensitive indicators of changing climate conditions (Smol and Douglas 2007). They can reflect shifts in temperature, precipitation, and even atmospheric CO_2 levels through species composition, distribution, and morphology changes. Therefore, understanding the climate change-induced responses of terrestrial diatoms can help in forecasting the broader impacts of global warming.

17.2 Potential of Terrestrial Diatoms as "Biological Indicators"

Diatoms have long been used as proxies in paleoclimate research, particularly in aquatic sediments, where they can reveal past environmental conditions through fossilized diatom assemblages. Recent studies have expanded this approach to terrestrial diatoms, suggesting their potential in monitoring both past and present climate changes (Finkelstein and Swaddle 2012). Terrestrial diatoms are also widely used in paleoclimatic reconstructions. Peat bogs and wetlands, which accumulate sediments over long periods, often contain well-preserved diatom frustules. Araujo et al. (2025) determined the influence of the environmental conditions of northern peatlands on the spatial and temporal variation in their diatom species composition. A key study by Carballeira and Pontevedra-Pombal (2020) focused on the use of diatom communities of peat bogs in northern Europe to reconstruct past climate conditions. The study found that certain diatom species, such as *Cymbella* spp. and

Fragilaria spp., were particularly abundant during colder periods, while others, such as *Navicula* spp., were more common in warmer, wetter conditions. This study demonstrated the utility of terrestrial diatoms as proxies for temperature and precipitation changes over long timescales, providing valuable insights into historical climate variability. In polar environments, cryoconite holes are habitats that act as refugia for the aquatic and terrestrial diatoms on glacier surfaces. Diatoms of this habitat have strong linkages between glaciers and the local basin characteristics of different temperature regimes and may promote speciation (Stanish et al. 2013). They observed that *Muelleria* and *Diadesmis* have an advantage in this habitat compared to other stream diatoms. Diatom assemblages are influenced by several factors, including temperature, precipitation, and CO_2 concentrations, all of which are impacted by climate change. For example, the abundance and diversity of diatom species in peat bogs and soils can indicate shifts in temperature and moisture regimes. In colder regions, warming temperatures may lead to an increase in terrestrial diatom species that prefer warmer conditions, while species that thrive in cooler climates may decline (Lavoie et al. 2004). Warmer temperatures can accelerate metabolic rates, leading to faster growth and reproduction in some species, while other species may be unable to tolerate the heat, resulting in shifts in community structure (Schneider and Root 1996). Similarly, alteration in the moisture availability in soils influences diatom distribution and species composition. Changes in precipitation patterns can lead to drier conditions in certain regions, favoring species that can withstand desiccation, while others may become dormant or die off (Peszek et al. 2021). Some species of terrestrial diatoms have developed adaptive strategies to cope with environmental extremes, such as the ability to enter dormancy during dry spells or the formation of resting spores that can withstand long periods of unfavorable conditions (Souffreau et al. 2010). The occurrence of diatoms is also best suited to the microhabitats within the tree trunks, though cyanobacteria and green algae are reported very often (Bhakta et al. 2014; Rybak et al. 2023).

Terrestrial diatoms offer several advantages as indicators of climate change. Their high sensitivity to environmental fluctuations allows for the detection of both short-term and long-term climate trends. One of the most notable examples of terrestrial diatoms being used as environmental indicators comes from studies in agricultural systems. A study by Zhang et al. (2020) investigated diatoms' role in assessing soil quality in agricultural fields. Diatom assemblages in surface soils were analyzed to determine their sensitivity to agricultural practices, such as fertilization and irrigation methods. The researchers found that changes in diatom communities, including shifts in species composition, were strongly correlated with soil disturbance (Vacht et al. 2014). They found that the abundance, absence, or presence of specific diatom species of riparian soils, e.g., *Hantzschia amphioxys*, *Fragilaria zeilleri* var. *elliptica*, and *Pinnularia lata,* could be potential indicators for anthropogenic disturbance levels. This suggested that diatoms could serve as sensitive indicators of agricultural impact, offering a tool for monitoring soil health in managed ecosystems. Additionally, diatom frustules are durable and can persist in soils and sediments for long periods, providing a historical record of past environmental conditions (Lund 1945). By analyzing changes in diatom communities over time, scientists can reconstruct past climate conditions with a high degree of resolution. Furthermore, because diatoms respond rapidly to environmental changes, they can provide early warnings of impending ecological shifts, offering valuable data for climate change mitigation and adaptation strategies.

Distributions of diatoms in aquatic habitats are controlled by many environmental and anthropogenic factors. Therefore, these diatoms are commonly studied the most for calculating different indices to monitor water quality (Soininen 2007). The anthropogenic forces affecting aquatic ecosystems, as well as species distribution influencing the lake ecosystem dynamics, were studied by Farqan et al. (2025). Variations in diatom assemblages were noted to differ between both core and surface sediment samples from Bosten Lake. The biological community is predominantly composed of oligotrophic taxa, including *Lindavia radiosa* and *Navicula peroblonga*. This community exhibited a shift toward meso-eutrophic species, such as *Pseudostaurosira*, *Staurosira*, and *Staurosirella* species. Ultimately, the lake became increasingly eutrophic, as evidenced by the rising abundance of *Fragilaria* cf. *crotonensis*. However, similar studies on terrestrial diatom communities are rare. A possible reason behind it may be a lack of adequate knowledge on the community structure of diatoms in terrestrial or aerial habitats and influential environmental parameters which affect species distribution (Antonelli et al. 2017). The systematic constraints in determining the identity of terrestrial diatoms are the small valve size and overlapping occurrence in any ecosystem (Johansen 2004). Several studies revealed that pennate diatoms dominate the terrestrial habitats over centric which can be attributed to their better adherence to the substratum for nutrient sequestration (Passy 2007; Zhang et al. 2020). The potentiality of these terrestrial diatoms in deciphering the impact of climate change can be interpreted from their behavior in response to any alteration in environmental parameters, which can range from phenotypic plasticity to changes in their community structure.

17.2.1 Soil Diatoms

Soil diatoms play an integral role in any agricultural ecosystem as they contribute significantly to nutrient cycling, augmentation of soil stability, and enhancement of soil productivity (Van de Vijver et al. 2002; Shein et al. 2016; Foets et al. 2021). Minaoui et al. (2021) identified 26 soil diatoms where *Nitzschia* was the dominant species for pollution tolerance and biomass production. It signified that diatom communities could be used as bioindicators of anthropogenic disturbances on urban soil. The community structure of these diatoms is regulated by soil moisture, pH, organic matter content, land use, and many anthropogenic factors like irrigation quality and pollution (Stanek-Tarkowska and Noga 2012; Vacht et al. 2014; Binoy and Ray 2016; Antonelli et al. 2017; Naz et al. 2023). Apart from the type of land use, the physical disturbances in the soil caused by different agricultural practices also play a key role in defining the soil diatom communities (Heger et al. 2012;

Foets et al. 2021). Hayek and Hulbary (1956) defined a relation between diatom abundance and soil textures and suggested rich diversity in sandy soils than the soils with clay. The species composition of soil diatoms was studied several times to assess edaphic quality and improvement of agricultural practices (Hahn and Neuhaus 1997; Zancan et al. 2006; Heger et al. 2012; Stanek-Tarkowska et al. 2015). Soil diatoms are also found vulnerable to distress caused by oil pollution and herbicide application (Bérard et al. 2004; Dorokhova 2007; Belnap et al. 2016; Maltsev et al. 2017). These diatoms, which are sensitive to any edaphic alterations and soil pollution can be considered as biological indicators.

Almost after a century since the preliminary documentations on soil diatoms (Petersen 1915, 1928, 1935), a comprehensive account on the region-based floristic of soil diatoms with detailed autecological information is still lacking, though there are bountiful of records of terrestrial diatoms from different parts of the globe (Van de Vijver and Beyens 1997; Bérard et al. 2004; Zancan et al. 2006; Moravcová et al. 2009; Lin et al. 2013; Vacht et al. 2014; Fazlutdinova and Sukhanova 2014; Ettl and Gärtner 2014; Chattová et al. 2022a). Despite the ubiquitous nature of diatoms, a qualitative uniqueness in their distribution in different ecosystems is remarkable. It was observed that the species of *Pinnularia microstauron*, *Nitzschia amphibia*, *Sellaphora nigri*, *Pinnularia borealis*, and *Humidophila contenta* are the indicators of yellow brown soil, *Luticola suecorum*, *Luticola cohnii*, *Luticola nivalis*, and *Luticola mutica* for Brown calcic soil, and *Stauroneis lundii*, *Luticola* species for yellow cinnamon soil (Zhang et al. 2020). However, several sensitive diatom species were also found to be adaptive to varied disturbed environmental conditions. Antonelli et al. (2017) observed similar behavior in *Caloneis lancettula* and *Stauroneis parathermicola* while deriving diatom indices to identify human-induced disturbances in soils. *Hantzschia amphioxys* is considered as a soil cosmopolitan species and is also found to be adaptive to anthropogenic stressed conditions (Olkova et al. 2017; Zhang et al. 2020). The adaptive occurrences were also recorded for *Pinnularia borealis* complex and some taxa of *Luticola* (Foets et al. 2021). Thus, it is believed that the development of diatom-based indices to monitor soil conditions can be an integrated mining of region-specific data as well as inclusion of appropriate eco-sensitive species.

17.2.2 Aerial Diatoms

Aerial diatoms inhabiting lichens, bryophytes, and tree barks are mostly pennate diatoms with smaller valves colonized with specific adaptive strategies and attachment mechanisms called stalks, constituting mucopolysaccharides, and sometimes with sticky mucopeptide substances (Szczepocka et al. 2024). Many diatom genera like *Humidophila*, *Luticola*, and *Boreogenicola* exclusively grow in aerial habitats (Radhakrishnan et al. 2022). The diversity of moss-inhabiting diatoms of the white poplar (*Populus alba* L.) bark revealed 47 diatom taxa, including three dominant species, e.g., *Achnanthes coarctata* (Brébisson) Grunow, *Luticola acidoclinata*

Lange-Bertalot, and *Stauroneis thermicola* (Petersen) Lund, and three species: *Luticola sparsipunctata* Levkov, Metzeltin, & Pavlov, *L. vanheurckii* Van de Vijver & Levkov, and *Hantzschia subrupestris* Lange-Bertalot, reported first time from Poland (Rybak et al. 2018). Some members of *Eunotia,* which generally prefer acidic aquatic environments, also colonize humid aerial habitats (Van Dam et al. 1994). However, the occurrence of some freshwater taxa in the aerial habitats cannot be repudiated, which could have been transferred from nearby aquatic habitats or absorbed through water vapors and ultimately splashed on the substratum through raindrops (Szczepocka et al. 2024). In the last couple of decades, many researchers have tried to understand the taxonomic composition in the aerial diatom communities around the globe and factors responsible for their distribution (Van de Vijver et al. 2002; Van de Vijver and Kopalova 2008; Furey et al. 2009; Kopalová et al. 2014; Chen et al. 2012; Pouličková et al. 2013; Das et al. 2020; Radhakrishnan et al. 2022).

The researchers so far have identified important community influencing environmental parameters, i.e., atmospheric humidity, temperature, and altitude (Borcard et al. 1992; Hájková et al. 2011; Chen et al. 2012; Das et al. 2020). Some common terrestrial diatom taxa are presented in Figs. 17.1 and 17.2. pH is also considered a critical factor when the substratum is exposed to extreme heat and desiccation, which causes high evaporation and reduction of moisture content in the substratum (Johansen 2004). High evaporation also leads to upscaled conductivity, which can be indicated by the appearance of halophilic species. Diatoms colonizing such extreme habitats are also excellent indicators of nutrient scarcity. Studies also revealed that many species of the genera *Gomphonema* and *Cymbella* found attached to moss patches are tolerant to elevated UV radiation (Bothwell et al. 1993; Johansen 2004). A study in the montane mires of China revealed that the moss-inhabited diatom assemblage composition is also regulated by the height of the substratum above the water table (Chen et al. 2012). In several ecosystems, the overlapping impact of multiple parameters was observed, instead of individual impact, proving their interrelatedness during impact (Goeyers et al. 2024). The potential of aerial diatoms as biological indicators is predominant (Kopalová et al. 2014; Das et al. 2020; Chattová et al. 2021; Goeyers et al. 2024). However, the main constraint in identifying such indicator species is the lack of adequate knowledge on their diversity and taxonomy.

17.2.3 Phenotypic Plasticity in Terrestrial Diatoms

Phenotypic plasticity is the concept of morphological alterations in diatom valves as an adaptive strategy to mitigate different environmental stressors (Vitug 2022). A wide range of morphological alterations are observed in diatom cells, e.g., valve size reduction, functional heterogeneity, Janus cells, innenschalen, and teratologies (Kociolek and Stoermer 2010). The principal causative factors behind the morphological plasticity are salinity, silica concentration, temperature, pH, heavy metals, and other pollutants (Falasco et al. 2009; Kociolek and Stoermer 2010). The impact

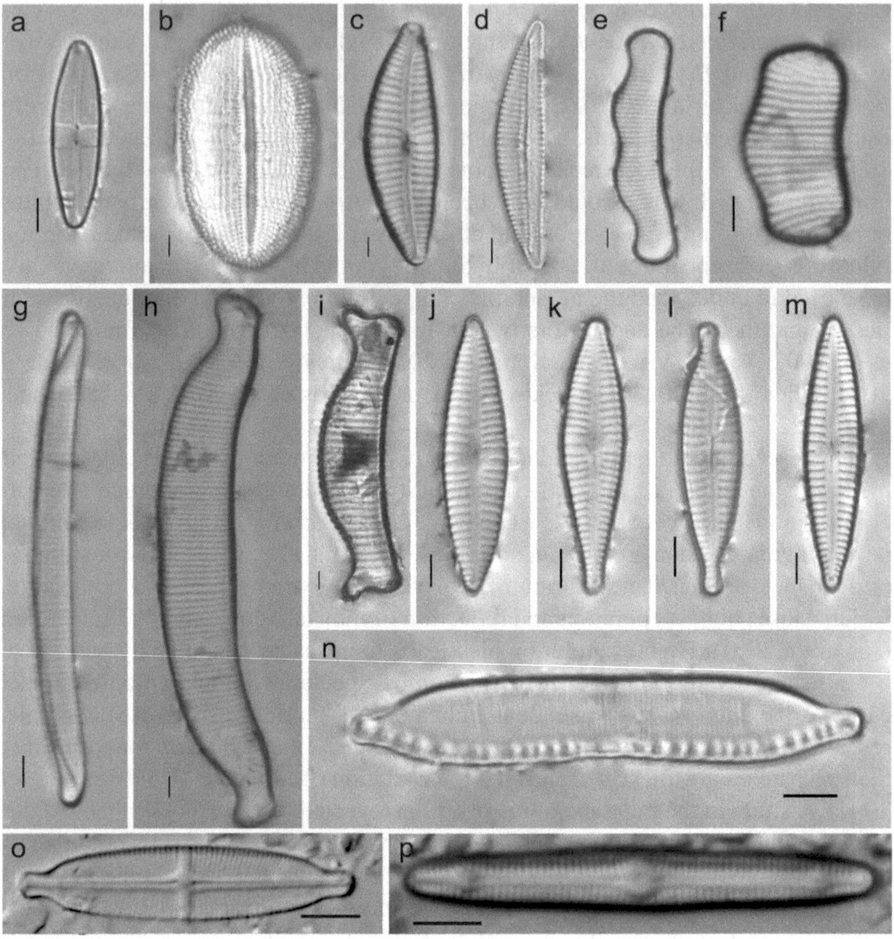

Fig. 17.1 Common terrestrial diatoms of the eastern Himalayas. (**a**) *Caloneis* cf. *clevei*; (**b**) *Cocconeis placentula* var. *euglypta*; (**c**) *Cymbella affinis*; (**d**) *Encyonema pergracile*; (**e**) *Eunotia bigibba*; (**f**) *E. curtagrunowii*; (**g**) *E. arcus*; (**h**) *E. paludosa*; (**i**) *Eunotia* sp.; (**j**) *Gomphonema amoenum*; (**k**) *G. insigne*; (**l**) *G. parvulum*; (**m**) *Gomphonema* sp.; (**n**) *Hantzschia amphioxys*; (**o**) *Stauroneis gracilis*; (**p**) *Humidophila platensis* (scale bar: a–p = 10 μm)

of atmospheric moisture on morphological abnormality was observed in the aerophilic taxon *Navicula gallica,* which showed a deformed valve contour in desiccated conditions (Lund 1945, 1946). So far, detections of morphological deformities are from aquatic diatoms. However, their occurrences in extreme conditions of terrestrial habitats are yet to be thoroughly explored.

Some of the recent studies on aerophilous diatoms inhabiting moss patches revealed valve size diminution in several diatoms to withstand drought conditions (Hayek and Hulbary 1956; Round 1957; Van de Vijver and Beyens 1997; Das et al. 2020). Valve size reduction in diatoms is, in fact, an adaptive strategy under depleted

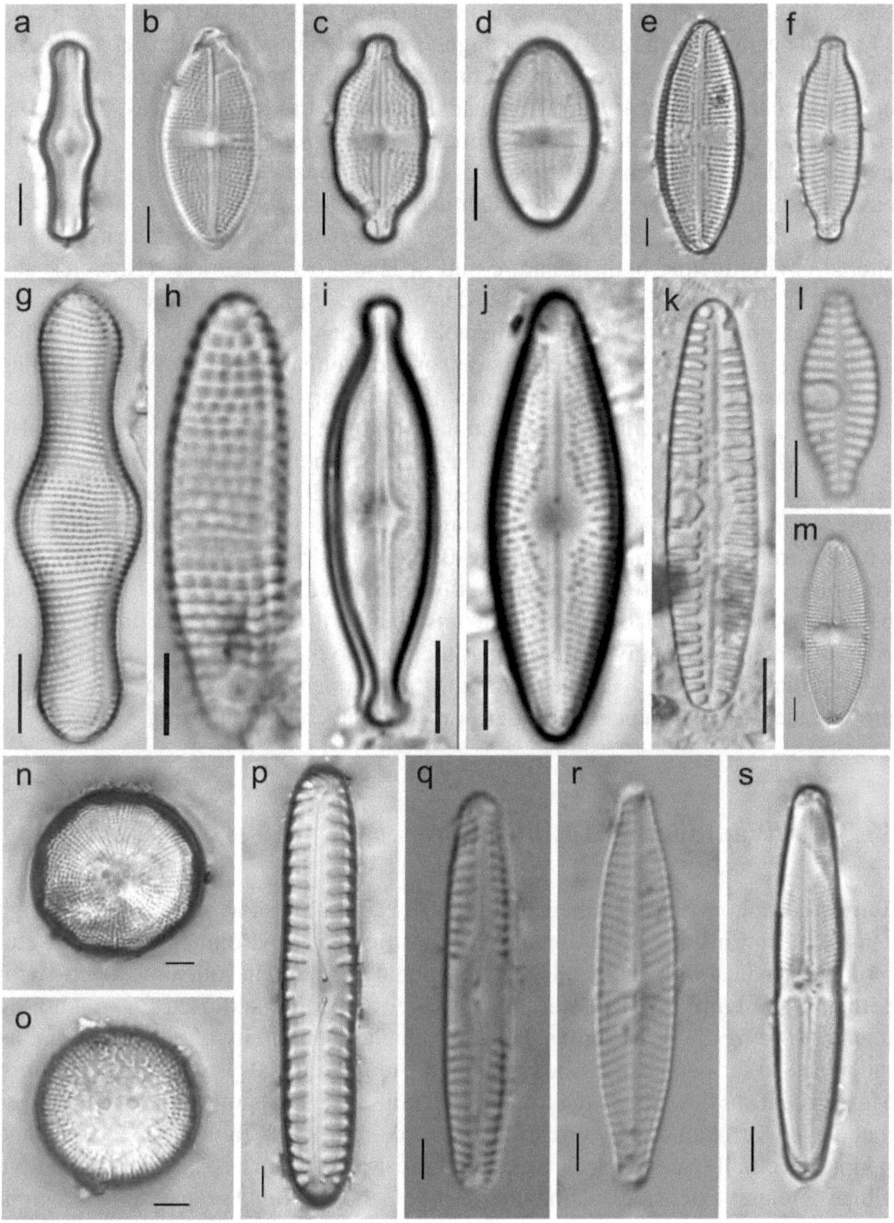

Fig. 17.2 Common terrestrial diatoms of the eastern Himalayas. (**a**) *Humidophila* sp.; (**b**) *Luticola goeppertiana*; (**c**) *L. muticoides*; (**d**) *L. nipkowii*; (**e**) *L. plausibilis*; (**f**) *Navicula elginensis*; (**g**) *Achnanthes inflata*; (**h**) *Achnanthes brevipes* var. *intermedia*; (**i**) *Brachysira microcephala*; (**j**) *Diadesmis confervacea*; (**k**) *Planothidium* sp.; (**l**) *Planothidium amphibium*; (**m**) *Psammothidium* cf. *acidoclinatum*; (**n**) *Orthoseira dendrophila*; (**o**) *O. roeseana*; (**p**) *Pinnularia borealis*; (**q**) *P. conica*; (**r**) *Navicula veneta*; (**s**) *Stauroneis* sp. (scale bar: a–s = 10 μm)

nutrients and moisture. It is a consequence of repeated vegetative reproduction with a gradual decrease in the size of the daughter cells (Drebes 1977). Earlier studies on the diatom size reduction from aquatic habitats revealed many environmental stressors (Cantonati et al. 2014; Barral-Fraga et al. 2016; Pandey and Bergey 2016). Das et al. (2020) identified *Eunotia paludosa*, *Eunotia bigibba,* and *Hantzschia amphioxys* as biological indicators for Neora Valley National Park, Eastern Himalayas, based on their size reductions in warmer and drier climates.

There are many historical instances of diatoms showing valve deformations in altered environments, both in nature and in laboratory conditions (Kociolek and Stoermer 2010; Andrejić et al. 2019). Many species gain or lose different morphological features at extreme environmental conditions (Villareal and Fryxell 1983; Tuchman et al. 1984). Some diatoms responded with strong heterovalvity, leading to the formation of Janus cells (Stoermer 1967; McBride and Edgar 1998; Thomas et al. 2009; Zhu et al. 2017). Apart from morphological changes, some diatoms also respond with irregular macromolecular content under changing environments, termed as "*ecophenotypic plasticity*" (Sackett et al. 2013). Innenschalen is a phenomenon in pennate diatoms, where a single valve or an entire frustule is developed within a frustule (Krammer and Lange-Bertalot 1991; Kociolek and Lowe 1993). Teratology or valve deformation is usually considered an indicator of heavy metal pollution, sometimes also caused due to unstable environmental conditions (Falasco et al. 2009). It is strongly believed that the phenotypic plasticity at the individual level of diatoms, along with the community level interpretations, can cumulatively provide ideal diatom indices for long-term environmental monitoring.

17.3 Advantages and Limitations

Terrestrial diatoms, due to their ubiquitous occurrence, short life cycle, and high sensitivity toward any alterations in the substrate and environment, can be useful biological indicators, specifically any nutritional alteration in soil, atmospheric temperature, and humidity. There are many indices prepared and used for monitoring the health of freshwater habitats. However, due to the scarcity of adequate data, no index has so far been established to monitor terrestrial ecosystems using diatoms. Many recent studies, in this regard, have identified the key biotic and abiotic factors that played a crucial role in terrestrial diatom composition (Table 17.1). Habitat and region specificity in the occurrence of terrestrial diatoms is another limitation, which demands extensive documentation across the globe and preparation of region-specific indices.

Table 17.1 Earlier studies on terrestrial diatoms across the globe and identification of factors influencing their community structure

Place	Type of terrestrial diatom	Factors influencing community structure	References
Edge Island, Svalbard	Aerial diatom (moss-associated)	Moisture	Beyens and De Bock (1989)
Zackenberg, Northeast Greenland	Soil diatom	Organic content, pH, permafrost depth, and moisture	van Kerckvoorde et al. (2000a)
Zackenberg, Northeast Greenland	Aerial diatom (moss-associated)	pH and moisture	van Kerckvoorde et al. (2000b)
Crozet Archipelago, sub-Antarctic	Soil diatom	Moisture and phosphate concentration	Van de Vijver et al. (2002)
Cambridge Bay area, Victoria Island, Canada	Aerial diatom (moss-associated)	pH and conductivity	Van de Vijver et al. (2003)
French Iles Kerguelen, sub-Antarctic	Aerial diatom (moss-associated)	Altitude	Gremmen et al. (2007)
Changbai Mountain, Northeast China	Aerial diatom (moss-associated)	Vegetation type, height above water table, pH, and moisture	Chen et al. (2012)
Jeseniky and Jizerské hory Mountains, Czech Republic	Aerial diatom (bryophyte-associated)	Conductivity, pH, and height above the water table	Poulíčková et al. (2013)
Livingston Island and James Ross Island, Maritime Antarctica	Aerial diatom (moss-associated)	Type of habitat, moisture, and biotic influences (salinity and nutrients)	Kopalová et al. (2014)
Attert River Basin, Luxembourg	Soil diatom	Seasonality, soil pH, and land use	Antonelli et al. (2017)
Windmill Islands, East Antarctica	Aerial diatom (moss- and lichen-associated)	Substrate vegetation type	Bishop et al. (2020)
Neora Valley National Park, Eastern Himalayas, India	Aerial diatom (moss-associated)	pH, conductivity, and relative humidity	Das et al. (2020)
Central-west China	Soil diatom	Magnesia, total phosphorus, organic phosphorus, and conductivity	Zhang et al. (2020)
Ile Amsterdam, sub-Antarctic	Aerial diatom (moss-associated)	Elevation, conductance, pH, and moisture	Chattová et al. (2021)
Luxembourg, Poland, Zackenberg (Greenland), Crozet Archipelago (sub-Antarctic)	Soil diatom	Anthropic disturbance	Foets et al. (2021)
Marrakesh city, Morocco	Soil diatom	Total organic carbon (TOC), phosphate, and nitrate content	Minaoui et al. (2021)

(continued)

Table 17.1 (continued)

Place	Type of terrestrial diatom	Factors influencing community structure	References
Argentine Islands, Maritime Antarctica	Aerial diatom (moss-associated)	Moisture	Chattová et al. (2022b)
James Ross Island, maritime Antarctica	Soil diatom	Vegetation coverage, moisture, conductivity, pH, and nutrient concentration	Chattová et al. (2022a, b)
Blue mountains, Mizoram, India	Aerial diatom (bryophyte-associated)	Altitude	Radhakrishnan et al. (2022)
Campbell Island, sub-Antarctic	Aerial diatom (bryophyte-associated)	Moisture and habitat type	Goeyers et al. (2022)
Darss-Zingst Peninsula, Germany	Aerial diatom (sand dunes)	Desiccation and temperature	Glaser et al. (2023)
Kalaallit Nunaat, Greenland	Aerial diatom (moss-associated)	Moisture, pH, conductivity, and temperature	Goeyers et al. (2024)
Zgierz, Poland	Aerial diatom (moss-associated)	Humidity	Szczepocka et al. (2024)

17.4 Conclusions

In the prevailing climate change scenario, a large sector of biodiversity is at stake, as the terrestrial environment houses more than 80% of the global biota. Mitigation toward this environmental concern needs extensive scientific scrutiny toward the identification of potential biological indicators. In this context, terrestrial diatoms are ideal because of their sensitivity to environmental alterations, both distributional as well as through morphological plasticity. The future research must head in the direction of the accumulation of information on region-specific diversity and autecology of diatoms. There is an urgent need for the preparation of different diatom indices for long-term monitoring of various ecosystems.

Acknowledgments The authors are thankful to Dr. A.A. Mao, Director, Botanical Survey of India, and Dr. Lal Ji Singh, Head of BSI-Andaman and Nicobar Regional Centre, for providing necessary support and facilities.

Declarations Informed Consent Statement Not applicable.

Data Availability Statement Not applicable.

Ethics Approval Not applicable.

Conflict of Interest There is no conflict of interest among the authors.

Funding There is no funding source for the presented study.

References

Andrejić JZ, Spaulding SA, Manoylov KM, Edlund MB (2019) Phenotypic plasticity in diatoms: Janus cells in four *Gomphonema* taxa. Diat Res 33(4):453–470. https://doi.org/10.1080/0269249X.2019.1572652

Antonelli M, Wetzel CE, Ector L, Teuling AJ, Pfister L (2017) On the potential for terrestrial diatom communities and diatom indices to identify anthropic disturbance in soils. Ecol Indic 75:73–81. https://doi.org/10.1016/j.ecolind.2016.12.003

Araujo CR, Wyatt KH, Rober AR (2025) For richer or poorer: diatoms as indicators of biological condition across a gradient of boreal fen types. Hydrobiologia 852:575–592. https://doi.org/10.1007/s10750-024-05581-8

Bardgett R (2005) The biology of soil – a community and ecosystems approach. Oxford University Press, Oxford

Barragán C, Wetzel CE, Ector L (2018) A standard method for the routine sampling of terrestrial diatom communities for soil quality assessment. J Appl Phycol 30:1095–1113. https://doi.org/10.1007/s10811-017-1336-7

Barral-Fraga L, Morin S, Rovira MDM, Urrea G, Magellan K, Guasch H (2016) Short-term arsenic exposure reduces diatom cell size in biofilm communities. Environ Sci Pollut Res 23:4257–4270. https://doi.org/10.1007/s11356-015-4894-8

Belnap J, Weber B, Büdel B (2016) Biological soil crusts as an organizing principle in drylands. In: Weber B, Büdel B, Belnap J (eds) Biological soil crusts: an organizing principle in drylands. Ecological Studies, vol 226. Springer, Cham, pp 3–13. https://doi.org/10.1007/978-3-319-30214-0_1

Bérard A, Rimet F, Capowiez Y, Leboulanger C (2004) Procedures for determining the pesticide sensitivity of indigenous soil algae: a possible bioindicator of soil contamination? Arch Environ Contam Toxicol 46:24–31. https://doi.org/10.1007/s00244-003-2147-1

Beyens L, De Bock P (1989) Moss dwelling diatom assemblages from Edgeøya (Svalbard). Polar Biol 9:423–430. https://doi.org/10.1007/BF00443228

Bhakta S, Pattanaik L, Dutta P, Sahu E, Bastia AK (2014) Diversity of corticolous algae from Similipal biosphere reserve, Mayurbhanj, Odisha. Phykos 44(1):9–16

Binoy TT, Ray JG (2016) Diversity and ecology of diatoms in oxic dystrustepts soils under three different vegetation of the Western Ghats, Kerala, South India. J Ecol 111:465–475

Bishop JM, Wasley J, Waterman MJ, Kohler TJ, Van de Vijver B, Robinson SA, Kopalová K (2020) Diatom communities differ among Antarctic moss and lichen vegetation types. Antarctic Sci 33(2):118–132. https://doi.org/10.1017/S0954102020000620

Borcard D, Legendre P, Drapeau P (1992) Partialling out the spatial component of ecological variation. Ecology 73:1045–1055. https://doi.org/10.2307/1940179

Both C, Bouwhuis S, Lessells CM, Visser ME (2006) Climate change and population declines in a long-distance migratory bird. Nature 441(4):81–83. https://doi.org/10.1038/nature04539

Bothwell ML, Sherbot D, Roberge AC, Daley RJ (1993) Influence of natural ultraviolet radiation on lotic periphytic diatom community growth, biomass accrual, and species composition: short-term versus long-term effects. J Phycol 29:24–35. https://doi.org/10.1111/j.1529-8817.1993.tb00276.x

Cantonati M, Angeli N, Virtanea L, Wojtal AZ, Gabriell J, Falasco E, Lavoie I, Morin S, Marchetto A, Fortin C, Smirnova S (2014) *Achnanthidium minutissimum* (Bacillariophyta) valve deformities as indicators of metal enrichment in diverse widely-distributed freshwater habitats. Sci Total Environ 475:201–215. https://doi.org/10.1016/j.scitotenv.2013.10.018

Carballeira R, Pontevedra-Pombal X (2020) Diatoms in paleoenvironmental studies of peatlands. Quaternary 3(2):10. https://doi.org/10.3390/quat3020010

Chattová B, Lebouvier M, Syrovátka V, Van de Vijver B (2021) Moss-inhabiting diatom communities from Ile Amsterdam (TAAF, southern Indian Ocean). Plant Ecol Evol 154:63–79. https://doi.org/10.5091/plecevo.2021.1767

Chattová B, Cahová T, Pinseel E, Kopalová K, Kohler TJ, Hrbáček F, Van de Vijver B, Nývlt D (2022a) Diversity, ecology and community structure of the terrestrial diatom flora from Ulu Peninsula (James Ross Island, NE Antarctic Peninsula). Polar Biol 45:873–894. https://doi.org/10.1007/s00300-022-03038-z

Chattová B, Zotov A, Parnikoza I (2022b) Moss inhabiting diatoms of Galindez Island, Argentine Islands (the maritime Antarctica) exhibit low diversity and pronounced differentiation. Czech Polar Rep 12:60–77. https://doi.org/10.5817/CPR2022-1-5

Chen X, Bu Z, Yang X, Wang S (2012) Epiphytic diatoms and their relation to moisture and moss composition in two montane mires, Northeast China. Fundam Appl Limnol 181:197–206. https://doi.org/10.1127/1863-9135/2012/0369

Das SK, Rawat DS, Dash SS, Banerjee A, Sinha BK, Singh P (2020) Moss-inhabiting diatoms as ecological indicators in Neora Valley National Park (Eastern Himalaya), India. Trop Ecol 61:226–237. https://doi.org/10.1007/s42965-020-00083-9

Dorokhova MF (2007) Diatoms as indicators of soil conditions in oil production regions. Oceanol Hydrobiol Stud 36(suppl 1):129–135

Drebes G (1977) Sexuality. In: Warner D (ed) The biology of diatoms. Botanical monograph 13. Blackwell Science Publications, Oxford, pp 250–283

Ettl H, Gärtner G (2014) Syllabus der Boden-, Luft-, und Flechtenalgen, 2, Auflage edn. Springer Spektrum, Berlin

Falasco E, Bona F, Badino G, Hoffmann L, Ector L (2009) Diatom teratological forms and environmental alterations: a review. Hydrobiologia 623:1–35. https://doi.org/10.1007/s10750-008-9687-3

Farqan M, Xiang L, Jiang L, Wang W, Zheng M, Mu X, Liu X, Song X, Zhu Z, Ayyamperumal R, Huang X (2025) Diatom assemblages indicated ecological shifts over the past two centuries in Bosten Lake, nothwest China: natural and anthropogenic forcings. Quart Int 716:109629. https://doi.org/10.1016/j.quaint.2024.109629

Fazlutdinova AI, Sukhanova NV (2014) Composition of soil diatoms in zones of impact from oil production complexes. Russ J Ecol 45:188–193. https://doi.org/10.1134/S1067413614030035

Finkelstein S, Swaddle J (2012) Diatoms as bioindicators of environmental change: a review of the methods and applications. Paleogeo Paleoclim Paleoecol 315:1–11

Foets J, Stanek-Tarkowska J, Teuling AJ, Van de Vijver B, Wetzel CE, Pfister L (2021) Autecology of terrestrial diatoms under anthropic disturbance and across climate zones. Ecol Indic 122:107248. https://doi.org/10.1016/j.ecolind.2020.107248

Furey PC, Lowe RL, Johansen JR (2009) Teratology in *Eunotia* taxa in the Great Smoky Mountains National Park and description of *Eunotia macroglossa* sp. nov. Diat Res 24(2):273–290. https://doi.org/10.1080/0269249X.2009.9705802

Glaser K, Kammann S, Plag N, Dressler M (2023) Ecophysiological performance of terrestrial diatoms isolated from biocrusts of coastal sand dunes. Frontiers Microbiol 14:1279151. https://doi.org/10.3389/fmicb.2023.1279151

Goeyers C, Vitt DH, Van de Vijver B (2022) Taxonomic and biogeographical analysis of diatom assemblages from historic bryophyte samples from Campbell Island (sub-Antarctic). Plant Ecol Evol 155(1):107–122. https://doi.org/10.5091/plecevo.84543

Goeyers C, Van de Vijver B, Kohler TJ, Verleyen E, Tytgat B, Gradstein SR, Sabbe K (2024) Local and regional, not latitudinal, variation in microclimate and bedrock shapes moss-associated diatom communities in Greenland. J Biogeogr 52(2):479–494. https://doi.org/10.1111/jbi.15050

Gremmen NJM, Van de Vijver B, Frenot Y, Lebouvier M (2007) Distribution of moss-inhabiting diatoms along and altitudinal gradient at sub-Antarctic Îles Kerguelen. Antarctic Sci 19(1):17–24. https://doi.org/10.1017/S0954102007000041

Hahn A, Neuhaus W (1997) Boden-diatomeen einer landwirtschaftlichennutzfläche bei potsdam, deutschland. Nova Hedwig 65:285–298

Hájková P, Bojková J, Fránková M, Opravilová V, Hájek M, Kintrová K, Horsák M (2011) Disentangling the effects of water chemistry and substratum structure on moss-dwelling unicellular and multicellular micro-organisms in spring-fens. J Limnol 70:54–64. https://doi.org/10.3274/JL11-70-S1-04

Hall RI, Smol JP (1999) Diatoms as indicators of lake eutrophication. In: Stoermer EF, Smol JP (eds) The diatoms: applications for the environmental and earth sciences. Cambridge University Press, Cambridge, pp 128–168

Hayek JMW, Hulbary RI (1956) A survey of soil diatoms. Proc Iowa Acad Sci 63:327–338

Heger TJ, Straub F, Mitchell EAD (2012) Impact of farming practices on soil diatoms and testate amoebae: a pilot study in the DOK-trial at Therwil, Switzerland. Eur J Soil Biol 49:31–36. https://doi.org/10.1016/j.ejsobi.2011.08.007

IPCC (2021) Climate change 2021: the physical science basis. contribution of working group I to the sixth assessment report. Cambridge University Press

Johansen JR (2004) Diatoms of aerial habitats. In: Stoermer EF, Smol JP (eds) The Diatoms – Applications for the environmental and earth sciences. Cambridge University Press, Cambridge, pp 264–273

Kociolek JP, Lowe RL (1993) Taxonomy and ultrastructure of *Meridion lineare* D.M. Williams (Bacillariophyceae) from North America. Nova Hedwigia 57:381–391

Kociolek JP, Stoermer EF (2010) Variation and polymorphism in diatoms: the triple helix of development, genetics and environment. a review of the literature. VIE MILIEU 60:75–87

Kooistra WHCF, Medlin LK (1996) Evolution of the diatoms (Bacillariophyta): IV. A reconstruction of their age from small subunit rRNA coding regions and the fossil record. Mol Phylogen Evol 6(3):391–407. https://doi.org/10.1006/mpev.1996.0088

Kooistra WHCF, Gersonde R, Medlin LK, Mann DG (2007) The origin and evolution of the diatoms: their adaptation to a planktonic existence. In: Falkowski PG, Knoll AH (eds) Evolution of primary producers in the sea. Elsevier, Boston, pp 207–249. https://doi.org/10.1016/B978-012370518-1/50012-6

Kopalová K, Ochyra R, Nedbalová L, Van de Vijver B (2014) Moss-inhabiting diatoms from two contrasting Maritime Antarctic islands. Plant Ecol Evol 147:67–84. https://doi.org/10.5091/plecevo.2014.896

Krammer K, Lange-Bertalot H (1991) Süsswasserflora von Mitteleuropa. Bacillariophyceae. Part 2/3. Centrales, Fragilariaceae, Eunotiaceae. Gustav Fisher, Stuttgart, p 576

Krammer K, Lange-Bertalot H (2000) Bacillariophyceae. In: Gerloff H, Heynig JH, Mollenhauer D (eds) Subwasserflora von Mitteleuropa, vol Band 2. Ettl., Spektrum Akademischer Verlag, Heidelberg/Berlin

Lavoie I, Vincent WF, Pienitz R, Painchaud J (2004) Benthic algae as bioindicators of agricultural pollution in the streams and rivers of southern Québec (Canada). Aquat. Ecosyst. Health Manage. 7(1):43–58. https://doi.org/10.1080/14634980490281236

Lin CH, Chou T, Wu J (2013) Biodiversity of soil algae in the farm lands of mid-Taiwan. Bot Stud 54:41. https://doi.org/10.1186/1999-3110-54-41

Lowe RL (2011) The importance of scale in understanding the natural history of diatom communities. In: Kociolek P (ed) Seckbach J. The Diatom World, Springer, pp 293–311

Lund JWG (1945) Observations on soil algae. I. The ecology, size and taxonomy of British soil diatoms. Part I. New Phytol 44:196–219. https://doi.org/10.1111/j.1469-8137.1945.tb05033.x

Lund JWG (1946) Observations on soil algae. I. The ecology, size and taxonomy of British soil diatoms. Part II. New Phytol 45:56–110

Malhi Y, Franklin J, Seddon N, Solan M, Turner MG, Field CB, Knowlton N (2020) Climate change and ecosystems: threats, opportunities and solutions. Phil. Trans. R. Soc. B 375:20190104. https://doi.org/10.1098/rstb.2019.0104

Maltsev YI, Maltseva IA, Solonenko AN, Bren AG (2017) Use of soil biota in the assessment of the ecological potential of urban soils. Biosyst Divers 25(4):257–262. https://doi.org/10.15421/011739

McBride SA, Edgar RK (1998) Janus cells unveiled: frustular morphometric variability in *Gomphonema angustatum*. Diatom Res 13:293–310. https://doi.org/10.1080/0269249X.1998.9705452

Menzel A, Sparks TH, Estrella N et al (2006) European phenological response to climate change matches the warming pattern. Global Change Biol 12(10):1969–1976. https://doi.org/10.1111/j.1365-2486.2006.01193.x

Minaoui F, Hakkoum Z, Douma M, Mouhri K, Loudiki M (2021) Diatom communities as bioindicators of human disturbances on suburban soil quality in arid Marrakesh area (Morocco). Water Air Soil Pollut 232:146. https://doi.org/10.1007/s11270-021-05094-3

Moravcová A, Beyens L, Van de Vijver B (2009) Diatom communities in soils influenced by the wandering albatross (*Diomedea exulans*). Polar Biol 33:241–255. https://doi.org/10.1007/s00300-009-0700-8

Naz S, Grover S, Khan AS, Verma J, Srivastava P (2023) Terrestrial diatoms and their potential for ecological monitoring. In: Srivastava P, Khan AS, Verma J, Dhyani S (eds) Insights into the world of diatoms: from essentials to applications, plant life and environmental dynamics. Springer, pp 131–141. https://doi.org/10.1007/978-981-19-5920-2_8

Olkova AS, Zimonina NM, Lyalina EI, Bobretsova VR (2017) Diagnostics of local pollution of urbanozem in the areas of petrol stations. Health Environ Res Online 1:579–222

Pandey LK, Bergey EA (2016) Exploring the status of motility, lipid bodies, deformities and size reduction in periphytic diatom community from Chronically metal (Cu, Zn) polluted water bodies as a biomonitoring tool. Sci Total Environ 550:372–381. https://doi.org/10.1016/j.scitotenv.2015.11.151

Parmesan C, Yohe G (2003) A globally coherent fingerprint of climate change impacts across natural systems. Nature 421(6918):37–42. https://doi.org/10.1038/nature01286

Parmesan C, Root TL, Willig MR (2000) Impacts of extreme weather and climate on terrestrial biota. Bull American Meteor Soc 81(3):443–450. https://doi.org/10.1175/1520-0477(2000)081<0443:IOEWAC>2.3.CO;2

Passy S (2007) Diatom ecological guilds display distinct and predictable behaviour along nutrient and disturbance gradients in running waters. Aquat Bot 86:171–178. https://doi.org/10.1016/j.aquabot.2006.09.018

Peszek L, Zgrundo A, Noga T, Kochman-Kędziora N, Poradowska A, Rybak M, Puchalski C, Lee J (2021) The influence of drought on diatom assemblages in a temperate climate zone: a case study from the Carpathian Mountains, Poland. Ecol Indic 125:107579. https://doi.org/10.1016/j.ecolind.2021.107579

Petersen JB (1915) Studier over danske aërofile alger. Kongel Danske Vidensk Selsk Skr Naturvidensk Math Afd 12:269–380

Petersen JB (1928) The aerial algae of Iceland. The botany of Iceland, In, pp 325–447

Petersen JB (1935) Studies on the biology and taxonomy of soil algae. C.A. Reitzels Forlag, Copenhagen

Poulíčková A, Bergová K, Hnilica R, Neustupa J (2013) Epibryic diatoms from ombrotrophic mires: diversity, gradients and indicating options. Nova Hedwigia 96:351–365. https://doi.org/10.1127/0029-5035/2013/0091

Radhakrishnan C, Yogeshwaran M, Karthick B (2022) Hanging in the air: tree moss diatoms from Indo-Burma biodiversity hot spot of India. Aerobiologica 38:557–566. https://doi.org/10.1007/s10453-022-09766-3

Raven JA, Waite AM (2004) The evolution of silicification in diatoms: inescapable sinking and sinking as escape? New Phytol 62(1):45–61. https://doi.org/10.1111/j.1469-8137.2004.01022.x

Round FE (1957) The diatom community of some bryophyta growing on sandstone. Bot J Linn Soc 55:657–661. https://doi.org/10.1111/j.1095-8339.1957.tb00028.x

Round FE, Crawford RM, Mann DG (1990) The diatoms biology and morphology of the genera. Cambridge University Press, Cambridge, pp 1–747

Rybak M, Noga T, Zubel R (2018) The aerophytic diatom assemblages developed on mosses covering the bark of Populus alba L. J Ecol Eng 19(6):113–123. https://doi.org/10.12911/22998993/92673

Rybak M, Czarnota P, Noga T (2023) Study of terrestrial diatoms in corticolous assemblages from deciduous trees in Central Europe with descriptions of two new *Luticola* D.G.Mann taxa. PhytoKeys 221:1–40. https://doi.org/10.3897/phytokeys.221.95248

Sackett O, Petrou K, Reedy B, de Grazia A, Hill R, Doblin M, Beardall J, Ralph P, Heraud P (2013) Phenotypic plasticity of Southern Ocean diatoms: key to success in the sea ice habitat? PLoS One 8(11):1–12, e81185. https://doi.org/10.1371/journal.pone.0081185

Schneider SH, Root TL (1996) Ecological implications of climate change will include surprises. Biodivers Conserv 5:1109–1119. https://doi.org/10.1007/BF00052720

Serieyssol KK (2012) Diatoms of Europe: diatoms of the European inland waters and comparable habitats. Diat Res 27(2):103. https://doi.org/10.1080/0269249X.2012.690210

Shein EV, Kharitonova GV, Milanovsky EY (2016) Aggregation of natural disperse formations: value of organic matter, soluble salts and diatoms. Biogeosyst Tech 1:77–86. https://doi.org/10.13187/bgt.2016.7.77

Sims PA, Mann DG, Medlin LK (2006) Evolution of the diatoms: insights from fossil, biological and molecular data. Phycologia 45(4):361–402. https://doi.org/10.2216/05-22.1

Smol JP, Douglas MS (2007) From controversy to consensus: making the case for recent climate change in the Arctic using lake sediments. Front Ecol Environ 5(9):466–474. https://doi.org/10.1890/060162

Soininen J (2007) Environmental and spatial control of freshwater diatoms – a review. Diatom Res. 22(2):473–490. https://doi.org/10.1080/0269249X.2007.9705724

Sorhannus U (2007) A nuclear-encoded small-subunit ribosomal RNA timescale for diatom evolution. Mar Micropaleontol 65(1):1–12. https://doi.org/10.1016/j.marmicro.2007.05.002

Souffreau C, Vanormelingen P, Verleyen E, Sabbe K, Vyverman W (2010) Tolerance of benthic diatoms from temperate aquatic and terrestrial habitats to experimental desiccation and temperature stress. Phycologia 49(4):309–324. https://doi.org/10.2216/09-30.1

Stanek-Tarkowska J, Noga T (2012) Diversity of diatoms (Bacillariophyceae) in the soil under traditional tillage and reduced tillage. Inź Ekol 30:287–296

Stanek-Tarkowska J, Noga T, Kochman-Kędziora N, Peszek Ł, Pajączek A, Kozak E (2015) The diversity of diatom assemblages developed on fallow soilin Pogórska Wola near Tarnów (southern Poland). Acta Agrobot 68(1):33–42. https://doi.org/10.5586/aa.2015.011

Stanish LF, Bagshaw EA, McKnight DM, Fountain AG, Tranter M (2013) Environmental factors imfluential in Antarctica cryoconite holes. Environ Res Lett 8(4):045006. https://doi.org/10.1088/1748-9326/8/4/045006

Stoermer EF (1967) Polymorphism in *Mastogloia*. J Phycol 3:73–77. https://doi.org/10.1111/j.1529-8817.1967.tb04633.x

Szczepocka E, Nowicka-Krawczyk P, Wolski GJ (2024) Terrestrial mosses as a substrate and potential host for cyanobacteria, green algae and diatoms. J Plant Res 137:847–861. https://doi.org/10.1007/s10265-024-01551-2

Thomas EW, Kociolek JP, Lowe RL, Johansen JR (2009) Taxonomy, ultrastructure and distribution of gomphonemoid diatoms (Bacillariophyceae) from Great Smoky Mountains National Park (USA). Diatom taxonomy, ultrastructure and ecology: modern methods and timeless questions. A tribute to Eugene F. Stoermer. Nova Hedwigia 135:201–237

Tuchman ML, Theriot E, Stoermer EF (1984) Effects of low-level salinity concentrations on the growth of *Cyclotella meneghiniana* Kütz. (Bacillariophyta). Arch Protistenk 128:319–326. https://doi.org/10.1016/S0003-9365(84)80003-2

Tylianakis JM, Didham RK, Bascompte J, Wardle DA (2008) Global change and species interactions in terrestrial ecosystems. Ecol Lett 11:1351–1363. https://doi.org/10.1111/j.1461-0248.2008.01250.x

Vacht P, Puusepp L, Koff T, Reitalu T (2014) Variability of riparian soil diatom communities and their potential as indicators of anthropogenic disturbances. Est J Ecol 63:168–184. https://doi.org/10.3176/eco.2014.3.04

Van Dam H, Martens A, Sinkeldam J (1994) A coded checklist and ecological indicator values of freshwater diatoms from the Netherlands. Netherlands J Aquat Ecol 28(1):117–133. https://doi.org/10.1007/BF02334251

Van de Vijver B, Beyens L (1997) The epiphytic diatom flora of mosses from Stromness Bay area, South Georgia. Polar Biol 17:492–501. https://doi.org/10.1007/s003000050148

Van de Vijver B, Kopalova K (2008) *Orthoseira gremmenii* sp. nov., a new aerophilic diatom from Gough Island (southern Atlantic Ocean). Cryptogam, Algol 29(2):105

Van de Vijver B, Ledeganck P, Beyens L (2002) Soil diatom communities from Ile de la Possession (Crozet, sub-Antarctica). Polar Biol 25:721–729. https://doi.org/10.1007/s00300-002-0392-9

Van de Vijver B, van Kerckvoorde A, Beyens L (2003) Freshwater and terrestrial moss diatom assemblages of the Cambridge Bay area, Victoria Island (Nunavut, Canda). Nova Hedwig 76:225–243. https://doi.org/10.1127/0029-5035/2003/0076-0225

van Kerckvoorde A, Trappeniers K, Nijs I, Beyens L (2000a) Terrestrial soil diatom assemblages from different vegetation types in Zackenberg (Northeast Greenland). Polar Biol 23:392–400. https://doi.org/10.1007/s003000050460

van Kerckvoorde A, Trappeniers K, Nijs I, Beyens L (2000b) The epiphytic diatom assemblages from terrestrial mosses in Zackenberg (Northeast Greenland). Syst Geogr Pl 70:301–314. https://doi.org/10.2307/3668649

Villareal TA, Fryxell GA (1983) Temperature effects on the valve structure of the bipolar diatoms *Thalassiosira antarctica* and *Porosira glacialis*. Polar Biol 2:163–169. https://doi.org/10.1007/BF00448966

Vitug LVD (2022) Diatom plasticity: trends, issues, and applications on modern and classical taxonomy, eco-evolutionary dynamics, and climate change. In: Annenkov V, Seckback J, Gordon R (eds.) Diatom morphogenesis, Scrivener Publishing LLC, pp. 261–268. doi:https://doi.org/10.1002/9781119488170.ch11

Wardle DA (2002) Communities and ecosystems: linking the above-ground and below ground components. Princeton University Press, Princeton

Zancan S, Trevisan R, Paoletti MG (2006) Soil algae composition under different agro-ecosystems in North-Eastern Italy. Agric Ecosyst Environ 112:1–12. https://doi.org/10.1016/j.agee.2005.06.018

Zhang Y, Ouyang S, Nie L, Chen X (2020) Soil diatom communities and their relation to environmental factors in three types of soil from four cities in central-west China. Eur J Soil Biol 98:103175. https://doi.org/10.1016/j.ejsobi.2020.103175

Zhu Z, Qu P, Fu F, Tennenbaum N, Tatters AO, Hutchins DA (2017) Understanding the blob bloom: Warming increases toxicity and abundance of the harmful bloom diatom *Pseudonitzschia* in California coastal waters. Harmful Algae 67:36–43. https://doi.org/10.1016/j.hal.2017.06.004

Chapter 18
Tiny Survivors in a Changing Climate: Modelling the Future Distribution of *Vallonia* (Mollusca, Gastropoda, Valloniidae) Snails in India

Sheikh Sajan, Basudev Tripathy, Krishnendu Mondal, and Paromit Chatterjee

Abstract The knowledge of accurate species distribution is a preliminary stage towards conservation in the current climate and land-use change scenario. It provides a preliminary understanding of the habitat and climate suitability of the species. However, in the modern scientific context, we can integrate various environmental, climatic, and anthropogenic factors with current species distribution data to accurately determine the suitable habitats for species occurrence and distribution. However, because of adequate data on species occurrences and distribution, the lesser invertebrates have attracted minimal attention on conservation (especially land snails) compared to vertebrates. The distribution of terrestrial molluscs, especially tiny snails such as *Vallonia* sp., *Alycaeus* sp., and *Diplomatina* sp., remains inadequately known due to their small size and the lack of sufficient ecological data. In this study, the possibility of the habitat of three species *Vallonia pulchella*, *Vallonia ladacensis*, and *Vallonia costata* are performed in relation to the climatic variable, human occurrence, and anthropogenic activities has been concluded by using maximum entropy modelling. The species distribution model results reveal a

S. Sajan (✉)
Zoological Survey of India, Kolkata, West Bengal, India

Gujarat Institute of Desert Ecology, Terrestrial Ecology Division, Bhuj, Kachchh, Gujarat, India

B. Tripathy
Zoological Survey of India, Western Regional Centre, Pune, Maharashtra, India

K. Mondal
Zoological Survey of India, Kolkata, West Bengal, India

Ministry of Environment, Forest and Climate Change, Government of India, New Delhi, India

P. Chatterjee
Zoological Survey of India, Kolkata, West Bengal, India

Habitat Lens Private Limited, Khardah, West Bengal, India

B. P. Barathan et al. (eds.), *Navigating Climate Change: Impacts on Biodiversity and Ecosystem Resilience*, Sustainable Development and Biodiversity 42,
https://doi.org/10.1007/978-981-95-0409-1_18

clear dependence of these species on climatic factors compared to other environmental variables. Notably, the endemic species *Vallonia ladacensis* shows a strong association with the mean diurnal temperature range. This chapter provides new insights into the distributional range and climatic requirements of these species, contributing to future climate change research, interspecific variation, and aiding in the formulation of effective management plans to limiting them before they become vermin.

Keywords Gastropod · Topography · Habitat · Distribution · *Vallonia* · Western Himalaya

18.1 Introduction

The maximum entropy modelling (MaxEnt) prediction probability on species distribution range based on geographical distribution data could give sufficient knowledge on the present and future distribution of land use and climate change (Pearson and Dawson 2003; Fortunato 2015). In the last decade, the use of species distribution modelling has been increasing subsequently to understand the requirement of environmental variables on the species distribution, occurrence and for the future prediction on habitat shifting, habitat shrinkage, extinction rate, and climatic viability in both flora and fauna (Ashraf et al. 2016; Merow et al. 2017; Rumi et al. 2017).

The *Vallonia* Risso (1826) is a type genus of the family Valloniidae of orthurethrous miniature air-breathing land snails, which are widely spread over the mountain ranges of Europe, North America, Australia, Northern Africa, Atlantic Islands, and the South Asian continent (Pilsbry 1948; Nevill 1878; Gude 1914; Sterki 1893; Rajagopal and Subba Rao 1972; Gerber 1996; Gerber and Bössneck 2009). In South Asia, it has predominantly been distributed along the western Himalaya of India, Nepal, East Tibet, Eastern China, and the mountain ranges of Kazakhstan (Gude 1914; Gerber and Bössneck 2009). The fossil records known date back to the Pleistocene, and apparently evolved in Mesozoic time, and have changed very little since the Eocene. In India, the genus *Vallonia* species is restricted to high altitudinal region of mountain range of Himalaya, mainly in Jammu and Kashmir, Punjab, Himachal Pradesh, and Uttarakhand state of western Himalayan landscape; among the genus, the species *V. costata* (Müller 1774), *V. ladacensis* (Nevill 1878), and *V. pulchella* (Müller 1774) are distributed throughout the valley. In contrast, *V. ladacensis* has been recorded from Sikkim of the central Himalaya (Gerber and Bössneck 2009). However, *V. miserrima* Gude, 1907 was described from the Anaimalai Hills of Southern India, considered as a junior synonym of *V. ladacensis* (Gude 1907, 1914; Raheem et al. 2014). In this current communication, we discuss the systematic revision of genus *Vallonia* and species distribution modelling (SDM), climatic requirements, current distributional range, and future scenario in India.

18.2 Material and Methods

18.2.1 Study Area

For the present study and this book chapter, the Western Himalayan region of India was considered, encompassing the Union Territories of Jammu and Kashmir and Ladakh, along with the states of Himachal Pradesh and Uttarakhand. This region represents a significant segment of the Indian Himalayas, characterized by complex topography, diverse microclimates, and rich ecological heterogeneity. Ranging from subtropical zones in the foothills to alpine and glacial environments at higher elevations, the Western Himalaya hosts a wide array of climatic conditions that influence species distribution. These varying climatic zones make the region an ideal landscape for examining the ecological preferences and climatic suitability of terrestrial molluscs such as *Vallonia pulchella*, *Vallonia ladacensis*, and *Vallonia costata*. The study area is also subject to increasing anthropogenic pressures, including land-use change, tourism, infrastructure development, and climate change, all of which pose potential threats to the native biodiversity. The selection of this region allows for a focused analysis of species–climate relationships under current conditions and projected future scenarios (Fig. 18.1).

18.2.2 Materials Examined and Data Collection

The molluscan materials used in the current manuscript are from the old collection in the National Zoological Collection of Zoological Survey of India (NZSI) and an extensive literature survey. The complete morphometric measurement and photographs have been extracted from Leica M205A multi-focus (Z-Stack) stereomicroscope in the Zoological Survey of India (Figs. 18.2 and 18.3). The species distribution data were obtained from the published paper and documented in grey literature, Nevill (1878), Gude (1914), Rajagopal and Subba Rao (1972), and Ramakrishna et al. (2010). The data on collection localities, species names, and geo coordinates were collected from 1878 to 2010 and used for species distribution and future prediction modelling.

18.2.3 Climatic Data

The various climatic variables, viz. minimum, maximum temperature, humidity, and monthly precipitation data of India, were obtained from the Indian Meteorological Department, Government of India. For the future distribution prediction, the data of annual mean temperature, mean diurnal range, temperature seasonality, mean temperature of wettest quarter, annual precipitation, the precipitation on driest month,

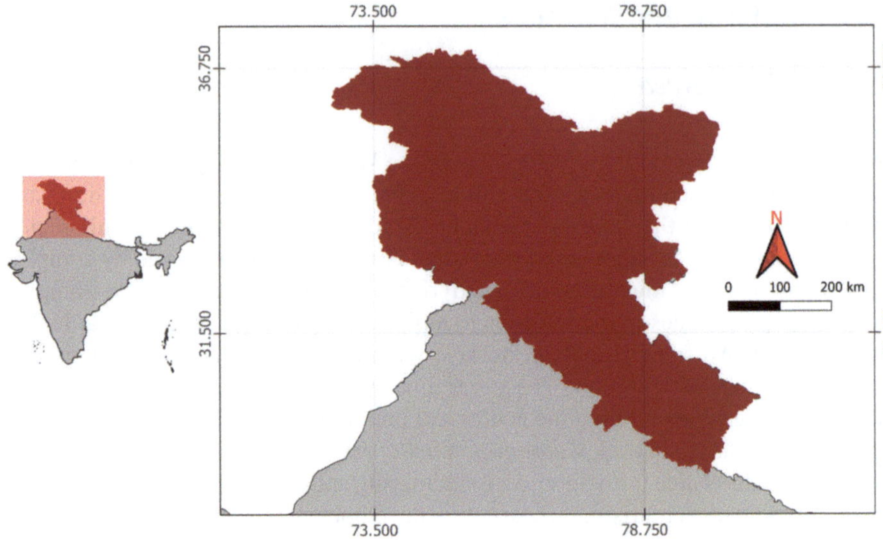

Fig. 18.1 Location map of the study area (Western Himalaya)

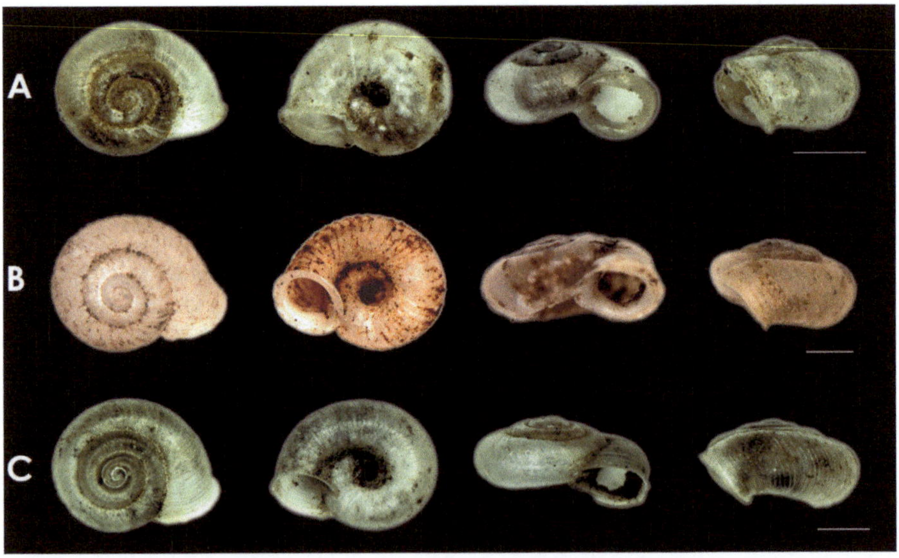

Fig. 18.2 Shells of *Vallonia*. (**a**) *Vallonia pulchella* (Müller 1774), (Scale 1 mm). (**b**) *Vallonia costata* (Müller 1774), (Scale 0.5 mm). (**c**) *Vallonia ladacensis* (Nevill 1878), (Scale: 1 mm)

Fig. 18.3 Key morphometric measurements of *Vallonia* used in this manuscript (see abbreviation in Table 18.2)

and precipitation seasonality were downloaded from WORLDCLIM (http://www.worldclim.org) 30 Arc second resolution. This climatic model was used for projecting species distribution and comparing the suitable change of habitat requirement of the *Vallonia* species about time and climate change (Table 18.1).

Table 18.1 List of variables used to find suitable habitat for Genus *Vallonia* in the western Himalaya

Variable	Variable type	Source
Habitat	LULC and vegetation pattern	Global Land Cover 2000
	Digital Elevation Model	United States Geological Survey
	Aspect	United States Geological Survey
	Slope	United States Geological Survey
	Hill shade	United States Geological Survey
	Human Population Density	NASA SEDAC project
Climatic	Annual Mean Temperature	WorldClim—Global Climate data
	Mean Diurnal Range	WorldClim—Global Climate data
	Temperature Seasonality	WorldClim—Global Climate data
	Mean Temperature of Wettest Quarter	WorldClim—Global Climate data
	Annual Precipitation	WorldClim—Global Climate data
	Precipitation of Driest Month	WorldClim—Global Climate data
	Precipitation Seasonality	WorldClim—Global Climate data

18.2.4 Habitat Variables and Data Source

To examine the relation between species and land use patterns, vegetation cover, human density, and different habitat variables were considered. The suitable habitat variables, viz. slope, aspect, hill shade, and the elevation data, were obtained from the US Geological Survey (https://www.usgs.gov/) to correlate the habitat requirement of *Vallonia* sp. in terms of distributional pattern. In addition to analysing the habitat changes, we downloaded the land use/land cover and vegetation data from Global Land Cover 2000 database (https://landcover.usgs.gov/glc/) with 30 m resolution. Moreover, the data on human population density of the western Himalayas were obtained from Socio-economic Data and Applications Center (NASA SEDAC) (http://sedac.ciesin.columbia.edu/) to calculate the human interference or involvement within the distribution of species (see Table 18.1).

18.2.5 Species Distribution Modelling

Species occurrence records were compiled from published literature, biodiversity databases such as GBIF, and validated field observations. Only precise location data were considered for modelling. To minimize spatial autocorrelation and overfitting—common issues in presence-only species distribution modelling—duplicate records and those located within 1 km^2 of each other were filtered out using a spatial thinning technique in ArcGIS. This ensured that the final dataset used for modelling represented spatially independent presence points for each species. Environmental variables were carefully selected to reflect a combination of bioclimatic, topographic, and anthropogenic influences on species distribution. A total of 19 bioclimatic variables were initially downloaded from the WorldClim v2.1 database, representing long-term average temperature and precipitation patterns from 1970 to 2000 at ~1 km^2 spatial resolution. To address multicollinearity among these variables, a Pearson correlation analysis was performed, and only non-correlated variables (correlation coefficient $r < |0.8|$) were retained for further analysis. In addition to climate data, topographic layers including elevation, slope, and aspect were derived from ASTER DEM. Anthropogenic influence was represented by human population density layer and land use/land cover data obtained from MODIS MCD12Q1 for the year 2020.

Species distribution models were developed using the Maximum Entropy (MaxEnt) algorithm (version 3.4.4), which is widely recognized for its effectiveness in predicting potential species distributions based on presence-only data. The models were built using a tenfold cross-validation method, with 75% of the data used for training and 25% for testing. Model performance was assessed using the Area Under the Receiver Operating Characteristic Curve (AUC) to evaluate predictive accuracy, and the True Skill Statistic (TSS) to measure the agreement between observed and predicted values. Additionally, jackknife analyses were conducted within MaxEnt to determine the relative importance of each environmental variable in predicting habitat suitability. Future climate projections were generated for the year 2050 using data from the CMIP5 climate model under the Representative Concentration Pathway (RCP) 4.5 scenario, which assumes stabilization of radiative forcing at 4.5 W/m^2 by the end of the century. The future bioclimatic variables were sourced from the WorldClim database, while topographic and anthropogenic layers were kept constant, allowing the models to isolate the effects of climate change on habitat suitability. The model outputs were generated as continuous probability maps, representing suitability values ranging from 0 (unsuitable) to 1 (highly suitable). These maps were classified into distinct suitability categories using natural breaks (Jenks method) for visualization and interpretation.

18.2.6 Morphological Analysis

To assess the morphological variation among the three *Vallonia* species, we employed a combination of descriptive, inferential, and multivariate statistical techniques. A total of nine morphometric traits were measured from shell specimens of *Vallonia costata*, *Vallonia pulchella*, and *Vallonia ladacensis*. These included whorl count, shell diameter (SD), apex height (APX), whorl measurements at two stages (Whorl_S1 and Whorl_S2), diameter of the whorl (DWL), aperture height (AH), aperture width (AW), and umbilicus width (UW). All measurements were first checked for consistency and completeness, and data were tabulated for statistical analysis.

Descriptive statistics were calculated for each species and trait, including mean, standard deviation, minimum, and maximum values. This provided an initial understanding of central tendencies and trait variability across species. To test whether the observed differences in morphometric traits were statistically significant among the species, we performed one-way Analysis of Variance (ANOVA) for each trait, with species as the categorical independent variable. Prior to ANOVA, the assumption of homogeneity of variances was tested using Levene's Test. Traits that passed this assumption were analysed using standard one-way ANOVA, and significance was evaluated at the 95% confidence level. Where p-values were less than 0.05, differences were considered statistically significant, and post-hoc Tukey tests were planned for pairwise comparisons if needed.

In addition to univariate analysis, we conducted Principal Component Analysis (PCA) to explore the overall morphometric variation and to visualize patterns of species separation in a reduced dimensional space. Prior to PCA, all morphometric variables were standardized (mean-centred and scaled to unit variance) to ensure comparability across traits with different units and ranges. PCA was carried out using the *prcomp()* function in R, which computes the principal components based on the covariance matrix of the standardized data. The variance explained by each principal component was examined through a scree plot, and the first two principal components, which together explained over 66% of the total variance, were used to generate a PCA scatter plot. This plot visualized individual specimens in the principal component space, with ellipses representing species-specific confidence areas, allowing a clear view of interspecific clustering. All statistical analyses and visualizations were performed using the R statistical programming environment, incorporating base functions along with the *ggplot2* and *ggpubr* packages for graphical output.

18.3 Results

18.3.1 Species Record and Distribution

A total of 42 occurrences of the *Vallonia* species were documented from the grey literature of three species found in the Indian Himalayan region. The *V. costata* (Müller 1774) limited to the Jammu & Kashmir state of India and *V. ladacensis* (Nevill 1878) has a broader distribution, being found throughout Jammu and Kashmir, Himachal Pradesh, and Uttarakhand state; the known localities are Kashmir, Mataian, Dras Valley, Leh, Ladak, Spiti, Punjab, and west Tibet. Whereas, *V. pulchella* (Müller 1774) is limited to southern Kashmir in India. All published records were used for distribution modelling and future prediction. However, the distribution of *V. miserrima* Gude, 1907 (=*V. ladacensis*), which is known only from a few localities of the Western Ghats, was excluded from the analysis.

18.3.2 Systematics Species Account

Family VALLONIIDAE Morse, 1864
Subfamily VALLONIINAE Morse, 1864
Genus *Vallonia* Risso, 1826
1826. *Vallonia* Risso, *Hist. Nat. Europemerid.* 4, 101.
Type species: *Helix pulchella* Müller, 1774 (type by monotypy) (=*Vallonia pulchella*) (Müller 1774).
Distribution: North America, Europe, North Africa, Northern and Central Asia (Gude 1914; Gerber 1996; Gerber and Bössneck 2009)

Vallonia costata (Müller 1774)
(Fig. 18.2a, Table. 18.3)
Vallonia costataeformis Jooss, 1912 (junior synonym)
1774. *Helix costata* Müller, *Verm. Terr. Flav. Hist.*, **2**: 30.
1878. *Vallonia costata*: Theobald, *J. Asiat. Soc. Beng.*, **47**: 144
Dimension: SD 2.878–3.238 (See Table 18.2)
Distribution: *India*: Kashmir, *Elsewhere*: Europe; Northern and Eastern Asia, Northern Africa, North America

Vallonia ladacensis (Nevill 1878)
(Fig. 18.2b, Table. 18.3)
1878. *Helix (Vallonia) ladacensis* Nevill, *Scient. Results Second Yorkund Mission*, Mollusca, p. 4.
1914. *Vallonia ladacensis*: Gude, *Fauna Brit. India, Mollusca*, **2**: 226.
Vallonia miserrima.—Gude, 1914 (junior synonym)
Dimension: SD 2.924–3.252 (See Table 18.2)

Table 18.2 Morphometric measurements of *Vallonia costata*, *V. pulchella*, and *V. ladacensis* ($n = 5$), measurements in mm, except the whorl count

Species	Whorl	SD	APX	Whorl (S1)	Whorl (S2)	DWL	AH	AW	UW
Vallonia costata	3.25	2.942	0.444	0.23	0.422	1.499	0.804	1.006	0.989
Vallonia costata	3.5	3.238	0.473	0.223	0.342	1.603	0.892	1.08	0.995
Vallonia costata	3.25	2.865	0.375	0.216	0.788	1.388	0.832	0.838	0.79
Vallonia costata	3.25	2.974	0.435	0.139	0.395	1.529	1.404	1.268	0.728
Vallonia costata	3.5	2.878	0.423	0.301	0.375	1.502	1.004	1.216	0.699
Vallonia pulchella	3.75	3.37	0.455	0.356	0.44	1.947	0.943	1.359	1.026
Vallonia pulchella	3.75	3.454	0.354	0.378	0.521	1.965	0.963	1.371	0.93
Vallonia pulchella	3.75	3.448	0.398	0.329	0.507	2.009	1.03	1.372	1.104
Vallonia pulchella	3.75	3.387	0.476	0.321	0.455	1.887	1.001	1.333	1.058
Vallonia pulchella	3.75	3.423	0.389	0.322	0.461	1.99	1.266	1.3	0.965
Vallonia ladacensis	3.5	3.001	0.462	0.329	0.389	1.667	0.999	1.272	0.75
Vallonia ladacensis	3.5	3.22	0.361	0.342	0.357	1.75	0.979	1.225	0.987
Vallonia ladacensis	3.5	3.252	0.357	0.284	0.476	1.841	0.853	1.43	0.988
Vallonia ladacensis	3.5	2.924	0.317	0.371	0.418	1.689	0.822	1.242	0.882
Vallonia ladacensis	3.5	3.069	0.288	0.292	0.392	1.642	0.977	1.31	0.805

SD diameter of the shell, *APX* apex, *S* width of the spire, *AH* height of the aperture, *AW* width of the aperture, *UW* width of the umbilicus, *DWL* diameter of the whorl

Table 18.3 Per cent contribution of the analysed habitat and environmental variables for predicting suitable habitat for three *Vallonia* species in the western Himalaya, India

Variable code	Variable name	Per cent contribution by variables while determining suitable habitat		
		V. costata	*V. pulchella*	*V. ladacensis*
wc8	Mean Temperature of Wettest Quarter	**56.6**	**75.9**	0
wc4	Temperature Seasonality	**15.5**	**14.0**	**26.1**
wc12	Annual Precipitation	**12.0**	0	0
slope1	Slope	**9.0**	0	0
wc1	Annual Mean Temperature	2.9	0	0
popden1	Human Population Density	2.0	0	**6.4**
dem1	Digital Elevation Model	1.5	**2.0**	1.0
lulc1	LULC and vegetation pattern	0.5	**3.6**	**3.0**
wc15	Precipitation Seasonality	0.1	0	**13.5**
wc14	Precipitation of Driest Month	0	**4.5**	**18.2**
wc2	Mean Diurnal Range	0	0	**30.6**
hshade1	Hill shade	0	0	0
aspect1	Aspect	0	0	1.2

Distribution: *India*: Kashmir, Mataian, Dras Valley, Leh, Ladak, Spiti, Punjab and west Tibet; Nepal

Vallonia pulchella (Müller 1774)
(Fig. 18.2c, Table. 18.3)
1774. *Helix pulchella* Müller, *Verm. Terr. Flav.*, **2**: 233.
1914. *Vallonia pulchella*: Gude, *Fauna Brit. India, Mollusca*, **2**: 223.
Dimension: SD 3.370–3.454 (See Table 18.2)
Distribution: *India*: Jammu & Kashmir; *Elsewhere*: North-East Asia continent, Europe, North Africa, North America.

18.3.3 Modelling and Influencing Factors

The maximum entropy modelling approach (Phillips et al. 2006) prediction based on high AUC (area under curve) of both training and test data of current and future scenarios is used for Vallonia species distribution modelling (SDM). The training AUC value of *Vallonia costata* is 0.998, and the test AUC is 0.993. Whereas, the main climatic variables influence the distribution of *Vallonia costata* are Mean Temperature of Wettest Quarter (wc8), Temperature Seasonality (wc4), Annual Precipitation (wc12), and Slope (Slope1). However, in *Vallonia pulchella* the training AUC is also 0.998, Test AUC is 0.993, with the standard deviation is 0.003 and the main bioclimatic factors affecting its distribution include the Mean Temperature of the Wet test Quarter (wc8), Temperature Seasonality (wc4), Precipitation of Driest Month (wc14), LULC and vegetation pattern (lulc1) and Digital Elevation Model (dem1) were influencing the species distribution, and for the endemic species *Vallonia ladacensis* the training AUC is 0.944 and Test AUC is 0.993, standard deviation is 0.004 and the main bioclimatic variables are Mean Diurnal Range (wc2), Temperature Seasonality (wc4), Precipitation of Driest Month (wc14), Precipitation Seasonality (wc15), Human Population Density (popden1), and LULC and vegetation pattern (lulc1) which are the major contribution on SDM (Table 18.3, Fig. 18.4).

18.3.4 Present Distributional Scenario of Vallonia *Species in India*

The MaxEnt model of *Valonia costata* predict the current distribution pattern of the species was restricted to the lower Himalayan landscape of the western Himalaya especially in Rajauri, Jammu, and Kathua (0.54–1 suitable score/>60%) districts of Jammu and Kashmir state, in Himachal the distribution of the species were restricted to the lower Himachal viz. Chamba, Kangra, Una, Hamirpur, Bilaspur, and Solan districts (0.38–0.69 suitable score/41–60%), and in Uttarakhand state the limited

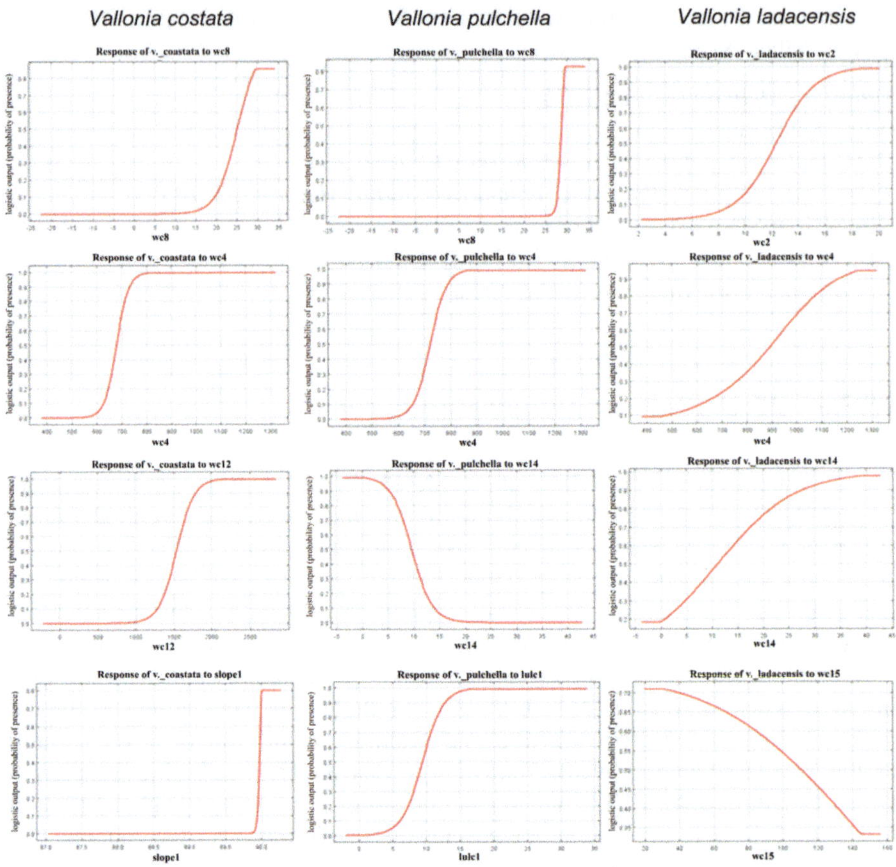

Fig. 18.4 Response curves of environmental variables and prediction of habitat suitability of *Vallonia costata* (first column), *V. pulchella* (second column), and *V. ladacensis* (third column) in western Himalaya

predictions were in Pauri Garhwal, Nainital and Champawat districts (0.23–0.38 suitable score/5–20%). However, the current prediction of *Vallonia pulchella* is limited to the Jammu district of Jammu and Kashmir (0.54–0.85 suitable score/>60%), and significantly less probability of occurrence has been identified in Himachal and Uttarakhand states of India. Moreover, the probability of endemic *Vallonia ladacensis* has been wider range in all three states of the western Himalaya, especially in the Trans-Himalaya region, viz. Leh, Ladakh, Kargil, etc. (Figs. 18.5 and 18.6).

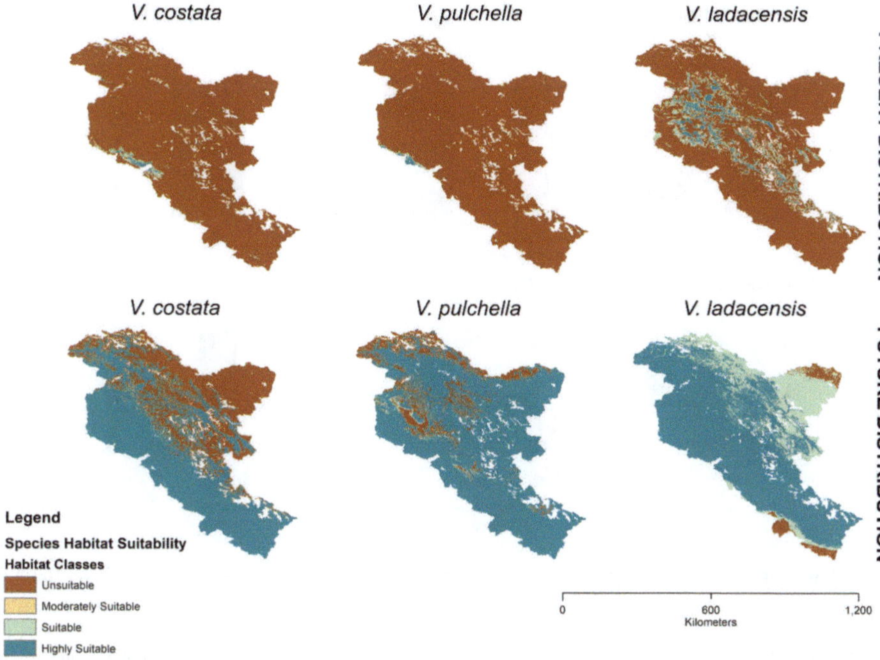

Fig. 18.5 Habitat suitability of the Genus *Vallonia* in the western Himalaya under present and future climatic conditions

18.3.5 Future Distributional Scenario of Vallonia *Species in India*

The MaxEnt model results offer an insightful projection of habitat suitability changes for *V. costata*, *V. pulchella*, and *V. ladacensis*, demonstrating significant habitat shifts between the present and future scenarios (Fig. 18.4). The True Skill Statistic (TSS) and Kappa values, which exceed 0.98 for all three species, confirm the high predictive accuracy of the model, minimizing classification errors and reinforcing the reliability of these habitat suitability projections. The key finding from this analysis is the dramatic reduction in unsuitable habitat and an increase in highly suitable habitats across all three species. This pattern strongly suggests that environmental changes—potentially climate-driven or influenced by landscape modifications—are reshaping habitat availability, making previously uninhabitable regions more conducive to mollusc populations (Fig. 18.5).

A closer look at habitat area change reveals intriguing trends. *V. costata* sees a 60.95% decrease in unsuitable habitat, with corresponding increases in moderately suitable (+1.85%), suitable (+2.53%), and highly suitable habitats (+56.57%). This shift suggests that large portions of its range will transform into more favourable conditions, potentially leading to a population increase or range expansion. The trend is even more pronounced for *V. ladacensis*, which experiences an 82.20%

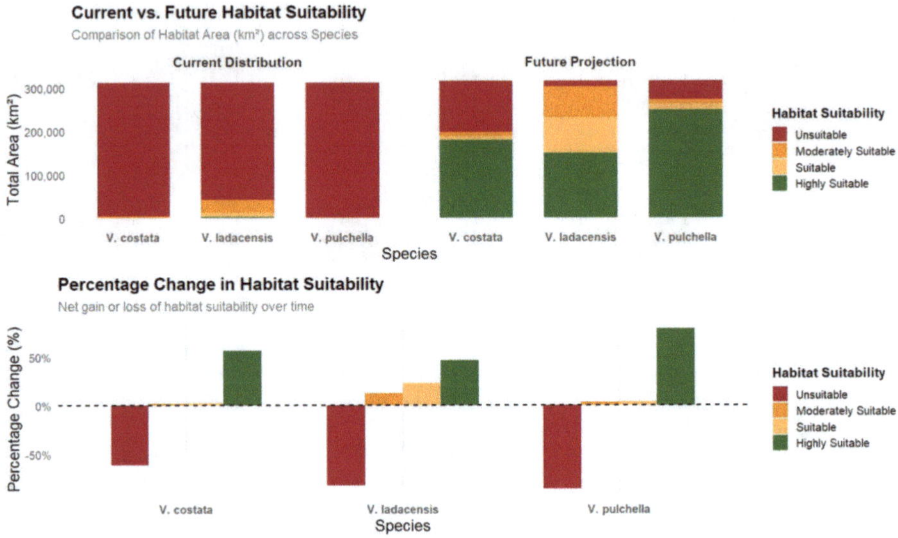

Fig. 18.6 Differences between current and future suitable areas in the western Himalaya, India

Table 18.4 Levene's test for equality of variances (Significant level ** - p = 0.005 Significant level *** - p = 0.001)

Trait	*F*_value	*p*_value	Significance	Levene_*F*	Levene_*p*	Equal_var_assumption
Whorl	32.67	0.000014	***	2.6667	0.1101	Yes
SD	17.51	0.000277	***	1.3937	0.2856	Yes
APX	2.724	0.106	ns	0.4232	0.6644	Yes
Whorl_S1	11.92	0.00141	**	0.4258	0.6627	Yes
Whorl_S2	0.576	0.577	ns	0.8506	0.4514	Yes
DWL	53.61	1.04E-06	***	0.2953	0.7496	Yes
AH	0.588	0.571	ns	0.6389	0.5449	Yes
AW	8.02	0.00614	**	3.3611	0.06934	Borderline
UW	3.478	0.0644		1.0562	0.378	Yes

** - p = 0.005
*** - p = 0.001

reduction in unsuitable habitat. In contrast, its moderately suitable habitat expands by 12.91%, suitable habitat grows by 23.02%, and highly suitable areas increase by 46.28%. However, the most striking transformation is seen in *V. pulchella*, where unsuitable habitat declines by 85.75%, while highly suitable habitat expands by an astonishing 78.66%. The sheer scale of these changes implies a significant ecological shift in how these species interact with their environment (Fig. 18.6).

18.3.6 Morphometric Variation in Vallonia *Species in India*

The summary statistics reveal distinct morphometric differences among the three *Vallonia* species. *Vallonia pulchella* exhibits the largest average shell diameter (mean SD = 3.42 mm) and aperture width (AW = 1.35 mm), suggesting a generally larger and more robust shell structure compared to the other two species. *Vallonia ladacensis*, although consistent in whorl count (3.5 across all individuals), shows intermediate shell diameter and relatively high values for AW and umbilicus width (UW), indicating a broader aperture and more open shell morphology. In contrast, *Vallonia costata* shows slightly smaller shell dimensions overall, with greater variability across traits such as aperture height and umbilicus width. These differences suggest that while all three species share similar whorl counts, they differ notably in shell breadth and aperture configuration, which may reflect adaptations to distinct microhabitats or ecological preferences within their overlapping Himalayan ranges.

The morphometric analysis using one-way ANOVA revealed significant interspecific differences among the three *Vallonia* species in several key traits. The number of shell whorls ($F = 32.67, p < 0.001$), shell diameter (SD; $F = 17.51, p < 0.001$), whorl measurement at stage 1 (Whorl_S1; $F = 11.92$, $p < 0.01$), diameter of the whorl (DWL; $F = 53.61$, $p < 0.001$), and aperture width (AW; $F = 8.02$, $p < 0.01$) showed highly significant variation between species. Among these, DWL and SD were particularly discriminative, with *V. pulchella* showing consistently larger values compared to the other two species.

In contrast, traits such as apex height (APX; $F = 2.72$, $p = 0.106$), whorl measurement at stage 2 (Whorl_S2; $F = 0.576$, $p = 0.577$), aperture height (AH; $F = 0.588, p = 0.571$), and umbilicus width (UW; $F = 3.478, p = 0.064$) did not show statistically significant variation, although UW approached marginal significance.

Levene's test confirmed the assumption of homogeneity of variances for all traits (Table 18.4), with the exception of aperture width (AW), which approached borderline significance ($F = 3.36$, $p = 0.069$). Nevertheless, variances were generally consistent across species, validating the robustness of the ANOVA results.

The Principal Component Analysis (PCA) effectively captured the major patterns of variation in the morphometric data of the three *Vallonia* species (Fig. 18.7). The scree plot indicates that the first two principal components (PC1 and PC2) explain a substantial proportion of the total variance—*48.8%* and *17.7%*, respectively, cumulatively accounting for *66.5%* of the morphometric variation across all measured traits. This justifies the use of a two-dimensional PCA plot to explore species-level differentiation (Fig. 18.8).

In the PCA scatter plot, individuals of *Vallonia pulchella* form a tight cluster along the positive axis of PC1, clearly separated from the other two species. This suggests that *V. pulchella* possesses a distinct set of morphometric traits, particularly contributing to the variation captured by PC1. *Vallonia costata* occupies a broader and more scattered region in the negative axis of PC1 and positive to neutral values along PC2, indicating greater intraspecific variability. In contrast, *Vallonia ladacensis* forms an intermediate cluster near the origin, overlapping slightly with

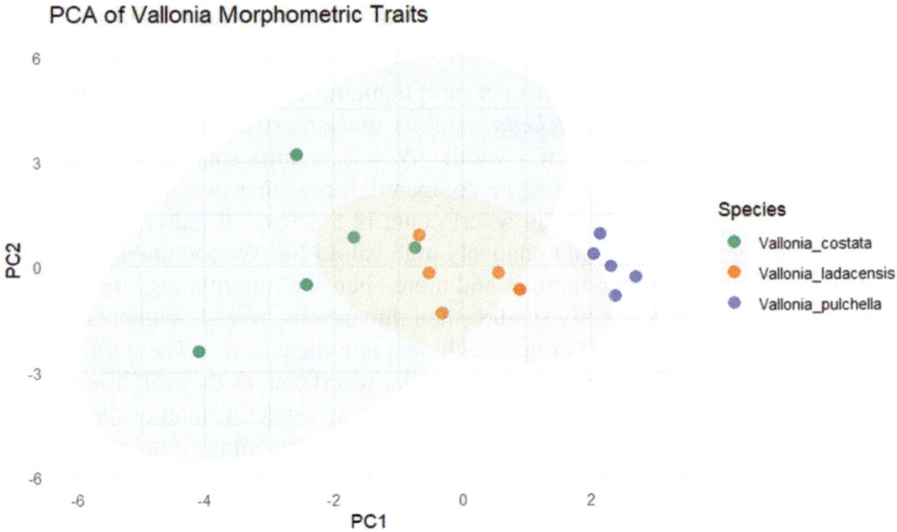

Fig. 18.7 Plot of components 1 and 2 resulting from the principal component analysis (PCA) of *Vallonia* species

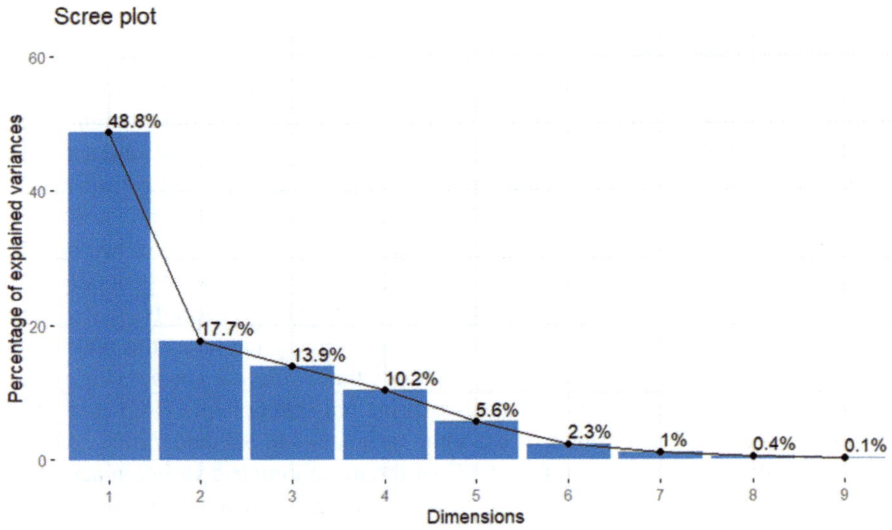

Fig. 18.8 Scree plot analysis of principal component analysis (PCA) contributing to morphometric analysis in *Vallonia* species

both *V. pulchella* and *V. costata*, but still maintaining a relatively confined distribution, suggesting it shares morphological characteristics with both species.

The confidence ellipses further support this observation, showing limited overlap between species and distinct morphospace occupancy, especially for *V. pulchella*. These results imply that species-level morphological differences, particularly in traits like shell diameter, aperture width, and whorl structure, are sufficient to differentiate the taxa and support their recognition as distinct biological units. Additionally, the observed variation along PC1 and PC2 may reflect underlying ecological or evolutionary divergence among the species in response to habitat or environmental pressures in the western Himalaya.

18.4 Discussion

The MaxEnt model demonstrated the clear effectiveness of the *Vallonia* sp. distribution in the western Himalaya and provided clear scenario for the suitable habitat of the species. However, regarding the climatic profile and other variables, the prediction of the future distribution will give sufficient information on the habitat shifting and extended the range of the species based on future climatic change. The mean Temperature of the Wettest Quarter, Temperature Seasonality, and annual precipitation were most significant variables in distribution of *V. costata* and *V. pulchella*, while, for *V. ladacensis* the mean diurnal range and temperature seasonality were acted as main climatic factors in future prediction and habitat shifting. Thus, the climatic variable will be the essential factor for species richness and geographical range extension and also habitat shifting in the *Vallonia* species. Unlike typical conservation concerns where species face habitat loss and fragmentation, these molluscs appear to be on the verge of significant range expansion, occupying new or previously marginally suitable habitats. The vast decline in unsuitable habitats, particularly for *V. pulchella* and *V. ladacensis*, indicates that these species may no longer be limited to small, specialized niches instead they could thrive in broader landscapes. This has two major ecological implications. First, it may reduce their vulnerability to extinction risks due to habitat degradation, as new suitable areas may provide alternative refugia. Second, this expansion could bring unforeseen ecological consequences, especially if these species begin to outcompete native mollusc populations or interact differently with their surroundings. While these species may not traditionally be considered invasive, the rapid and large-scale habitat gains could allow them to be established in regions where they were previously rare or absent, potentially altering community structures and ecosystem dynamics.

Despite the findings, it is crucial to reassess the ecological roles of these molluscs and their potential impact on new environments. If they expand into areas beyond their historical range, their population dynamics could shift, influencing factors such as resource competition, predator–prey interactions, and overall biodiversity balance. Although the direct risk of invasion remains speculative, a scenario where these species outcompete local mollusc populations or disrupt established trophic relationships cannot be ruled out. Additionally, the shift in habitat suitability could create novel ecological interactions, requiring continuous monitoring to assess whether these species integrate harmoniously into new habitats or trigger cascading effects on ecosystem processes.

This study also highlights significant morphometric differences among three species of the genus *Vallonia*—*V. costata*, *V. pulchella*, and *V. ladacensis*—inhabiting the western Himalayan region. Using a combination of descriptive statistics, inferential testing, and multivariate analysis, we were able to delineate interspecific variation and morphospace occupation, shedding light on species-level distinctions that are critical for ecological interpretation and taxonomic clarity.

The results of the one-way ANOVA indicate that several shell traits exhibit statistically significant interspecific variation. Notably, shell diameter (SD), diameter of the whorl (DWL), and aperture width (AW) were significantly different across species, with *V. pulchella* consistently exhibiting larger values. The number of whorls and penultimate whorl measurements at stage 1 (Whorl_S1) also differed significantly, supporting the idea that *V. pulchella* is morphologically distinct. In contrast, traits such as apex height (APX), aperture height (AH), and penultimate whorl measurement at stage 2 (Whorl_S2) did not differ significantly, indicating these may not contribute substantially to species separation. Interestingly, umbilicus width (UW) approached marginal significance, suggesting it may be a weakly differentiating trait that requires further sampling to clarify its role.

These statistical findings are further substantiated by Principal Component Analysis (PCA), which visually illustrates the morphospace partitioning among species. The first two principal components explained approximately 66.5% of the total variance, with PC1 alone accounting for nearly half. *V. pulchella* formed a clearly separated cluster, suggesting strong differentiation in shell morphology, likely driven by traits heavily loaded on PC1 such as shell diameter and aperture width. *V. ladacensis* and *V. costata* exhibited some degree of overlap, although the former occupied a more central and constrained area of the morphospace. This pattern may suggest morphological convergence or intermediate character states, potentially due to overlapping ecological niches or shared evolutionary ancestry.

The relatively higher morphological variation observed in *V. costata* compared to the other two species could point to either ecological plasticity or cryptic diversity, especially considering its broader spread in PCA space. On the other hand, the tight clustering of *V. pulchella* may reflect ecological specialization or stabilizing selection in its habitat. While *V. ladacensis* maintained consistent whorl numbers across individuals, it exhibited intermediate values for several traits, reinforcing its potential status as a morphologically conserved but ecologically distinct species.

From an ecological perspective, the observed interspecific differences in shell and aperture dimensions may reflect adaptation to microhabitat conditions such as moisture retention, soil substrate, or predation pressure. The robust shell and wide aperture of *V. pulchella* could facilitate greater mobility or moisture conservation, allowing it to occupy a broader ecological niche. Conversely, the more compact shell structure of *V. costata* and the intermediate morphology of *V. ladacensis* may be suited to more specific or less disturbed habitats.

Overall, the integrative approach combining univariate and multivariate statistical tools has proven effective in characterizing morphometric variation among closely related snail species. These findings are valuable not only for refining species identification but also for understanding ecological divergence and potential conservation needs. Given that molluscs often remain understudied in biodiversity assessments, especially in montane systems like the western Himalayas, studies like this provide essential baselines for future taxonomic and ecological research. Further investigation using larger sample sizes, molecular data, and environmental correlation models could enhance our understanding of species boundaries, cryptic diversity, and habitat specificity within the genus *Vallonia*.

From a conservation and management perspective, these results emphasize the need for long-term ecological monitoring to track population trends, habitat utilization, and interspecies interactions. While this transformation may seem beneficial for species persistence, it also raises questions about ecosystem resilience and the adaptive responses of other organisms within the changing landscapes. Conservation efforts should not only focus on protecting suitable habitats but also consider the potential consequences of species expansion, ensuring that biodiversity remains balanced and ecosystem functions are not disrupted by sudden demographic shifts. Future research should explore how these species respond to competition in newly available habitats, whether they alter food web structures, and if they trigger any secondary effects on native fauna and flora.

Ultimately, the study reveals that *V. costata*, *V. pulchella*, and *V. ladacensis* are not facing habitat loss but rather an unprecedented range transformation, making their future conservation and ecological monitoring more critical than ever. Their ability to adapt to new environments, coexist with other species, and sustain populations in emerging habitats will be key determinants of their long-term survival and ecological impact.

18.5 Conclusions

This study presents a comprehensive assessment of the current and future habitat suitability of three micro snail species—*Vallonia costata*, *Vallonia pulchella*, and *Vallonia ladacensis*—in the Western Himalayan region of India. Utilizing a robust species distribution modelling approach through MaxEnt, we were able to identify the bioclimatic and anthropogenic variables most influencing the distribution of these species and project how their suitable habitats may shift under future climatic conditions. The high predictive accuracy of the models, reflected in AUC and TSS values, strengthens the reliability of our findings and provides critical insights into the climatic dependencies and potential expansion trajectories of these poorly studied molluscan taxa.

Contrary to the general conservation narrative for invertebrates, especially terrestrial molluscs, which often emphasizes range contraction and habitat fragmentation, the findings of this study reveal a substantial positive shift in suitable habitat areas for all three *Vallonia* species in future climate scenarios. This is particularly significant in the case of *V. ladacensis*, an endemic species, which exhibits a notable increase in habitat suitability, primarily influenced by the mean diurnal temperature range and temperature seasonality. These results suggest that climate change, while posing severe threats to many taxa, may simultaneously create new ecological opportunities for certain adaptable invertebrates.

The projected expansion in suitable habitats raises important ecological and conservation considerations. On one hand, the anticipated increase in highly suitable environments may reduce extinction vulnerability and allow populations to thrive in broader ecological niches. On the other hand, such expansion could alter local community dynamics, especially if these *Vallonia* species begin to colonize areas beyond their historical ranges, potentially affecting existing mollusc assemblages and interacting differently within newly occupied ecosystems. Although these species are not traditionally invasive, the scale of future habitat gains necessitates a precautionary approach to ecological monitoring. Our study highlights the importance of incorporating lesser-known taxa like micro-snails into climate change impact assessments and conservation planning. The role of invertebrates in ecosystem functioning—such as nutrient cycling, soil health, and food web dynamics—should not be underestimated. By identifying key climatic and topographic drivers of distribution, this work provides a baseline for understanding how small-bodied terrestrial molluscs respond to environmental changes, especially in climatically sensitive mountain systems like the Himalayas. Looking ahead, there is an urgent need for expanded field surveys, long-term monitoring programmes, and experimental studies to validate these model projections and explore the biological and ecological traits that confer adaptability or vulnerability to climate shifts. Future research should also investigate potential interspecific interactions, range overlap with native species, and the implications for local biodiversity and ecosystem resilience. Conservation strategies should therefore be dynamic, forward-looking, and inclusive of such taxonomic groups that have historically received minimal attention.

In connection with the morphometric analysis, this study highlights significant differences among three species of the genus *Vallonia*—*V. costata*, *V. pulchella*, and *V. ladacensis*—inhabiting the western Himalayan region. Using a combination of descriptive statistics, inferential testing, and multivariate analysis, we were able to delineate interspecific variation and morphospace occupation, shedding light on species-level distinctions that are critical for ecological interpretation and taxonomic clarity.

The results of the one-way ANOVA indicate that shell traits exhibit statistically significant interspecific variation. Notably, shell diameter (SD), diameter of the whorl (DWL), and aperture width (AW) were significantly different across species, with *V. pulchella* consistently exhibiting larger values. The number of whorls and whorl measurements at stage 1 (Whorl_S1) also differed significantly, supporting the idea that *V. pulchella* is morphologically distinct. In contrast, traits such as apex height (APX), aperture height (AH), and whorl measurement at stage 2 (Whorl_S2) did not differ significantly, indicating these may not contribute substantially to species separation. Interestingly, umbilicus width (UW) approached marginal significance, suggesting it may be a weakly differentiating trait that requires further sampling to clarify its role.

These statistical findings are further substantiated by Principal Component Analysis (PCA), which visually illustrates the morphospace partitioning among species. The first two principal components explained approximately 66.5% of the total variance, with PC1 alone accounting for nearly half. *V. pulchella* formed a clearly separated cluster, suggesting strong differentiation in shell morphology, likely driven by traits heavily loaded on PC1 such as shell diameter and aperture width. *V. ladacensis* and *V. costata* exhibited some degree of overlap, although the former occupied a more central and constrained area of the morphospace. This pattern may suggest morphological convergence or intermediate character states, potentially due to overlapping ecological niches or shared evolutionary ancestry.

The relatively higher morphological variation observed in *V. costata* compared to the other two species could point to either ecological plasticity or cryptic diversity, especially considering its broader spread in PCA space. On the other hand, the tight clustering of *V. pulchella* may reflect ecological specialization or stabilizing selection in its habitat. While *V. ladacensis* maintained consistent whorl numbers across individuals, it exhibited intermediate values for several traits, reinforcing its potential status as a morphologically conserved but ecologically distinct species.

From an ecological perspective, the observed interspecific differences in shell and aperture dimensions may reflect adaptation to microhabitat conditions such as moisture retention, soil substrate, or predation pressure. The robust shell and wide aperture of *V. pulchella* could facilitate greater mobility or moisture conservation, allowing it to occupy a broader ecological niche. Conversely, the more compact shell structure of *V. costata* and the intermediate morphology of *V. ladacensis* may be suited to more specific or less disturbed habitats.

Overall, the integrative approach combining univariate and multivariate statistical tools has proven effective in characterizing morphometric variation among closely related snail species. These findings are valuable not only for refining species identification but also for understanding ecological divergence and potential conservation needs. Given that molluscs often remain understudied in biodiversity assessments, especially in montane systems like the western Himalayas, studies like this provide essential baselines for future taxonomic and ecological research. Further investigation using larger sample sizes, molecular data, and environmental correlation models could enhance our understanding of species boundaries, cryptic diversity, and habitat specificity within the genus *Vallonia*.

In conclusion, this study not only advances our understanding of the ecological requirements and biogeography of *Vallonia* species in India but also contributes to broader discussions on species resilience under climate change. By integrating ecological modelling with climatic and anthropogenic datasets, it offers a predictive framework that can inform adaptive conservation strategies for micro-snails and similar taxa in mountainous biodiversity hotspots.

Acknowledgements The authors are grateful to the Director, Zoological Survey of India, Kolkata, and Gujarat Institute of Desert Ecology, Bhuj, for providing necessary facilities and encouragement for the study. The author SS is thankful to Dr. Jochen Gerber, Field Museum of Natural History, Chicago, USA; Dr. Małgorzata Proćków, Museum of Natural History, University of Wroclaw, Poland, and the Biodiversity Heritage Library (BHL) for the literature support. We are thankful to the anonymous reviewers and the editor for their constructive comments and suggestions in improving the chapter.

Funding Statement No external funds, grants, or other support was received to carry out this work.

Data Availability Statement Data are available from the authors on request.

Conflict of Interest The authors declare that they have no known competing financial interests or personal relationships that could have appeared to influence the work reported in this paper.

Ethics Approval This study does not involve humans or animals requiring ethical approval.

References

Ashraf U, Hussan A, Chaudry MN, Ashraf I, Botool A, Saqib Z (2016) Predicting the potential distribution of *Olea ferruginea* in Pakistan incorporating climate change by using Maxent model. Sustainability 8:722. https://doi.org/10.3390/su8080722

Fortunato H (2015) Mollusks: tools in environmental and climate research. Am Malacol Bull 33:1–15

Gerber J (1996) Revision der Gattung *Vallonia* Risso, 1826 (Mollusca: Gastropoda: Valloniidae). Schriften zur Malakozoologie 8:1–227

Gerber J, Bössneck U (2009) The genus *Vallonia* in Nepal (Gastropoda: Pulmonata: Valloniidae). Arch Molluskenkd 138:43–52

Gude GK (1907) Description of a new species of *Vallonia* from South India. Proc Malacol Soc Lond 7:306–307

Gude GK (1914) The fauna of British India, including Ceylon and Burma Mollusca: II (Trochomorphidae-Janellidae). Today Tomorrow's Printers & Publishers, New Delhi. 520 pp

Jooss CH (1912) Neue Landschnecken aus dem Obermiocän von Steinheim am Aalbuch in Württemberg. Nachrichtsblatt der Deutschen Malakozoologischen Gesellschaft. 44(1): 30-45.https://www.biodiversitylibrary.org/page/15107739

Merow C, Bois ST, Allen JM, Xie Y, Silander JA (2017) Climate change both facilitates and inhibits invasive plant ranges in New England. Proc Natl Acad Sci 114:E3276–E3284

Müller OF (1774) Vermium terrestrium et fluviatilium, seu animalium infusorium, Helminthicorum, et testaceorum, non marinorum, succincta historia. vol 2. Havniae et Lipsiae, apud Heineck et Faber, ex officina Mölleriana. IXXXVI, 1-214 pp, 10.https://www.biodiversitylibrary.org/page/12893715

Nevill G (1878) Mollusca. Scientific Results of the Second Yarkand Mission; based upon the collections and notes of the late Ferdinand Stoliczka Ph.D.: 1–21, 1 pl. Office of the Superintendent of Government Printing, Calcutta

Pearson RG, Dawson TP (2003) Predicting the impacts of climate change on the distribution of species: are bioclimate envelope models useful? Glob Ecol Biogeogr 12:361–371

Phillips SJ, Anderson RP, Schapire RE (2006) Maximum entropy modeling of species geographic distributions. Ecol Model 190:231–259

Pilsbry HA (1948) Land Mollusca of North America (north of Mexico), Volume II, Part 2. The Academy of Natural Sciences of Philadelphia, Monographs 3. Wickersham Printing Company, Lancaster, pp i–xlvii + 521–1113

Raheem DC, Taylor H, Ablett J, Preece RC, Aravind NA, Naggs F (2014) A systematic revision of the land snails of the Western Ghats of India. Trop Nat Hist Suppl 4:1–308

Rajagopal AS, Subba Rao NV (1972) Some land molluscs of Kashmir, India. Rec Zool Surv India 66:197–212

Ramakrishna G, Mitra SC, Dey A (2010) Annotated check list of Indian land molluscs. Rec Zool Surv India 306:1–359. https://archive.org/details/dli.zoological.occpapers.306/mode/2up?view=theater

Risso A (1826) Histoire naturelle des principales productions de l'Europe méridionale et particulièrement de celles des environs de Nice et des Alpes Maritimes. Tome Quatrième 4:VII+439 S., 12T; Paris

Rumi A, Vogler RE, Beltramino AA (2017) The South-American distribution and southernmost record of *Biomphalaria peregrina*—a potential intermediate host of schistosomiasis. PeerJ 5:e3401

Sterki V (1893) Observations on *Vallonia*. Proc Acad Natl Sci Phila 45:234–279

Chapter 19
Restoration Efforts in Mitigation of Climate Change: Case Study from Guneri Village, Gujarat

Deepa Gavali, Jayendra Lakhmapurkar, and Pankaj N. Joshi

Abstract Guneri village is located in the western part of the Kachchh district in Gujarat, with low rainfall and arid conditions. The local pastoral community is dependent on livestock rearing and rainfed agriculture for livelihood. The village grasslands are suffering from low grass productivity due to overexploitation, salinity, and invasion of exotic species *Prosopis juliflora*. The invasive *Prosopis juliflora* was removed with the help of the local community in the first year. Following the removal of *Prosopis* plants and treatment with manure, propagation of the native grass seeds was done in the form of seed balls. Continuous interventions were done for soil moisture conservation and grass seed propagation. The areas showed positive results in the third year with the growth of good ground vegetation cover, and the soil pH declined to 7.6. The trend observed in the soil pH and salinity were significant at $p = 0.02$ level and indicates substantial improvement in soil fertility. With the decline in soil salinity, certain species that were reported in the first year diminished and gradually replaced by other species in the third year. In the first year, *Cyperus arenarius* was the dominant species followed by *Cressa cretica, Aristida funiculata, Sporobolus marginatus,* and *Eleusine indica*. The top five species in the third year recorded include *Aristida adscensionis, Indigofera linifolia, Indigofera cordifolia, Goniogyna hirta,* and *Melanocenchris jacquemontii*. Herbaceous layer species diversity increased from 19 at the beginning of the restoration (first year) to 33 species at the end of the second year and 49 species at the end of the third year. The grass density increased by about 23 times in the third year as compared to the first year. PCA analysis points out that positive impacts were visible with improvement in the soil properties and vegetation.

Keywords Restoration · Ground vegetation · Soil improvement · Community involvement

---

D. Gavali (✉) · J. Lakhmapurkar
Gujarat Ecology Society, Vadodara, Gujarat, India

P. N. Joshi
Ecologist, Bhuj Kutch, Gujarat, India

B. P. Barathan et al. (eds.), *Navigating Climate Change: Impacts on Biodiversity and Ecosystem Resilience*, Sustainable Development and Biodiversity 42,
https://doi.org/10.1007/978-981-95-0409-1_19

19.1 Introduction

Global efforts to combat Green house gases (GSH) emissions and enhance carbon sequestration through improved protection and restoration of ecosystems are being implemented. These interventions are referred to as nature-based solution and result in changes for social resilience, livelihood and adaptive capacity, biodiversity management, and ecosystem services. Interventions are rarely implemented in isolation; integrated conservation and development approaches are often used to address social and ecological drivers that aim to reduce soil disturbance, maintain soil cover, and encourage crop rotation. Moreover, interventions take place within broader socio-ecological systems with different interventions often implemented within the same landscape. These complexities make it challenging to disentangle attribution of effects at scale and the interactions between effects. Restoration of 15% of degraded ecosystems around the world has been the priority goal targeted by UN convention on biodiversity strategic plan for Biodiversity 2011–20 (CBD 2015, 2016). Capacity to regenerate is limited and restoring of degraded lands is becoming essential to repair the integrity of impacts on forests or grasslands or mine affected areas (Ockendon et al. 2018). The process involves establishing the linkages between the ecosystem services and society. Abbandonato et al. (2018) suggested the need for participatory approach for successful restoration whenever multiple stakeholders are involved.

Restoration strategies vary according to the affected ecosystem and the cause of degradation. Re-forestation within natural or plantation sites can restore forest ecosystems (Gastauer et al. 2020). Re-seeding can provide vegetation cover and prevent the erosion and desertification of rangelands (Stock et al. 2020). Conservation tillage, nutrient, and organic replenishment can restore poor soils, while wetland restoration aims to recover the natural hydrology of rivers and lagoons (Santini et al. 2019). It is essential to examine the effects of vegetation restoration and environmental factors on soil nutrients to optimize restoration approaches. Coupling the goals of ecological restoration with the participation of local communities is needed for successful, long-term restoration of ecosystems (Robinson et al. 2021). Therefore, restoring degraded habitats is often critical for the well-being of these communities, who are well-positioned with an interest in restoring the land upon which their livelihood depends. The knowledge of indigenous and local communities can inform and guide effective restoration. Traditional knowledge can identify keystone species or sites that are foundational to restoration efforts. Local communities also have experience with successional regeneration processes that occur during habitat restoration. The land management practices of local communities, such as rotational farming, agroforestry, and enclosures have been developed over thousands of years and are effective strategies to prevent environmental degradation (Santini and Miquelajauregui 2022). Local communities can also actively monitor ecosystem health and restoration success, and collect data on species and ecological trends, particularly in remote sites and over long time periods (Eicken et al. 2021). The present restoration efforts were undertaken at Guneri village near Lakhpat area of Gujarat. Historically, the Sindhu River nourished Lakhpat's plains, supporting agriculture and trade. However, the 1819 earthquake and subsequent formation of the Allah Bund significantly altered the region's hydrology, geography, and socio-economic, leading to arid conditions and salinity issues that persist to this day (CGWB 2013). Despite harsh climatic conditions, Lakhpat supports endemic and threatened plant species, emphasizing the region's ecological importance. The region continues to face challenges such as salinity, drought, invasive species, and human pressure. However, ongoing research and conservation efforts emphasize the importance of preserving this unique ecosystem through sustainable practices and community engagement.

19.2 Materials and Methods

Guneri village lies on the western most part of Kachchh district in Gujarat. Guneri village is situated north of Lakhpat fort with a population of 967 as per the 2011 census. The region falls under an arid climatic zone with an average rainfall of around 350 mm (2001–2020). There is a temporal variation in rainfall with the highest rainfall of 843 mm (2003) and 810 mm (2020) and a minimum of 18 mm (2018) and 58 mm (2002) rainfall. The general slope of the Guneri's ground is from south to north (towards Rann). The highest elevation is around 67 m in the south and the lowest 4 m. The general gradient of the area is 1.6 m. Several streams originate in the hilly area and drain into the Rann flowing north and northwest direction. Around 59% of the village area is barren land dominantly hilly rocky terrain (34%) followed by saline waste (25%). About 21% of area is under agriculture and 3% forms the pastureland (Fig. 19.1).

The village is afflicted with invasion of *Prosopis juliflora*, which has taken over natural grasslands and vegetation found in the area. The village is also known for the presence of inland mangroves and close to being declared as Biodiversity Heritage site. The local community are largely pastoral community and undertake rainfed agriculture. The community is largely represented by Darbar community, who claim to have migrated from Sindh area of Pakistan, long before the partition. The village is dominated by the Jadeja clan of Rajputs. The village population is 967 with 200 (21%) belonging to Scheduled Caste. The village has 127 total workers with only 13% employment. The majority (80%) of workers are dependent on agriculture and agriculture-based industries. The livestock data of the Guneri suggest the dominance of browsers mainly sheep and Goat (52%) indicative of an arid system followed by buffalo (39%). *Prosopis* species is used by the locals as source of firewood, collected from the open areas. Buffaloes are taken out for grazing in the pastureland; however, post January there is negligible grasses present in the open areas and purchase of dry grass is the main source of fodder. Goats and sheep are left to browse on the open areas, with whatever grass blade is left and move to the saline wastelands.

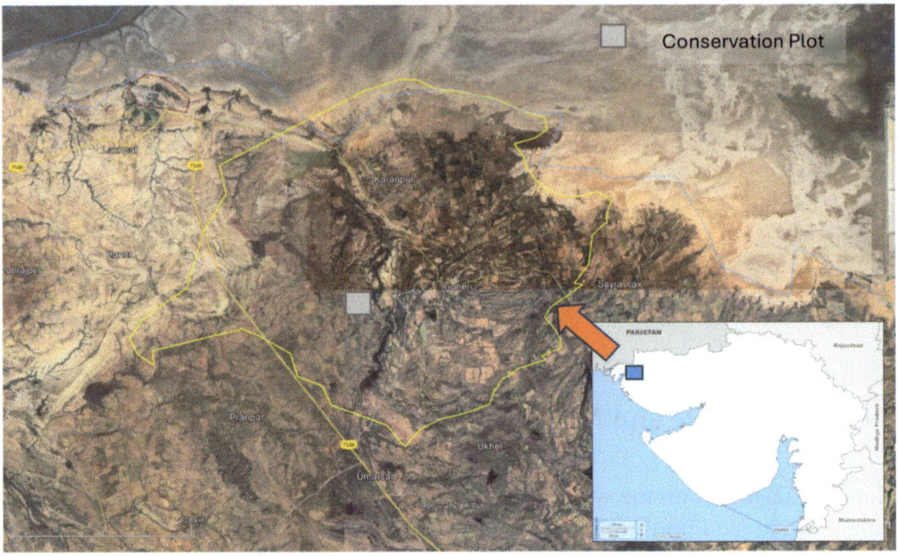

Fig. 19.1 Location of the implementation site

The restoration efforts were initiated at a small pilot stage of the common area infested with *Prosopis* and was unproductive. Restoration process was initiated, and the following steps were followed. Interaction with the local community to own the restoration process and enhance their participation were undertaken and one person was appointed from the village, who would be involved in the restoration process. The second step was to record the existing vegetation and soil quality of the study area. The third step involved removing *Prosopis juliflora*, an exotic and invasive species of the region. The area was tilled and fenced to prevent entry of grazing natives. Thereafter, the native grass seeds belonging to the region was procured from Central Arid Zone Research Institute, Bhuj and broadcasted on the first year. Organic manure was added to the study area to improve the soil properties and post rain, the growth of the grass and other vegetation was recorded. The observation was recorded for the second and third years of intervention. The local community were trained on collection of grass seeds and preservation. For effective documentation of the vegetation, the area was gridded into 50 × 50m (Fig. 19.2). The primary vegetation data of the sites for restoration were initiated. The delay in rainfall has postponed the primary survey. 5 × 5 m quadrate for estimation of shrub and tree layer and 1 × 1 m for ground vegetation were laid (Fig. 19.3).

Standard method of soil analysis as prescribed by American Public Health Association (APHA) was followed. Vegetation recording was done following stratified quadrate method.

Community involvement in the restoration process involved the following steps:

1. Awareness and orientation sessions were conducted with the local community on the process of restoration, benefits, and their participation.

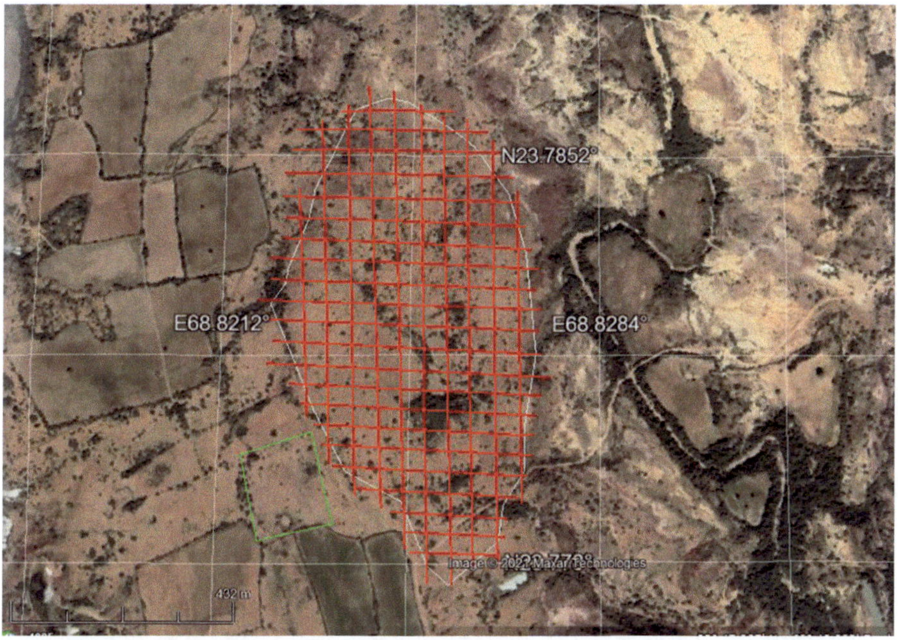

Fig. 19.2 Grids created for vegetation data collection

2. The local community were trained in preparation of seed balls using local manure and grass seeds and the time of broadcast for effective regeneration (Fig. 19.4). Rainfall being scanty in the region, broadcasting of seed balls was done after first two showers for effective results.
3. The local community were trained in seed collection and storage of grass seeds for use in the subsequent years (Fig. 19.5).
4. The monitoring of vegetation was undertaken with the help of local community to enhance their knowledge on the changes taking place due to restoration process.
5. School children were also involved in the seed ball preparation.

19.3 Results and Discussion

P. juliflora was introduced in India way back in 1878 during the British regime for afforesting deserts of Sindh and Southern Punjab. In Gujarat, the forest department planted this species to check desertification in mid-1950s. *P. juliflora* was planted largely at fringes of Rann to stop desertification. Since then, it has invaded almost the entire coastal area and invaded areas with high ground table throughout the state. Study conducted by Gavali and Lakhmapurkar (2009) on impact of *Prosopis juliflora* invasion in Banni grasslands have reported replacement of native species

Fig. 19.3 Laying of quadrates for vegetation data collection

like *Capparis* spp., *Salvadora* spp., *P. cineraria,* and *Acacia nilotica. Prosopis juliflora* has largely invaded into the ravine and wastelands in Gujarat. Thus, *Prosopis juliflora* is posing threat to the biodiversity especially birds like Lesser Florican and Houbara Bustard (Gadhvi 2003; Rahamni 1998). Temporal analysis of land use and vegetation conducted by Das et al. (2019) revealed a decline in native vegetation cover in Lakhpat with an alarming increase in invasive *Prosopis juliflora* between 2009 and 2017. Satellite imagery comparisons from 2009 to 2017 indicate a decline in native vegetation, particularly in the western parts of the taluka. Rastogi et al. (2016) further analysed rainfall data, identifying Lakhpat as the region receiving the least rainfall in Kachchh.

Soil is the primary indicator of the health of the area and indicator of the direction of the restoration efforts undertaken. During the present restoration efforts, the pH of the soil in the beginning and second year was in the range of 8 and above. The soil characteristics of Lakhpat Taluka reflect its arid and saline environment, presenting challenges to agriculture and vegetation growth. Chaudhary et al. (2018) assessed the nutrient status of soils in Kachchh, reporting high potassium (2.6 kg/ha), marginal phosphorus (1.77 kg/ha), and low nitrogen (1.08 kg/ha). The study

Fig. 19.4 Local preparing the seed ball

noted an electrical conductivity value of 6.05 dS/m, which inhibit seed germination and crop productivity, indicating severe salinity. Further studies on irrigation water quality highlight high concentrations of sodium, calcium, and magnesium ions, contributing to soil degradation and poor biomass production (Sipai et al. 2022). In the first two years, the soil properties in relation to the pH did not differ much; however, in the third year, post growth of good ground vegetation cover, the soil pH declined to 7.6 (Fig. 19.6). The trend observed in the soil pH is significant at $p = 0.02$ level and indicates substantial improvement in the soil fertility. Besides pH, organic matter content could also have contributed to the greater concentrations of Ca, Mg, and K. This is because of the influence of organic matter on cation exchange capacity (CEC) and consequently soil fertility (Liu et al. 2020).

Fig. 19.5 Local collecting the grass seeds

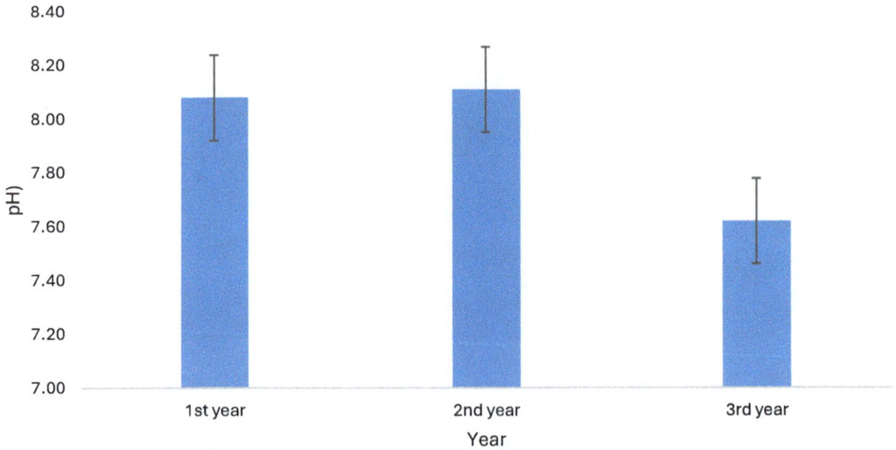

Fig. 19.6 Soil pH variation across the three years of restoration

The conductivity of the soil determines the quantum of exchangeable ions available to the plants for growth. Higher conductivity restricts the movement of the ions and for effective movement of essential nutrients across the soil root zones, low conductivity is preferred. In the present study, the conductivity declined from 380 mS to 120 mS in the third year and the changes recorded are significant at 0.02 level (Fig. 19.7).

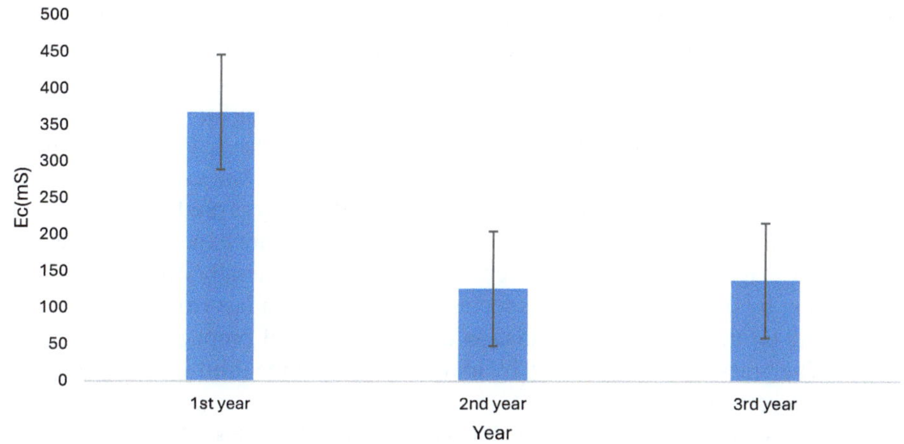

Fig. 19.7 Soil Conductivity (mS) variation across the three years of restoration

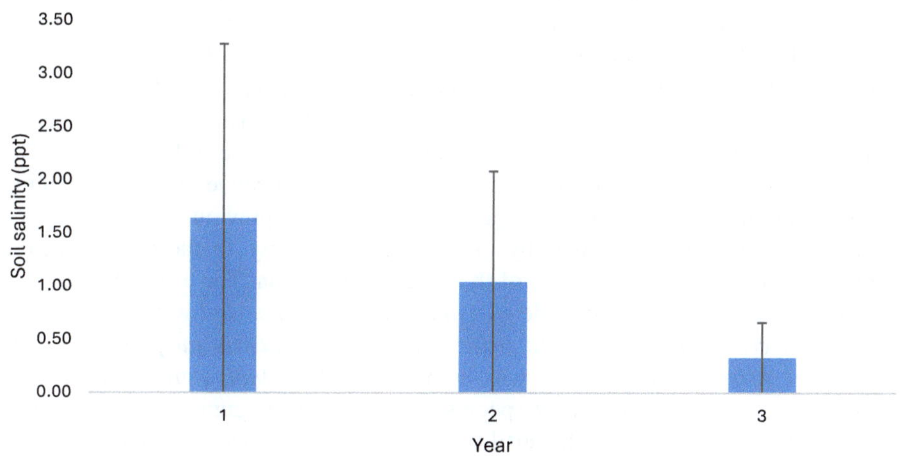

Fig. 19.8 Soil salinity (ppt) variation across the three years of restoration

The soil salinity of the area prior to the restoration initiation was around 2 ppt, showing typical arid condition with salinity ingress related land degradation issues. The growth of *Prosopis juliflora* in high density marks the presence of high salinity. However, as the restoration process was initiated, in the second year itself considerable reduction in the soil salinity was recorded and reduced to almost half as compared to first year (Fig. 19.8). By the third year, the soil salinity reduced further, and the difference observed is significant at ($p = 0.02$). Reduction in soil salinity ensures movement of nutrients across the system and indicates improvement in soil fertility. With the decline in soil salinity certain species that were reported in the first year diminished and gradually replaced by other species in the third year. Organic amendments such as biosolids, composted material and mulches may increase soil microbial activity, which favour organic matter decomposition and mineralization, and generally increases plant productivity and carbon sequestration in the medium or long-term, as highlighted by Valdecantos and Fuentes (2018).

Herbaceous layer species diversity increased from 19 in the beginning of the restoration (first year) to 33 species at the end of second year and 49 species at the end of third year. Low Shannon diversity index indicates low evenness and high ecological dominance. This shows that low species diversity is predominately occupied by many individuals of a few species. These results are expected for areas in the initial stage of ecological succession, which are generally characterized by occupation by only a few pioneer species, a phase that can last from 5 to 10 years, depending on the degree of degradation and/or scarcity of propagules (Neto et al. 2021). The grass density increased by about 23 times in the third year as compared to the first year (Table 19.1).

In the first year, *Cyperus arenarius* was the dominant species followed by *Cressa cretica, Aristida funiculata, Sporobolus marginatus,* and *Eleusine indica*. These can be considered as the primary species present in the area before restoration work was initiated. The species composition indicated the presence of saline tolerant species and higher degraded state. In the second year, post-fencing and control of grazing along with soil remediation measures increased the diversity to 33. At this stage, the sequence of dominant species was *Aristida funiculate, Goniogyna hirta, Eleusine compressa, Triumfetta rotundifolia,* and *Cressa cretica.* These pioneer species played an important role in nutrient recycling and enriching the soil conditions. The resultant was a decline in the conductivity and salinity indicative of improvement in the soil conditions. The generation of biomass from the previous year also added organic carbon to the soils thereby improving the fertility. In the third year, the number of species increased to 50 of the ground vegetation. The top five species in the third year recorded include *Aristida adscensionis, Indigoferea linifolia, Indigofera cordifolia, Goniogyna hirta,* and *Melanocenchris jacquemontii.* The density of *Aristida adscensionis* was high followed by *Indigoferea linifolia* and *Indigofera cordifolia. Indigofera* species belong to the Leguminosae family and could fix nitrogen and enrich the fertility of the soils. *Melanocenchris jacquemontii,* a palatable grass species has reported an increase in density. The species *Goniogyna hirta* grows close to the soil and has high mulching properties and retains soil moisture. Another species that showed promising growth in the restored plot was *Enicostemma hyssopifolium*, which has high medicinal value. Thus, in the restoration process, the coming up of the top five species indicates improvement in the

Table 19.1 Species diversity indices at the restoration plot

	2021	2022	2023
No. of species	19	33	49
Simpson_1-D	0.2764	0.6674	0.7656
Shannon_H	0.7334	1.576	1.833
Evenness_e^H/S	0.1157	0.1612	0.1787
Grass density (No./Ha)	32535.71	119785.71	749333.66

habitat and the recovery of the degraded system. Restoration increased the soil nutrients and vegetation biomass. Grass diversity mainly affected the organic matter variance in the semi-arid Naiman desert (Zhou et al. 2023).

The restoration process was initiated with the involvement of the local community. The community removed the invasive *Prosopis juliflora* in the first year and collected the grass seeds produced in the first year for use in the second year. This initiative improved the quality of grass density in the second year and subsequently in the third year (Fig. 19.9). Along with this, the community enriched the soils with organic manure produced locally as they are the pastoral community and rear livestock. This step ensured improvement in the soil properties and good green cover (Table 19.1). Revegetation efforts generally result in improvement of soil fertility and enhancement of ecosystem services and functions such as carbon sequestration (Gao et al. 2018), nutrient cycling (Barliza et al. 2018; Hu et al. 2018) and soil microbial diversity and activity (García-Palacios et al. 2018). Similar phenomenon has taken place in the present restoration efforts. Abbandonato et al. (2018) stressed the need for use of native seed species in restoration efforts.

Vegetation restoration significantly increased the proportion of macroaggregates and aggregate stability. Vegetation restoration increased the aggregate-associated Organic carbon (OC) content and bulk Soil Organic carbon (SOC) content, and macroaggregates played an important role in soil carbon sequestration (Zhuoxia et al. 2025). Rosenzweig et al. (2016) suggested that soil properties related to nutrient retention and regulation of nutrient availability can be reestablished within the first few decades of restoration, while C sequestration will continue to scale.

PCA analysis points out that the restoration process in all the 3 years behaved differently (Fig. 19.10). In the first year, all the parameters studied were not in so good condition. In the second year, vegetation improved; however, the soil salinity was on higher scale, though reduced as compared to first year. In the third year, all the positive impacts were visible with improvement in the soil properties and vegetation. Feng et al. (2011) found that revegetation had a positive effect on soil bulk density decrease, and total porosity and capillary porosity increase in the surface soil layers but not in the subsurface layer. Additionally, soil organic matter, total nitrogen, available nitrogen, and available potassium were greater at shrubland and woodland sites compared with other land uses.

Zhao et al. (2022) studied the secondary forest restoration effects in southern China particularly soil physico-chemical and microbiological properties. As the forest restoration proceeded, the soil properties, Bulk Density (BD), porosity, and water holding capacity were significantly improved. The soil structure, especially the topsoil was also largely optimized. Total soil nutrients and soil microbial biomass also gradually increased.

On the invasion of *Prosopis juliflora* in protected areas of India under climate change have stressed the need to manage its spread protect and conserve the threatened vertebrates' habitats (Singh et al. 2024). The study further highlighted that overall area of *P. juliflora* suitability in the protected areas will increase by 2050 and higher vertebrate diversity are in the area suitable for *P. juliflora.* The present

Fig. 19.9 The growth of ground vegetation for three years

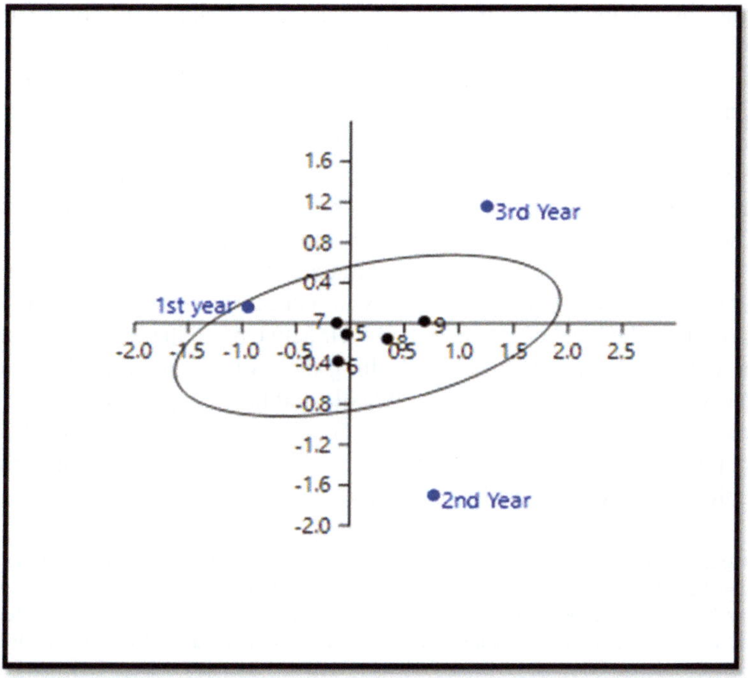

Fig. 19.10 Graph representing the PCA analysis between the sites

approach of restoration involving community participation will be useful in controlling the invasion of *Prosopis* and its control.

Effective restoration strategies are often specific to a site, and successes in one region or ecosystem may not be easily transferred to another location. Restoration requires committed and concerted efforts over many years. Zhuoxia et al. (2025) provided an important example of building the ecological restoration roadmap and forecasting Green House Gases (GHG) emissions caused by vegetation restoration from a broad landscape of fragile systems, supporting climate mitigation policies. Abbandonato et al. (2018) proposed native seed policy for European countries. Through the policy, it is envisaged that there will be availability of large number of herbaceous native seed for restoration.

19.4 Conclusions and Recommendations

In the arid zone, where soil salinity is a challenge, the present model, where ~~in~~ restoration is undertaking by simply fencing the area and addition of organic manure proves a low-cost technique to improve degraded lands in arid climatic zones. This substantiated with collection of productive seeds by the locals and broadcasting before the onset of the monsoon will ensure long term sustainability of the area. The improvement in the plant diversity will ensure availability of fodder to the local community and aid to fight climate inducted challenges. This could prove to be an important tool to restore vast tracts of degraded lands by creating smaller enclosures/restoration plots and ensure to achieve climate mitigation strategies. IUCN stresses the urgent need to link all climate change mitigation and adaptation activities, including restoration, for improving biodiversity, ecosystem integrity, and human well-being. Every restoration site is embedded within a larger ecological, cultural, and socio-economic landscape or seascape, and conditions and activities within the broader geographic area greatly influence trajectories of recovery.

Empowering the local community and their active participation in the restoration process will prove to be useful in controlling the invasion of exotic species. Small level intervention plots will be easy to handle and have long term benefits as well. The local community use the common land for grazing and involving the herders in the restoration process will enable control of *Prosopis* and at the same time, the community will reap the benefits of reintroduction of the native biodiversity.

Acknowledgement The authors are indebted to the local community of Guneri Village, especially Mansinh Jadeja, who was the focal point of contact between us and the community. The support and assistance of the local community at each phase of the restoration is worth mentioning. The study was undertaken with financial support from Adani Ports and SEZ Ltd, Mundra.

Declarations Informed Consent Statement: Not applicable.

Data Availability Statement: Data are available from the authors on request.

Ethics Approval: Not applicable.

Conflict of interest: The authors declare no conflict of interest.

References

Abbandonato H, Pedrini S, Pritchard HW, De Vitis M, Bonomi C (2018) Native seed trade of herbaceous species for restoration: a European policy perspective with global implications. Restor Ecol 26(5):820–826

Barliza JC, Peláez JDL, Campo J (2018) Recovery of biogeochemical processes in restored tropical dry forest on a coal mine spoil in La Guajira, Colombia. Land Degrad Dev 29:3174–3183. https://doi.org/10.1002/ldr.3069

CBD (Convention on Biological Diversity) (2015) History of the convention, introduction. http://www.cbd.int/history/

CBD (Convention on Biological Diversity) (2016) CBD COP13 decision XIII/23. http://www.cbd.int/doc/decisions/cop-13/cop-13-dec-23-en.pdf

CGWB (2013) Groundwater brochure Kachchh district, technical report series. Central Water Board, Government. http://cgwb.gov.in/District_Profile/Gujarat/Kachchh.pdf

Chaudhary JP, Sipai AH, Chaudhary PP, Vaghela ID (2018) Major-nutrients status in soils of Kachchh district of Gujarat

Das A, Shukla SH, Joshi PN, Prajapati PP (2019) Wetland ecosystem sustainability over human wildlife conflict in and around Lakhpat taluka, Kachchh district, India. Themat J Geogr 8(4)

Eicken H, Danielsen F, Sam J-M, Fidel M, Johnson N, Poulsen MK, Lee OA, Spellman KV, Iversen L, Pulsifer P, Enghoff M (2021) Connecting top-down and bottom-up approaches in environmental observing. Bioscience 71. https://doi.org/10.1093/biosci/biab018

Feng J, Wen Z-M, Shao-Shan (2011) Changes in soil properties across a chrono sequence of vegetation restoration on the Loess Plateau of China. Catena 86(2):110–116., ISSN:0341-8162. https://doi.org/10.1016/j.catena.2011.03.001

Gadhvi JIR (2003) Monitoring nesting of lesser florican (*Sypheotides indica*) in and around blackbuck National Park, Gujarat. Zoos Print J 18(5):1135–1142

Gao Y, Dang P, Zhao Q, Liu J, Liu J (2018) Effects of vegetation rehabilitation on soil organic and inorganic carbon stocks in the Mu Us Desert, northwest China. Land Degrad Dev 29(4):1031–1040. https://doi.org/10.1002/ldr.283

García-Palacios P, Gross N, Gaitán J, Maestre FT (2018) Climate mediates the biodiversity-ecosystem stability relationship globally. Proc Natl Acad Sci U S A. 115(33):8400–8405

Gastauer M, Cavalcante RBL, Caldeira CF, Nunes S (2020) Structural hurdles to large-scale forest restoration in the Brazilian Amazon. Front Ecol Evol 8:1–6. https://doi.org/10.3389/fevo.2020.593557

Gavali DJ, Lakhmapurkar JJ (2009) Soil salinity versus *Prosopis juliflora* growth – a case of Bani Grasslands. In: Proceedings of the national symposium on Prosopis: ecological, economic significance and management challenges, pp 33–36

Hu YF, Shu XY, He J, Zhang YL, Xiao HH, Tang XY, Gu Y-F, Lan T, Xia J-G, Ling J, Chen GD, Wang C-Q, Deng L-J, Yuan S (2018) Storage of C, N, and P affected by afforestation with *Salix cupularis* in an alpine semiarid desert ecosystem. Land Degrad Dev 29:188–198. https://doi.org/10.1002/ldr.2862

Liu H, Ye Q, Wiens JJ (2020) Climatic-niche evolution follows similar rules in plants and animals. Nat Ecol Evol 4:753–763

Neto ABB, Schwartz G, Noronha NC, Andr'e Piedade Gama M, Ferreira GC (2021) Natural regeneration for restoration of degraded areas after bauxite mining: a case study in the Eastern Amazon. Ecol Eng 171(2021):106392

Ockendon N, Thomas DH, Cortina J, Adams WM, Aykroyd T, Barov B, Boitani L, Bonn A, Branquinho C, Brombacher M, Burrell C, Carver S, Crick HQP, Duguy B, Everett S, Fokkens B, Fuller RJ, Gibbons DW, Sutherlansald WJ (2018) One hundred priority questions for landscape restoration in Europe. Biol Conserv 221:198–208. https://doi.org/10.1016/j.biocon

Rahamni AR (1998) The Banni grasslands: natural resource under siege. Sanctuary Asia XVIII(2):40–49

Rastogi BK, Kothyari GC, Sairam B, Kumar GP, Patel V, Goswami OP, Prajapati N (2016) Geophysical evidence for paleochannels and possible sources of groundwater: a case study from Kachchh region, western peninsular India. J Earthq Sci Eng 3:1–17

Robinson JM, Gellie N, MacCarthy D, Mills JG, O'Donnell K, Redvers N (2021) Traditional ecological knowledge in restoration ecology: a call to listen deeply, to engage with, and respect Indigenous voices. Restor Ecol 29. https://doi.org/10.1111/rec.13381

Rosenzweig ST, Carson MA, Baer SG, Blair JM (2016) Changes in soil properties, microbial biomass, and fluxes of C and N in soil following post-agricultural grassland restoration. Appl Soil Ecol 100:186–194. ISSN:0929-1393

Santini NS, Miquelajauregui Y (2022) The restoration of degraded lands by local communities and indigenous peoples. Min review. Front Conserv Sci 3:873659. https://doi.org/10.3389/fcosc.2022.873659

Santini NS, Lovelock CE, Hua Q, Zawadzki A, Mazumder D, Mercer TR (2019) Natural and regenerated saltmarshes exhibit similar soil and belowground organic carbon stocks, root production and soil respiration. Ecosystems 22:1803–1822. https://doi.org/10.1007/s10021-019-0037

Singh M, Kumar L, Rajasekaran A, Amaladas P, Anuja H (2024) *Prosopis juliflora* (Swartz) DC. invasion in protected areas of India under climate change. J Nat Conserv 79:126622., ISSN:1617-1381. https://doi.org/10.1016/j.jnc.2024.126622

Sipai AH, Chaudhary JP, Damor NN, Bhavik R, Chauhan VA (2022) Major nutrients status in soil and assessment of quality of irrigation water of Kachchh district of Gujarat, India. Plant Arch (09725210) 22(2)

Stock E, Standish RJ, Muñoz-Rojas M, Bell RW, Erickson TE (2020) Field-deployed extruded seed pellets show promise for perennial grass establishment in arid zone mine rehabilitation. Front Ecol Evol 8:1–15. https://doi.org/10.3389/fevo.2020.576125

Valdecantos A, Fuentes D (2018) Carbon balance as affected by biosolid application in reforestations. Land Degrad Dev 29:1442–1452. https://doi.org/10.1002/ldr.2897

Zhao X, Peiling L, Yingjie F, Weiqiang Z, Brian N, Fengling L, Zhou Qing Q, Chao GX, Xiaodong L (2022) Changes in soil physico-chemical and microbiological properties during natural succession: a case study in lower subtropical China. Front Plant Sci 13. https://doi.org/10.3389/fpls.2022.878908

Zhou W, Li C, Wang S, Ren Z, Stringer LC (2023) Effects of vegetation restoration on soil properties and vegetation attributes in the arid and semi-arid regions of China. J Environ Manag 343:118186. ISSN:0301-4797

Zhuoxia S, Zhu S, Wei Z, He Y, Bingqian S, Zhang K, Ma X, Shangguan Z (2025) Vegetation restoration changed the soil aggregate stability and aggregate carbon stabilization pathway according to δ13C signatures. Agric Ecosyst Environ 378:10931. ISSN:0167-8809

Chapter 20
Climate Change and the Dynamics of Coastal Pollution: A Plankton-Based Assessment of Chennai's Urban Creeks

Radha Vijayaraj, Kareem Altaff, Prem Kumar Duraivel, Philip Theodore Reginald, and Raju Piliyan

Abstract The coastal waters of Chennai face increasing threats from pollution caused by untreated sewage, industrial effluents, and urban runoff, particularly through creeks such as Ennore, Napier, and Adyar. This study assesses the extent of coastal pollution using phytoplankton and zooplankton communities as bioindicators. Environmental parameters including temperature, pH, salinity, total dissolved solids (TDS), dissolved oxygen (DO), nitrate, inorganic phosphate, and total chlorophyll were measured at each site to evaluate water quality. Results revealed significant spatial variability in physicochemical parameters, with Napier Creek exhibiting the highest pollution load, characterized by low DO levels (<2 mg/L) and reduced plankton densities. In contrast, Adyar Creek showed elevated plankton densities (>500 individuals/L), indicative of eutrophic conditions driven by nutrient enrichment. Regression analysis demonstrated strong correlations between plankton densities and environmental parameters, underscoring the sensitivity of plankton communities to changes in water quality. Phytoplankton diversity included genera such as *Chlorella* spp., *Nitzschia* spp., and *Skeletonema* spp., while zooplankton communities were dominated by copepods such as *Paracalanus crassirostris*, *Tortanus discaudatus*, and *Oithona* spp. The diversity and density of these bioindicators provided valuable insights into the ecological status of the creeks. Moreover, climate change is expected to intensify pollution dynamics and biodiversity shifts by altering temperature, salinity, and nutrient availability factors that directly affect planktonic community structures and overall ecosystem health. This study demonstrates the utility of plankton-based biomonitoring in assessing coastal pollution and underscores the urgent need for improved wastewater treatment and ongoing environmental surveillance to safeguard Chennai's coastal ecosystems.

R. Vijayaraj (✉) · K. Altaff · P. K. Duraivel · P. T. Reginald · R. Piliyan
Aquaculture Research Laboratory, Department of Marine Biotechnology, AMET University, Chennai, India
e-mail: vijayarajr@ametuniv.ac.in

B. P. Barathan et al. (eds.), *Navigating Climate Change: Impacts on Biodiversity and Ecosystem Resilience*, Sustainable Development and Biodiversity 42,
https://doi.org/10.1007/978-981-95-0409-1_20

Keywords Planktonic bioindicators · Environmental monitoring · Coastal pollution · Climate change · Biodiversity

20.1 Introduction

The city of Chennai, situated on the southeastern coast of India, stands as the fourth largest metropolis in the country (Arun Kumar and Kunte 2012; Krishnamurthy and Desouza 2015; Mani and Jayalakshmi 2024). Its coastal zone spans approximately 30 km from the Adyar River in the south to the Ennore creek in the north (Neelavannan et al. 2021). Unfortunately, this coastal region exemplifies uncontrolled disposal of wastewater and severe pollution levels (Shanmugam et al. 2007; Sudha Rani et al. 2015). The degradation is largely attributed to various anthropogenic factors such as rapid population growth (Padmanaban et al. 2017), the development of both small and large-scale industries (Saravanan et al. 2019), expansion of harbors (Ramasamy et al. 2018), tourism-related activities (VishnuRadhan et al. 2015), inadequate waste management (Shiam Babu et al. 2025), and a plethora of recreational and commercial activities. These not only diminish the quality of coastal water but also pose substantial health risks to both marine life and humans (Shanmugam et al. 2007). The primary sources of coastal pollution in Chennai include the Cooum and Adyar rivers, outlets of the Buckingham canal, industrial effluent discharge points, and industrial and domestic sewage discharge locations along the coastal stretch, particularly prominent in the northern Chennai region (Shanmugam et al. 2007; Dhandapani 2018; Basheer Ahamed 2023). This pollution significantly impacts the marine environment, necessitating a comprehensive biomonitoring and bioindicators study.

Biomonitoring and bioindicators play a crucial role in assessing the health and ecological condition of an environment, especially in the case of polluted coastal areas (Manzoor et al. 2021; Katkova-Zhukotskaya and Kalyujnaya 2023). Biomonitoring involves the use of organisms to evaluate the quality of an environment by analyzing their responses to various environmental stressors (Ogidi et al. 2024). Bioindicators are specific organisms or biological responses that signify particular environmental conditions or stressors (Dahiya et al. 2022). Both biomonitoring and bioindicators serve as valuable tools in environmental science, aiding in the assessment, management, and preservation of ecosystems.

In the context of Chennai's coastal pollution, conducting a biomonitoring and bioindicators study is of essential. This study aims to comprehensively investigate different coastal areas, considering physicochemical parameters and the plankton community. Moreover, climate change is likely to exacerbate the dynamics of coastal pollution and biodiversity. Variability in ocean temperature and dissolved oxygen levels, driven by climate change, can affect primary productivity and plankton communities. Warming waters can alter phytoplankton growth rates and distribution, while reduced oxygen availability can lead to hypoxic conditions that disrupt plankton populations, impacting the entire marine food web. These factors further highlight the importance of monitoring planktonic communities as bioindicators of both pollution and climate-induced environmental changes. To comprehensively address this issue and safeguard the marine ecosystem and human health, a biomonitoring and bioindicators study focusing on physicochemical parameters and the plankton community is essential. This study will provide critical data to assess the extent of pollution, guide conservation efforts, and ultimately work toward restoring the health and balance of Chennai's coastal environment.

20.2 Materials and Methods

20.2.1 Sampling Site and Sampling Method

Samples were collected from three coastal regions of Chennai (Ennore creek, Napier creek, Adyar creek) along the Bay of Bengal, known for fishing, industrial activities, and tourism. The Ennore station is affected by industrial effluents, extensive fish landings and harbor activities. Adyar station is in the estuarine region which is the breeding ground of commercially important finfish and shellfish. This station is highly polluted with domestic discharge having organic content. Standard plankton nets were used to collect samples: a phytoplankton net with a 20-micron mesh size and a zooplankton net made of Bolton silk with a 150-micron mesh size.

Phytoplankton samples were immediately preserved in Lugol's iodine solution, and zooplankton samples were preserved in 5% formalin buffered with seawater, then transported to the laboratory for further analysis (Altaff 2004; Sivakumar et al. 2021). Phytoplankton species were identified using a high-magnification compound microscope, following the methods of Gopinathan et al. (2007) and Verlecar and Desai (2004). Zooplankton species were identified using standard taxonomic keys (Kasturirangan 1963; Boxshall and Halsey 2004; Altaff 2004; Razouls et al. 2022).

Quantitative analysis of phytoplankton was performed using a Sedgewick-Rafter counting chamber. One milliliter of subsample was taken and transferred to the counting chamber, where cells in 10 randomly selected squares were counted. For zooplankton, samples were mixed thoroughly, and 1 mL (equivalent to 20 drops) was extracted using a wide-mouthed pipette and transferred to the Sedgewick-Rafter counting chamber for counting under a compound microscope. Mean values were calculated from three subsamples (Perry 2010; Sivakumar et al. 2021).

20.2.2 Analysis of Physicochemical Parameters

The physicochemical parameters, including temperature, pH, total dissolved solids (TDS), salinity, total alkalinity, dissolved oxygen, nitrate, inorganic phosphate, and total chlorophyll, were analyzed during April 2024. The water temperature was measured using a standard mercuric thermometer. Salinity was determined using Mohr's method (Belcher et al. 1957). pH was assessed using a standard pH meter, while dissolved oxygen levels were estimated using a modified Winkler's method (Golterman 1983). Nitrate in the water was converted to nitrite by passing a sample through a column containing cadmium filings lightly coated with metallic copper. An adapted colorimetric method was employed to analyze inorganic phosphate in seawater, generating a highly colored blue Phosphomolybdate complex (Murphy and Riley 1962). All statistical analysis was carried out using PAST 4.09.

20.3 Results and Discussion

Chennai's marine environment faces significant pollution challenges, primarily due to the discharge of untreated sewage from the Ennore, Napier, and Adyar Creeks. This leads to elevated levels of pathogens, harmful algal blooms, eutrophication, and oxygen depletion, particularly near river mouths and creeks. Biomonitoring studies using plankton communities, especially phytoplankton and zooplankton, are ideal for assessing pollution levels in marine ecosystems. Therefore, this study focuses on biomonitoring coastal pollution in Chennai using planktonic bioindicators, with particular emphasis on the Ennore, Napier, and Adyar Creeks. In addition to anthropogenic pollution, climate change has further exacerbated the challenges to Chennai's coastal ecosystem. The rising ocean temperatures, coupled with altered oxygen dynamics, have influenced plankton populations. Warming waters have been shown to accelerate the growth rates of certain phytoplankton species, but also cause a shift in species composition, favoring heat-tolerant species that may not contribute as effectively to marine food webs. Moreover, reduced oxygen levels, driven by both pollution and climate change, have created hypoxic conditions in some areas, significantly impacting zooplankton populations, particularly in the Ennore and Napier Creeks. Zooplankton, which serve as critical food sources for fish larvae, are especially vulnerable to these changes. The decrease in zooplankton densities, along with changes in phytoplankton species composition, underscores the need for a comprehensive understanding of how climate variability interacts with pollution stressors in coastal regions. This study highlights how the interactions between climate change and pollution are altering marine biodiversity, particularly the delicate plankton communities that serve as bioindicators of ecosystem health. The data presented in this study reveal significant spatial variability in plankton diversity and density, with areas impacted by both sewage discharge and warming waters showing reduced plankton populations and altered community structures.

These findings reinforce the need for integrated environmental monitoring that takes into account both pollution and climate change to develop effective conservation strategies for Chennai's coastal ecosystems.

20.3.1 Environmental Parameters of Ennore, Napier, and Adyar Creeks

Chennai's coastal waters, specifically Ennore, Napier, and Adyar Creeks, are increasingly affected by pollution due to uncontrolled waste disposal and industrial discharges, which have significant impacts on the marine environment and plankton communities (Waters 2004; Reddy and Nagendra 2011; Begum et al. 2022). Elevated concentrations of heavy metals such as lead, nickel, cobalt, and cadmium have been detected in certain locations (Waters 2004; Reddy and Nagendra 2011; Begum et al. 2022). This contamination can influence the growth and abundance of phytoplankton, the primary producers in the marine food web, and may also affect zooplankton populations (Shanmugam et al. 2007). Abiotic factors in the marine environment, which are the non-living components, play a critical role in shaping the physical and chemical characteristics of the ocean, thereby influencing the distribution and interactions of living organisms (Ali 2023; Sumuki 2023). These factors include temperature, salinity, sunlight, depth, dissolved gases, and substrate type (Gatamaneni et al. 2018; Sumuki 2023). With the ongoing impacts of climate change, ocean temperatures are rising, further altering the distribution of plankton species and their physiological responses. Additionally, the decreasing oxygen content in coastal waters is an emerging challenge, likely driven by both pollution and climate change. Understanding these abiotic parameters is essential for assessing the relationship between plankton communities and the extent of pollution in marine ecosystems (Häder and Gao 2015; Gunathilaka et al. 2023).

20.3.2 Pollution Assessment Based on Environmental Parameters

The measured environmental parameters reflect varying levels of anthropogenic impact across the three creeks (Table 20.1). The high TDS levels observed in Ennore Creek (26,123.33 mg/L) compared to Napier and Adyar Creeks may be attributed to the discharge of untreated industrial effluents and thermal power plant activities in the region, which are known to elevate salinity and dissolved solids through chemical runoff and heated water discharge (Pandey et al. 2020; Lemessa et al. 2023). Similarly, elevated nitrate and phosphate concentrations, particularly in Napier Creek (nitrate: 4.6 mg/L; phosphate: 1.82 mg/L), indicate nutrient enrichment

Table 20.1 Environmental parameters of Ennore, Napier, and Adyar Creeks

Parameter	Ennore creek	Napier creek	Adyar creek
Temperature	26.1 ± 0.95	26.2 ± 1.03	25.95 ± 0.82
pH	7.6 ± 0.51	7.46 ± 0.39	7.5 ± 0.34
Total dissolved solids mg/L	26123.33 ± 20.20	2765.33 ± 134.52	2234 ± 76.2
Salinity (PSU)	24.33 ± 5.15	23 ± 6.9	22.33 ± 4.57
Total alkalinity (mg/L)	246.66 ± 5.77	253.33 ± 11.54	241.33 ± 1.15
Dissolved oxygen mg/L	3.9 ± 0.57	3.2 ± 2.5	5.2 ± 0.51
Nitrate (mg/L)	3.11 ± 0.04	4.6 ± 0.99	2.25 ± 1.11
Inorganic phosphate (mg/L)	1.5 ± 0.15	1.82 ± 0.61	1.23 ± 0.83
Total chlorophyll	1.82 ± 0.08	1.21 ± 1.03	2.31 ± 1.91

The values are represented as mean ± standard deviation

(eutrophication) from domestic sewage inflow and urban runoff (Sheffield 2019; Nethavhani 2023). This can lead to algal blooms, followed by hypoxic conditions, negatively affecting aquatic organisms, especially sensitive zooplankton species (Sheffield 2019; Nethavhani 2023). Dissolved oxygen levels were lowest in Napier Creek (3.2 ± 2.5 mg/L), suggesting possible organic pollution and microbial degradation processes that consume oxygen, further stressing aquatic life (Kannan et al. 2022). In contrast, Adyar Creek, with relatively higher DO (5.2 mg/L), may receive more freshwater inflow or experience better tidal flushing, which helps maintain oxygen levels (O'Connor et al. 2022). The slightly acidic pH values across all sites (pH ~7.4–7.6) could reflect increased CO_2 levels from microbial activity associated with organic waste decomposition (Freitas et al. 2022). These changes, combined with climate-induced temperature fluctuations and reduced oxygen solubility in warmer waters, exacerbate the effects of pollution. Rising temperatures may also enhance the breakdown of organic material, leading to further depletion of oxygen and promoting harmful algal blooms.

Among the three creeks, Ennore Creek exhibited the highest pollution levels, as indicated by the exceptionally high concentration of total dissolved solids (26,123.33 mg/L) and relatively low dissolved oxygen (3.9 mg/L). Elevated TDS often reflects a higher load of dissolved salts, heavy metals, and organic pollutants, while low DO indicates poor water quality and stress on aquatic life (Scannell and Jacobs 2001; Adjovu et al. 2023). The moderate levels of nutrients (nitrate and phosphate) and chlorophyll also suggest eutrophic conditions potentially driven by industrial and sewage discharges (Jarvie et al. 2006; Silva et al. 2013). In contrast, Adyar Creek displayed the lowest pollution levels, with lower TDS (2234 mg/L) and the highest DO (5.2 mg/L) among the sites, suggesting comparatively better water quality (Dhayachandhran and Jothilakshmi 2021). Although it exhibited slightly higher total chlorophyll, indicating productive conditions, the lower nutrient concentrations and better oxygenation suggest a relatively less polluted environment (Badran 2001; Häder and Gao 2015). Napier Creek falls in an intermediate pollution category, showing elevated nutrient concentrations (nitrate: 4.6 mg/L, phosphate: 1.82 mg/L) and variable DO (3.2 mg/L), which may point to organic pollution and nutrient loading from urban runoff or domestic sewage (Nethavhani 2023). These insights help clarify the spatial variation in pollution across Chennai's coastal systems and underline the urgent need for targeted environmental management, especially in the Ennore region.

20.3.3 Total Density of Plankton Communities in Ennore, Napier, and Adyar Creeks

Phytoplankton density serves as a critical biomonitoring parameter, offering insights into the ecological status of aquatic ecosystems (Kashyap et al. 2024; Hu et al. 2025). As the foundation of the aquatic food web, phytoplankton populations are particularly sensitive to changes in water quality and can reflect the presence of environmental stressors, such as nutrient enrichment, chemical contamination, or reduced oxygen levels (Häder and Gao 2015; Salmaso and Tolotti 2021). Climate change, especially rising ocean temperatures, can further stress plankton communities by altering nutrient cycling and promoting the dominance of harmful algal species (Burford et al. 2018). Warmer waters can accelerate algal blooms and exacerbate hypoxic conditions, further impeding plankton survival and community structure. In the present study, the total plankton density (phytoplankton and zooplankton) was assessed in Ennore, Napier, and Adyar Creeks, each experiencing varying degrees of anthropogenic pollution and climatic stress (Table 20.2).

In Ennore Creek, the phytoplankton density was 2012.2 ± 172 cells/L, indicating a moderately high primary productivity level. However, this could be associated with organic pollution and nutrient enrichment from industrial effluents and thermal discharges from nearby power plants (Begum et al. 2022). Zooplankton density in this region was also high (28.64 ± 1.4 ind/L), which may reflect a stable trophic response to increased phytoplankton availability, although such responses may also indicate an imbalance caused by eutrophication (Striebel et al. 2012; Ger et al. 2016). Napier Creek exhibited the lowest phytoplankton density among the three sites (723 ± 31 cells/L), suggesting either poor nutrient availability or high levels of pollution that inhibit algal growth (Zhu et al. 2010). The low zooplankton density (8.14 ± 0.405 ind/L) further indicates ecological degradation, likely due to high levels of organic and chemical pollutants entering the creek through untreated sewage and urban runoff (de Paggi Susana et al. 2008; Xiong et al. 2019). This site appears to be under significant ecological stress, leading to reduced biological productivity and biodiversity (de Paggi Susana et al. 2008; Xiong et al. 2019). In contrast, Adyar Creek showed the highest phytoplankton density (2218.2 ± 182 cells/L),

Table 20.2 Total plankton density at Ennore, Napier, and Adyar Creeks

Total density	Ennore creek	Napier creek	Adyar creek
Phytoplankton density (cells/L)	2012.2 ± 172^{a}	723 ± 31^{b}	2218.2 ± 182^{a}
Zooplankton density (ind/L)	28.64 ± 1.4^{a}	8.14 ± 0.405^{b}	36.5 ± 2.16^{a}

potentially indicating nutrient enrichment from domestic wastewater inflows (Skejić et al. 2014). The accompanying high zooplankton density (36.5 ± 2.16 ind/L) suggests that despite pollution, the system supports a productive food web (Bettinetti and Manca 2013). However, elevated plankton densities in urban waterways are often symptomatic of eutrophication, driven by nutrient loading and decreased water quality (Oliver et al. 2019; Freeman et al. 2019). Overall, the variation in plankton density among the creeks highlights the differing pollution loads and ecological impacts. While Ennore and Adyar Creeks show signs of eutrophication due to nutrient enrichment, Napier Creek's low productivity reflects possible acute toxicity or oxygen depletion, severely impairing the plankton community. These observations underscore the need for continuous monitoring and pollution mitigation strategies, especially in light of the increasing pressures from climate change, to restore ecological balance in Chennai's coastal water bodies.

The values are represented as mean ± standard deviation; Values with different alphabet superscripts in column denote they show significance with each other in ANOVA followed by Tukey's test ($p < 0.05$).

A comprehensive regression analysis was conducted to examine the relationships between phytoplankton and zooplankton densities and various environmental parameters. The regression coefficients (R) offer valuable insights into the strength and direction of these associations. For phytoplankton density, negative correlations were observed with temperature ($R = -0.87$), total alkalinity ($R = -0.94$), nitrate ($R = -0.95$), and inorganic phosphate ($R = -0.94$), indicating a decline in phytoplankton density with increasing levels of these parameters. In contrast, positive correlations were found with pH ($R = 0.63$), total dissolved solids ($R = 0.37$), dissolved oxygen ($R = 0.75$), and total chlorophyll ($R = 0.94$), suggesting that higher values of these factors support increased phytoplankton density (Table 20.3).

Similarly, zooplankton density showed negative correlations with temperature($R = -0.89$), total alkalinity ($R = -0.957$), nitrate ($R = -0.97$), and inorganic phosphate ($R = -0.95$). Positive correlations were observed with pH ($R = 0.60$), total dissolved solids ($R = 0.34$), dissolved oxygen ($R = 0.77$), and total chlorophyll ($R = 0.95$). These regression results highlight the complex interactions between plankton populations and environmental parameters. Understanding these relationships is essential for interpreting the dynamics of plankton communities and their responses to changing environmental conditions, thereby enhancing our comprehension of the marine ecosystem.

20.3.4 Diversity of Phytoplankton in Ennore, Napier, and Adyar Creeks

Phytoplankton are crucial bioindicators in marine ecosystems due to their rapid response to environmental changes, providing valuable insights into water quality and the presence of pollutants (Chandel et al. 2024; Gouda et al. 2024). Their

Table 20.3 Regression analysis between phytoplankton and zooplankton with environmental parameters

Regression coefficient (R)	Phytoplankton density	Zooplankton density
Temperature	−0.87	−0.89
pH	0.63	0.6
Total dissolved solids	0.37	0.34
Salinity	0.1	0.02
Total alkalinity	−0.94	−0.957
Dissolved oxygen	0.75	0.77
Nitrate	−0.95	−0.97
Inorganic phosphate	−0.94	−0.95
Total chlorophyll	0.94	0.95

Table 20.4 Diversity of phytoplankton observed at Ennore, Napier, and Adyar Creeks

Species	Ennore creek	Napier creek	Adyar creek
Cymbella marina	++	−	−
Chlorella sp.	++	+	+
Cymbella marina	++	+	+
Nanochloropsis	+	+	+
Planktoniella sol	−	−	+
Nitzschia longissima	−	−	+
Spirulina sp	+	−	++
Skeletonema costatum	−	−	+
Stephanophyxix palmeriana	−	−	−

+ denotes presence of the species in the station; ++ denotes the dominance of the species in the station; − denotes absence of the species in the station

sensitivity to shifts in the environment makes them indispensable tools for monitoring and managing marine health. In this study, the diversity of phytoplankton species across Ennore, Napier, and Adyar Creeks offers significant insights into the physicochemical composition and overall water quality of these coastal areas (Table 20.4). In Ennore Creek, the presence of *Cymbella marina* and *Chlorella* sp. suggests an environment that supports their growth, likely due to adequate nutrient availability. The occurrence of *Stephanopyxis palmeriana* further indicates favorable environmental conditions, while the presence of *Nitzschia longissima* reflects moderate nutrient concentrations. These species are generally associated with environments that are nutrient-rich and exhibit a balanced ecosystem conducive to phytoplankton proliferation. At Napier Creek, although specific species were not highlighted in this section, the lower planktonic density observed earlier may point to a disruption in the normal phytoplankton community due to elevated pollution levels. This suggests that nutrient overloads from industrial and domestic runoff could be inhibiting the growth of certain phytoplankton species. In contrast, Adyar Creek exhibits a more diverse phytoplankton community, with species such as *Skeletonema costatum*, *Nitzschia longissima*, *Cymbella marina*, and *Planktoniella sol*. These species indicate a range of nutrient levels and ecological conditions, pointing to a balanced ecosystem with moderate nutrient enrichment. The presence of *Stephanopyxis palmeriana* and *Spirulina* sp. further supports the observation of favorable growth conditions, suggesting a system that, despite pollution stress, supports a productive phytoplankton community. Together, the phytoplankton compositions across the three creeks highlight the dynamic interplay between nutrient availability, salinity, dissolved oxygen, and other environmental factors. These findings underscore the importance of monitoring these bioindicators to assess the health of coastal ecosystems and the impacts of pollution on marine biodiversity.

Table 20.5 Diversity of zooplanktons in Ennore, Napier, and Adyar Creeks

Order	Species	Ennore creek	Napier creek	Adyar creek
Decapoda	*Lucifer* spp.	+	+	–
Calanoida	*Acartia* sp.	+	–	+
	Acartia erythrea	–	–	++
	Acartia tonsa	–	–	+
	Acrocalanus gracilis	+	–	+
	Acrocalanus gibber	+	+	++
	Paracalanus aculeatus	–	–	+
	Parvocalanus crassirostris	+	–	+
	Clausocalanus minor	–	–	+
	Centropages furcatus.	–	–	+
	Subeucalanus flemingeri	+	–	–
	Pseudodiaptomus serricaudatus	+	+	+
	Temora turbinata	+	+	+
	Temora discaudata.	+	–	–
Cyclopoida	*Apocyclops royi*	++	++	++
	Apocyclops sp.	+	+	+
	Dioithona rigida	++	+	++
	Oithona similis	–	–	+
	Oithona attenuata	–	–	+
	Oithona sp.	+	–	+
	Corycaeus sp.	+	+	–
Harpacticoida	*Euterpina acutifrons*	+	+	+
	Onychocamptus bengalensis	–	–	+
Copepod nauplius		–	–	+
Rotifers		+	–	+
Cladocerans		+	–	+
Ostracods		+	+	+
Polychaete larvae		–	–	+
Fish eggs		+	–	+

+ denotes presence of the species in the station; ++ denotes the dominance of the species in the station; – denotes absence of the species in the station

20.3.5 *Diversity of Zooplankton in Ennore, Napier, and Adyar Creeks*

Zooplankton, particularly copepods, are widely recognized as effective bioindicators in marine environments due to their rapid response to changes in water quality (Bakhtiyar et al. 2020). They play a critical role in assessing ecosystem health, as many copepod species are highly sensitive to environmental stressors such as pollution and climate change (Vehmaa et al. 2013; Wang et al. 2018; Hussain et al. 2020). In this study, the composition of zooplankton across three coastal locations Ennore, Napier, and Adyar Creeks serves as a key indicator of the ecological health and physicochemical characteristics of these marine ecosystems (Table 20.5).

At Ennore, the zooplankton community comprises *Lucifer* spp., *Parvocalanus crassirostris*, *Subeucalanus flemingeri*, *Acrocalanus gibber*, and *Temora discaudata*, indicating a high species diversity and suggesting favorable environmental conditions that support zooplankton proliferation. At Marina Beach, the zooplankton assemblage includes *Lucifer* spp., *Oithona* sp., *Pseudodiaptomus serricaudatus*, and *Paracalanus aculeatus*, reflecting a dynamic and balanced marine ecosystem. The occurrence of *Lucifer* spp., a genus of bioluminescent zooplankton (Takenaka et al. 2012), may be influenced by physicochemical parameters such as temperature and dissolved oxygen levels. Across all study sites, the zooplankton community is notably diverse, featuring taxa such as *Centropages furcatus*, *Pseudodiaptomus serricaudatus*, *Euterpina acutifrons*, *Acrocalanus gibber*, *Clausocalanus minor*, *Paracalanus aculeatus*, *Oithona similis*, and *Oithona attenuata*. This high biodiversity points to a complex and dynamic ecosystem shaped by varying levels of nutrients, salinity, and temperature (Hooper et al. 2005; Jin 2008; Telesh et al. 2013). Among the study areas, the Adyar Creek Estuary stands out for its exceptional zooplankton diversity. The presence of multiple indicator species highlights the ecological richness and vitality of this estuarine system. Species such as *Parvocalanus crassirostris* and *Euterpina acutifrons* are particularly sensitive to fluctuations in temperature, dissolved oxygen, and nutrient concentrations, making them valuable bioindicators of environmental health.

20.3.6 *Biomonitoring and Bioindicators: Relationship Between Physicochemical Parameters and Plankton Communities*

Plankton communities comprising both phytoplankton and zooplankton serve as vital bioindicators for assessing the ecological health of aquatic ecosystems (Jakhar 2013). Their diversity, abundance, and composition are closely regulated by key physicochemical parameters such as temperature, pH, dissolved oxygen (DO), salinity, total dissolved solids (TDS), and nutrient concentrations (nitrate and phosphate) (Inyang and Wang 2020; El Nahhal et al. 2021). Through biomonitoring, the observed plankton responses to these variables can be effectively used to evaluate the level of pollution and environmental stress in marine and estuarine habitats (Hemraj et al. 2017; Sumudumali and Jayawardana 2021). Additionally, climate change significantly impacts plankton populations, altering their distribution, abundance, and seasonal dynamics. Changes in temperature, ocean acidification, and altered nutrient cycles directly affect plankton communities, modifying their roles as bioindicators and influencing their ability to reflect environmental conditions. This makes them even more critical for monitoring shifts in marine ecosystem health due to changing climatic conditions.

Temperature is a critical determinant of metabolic rates and reproductive cycles in plankton. In this study, a negative correlation was observed between temperature and both phytoplankton ($R = -0.87$) and zooplankton densities ($R = -0.89$), indicating that higher temperatures may inhibit plankton growth, possibly due to thermal stress or its effect on dissolved oxygen availability. Elevated temperatures may also reflect thermal pollution, potentially linked to industrial discharges. Moderate pH values around neutrality favor the survival of most plankton species. The positive correlation between pH and plankton densities (phytoplankton: $R = 0.63$; zooplankton: $R = 0.6$) suggests that slightly alkaline conditions support a more stable and productive planktonic community. Extreme pH values, often a result of chemical pollution, can disrupt cellular functions and enzyme activity, making pH a crucial bioindicator of pollution. Dissolved oxygen is a primary determinant of water quality and ecosystem productivity. High DO levels typically indicate well-oxygenated environments supportive of aerobic organisms. The strong positive correlation of DO with phytoplankton ($R = 0.75$) and zooplankton ($R = 0.77$) densities emphasizes its role in sustaining biological productivity. Low DO can indicate organic pollution and eutrophication, leading to hypoxia and reduced biodiversity. Nutrients are essential for primary productivity, but excessive levels can lead to eutrophication. The negative correlations between nitrate and plankton densities (phytoplankton: $R = -0.95$; zooplankton: $R = -0.97$), as well as phosphate (phytoplankton: $R = -0.94$; zooplankton: $R = -0.95$), suggest nutrient overloading in certain areas. These findings reflect a state of nutrient imbalance, where excessive inputs—likely from domestic or industrial wastewater can initially promote algal blooms but eventually destabilize the ecosystem, reduce biodiversity, and increase the risk of harmful algal blooms. Total alkalinity, which reflects the water's buffering capacity, also showed a strong negative relationship with both phytoplankton ($R = -0.94$) and zooplankton ($R = -0.957$). High alkalinity levels may be linked with industrial effluents or carbonate-rich discharges, which can disrupt ionic balance and affect organismal physiology. These parameters influence the osmotic balance and habitat preference of plankton. Moderate positive correlations with TDS (phytoplankton: $R = 0.37$; zooplankton: $R = 0.34$) suggest that mineral-rich environments support plankton growth to some extent. However, very high TDS or

Table 20.6 Plankton communities as bioindicators

Species	Environmental indication	Site observations
Phytoplankton as bioindicators		
C. marina	Indicates moderate nutrient conditions and stable environmental quality	Presence in Ennore Creek suggests relatively balanced conditions
Chlorella sp.	Tolerant, found in moderate to eutrophic waters; may indicate organic pollution when abundant	Widespread presence suggests nutrient availability
S. costatum	Indicator of eutrophication; linked to high nitrogen and phosphorus levels	Dominance in Adyar Creek points to nutrient-rich waters
Spirulina sp.	Needs sunlight and nutrients; high levels may imply imbalance	Dominance in Adyar suggests high productivity
N. longissima	Tolerates polluted or organically enriched environments	Presence in Napier and Adyar Creeks may reflect pollution
Nanochloropsis	Common in eutrophic and aquaculture systems; potential for algal blooms	Suggests high productivity and nutrient-rich conditions
P. sol	Found in nutrient-rich estuarine systems; indicates complex trophic dynamics	Suggests diverse and dynamic water quality
Zooplankton as bioindicators		
P.crassirostris, A. gibber	Found in moderately clean to eutrophic waters; adaptable	Presence across multiple sites indicates moderate productivity
O. similis, O. attenuata	Thrive in brackish, nutrient-rich environments	Presence in Adyar Creek suggests eutrophication and estuarine mixing
C. furcatus, C. minor	Sensitive calanoids; prefer diverse, well-oxygenated marine systems	Presence in Adyar indicates high diversity and ecological complexity
A. royi, D. rigida	Resilient and widely distributed; tolerant to fluctuations	Dominance across all sites indicates robustness in variable water quality
P. serricaudatus	Indicator of nutrient-rich coastal waters and moderate organic enrichment	Presence suggests organic enrichment
Rotifers, Cladocerans, Fish eggs	Indicators of secondary productivity and trophic function	Presence in Adyar reflects a productive and functional food web

salinity can stress freshwater species and reduce diversity, as seen in areas with industrial contamination or tidal intrusion.

20.3.7 Pollution Assessment Based on Plankton Bioindicators

Plankton communities serve as effective bioindicators for assessing pollution levels across the study sites. Ennore Creek exhibits moderate nutrient levels and stable plankton populations, indicating low to moderate pollution, likely influenced by industrial activity, yet maintaining ecological balance. Napier Creek is characterized by low plankton densities and high variability in dissolved oxygen and nutrient concentrations, reflecting a more polluted environment probably due to anthropogenic discharges and nutrient imbalances. In contrast, Adyar Creek shows high diversity and density of plankton. Despite elevated nutrient levels, the system appears highly productive, suggesting potential eutrophication mitigated by robust biological activity. Phytoplankton, in particular, respond rapidly to fluctuations in nutrient availability, light penetration, and water chemistry, making them reliable indicators of primary productivity and nutrient enrichment (eutrophication). Based on these observations, environmental indicators and site-specific conditions are summarized in Table 20.6.

20.3.8 Comprehensive Analysis of Physicochemical Parameters and Plankton Community Dynamics

The comprehensive analysis of physicochemical parameters and plankton community dynamics across Ennore Creek, Napier Creek, and Adyar Creek Estuary offers a reliable assessment of coastal water quality through biomonitoring, using phytoplankton and zooplankton as key bioindicators. Adyar Creek supports a rich and diverse planktonic community, characterized by the presence of dominant phytoplankton such as *Spirulina* sp., *Skeletonema costatum*, and *Planktoniella sol*, alongside zooplankton species like *Centropages furcatus*, *Clausocalanus minor*, and *Oithona similis*. Elevated levels of dissolved oxygen and chlorophyll indicate high primary and secondary productivity. However, the relatively high nutrient concentrations suggest a moderately eutrophic system, where increased productivity may be driven by domestic and urban runoff, raising concerns about the onset of eutrophication. Napier Creek shows signs of environmental degradation. Despite similar temperature and salinity conditions compared to the other sites, the plankton densities are markedly lower. High concentrations of nutrients such as nitrate and phosphate, combined with the presence of pollution-tolerant species like *Nitzschia longissima*, and fluctuating dissolved oxygen levels, point toward nutrient enrichment and possible organic pollution, indicating a stressed and potentially eutrophic environment. Ennore Creek presents a more balanced ecological profile. Moderate nutrient levels and stable dissolved oxygen concentrations support relatively high phytoplankton and zooplankton densities. The occurrence of indicator species such as *Cymbella marina*, *Chlorella* sp., and *Temora discaudata* suggests moderate productivity and ecological stability, with conditions conducive to sustaining a balanced aquatic ecosystem. Regression coefficient analysis further underscores these findings. Plankton densities show strong negative correlations with increasing nutrient and alkalinity levels, while positive correlations are observed with dissolved oxygen and total chlorophyll. These patterns reflect the close interdependence between water quality parameters and plankton community structure. Overall, the study reveals a gradient of ecological quality across the three sites. Adyar Creek emerges as biologically rich but nearing eutrophic stress; Ennore Creek appears moderately healthy with balanced productivity; while Napier Creek is the most impacted, displaying low biological diversity and elevated nutrient loads indicative of significant anthropogenic pollution.

20.3.9 Recovery of Marine Ecosystems from Oxygen Depletion and Resilience: Insights from Ennore, Napier, and Adyar Creeks

Marine ecosystems are highly susceptible to oxygen depletion, a phenomenon often exacerbated by human activities such as industrial effluents, sewage discharge, and nutrient loading. In areas where oxygen depletion occurs, the ability of ecosystems to recover is intrinsically linked to their resilience, which is defined by their capacity to absorb, adapt to, and reorganize in response to disturbances. The study of planktonic communities in Ennore, Napier, and Adyar Creeks offers valuable insights into how these ecosystems respond to and recover from oxygen depletion, highlighting the importance of resilience in sustaining biodiversity and ecosystem functionality.

20.3.10 Impact of Oxygen Depletion and Resilience in Ennore, Napier, and Adyar Creeks

The study observed varying levels of plankton density across the three study sites, with distinct differences in ecological health and potential for recovery. In Ennore Creek, which exhibited moderately high phytoplankton and zooplankton densities (2012.2 ± 172 cells/L and 28.64 ± 1.4 ind/L, respectively), the presence of significant pollution from industrial effluents and thermal discharges suggested a scenario where oxygen depletion could have occurred. However, the high plankton densities indicate a resilient ecosystem that may still support a productive food web despite environmental stressors, potentially indicating the system's capacity for recovery under optimal conditions. As oxygen levels gradually improve, this ecosystem can regenerate, as seen in the strong presence of copepods and other zooplankton species that are sensitive to dissolved oxygen levels (e.g., *Temora discaudata*, *Acrocalanus gibber*).

In contrast, Napier Creek exhibited the lowest plankton densities (723 ± 31 cells/L for phytoplankton and 8.14 ± 0.405 ind/L for zooplankton), suggesting a more stressed environment, potentially due to long-term organic and chemical pollution. The low diversity of plankton here may reflect diminished resilience, with the system struggling to recover from oxygen depletion and other environmental stressors. Without intervention to reduce nutrient loading and pollution, Napier

Creek may struggle to rebuild its ecosystem. Adyar Creek, showing the highest plankton densities (2218.2 ± 182 cells/L for phytoplankton and 36.5 ± 2.16 ind/L for zooplankton), suggests a more resilient system that can support a productive food web despite nutrient enrichment from domestic wastewater inflows. The resilience in Adyar Creek could be attributed to the diverse community of plankton species present, such as *S. costatum* and *N. longissima*, which are more capable of adjusting to fluctuations in environmental conditions, including oxygen levels. The presence of these diverse species highlights the system's capacity to adapt to changes in oxygen and nutrient availability, further supporting the idea that biodiversity enhances resilience in marine ecosystems.

20.3.11 Role of Resilience in Recovery: A Bioindicator Perspective

The resilience of these marine ecosystems is closely linked to the biodiversity of planktonic communities. In ecosystems like Adyar Creek, the diverse zooplankton species (e.g., *P. crassirostris*, *E. acutifrons*) reflect an ecosystem capable of withstanding periodic disturbances, such as oxygen depletion. These species are valuable bioindicators of ecological health and resilience, as they can adapt to fluctuating environmental conditions, including low oxygen levels. Their ability to survive in hypoxic conditions suggests that the system can recover once conditions improve, particularly if the stressors like nutrient overload and organic pollution are reduced. Conversely, the limited diversity and low abundance of plankton in Napier Creek indicate reduced resilience. The ecosystem may have passed a tipping point where oxygen depletion, compounded by high pollutant levels, prevents recovery. Such ecosystems may experience prolonged periods of low productivity, reduced biodiversity, and impaired ecosystem functions, which further delay the recovery process.

20.3.12 Recovery Process and the Importance of Biodiversity

The recovery of marine ecosystems from oxygen depletion requires a multifaceted approach. As dissolved oxygen levels rise, plankton populations, particularly phytoplankton and zooplankton, will gradually repopulate the area, helping to restore the biological productivity of the system. However, the speed and success of this recovery are directly linked to the resilience of the ecosystem, which is supported by biodiversity. High species diversity allows ecosystems to function effectively under stress, as different species respond to varying environmental conditions, thus ensuring ecosystem stability and productivity. In the case of Ennore and Adyar Creeks, the observed plankton community suggests that both sites have some degree of resilience, with species capable of adapting to environmental changes. This is crucial for long-term recovery, as resilient ecosystems can regenerate more quickly, restore biodiversity, and maintain ecosystem functions such as nutrient cycling and primary production.

20.4 Conservation Strategies for Chennai's Polluted Creeks

Based on the biomonitoring study and the observed environmental conditions in Ennore, Napier, and Adyar Creeks, the following recommendations are proposed to address coastal pollution and conserve the marine ecosystem:

- Reduce nutrient loading from domestic, industrial, and agricultural runoff by enforcing stricter regulations on wastewater treatment and improving the management of stormwater. This would help to reduce the risk of eutrophication and harmful algal blooms in the creeks.
- Upgrade the existing wastewater treatment infrastructure to ensure proper treatment of both domestic and industrial effluents before they are discharged into the creeks. Establish sewage treatment plants and improve the treatment processes to reduce the levels of pollutants like nitrates, phosphates, and organic matter, which contribute to water pollution and disrupt the planktonic community.
- Develop and protect riparian buffer zones along the creeks to act as natural filtration systems. These buffer zones can help to absorb excess nutrients, reduce sediment runoff, and filter pollutants before they reach the water.
- Implement continuous monitoring of physicochemical parameters such as temperature, pH, dissolved oxygen, salinity, nutrients (nitrates and phosphates), and TDS to identify early signs of pollution and take proactive measures.
- Establish regular monitoring of plankton communities as bioindicators to track changes in ecosystem health and detect pollution stress before it significantly impacts the marine biodiversity.
- Launch public awareness campaigns targeting local communities, industries, and fishermen about the importance of protecting the marine ecosystem and reducing pollution. And also involve local communities in conservation efforts by promoting sustainable fishing practices, reducing plastic waste, and encouraging eco-friendly disposal of waste.
- Focus on restoring damaged ecosystems in Napier Creek, where pollution levels are highest. Habitat restoration projects such as planting mangroves can enhance water quality, increase biodiversity, and stabilize the shoreline.
- Collaborate with local authorities, environmental NGOs, academic institutions, and industrial sectors to create a unified approach to pollution management and conservation in the creeks. Promote interdisciplinary research to develop sustainable technologies and solutions for pollution control and ecosystem management in the region.
- Encourage sustainable development practices that consider the long-term health of the creeks, including responsible waste disposal and environmental impact assessments for new projects.

By implementing these measures, it is possible to mitigate the impact of pollution on Ennore, Napier, and Adyar Creeks and contribute to the long-term conservation of the marine ecosystem. These efforts should be sustained and periodically evaluated to ensure the protection and restoration of water quality and biodiversity.

20.5 Conclusion

The ability of marine ecosystems to recover from oxygen depletion is inherently tied to their resilience. Ecosystems with high biodiversity, such as Adyar Creek, demonstrate greater resilience, allowing for faster recovery and a return to a more balanced state following disturbances. In contrast, ecosystems with low biodiversity, such as Napier Creek, may face prolonged recovery times and difficulty regaining ecological health. The presence of diverse planktonic species in the study areas serves as a key indicator of the system's ability to adapt to changes in environmental conditions, including oxygen depletion. To enhance resilience and promote the recovery of these ecosystems, it is crucial to manage nutrient loading, control pollution, and mitigate climate change impacts.

Acknowledgements The authors express their gratitude to the Management of AMET University for providing research facilities to carry out this work.

Ethical Approval *Ethical* review and approval were waived for this study as it complied with the regulations of the Committee for the Purpose of Control and Supervision of Experiments on Animals (CPCSEA), Department of Animal Husbandry and Dairying, Ministry of Fisheries, Animal Husbandry and Dairying, Government of India (2021). Additionally, the guidelines of the Indian National Science Academy were followed. Ethical approval by an Animal Welfare Committee is not required for studies involving Planktons.

Author Contributions All the authors contributed equally to this research work.

Conflict of Interest Statement The authors have no conflict of interest.

Funding This research was funded by Anusandhan National Research Foundation (***Formerly, Science and Engineering Research Board (SERB)***), Government of India for funding a major research project (EEQ/**2023/000185**).

References

Adjovu GE, Stephen H, James D, Ahmad S (2023) Measurement of total dissolved solids and total suspended solids in water systems: a review of the issues, conventional, and remote sensing techniques. Remote Sens 15(14):3534

Ali EM (2023) And abiotic components of marine ecosystem. In: Marine ecosystems: a unique source of valuable bioactive compounds, vol 3, pp 80–108

Altaff K (2004) A manual of zooplankton. University Grants Commission, New Delhi, pp 1–155

Arun Kumar A, Kunte PD (2012) Coastal vulnerability assessment for Chennai, east coast of India using geospatial techniques. Nat Hazards 64:853–872

Badran MI (2001) Dissolved oxygen, chlorophyll a and nutrients: seasonal cycles in waters of the Gulf of Aquaba, Red Sea. Aquat Ecosyst Health Manage 4(2):139–150

Bakhtiyar Y, Arafat MY, Andrabi S, Tak HI (2020) Zooplankton: the significant ecosystem service provider in aquatic environment. In: Bioremediation and biotechnology, Vol 3: persistent and recalcitrant toxic substances, pp 227–244

Belcher R, Macdonald AMG, Parry E (1957) On mohr's method for the determination of chlorides. Analytica chimica acta, 16:524–529

Basheer Ahamed SMZ (2023) Restoration of Buckingham Canal Chennai as a case study

Begum M, Naik S, Pradhan UK, Ezhilarasan P, Sura AN, Bharathi MD et al (2022) Assessment of Ennore oil spill 2017 on Chennai coastal water and biota. In: Oceans 2022-Chennai. IEEE, pp 1–7

Bettinetti R, Manca M (2013) Understanding the role of zooplankton in transfer of pollutants through trophic food webs. In: Kehayias G (ed) Zooplankton: species diversity, distribution and seasonal dynamics, pp 1–18

Boxshall G, Halsey S (2004) An introduction to copepod diversity. The Ray Society

Chandel P, Mahajan D, Thakur K, Kumar R, Kumar S, Brar B et al (2024) A review on plankton as a bioindicator: a promising tool for monitoring water quality. World Water Policy 10(1):213–232

Dahiya UR, Das J, Bano S (2022) Biological indicators of soil health and biomonitoring. In: Advances in bioremediation and phytoremediation for sustainable soil management: principles, monitoring and remediation. Springer, Cham, pp 327–347

de Paggi Susana J, Juan P, Pablo C, Jorge C, Bernal G (2008) Water quality and zooplankton composition in a receiving pond of the stormwater runoff from an urban catchment. J Environ Biol 29(5):693–700

Dhandapani D (2018) Urbanization and its influences along with cooum. For the Situation of Chennai Flood, River

Dhayachandhran KS, Jothilakshmi M (2021) Quality assessment of ground water along the banks of Adyar river using GIS. Mater Today Proc 45:6234–6241

El Nahhal D, El-Nahhal I, Al Najar H, Al-Agha M, El-Nahhal Y (2021) Acidity, electric conductivity, dissolved oxygen total dissolved solid and salinity profiles of marine water in Gaza: influence of wastewater discharge. Am J Anal Chem 12(11):408–428

Freeman LA, Corbett DR, Fitzgerald AM, Lemley DA, Quigg A, Steppe CN (2019) Impacts of urbanization and development on estuarine ecosystems and water quality. Estuar Coasts 42:1821–1838

Freitas FS, Arndt S, Hendry KR, Faust JC, Tessin AC, März C (2022) Benthic organic matter transformation drives pH and carbonate chemistry in Arctic marine sediments. Glob Biogeochem Cycles 36(7):e2021GB007187

Gatamaneni BL, Orsat V, Lefsrud M (2018) Factors affecting growth of various microalgal species. Environ Eng Sci 35(10):1037–1048

Ger KA, Urrutia-Cordero P, Frost PC, Hansson LA, Sarnelle O, Wilson AE, Lürling M (2016) The interaction between cyanobacteria and zooplankton in a more eutrophic world. Harmful Algae 54:128–144

Golterman HL (1983) Algal bioassays and algal growth controlling factors in eutrophic shallow lakes. Hydrobiologia, 100(1):59–64

Gopinathan CP, Rajagopalan M, Kaladharan P, Prema D (2007) Training manual on phytoplankton identification/taxonomy. Central Marine Fisheries Research Institute

Gouda SA, Hesham H, Taha A (2024) Bioindicators: a promising tool for detecting and evaluating water pollution. Egypt J Aquat Biol Fish 28(4):1209–1235

Gunathilaka ML, Bao S, Liu X, Li Y, Pan Y (2023) Antibiotic pollution of planktonic ecosystems: a review focused on community analysis and the causal chain linking individual-and community-level responses. Environ Sci Technol 57(3):1199–1213

Häder DP, Gao K (2015) Interactions of anthropogenic stress factors on marine phytoplankton. Front Environ Sci 3:14
Hemraj DA, Hossain MA, Ye Q, Qin JG, Leterme SC (2017) Plankton bioindicators of environmental conditions in coastal lagoons. Estuar Coast Shelf Sci 184:102–114
Hooper DU, Chapin FS III, Ewel JJ, Hector A, Inchausti P, Lavorel S et al (2005) Effects of biodiversity on ecosystem functioning: a consensus of current knowledge. Ecol Monogr 75(1):3–35
Hu TH, Hu CQ, Liu QF, Mai YZ, Sun JH, Wang C (2025) Trend analysis of research on phytoplankton traits as ecological indicators. Appl Ecol Environ Res 23(2):2343–2363
Hussain MB, Laabir M, Yahia MND (2020) A novel index based on planktonic copepod reproductive traits as a tool for marine ecotoxicology studies. Sci Total Environ 727:138621
Inyang AI, Wang YS (2020) Phytoplankton diversity and community responses to physicochemical variables in mangrove zones of Guangzhou Province, China. Ecotoxicology 29(6):650–668
Jakhar P (2013) Role of phytoplankton and zooplankton as health indicators of aquatic ecosystem: a review. Int J Innov Res Stud 2(12):489–500
Jarvie HP, Neal C, Withers PJ (2006) Sewage-effluent phosphorus: a greater risk to river eutrophication than agricultural phosphorus? Sci Total Environ 360(1–3):246–253
Jin C (2008) Biodiversity dynamics of freshwater wetland ecosystems affected by secondary salinisation and seasonal hydrology variation: a model-based study. Hydrobiologia 598:257–270
Kannan N, Venkataraman K, Stroebel J, Taylor L (2022) Investigating the causes of reduced dissolved oxygen concentrations in Kickapoo Creek, TX. Environ Monit Assess 194(8):574
Kashyap NK, Hait M, Bhardwaj AK (2024) Planktons as a sustainable biomonitoring tool of aquatic ecosystem. In: Biomonitoring of pollutants in the global south. Singapore, Springer, pp 275–319
Kasturirangan LR (ed) (1963) A key for identification of the most common planktonic copepod of Indian coastal water. Council of Scientific and Industrial Research
Katkova-Zhukotskaya O, Kalyujnaya S (2023) Bioindication using fish and its role in an integrated assessment of the condition of coastal ecosystems: a review. J Stress Physiol Biochem 19(4):203–216
Krishnamurthy R, Desouza KC (2015) Chennai, India. Cities 42:118–129
Lemessa F, Simane B, Seyoum A, Gebresenbet G (2023) Assessment of the impact of industrial wastewater on the water quality of rivers around the Bole Lemi Industrial Park (BLIP), Ethiopia. Sustainability 15(5):4290
Mani S, Jayalakshmi S (2024) Urban contrasts: a comprehensive review of Chennai and Coimbatore. Int J Res Publ Rev 5(1):2450
Manzoor M, Bhat KA, Khurshid N, Yatoo AM, Zaheen Z, Ali S, Rehman MU (2021) Bio-indicator species and their role in monitoring water pollution. In: Freshwater pollution and aquatic ecosystems. Apple Academic Press, pp 321–347
Murphy J. A. M. E. S, Riley JP (1962) A modified single solution method for the determination of phosphate in natural waters. Analytica chimica acta, 27:31–36
Neelavannan K, Narasimhan CL, Sivaraj K, Nisha V, Sekar S (2021) Environmental magnetic and textural characteristics of two estuarine core sediments from Bay of Bengal, India. Arab J Geosci 14(18):1934
Nethavhani EC (2023) Seasonal variations and trends of nitrates and phosphates concentration along the Boesmanspruit River in Mpumalanga Province. University of Johannesburg (South Africa)
O'Connor JA, Erler DV, Ferguson A, Maher DT (2022) The tidal freshwater river zone: physical properties and biogeochemical contribution to estuarine hypoxia and acidification-the "hydrologic switch". Estuar Coast Shelf Sci 268:107786
Ogidi OI, Onwuagba CG, Richard-Nwachukwu N (2024) Biomonitoring tools, techniques and approaches for environmental assessments. In: Biomonitoring of pollutants in the global south. Singapore, Springer, pp 243–273
Oliver S, Corburn J, Ribeiro H (2019) Challenges regarding water quality of eutrophic reservoirs in urban landscapes: a mapping literature review. Int J Environ Res Public Health 16(1):40

Padmanaban R, Bhowmik AK, Cabral P, Zamyatin A, Almegdadi O, Wang S (2017) Modelling urban sprawl using remotely sensed data: a case study of Chennai city, Tamilnadu. Entropy 19(4):163

Pandey V, Ray M, Kumar V (2020) Assessment of water-quality parameters of groundwater contaminated by fly ash leachate near Koradi thermal power plant, Nagpur. Environ Sci Pollut Res 27:27422–27434

Perry R (2010) Guide to the common inshore marine plankton of Southern California. UCLA Ocean GLOBE & Malibu High School

Ramasamy SM, Vijay A, Dhinesh S (2018) Geo-anthropogenic aberrations and Chennai floods: 2015, India. Nat Hazards 92:443–477

Razouls C, Desreuma N, Kouwenberg J, Bovée F (2022) Biodiversity of marine planktonic copepod: morphology, geographical distribution, and biological data. Sorbonne University, CNRS

Reddy AN, Nagendra R (2011) Assessment of environmental pollution along the Chennai coast-a review. In: Proceedings of the national seminar on environment and public participation, pp 19–25

Salmaso N, Tolotti M (2021) Phytoplankton and anthropogenic changes in pelagic environments. Hydrobiologia 848(1):251–284

Saravanan SP, Desmet M, Neelakanta Pillai Kanniperumal A, Ramasamy S, Shumskikh N, Grosbois C (2019) Geochemical footprint of megacities on river sediments: a case study of the fourth most populous area in India, Chennai. Minerals 9(11):688

Scannell PW, Jacobs LL (2001) Effects of total dissolved solids on aquatic organisms. In: Alaska department of fish and game: division of habitat and restoration, vol 1, p 4

Shanmugam P, Neelamani S, Ahn YH, Philip L, Hong GH (2007) Assessment of the levels of coastal marine pollution of Chennai city, Southern India. Water Resour Manag 21:1187–1206

Sheffield NA (2019) Nitrogen and phosphorus uptake in an urban stream ecosystem. Illinois State University

Shiam Babu R, Prasanna K, Senthil Kumar P, Rangasamy G (2025) Comprehensive assessment of current municipal solid waste management in Chennai, India: a critical case study with real-time analysis. Int J Environ Sci Technol 22(7):5729–5746

Silva MAM, Eça GF, Santos DF, Guimarães AG, Lima MC, de Souza MFL (2013) Dissolved inorganic nutrients and chlorophyll a in an estuary receiving sewage treatment plant effluents: Cachoeira River estuary (NE Brazil). Environ Monit Assess 185:5387–5399

Sivakumar K, Nawaz MA, Saboor A (2021) Population composition of calanoid copepod of the Chennai coast, Tamil Nadu. Indian J Geo-Mar Sci 50(9):693–700

Skejić S, Bojanić N, Matijević S, Vidjak O, Grbec B, Ninčević ŽG et al (2014) Analysis of phytoplankton community in the vicinity of domestic sewage outflow during stratified conditions. Mediterr Mar Sci 15:574–586

Striebel M, Singer G, Stibor H, Andersen T (2012) "Trophic overyielding": phytoplankton diversity promotes zooplankton productivity. Ecology 93(12):2719–2727

Sudha Rani NNV, Satyanarayana ANV, Bhaskaran PK (2015) Coastal vulnerability assessment studies over India: a review. Nat Hazards 77:405–428

Sumudumali RGI, Jayawardana JMCK (2021) A review of biological monitoring of aquatic ecosystems approaches: with special reference to macroinvertebrates and pesticide pollution. Environ Manag 67(2):263–276

Sumuki S (2023) Abiotic: exploring the non-living components of our ecosystems. Ukr J Ecol 13(3):58–59

Takenaka Y, Yamaguchi A, Tsuruoka N, Torimura M, Gojobori T, Shigeri Y (2012) Evolution of bioluminescence in marine planktonic copepods. Mol Biol Evol 29(6):1669–1681

Telesh I, Schubert H, Skarlato S (2013) Life in the salinity gradient: discovering mechanisms behind a new biodiversity pattern. Estuar Coast Shelf Sci 135:317–327

Vehmaa A, Hogfors H, Gorokhova E, Brutemark A, Holmborn T, Engström-Öst J (2013) Projected marine climate change: effects on copepod oxidative status and reproduction. Ecol Evol 3(13):4548–4557

Verlecar XN, Desai SR (2004) Phytoplankton identification manual. National Institute of Oceanography

VishnuRadhan R, David Thresyamma D, Sarma K, George G, Shirodkar P, Vethamony P (2015) Influence of natural and anthropogenic factors on the water quality of the coastal waters around the South Andaman in the Bay of Bengal. Nat Hazards 78:309–331

Wang M, Jeong CB, Lee YH, Lee JS (2018) Effects of ocean acidification on copepods. Aquat Toxicol 196:17–24

Waters NCC (2004) Waste load allocation & waste assimilative capacity studies for Ennore. Creek & North Chennai Coastal Waters

Xiong W, Ni P, Chen Y, Gao Y, Li S, Zhan A (2019) Biological consequences of environmental pollution in running water ecosystems: a case study in zooplankton. Environ Pollut 252:1483–1490

Zhu W, Wan L, Zhao L (2010) Effect of nutrient level on phytoplankton community structure in different water bodies. J Environ Sci 22(1):32–39

Chapter 21
Evaluating CO_2 Removal Technologies: A Life Cycle Assessment for Effective Climate Change Mitigation

Divya Meril, Piliyan Raju, Felix Suganthan, and Santhanam Perumal

Abstract Climate change is an urgent global crisis, driven by rising atmospheric CO_2 levels that threaten both ecosystems and human societies. While reducing emissions remains essential, carbon dioxide removal (CDR) technologies are increasingly vital for achieving net-zero and net-negative targets, especially in hard-to-abate sectors. This review critically assesses the environmental sustainability of key CDR approaches—Direct Air Capture (DAC), bioenergy with carbon capture and storage (BECCS), afforestation/reforestation, ocean alkalinity enhancement, and soil carbon sequestration—using the Life Cycle Assessment (LCA) framework. Findings reveal wide-ranging environmental trade-offs. BECCS has high carbon removal potential, but poses risks related to land-use change, water consumption, and biodiversity. DAC is effective but energy-intensive, with its sustainability hinging on the energy source. Nature-based solutions offer ecological co-benefits but are constrained by land availability and risks of carbon reversal. Ocean-based methods are still experimental and could disrupt marine systems. To ensure CDR supports climate goals without creating new environmental problems, policies must prioritize renewable energy use, mitigate land-use impacts, and improve monitoring of carbon permanence. Immediate, informed action is crucial to deploy sustainable

D. Meril (✉)
Department of Aquatic Environment Management, St. Devasahayam Institute of Fisheries Science and Technology, Kanniyakumari, Tamil Nadu, India

Dr. J. Jayalalithaa Fisheries University, Kanniyakumari, Tamil Nadu, India

P. Raju
Department of Marine Biotechnology, AMET University, Chennai, Tamil Nadu, India

F. Suganthan
Dr. J. Jayalalithaa Fisheries University, Kanniyakumari, Tamil Nadu, India

Department of Aquaculture, St. Devasahayam Institute of Fisheries Science and Technology, Kanniyakumari, Tamil Nadu, India

S. Perumal
Marine Planktonology and Aquaculture Laboratory, Department of Marine Science, Bharathidasan University, Tiruchrappalli, Tamil Nadu, India

B. P. Barathan et al. (eds.), *Navigating Climate Change: Impacts on Biodiversity and Ecosystem Resilience*, Sustainable Development and Biodiversity 42,
https://doi.org/10.1007/978-981-95-0409-1_21

CDR solutions at scale and help stabilize atmospheric CO_2 levels in time to avoid the worst effects of climate change.

Keywords Carbon dioxide removal · Direct air capture · Life Cycle Assessment · CO_2

21.1 Introduction

The Intergovernmental Panel on Climate Change (IPCC 2021) identifies anthropogenic CO_2 emissions as the primary driver of climate change, contributing to global temperature rise, extreme weather events, and sea-level rise. According to the IPCC Sixth Assessment Report, limiting global warming to well below 2 °C preferably to 1.5 °C requires deep and sustained reductions in greenhouse gas emissions, along with the deployment of carbon dioxide removal (CDR) strategies. To meet this target, it is not only critical to achieve net-zero emissions, but also necessary to actively remove excess atmospheric CO_2. While net-zero emissions stabilize atmospheric carbon levels, they do not address historical emissions. Therefore, CDR approaches, including Negative Emissions Technologies (NETs), are essential to bridge this gap and ensure long-term climate stability (Rogelj et al. 2018; IPCC 2021).

CDR technologies can be broadly categorized into nature-based and technological approaches. Nature-based solutions such as afforestation, reforestation, soil carbon sequestration, and blue carbon leverage natural ecosystems to sequester atmospheric carbon (Griscom et al. 2017). In India, large-scale afforestation and reforestation are carried out under programs like the National Afforestation Programme (NAP) and Green India Mission, which aim to enhance carbon sinks through forest restoration. Mangrove afforestation projects in the Sundarbans and coastal Andhra Pradesh contribute to blue carbon capture. Additionally, soil carbon sequestration is promoted under Paramparagat Krishi Vikas Yojana (PKVY) and other organic/natural farming schemes, improving soil health and carbon storage. Technological approaches—including direct air capture (DAC), bioenergy with carbon capture and storage (BECCS), enhanced weathering, ocean alkalinization, and mineralization—are still in early stages in India but are gaining attention. For instance, Indian Oil Corporation and Bharat Petroleum are exploring carbon capture and utilization (CCU) technologies in their refineries. NETRA (NTPC Energy Technology Research Alliance) is conducting pilot studies on CO_2 capture from thermal power plant emissions. Research on biochar application—a form of carbon-negative technology derived from biomass pyrolysis—is being supported by agricultural universities and ICAR institutes as a scalable solution for both carbon sequestration and soil improvement. Preliminary investigations into enhanced weathering using basalt rock dust in Indian croplands are also underway, particularly in Karnataka and Maharashtra. Each of these approaches presents distinct principles, scalability, and environmental trade-offs, necessitating thorough evaluation through LCA to assess their feasibility and sustainability (Fuss et al. 2018).

LCA is a comprehensive method for evaluating the environmental impacts of a technology across its entire life cycle, from resource extraction to disposal (ISO 2006a). In the context of CDR technologies, LCA is essential for understanding broader environmental impacts, including energy consumption, resource utilization, and emissions, beyond just CO_2 removal. For example, while DAC facilitates the removal of atmospheric CO_2, its energy-intensive operations can lead to substantial indirect emissions if powered by nonrenewable energy sources (Tanzer and Ramírez 2019). LCA also helps identify unintended consequences, such as land-use conflicts associated with afforestation or potential biodiversity loss in blue carbon initiatives, ensuring these technologies align with overarching sustainability goals. The LCA process follows four stages (Fig. 21.1)—goal and scope definition, inventory analysis, impact assessment, and interpretation—according to ISO 14040/14044 standards to ensure consistency and reliability (ISO 2006a, b).

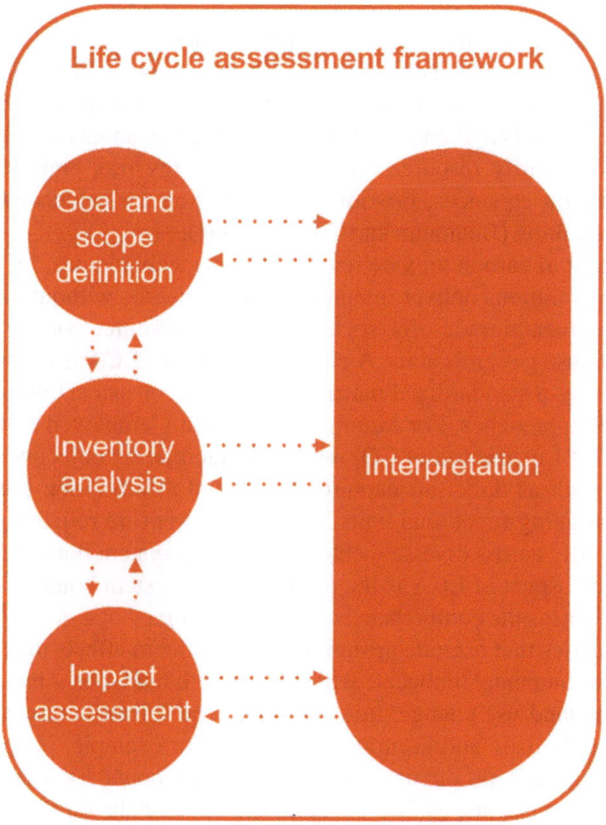

Fig. 21.1 Different stages of LCA

This chapter seeks to provide a comprehensive LCA analysis of CDR technologies, exploring their environmental, social, and economic dimensions. It evaluates both nature-based and technological solutions, such as afforestation, soil carbon sequestration, DAC, and BECCS, with a focus on scalability and climate mitigation potential (Minx et al. 2018). The analysis adopts global geographic boundaries and cradle-to-grave system boundaries, with a temporal focus on present and near-future applications. Key limitations include data availability, regional variations, and long-term uncertainties. The review covers critical topics such as climate urgency, LCA methodologies, CDR technologies, and policy recommendations, offering a holistic perspective on the subject.

21.2 Methodological Framework of LCA

21.2.1 Definition and Function of LCA in Environmental Analysis

LCA is a systematic and standardized framework designed to evaluate the environmental impacts associated with products, processes, or technologies across their entire life cycle, from raw material extraction to final disposal (ISO 2006a, b). Defined by the ISO 14040 and 14,044 standards, LCA adopts a "cradle-to-grave" perspective to quantify resource consumption, energy use, emissions, and other environmental consequences, providing a foundation for evidence-based sustainable decision-making (Baumann and Tillman 2004).

In the context of carbon dioxide removal (CDR) technologies, LCA is critical to ensuring these solutions deliver intended climate benefits without introducing significant environmental trade-offs, such as excessive resource consumption or unintended greenhouse gas emissions. A core component of LCA is the functional unit, which establishes a standardized metric for fair and consistent comparisons across different CDR approaches. For example, a commonly employed functional unit in CDR studies is "1 metric ton of CO_2 removed," facilitating objective evaluations of technologies such as direct air capture (DAC) and afforestation (Hauschild et al. 2018). Standardizing functional units is essential to ensure reliability, consistency, and comparability across diverse methodologies and system boundaries.

Another key aspect of LCA is the delineation of system boundaries, which significantly influences the comprehensiveness and accuracy of assessments. Narrowly defined boundaries that exclude upstream or downstream effects may lead to underestimated environmental impacts, while broader boundaries—including indirect effects such as land-use change, transportation, and infrastructure requirements—provide a more holistic and accurate evaluation. For example, in aquaculture systems, excluding feed production from the system boundary can underestimate greenhouse gas emissions, as feed accounts for a major share of the environmental burden. Including processes such as crop cultivation, processing, and transport of feed ingredients reveals a more realistic footprint (Henriksson et al. 2012). Similarly, conducting an LCA of bioenergy with carbon capture and storage (BECCS) requires the inclusion of processes such as CO_2 capture and storage, biomass cultivation, land-use changes, transportation, and associated energy consumption to reflect the full environmental impact (Fuss et al. 2018).

By encompassing both direct and indirect environmental impacts, LCA offers valuable insights into the sustainability of CDR technologies. Adherence to internationally recognized guidelines, such as those outlined in ISO standards, ensures methodological rigor, transparency, and comparability, enabling policymakers and stakeholders to make well-informed decisions (Guinée 2002).

LCA is a comprehensive methodology for evaluating the environmental impacts associated with all stages of a product or process, from raw material extraction to end-of-life disposal. It plays a vital role in identifying resource consumption and environmental outputs, thereby enabling decision-makers to develop more sustainable strategies (ISO 2006a; Goedkoop et al. 2013). By uncovering hidden environmental costs, LCA supports informed and responsible choices across various sectors. For instance, in aquaculture, LCA helps evaluate the trade-offs of alternative feed ingredients. Replacing fishmeal with plant-based options like soybean meal can reduce pressure on marine ecosystems, but may increase land use and nutrient pollution (Henriksson et al. 2012). In the renewable energy sector, LCA is instrumental in comparing technologies such as solar PV and wind power—noting their lower life-cycle emissions relative to fossil fuels while highlighting issues like material sourcing and waste management (Bhandari et al. 2015). Similarly, for carbon dioxide removal (CDR) technologies like direct air capture (DAC), LCA is essential for assessing overall climate benefits by accounting for energy use, infrastructure, and operational emissions (Deutz and Bardow 2021). This paper provides an in-depth analysis of the key stages of LCA, with a particular focus on their methodologies, significance, and applications within the realm of emerging carbon dioxide removal (CDR) technologies (Fig. 21.2).

21.3 Key Stages of Life Cycle Assessment (LCA): A Comprehensive Review

21.3.1 Goal and Scope Definition

The goal and scope definition phase establishes the foundational framework for the entire LCA process. This phase involves clearly defining the study's objectives, setting system boundaries, and selecting an appropriate functional unit for comparison (ISO 2006a). Clearly articulated objectives are essential to ensure that the study aligns with its intended purpose, which may range from evaluating environmental performance to informing policy and regulatory decision-making (Ness and Andersson 2020). For instance, in the context of carbon dioxide removal (CDR)

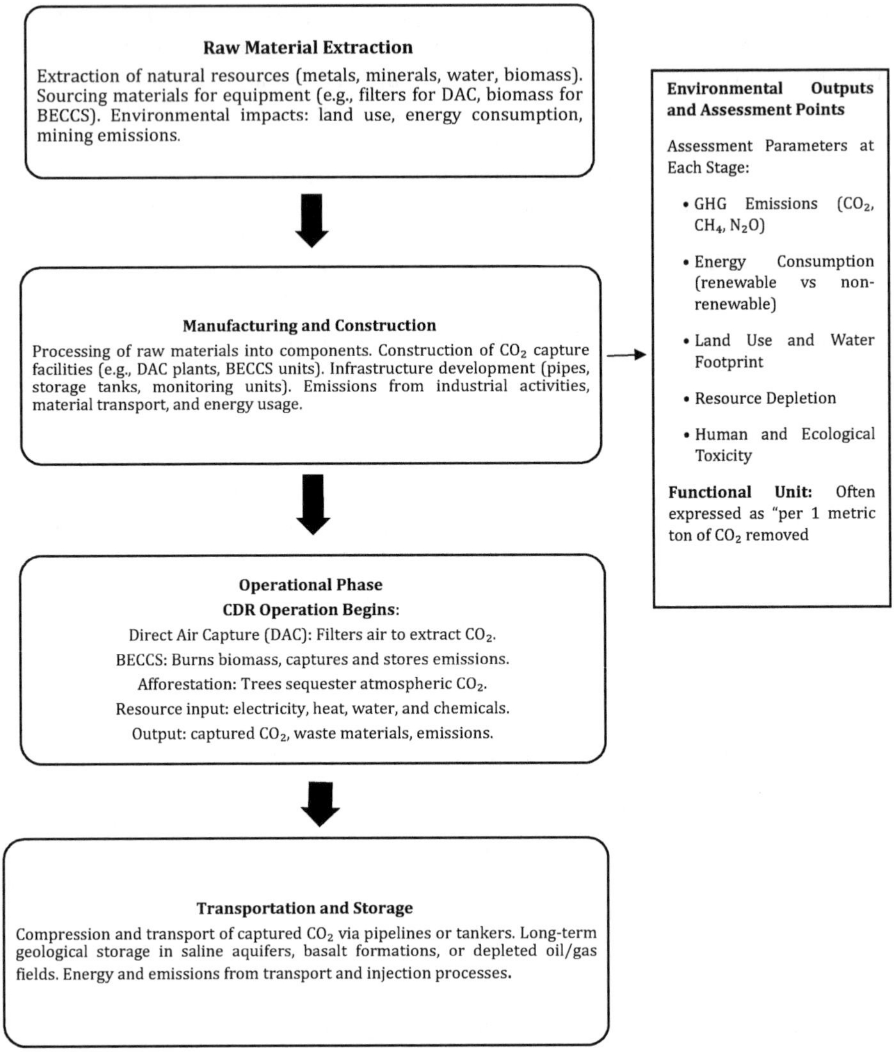

Fig. 21.2 Life cycle assessment (cradle-to-grave) of carbon dioxide removal technologies

technologies, this phase often focuses on determining whether the technology achieves net negative carbon emissions, accounting for the life cycle emissions of all associated processes (Chelvam and Hanafiah 2023).

The definition of system boundaries is equally critical, as it delineates the life cycle stages to be considered (e.g., cradle-to-grave). The selection of these boundaries must reflect the study's objectives and address the concerns of relevant stakeholders, including environmental organizations, policymakers, and industry representatives (Muller et al. 2018). Sensitivity analysis is a vital component of this phase, as it evaluates the impact of varying assumptions and data uncertainties on the LCA results (Roy et al. 2020). Additionally, a thorough assessment of data quality is essential to ensure the reliability and validity of the study's outcomes (Hauschild et al. 2018).

21.3.2 Inventory Analysis

The inventory analysis phase involves identifying and quantifying all inputs and outputs associated with the system under study. Inputs generally include raw materials, energy, and water, while outputs encompass emissions, by-products, and waste (Guinea 2019). This comprehensive quantification of inputs and outputs is vital for a detailed environmental analysis.

Primary data, gathered through direct measurements or site-specific observations, are ideal because they reflect the actual performance of the studied processes. However, primary data collection can be resource-intensive and time-consuming. Secondary data, often sourced from life cycle inventory databases such as Ecoinvent or GaBi, are commonly used in the absence of primary data (Bakker et al. 2022). Secondary data are more accessible and comprehensive, but may introduce uncertainties if they do not reflect the specific context of the study. Life cycle inventory databases, such as Ecoinvent, remain essential tools for aggregating large amounts of data across various industries. However, the use of secondary data must be handled with care, as it can introduce bias or inaccuracies, particularly if the data are outdated or not region-specific (Vázquez-Rowe et al. 2020a, b).

21.3.3 Impact Assessment

Impact assessment is the stage where the inventory results are translated into environmental impacts. The aim is to quantify the potential environmental consequences of the inputs and outputs identified in the inventory phase (ISO 2006b). This stage often employs impact categories such as global warming potential (GWP), eutrophication, acidification, and resource depletion, among others (Hauschild et al. 2018).

Several impact assessment methods are available, each with its strengths and limitations. The ReCiPe method, for instance, combines midpoint and endpoint indicators, allowing for a comprehensive evaluation of environmental impacts (Goedkoop et al. 2013). Other methods such as CML or TRACI may be more appropriate in specific contexts. For example, TRACI has been particularly useful in the United States for addressing region-specific impacts, such as human health impacts due to air pollution (Bare 2019). For technologies like CDR, key impact categories include GWP, land use change, water consumption, and resource depletion, all of which are critical when evaluating the potential trade-offs of different CO2 sequestration technologies (Chelvam and Hanafiah 2023).

21.3.4 Data Quality Assessment

Data quality assessment ensures that the results of the LCA are based on reliable, accurate, and relevant data. This step involves evaluating the precision, completeness, consistency, and temporal relevance of the data used (Weidema and Wesnaes 1996; Vázquez-Rowe et al. 2020a, b). The quality of data directly impacts the credibility of the LCA findings and ultimately informs decision-making processes.

Data transparency is also essential. Detailed documentation of the data sources, methodologies, and assumptions enhances the transparency of the LCA process and allows other researchers or stakeholders to replicate the study (Baumann and Tillman 2004). Transparent reporting ensures that uncertainties and assumptions are clearly communicated, providing a basis for the validity and reproducibility of the results (Hauschild et al. 2018).

21.3.5 Interpretation

The interpretation phase synthesizes the findings from the inventory and impact assessment stages to derive conclusions and recommendations. This phase is essential for identifying environmental hotspots—areas of the life cycle that contribute disproportionately to overall environmental impacts (Ness and Andersson 2020). For instance, in a CDR technology such as DAC, energy-intensive capture processes may emerge as hotspots, leading to recommendations for technological optimization or the use of cleaner energy sources (Chelvam and Hanafiah 2023).

Uncertainty and limitations must be thoroughly discussed in this phase. Scenario analysis and Monte Carlo simulations are common methods for quantifying uncertainty, helping to identify which variables most affect the outcomes of the LCA (Roy et al. 2020). The results of sensitivity analysis are invaluable in guiding decisions, as they pinpoint the factors that require further investigation or improvement to optimize the environmental performance of the system under study (Bare 2019).

21.4 Challenges and Considerations in LCA of CDR Technologies

LCA is increasingly recognized as a critical tool for assessing the environmental impact of CDR technologies, which are seen as essential in addressing climate change. However, LCA applied to CDR technologies is faced with a range of challenges and considerations that must be effectively addressed to ensure that the findings are meaningful and actionable.

One of the foremost challenges in applying LCA to CDR technologies is the availability and reliability of data. Due to the early-stage development of many

CDR methods, including direct air capture (DAC) and bioenergy with carbon capture and storage (BECCS), reliable data are often scarce, making it difficult to perform a comprehensive analysis (Fajardy et al. 2019). Additionally, uncertainties in data, especially related to the performance of these technologies over long time periods, further complicate LCA. The future performance of CDR technologies, particularly with respect to long-term carbon storage, is highly uncertain, making the assessment of potential environmental impacts inherently speculative (Finkbeiner et al. 2020; Chelvam and Hanafiah 2023).

Modeling complex biological processes presents substantial challenges, particularly in the context of carbon removal strategies such as afforestation and soil carbon sequestration. These approaches necessitate a comprehensive understanding of biological and ecological systems, which remain inadequately characterized and are often difficult to represent through existing modelling frameworks. (Jones et al. 2016; Keith et al. 2021). The intrinsic variability of biological systems, influenced by factors such as climate conditions, soil health, and biodiversity, further complicates the precise modelling of environmental outcomes (Zhao et al. 2021). Moreover, the establishment of a robust reference system proves challenging, especially given the critical role of spatial context in shaping the impacts of carbon dioxide removal (CDR) technologies (Muller et al. 2018). For instance, land-use changes associated with nature-based solutions frequently result in region-specific environmental consequences that elude quantification through generalized modelling approaches (Chelvam and Hanafiah 2023).

Another significant challenge lies in scaling up CDR technologies from pilot projects to large-scale implementations. The behaviour of these technologies at larger scales is often unpredictable, as evidenced by the disparity in energy requirements, efficiency, and infrastructure needs between small and large-scale operations. These variations introduce additional uncertainty into LCAs and require careful modelling to account for the potential differences in operational performance at larger scales (Fajardy et al. 2019; Heath et al. 2020).

21.5 Considerations in LCA of CDR Technologies

Addressing the impacts of land-use change (LUC) is crucial when evaluating nature-based carbon dioxide removal (CDR) strategies. Methods such as afforestation, reforestation, and soil carbon sequestration involve significant land-use alterations, which can yield both positive and negative environmental outcomes. While these approaches have the potential to sequester carbon on a large scale, they may also lead to unintended consequences, including adverse effects on biodiversity, water resources, and local communities (Zhao et al. 2021). Therefore, a thorough assessment of their net benefits requires an in-depth evaluation of both direct and indirect environmental impacts associated with LUC.

Similarly, the environmental costs of materials, energy, and infrastructure required for technological CDR methods, such as direct air capture (DAC) and bioenergy with carbon capture and storage (BECCS), are of critical importance. Despite their promising potential for carbon removal, these technologies demand significant energy inputs and infrastructure development, which contribute to their overall environmental footprint. Energy use is particularly critical, as the energy-intensive processes underpinning many CDR technologies can partially offset their carbon removal gains (Muller et al. 2018; Heath et al. 2020). Evaluating these technologies within a life cycle framework—including resource use, infrastructure demands, and material requirements—provides a more comprehensive understanding of their environmental implications.

The temporal aspect is another key factor in assessing CDR technologies. Nature-based solutions, in particular, often require decades or even centuries to realize their full carbon sequestration potential. LCA must account for the duration of carbon storage, the risk of carbon leakage, and the influence of future technological advancements that could either enhance or undermine the long-term effectiveness of these methods (Zhao et al. 2021).

Finally, multi-criteria decision-making (MCDM) is essential for selecting optimal CDR approaches, especially when trade-offs arise among environmental, social, and economic impacts. For instance, a technology may excel in carbon removal but impose significant burdens in terms of water use or land-use change. MCDM enables the evaluation of multiple criteria simultaneously, facilitating the identification of solutions that achieve a balanced trade-off between competing priorities (Ness and Andersson 2020). Incorporating stakeholder input and societal values into these processes is equally important to ensure alignment with broader sustainability goals (Muller et al. 2018).

21.6 Specific LCA Software and Databases

The use of specialized LCA software and databases is pivotal in overcoming some of the challenges discussed above and ensuring that the LCA process is both accurate and efficient. Popular LCA software tools include SimaPro, GaBi, and Open LCA. SimaPro, for example, is known for its extensive database and advanced modelling capabilities, allowing users to assess the environmental impacts of various technologies across their life cycles (Goedkoop et al. 2013). GaBi, another widely used tool, offers a comprehensive database that covers a broad range of industries and processes, making it particularly useful for conducting LCAs of industrial-scale technologies like DAC and BECCS (Weidema and Wesnæs 2021).

Open LCA, an open-source LCA tool, provides flexibility in terms of impact assessment methods, which can be tailored to specific study needs. It supports a variety of impact assessment methodologies, including ReCiPe, CML, and TRACI, providing users with a diverse range of options for evaluating the environmental impacts of CDR technologies (Frischknecht et al. 2020). The ability to integrate different impact categories and databases makes Open LCA a versatile tool for conducting comprehensive assessments of emerging technologies.

Table 21.1 Key features of common LCA software and databases

Tool/ database	Type	Key features/uses	Reference
SimaPro	Software	Advanced modelling capabilities; supports multiple impact assessment methods; includes extensive Ecoinvent data	Goedkoop et al. (2013)
GaBi	Software	Comprehensive industrial database; widely used for complex industrial LCAs (e.g., BECCS, DAC)	Weidema and Wesnæs (2021)
Open LCA	Software	Open-source; supports various impact methods (e.g., ReCiPe, TRACI); highly customizable	Frischknecht et al. (2020)
Ecoinvent	Database	High-quality, peer-reviewed life cycle inventory data across multiple sectors	Wernet et al. (2016)
GaBi Database	Database	Detailed LCI data for energy, materials, and manufacturing processes; integrated with GaBi software	Sweeney et al. (2021)
USLCI	Database	Focused on US-specific inventory data; suitable for regionalized impact assessments	Sweeney et al. (2021)

In terms of LCA databases, Eco invent, GaBi data, and the USLCI database are among the most used. Eco invent is renowned for its detailed and reliable life cycle inventory data, which cover a wide range of sectors and processes, making it a vital resource for LCA practitioners (Wernet et al. 2016). GaBi data, which are similarly comprehensive, offer detailed life cycle inventory data for various sectors, facilitating accurate environmental impact assessments (Sweeney et al. 2021). The USLCI database, which focuses specifically on US-based data, is especially useful for assessing region-specific impacts and integrating local data into global life cycle assessments (Sweeney et al. 2021). Key features of common LCA software and databases are shown in Table 21.1.

21.7 LCA of Nature-Based CDR Technologies

21.7.1 Afforestation and Reforestation

Afforestation and reforestation are prominent nature-based carbon dioxide removal (CDR) strategies that involve either planting trees on land that was previously non-forested (afforestation) or restoring forests that have been degraded or deforested (reforestation). These strategies are critical because they not only sequester carbon but also contribute to biodiversity enhancement, water regulation, and soil protection (Cipolla et al. 2021). The effectiveness of these strategies is largely contingent on the carbon sequestration rates over the lifecycle of the trees and the longevity of the carbon stored in forest biomass and soil (Davidson et al. 2020a, b).

An important aspect of LCA in afforestation and reforestation is accounting for carbon sequestration rates, which vary significantly depending on factors such as tree species, soil type, climate, and forest management practices (Chadwick et al. 2019). These factors influence the rate at which forests absorb CO_2 and how much of this carbon remains sequestered over time. While afforestation has the potential to sequester significant amounts of carbon, it may also involve land-use changes that can result in the displacement of agricultural land, leading to indirect emissions, especially if high-carbon ecosystems such as peatlands or grasslands are converted to forests (Houghton 2021).

Water demand is another critical factor, as forests require substantial amounts of water, which may lead to water scarcity in regions where water resources are already stressed (Cipolla et al. 2021). Nutrient runoff from fertilization practices can also lead to water quality degradation, potentially harming aquatic ecosystems (Davidson et al. 2020a, b). Furthermore, the maintenance of plantations, including fertilization, pest control, and harvesting operations, adds to the overall environmental impact of afforestation and reforestation projects. These activities contribute to greenhouse gas emissions using machinery, transportation, and processing (Lenton et al. 2020).

Case studies, such as those conducted in Southeast Asia and the Amazon, demonstrate both the successes and challenges of afforestation and reforestation efforts. While these projects show promise in terms of carbon sequestration, challenges such as biodiversity loss, the disruption of local livelihoods, and long-term sustainability issues remain (Cipolla et al. 2021). Additionally, trade-offs between carbon sequestration and the preservation of other ecosystem services like water regulation and biodiversity conservation need to be carefully balanced (Duarte et al. 2020).

21.7.2 Soil Carbon Sequestration

Soil carbon sequestration, the process of capturing and storing carbon in soils through strategic land management practices, is a key approach to mitigating climate change. Practices such as no-till farming, agroforestry, cover cropping, and reduced fertilizer application have shown considerable potential to enhance soil organic carbon, thereby removing CO_2 from the atmosphere (Davidson et al. 2020a, b). While soils are recognized as effective carbon sinks, their sequestration potential is highly dependent on the specific management practices adopted and the characteristics of the soil type (Lal 2021).

A comprehensive LCA of soil carbon sequestration involves quantifying carbon storage rates across different practices and evaluating their broader impacts on soil health, water usage, and greenhouse gas (GHG) emissions. For example, no-till farming has been shown to increase soil organic matter, enhance carbon retention, reduce erosion, and improve water conservation (Lal 2021). However, improper management, such as excessive fertilizer use, can result in significant nitrous oxide (N_2O) emissions—a potent GHG—thereby offsetting the benefits of carbon sequestration (Smith et al. 2016).

Land-use changes are another critical factor influencing soil carbon sequestration. The conversion of natural ecosystems such as grasslands and wetlands to agricultural use can result in substantial carbon losses from soils (MacDougall et al. 2019). Additionally, conventional tillage practices can disrupt soil carbon stocks, releasing stored carbon back into the atmosphere. As a result, the overall net benefits of soil carbon sequestration depend on how these practices interact with broader environmental processes, including nutrient cycling and water quality (Chadwick et al. 2019).

Case studies from Europe, North America, and Africa highlight the significant potential of soil carbon sequestration within agricultural systems. Nonetheless, challenges such as the need for sustained management and uncertainties surrounding long-term carbon storage remain pressing concerns (Davidson et al. 2020a, b). Furthermore, balancing carbon sequestration with other ecosystem services, such as food production and water resource management, requires careful navigation of the inherent trade-offs (Trost et al. 2018).

21.7.3 Coastal and Ocean-Based Carbon Sequestration (Blue Carbon)

Coastal ecosystems, including mangroves, seagrass beds, and salt marshes, are highly effective at sequestering carbon in both vegetation and sediments. These ecosystems, collectively known as blue carbon ecosystems, play a crucial role in global carbon cycling and are considered one of the most efficient carbon sinks on the planet (Duarte et al. 2020). The LCA of blue carbon sequestration involves evaluating the carbon sequestration rates of these ecosystems, as well as their potential for long-term carbon storage in sediments, which can remain stable for centuries (Shackley et al. 2018).

Mangroves, seagrass beds, and salt marshes are particularly valuable due to their high productivity and their ability to trap carbon in waterlogged soils, which prevents decomposition. However, the effectiveness of blue carbon ecosystems is dependent on factors such as water quality, salinity, and tidal conditions, which can vary significantly across different regions (Zhang et al. 2021). Additionally, nutrient runoff from agricultural activities can degrade these ecosystems, leading to a decrease in their carbon sequestration potential and harming biodiversity (Duarte et al. 2020).

Restoration and conservation efforts for blue carbon ecosystems have been shown to increase both carbon storage and biodiversity. For example, the restoration of mangrove forests in Southeast Asia and the Caribbean has led to substantial increases in carbon sequestration, while also enhancing fishery productivity and coastal protection (Duarte et al. 2020). However, challenges remain in terms of scaling up these efforts and addressing the potential impacts of climate change, such as rising sea levels, which could affect the stability and carbon storage capacity of coastal ecosystems (Fletcher et al. 2020).

21.7.4 Enhanced Weathering

Enhanced weathering is a geoengineering technique that accelerates the natural process of CO_2 absorption by silicate rocks. When these rocks, such as basalt and olivine, are weathered, they react with CO_2 in the atmosphere, forming stable carbonates and effectively removing CO_2 from the air (Kong et al. 2020). The LCA of enhanced weathering involves evaluating the environmental impacts associated with the mining, transportation, crushing, and spreading of these rocks, as well as potential environmental risks such as nutrient leaching and trace metal mobilization (Shackley et al. 2018).

The potential of enhanced weathering to sequester carbon at a large scale is promising, but the process requires significant energy inputs for mining and processing the rocks. Additionally, the long-term stability of the carbon stored in the form of carbonates is still uncertain (Kong et al. 2020). Nutrient leaching, particularly the release of phosphorus and other trace elements, is another concern, as it could lead to water quality issues and ecosystem imbalances in the areas where the rocks are spread (Shackley et al. 2018).

Case studies and pilot projects conducted in Iceland and Australia have demonstrated the technical feasibility of enhanced weathering. However, challenges related to scalability, infrastructure demands, and the environmental impacts associated with large-scale mining must be resolved before this technology can achieve widespread implementation (Shackley et al. 2018). Enhanced weathering involves trade-offs, including potential land-use conflicts and environmental risks arising from the extraction and processing of mineral resources (Lenton et al. 2020). Advancing this technology will require further research focused on optimizing its efficiency and minimizing its environmental footprint (MacDougall et al. 2019).

21.8 Life Cycle Assessment of Technological Carbon Dioxide Removal Methods

21.8.1 Direct Air Capture (DAC)

Direct air capture (DAC) is an advanced technology developed to extract carbon dioxide (CO_2) directly from ambient air, representing a pivotal strategy for addressing climate change. DAC systems are primarily categorized into solvent-based and solid sorbent systems. Solvent-based systems utilize liquid chemicals, such as amines, to absorb CO2, which is subsequently released through thermal regeneration for storage or utilization (Wei et al. 2020). In contrast, solid sorbent systems employ materials like metal-organic frameworks to adsorb CO2 at lower pressures and temperatures, with desorption occurring at elevated temperatures (Zhao et al. 2021).

Both approaches are energy-intensive, particularly during the regeneration of capture media and the compression of CO_2 for transportation and storage (Soria et al. 2022). Energy requirements are typically met through electricity or heat, which, if sourced from fossil fuels, can offset the net carbon removal benefits (Bui et al. 2020). Consequently, integrating DAC technologies with renewable energy sources is critical to enhancing their sustainability (Azar et al. 2021). Additionally, water consumption is a notable consideration, particularly in solvent-based systems where cooling and regeneration processes demand significant water inputs (Matsumoto et al. 2020).

The transportation of captured CO_2 adds another layer of complexity, as compression and conveyance via pipelines or shipping can introduce additional emissions (Luderer et al. 2021). Despite these challenges, DAC is recognized as a promising and scalable solution for CO_2 removal, especially when combined with renewable energy and sustainable storage options such as deep geological formations (Lackner et al. 2018; Ruth and Pacala 2020).

21.8.2 Bioenergy with Carbon Capture and Storage (BECCS)

Bioenergy with carbon capture and storage (BECCS) is a carbon-negative technology that integrates bioenergy production with CO_2 capture and storage, achieving net-negative emissions. The process begins with the cultivation of biomass—such as trees, agricultural residues, or dedicated energy crops—which absorbs atmospheric CO_2 through photosynthesis. When the biomass is combusted or processed to generate energy, the resulting CO_2 emissions are captured and stored in geological formations, effectively preventing their release into the atmosphere (Fajardy and Mac Dowell 2017). BECCS systems are adaptable, with variations depending on the type of biomass utilized and the CO_2 capture method employed, such as post-combustion or precombustion techniques (Heath et al. 2020). However, LCA of BECCS reveals several environmental challenges. Large-scale biomass production can trigger land-use changes that result in deforestation, biodiversity loss, and additional CO_2 emissions (Searchinger et al. 2015; Smith et al. 2016). Furthermore, BECCS operations require substantial energy inputs for biomass cultivation, transportation, conversion, and the subsequent capture, compression, and storage of CO_2 (Azar et al. 2021).

While BECCS offers significant potential for climate change mitigation, its scalability is constrained by these energy demands, land-use implications, and the necessity for reliable, long-term CO_2 storage solutions (Smith et al. 2016; Fuss et al. 2018). Addressing these limitations is crucial for maximizing the viability and sustainability of BECCS as a carbon removal strategy.

21.8.3 Mineralization

Mineralization, or mineral carbonation, is the process of accelerating the natural chemical reaction between CO_2 and abundant minerals to form stable carbonates, representing a promising strategy for carbon sequestration (Schroeder et al. 2019). This natural process occurs when CO_2 reacts with minerals such as olivine and serpentine, resulting in the formation of carbonates that can sequester CO_2 for millennia. The accelerated carbonation process, which enhances the natural reaction through elevated temperatures, pressures, or chemical catalysts, requires significant energy inputs (Krause et al. 2019). Mining and transporting raw minerals like olivine are energy-intensive and generate CO_2 emissions (Schroeder et al. 2019). Once extracted, the minerals must be crushed and processed, further increasing the energy demand (Baciocchi et al. 2021).

Despite these challenges, the primary advantage of mineral carbonation is the permanent sequestration of CO_2 in solid form, which mitigates the risk of leakage associated with other storage methods, such as geological sequestration (Wilke et al. 2021). Case studies, such as the CarbFix project in Iceland, have demonstrated the feasibility of this technology, wherein CO_2 is injected into basalt rock formations to form stable carbonates (Wilke et al. 2021). However, the scalability of mineral carbonation is hindered by the high energy demands and costs related to mineral extraction, processing, and carbonation, necessitating continued technological advancements to enhance efficiency and cost-effectiveness (Lackner et al. 2018).

21.8.4 Ocean Alkalinization

Ocean alkalinization is an emerging geoengineering approach aimed at increasing the alkalinity of seawater to enhance the ocean's ability to absorb CO_2 from the atmosphere. This process involves adding alkaline substances, such as lime or other minerals, to seawater, which reacts with CO_2 to form stable bicarbonate ions. By increasing the ocean's buffering capacity, this approach could potentially amplify the ocean's natural carbon sequestration processes (Kohler et al. 2013).

However, the large-scale implementation of ocean alkalinization presents significant challenges. The mining and processing of the necessary materials, such as limestone, require considerable energy, and the transportation of these materials to ocean sites further contributes to environmental costs (Lawania and Biswas 2018). Additionally, the ecological impacts of such interventions on marine ecosystems are not fully understood, raising concerns about potential adverse effects on marine life and biodiversity (Koh et al. 2020). Despite these uncertainties, ocean alkalinization holds potential as a climate change mitigation strategy if carefully managed. The main trade-offs involve balancing the energy and material inputs required for the process with the possible ecological consequences, which will need to be thoroughly assessed through pilot studies and long-term monitoring (Lawania and Biswas 2018).

21.9 Comparative Analysis of LCA Results

21.9.1 Cross-Technology Comparison

A comparative analysis of LCA results across various carbon dioxide removal (CDR) technologies offers valuable insights into their relative effectiveness in mitigating global carbon emissions. Technologies such as direct air capture (DAC), bioenergy with carbon capture and storage (BECCS), mineral carbonation, and ocean alkalinization demonstrate significant potential for CO_2 sequestration, yet exhibit diverse impacts across different stages of their life cycle. DAC, for example, necessitates substantial energy inputs, primarily in the form of electricity and heat, making it highly sensitive to the energy mix used during its operation (Ruth and Pacala 2020). While DAC is efficient in capturing CO_2 directly from the atmosphere, its carbon footprint remains relatively high unless powered by renewable energy sources (Fajardy and MacDowell 2017). In contrast, BECCS relies on biomass cultivation, processing, and combustion, which introduces land-use change effects and emissions from biomass production (Searchinger et al. 2015). Mineral carbonation and ocean alkalinization represent more natural processes but encounter challenges related to material requirements, energy consumption, and potential ecological impacts. In terms of carbon removal potential, BECCS typically offers the highest sequestration rates due to the large quantities of biomass available for capture, whereas DAC is more effective at capturing lower concentrations of CO_2 directly from ambient air (Smith et al. 2016). A comprehensive understanding of the strengths and limitations of each technology is essential for informing climate policies and strategies aimed at achieving global carbon removal targets.

21.9.2 Identification of Environmental Hotspots

LCA highlights that the most environmentally impactful stages of carbon dioxide removal (CDR) technologies typically occur during the production, transportation, and operational phases. In the case of direct air capture (DAC), energy consumption during the CO_2 capture and regeneration processes represents a significant environmental hotspot, particularly when fossil fuels are used, leading to increased CO_2 emissions (Lackner et al. 2018). BECCS faces comparable challenges, primarily arising from land-use changes associated with biomass production, which can result in deforestation and loss of biodiversity (Searchinger et al. 2015). For mineral carbonation, the most notable environmental impact stems from the energy-intensive processes involved in mineral extraction, transportation, and crushing of minerals such as olivine (Schroeder et al. 2019). Ocean alkalinization also carries environmental risks, particularly about the potential disruption of marine ecosystems due to the large-scale application of alkaline substances (Koh et al. 2020). Addressing these hotspots requires the adoption of renewable energy sources for DAC and

BECCS operations, optimization of land management practices for biomass cultivation, and improvements in the efficiency of material extraction and processing for mineral carbonation and ocean alkalinization. Moreover, implementing robust environmental monitoring and management strategies can help mitigate the associated risks of these technologies.

21.9.3 Sensitivity Analysis and Uncertainty Assessment

Sensitivity analysis and uncertainty assessment play a critical role in understanding the robustness of LCA results, especially in technologies like DAC, BECCS, and mineral carbonation. Variations in input data, such as energy requirements, land-use assumptions, and CO_2 transport distances, can lead to significant differences in the overall environmental impact of CDR technologies (Kohler et al. 2013). The choice of impact assessment methods also influences the interpretation of results, with some methods emphasizing energy use and others focusing on emissions or ecosystem impacts (Owsianiak et al. 2022). Long-term uncertainties, particularly regarding the durability of carbon sequestration in geological storage or mineral carbonation, require careful consideration in LCA studies. For example, the long-term permanence of CO_2 storage in BECCS and mineral carbonation systems remains uncertain, with potential risks of leakage introducing significant reliability concerns (Wilke et al. 2021). Geographical location also plays a significant role, influencing factors like energy sources, transportation distances, and land availability for biomass production or mineral extraction, thereby impacting the results of the LCA (Fajardy and MacDowell 2017). Incorporating these uncertainties into the decision-making process is essential for selecting the most effective and sustainable CDR technology.

21.9.4 Trade-Offs and Synergies

Recent studies have further explored the trade-offs and synergies among various carbon dioxide removal (CDR) technologies, emphasizing the importance of integrated approaches to optimize their effectiveness. For instance, a study by Gautam et al. (2025) discusses the convergence of carbon capture within the circular economy paradigm, highlighting the potential for integrating carbon capture and storage (CCS) with renewable energy sources to achieve multifaceted sustainable development goals. Additionally, Kim et al. (2024) assessed the economic viability of integrating solid direct air capture (DAC) systems with green hydrogen production. Their findings indicate that combining these technologies can yield significant economic advantages, enhancing both efficiency and sustainability. Furthermore, Nguyen et al. (2023) proposed a linear optimization modelling approach to identify the optimal contributions of renewables and CCS toward maximizing carbon reduction. Their results suggest that a combination of these technologies could enable a substantial reduction in CO_2 emissions in the energy sector by 2050. These studies underscore the critical need for policymakers to consider integrated, multi-criteria decision-making frameworks that balance factors such as cost, efficiency, land use, and social acceptance. By analyzing the synergies and trade-offs among various CDR technologies, optimal portfolios can be identified that align with both sustainability and climate objectives.

21.10 Policy Implications and Recommendations

21.10.1 Role of LCA in Policymaking

LCA is an essential tool for guiding policy decisions regarding carbon dioxide removal (CDR) technologies. It offers a comprehensive evaluation of the environmental, social, and economic impacts of various technologies, providing policymakers with the data needed to make well-informed decisions. By assessing technologies such as direct air capture (DAC), bioenergy with carbon capture and storage (BECCS), and ocean alkalinization, LCA can inform the allocation of public funds, the design of subsidies, and the implementation of regulations aimed at addressing climate change. Moreover, incorporating LCA into carbon accounting frameworks ensures that policymakers can accurately assess and track the effectiveness of different CDR technologies in reducing atmospheric CO_2 (Nemet et al. 2020; Terlouw et al. 2021; Butnar and Lynch 2021). LCA also helps identify the full range of potential impacts and benefits, enabling the development of policies that support the most sustainable, efficient, and scalable technologies. By integrating LCA findings into the decision-making process, policymakers can more effectively prioritize strategies and technologies that align with global climate goals.

21.10.2 Recommendations for Technology Development

LCA plays a pivotal role in identifying opportunities for technological improvements and more environmentally sustainable alternatives within carbon dioxide removal (CDR) technologies. By thoroughly analyzing the environmental impacts of various technologies, LCA can highlight energy-intensive processes, resource inefficiencies, and potential adverse environmental effects (Gagnon et al. 2021; Terlouw et al. 2021). For instance, DAC technologies, which are energy-intensive, can be optimized by enhancing the efficiency of CO_2 capture materials and processes. Furthermore, LCA can direct efforts toward the development of more sustainable biomass production methods in BECCS, mitigating land-use impacts and reducing associated carbon emissions (Fajardy and MacDowell 2017). To address uncertainties in LCA studies, especially concerning the long-term durability and scalability of carbon sequestration technologies, continued research is essential. As these technologies evolve, ongoing monitoring through LCA will be critical for refining models and assumptions, thereby improving their overall environmental performance (Walworth et al. 2024). Additionally, public engagement and education are vital to the successful development and deployment of these technologies, as public support is integral to their widespread acceptance and integration into climate policy.

21.10.3 Strategies for Effective Climate Change Mitigation

Integrating carbon dioxide removal (CDR) technologies into a comprehensive climate change mitigation strategy necessitates meticulous planning and coordination. CDR serves as a critical complement to emission reduction efforts, enabling the removal of excess CO_2 from the atmosphere that cannot be fully addressed through reductions alone (Fuss et al. 2018). Effective integration involves scaling up technologies such as BECCS, DAC, and mineral carbonation, while concurrently advancing efforts to reduce emissions through renewable energy, energy efficiency, and carbon-neutral technologies. International cooperation is essential for the global deployment of these technologies, as regions encounter distinct challenges related to infrastructure, energy access, and land use (Pillai et al. 2021). A holistic approach to climate mitigation must encompass both emission reductions and carbon removals to achieve net-zero emission targets. Collaborative frameworks, coupled with robust research and development initiatives, are crucial to ensuring the long-term success of CDR technologies in combating climate change. By synergizing CDR strategies with emissions reduction efforts, the transition to a net-zero world can be expedited.

21.11 Conclusion

The key role of LCA is to assess various carbon dioxide removal (CDR) technologies such as direct air capture(DAC), bioenergy with carbon capture and storage (BECCS), and ocean alkalization. While each technology offers significant potential for CO2 sequestration, they each face challenges related to energy requirements, land use, and scalability. Future efforts should focus on refining LCA models to reduce uncertainties around long-term performance and geographic variations, exploring alternative feedstocks, and improving the efficiency of these technologies. Additionally, integrating LCA into policy development and fostering international collaboration will help identify the most sustainable solutions for achieving net-zero emissions and support the global transition to a carbon-neutral future.

Future Prospects

The future of carbon dioxide removal (CDR) technologies depends on smart innovations and careful environmental assessment. New approaches such as engineered microbes to absorb CO_2, compact and energy-efficient DAC units, and ocean-based methods offer strong potential. To ensure long-term sustainability, improved Life Cycle Assessment (LCA) tools are needed to capture both direct and indirect environmental impacts. Technologies like artificial intelligence can make LCA faster and more accurate. Global collaboration and open data sharing will help identify the most effective and environmentally friendly CDR solutions. With the right strategies, CDR can become a key contributor to climate change mitigation and a cleaner future.

Conflicts of Interest The authors declare that they have no conflict of interest.

Availability of Data and Material Not Applicable.

Ethics Approval Not applicable—No human or animal subjects were involved.

References

Azar C, Lindgren K, Mollersten K (2021) Direct air capture of CO_2: a technology review and potential for deployment. Renew Sust Energ Rev 135:110234. https://doi.org/10.1016/j.rser.2020.110234

Baciocchi R, Allwood JM, Lopez-Amo J (2021) Review of the challenges of scaling up carbon mineralization for industrial carbon capture. J CO_2 Util 42:101308. https://doi.org/10.1016/j.jcou.2020.101308

Bakker E, Tuit S, Pfauder E (2022) Using secondary data in life cycle assessments: the pitfalls and potential. J Ind Ecol 26(5):1420–1432. https://doi.org/10.1111/jiec.13145

Bare JC (2019) TRACI 2.1: the tool for the reduction and assessment of chemical and environmental impacts. Environ Impact Assess Rev 78:55–58. https://doi.org/10.1016/j.eiar.2019.01.001

Baumann H, Tillman AM (2004) The hitch hiker's guide to LCA. Studentlitteratur, Lund. https://doi.org/10.1065/lca2006.02.008

Bhandari KP, Collier JM, Ellingson RJ, Apul DS (2015) Energy payback time (EPBT) and energy return on energy invested (EROI) of solar photovoltaic systems: A systematic review and meta-analysis. Renewable and Sustainable Energy Reviews 47:133–141

Bui M, Adjiman CS, Bardow A, Anthony EJ, Boston A, Brown S, Fennell PS (2020) Carbon capture and storage (CCS): the way forward. Energy Environ Sci 13(5):1344–1370. https://doi.org/10.1039/C9EE03029F

Butnar I, Lynch J (2021) A review of life cycle assessment methods to inform the scale-up of carbon dioxide removal interventions. Wiley Interdiscip Rev Energy Environ 10(6):e540. https://wires.onlinelibrary.wiley.com/doi/epdf/10.1002/wene.540

Chadwick R, Good P, Martin G, Rowell DP (2019) Large rainfall changes consistently projected over substantial areas of tropical land. Nat Clim Chang 6(2):177–181. https://doi.org/10.1038/nclimate2805

Chelvam K, Hanafiah MM (2023, May) Life cycle assessment of carbon capture, utilisation and storage technologies: an analytical review. In: IOP conference series: earth and environmental science (vol 1167, no 1, p 012036). IOP Publishing. https://doi.org/10.1088/1755-1315/1167/1/012036

Cipolla M, Pischke CR, Nabuurs G-J, Leip A, Briner G (2021) A critical review of carbon sequestration potential of afforestation and reforestation strategies. For Ecol Manag 491:119164. https://doi.org/10.1016/j.foreco.2021.119164

Davidson EA, Savage KE, Finzi AC, Gartner TR (2020a) Belowground carbon allocation in forests depends on soil moisture. Nature 408(6809):789–792. https://doi.org/10.1038/35048563

Davidson EA, Savory DJ, Suddick EC, Hartman MD (2020b) Soil carbon sequestration through land management: global trends and regional patterns. Glob Chang Biol 26(2):501–513. https://doi.org/10.1111/gcb.14876

Deutz S, Bardow A (2021) Life-cycle assessment of an industrial direct air capture process based on temperature–vacuum swing adsorption. Nature Energy 6(2):203–213

Duarte CM, Middelburg JJ, Caraco N (2020) The potential of coastal ecosystem-based carbon sequestration in a changing climate. Glob Environ Chang 63:102119. https://doi.org/10.1016/j.gloenvcha.2020.102119

Fajardy M, Mac Dowell N (2017) Can BECCS deliver sustainable and resource-efficient negative emissions?. Energy and Environmental Science 10(6):1389–1426

Fajardy M, MacDowell N, Kythreotis A (2019) The role of life cycle assessment in evaluating the feasibility of large-scale carbon capture and storage. J Clean Prod 239:118033. https://doi.org/10.1016/j.jclepro.2019.118033

Finkbeiner M, Reuter G, Herrmann C (2020) Uncertainty in life cycle assessments: sources and methods. Int J Life Cycle Assess 25(2):222–235. https://doi.org/10.1007/s11367-019-01653-2

Fletcher M, Sarah E, Gruber N, Jacobson AR, Gloor M, Doney SC (2020) Carbon sequestration potential of coastal ecosystems and challenges in their implementation. Nat Sustain 3(12):968–977. https://doi.org/10.1038/s41893-020-0571-4

Frischknecht R, Baumann M, Schryver D, Alexander. (2020) OpenLCA: a new approach to life cycle assessment for sustainable development. J Ind Ecol 24(5):1091–1100. https://doi.org/10.1111/jiec.13048

Fuss S, Lamb WF, Callaghan MW, Hilaire J, Creutzig F, Amann T et al (2018) Negative emissions—Part 2: costs, potentials and side effects. Environ Res Lett 13(6):063002. https://doi.org/10.1088/1748-9326/aabf9f

Gagnon B, Norris G, Ayer N (2021) Addressing resource efficiency in carbon dioxide removal through life cycle assessment. J Ind Ecol 25(3):560–575. https://doi.org/10.1111/jiec.13101

Gautam P, Salunke D, Lad D, Gautam A (2025) Convergent synergy of carbon capture within the circular economy paradigm: a nexus for realizing multifaceted sustainable development goals. Carbon Res 4:3. https://doi.org/10.1007/s44246-024-00178-1

Goedkoop, Mark, Heijungs, Reinout, Huijbregts, Mark, & de Schryver, An. (2013). ReCiPe 2008: a life cycle impact assessment method, 2013 update. Ministry of Infrastructure and Water Management, Netherlands doi:https://doi.org/10.1016/j.jclepro.2019.01.022

Griscom BW, Adams J, Ellis PW, Houghton RA, Lomax G, Miteva DA et al (2017) Natural climate solutions. Proc Natl Acad Sci 114(44):11645–11650. https://doi.org/10.1073/pnas.1710465114

Guinea F, Walet NR (2019) Continuum models for twisted bilayer graphene: Effect of lattice deformation and hopping parameters. Physical Review B 99(20):205134

Guinée JB (2002) Handbook on life cycle assessment: operational guide to the ISO standards (vol. 7). Springer Science & Business Media

Hauschild MZ, Rosenbaum RK, Olsen SI (2018) Life cycle assessment: theory and practice. Springer. https://doi.org/10.1007/978-3-319-56475-8

Heath G, McLaren J, Nemet G (2020) Scaling carbon capture and storage: environmental, technical, and economic challenges. Environ Sci Technol 54(12):7560–7569. https://doi.org/10.1021/acs.est.0c01545

Henriksson PJG, Guinée JB, Kleijn R, de Snoo GR (2012) Life cycle assessment of aquaculture systems—a review of methodologies. Int J Life Cycle Assess 17(3):304–313. https://doi.org/10.1007/s11367-011-0369-4

Houghton RA (2021) Emissions of greenhouse gases from land-use changes in tropical regions: a critical review of methodologies. Glob Chang Biol 27(5):1520–1538. https://doi.org/10.1111/gcb.15532

International Organization for Standardization (ISO) (2006a) Environmental management-life cycle assessment-principles and framework (ISO 14040:2006). International Organization for Standardization. https://www.iso.org/standard/37456.html

International Organization for Standardization (ISO) (2006b) ISO 14040: environmental management-life cycle assessment-principles and framework. International Organization for Standardization

IPCC (2021) Climate change 2021: the physical science basis. Contribution of working group I to the sixth assessment report of the intergovernmental panel on climate change. Cambridge University Press. https://doi.org/10.1017/9781009157896

Jones CD, Ciais P, Davis SJ, Friedlingstein P, Gasser T, Peters GP, van Vuuren DP (2016) Simulating the Earth system response to negative emissions. Environ Res Lett 11(9):095012. https://doi.org/10.1088/1748-9326/11/9/095012

Keith DW, Holmes G, St. Angelo D, Heidel K (2021) A process for capturing CO_2 from the atmosphere. Joule 5(1):1–22. https://doi.org/10.1016/j.joule.2020.11.006

Kim S, Park J, Lee JH (2024) Integrating solid direct air capture systems with green hydrogen production: economic synergy of sector coupling. arXiv preprint arXiv:2406.00665

Koh CP, Zhan Z, Wong M (2020) Potential risks of ocean alkalinization to marine ecosystems. Nat Sustain 3(7):461–469. https://doi.org/10.1038/s41893-020-0478-1

Kohler P, Doney SC, Mauzerall DL (2013) Ocean alkalinization as a method for carbon dioxide removal: a systematic review. Environ Sci Technol 47(9):4642–4650. https://doi.org/10.1021/es400617m

Kong AR, Maher DT, Santos IR (2020) Enhanced weathering and its role in mitigating global warming. Environ Res Lett 15(3):034024. https://doi.org/10.1088/1748-9326/ab6fa4

Krause A, Greiner L, Götz C (2019) Mineral carbonation for permanent CO2 storage: a systematic review and outlook. Environ Sci Technol 53(5):2695–2703. https://doi.org/10.1021/acs.est.8b05753

Lackner KS, Brennan S, Park A-HA (2018) The potential of direct air capture of CO2. Proc Natl Acad Sci 115(5):1917–1922. https://doi.org/10.1073/pnas.1719188115

Lal R (2021) Soil organic carbon sequestration as a key to mitigating climate change. In: Johnson JR (ed) Global climate change and the role of soil carbon. Springer, pp 179–201. https://doi.org/10.1007/978-3-030-39461-9_10

Lawania K, Biswas WK (2018) Application of life cycle assessment approach to deliver low carbon houses at regional level in Western Australia. Int J Life Cycle Assess 23:204–224. https://doi.org/10.1007/s11367-017-1314-y

Lenton TM, Rockström J, Gaffney O, Rahmstorf S, Richardson K, Steffen W, Schellnhuber HJ (2020) Accelerating the global implementation of carbon capture and storage technologies: opportunities and challenges. Nat Commun 11(1):1–13. https://doi.org/10.1038/s41467-020-19493-w

Luderer G, McJeon H, Hilaire J (2021) Climate change mitigation potential of direct air capture technologies. Nat Commun 12(1):2442. https://doi.org/10.1038/s41467-021-22696-6

MacDougall AS, McCann KS, Gellner G, Turkington R, Levine JM (2019) The impact of land-use changes on carbon sequestration and greenhouse gas emissions: implications for land management strategies. Environ Manag 63(5):876–890. https://doi.org/10.1007/s00267-019-01156-x

Matsumoto M, Kawano Y, Tsutsumi A (2020) Water usage and environmental considerations for direct air capture technologies. Environ Sci Technol 54(6):3709–3717. https://doi.org/10.1021/acs.est.0c00598

Minx JC, Lamb WF, Callaghan MW, Fuss S, Hilaire J, Creutzig F et al (2018) Negative emissions-Part 1: research landscape and synthesis. Environ Res Lett 13(6):063001. https://doi.org/10.1088/1748-9326/aabf9b

Muller W, Frischknecht R, Weidema BP (2018) Life cycle assessment of products: theory and practice. Springer. https://doi.org/10.1007/978-3-319-56475-8

Nemet GF, Callaghan MW, Creutzig F, Fuss S, Hartmann J, Hilaire J et al (2020) Negative emissions—Part 3: innovation and upscaling. Environ Res Lett 15(3):033003. https://iopscience.iop.org/article/10.1088/1748-9326/ab4cf9

Ness B, Andersson P (2020) The importance of goal and scope definition in life cycle assessments for sustainability assessments of new technologies. J Clean Prod 272:122711. https://doi.org/10.1016/j.jclepro.2020.122711

Nguyen DH, Chapman A, Tsuji T (2023) Assessing the optimal contributions of renewables and carbon capture and storage toward carbon neutrality by 2050. arXiv preprint arXiv:2305.05165

Owsianiak M, van Oers L, Drielsma J, Laurent A, Hauschild MZ (2022) Identification of dissipative emissions for improved assessment of metal resources in life cycle assessment. J Ind Ecol 26(2):406–420

Pillai R, Jain A, Lickley M, Sgouridis S (2021) International cooperation and policy frameworks for scaling up carbon dioxide removal technologies. Nat Commun 12(1):2203. https://doi.org/10.1038/s41467-021-22476-5

Rogelj J, Popp A, Calvin KV, Luderer G, Emmerling J, Gernaat D, Tavoni M (2018) Scenarios towards limiting global mean temperature increase below 1.5 C. Nat Clim Chang 8(4):325–332. https://doi.org/10.1038/s41558-018-0091-3

Roy P, Labadie J, Mazijn B (2020) Uncertainty analysis in LCA: methods and applications. Environ Impact Assess Rev 82:106405. https://doi.org/10.1016/j.eiar.2020.106405

Ruth M, Pacala S (2020) Global assessment of direct air capture technology. Nat Clim Chang 10(3):234–239. https://doi.org/10.1038/s41558-020-0693-2

Schroeder D, Seitz S, Schlüter L (2019) Mineral carbonation technologies for CO2 sequestration: from fundamental principles to industrial applications. Nat Commun 10(1):4289. https://doi.org/10.1038/s41467-019-12288-5

Searchinger T, Heimlich R, Houghton R (2015) Avoiding bioenergy pitfalls. Science 350(6258):1–3. https://doi.org/10.1126/science.aac8365

Shackley S, Sohi SP, Ibarrola R, Hammond J, Fawcett T (2018) Enhanced weathering for CO2 removal: a critical review of environmental impacts and economic potential. Environ Sci Technol 52(9):5160–5171. https://doi.org/10.1021/acs.est.7b06193

Smith P, Bustamante M, Ahammad H, Clark H, Dong H, Elsiddig E, Haberl H, Harper R, House J, Jafari M, Masera O, Mbow C, Ravindranath NH, Rice C, Robledo-Abad C, Romanovskaya A, Sperling F, Tubiello F (2016) Biophysical and economic limits to global carbon dioxide removal. Nat Clim Chang 6(10):993–1002. https://doi.org/10.1038/nclimate3024

Soria A, Wang T, Shah N (2022) Energy and economic implications of solvent-based and solid sorbent direct air capture systems. Energy Convers Manag 239:114280. https://doi.org/10.1016/j.enconman.2021.114280

Sweeney J, Nunes B, Parker M (2021) Life cycle inventory database for the United States: USLCI database update. J Ind Ecol 25(6):1398–1410. https://doi.org/10.1111/jiec.13095

Tanzer SE, Ramírez A (2019) When are negative emissions negative emissions? Energy Environ Sci 12(4):1210–1218. https://doi.org/10.1039/C8EE03338B

Terlouw T, Bauer C, Rosa L, Mazzotti M (2021) Life cycle assessment of carbon dioxide removal technologies: a critical review. Energy Environ Sci 14(4):1701–1721. https://pubs.rsc.org/en/content/articlehtml/2021/ee/d0ee03757e

Trost B, Prochnow A, Drastig K, Meyer-Aurich A, Ellmer F, Baumecker M (2018) Irrigation, soil organic carbon and N_2O emissions. A review. Agron Sustain Dev 33(4):733–749. https://doi.org/10.1007/s13593-013-0134-0

Vázquez-Rowe I, Marvuglia A, Rege S, Benetto E (2020a) Applying consequential LCA to support energy policy: land use change effects of bioenergy production. Sci Total Environ 472:78–89. https://doi.org/10.1016/j.scitotenv.2013.10.097

Vázquez-Rowe I, García-Álvarez J, Moreira MT (2020b) Secondary data in life cycle assessments: challenges and opportunities. J Ind Ecol 24(6):1215–1227. https://doi.org/10.1111/jiec.13053

Walworth CM, Gannon KE, Smith SM (2024) Public engagement and collaboration for carbon dioxide removal: lessons from a coastal enhanced weathering project. Front Clim 2:1290999. https://www.frontiersin.org/articles/10.3389/fclim.2024.1290999/full

Wei M, Patel M, Li Y (2020) Solvent-based direct air capture: a comprehensive review of operational and environmental impact. Appl Energy 269:115135. https://doi.org/10.1016/j.apenergy.2020.115135

Weidema BP, Wesnæs MS (1996) Data quality management for life cycle inventories—an example of using data quality indicators. J of Cleaner Production 4(3–4):167–174

Weidema BP, Wesnæs MS (2021) Data quality management for life cycle inventories: new methods and approaches. Environ Sci Technol 55(7):4135–4147. https://doi.org/10.1021/acs.est.0c06041

Wernet G, Bauer C, Steubing B (2016) The ecoinvent database: overview and applications in LCA. Int J Life Cycle Assess 21(5):950–960. https://doi.org/10.1007/s11367-016-1042-2

Wilke J, Smit F, Hülsebusch M (2021) CarbFix: a scalable solution to permanent carbon dioxide sequestration. Nat Commun 12(1):1189. https://doi.org/10.1038/s41467-021-21398-1

Zhang H, Li Y, Liu X, Wang X (2021) The role of coastal ecosystems in carbon sequestration and climate change mitigation. Sci Total Environ 746:141090. https://doi.org/10.1016/j.scitotenv.2020.141090

Zhao Z, Wang H, Liu M (2021) Land-use change impact in nature-based carbon dioxide removal methods: a critical review. J Clean Prod 286:125490. https://doi.org/10.1016/j.jclepro.2020.125490

Chapter 22
Synergies Between Algae and Aquaponics for Ecosystem Restoration and Climate Change Adaptation

P. Santhanam, G. Roseline Jebapriya, A. Gowthami, M. Hema, M. Saranya, S. Dhanasundaram, R. Anburaj, and P. Perumal

Abstract Environmental sustainability and global food system security are at risk due to climate change and ecosystem degradation. Innovative integrated algae-based aquaponics systems promise a potential pathway for improved resource use, adaptation to climate change, and ecosystem restoration. This chapter delves into the synergies between aquaponics and algae in advancing sustainable efforts in aquaculture and agriculture. Algae play several crucial roles in ecosystem restoration, including carbon sequestration, bioremediation, and biodiversity conservation. It supports environmental restoration through cutting-edge methods that may help absorb nutrient run-off from farming and mitigate climate change. Water efficiency is further enhanced by incorporating algae into aquaponic systems, along with nutrient recycling and climate resilience. Aquaponics is the closed-loop farming of aquaculture and hydroponics, which minimizes waste and water use to produce food sustainably. Incorporating algae into these systems not only improves nutrient availability and acts as a natural biofilter but also supports the health of both fish and crops while delivering essential ecosystem services. Combining algae with aquaponics offers a viable and scalable option to adapt to climate change by lowering environmental footprints, boosting ecosystem resilience, and safeguarding food and energy security. However, financial constraints, potential environmental risks, and technological obstacles limit the large-scale implementation of algae-aquaponics systems. Overcoming these hurdles will require coordinated multidisciplinary research, policy support, and technological innovation. This chapter concludes by

P. Santhanam (✉) · G. R. Jebapriya · A. Gowthami · M. Hema · M. Saranya · S. Dhanasundaram · P. Perumal
Marine Planktonology & Aquaculture Laboratory, Department of Marine Science, School of Marine Sciences, Bharathidasan University, Tiruchirappalli, Tamil Nadu, India

R. Anburaj
Department of Microbiology, Thanthai Hans Roever College (Autonomous), Perambalur, Tamil Nadu, India

B. P. Barathan et al. (eds.), *Navigating Climate Change: Impacts on Biodiversity and Ecosystem Resilience*, Sustainable Development and Biodiversity 42,
https://doi.org/10.1007/978-981-95-0409-1_22

outlining future research directions and emphasizing the urgent need for legislative initiatives to scale algae-aquaponics systems as a sustainable solution to pressing global environmental and agricultural challenges.

Keywords Aquaponics · Algae · Ecosystem restoration · Climate change adaptation · Carbon sequestration · Bioremediation · Sustainable agriculture · Food security · Environmental sustainability

22.1 Introduction

Global environmental problems are a result of ecosystem degradation caused by both natural and man-made influences. The degradation of ecosystem services, biodiversity, and the resilience of natural systems has resulted in climate change, habitat destruction, pollution, overexploitation, and unsustainable land-use practices (Hüttl and Schneider 1998; Glavovic et al. 2015; Sharma and Birman 2024). The undervaluation of ecosystem services, including flood protection, sediment retention, biodiversity conservation, maintenance of coastal and marine food systems, and water quality management, is a significant contributing element to this deterioration (Crooks 1999). Environmental instability is made terrible by these disturbances and climate change, which can have an impact on socioeconomic systems and result in human relocation (Warner et al. 2010; Malhi et al. 2020). In addition, long-lived greenhouse gases, such as CO_2, CH_2, and N_2O, contribute to rising global temperatures, while aerosols may induce localized warming or cooling effects (Doney et al. 2012). The impacts of climate change alterations in precipitation patterns, ocean chemistry, water balance, and extreme weather events are rapidly transforming ecosystems and further intensifying environmental stressors (Day and Rybczyk 2019). Addressing these challenges requires innovative, sustainable solutions that integrate ecological, technological, and economic benefits.

Microalgae incorporation into aquaponic systems is one such potential strategy. Microalgae are photosynthetic microorganisms that are found in both fresh and marine environments. They are essential for wastewater treatment, atmospheric CO_2 mitigation, and the synthesis of valuable bioproducts such as proteins, lipids, pigments, polysaccharides, vitamins, and bioactive compounds (Khan et al. 2018). They are a crucial part of climate change adaptation plans because of their capacity to absorb CO_2 and enhance water quality. Combining hydroponics and aquaculture, aquaponics is a sustainable system that has gained popularity as a climate-resilient food production technique. A closed-loop system that reduces resource consumption and environmental impact is produced in aquaponic systems by turning fish waste into nutrients that promote plant development (Love et al. 2015). A growing number of researchers are concentrating on cutting-edge aquaponic methods that improve system effectiveness and climatic variability resistance (Vasdravanidis et al. 2022). According to Goddek et al. (2019) and Knaus and Palm (2017), aquaponics maximizes nutrient recycling while reducing environmental volatility and guaranteeing sustainable aquaculture output by combining freshwater fish farming with plant cocultivation.

The combination of aquaponics and algae offers a fresh strategy for mitigating climate change and restoring ecosystems. Enhanced nutrient removal, better water quality, pH stability, dissolved oxygen generation, and the synthesis of polyunsaturated fatty acids for fish feed are only a few advantages of carefully controlled algal integration in aquaponic systems (Addy et al. 2017). Fish health and overall system productivity can be improved by using microalgae, which are high in lipids and omega-3 fatty acids, as a sustainable and nutritious feed ingredient (Li et al. 2011; Zhou et al. 2012; Cheunbarn and Cheunbarn 2015a, b; Tocher 2010). Algae also contribute to system stability by balancing pH levels, regulating ammonia concentrations, and efficiently removing nitrogen, often outperforming traditional vegetable crops in nitrogen assimilation (Addy 2017). There are still difficulties, though, especially when it comes to maximizing algae production for aquaponic systems' financial sustainability. To fully realize the promise of algae-aquaponic integration, regulated growing systems that optimize biomass increase while preserving system balance must be developed.

This review explores the synergistic potential of algae and aquaponics as a sustainable strategy for ecosystem restoration and climate change adaptation. It examines the role of microalgae in enhancing aquaponic system efficiency, mitigating environmental stressors, and contributing to carbon sequestration. Additionally, the review highlights the challenges and future prospects of integrating algae within aquaponic frameworks, emphasizing its potential as an eco-friendly and cost-effective approach to sustainable agriculture and aquaculture.

22.2 Role of Algae in Ecosystem Restoration

22.2.1 Diversity and Functionality of Algae

Algae are a remarkably diverse group of organisms that are present in almost every habitat on Earth. This diversity ranges from microscopic microalgae to complex macroalgae. One of the oldest living things on Earth is microalgae, which are small, unicellular, or multicellular creatures that include eukaryotic diatoms and green algae, and prokaryotic cyanobacteria (Mishra et al. 2019a, b). Algae have the capability of rapid growth and remarkable adaptability to diverse environments. Algae can thrive in freshwater, brackish water, marine environments, and even extreme habitats like hot springs and polar regions. Algae are classified into nine major phyla based on their pigments, cellular structures, and evolutionary origins. There are one prokaryotic phyla chloroxybacteria (cyanobacteria) and the eukaryotic phyla Chlorophyta (green algae), Cryptophyta (cryptomonads), Glaucophyta, (glaucophytes), Euglenophyta (euglenoids), Ochrophyta (chrysophyceans, tiny flagellates, diatoms, and brown algae), Haptophyta (haptophytes), Rhodophyta (red algae), and Dinophyta (dinoflagellates) (Sahoo and Seckbach 2015). Cyanobacteria are the oldest life forms on Earth, responsible for oxygenating the early atmosphere billions of years ago (Mobin and Alam 2017). Eukaryotic algae contain membrane-bound organelles, including chloroplasts, which are essential for photosynthesis. These chloroplasts originated from ancient endosymbiotic events, where a eukaryotic cell engulfed a photosynthetic bacterium, leading to the evolution of algae. The diversity of algal groups is further expanded by secondary endosymbiosis, where one eukaryotic alga was engulfed by another, giving rise to complex groups like dinoflagellates and haptophytes (Bolton 2016). Algae exhibit remarkable metabolic flexibility, which enables them to thrive in various environmental conditions. While many algae are autotrophic, producing their own food via photosynthesis, some, such as dinoflagellates and euglenoids, are mixotrophic, combining photosynthesis with the consumption of organic matter. Others are heterotrophic, entirely relying on organic carbon sources (Perin et al. 2019).

The ecological roles of algae are vital and responsible for a significant portion of global carbon fixation, oxygen production, and a critical role in marine food webs (Zhu et al. 2017). Algal blooms can sometimes be harmful, also influence the marine ecosystems by altering nutrient cycles. Furthermore, macroalgae like kelp forests provide important habitats for a wide range of marine species, offering food and shelter (Peteiro et al. 2016). Beyond their ecological importance, algae have immense economic and commercial value. Algae's adaptability to different environmental stressors, including changes in light, salinity, and temperature, has further increased interest in them for biotechnological applications. Their ability to produce high-value secondary metabolites under stress conditions can be harnessed for the production of nutraceuticals, cosmetics, and biofertilizers. Generally, macroalgae are harvested for their phycocolloids, such as agar, carrageenan, and alginate, which are used in food, pharmaceuticals, and industrial applications. In many parts of the world, algae are also consumed as food, such as nori, used in sushi, and wakame, popular in soups and salads. Microalgae have gained attention for their potential in biofuel production, as they can produce bio-oils, biodiesel, and other renewable energy sources (Abu-Ghosh et al. 2021). Additionally, algae are rich in bioactive compounds with antiviral, antibacterial, and anticancer properties, making them a promising resource for pharmaceuticals and nutraceuticals (Gauthier et al. 2020).

22.2.2 Role of Algae in Aquatic Biodiversity

Algae play a pivotal role in supporting aquatic biodiversity through various ecological functions. They are primary producers, forming the foundation of aquatic food webs by converting sunlight into energy via photosynthesis, which sustains herbivores and higher trophic levels (Zhou et al. 2023). Algae contribute significantly to oxygen production, releasing oxygen into the water, which is essential for the survival of fish, invertebrates, and other aquatic organisms (Cuellar-Bermudez et al.

2015). They also regulate nutrient cycles by absorbing excess nitrogen and phosphorus from the water. This prevents nutrient over-enrichment and reduces the risk of harmful algal blooms, maintaining ecosystem balance (Aketo et al. 2020). Algae provide food for a diverse range of organisms, from microscopic zooplankton to larger species like fish. Additionally, their presence supports microbial communities, which further enhance nutrient recycling and decomposition processes (Zhou et al. 2023).

22.2.3 Algal Bioremediation

The growing demand for clean water and the escalating pollution crisis necessitate sustainable remediation strategies (Javan et al. 2024). Algal bioremediation, utilizing the unique capabilities of algae, offers a promising solution to water pollution and resource scarcity. Algae efficiently remove a variety of pollutants from wastewater, including heavy metals, hydrocarbons, and nutrients (Iftehimul et al. 2023).

Microalgae degrade organic contaminants such as pesticides, hydrocarbons, and phenols through enzymatic degradation, bioadsorption, and assimilation (Mustafa et al. 2021). *Chlorella vulgaris* and *Scenedesmus obliquus* have shown the ability to degrade organophosphate and carbamate pesticides via hydrolases and oxidases, achieving degradation rates up to 96% for diazinon and 100% for methamidophos (Lanelly and Elsa 2024). Hydrocarbons from petroleum products are metabolized by microalgae like *Botryococcus braunii*, significantly reducing crude oil concentrations in aquatic systems (Nazloo et al. 2024). For phenolic compounds, *Isochrysis galbana* and *Phaeodactylum tricornutum* have demonstrated notable degradation capabilities (Wang et al. 2019).

Heavy metals such as cadmium, lead, and mercury pose serious environmental risks. Microalgae like *Chlamydomonas reinhardtii* and *Chlorella* sp. bioaccumulate these metals via their polysaccharide- and protein-rich cell walls. *Chlorella vulgaris* adsorbed 17.49 mg/g of mercury within 2 hours (Solisio et al. 2019). Nutrients like nitrates and phosphates, responsible for eutrophication, are also assimilated by algae. *Chlorella* sp. removed 31.26 mg/g of nitrate and 11.07 mg/g of phosphate within 7 days in wastewater treatment setups (Fazal et al. 2021).

Pharmaceuticals, personal care products, and microplastics are rising concerns. Microalgae have shown capacity to degrade or adsorb these pollutants. *Haematococcus pluvialis* exhibited the highest efficiency in degrading various antibiotics under controlled conditions (Kiki et al. 2020). Endocrine disruptors like parabens and triclosan are effectively removed by *Scenedesmus* and *Tetraselmis* species through bioadsorption and enzymatic transformation (Dai et al. 2021). Additionally, *Picochlorum maculatum* demonstrated a 20% reduction of LDPE microplastics by altering their physical and chemical properties (Gowthami et al. 2023).

22.2.4 Carbon Sequestration

Microalgae contribute approximately 50% oxygen production to earth by photosynthesis, in which large portion of carbon-di-oxide is also fixed (Banerjee et al. 2020). Microalgae are also one of the carbon sinks because they have high photosynthetic efficiency and capable of removing up to 273.75 g/L of CO_2 (Koukoumaki et al., 2024). In addition, ubiquitous nature of microalgae makes them a potential candidate for the bio-fixation of CO_2 and for bioenergy production (Thanigaivel et al. 2022). The integration of microalgae with wastewater treatment offers multiple benefits (Rame et al. 2023).

Microalgae like *Botryococcus braunii, Chlorella kessleri, Dunaliella tertiolecta, Isochrysis galbana, Chlorococcum littorale, Spirulina platensis, Nannochloropsis, Dunaliella salina, Scenedesmus obliquus*, and *Chlorella* sp. have been extensively studied for their role in carbon sequestration. Freshwater species, such as *Scenedesmus, Spirulina platensis*, and *Chlorella* sp., are commonly used for carbon capture (Alami et al. 2020; Mondal et al. 2016; Wang et al. 2018). Previous report indicated that microalgae can fix up to 1.83 kg of CO_2 per kg of biomass, highlighting their high potential for CO_2 sequestration (Cheah et al. 2015). Atmospheric CO_2 concentration typically ranges between 0.03% and 0.06% (v/v), while flue gases from industrial sources can contain 6% to 20% CO_2 (Brilman et al. 2013). For instance, *Chlorella vulgaris* and *Anabaena* sp. can fix CO_2 at a rate of 6.24 g/L/day (Ghorbani et al. 2014). Therefore, microalgae cultivation is often located near large CO_2 emitters, such as power plants and industrial facilities, to optimize CO_2 capture. CO_2 assimilation efficiency of microalgae can be enhanced through the regulation of phytohormones, genetic engineering, and using advanced cultivation technologies (Chen et al. 2023). Incorporation of advanced technologies such as self-monitoring analysis, sensors, artificial intelligence, and bioengineering can enhance microalgal CO_2 sequestration and CO_2 valorization (Ma et al. 2022).

Macroalgae are also recognized as a highly efficient natural carbon sink due to their rapid growth and high productivity. Certain species, such as *Macrocystis pyrifera, Sargassum* sp., and *Laminaria hyperborea*, can fix carbon at rates exceeding 3000 g C/m^2 annually (Paine et al. 2021). Approximately 30–50% of seaweed's fixed carbon is released as dissolved organic carbon (DOC), of which a significant fraction can be refractory, persisting for centuries in the deep ocean (Krause-Jensen and Duarte 2016). Seaweed farming further enhances the carbon capturing efficiency. By optimizing species selection, spacing, and harvest cycles, farms can increase productivity and carbon capture. Studies suggest that farming a mere 0.2% of the ocean's suitable area could sequester 0.1 Gt CO_2 annually. Moreover, seaweed's adaptability to varying marine conditions allows it to function in diverse environments, enhancing its potential for scalable carbon sequestration. This positions seaweed farming as a critical tool in global carbon mitigation efforts (Froehlich et al. 2019).

22.3 Aquaponics

Aquaponics is an integrated system that combines the principles of hydroponics and recirculating aquaculture. In this system, nutrient-rich water derived from fish waste serves as a natural fertilizer for plant growth, contrasting with conventional hydroponics, which relies on the addition of mineral fertilizers to supply essential nutrients (Rajalakshmi et al. 2022). A major advantage of aquaponics is the reduced need for water exchange; unlike in traditional aquaculture, excess nutrients are utilized by plants rather than requiring periodic flushing with fresh water. This closed-loop approach fosters the sustainable use of both water and nutrients, promoting a symbiotic relationship among fish, microbes, and plants (Gichana et al. 2018), and it was clearly illustrated in Fig. 22.1.

Importantly, the ecological limitations associated with hydroponics and aquaculture when practiced independently are transformed into complementary strengths within aquaponics systems (Schneider et al. 2021). As a result, the combined system significantly lowers both waste production and the need for external fertilizer inputs (Jouhara et al. 2017). Plants require a range of macronutrients such as carbon (C), hydrogen (H), oxygen (O), nitrogen (N), phosphorus (P), potassium (K), calcium (Ca), sulfur (S), and magnesium (Mg) and micronutrients including iron (Fe), chlorine (Cl), manganese (Mn), boron (B), zinc (Zn), copper (Cu), and molybdenum (Mo) to support healthy growth (Ahmed et al. 2020). Except for C, H, and O, which are naturally obtained from air and water, all essential nutrients must be provided to plants in ionic forms, particularly in hydroponic and aquaponic systems (Silber and Bar-Tal 2019).

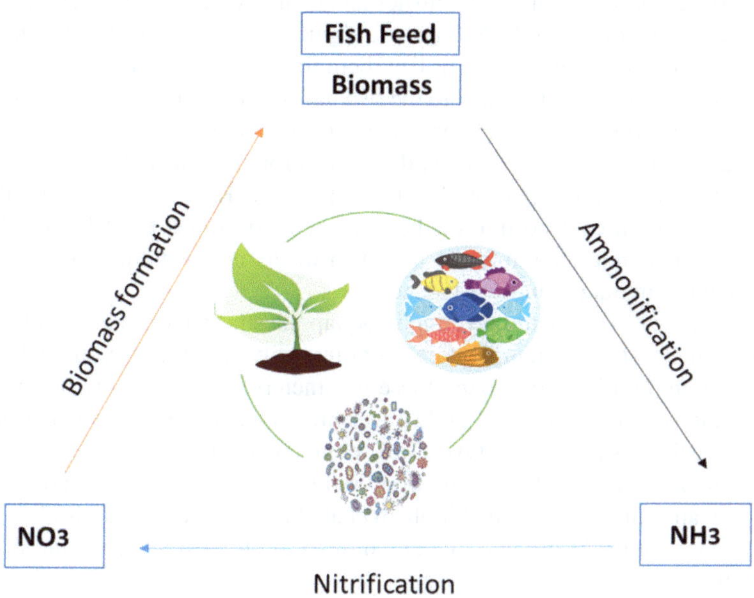

Fig. 22.1 Nutrient flow in the aquaponics system

In aquaponics, the dissolved fish waste supplements plant nutrient inputs by providing a mix of both organic and inorganic compounds, many of which are broken down into absorbable ionic forms (Amin 2022). Regular monitoring of macro- and micronutrient concentrations is critical to ensure optimal plant development (Nadeem et al. 2018). Occasionally, external supplementation may be necessary; for instance, fish waste alone may not supply adequate levels of iron, a key micronutrient for plant health. Microbial communities play a vital role in managing fish waste and ensuring nutrient availability. Aquaponics systems must support diverse microbial populations capable of processing organic matter and facilitating nutrient cycling (Gichana et al. 2018). A key microbial process involves the conversion of toxic ammonia (NH_4^+) excreted by fish via urine and gill discharge into less harmful nitrate through nitrification. This transformation is driven by nitrifying autotrophic bacteria, particularly species like *Nitrosomonas* (ammonia-oxidizing bacteria) and *Nitrospira* or *Nitrobacter* (nitrite-oxidizing bacteria), which form biofilms on solid surfaces within the system (Goddek et al. 2015).

22.3.1 Nutrient Recycling in the Aquaponics System

Nutrient recycling in aquaponics is a sophisticated process that integrates aquaculture and hydroponics into a sustainable model for food production. Within this closed-loop system, fish waste primarily in the form of ammonia serves as the main nutrient source. However, ammonia is toxic to fish at elevated concentrations (Goddek et al. 2016). Beneficial bacteria, particularly *Nitrosomonas* and *Nitrobacter* species, convert ammonia first into nitrites and then into nitrates through the process of nitrification (Gerardi 2002). Nitrates are significantly less harmful to fish and are readily absorbed by plants, which utilize them for growth, effectively purifying the water before it is recirculated back to the fish tanks (Gichana et al. 2018). The nutrient dynamics in aquaponics can be visualized through the flow of inputs and outputs (Fig. 22.2). Fish are fed a formulated diet, and through their metabolism, nutrients such as nitrogen (N), phosphorus (P), and potassium (K) are released into the system. While a significant portion of these nutrients becomes available to plants, a fraction remains unutilized in the form of fish waste, necessitating efficient system management to maintain nutrient balance.

The efficiency of nutrient recycling in aquaponics is influenced by multiple factors, including water pH, temperature, light intensity, and the overall health of the microbial communities; optimizing these parameters enhances nutrient uptake and plant productivity (Kumar et al. 2022). Despite its advantages, challenges such as nutrient imbalances can occur. Over time, deficiencies in elements like potassium, calcium, magnesium, and trace micronutrients may arise, necessitating regular monitoring and supplementation (Graham et al. 2001). Moreover, biofouling caused by the accumulation of organic matter can lead to clogged systems and disrupted water flow.

Fig. 22.2 Nutrient flow in aquaponics systems: transformation of fish feed into plant-available nutrients and the occurrence of unutilized nitrogen (N), phosphorus (P), and potassium (K) in fish waste

To optimize nutrient recycling, practitioners employ strategies such as frequent water quality testing, implementation of solid waste treatment systems to convert waste into plant-available fertilizers, and tailoring system designs to meet specific crop requirements (Umeh 2024). By understanding and effectively managing these dynamics, aquaponics systems can achieve high productivity while minimizing environmental impacts through efficient resource utilization and waste reduction. This innovative approach not only addresses critical food security issues but also promotes ecological sustainability by creating a balanced ecosystem where aquatic life and plant cultivation coexist harmoniously.

22.3.2 Water Efficiency in Aquaponics

Aquaponics is a sustainable agricultural practice that merges aquaculture (fish farming) with hydroponics (soilless plant cultivation), achieving remarkable water efficiency through a closed-loop system (Ibrahim et al. 2023). In this integrated approach, fish waste provides essential nutrients for plants, while beneficial bacteria convert toxic ammonia into plant-available nitrates (Yavuzcan Yildiz et al. 2017). The continuous recirculation of water minimizes evaporation and runoff, enabling aquaponics to use up to 90–99% less water than conventional farming systems. For example, producing one kilogram of vegetables in an aquaponics setup may require only 0.49 cubic meters of water, a significant reduction compared to traditional agriculture. Additionally, aquaponics eliminates soil erosion and greatly reduces the need for chemical fertilizers and pesticides, lessening environmental impacts and fostering healthier ecosystems (Joyce et al. 2019). Its suitability for urban farming makes it a promising option for local food production, reducing food miles and further enhancing sustainability.

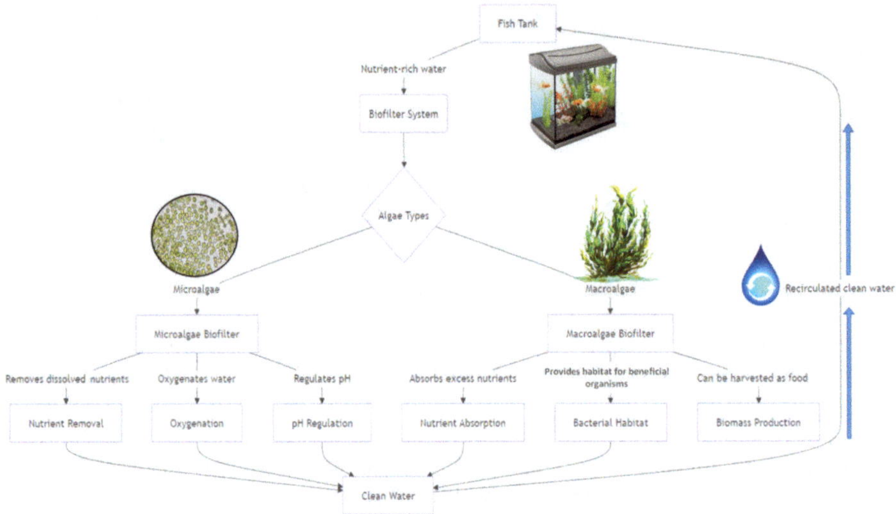

Fig. 22.3 Flowchart of algal-associated biofiltration system in aquaponics

22.4 Algae as a Biofilter in Aquaponics

Algae act as an efficient biofilter in aquaponic systems by removing excess nutrients and pollutants from the water (Fang et al. 2017). They play a vital role in the nitrogen cycle by transforming toxic ammonia from fish waste into safer compounds like nitrate, which plants can absorb as nutrients. This not only improves water quality but also boosts plant growth. Additionally, algae help purify water by absorbing heavy metals and organic pollutants, while their interaction with microbes enhances organic matter decomposition, maintaining a balanced ecosystem (Kushwaha et al. 2023). The efficiency of algae in biofiltration is influenced by factors such as light, temperature, and photoperiod, making it crucial to optimize these conditions for optimal performance (Fig. 22.3) (Calderini et al. 2021).

22.4.1 Microalgae's Role as Biofilter

In a study conducted by researchers, the microalga, *Tetradesmus dimorphus*, was combined with nitrifying bacteria to form a photo-biofilter, which was integrated into a Recirculating Aquaculture System (RAS) for rearing rainbow trout. The results demonstrated that the algal–bacterial community effectively removed ammonium, total ammonium nitrogen (TAN), nitrate, and phosphate simultaneously. Additionally, the system achieved an average daily TAN removal rate of 1.18 kg per m^3 of biofilter beads (Rodríguez-Leal et al. 2023). In aquaponics systems, algae not only remove excess nutrients, but also promote ecosystem stability by interacting with microbial communities. These interactions accelerate the decomposition of organic matter and nutrient mineralization, improving nutrient availability for plants (Munguia-Fragozo et al. 2015). Algae can also suppress harmful bacteria and pathogens, ensuring a safer environment for both fish and plants (Kasozi et al. 2021). The effectiveness of algae as biofilters is shaped by factors like light, temperature, and nutrient levels, which must be properly managed for optimal results. Recent research shows that integrating algae with bacterial communities can enhance the removal of nitrogenous waste and other pollutants, boosting the sustainability and efficiency of aquaponics systems (Vo et al. 2024; Ji et al. 2024). By using the algae, absorption of different nitrogen compounds, an algal–bacterial-based aquaponic system can decrease the production of waste and boost nitrogen utilization efficiency (Fang et al. 2018). In aquaculture, microalga like *Tetraselmis suecica* has shown great promise as biofilters, particularly when it comes to removing nitrogenous waste from water (Michels et al. 2014). Absorption of ammonium and other nutrients including phosphate of these autotrophic organisms lower hazardous nitrogen levels (Thakur et al. 2023). Besides, microalgae enhance nutrient cycling in aquaponics systems by aiding the creation of oxygen. Their application offers a strong substitute for conventional bacterial biofilters, which can result in the production of off-flavor chemicals like geosmin that lower fish quality. According to studies, microalgae can efficiently extract significant amounts of phosphorus and nitrogen from aquaculture wastewater. *Tetraselmis suecica*, for example, was able to remove 99% of the phosphorus and 49.4% of the nitrogen from saltwater aquaculture effluents (Michels et al. 2014). Depending on the nitrogen concentration of the algal biomass, a photobioreactor employing microalgae for ammonium removal was calculated to need 0.4 to 1.26 hectares on a system scale in order to completely substitute the ammonium-removing capability of a conventional biofilter.

22.4.2 *Macroalgae's Role as Biofilter*

Macroalgae are feasible substitutes for ammonium removal in addition to microalgae, albeit with lower overall efficiency (Ojanen et al. 2017). In this regard, due to their bigger size and propensity to create floating mats, macroalgae are simpler to collect. Likewise, they provide possible by-products for sectors like biofuels and food (Ganesan et al. 2024). Even though macroalgae might not be as effective as microalgae at removing nutrients, they are still a good choice for integrated aquaponic systems because of their economic potential and simplicity of harvesting. The potential of marine macroalgae as alternatives for conventional biofilters in RAS for ammonium removal has also been investigated. In comparison to traditional biofiltration techniques, studies by Cahill et al. (2010) showed that marine macroalgae—specifically, *Ulva lactuca* and *Undaria pinnatifida*—effectively maintained reduced ammonium concentrations in marine farming settings. Similarly, Azman et al. (2014) looked at the combined cultivation of *Ulva lactuca* and *Gracilaria edulis*, two macroalgal species, in an outdoor recirculating system intended for the treatment of shrimp wastewater. According to their research, both species considerably improved the overall quality of the water by lowering the wastewater's ammonium levels.

22.4.3 Integrated Aquaponics Systems with Algae

Algae-based integrated aquaponics systems provide an effective and sustainable way to control water quality and nutrient cycles (Ji et al. 2022). Since they absorb inorganic nutrients from the water, including phosphate, nitrate, and ammonium, algae especially microalgae play an essential role in these systems. This procedure lessens the requirement for outside nutrient inputs while also aiding in maintaining ideal water quality. By adding another source of oxygen through photosynthesis process, which is vital for the health of both fish and plants, algae improve the ponics-system's overall production (Zhang et al. 2022). Additionally, algae can be collected and used as a valuable resource to produce animal feed, biofuels, and other bioproducts, which would add a financial component to the environmental advantages (Ajeng et al. 2022) (Fig. 22.4).

By incorporating algae into aquaponics systems helps overcome some of the disadvantages of traditional biofilters (Ramli et al. 2020). Algae are highly effective in removing a variety of contaminants, including heavy metals and organic pollutants, which can be difficult to manage through standard filtration methods (Kumar et al. 2015). Likewise, algae foster a synergistic interaction with microbial communities, promoting the decomposition of organic materials and enhancing the availability of nutrients for plant growth (Renganathan et al. 2024). This holistic approach not only optimizes nutrient recycling, but also encourages the development of a diverse and resilient ecosystem. Emerging research has underscored the potential of algae to significantly boost the sustainability and productivity of aquaponics, positioning them as a key solution for future food production challenges. Algae serve as effective biostimulants in aquaponics systems, encompassing a wide array of organic compounds and beneficial microorganisms, such as bacteria, fungi, and algae (Braun and Colla 2023). These biostimulants are directly applied to plants

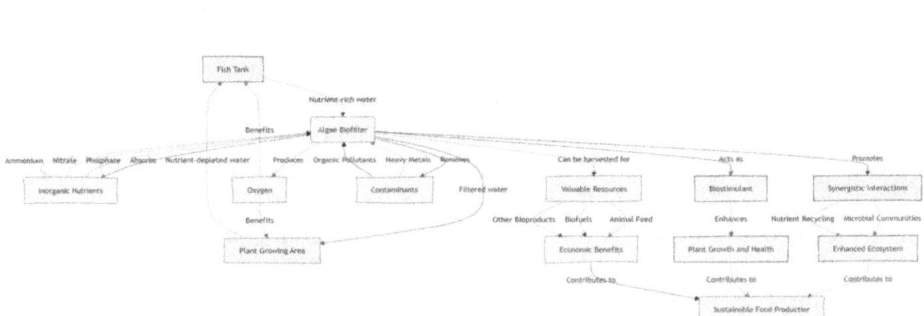

Fig. 22.4 Integrated aquaponics systems with algae for sustainable food productions

during different growth stages to enhance and regulate their physiological and developmental processes. By doing so, algae not only support plant health and growth, but also contribute to optimizing the overall productivity of the aquaponics ecosystem.

22.4.4 Co-benefits for Crop and Fish Health

Integrated aquaponics systems that incorporate algae offer significant co-benefits for both crop and fish health, enhancing the overall sustainability and productivity of the system. One of the primary advantages is the efficient nutrient cycling facilitated by the symbiotic relationship between fish and plants (Wongkiew et al. 2021). Fish waste provides essential nutrients, such as nitrogen and phosphorus, which are crucial for plant growth. In turn, plants help to filter and purify the water, creating a healthier environment for the fish. This natural filtration process reduces the need for chemical fertilizers and water treatments, promoting a more organic and sustainable method of food production (Chiquito-Contreras et al. 2022). Moreover, the presence of algae in these systems can further enhance these benefits (Addy et al. 2017). Algae not only contribute to nutrient removal but also produce oxygen through photosynthesis, which is vital for maintaining dissolved oxygen levels in the water (Fang et al. 2018). This is particularly beneficial for fish health, as adequate oxygen levels are essential for their growth and well-being. Additionally, algae can help control harmful bacteria and pathogens, reducing the incidence of diseases in both fish and plants (Stouvenakers et al. 2019). The integration of algae thus supports a more balanced and resilient ecosystem, improving the overall health and productivity of the aquaponics system. Further research highlights that algae can also play a role in mitigating greenhouse gas emissions within aquaponics systems. By assimilating carbon dioxide during photosynthesis, algae contribute to carbon sequestration, which can help reduce the overall carbon footprint of the system (Kotzen et al. 2019). This aspect is particularly important in the context of climate change and the need for sustainable agricultural practices. Additionally, the biomass produced by algae can be harvested and utilized for various high-value applications, such as biofuels, animal feed, and pharmaceuticals, adding an economic dimension to their environmental benefits (Table 22.1).

22.4.5 Enhancing Resilience Against Climate Variability

Climate change affects the aquaculture sector directly as well as indirectly, having an immediate and long-term effect on the fishing sector. The impacts of climate change, including rising temperatures, ocean acidification, and harmful algal blooms, can directly influence the physical and physiological traits of finfish and shellfish (Mitra et al. 2023). Fish have elevated levels of catecholamines,

Table 22.1 Co-benefits of integrated aquaponics systems incorporating algae for crop and fish production

Microalgae species name	Fish types grown in aquaponics system	Crop types growing in aquaponics system	Productivity from the aquaponics system	References
Chlorella vulgaris	Nile Tilapia (*Oreochromis niloticus*)	Lettuce (*Lactuca sativa*)	High nitrogen removal efficiency, leading to improved water quality for tilapia and high lettuce yield. Enhanced sustainability by recovering up to 80.84% of nitrogen in the system	Kankia et al. (2024) and Sánchez-Morales et al. (2024)
Spirulina platensis	Catfish (*Clarias gariepinus*)	Basil (*Ocimum basilicum*)	Effective at improving water quality and promoting higher basil yield while reducing nutrient waste in the system	Knaus et al. (2020)
Scenedesmus obliquus	Common Carp (*Cyprinus carpio*)	Spinach (*Spinacia oleracea*)	Demonstrated significant ammonia reduction, promoting better growth rates for both fish and spinach in the system	Mercado et al. (2020)
Tetraselmis suecica	Rainbow Trout (*Oncorhynchus mykiss*)	Tomatoes (*Solanum lycopersicum*)	Highly effective in ammonium removal, leading to optimized conditions for fish rearing and tomato cultivation. Reduced environmental impact and improved water recycling efficiency	Andreotti et al. (2019)
			Tetraselmis suecica can remove up to 49.4% of nitrogen and 99% of phosphorus from aquaculture wastewater, which enhances conditions for both rainbow trout and tomato cultivation	

corticosteroids, cortisol, glycogen, and lactic acid when they are subjected to many stressors, including the effects of climate change (Islam et al. 2022; Abdel-Tawwab et al. 2019). Hormones released into the blood and tissues because of thermal stress reactions alter the immunological responses, energy mobilization, heart activity, ionic balance, physical performance, growth, and their reproduction rate (Reid et al. 1998; Barton 2002). Thus, the aquaponics-algal integration robust solution to enhance resilience against climate variability and also mitigate their fish-associated diseases. Aquaponics systems capitalize on the natural symbiosis between fish, plants, and algae to establish a self-regulating ecosystem capable of adapting to environmental changes (Mathew et al. 2024). Algae play an essential role in this integration by absorbing excess nutrients like nitrogen and phosphorus from fish waste, thus maintaining water quality and minimizing the need for chemical interventions (Nie et al. 2020). This nutrient absorption not only promotes plant growth, but also stabilizes the system against fluctuations in nutrient levels, which may arise due to climate variability (Fig. 22.5).

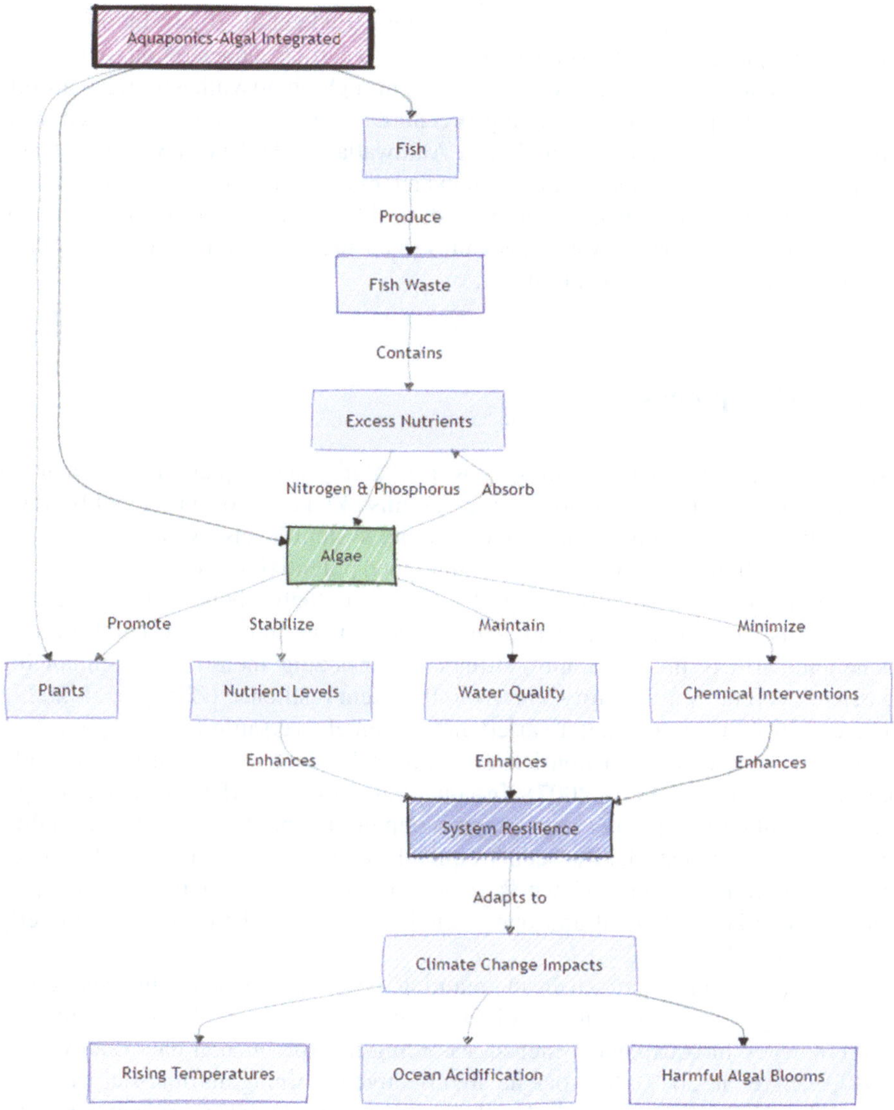

Fig. 22.5 Enhancing resilience against climate variability in aquaponics-algal integrated systems

Algae enhance the resilience of aquaponics systems by producing oxygen through photosynthesis, which is crucial for sustaining adequate dissolved oxygen levels necessary for fish health (Omweno et al. 2024). This function becomes increasingly important as rising temperatures and other climate-related stressors threaten to deplete oxygen levels in aquatic environments. The algae such as flagellates, dinoflagellates, and other harmful species within these taxonomic groups are potentially toxic and can lead to significant stress or even mortality in finfish and shellfish populations (Oh et al. 2023). Additionally, algae can reduce the risk of harmful algal blooms by outcompeting undesirable species for nutrients and light, ensuring a more balanced and robust ecosystem (Ende et al. 2024). Incorporating algae into aquaponics systems also contributes to carbon sequestration, helping to lower the system's overall carbon footprint. Through photosynthesis, algae absorb carbon dioxide, potentially offsetting greenhouse gas emissions associated with aquaculture and agriculture (Singh and Ahluwalia 2013). The development and adoption of innovative, sustainable systems are essential to mitigate environmental fluctuations that could negatively affect fish health and reduce productivity (Yadav et al. 2024). These ponics-system also aims to ensure efficient aquaculture production with minimal environment impact.

22.5 Algaeponics

Algaeponics, a term that combines the words "algae" and "aquaponics," is a novel idea that uses microalgae in aquaponic systems to take use of its special biological qualities for benefits at the ecosystem and system levels. Algaeponics introduces a new dimension by utilizing the untapped potential of microalgae, whereas aquaponics has historically focused on the mutually beneficial connection between fish and plants. Microalgae have been shown in recent research to have beneficial effects in aquaponic settings, emphasizing their role in enhancing nutrient cycling, water quality, and overall system resilience (Zhang et al. 2022). The capacity of microalgae, a varied and extremely versatile class of photosynthetic microorganisms, to flourish even under low-nutrient conditions is well-known (Gordon and Polle 2007). In aquaponic systems, when unchecked algal growth resulted in pipe blockage, oxygen depletion, and declining water quality, microalgae have historically been frequently considered a nuisance. However, new research has changed viewpoints, acknowledging that microalgae may provide substantial ecological and operational advantages when managed properly (Addy et al. 2017).

Microalgae can be crucial in maintaining system pH balance, increasing dissolved oxygen concentrations, regulating ammonia levels, and boosting ammonia use efficiency in aquaponic systems, according to experimental data (Addy et al. 2017; Fang et al. 2017). Microalgae are effective in turning nutrient-rich aquaculture effluent into useful biomass in addition to stabilizing system parameters. This biomass is a promising feedstock for aquatic animals and a possible source of high-value bioproducts because it is abundant in essential proteins, amino acids, lipid-rich polyunsaturated fatty acids (PUFAs), carbohydrates, pigments, and a range of bioactive compounds (Ansari et al. 2021; Siddiki et al. 2022).

Cultivating microalgae as a sustainable fish feed ingredient is one creative algaeponics technique, maybe accomplished through a special cultivation loop in the aquaponic system. By adding PUFAs to fish diets, this method not only provides a sustainable substitute for traditional fishmeal but also improves their nutritional profile. PUFAs are essential for fish development, health, and disease resistance (Cheunbarn and Cheunbarn 2015a, b).

Microalgae are still not a common element of popular aquaponic systems, despite the rising interest in them. There are still obstacles to overcome, such as the possibility that microalgae would outcompete vegetables for nutrients, light, and space, all of which might affect plant output. Additionally, while first results are encouraging, more thorough investigation and optimization are needed to determine the specific ecological functions and long-term effects of microalgae in aquaponic systems (Ansari et al. 2021). A more thorough comprehension of system dynamics, nutrient flow management, and algal biomass control will be essential as algaeponics advances. Algaeponics can improve aquaponic systems' sustainability, resilience, and productivity with thoughtful system design and integrated management techniques, making a significant contribution to upcoming climate-smart food production and ecosystem restoration initiatives.

22.5.1 Algaeponics' Role in Climate Change and Sustainability

Traditional food production techniques, biodiversity, and water resources might all be impacted by climate change. An inventive way to handle several environmental and agricultural issues at once is using algae-integrated aquaponic systems. These systems provide a closed-loop environment where waste products are reduced and resources are continuously recycled, in addition to encouraging more effective use of water and nutrients. Due to their ability to absorb carbon dioxide (CO_2) from the atmosphere, microalgae are crucial in the fight against climate change (Merlo et al. 2021; Olabi et al. 2023). They actively contribute to climate action (SDG 13) by absorbing CO_2 during photosynthesis, which lessens the greenhouse gas burden in aquaponic systems. Additionally, the addition of microalgae to aquaponics systems improves water quality by eliminating surplus nutrients like phosphorus and nitrogen, which are frequently contaminants in conventional aquaculture systems (Nair et al. 2025). By supporting clean water and sanitation (SDG 6), this feature makes sure aquaponic systems run sustainably and do not harm the environment.

By providing a robust and sustainable approach to food production, algae-integrated aquaponics systems also closely correspond with the SDG 2 aim of eradicating hunger (Ibrahim et al. 2023). A very effective method of producing both fish and vegetables, aquaponics combines hydroponic plant culture with fish farming to generate food year-round with little land usage. When added to these

Table 22.2 Roles of aquaponics and algaeponics in sustainability and climate action

	Aquaponics	Algaeponics	SDG goals
Nutrient Recycling	Recycles fish waste into nutrients for plant growth	Converts excess nutrients (N, P) into valuable algal biomass	SDG 6 (Clean Water and Sanitation), SDG 12 (Responsible Consumption and Production)
Water Conservation	Uses 90% less water compared to traditional agriculture	Enhances water purification through nutrient absorption	SDG 6 (Clean Water and Sanitation)
Carbon Sequestration	Indirect CO_2 mitigation via sustainable practices	Direct CO_2 absorption during photosynthesis	SDG 13 (Climate Action)
Food Production and Security	Produces fish and vegetables simultaneously year-round	Enhances fish nutrition via PUFA-enriched algal feed, improving growth and health	SDG 2 (Zero Hunger), SDG 12 (Responsible Consumption and Production)
System Stability and Resilience	Creates a closed-loop ecosystem minimizing external inputs	Improves dissolved oxygen levels, regulates pH, and controls ammonia	SDG 2 (Zero Hunger), SDG 13 (Climate Action)
Waste Minimization	Reduces discharge into the environment	Further utilizes waste nutrients, minimizing effluent discharge	SDG 12 (Responsible Consumption and Production)
Marine Resource Conservation	Reduces reliance on traditional fishmeal	Offers sustainable algae-based fish feed alternative to overfished marine resources	SDG 14 (Life Below Water), SDG 12 (Responsible Consumption and Production)

systems, microalgae enhance the nutritional content of fish feed, promoting fish health and increasing system production (Olabi et al. 2023). Particularly in regions where traditional agriculture is impeded by resource limitations or climatic extremes, this might offer a remedy for people at risk of food poverty. Additionally, fishmeal, which is frequently obtained from overfished marine resources, might be lessened by using algae-based feed. Long-term sustainability of aquaponic systems is increased by this move toward more sustainable feed sources, which also encourages responsible consumption and production (SDG 12) and protects marine habitats. Table 22.2 incorporates both Aquaponics and Algaeponics, showing their combined roles and how each component contributes to sustainability and SDGs.

22.6 Challenges and Limitations

The integration of algae cultivation within aquaponic systems demands precise environmental control, including optimal light, temperature, and nutrient conditions, making large-scale adoption challenging, particularly in areas with limited infrastructure. Advanced technologies for water filtration, nutrient monitoring, and system automation are needed to enhance efficiency. Risks such as power outages, equipment failures, and high initial setup and operational costs further hinder scalability, especially for small-scale farmers. Specialized knowledge and continuous monitoring add to these barriers. The economic viability of algae-derived products like biofuels, nutraceuticals, and fish feed remains dependent on technological innovations, market demand, regulatory approvals, and consumer acceptance. Uncontrolled algae growth may cause system imbalances, oxygen depletion, and nutrient competition, while introducing non-native species could disrupt local ecosystems. Proper waste management and biosecurity measures are essential to mitigate ecological risks.

Moreover, limited public awareness, technical expertise, and unclear regulatory frameworks slow down the sector's growth. Policy support, education initiatives, and sustainable production guidelines are urgently needed. Large-scale implementation is also constrained by logistical factors such as the availability of land, water, energy, and efficient supply chains. Strategic collaborations between research institutions, industries, and policymakers are crucial to overcome these hurdles and advance algae-integrated aquaponics as a climate adaptation strategy.

22.7 Future Perspective and Conclusion

Aquaponics offers a revolutionary model for sustainable food production by addressing water conservation and environmental inefficiencies inherent in traditional farming. Integrating microalgae into aquaponic systems amplifies these benefits, as algae naturally regulate pH, enhance oxygen levels, and control ammonia concentrations by assimilating nitrogen and phosphorus. However, the selection of appropriate algae species is critical for system success, necessitating further research.

Beyond improving system resilience, cultivated microalgae present versatile applications in biofuel, bioethanol, aquaculture, and nutraceutical industries, enhancing the economic sustainability of algaeponics. Cocultivated algae also contribute to atmospheric CO_2 regulation, positioning algaeponics as a valuable tool in combating climate change. Overall, algae-integrated aquaponics represents a promising convergence of sustainable agriculture and climate change adaptation. While technical, economic, and policy challenges persist, continued research, innovation, and supportive frameworks will be vital to fully realize its potential. As global climate pressures intensify, such integrated systems offer a resilient and sustainable pathway toward securing future food and energy needs.

Conflicts of Interest The authors declare that they have no conflict of interest.

Availability of Data and Material Not Applicable.

Ethics Approval Not applicable—No human or animal subjects were involved.

References

Abdel-Tawwab M, Monier MN, Hoseinifar SH, Faggio C (2019) Fish response to hypoxia stress: growth, physiological, and immunological biomarkers. Fish Physiol Biochem 45:997–1013

Abu-Ghosh S, Dubinsky Z, Verdelho V, Iluz D (2021) Unconventional high-value products from microalgae: a review. Bioresour Technol 329:124895

Addy MM, Kabir F, Zhang R, Lu Q, Deng X, Current D, Griffith R, Ma Y, Zhou W, Chen P, Ruan R (2017) Co-cultivation of microalgae in aquaponic systems. Bioresour Technol 245:27–34

Ahmed M, Hasanuzzaman M, Raza MA, Malik A, Ahmad S (2020) Plant nutrients for crop growth, development and stress tolerance. Sustain Agric Era Clim Change:43–92

Ajeng AA, Rosli NSM, Abdullah R, Yaacob JS, Qi NC, Loke SP (2022) Resource recovery from hydroponic wastewaters using microalgae-based biorefineries: a circular bioeconomy perspective. J Biotechnol 360:11–22

Aketo T, Hoshikawa Y, Nojima D, Yabu Y, Maeda Y, Yoshino T, Takano H, Tanaka T (2020) Selection and characterization of microalgae with potential for nutrient removal from municipal wastewater and simultaneous lipid production. J Biosci Bioeng 129(5):565–572

Alami AH, Faraj M, Aokal K, Abu Hawili A, Tawalbeh M, Zhang D (2020) Investigating various permutations of copper iodide/FeCu tandem materials as electrodes for dye-sensitized solar cells with a natural dye. Nano 10(4):784

Amin AANM (2022) Utilization of fish farms water for vegetables production. Doctoral dissertation, Benha University

Andreotti V, Solimeno A, Chindris A, Marazzi F, García J (2019) Growth of Tetraselmis suecica and Dunaliella tertiolecta in aquaculture wastewater: numerical simulation with the BIO_ALGAE model. Water Air Soil Pollut 230:1–14

Ansari F, Guldhe A, Gupta S, Rawat I, Bux F (2021) Improving the feasibility of aquaculture feed by using microalgae. Environ Sci Pollut Res 28:43234–43257

Azman S, Said MIM, Ahmad F, Mohamad M (2014) Biofiltration potential of macroalgae for ammonium removal in outdoor tank shrimp wastewater recirculation system. Biomass Bioenergy 66:103–109

Banerjee C, Dubey KK, Shukla P (2020) Carbon sequestration and bioenergy production by microalgae: opportunities and challenges. Energy Convers Manag 205:112378. https://doi.org/10.1016/j.enconman.2019.112378

Barton BA (2002) Stress in fishes: a diversity of responses with particular reference to changes in circulating corticosteroids. Integr Comp Biol 42(3):517–525

Bolton JJ (2016) What is aquatic botany? And why algae are plants: The importance of non-taxonomic terms for groups of organisms. Aquat Bot 132:1–4

Braun JC, Colla LM (2023) Use of microalgae for the development of biofertilizers and biostimulants. Bioenergy Res 16(1):289–310

Brilman W, Garcia Alba L, Veneman R (2013) Capturing atmospheric CO2 using supported amine sorbents for microalgae cultivation. Biomass Bioenergy 53:39–47

Cahill PL, Hurd CL, Lokman M (2010) Keeping the water clean – seaweed biofiltration outperforms traditional bacterial biofilms in recirculating aquaculture. Aquaculture 306(1–4):153–159
Calderini ML, Stevčić Č, Taipale S, Pulkkinen K (2021) Filtration of Nordic recirculating aquaculture system wastewater: effects on microalgal growth, nutrient removal, and nutritional value. Algal Res 60:102486
Cheah WY, Show PL, Chang J-S, Ling TC, Juan JC (2015) Biosequestration of atmospheric CO2 and flue gas-containing CO2 by microalgae. Bioresour Technol 184:190–201
Chen H, He C, Li Y, Hu Z (2023) Enhancing carbon fixation of microalgae through phytohormone regulation and bioengineering. Algal Res 72:102911
Cheunbarn T, Cheunbarn S (2015a) Cultivation of algae in vegetable and fruit canning industrial wastewater treatment effluent for tilapia (Oreochromis niloticus) feed supplement. Int J Agric Biol 17(3):653–657
Cheunbarn T, Cheunbarn S (2015b) Cultivation of algae in vegetable and fruit canning industrial wastewater treatment effluent for tilapia (Oreochromis niloticus) feed. Survival 1(F2):100
Chiquito-Contreras RG, Hernandez-Adame L, Alvarado-Castillo G, Martínez-Hernández MDJ, Sánchez-Viveros G, Chiquito-Contreras CJ, Hernandez-Montiel LG (2022) Aquaculture – production system and waste management for agriculture fertilization – a review. Sustainability 14(12):7257
Crooks SRK (1999) Turner. In: Advances in ecological research 1999
Cuellar-Bermudez SP, Garcia-Perez JS, Rittmann BE, Parra-Saldivar R (2015) Photosynthetic bioenergy utilizing CO2: an approach on flue gases utilization for third generation biofuels. J Clean Prod 98:53–65
Dai Z, Luo X, Yang A, Wang J, Fu H, Wu Y (2021) The effects of Triclosan on physiological and photosynthetic characteristics of Chlorella vulgaris. Water 13(10):1355
Day JW, Rybczyk JM (2019) Global change impacts on the future of coastal systems: perverse interactions among climate change, ecosystem degradation, energy scarcity, and population. In: Coasts and Estuaries. Elsevier, pp 621–639
Doney SC, Ruckelshaus M, Emmett Duffy J, Barry JP, Chan F, English CA, Talley LD (2012) Climate change impacts on marine ecosystems. Annu Rev Mar Sci 4(1):11–37
Ende S, Henjes J, Spiller M, Elshobary M, Hanelt D, Abomohra A (2024) Recent advances in recirculating aquaculture systems and role of microalgae to close system loop. Bioresour Technol:131107
Fang Y, Hu Z, Zou Y, Zhang J, Zhu Z, Zhang J, Nie L (2017) Improving nitrogen utilization efficiency of aquaponics by introducing algal-bacterial consortia. Bioresour Technol 245:358–364
Fang Y, Chen X, Hu Z, Liu D, Gao H, Nie L (2018) Effects of hydraulic retention time on the performance of algal-bacterial-based aquaponics (AA): focusing on nitrogen and oxygen distribution. Appl Microbiol Biotechnol 102:9843–9855
Fazal T, Rehman MSU, Javed F, Akhtar M, Mushtaq A, Hafeez A, Din AA, Iqbal J, Rashid N, Rehman F (2021) Integrating bioremediation of textile wastewater with biodiesel production using microalgae (Chlorella vulgaris). Chemosphere, 281:130758
Froehlich HE, Afflerbach JC, Frazier M, Halpern BS (2019) Blue growth potential to mitigate climate change through seaweed offsetting. Curr Biol 29(18):3087–3093
Ganesan JJ, Nagarajan D, Rathinam B, Chien JRC (2024) Low-carbon biofuels from macroalgae towards a sustainable circular bioeconomy and green future. Biomass Bioenergy 190:107389
Gauthier A, Araya-Cloutier C, Benoit L, Weidmann S (2020) Microalgae as natural sources of bioactive compounds: strategies and prospects. J Appl Phycol 32(6):3807–3823
Gerardi MH (2002) Nitrification and denitrification in the activated sludge process. Wiley
Ghorbani R, Yeganeh B, Hosseini M, Khosravi-Darani K (2014) Carbon dioxide biofixation by microalgae: an overview of carbon capture and storage strategies. Environ Eng Manag J 13(8):1905–1915. https://doi.org/10.30638/eemj.2014.207
Gichana ZM, Liti D, Waidbacher H, Zollitsch W, Drexler S, Waikibia J (2018) Waste management in recirculating aquaculture system through bacteria dissimilation and plant assimilation. Aquac Int 26:1541–1572

Glavovic BC, Limburg K, Liu KK, Emeis KC, Thomas H, Kremer H, Swaney DP (2015) Living on the Margin in the Anthropocene: engagement arenas for sustainability research and action at the ocean–land interface. Curr Opin Environ Sustain 14:232–238

Goddek S, Delaide B, Mankasingh U, Ragnarsdottir KV, Jijakli H, Thorarinsdottir R (2015) Challenges of sustainable and commercial aquaponics. Sustainability 7(4):4199–4224

Goddek S, Espinal CA, Delaide B, Jijakli MH, Schmautz Z, Wuertz S, Keesman KJ (2016) Navigating towards decoupled aquaponic systems: a system dynamics design approach. Water 8(7):303

Goddek S, Joyce A, Kotzen B, Burnell GM (2019) Aquaponics food production systems: combined aquaculture and hydroponic production technologies for the future. Springer, p 619

Gordon JM, Polle JE (2007) Ultra-high bioproductivity from algae. Appl Microbiol Biotechnol 76(5):969–975

Gowthami R, Kumari A, Shalini K, Anantharaman P (2023) Algae-mediated degradation of microplastics: current research and future perspectives. Sci Total Environ 858:159932

Graham RD, Welch RM, Bouis HE (2001) Addressing micronutrient malnutrition through enhancing the nutritional quality of staple foods: principles, perspectives and knowledge gaps

Hüttl RF, Schneider BU (1998) Forest ecosystem degradation and rehabilitation. Ecol Eng 10(1):19–31

Ibrahim LA, Shaghaleh H, El-Kassar GM, Abu-Hashim M, Elsadek EA, Alhaj Hamoud Y (2023) Aquaponics: a sustainable path to food sovereignty and enhanced water use efficiency. Water 15(24):4310

Iftehimul I, Rahman MR, Hossain MZ, Hasan M (2023) Microalgae as a sustainable tool for bioremediation and wastewater treatment. Environ Res 216:114615

Islam MJ, Kunzmann A, Slater MJ (2022) Responses of aquaculture fish to climate change-induced extreme temperatures: a review. J World Aquacult Soc 53(2):314–366

Javan T, Mohammadi A, Gharibi D (2024) Algae-based bioremediation: an eco-friendly approach for wastewater treatment. Chemosphere 339:139634

Ji M, Gao H, Diao L, Zhang J, Liang S, Hu Z (2022) Environmental impacts of antibiotics addition to algal-bacterial-based aquaponic system. Appl Microbiol Biotechnol 106(9):3777–3786

Ji M, Gao H, Zhang J, Hu Z, Liang S (2024) Environmental impacts on algal–bacterial-based aquaponics system by different types of carbon source addition: water quality and greenhouse gas emission. Environ Sci Pollut Res 31(18):26665–26674

Jouhara H, Czajczyńska D, Ghazal H, Krzyżyńska R, Anguilano L, Reynolds AJ, Spencer N (2017) Municipal waste management systems for domestic use. Energy 139:485–506

Joyce A, Goddek S, Kotzen B, Wuertz S (2019) Aquaponics: closing the cycle on limited water, land and nutrient resources. Aquapon Food Prod Syst 19

Kankia AI, Manoj VR, Orugba HO, Owamah HI (2024) Enhancing the sustainability of aquaculture systems via nutrient recycling using aquaponics system. Water Air Soil Pollut 235(6):397

Kasozi N, Abraham B, Kaiser H, Wilhelmi B (2021) The complex microbiome in aquaponics: significance of the bacterial ecosystem. Ann Microbiol 71:1–13

Kiki C, Rashid A, Wang Y, Li Y, Zeng Q, Yu CP, Sun Q (2020) Dissipation of antibiotics by microalgae: kinetics, identification of transformation products and pathways. J of hazardous materials, 387:121985

Khan MI, Shin JH, Kim JD (2018) The promising future of microalgae: current status, challenges, and optimization of a sustainable and renewable industry for biofuels, feed, and other products. Microb Cell Factories 17:36

Knaus U, Palm HW (2017) Effects of the fish species choice on vegetables in aquaponics under spring-summer conditions in northern Germany (Mecklenburg Western Pomerania). Aquaculture 473:62–73

Knaus U, Pribbernow M, Xu L, Appelbaum S, Palm HW (2020) Basil (Ocimum basilicum) cultivation in decoupled aquaponics with three hydro-components (grow pipes, raft, gravel) and African catfish (Clarias gariepinus) production in Northern Germany. Sustainability 12(20):8745

Kotzen B, Emerenciano MGC, Moheimani N, Burnell GM (2019) Aquaponics: alternative types and approaches. In: Aquaponics food production systems: combined aquaculture and hydroponic production technologies for the future, pp 301–330

Koukoumaki DI, Tsouko E, Papanikolaou S, Ioannou Z, Diamantopoulou P, Sarris D (2024) Recent advances in the production of single cell protein from renewable resources and applications. Carbon Resources Conversion, 7(2):100195
Krause-Jensen D, Duarte CM (2016) Substantial role of macroalgae in marine carbon sequestration. Nat Geosci 9(10):737–742
Kumar KS, Dahms HU, Won EJ, Lee JS, Shin KH (2015) Microalgae – a promising tool for heavy metal remediation. Ecotoxicol Environ Saf 113:329–352
Kumar S, Sindhu SS, Kumar R (2022) Biofertilizers: an ecofriendly technology for nutrient recycling and environmental sustainability. Curr Res Microb Sci 3:100094
Kushwaha J, Priyadarsini M, Rani J, Pandey KP, Dhoble AS (2023) Aquaponic trends, configurations, operational parameters, and microbial dynamics: a concise review. Environ Dev Sustain:1–34
Lanelly BC, Elsa A (2024) Microalgae-based bioremediation of pesticide-contaminated environments. Ecotoxicol Environ Saf 260:115035
Li Y, Zhou W, Hu B, Min M, Chen P, Ruan RR (2011) Integration of algae cultivation as biodiesel production feedstock with municipal wastewater treatment: strains screening and significance evaluation of environmental factors. Bioresour Technol 102(23):10861–10867
Love DC, Fry JP, Li X, Hill ES, Genello L, Semmens K, Thompson RE (2015) Commercial aquaponics production and profitability: findings from an international survey. Aquaculture 435:67–74
Ma X, Jiang L, Zhu C, Chen X, Zhang J (2022) Smart cultivation and harvesting technologies for microalgae: recent advances and future perspectives. Bioresour Technol 351:127006
Malhi Y, Franklin J, Seddon N, Solan M, Turner MG, Field CB, Knowlton N (2020) Climate change and ecosystems: threats, opportunities and solutions. Philos Trans R Soc B 375(1794):20190104
Mathew JT, Inobeme A, Musah M, Azeh Y, Abdullahi A, Shaba EY, Ndamitso MM (2024) A critical review of green approach on wastewater treatment strategies. J Appl Sci Environ Manag 28(2):363–391
Mercado I, Álvarez X, Verduga ME, Cruz A (2020) Enhancement of biomass and lipid productivities of Scenedesmus sp. cultivated in the wastewater of the dairy industry. Processes 8(11):1458
Merlo S, Gabarrell Durany X, Pedroso Tonon A, Rossi S (2021) Marine microalgae contribution to sustainable development. Water 13(10):1373
Michels MH, Vaskoska M, Vermuë MH, Wijffels RH (2014) Growth of Tetraselmis suecica in a tubular photobioreactor on wastewater from a fish farm. Water Res 65:290–296
Mishra N, Prasad SM, Mishra N (2019a) Influence of high light intensity and nitrate deprivation on growth and biochemical composition of the marine microalgae Isochrysis galbana. Braz Arch Biol Technol 62:e19180398
Mishra SK, Shrivastav A, Dubey S (2019b) Algae: an efficient tool for sustainability, bioenergy and climate change mitigation. Renew Sust Energ Rev 112:54–62
Mitra A, Abdel-Gawad FK, Bassem S, Barua P, Assisi L, Parisi C, Guerriero G (2023) Climate change and reproductive biocomplexity in fishes: innovative management approaches towards sustainability of fisheries and aquaculture. Water 15(4):725
Mobin S, Alam F (2017) Some promising microalgal species for commercial applications: a review. Energy Procedia 110:510–517. https://doi.org/10.1016/j.egypro.2017.03.178
Mondal M, Ghosh A, Tiwari ON, Gayen K, Das P (2016) Influence of carbon dioxide, light and nutrients on algal productivity: a review. Renew Sust Energ Rev 65:902–909. https://doi.org/10.1016/j.rser.2016.07.065
Munguia-Fragozo P, Alatorre-Jacome O, Rico-Garcia E, Torres-Pacheco I, Cruz-Hernandez A, Ocampo-Velazquez RV, Guevara-Gonzalez RG (2015) Perspective for aquaponic systems: "omic" technologies for microbial community analysis. Biomed Res Int 2015(1):480386
Mustafa AM, Shou-Qing L, Khan SA (2021) Recent advancements in microalgae-mediated bioremediation strategies: a review. Environ Res 201:111531. https://doi.org/10.1016/j.envres.2021.111531
Nadeem F, Hanif MA, Majeed MI, Mushtaq Z (2018) Role of macronutrients and micronutrients in the growth and development of plants and prevention of deleterious plant diseases-a comprehensive review. Int J Chem Biochem Sci 13:31–52

Nair CS, Manoharan R, Nishanth D, Subramanian R, Neumann E, Jaleel A (2025) Recent advancements in aquaponics with special emphasis on its sustainability. J World Aquacult Soc 56(1):e13116

Nazloo EK, Danesh M, Sarrafzadeh MH, Moheimani NR, Ennaceri H (2024) Biomass and hydrocarbon production from Botryococcus braunii: A review focusing on cultivation methods. Science of the Total Environment, 926:171734

Nie X, Mubashar M, Zhang S, Qin Y, Zhang X (2020) Current progress, challenges and perspectives in microalgae-based nutrient removal for aquaculture waste: a comprehensive review. J Clean Prod 277:124209

Oh JW, Pushparaj SSC, Muthu M, Gopal J (2023) Review of harmful algal blooms (HABs) causing marine fish kills: toxicity and mitigation. Plants 12(23):3936

Ojanen S, Tyystjärvi E, Holmberg H, Kouhia M, Ahtila P (2017) Can bacterial biofiltration be replaced by autotrophic organisms in recirculating fresh water aquaculture? Aquac Int 25:1427–1440

Olabi AG, Shehata N, Sayed ET, Rodriguez C, Anyanwu RC, Russell C, Abdelkareem MA (2023) Role of microalgae in achieving sustainable development goals and circular economy. Sci Total Environ 854:158689

Omweno JO, Argwings O, Sieberi BM, Angima M, Ondieki PM (2024) Utilization of aquatic plants and microalgae for sustainable aquaculture production and potential biotechnological applications. Trop Environ Biol Technol 2(2):57–71

Paine RT, Tegner MJ, Johnson EA (2021) Compounded perturbations yield ecological surprises. Ecosystems 24(2):301–315. https://doi.org/10.1007/s10021-020-00554-4

Perin S, Casabianca S, Battocchi C, Accoroni S, Totti C (2019) Mixotrophy and harmful algal blooms: a review. Harmful Algae 87:101632. https://doi.org/10.1016/j.hal.2019.101632

Peteiro C, Sánchez N, Martínez B (2016) Cultivation of the kelp Saccharina latissima in the Atlantic coast of Spain: production and biochemical composition. J Appl Phycol 28(2):1109–1117. https://doi.org/10.1007/s10811-015-0641-7

Rajalakshmi M, Manoj VR, Manoj H (2022) Comprehensive review of aquaponic, hydroponic, and recirculating aquaculture systems

Rame JL, Ng KK, Show PL, Ling TC (2023) Microalgae-based integrated systems for CO2 capture and wastewater treatment: advances and prospects. Sci Total Environ 869:161703. https://doi.org/10.1016/j.scitotenv.2023.161703

Ramli NM, Verreth JAJ, Yusoff FM, Nurulhuda K, Nagao N, Verdegem MC (2020) Integration of algae to improve nitrogenous waste management in recirculating aquaculture systems: a review. Front Bioeng Biotechnol 8:1004

Reid SG, Bernier NJ, Perry SF (1998) The adrenergic stress response in fish: control of catecholamine storage and release. Comp Biochem Physiol C: Pharmacol Toxicol Endocrinol 120(1):1–27

Renganathan P, Puente EOR, Sukhanova NV, Gaysina LA (2024) Hydroponics with microalgae and cyanobacteria: emerging trends and opportunities in modern agriculture. Biotech 13(3):27

Rodríguez-Leal S, Silva-Acosta J, Marzialetti T, Gallardo-Rodríguez JJ (2023) Lab-and pilot-scale photo-biofilter performance with algal–bacterial beads in a recirculation aquaculture system for rearing rainbow trout. J Appl Phycol 35(4):1673–1683

Sahoo D, Seckbach J (2015) The algae world. Springer. https://doi.org/10.1007/978-94-017-7321-8

Sánchez-Morales DJ, Rubiños-Panta JE, Crespo-Pichardo G, Mendoza-Pérez C, de Lourdes Pérez-Ávila M, Peralta-Inga M, Rubiños-Hernandez CE (2024) Evaluación del Sistema Acuapónico de Pequeña Escala para la Producción de Tilapia y Lechuga. Revista Terra Latinoamericana 42

Schneider P, Rochell V, Plat K, Jaworski A (2021) Circular approaches in small-scale food production. Circ Econ Sustain 1:1–25

Sharma I, Birman S (2024) Biodiversity loss, ecosystem services, and their role in promoting sustainable health. In: The climate-health-sustainability nexus: understanding the interconnected impact on populations and the environment. Springer, Cham, pp 163–188

Siddiki SYA, Mofijur M, Kumar PS, Ahmed SF, Inayat A, Kusumo F, Mahlia TMI (2022) Microalgae biomass as a sustainable source for biofuel, biochemical and biobased value-added products: an integrated biorefinery concept. Fuel 307:121782

Silber A, Bar-Tal A (2019) Nutrition of substrate-grown plants. Soilless Culture:291–339

Singh UB, Ahluwalia AS (2013) Microalgae: a promising tool for carbon sequestration. Mitig Adapt Strateg Glob Chang 18(1):73–95

Solisio C, Lodi A, Veglio F, Converti A (2019) Heavy metal removal by microalgae. Biomass Bioenergy 122:363–371. https://doi.org/10.1016/j.biombioe.2019.01.015

Stouvenakers G, Dapprich P, Massart S, Jijakli MH (2019) Plant pathogens and control strategies in aquaponics. Aquapon Food Prod Syst 14:353–378

Thakur K, Kuthiala T, Singh G, Arya SK, Iwai CB, Ravindran B, Awasthi MK (2023) An alternative approach towards nitrification and bioremediation of wastewater from aquaponics using biofilm-based bioreactors: a review. Chemosphere 316:137849

Thanigaivel S, Vijayanand N, Jeyaraj S (2022) Microalgae carbon sequestration potential: a sustainable approach towards climate change mitigation. Environ Res 208:112741

Tocher DR (2010) Fatty acid requirements in ontogeny of marine and freshwater fish. Aquac Res 41(5):717–732

Umeh CA (2024) Institutional and policy barriers to the adoption of urine recycling as a sustainable solution for nutrient management and agricultural productivity in Ethiopia. Master's thesis, Norwegian University of Life Sciences

Vasdravanidis C, Alvanou MV, Lattos A, Papadopoulos DK, Chatzigeorgiou I, Ravani M, Giantsis IA (2022) Aquaponics as a promising strategy to mitigate impacts of climate change on rainbow trout culture. Animals 12(19):2523

Vo TP, Danaee S, Chaiwong C, Pham BT, Kim M, Kuzhiumparambil U, Vo PH (2024) Microalgae-bacteria consortia for organic pollutants remediation from wastewater: a critical review. J Environ Chem Eng:114213

Wang H, Wang S, Xu X, Zhang Y, Ma L (2018) Research progress in microalgae-based CO2 capture and utilization. Renew Sust Energ Rev 96:505–517

Wang J, Wang X, Xiong Z, Chen H, Li X (2019) Microalgae-mediated removal of phenolic compounds from water: a review. Algal Res 40:101485

Warner K, Hamza M, Oliver-Smith A, Renaud F, Julca A (2010) Climate change, environmental degradation and migration. Nat Hazards 55:689–715

Wongkiew S, Hu Z, Lee JW, Chandran K, Nhan HT, Marcelino KR, Khanal SK (2021) Nitrogen recovery via aquaponics–bioponics: engineering considerations and perspectives. ACS ES&T Eng 1(3):326–339

Yadav NK, Patel AB, Singh SK, Mehta NK, Anand V, Lal J, Devi NC (2024) Climate change effects on aquaculture production and its sustainable management through climate-resilient adaptation strategies: a review. Environ Sci Pollut Res 31:1–21

Yavuzcan Yildiz H, Robaina L, Pirhonen J, Mente E, Domínguez D, Parisi G (2017) Fish welfare in aquaponic systems: its relation to water quality with an emphasis on feed and faeces – a review. Water 9(1):13

Zhang Y, Zhang YK, Li Z (2022) A new and improved aquaponics system model for food production patterns for urban architecture. J Clean Prod 342:130867

Zhou W, Hu B, Li Y, Min M, Mohr M, Du Z, Ruan R (2012) Mass cultivation of microalgae on animal wastewater: a sequential two-stage cultivation process for energy crop and omega-3-rich animal feed production. Appl Biochem Biotechnol 168(2):348–363

Zhou W, Wang J, Chen P, Min M (2023) Harnessing algal biodiversity for sustainable ecosystem management: insights and perspectives. Sci Total Environ 860:160495

Zhu B, Chen G, Cao X, Wei D (2017) Molecular characterization of CO2 sequestration and assimilation in microalgae and its biotechnological applications. Bioresource Technology, 244:1207–1215

Chapter 23
Remote Sensing Observation of Sea Surface Temperature (SST) and pCO_2 over the Bay of Bengal and Arabian Sea and Its Relation with Chlorophyll Variability

R. K. Sarangi

Abstract The study is carried out to estimate the satellite-derived partial pressure of carbon dioxide (pCO_2) in the Bay of Bengal (BoB) and Arabian Sea (AS) using sea surface temperature (SST)-based algorithm. The relationship of satellite-derived pCO_2 with SST and chlorophyll has been understood in different seasonal months and years. The SST images are generated for the Bay of Bengal and Arabian Sea during two distinct seasonal months, December 2013 and 2014 and April 2014 and 2015. The daily and 8 days, monthly composite SST images are generated using INSAT-3D, MODIS-Aqua, and GHRSST datasets. The corresponding overpass time of MODIS-Aqua and INSAT-3D 13:30Hrs SST data has been archived. The SST is observed in the range of 24–32 °C. The SST-based pCO_2 algorithm is applied over the northern Indian ocean and the pCO_2 variability in two different seasons monitored. The pCO_2 ranged around 350–750 μatm. The INSAT-3D derived 30-min time interval images processed on intra-day basis having 48 passes per day. The pCO_2 images observed directly proportional relationship with the SST images during summer and inverse trend during winter. With the increase in SST by 1–2 °C, there has been increase in pCO_2 by 2–5% during summer. The comparison of pCO_2 on weekly and monthly time scales using the INSAT-3D, MODIS and GHRSST data has been observed to be interesting and showed matching trend. This exemplifies the preliminary study to understand the hourly, daily, weekly, monthly, and seasonal trend of SST and pCO_2 variability in the northern Indian Ocean basins using satellite datasets. The MODIS-Aqua monthly composite chlorophyll images indicate that the high chlorophyll (0.8–1.4 mg m^{-3}) patches are matching well with the high pCO_2 concentration (400–450 μatm) patches during winter month and similar trend is not observed during summer month. Main findings of the paper are to have the pCO_2 estimation using SST data in Indian scenario using multiple satellite datasets from MODIS-Aqua, INSAT, and GHRSST datasets and the comparison

R. K. Sarangi (✉)
Marine Ecosystem Division, BPSG/EPSA, Space Applications Centre (ISRO), Ahmedabad, Gujarat, India

B. P. Barathan et al. (eds.), *Navigating Climate Change: Impacts on Biodiversity and Ecosystem Resilience*, Sustainable Development and Biodiversity 42,
https://doi.org/10.1007/978-981-95-0409-1_23

with ocean productivity using satellite-derived chlorophyll data. This study has a strong relevance in terms of ocean acidification monitoring using satellite data- and model-based time-series map generation. The study is important from the point of view of air-sea interaction, ocean acidification, and ocean biogeochemistry. The in situ pCO_2 measurements, data validation, and fine-tuning would rely on the scope for regional algorithm development as future study and trend analysis from climate change perspective.

Keywords SST · pCO_2 · Chlorophyll · Satellite data · Bay of Bengal · Arabian Sea

23.1 Introduction

The carbon dioxide is one such important atmospheric gas that is exchanged through the air-sea interface, and hence, oceanic fluxes of carbon dioxide are significant in the global carbon cycle (Chester 2000). Carbon dioxide, being an efficient absorber of earth-emitted radiation, has vital contribution to the heating of atmosphere (Lutgens and Tarbuck 2004). The biogeochemical cycles of the atmospheric gases play important roles in the global ecosystem, and the air-sea exchange of these gases is noteworthy to study. The atmospheric concentration of carbon dioxide (CO_2) is being controlled by spontaneous exchange of oceanic CO_2 at the sea surface via carbon cycle. Due to upwelling event, nutrient enrichment takes place up to the upper layer of sea surface and the primary productivity gets enhanced. This also brings the subsurface CO_2-rich water to the sea surface, resulting in the increased rate of CO_2 emissions to the atmosphere (Chester 2000). The partial pressure of CO_2 in sea water is mostly regulated by SST and by phytoplankton photosynthesis and respiration, biologically. Although both these parameters show similar magnitude of effect on pCO_2, their effects are often found to be compensating, thus the net effect causes a balance between uptake and release of CO_2 by sea surface (Feely 2004). During bloom conditions, the phytoplankton affects the CO_2 absorption due to the increased rate of photosynthetic activity leading to increased uptake of CO_2 by the sea water (Chester 2000). Studies have been conducted worldwide to determine the effect of sea surface temperature (Andrié et al. 1986; Takahashi et al. 1993; Kumamoto et al. 1995; Sarma et al. 2000a; Takahashi et al. 2002; Lefèvre and Taylor 2002; Chen and Gao 2007; Zhu et al. 2009) and biological processes (Trela et al. 1995; Kumamoto et al. 1995; Sarma et al. 2000a; Takahashi et al. 2002; Zhu et al. 2009) on sea water CO_2 partial pressure.

The CO_2 emissions from Arabian Sea during the SW monsoon were studied by Körtzinger et al. (1997) which showed a pCO_2 super-saturation condition during this season resulting in strong emissions to the atmosphere. The coastal region around Oman showed super-saturation of surface water pCO_2 reaching up to 750 μatm. Elevated pCO_2 levels up to 525 μatm are observed off the Oman coast during southwest monsoon featuring cold upwelled waters and also exhibited by Ekman pumping. Ekman pumping is basically the vertical movement of water caused by the wind stress. The Coriolis force pulls the surface water to the left side of the wind vector. With this effect, the water converged and pushed downwards. Ekman pumping occurs in coastal water and in open ocean waters. Hence, it has a relevance on the *p*CO2 variability in the ocean water column.

Temporal variations of pCO_2 in the Arabian Sea were analysed by Goet et al. (1991) during 1995, which involved continuous measurements of sea surface pCO_2 and atmospheric pCO_2. The coastal values in the western region showed 260 μatm, Sarma et al. (2000b) studied the seasonal and inter-annual variations in the total carbon dioxide (TCO_2) and carbon dioxide partial pressure (pCO_2) in the central and eastern Arabian Sea during the JGOFS programme. The seasonal changes in pCO_2 concentration observed the high range of 520–685 μatm and low range around 266 μatm in the southwest coast of India during southwest monsoon. The temporal variation of surface water pCO_2 was studied by Kumamoto et al. (1995) in Iyo Nada in the Seto Inland Sea in Japan using infrared absorption method. There has been low variability during January to May end ranging 270–340 μatm and during September it ranged 200–450 μatm. Zhu et al. (2009) estimated the sea surface CO_2 partial pressure and CO_2 fluxes in the northern South China Sea using satellite data-derived SST.

The pCO_2 is driven by coupled bio-physical system as a consequence of air-sea interaction processes. The pCO_2 varies on a range of timescales, from days to weeks and months to decades. Hence, attempt has been made to understand and determine the driving forces behind the changes and its monitoring ocean's ability to absorb atmospheric CO_2 and effect on climate (Takahashi et al. 2009). Unlike the atmospheric pCO_2, oceanic pCO_2 varies both spatially and temporally (Mahadevan et al. 2011). Physical and biogeochemical factors, including surface temperature-driven solubility, biological processes, fall-to-winter vertical mixing, ocean circulation, river runoff, and shelf-ocean exchange, will influence the temporal and spatial variability of the surface-ocean pCO_2 on coastal areas (Signorini et al. 2013). The study contributes to estimate the coastal water carbon flux, quantify the role of coastal regions in carbon cycle, and provide important information to understand the ocean acidification mechanisms. Due to ever increasing anthropogenic emissions, atmospheric CO_2 concentrations continue to rise and trigger climate change (Solomon 2007; Le Quéré et al. 2009; Peters et al. 2011). In this context, the trends in the physical and biogeochemical characteristics of the Indian Ocean, such as surface ocean warming (Swapna et al. 2014; Roxy et al. 2014; Roxy et al. 2016), ocean carbon fluxes (Valsala and Maksyutov 2013; Valsala and Murtugudde 2015; Chakraborty et al. 2021), ocean acidification (Sreeush et al. 2019), and carbon sequestration (Gruber et al. 2019), deserve special attention.

The marine science community now regularly uses climate change projections released by the Intergovernmental Panel on Climate Change (IPCC; IPCC 2007) to make qualitative and quantitative projections of marine ecosystem responses to environmental changes associated with the accumulation of greenhouse gases in the atmosphere (e.g. climate change and ocean acidification). These projections indicate that climate change will affect fish, fisheries, and fisheries-based economies around the globe as well as broader components of marine ecosystems (Allison et al. 2009; Cochrane et al. 2009; Drinkwater et al. 2010; Doney et al. 2012; Merino et al. 2012). The accumulation of CO_2 in the atmosphere and associated climate changes is expected to cause ocean acidification and expansion of oligotrophic gyres (Doney et al. 2012). These physical and chemical changes are expected to result in shifts in the timing, species composition, and magnitude of seasonal phytoplankton production (Yoder, 2000; Cochrane et al. 2009; Doney et al. 2012). These physical, and resulting biological, changes will occur at different spatial and temporal scales throughout the world's oceans (Burrows et al. 2011; Gnanadesikan et al. 2011; King et al. 2011). Climate-driven changes in the environment may affect the physiology, phenology, and behaviour of marine fish and shellfish at any life-history stage, and any of these effects may drive population-level changes in distribution and abundance (Drinkwater et al. 2010; Jorgensen et al. 2012). The current study has been carried out in the northern Indian Ocean basins covering the Bay of Bengal and Arabian Sea regions for retrieving sea surface temperature (SST) and derived pCO_2 variability in two different seasons and its relationship with biological productivity.

23.2 Data and Methodology

23.2.1 Study Area

Bay of Bengal (BoB) system is controlled by seasonal reversing monsoon winds. As a result of this, the ocean productivity has been effective during both the southwest and the northeast monsoon timeframe. Bay of Bengal is fed with large volume of fresh water from rainfall and from the rivers like Hoogly, Ganges, Brahmaputra, Airawati, Mahandi, Godavari, Krishna, Kaveri, and so on (Subramanian 1993). The important nutrients such as nitrate, phosphate, and silicate and their enrichment have been on record as major sources from above rivers and riverine confluences and via estuaries falling into BoB. Also, one of the major suppliers of the nutrients to the surface in BoB is mixing due to cyclones, frequent during post-monsoon season (Madhu et al. 2002; Nayak et al. 2002; Sarangi et al. 2015). The Arabian Sea is recognised as low-latitude Sea; terminates at latitude around 25°N and marked with continental influence. It is the region of monsoon and its biological production is affected by the physical processes altering the vertical flux of nutrients in the mixed layer (Banse and English 2000).

23.2.2 Processing of INSAT-3D Level-3 SST Datasets

Indian National satellite (INSAT-3D) Level-3 4 km resolution sea surface temperature datasets have been archived from (http://www.mosdac.gov.in) and processed to generate SST images for the months December 2013 and 2014 and April 2014 and

2015 covering Bay of Bengal (BoB) and Arabian Sea (AS). The INSAT-3D daily data of 13:30 h over pass has been averaged to 8-day and monthly composites taking into account the same overpass time of MODIS-Aqua, i.e. 13:30 IST.

23.2.3 Processing of MODIS-Aqua Level-3 SST and Chlorophyll Datasets

The chlorophyll and SST Level-3 datasets have been retrieved from the MODIS-Aqua satellite sensor during two different seasonal months 8-day composite and monthly datasets (December 2013 and 2014 and April 2014 and 2015). MODIS-Aqua data of 4-km resolutions covering the study region (Latitude 5–24°N and Longitude 65–95°E) over the Bay of Bengal basin and Arabian Sea has been processed (http://modis.gsfc.nasa.gov/Oceancolor).

23.2.4 Processing of GHRSST Level-3 Data

Sea surface temperature (SST) high resolution (1 km) reanalysis global data has been retrieved from GHRSST link (http://noaa.gov.in/GHRSST) and subsets for northern Indian Ocean region have been collected for the current study.

23.2.5 Processing of pCO_2 Images Based on SST Data

The opposite seasons (summer and winter) based on temperature variability and dynamics have been chosen. These two seasons exhibit different phytoplankton growth and biomass patterns in the Indian coastal ocean water-mass and the observance of two extreme weather conditions, which is relevant to SST variability study and hence processed these two different seasonal months data for model and satellite-based studies. The satellite datasets were retrieved from MODIS-Aqua satellite and INSAT and GHRSST data from above respective websites archived and processed using ENVI and SEADAS softwares to generate colour maps with respective data range and scales. The monthly, daily, 8-day, and intra-day data for study area and location wise time series plots generated and interpreted over two different seasonal months and compared. INSAT 3D is a geostationary orbit-based sensor (around 36,000 km height orbit) and has a repeat cycle of every 30 min. But MODIS-Aqua is polar orbiting satellite which is above 600 km orbit above earth surface, with a daily repeat cycle. Also, the bands/channels are different in case of INSAT (Thermal Infra-red TIR-1, 10.2–11.2 and TIR-2, 11.5–12.5 μm) and Modis satellite (Thermal Infra-red TIR-1, 10.78–11.28 and TIR-2, 11.77–12.27, and

TIR-3, 13.185–13.485 μm). The radiometry and imaging mechanism are also different for both satellites. GHRSST is a combined satellite- and model-derived product with 1 km resolution like MODIS and INSAT.

The SST-based pCO_2 algorithm (Zhu et al. 2009) has been applied to INSAT-3D (1 km resolution), MODIS-Aqua (4 km resolution), and GHRSST (1 km resolution) SST data sets to retrieve pCO_2 images for two different seasonal months for 2 years, December 2013 and 2014 and April 2014 and 2015.

The algorithm has the following form,

$$pCO_2 = 6.34T^2 - 366.65T + 5678.53 \quad (23.1)$$

Where, T is the sea surface temperature (SST).

23.3 Results and Discussion

The SST has been interpreted using 3-sensor datasets for two different seasonal months and years (Fig. 23.1) and similarly the pCO_2 monthly scale images have been generated using 3-data sources (Fig. 23.2). GHRSST-derived monthly pCO_2 and Modis monthly chlorophyll images have been interpreted during December 2013–14 and April 2014–15 (Fig. 23.3). The GHRSST April and December months indicate similar trend during both the years, for SST and pCO_2 parameters. But, 2015 April SST was higher in the southern Indian waters. Chlorophyll has been observed to be high in the month of December in the western BoB and northern AS. The high chlorophyll concentration was observed in the range of 0.4–1.5 mg m^{-3}. The seasonal variation study is always interesting in terms of the ocean biomass accumulation, phytoplankton blooms, river fluxes, seasonal upwelling, convection, advection, eddies, seasonal ocean circulation, etc.; these processes and phenomena regulate the chlorophyll, SST, and pCO_2 dynamics in temporal and spatial scales over the seasons. The low SST is one of the regulating parameter to induce phytoplankton biomass and blooming events in the coastal water of India. The phytoplankton and their growth and phenology is SST-limited. The upwelling brings cooler water enriched with nutrients from bottom water and water column to surface water. Hence, the change in SST due to upwelling other processes like eddies, convection, advection, eddies, etc. play vital role for the marine life and the linkage to biogeochemistry and biological cycle in ocean water column. The wind-induced upwelling and convection/convective eddies play crucial role to add the CO_2 from water column and bottom to surface and also have been of interest to researchers. This is a study of further research to understand this concept. Also the phytoplankton bloom absorbs more CO_2 and hence there is observation of more pCO_2 drawdown and has shown better correlation with chlorophyll. In summer typically high pCO_2 content observed in sea water incurring with more wind stress but chlorophyll and phytoplankton biomass is less, hence less resemblance observed (Fig. 23.3).

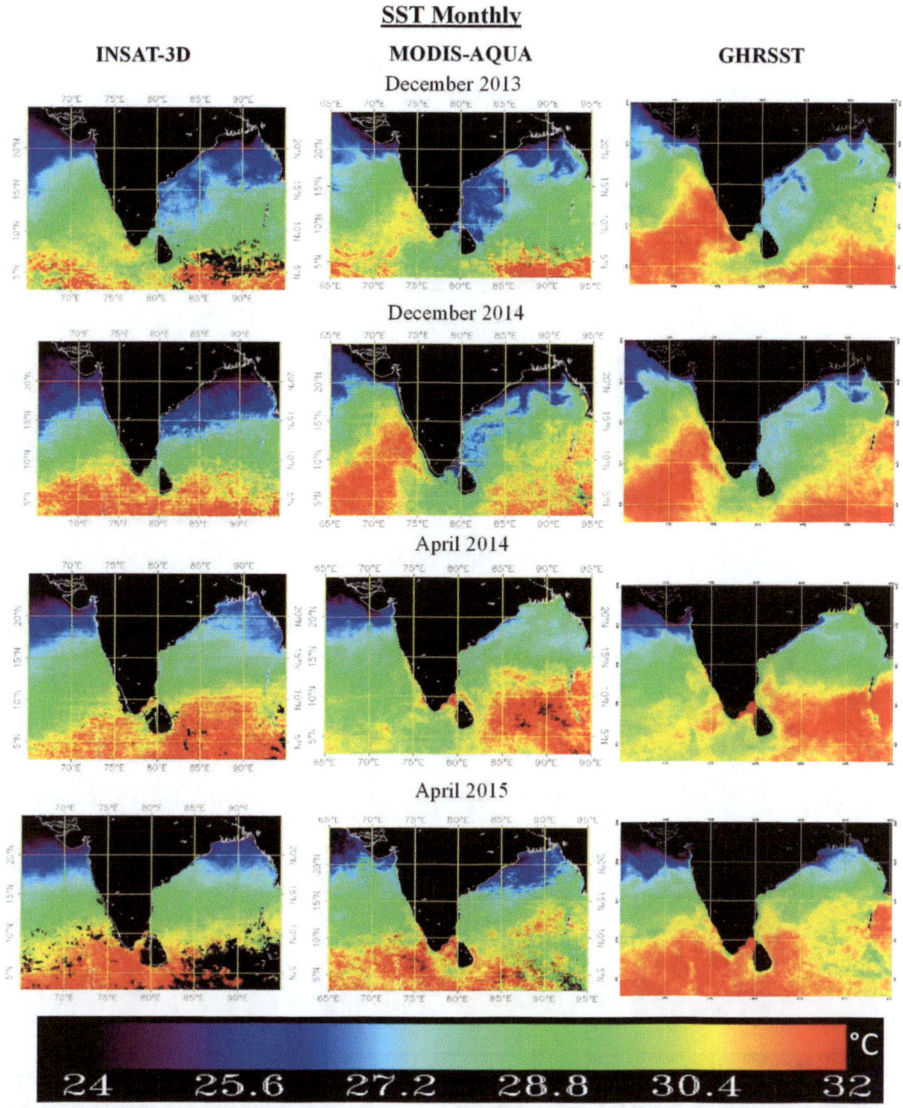

Fig. 23.1 Monthly composite GHRSST and derived pCO_2 and MODIS-Aqua-derived chlorophyll during December 2013–14 and April 2014–15

The relationship has been established between SST and pCO_2 parameters trends, observed using Modis and INSAT weekly datasets covering two different seasonal/months data points; December 2013 and 2014 and April 2014 and 2015 (Fig. 23.4). The figure indicates the variability of pCO_2 on weekly scale during April (2014–15) and during December (2013–14) with the mean and standard deviation plots. The overall range observed 390–430 μatm for INSAT and Modis (Fig. 23.5). The SST weekly mean trend during April and December has been observed using INSAT and

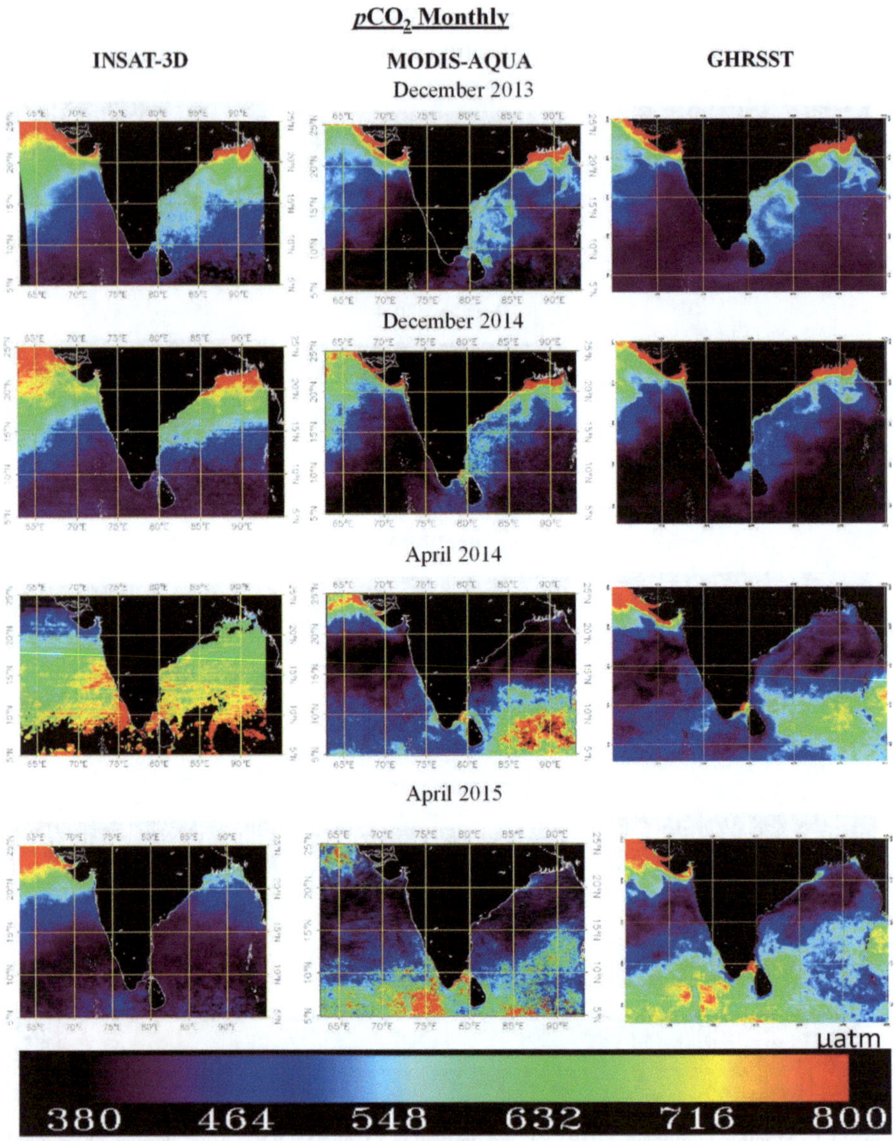

Fig. 23.2 INSAT-3D, MODIS-Aqua, and GHRSST-derived monthly composite SST images during December 2013–14 and April 2014–15

Modis data. The Modis SST showed higher values (1 °C) compared to INSAT, Modis range observed to be 27–30.4 °C, and INSAT SST range observed to be 26.5–29.8 °C. During December, SST was lower as compared to April month SST. The pCO_2 weekly mean trend has been observed using INSAT and Modis data. pCO_2 is observed to be higher during the month of December (INSAT range,

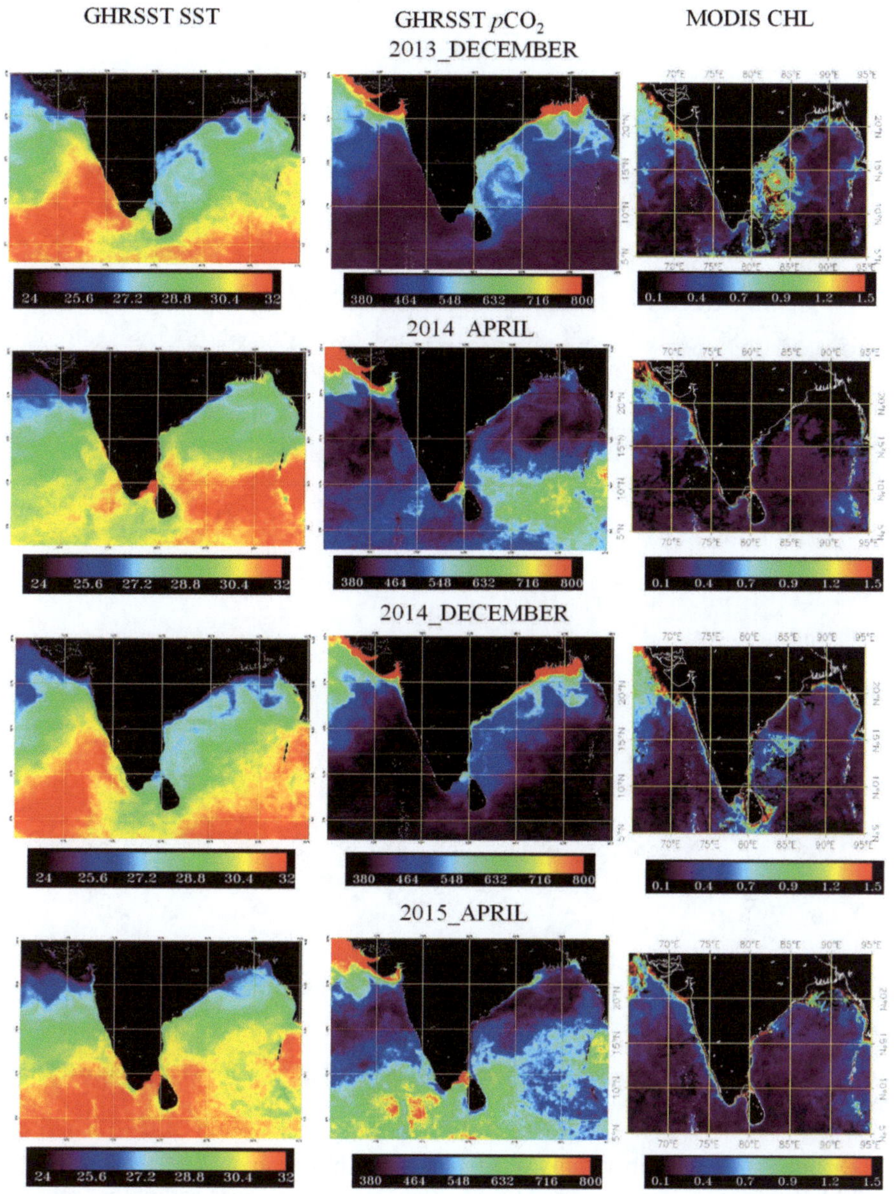

Fig. 23.3 INSAT-3D, MODIS-Aqua, and GHRSST-derived monthly composite pCO_2 images during December 2013–14 and April 2014–15

400–430 μatm, and Modis range 385–415 μatm). During April, range for INSAT was 390–410 μatm and for Modis, 390–405 μatm. December trend is observed to be higher than April month pCO_2 (Figs. 23.4 and 23.5).

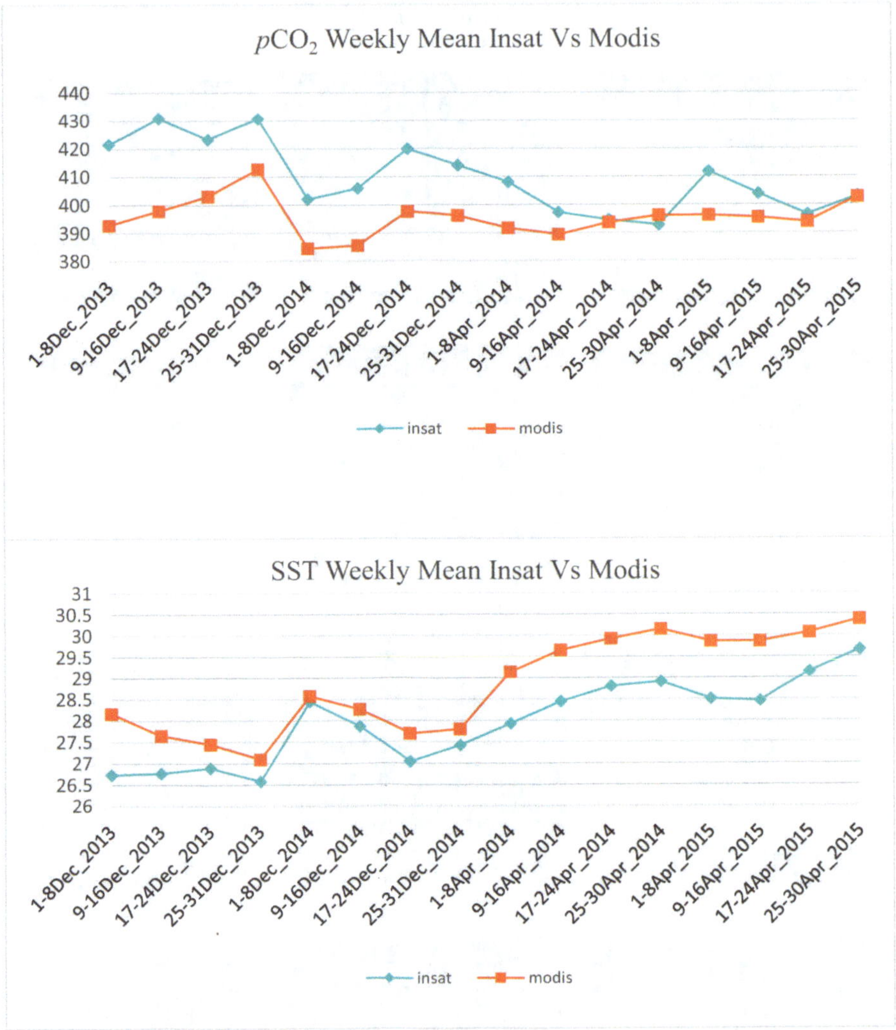

Fig. 23.4 Line plots showing the weekly scale variability of pCO_2 and SST mean derived from INSAT and Modis datasets

During 2013 December, daily pCO_2 variability has been observed using INSAT- and Modis-derived images. The INSAT is observed to be over-estimating than the Modis daily pCO_2 datasets (Fig. 23.6). INSAT range is observed to be 420–470 μatm and the Modis range observed to be 395–450 μatm. During December 2014, the daily pCO_2 images have been observed and interpreted. The INSAT and Modis trend appeared similar with little over-estimating trend seen from INSAT data points. The overall pCO_2 range observed is 390–445 μatm. During April 2014, daily pCO_2 has been observed using INSAT and Modis data. The mean INSAT observed to be over-estimating as compared to Modis data. INSAT range found to be

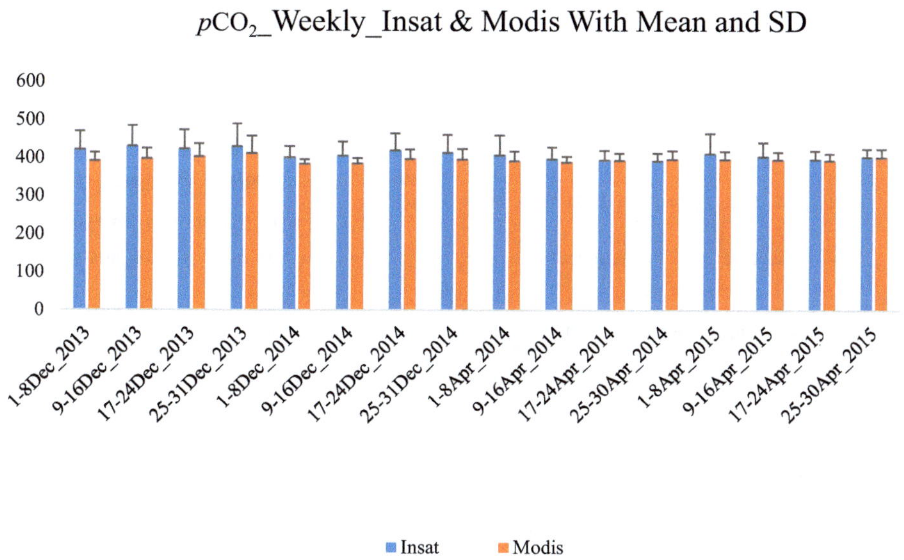

Fig. 23.5 Histogram plots showing the weekly scale mean and standard deviation variability of pCO_2 derived from INSAT and Modis datasets

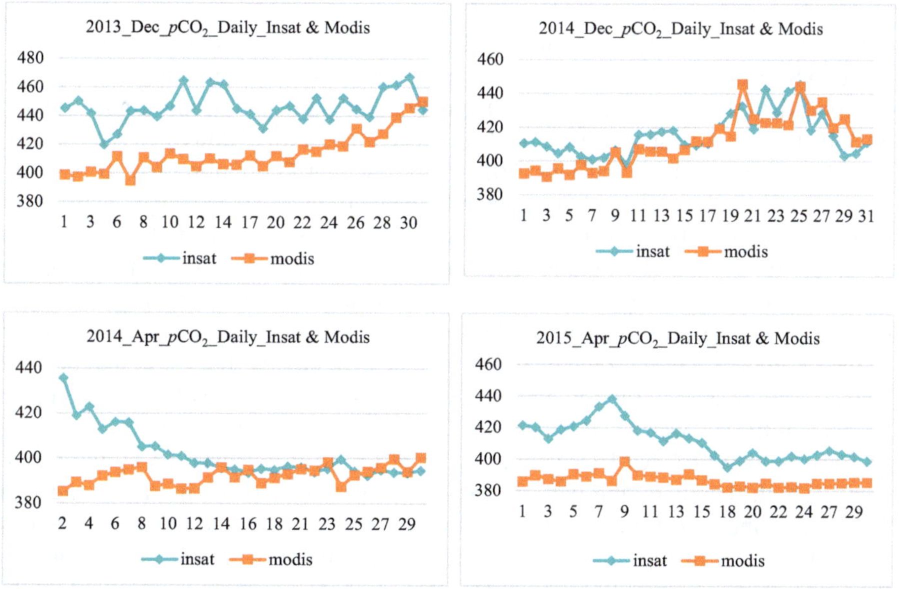

Fig. 23.6 Line plots showing the daily scale variability of pCO_2 derived from INSAT and Modis datasets in seasonal months and annual scale

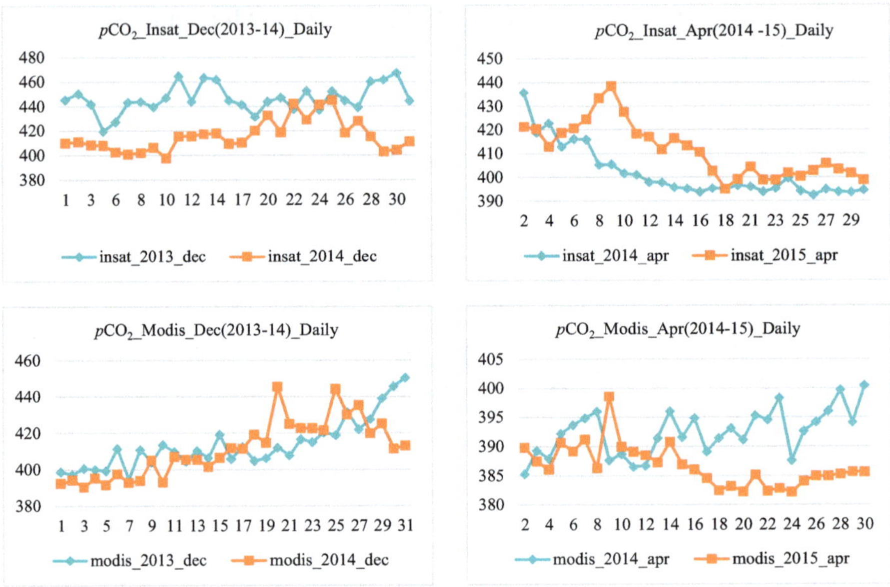

Fig. 23.7 Line plots showing the daily variability of pCO_2 derived from INSAT and Modis datasets in seasonal months and annual scale

395–400 μatm (Fig. 23.7). During April 2015, daily pCO_2 mean has been observed using INSAT and Modis data. INSAT is observed to be overestimating as compared to Modis data. INSAT range is found to be 395–440 μatm and Modis range found to be 380–400 μatm. During 2013 and 2014 December, INSAT daily pCO_2 mean data have been observed. In the year 2013 December, pCO_2 is observed to be over-estimating compared to 2014 December daily pCO_2. In 2013 December, daily pCO_2 range is found to be 420–465 μatm and for 2014 December, daily pCO_2 range observed to be 395–450 μatm. The variation of time to time daily insolation intensity-based SST changes with the sun zenith angle and viewing angle variation plays crucial role on the changes observed with the modelled pCO_2 in the ocean water. Of course, the seasonal changes, wind speed, clouds' presence, and also the changes on air-sea interaction and ocean currents and processes lead to change in the SST of the sea water and hence the pCO_2 is more connected to SST.

During 2013 and 2014 December, daily pCO_2 mean is observed by the Modis data. In both the years, the daily variability trend is observed similar. In the year 2013 December, daily pCO_2 range is found to be 395–450 μatm and during 2014 December, daily pCO_2 range found to be 390–450 μatm, which is similar for both the years' data. During 2014 and 2015 April, daily pCO_2 mean is observed by the INSAT data. The April 2015 daily pCO_2 mean data are observed to be over-estimating compared to 2014 April pCO_2 daily datasets. During 2014 April, pCO_2 daily data range is found to be 390–435 μatm and during 2015 April, daily pCO_2 range found to be 395–440 μatm. During April 2014–15, daily pCO_2 mean is observed by Modis. The April 2014 pCO_2 daily data are observed over-estimating

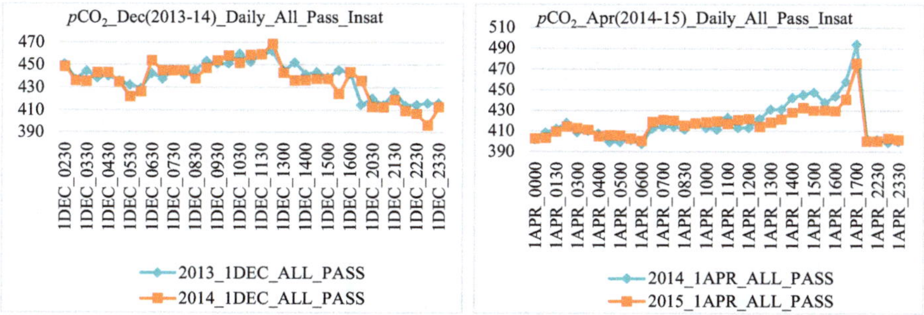

Fig. 23.8 Line plots showing the intra-day variability of pCO_2 derived from INSAT datasets in seasonal months and annual scale

compared to 2015 April daily data. During 2014 April, daily pCO_2 range is found to be 385–400 μatm and during 2015 April, daily pCO_2 range found to be 382–398 μatm.

During 2013–14, the intra-day hourly interval INSAT-3D-derived mean pCO_2 has been observed for the AS and BoB (Fig. 23.8). The trend for both the years' data points observations has been found to be similar for this single day datasets derived from December 1, 2013 and 2014. It has been observed that the passes during 2:30–11:30 h have shown higher pCO_2 as compared to next 12 h pCO_2 observation. It shows there is the impact of solar insolation on the pCO_2 variability in intra-day time scale. The range has been observed 390–470 μatm during 2013–14. Similarly, the intra-day pCO_2 variability has been plotted for April 2014–15. For April 1 date hourly interval data points, the trend for both the years' single day data has been found similar. Range is observed to be 395–495 μatm. It is observed that from morning to afternoon hours, the pCO_2 is observed to be increasing as seen during 2014–15 April intra-day datasets.

Modis chlorophyll mean has been correlated with 8-day pCO_2 mean derived from three sensors datasets, i.e. Modis, INSAT, and GHRSST. The reverse trend has been observed from the regression; like with increasing chlorophyll, pCO_2 has been observed to be decreasing (Figs. 23.9, 23.10, and 23.11). The Table 23.1 shows the monthly mean SST and pCO_2 from three different sources and, in synoptic scale, there are not more deviation in the mean SST and pCO_2 values. The GHRSST and Modis SST are observed to be higher than INSAT data. The INSAT-derived pCO_2 showed higher values as compared to Modis and GHRSST-derived data points. The phytoplankton bloom and their photosynthetic process take care of CO_2 absorption. The diatoms are mostly dominant and absorb the CO_2 in the northern Indian Ocean context. As the other groups of phytoplankton, dinoflagellate, cyanophyte, etc. composition is less. The optimal temperature is the main reason for the phytoplankton types and the suitable distribution regionally. Also, the coastal fluxes, riverine fluxes, and induced water quality and nutrient enrichment lead to enhance phytoplankton biomass, and in suitable ambient temperature conditions, the phytoplankton absorbs the CO_2 with different rates in different sunlight intensities as well. The increased chlorophyll is linked to the food chain and food web in terms of intake by

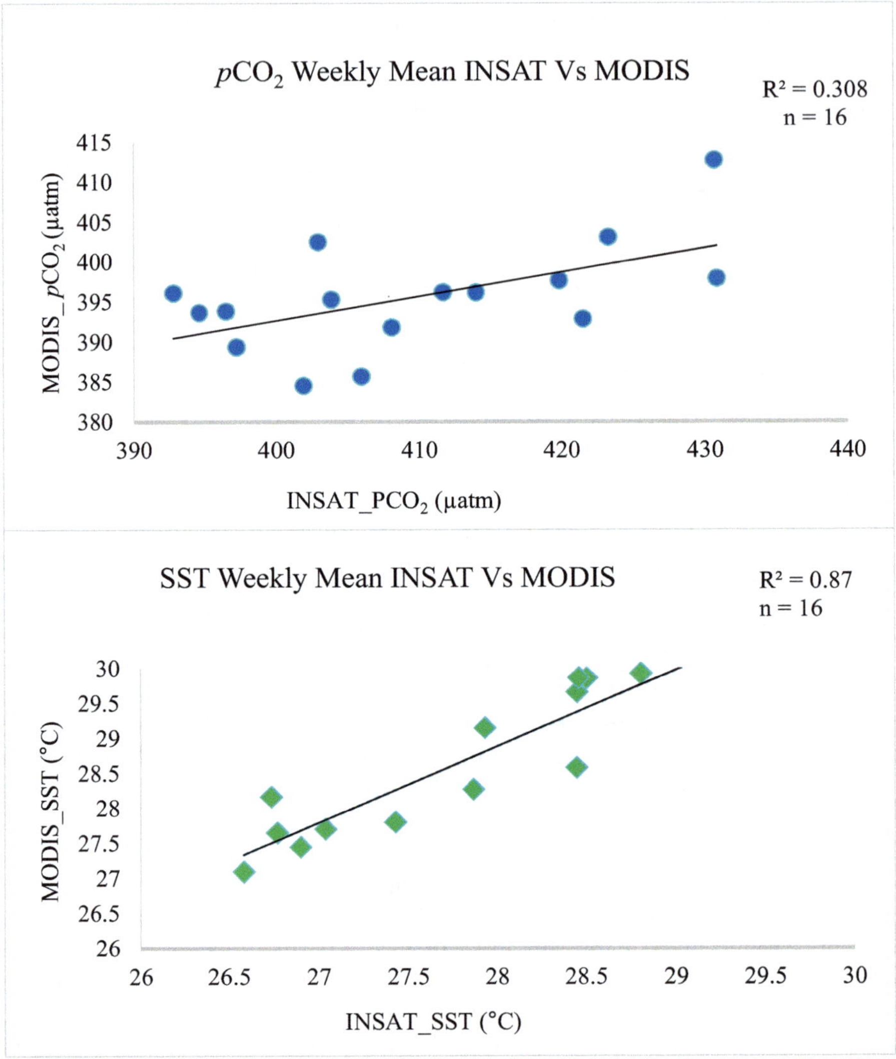

Fig. 23.9 Inter-relationship of SST and pCO_2 derived from weekly mean datasets from INSAT and Modis

zooplankton and smaller fishes and larger fishes and nektons and also degradation and decay as part of microbial loop as well. The seasonal cycles of chlorophyll concentration are related to the growth phase, blooming, and phytoplankton phenology in the marine ecosystem. The chlorophyll increases with phytoplankton bloom cycles and impacts zooplankton biomass and ecological succession as well. The increase in SST is linked to increased CO_2 level and mostly reduces the chlorophyll concentration in seawater and there is decrease in phytoplankton biomass which is represented by reduction in chlorophyll concentration. Hence, understanding the

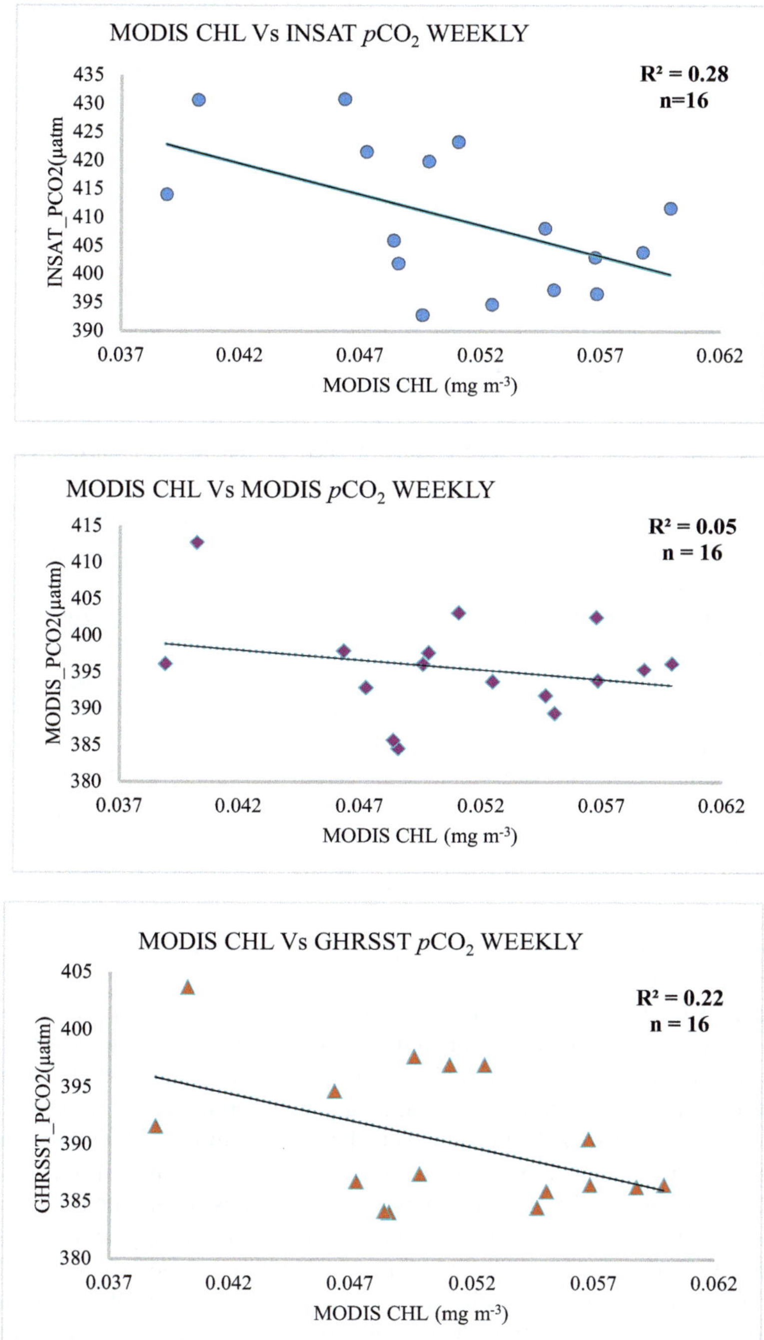

Fig. 23.10 Inter-relationship of chlorophyll with pCO_2 derived from weekly mean datasets

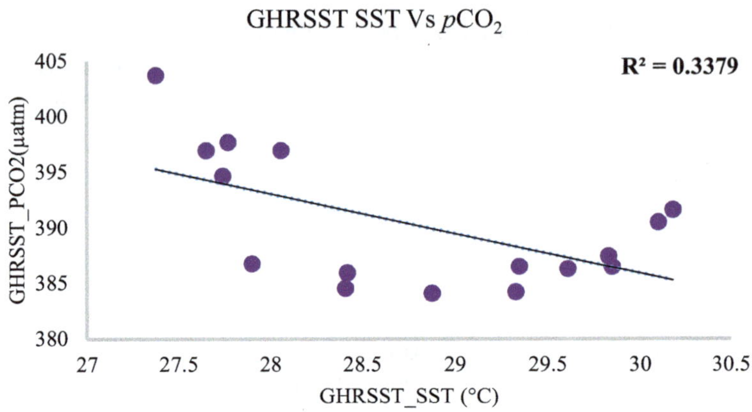

Fig. 23.11 Inter-relationship of GHRSST with pCO_2 derived from weekly mean datasets

Table 23.1 SST and pCO_2 monthly mean for two seasonal months retrived from three satellites data

	INSAT-3D	MODIS-Aqua	GHRSST
SST monthly mean			
December-2013	27.06	27.69	27.63
December-2014	27.62	27.71	29.53
April-2014	28.43	29.83	28.12
April-2015	28.67	29.89	29.70
pCO_2 monthly mean			
December-2013	417.82	396.46	395.51
December-2014	406.62	392.03	386.82
April-2014	398.01	392.71	390
April-2015	403.34	392.63	387.43

biological variability in terms of chlorophyll during different pCO_2 conditions and seasons is essential.

These pCO_2 levels range varies in coastal to offshore regions and has been observed by many authors with such ranges (Sarma et al. 2000a, b; Joshi and Warrior 2022). These similar ranges (350–750 µmol) are monitored and understood with few studies in northern Indian Ocean and vary in different oceans (Lekshmi 2013; Joshi and Warrior 2022). As the subtropical region is influenced well by SST and CO_2 dissociation with the process via solubility pump and air-sea interaction. In some cases, when the CO_2 absorption exceeds in the ecosystem even if there is high chlorophyll observed, there is chance of ocean acidification happening with saturation of pCO_2 and possibly due to river fluxes and inorganic load increasing. There is need for forecasting on ocean acidification status based on water stoichiometry and imbalances in the water column. This seems future challenge of research on pCO_2 dynamics.

This study is important from the point of view of climate change perspective in connection with the process of global warming and connected ocean acidification process in different regional seas in temporal scale. Hence, monitoring SST from satellite and its time series and applying the modelled pCO_2 data and image on the regional scale has been an important aspect of research topic in the current frontiers of ocean research. The study is linked to biogeochemistry as the source and sink of CO_2 is related to the ocean carbon cycle and biological pump. The more in situ data and modelling the ocean biogeochemistry have been the current challenge for ongoing and future research to study the carbon fluxes and its rate of dissipation via the air-sea interaction process.

23.4 Conclusion

The SST monthly composite images derived from three different data sources, INSAT-3D, Modis-Aqua, and GHRSST, have been interpreted for two different seasonal months, December 2013 and 2014 and April 2014 and 2015. There has been observation of similar trend of SST for three different data sources. But, the GHRSST has shown higher range compared to INSAT-3D and Modis-Aqua. The INSAT-3D-derived SST has observed SST within the range (24–32 °C) and picked up the ocean features and trends as compared to Modis and GHRSST datasets. The pCO_2 derived from three SST data sources has been processed and the images have been interpreted for the first time using three different data sources for the Indian water. The coastal water in BoB has shown high pCO_2 value (500–650 µatm) in December month, but the INSAT has over-estimated the pCO_2 in the coastal water. During April month, the pCO_2 has shown low pCO_2 range (380–460 µatm). And in the southern Indian water, there has been observation of high pCO_2 range (600–750 µatm) during April month. INSAT pCO_2 range has shown over-estimation during April 2014. Overall, the high chlorophyll trend coastal water in BoB and AS has shown higher pCO_2 trend and low SST trend, as evident during December 2013–14 monthly images. During winter, there has been observation of low SST (27–28 °C) in the coastal BoB, where there is high pCO_2 (550–650 µatm) and observation of high chlorophyll (0.7–1.5 mg/m^3). So during the winter month, low SST is linked to high pCO_2 and high chlorophyll in the coastal BoB. But in Arabian Sea, during December, there is higher SST (29-32 °C), low pCO_2 (400–550 µatm), and moderate chlorophyll (0.1–0.8 mg/m^3). So, it indicated the higher SST in Arabian Sea during December, has reduced the pCO_2 and chlorophyll concentration as compared to BoB. The understanding of water column structure with pCO_2 in situ measurement would convince understanding the processes.

During summer month April, there has been change in the SST trend as compared to December in the BoB. The high SST (28–31 °C) has been observed and pCO_2 is low (400–550 µatm) and chlorophyll is also low (0.1–0.4 mg/m^3). In coastal Arabian Sea, SST during April is moderate and high, pCO_2 is low (400–550 µatm), and chlorophyll is low (0.1–0.4 mg/m^3). So, overall it indicates that the ocean water column and bottom flux to surface play a greater role in regulating the surface productivity and the increase in pCO_2 concentration and the decrease in SST. There has been inverse trend irrespective of season and region exemplified by BoB and AS. It infers that due to possible upwelling and convection process, the bottom water possibly brings up CO_2 to the surface water in addition to the gaseous transfer from the air to sea. But during summer, with high SST the pCO_2 has been observed higher as evidenced in coastal and offshore waters and there has been low chlorophyll. Thorough study with in situ measurements would confirm our inference and also the seasonal variability. This is an important piece of study to supplement on the climate change processes as the pCO_2 variability plays crucial role in oceans state and impact due to climate change in relation with ocean acidification processes. The study on pCO_2 measurements in different timeframes has its role on studying seasonal and inter-annual observations in the perspective view of climate change in regional and global oceans. Hence, the warming events in oceans have been linked to CO_2 fluxes and its sequestration and pCO_2 levels in oceans. So the satellite technology is having advantage to map and monitor in a synoptic scale and interpret efficiently and effectively our oceans. The development of more robust models with inclusion of more parameters and also the in situ validation has the future scope and the emerging trend of new instruments would justify the models and accuracy achievements.

Acknowledgements The author is thankful to Dr. B.K.Bhattacharya, Group Director, BPSG, Dr. I.M.Bahuguna, Deputy Director, EPSA/SAC and Director, Space Applications Centre, Ahmedabad, for necessary guidance and facilities for carrying out the work. A note of thanks goes to Ms. Megha Pandya, Software Engineer from Globaltech (I) Pvt. Ltd. for assistance in data processing. Author acknowledge the data archival supports rendered from NASA Ocean color, GHRSST and MOSDAC archival websites.

References

Allison EH, Perry AL, Badjeck M-C, Adger WN, Brown K, Conway D, Halls AS et al (2009) Vulnerability of national economies to potential impacts of climate change on fisheries. Fish Fish 10:173–196

Andrié C, Oudot C, Genthon C, Merlivat L (1986) CO_2 fluxes in the tropical Atlantic during FOCAL cruises. J Geophys Res Oceans 91(C10):11741–11755

Banse K, English DC (2000) Geographical differences in seasonality of CZCS-derived phytoplankton pigment in the Arabian Sea for 1978-1986. Deep Sea Res 47:1623–1677

Burrows MT, Schoeman DS, Buckley LB, Moore P, Poloczanska ES, Brander KM, Brown C et al (2011) The pace of shifting climate in marine and terrestrial ecosystems. Science 334:652–655

Chakraborty K, Valsala V, Bhattacharya T, Ghosh J (2021) Seasonal cycle of surface ocean pCO2 and pH in the northern Indian Ocean and their controlling factors. Prog Oceanogr 198:102683. https://doi.org/10.1016/j.pocean.2021.102683

Chen L, Gao Z (2007) Spatial variability in the partial pressures of CO_2 in the northern Bering and Chukchi seas. Deep-Sea Res II Top Stud Oceanogr 54(23):2619–2629

Chester R (2000) Marine geochemistry, 2nd edn. Blackwell Science Ltd, Oxford, 506 pp

Cochrane K, DeYoung C, Soto D, Bahri T (2009) Climate change implications for fisheries and aquaculture, FAO fisheries and aquaculture technical paper, vol 530. Food and Agriculture Organization of the United Nations, Rome, 212 pp

Doney SC, Ruckelshaus M, Duffy JE, Barry JP, Chan F, English CA, Galindo HM et al (2012) Climate change impacts on marine ecosystems. Annu Rev Mar Sci 4:1–27

Drinkwater KF, Beaugrand G, Kaeriyama M, Kim S, Ottersen G, Perry RI et al (2010) On the processes linking climate to ecosystem change. J Mar Syst 79:374–388

Feely RA (2004) Impact of anthropogenic CO2 on the CaCO3 system in the oceans. Science 305(5682):362–366

Gnanadesikan A, Dunne JP, John J (2011) What ocean biogeochemical models can tell us about bottom-up control of ecosystem variability. ICES J Mar Sci 68:1030–1044

Goet C, Beauverger C, Brunet C, Poisson A (1991) Distribution of carbon dioxide partial pressure in surface waters of the Southwest Indian Ocean. Tellus B 43(1):1–11

Gruber N, Clement D, Carter BR, Feely RA, van Heuven S, Hoppema M, Ishii M, Key RM, Kozyr A, Lauvset SK, Lo Monaco C, Mathis JT, Murata A, Olsen A, Perez FF, Sabine CL, Tanhua T, Wanninkhof R (2019) The oceanic sink for anthropogenic CO_2 from 1994 to 2007. Science 363(6432):1193–1199. https://doi.org/10.1126/science.aau5153. PMID: 30872519

IPCC (2007) Climate change 2007: synthesis report. In Contribution of Working Groups I, II and III to the fourth assessment report of the Intergovernmental Panel on Climate Change. Ed. by R. K. Pachauri, A. Reisinger, and Core Writing Team. IPCC, Geneva, 104 pp

Jorgensen C, Peck MA, Antognarelli F, Azzurro E, Burrows MT, Cheung WWL, Cucco A et al (2012) Conservation physiology of marine fishes: advancing the predictive capacity of models. Biol Lett 8:900–903

Joshi AP, Warrior HV (2022) Comprehending the role of different mechanisms and drivers affecting the sea-surface pCO2 and the air-sea CO2 fluxes in the Bay of Bengal: a modeling study. Mar Chem 243:104120

King JR, Agostini VN, Harvey CJ, McFarlane GA, Foreman MGG, Overland JE, Di Lorenzo E et al (2011) Climate forcing and the California Current ecosystem. ICES J Mar Sci 68:1199–1216

Körtzinger A, Duinker JC, Mintrop L (1997) Strong CO_2 emissions from the Arabian Sea during south-west monsoon. Geophys Res Lett 24(14):1763–1766

Kumamoto YI, Fujiwara K, Tsubota H (1995) Temporal variation of the partial pressure of CO_2 in surface seawater of Iyo Nada in the Seto Inland Sea, Japan. J Oceanogr 51(6):681–697

Le Quéré C, Raupach MR, Canadell JG, Marland G, Bopp L, Ciais P, Conway TJ et al (2009) Trends in the sources and sinks of carbon dioxide. Nat Geosci 12:831–836

Lefèvre N, Taylor A (2002) Estimating pCO_2 from sea surface temperatures in the Atlantic gyres. Deep-Sea Res I Oceanogr Res Pap 49(3):539–554

Lekshmi K (2013) Measurement of carbon dioxide fluxes in the Arabian Sea using Remote Sensing data, Masters degree thesis, Andhra University

Lutgens FK, Tarbuck EJ (2004) The atmosphere: an introduction to meteorology, 9th edn. Prentice Hall, New Jersey, p 14

Madhu NV, Maheswaran PA, Jyotibabu R, Sunil V, Ravichandran C, Balasubramaniam T, Gopalakrsishnaan TC, Nair KKC (2002) Enhanced biological production off Chennai triggered by October 1999 super cyclone (Orissa). Curr Sci 82(12):1472–1479

Mahadevan A, Tagliabue A, Bopp L, Lenton A, Mémery L, Lévy M (2011) Impact of episodic vertical fluxes on sea surface pCO_2. Philos Trans R Soc A Math Phys Eng Sci 369(1943):2009–2025

Merino G, Barange M, Blanchard J, Harle J, Holmes R, Allen I, Allison EH et al (2012) Can marine fisheries and aquaculture meet fish demand from a growing human population in a changing climate? Glob Environ Chang 22:795–806

Nayak SR, Sarangi RK, Rajawat AS (2002) Application of IRS-P4 OCM data to study the impact of cyclone on coastal environment of Orissa. Curr Sci 80:1208–1213

Peters K, Johannes Q, Nicolas B (2011) Effects of absorbing aerosols in cloudy skies: a satellite study over the Atlantic Ocean. Atmos Chem Phys 11(4):1393–1404

Roxy MK, Kapoor R, Pascal T, Sébastien M (2014) The curious case of Indian Ocean warming. J Clim 27(22):8501–8509
Roxy MK, Modi A, Murtugudde R, Valsala V, Swapna P, Prasanna Kumar S, Ravichandran M, Vichi M, Lévy M (2016) A reduction in marine primary productivity driven by rapid warming over the tropical Indian Ocean. Geophys Res Lett 43(2):826–833
Sarangi RK, Mishra M, Chauhan P (2015) Remote sensing observations on impact of Phailin cyclone on phytoplankton distribution in northern Bay of Bengal. IEEE J Sel Top Appl Earth Obs Remote Sens 8(2):539–549
Sarma VVSS, Dileep Kumar M, Gauns M, Madhupratap M (2000a) Seasonal controls on surface pCO_2 in the central and eastern Arabian Sea. Indian Academy of Sciences
Sarma VVSS, Kumar MD, Gauns M, Madhupratap M (2000b) Seasonal controls on surface pCO2 in the central and eastern Arabian Sea. J Earth Syst Sci 109:471–479
Signorini SR, Mannino A, Raymond G, Marjorie N, Friedrichs AM, Cai WJ, Salisbury J, Wang ZA, Thomas H, Shadwick E (2013) Surface ocean pCO2 seasonality and sea-air CO2 flux estimates for the North American east coast. J Geophys Res Oceans 118(10):5439–5460
Solomon S (ed) (2007) Climate change 2007 – the physical science basis: Working Group I contribution to the fourth assessment report of the IPCC, vol 4. Cambridge University Press
Sreeush MG, Rajendran S, Valsala V, Sreenivas P, Prasad KVSR, Murtugudde R (2019) Variability, trend and controlling factors of ocean acidification over Western Arabian Sea upwelling region. Mar Chem 209:14–24
Subramanian V (1993) Sediment load of Indian rivers. Curr Sci 64:928–930
Swapna P, Krishnan R, Wallace JM (2014) Indian Ocean and monsoon coupled interactions in a warming environment. Clim Dyn 42:2439–2454
Takahashi T, Olafsson J, Goddard JG, Chipman DW, Sutherland SC (1993) Seasonal variation of CO_2 and nutrients in the high-latitude surface oceans: a comparative study. Glob Biogeochem Cycles 7(4):843–878
Takahashi T, Sutherland SC, Sweeney C, Poisson A, Metzl N, Tilbrook B, Bates N, Wanninkhof R, Feely RF, Sabine C, Olafsson J, Nojiri Y (2002) Global sea-air CO_2 flux based on climatological surface ocean pCO_2 and seasonal biological and temperature effects. Deep-Sea Res II Top Stud Oceanogr 49(9):1601–1622
Takahashi T, Sutherland SC, Wanninkhof R, Sweeney C, Feely RA, Chipman DW, de Baar HJ (2009) Climatological mean and decadal change in surface ocean pCO2, and net sea–air CO2 flux over the global oceans. Deep-Sea Res II Top Stud Oceanogr 56(8–10):554–577
Trela P, Sathyendranath S, Moore RM, Kelley DE (1995) Effect of the nonlinearity of the carbonate system on partial pressure of carbon dioxide in the oceans. J Geophys Res Oceans 100(C4):6829–6844
Valsala V, Maksyutov S (2013) Interannual variability of the air–sea CO2 flux in the north Indian Ocean. Ocean Dyn 63(2):165–178
Valsala V, Murtugudde R (2015) Mesoscale and intraseasonal air–sea CO2 exchanges in the western Arabian Sea during boreal summer. Deep-Sea Res I Oceanogr Res Pap 103:101–113
Yoder JA (2000) An overview of temporal and spatial patterns in satellite-derived chlorophyll imagery and their relation to ocean processes. In: Halpern D (ed) Satellites, oceanography and society. Elsevier Science BV, Amsterdam, pp 225–238
Zhu Y, Shang S, Zhai W, Dai M (2009) Satellite-derived surface water pCO_2 and air-sea CO_2 fluxes in the northern South China Sea in summer. Prog Nat Sci 19(6):775–779

Chapter 24
Warming Seas: The Impact of Climate Change on Ocean Life and Ecosystems

Soma Mishra

Abstract Ocean ecosystems are crucial for life on Earth, providing food, regulating the climate, and supporting biodiversity. However, the ongoing effects of climate change, such as rising ocean temperatures, ocean acidification, and rising sea levels, are significantly altering marine environments. The impact of climate change on ocean ecosystems is an urgent issue with far-reaching consequences for marine life and human societies. This chapter comprehensively examines the causes and mechanisms driving ocean warming, including the role of greenhouse gases and the heat-buffering capacity of oceans. It discusses the physical and chemical changes occurring in the marine environment, such as increased sea surface temperatures, ocean acidification, deoxygenation, and shifts in ocean currents. The chapter also explores the repercussions for marine life, highlighting the vulnerability of plankton, fish species, coral reefs, and marine megafauna. In addition, the decline in ecosystem services, such as fisheries and biodiversity, and the socioeconomic impacts on coastal communities, food security, and global economies are addressed.

Keywords Climate Change · Ocean Warming · Marine Ecosystems · Ocean Acidification · Coral Bleaching · Fisheries · Biodiversity Loss · Socioeconomic Impacts

S. Mishra (✉)
Dr. Vishwanath Karad MIT World Peace University, Pune, Maharashtra, India
e-mail: Soma.mishra@mitwpu.edu.in

B. P. Barathan et al. (eds.), *Navigating Climate Change: Impacts on Biodiversity and Ecosystem Resilience*, Sustainable Development and Biodiversity 42,
https://doi.org/10.1007/978-981-95-0409-1_24

24.1 Introduction

24.1.1 Overview of Climate Change and Its Impact on the Ocean

The oceans are essential to life on Earth. They cover more than two-thirds of the planet's surface and provide many important services. Oceans help regulate the Earth's climate by absorbing heat and storing large amounts of carbon dioxide (IPCC 2021a, b). They support a wide variety of marine species and supply food, oxygen, and income to billions of people around the world (FAO 2022). However, because of human activities, especially the burning of fossil fuels, the oceans are warming at an alarming rate (Cheng et al. 2019). This increase in ocean temperatures is leading to major environmental changes that affect marine life, ecosystems, and people who depend on the ocean.

Over the past few decades, scientists have recorded a steady rise in sea surface temperatures, with the upper ocean absorbing more than 90% of the heat from global warming (Cheng et al. 2020). This warming is causing serious impacts, including coral bleaching, melting polar ice, rising sea levels, and stronger and more frequent storms (Hoegh-Guldberg et al. 2014; IPCC 2019). Many marine animals, such as fish, plankton, and seabirds, are moving toward cooler waters or deeper ocean zones to survive, which is changing food webs and affecting fisheries (Poloczanska et al. 2013; Pinsky et al. 2020).

Warming oceans also affect the chemistry of seawater. As more carbon dioxide is absorbed by the ocean, the water becomes more acidic, which is harmful to species like corals, oysters, and other shell-forming animals (Doney et al. 2009; Gattuso et al. 2015). At the same time, ocean warming reduces the amount of oxygen in the water, which creates low-oxygen zones where many species cannot survive (Breitburg et al. 2018). These changes together—warming, acidification, and deoxygenation—are sometimes called the "triple threat" to ocean health. Figure 24.1 shows the Climate Change and its impact on Ocean.

The social and economic impacts of warming oceans are just as serious. Fisheries around the world are becoming less predictable and less productive, especially in tropical regions where many communities rely on fish for food and income (Barange et al. 2018; FAO 2022). Coastal areas are experiencing more flooding, erosion, and damage from extreme weather events, putting millions of people at risk (Nurse et al. 2014; Oppenheimer et al. 2019). Small island nations and low-lying coastal countries are especially vulnerable to sea level rise and storm surges.

Because of these wide-ranging and interconnected effects, studying ocean changes caused by warming seas is extremely important. This chapter aims to explore how ocean temperatures are changing in different regions, how marine ecosystems are responding, and how these changes are affecting human communities. The main objectives of this chapter are:

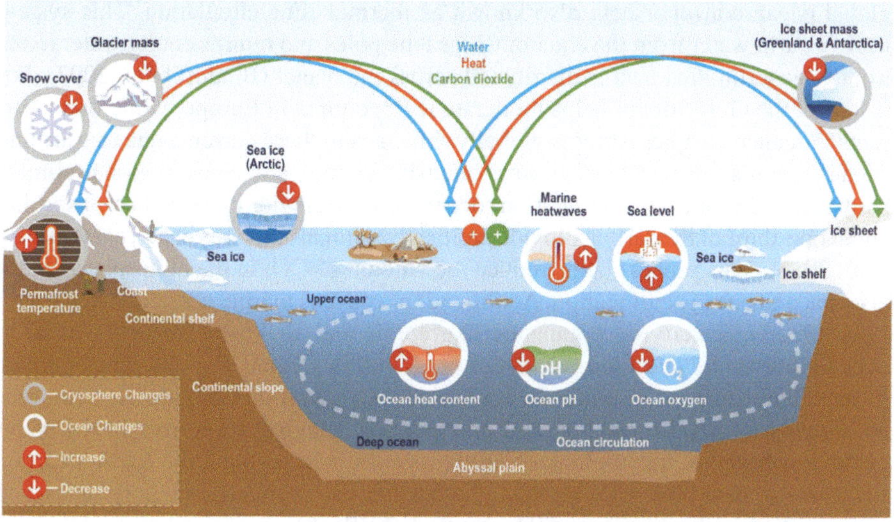

Fig. 24.1 Climate Change and its impact on Ocean. (Source: https://www.ipcc.ch/srocc/chapter/technical-summary)

- To explain the scientific reasons behind ocean warming and identify global trends
- To describe how warming affects marine biodiversity, ocean chemistry, and ecosystem functioning
- To assess how these changes impact coastal populations, fisheries, and economies
- To review global and local efforts to adapt to and reduce the impacts of warming seas

The scope of this chapter is broad and interdisciplinary. It brings together knowledge from oceanography, marine biology, climate science, and social science. By presenting a clear overview of the science and impacts of ocean warming, this chapter hopes to raise awareness of the urgent need for action. It encourages stronger cooperation between scientists, governments, and local communities to build solutions that protect ocean life and human well-being in a warming world.

24.1.2 *Importance of Oceans in Regulating Earth's Climate*

The oceans play a vital role in regulating Earth's climate by acting as a vast reservoir for heat and carbon dioxide. Covering about 71% of the planet's surface, they absorb and store solar energy, distributing heat through currents and maintaining temperature balance between different regions. According to the IPCC (2019), oceans have absorbed more than 90% of the excess heat generated by greenhouse gas emissions since the mid-twentieth century, significantly slowing the pace of global warming on land. One critical way oceans regulate the climate is through the global ocean conveyor belt, also known as thermohaline circulation. This system moves warm water from the equator toward the poles and returns cooler water to the tropics, ensuring that heat is distributed across the planet (Bindoff et al. 2007). For instance, the Gulf Stream helps moderate temperatures in Europe, while other currents influence weather patterns globally. Changes in these currents due to warming oceans can disrupt weather systems, intensify storms, and cause regional climate shifts. In addition to heat regulation, oceans act as a major carbon sink, absorbing about one-third of the carbon dioxide emitted by human activities each year (Gattuso et al. 2015). This process helps reduce the amount of CO_2 in the atmosphere, but it also leads to ocean acidification. Acidification weakens marine ecosystems, particularly affecting calcifying organisms like corals and shellfish, which are vital to ocean biodiversity and food webs (Orr et al. 2005). Oceans also influence climate through their interactions with the atmosphere. Evaporation from the ocean surface contributes to cloud formation and precipitation, playing a key role in the water cycle. Furthermore, phytoplankton in the oceans help regulate the Earth's carbon balance by absorbing CO_2 during photosynthesis, indirectly supporting carbon sequestration in the deep ocean (Doney et al. 2012a, b).

However, as the oceans continue to warm and absorb CO_2, their capacity to buffer climate change diminishes. This poses risks to their ability to regulate temperature and carbon levels effectively, exacerbating climate impacts. Protecting and understanding the role of oceans is critical to mitigating climate change and ensuring the stability of Earth's climate systems.

24.2 Ocean Warming: Causes and Mechanisms

The warming of the world's oceans is a central aspect of climate change, directly linked to the accumulation of greenhouse gases in the atmosphere. As the Earth's largest heat reservoir, the oceans have absorbed the majority of the excess energy trapped by greenhouse gases, moderating atmospheric warming but altering marine ecosystems and global climate patterns. This complex process involves heat absorption mechanisms, historical trends in temperature increases, regional differences in warming, and a clear connection to greenhouse gas emissions.

24.2.1 Greenhouse Gases and Their Role in Trapping Heat

Greenhouse gas emissions drive ocean warming by trapping heat, much of which is absorbed by the oceans, altering temperature, density, and chemistry. This process has slowed atmospheric warming but led to acidification and disrupted ocean circulation, such as the weakening of the Atlantic Meridional Overturning Circulation (AMOC), which can trigger extreme weather and rainfall shifts (Gattuso et al. 2015).

Warming oceans also intensify marine heatwaves, devastating ecosystems like coral reefs and kelp forests, with severe consequences for biodiversity, fisheries, and coastal economies (Frölicher et al. 2018). While the oceans act as a heat sink, this buffer comes at a high cost to marine life and global climate stability. Addressing ocean warming requires reducing greenhouse gas emissions, protecting marine ecosystems, and advancing climate research to safeguard both biodiversity and planetary health.

24.2.2 Heat Absorption by Oceans: The Buffer Effect

The oceans play a critical role in regulating the Earth's climate by absorbing solar energy and redistributing it across the planet. This heat absorption is governed by several mechanisms. Sunlight penetrates the ocean's surface, and the top layers absorb much of this energy. Water's high heat capacity allows it to store large amounts of energy without a significant temperature change, making the ocean an effective heat sink (Bindoff et al. 2007).

The ocean's ability to absorb heat is influenced by circulation patterns, such as the global thermohaline circulation, also known as the "ocean conveyor belt." This system transports warm surface water to higher latitudes and colder, denser water to the depths, enabling the distribution of heat across various layers and regions. Additionally, phenomena like the El Niño-Southern Oscillation (ENSO) and the Pacific Decadal Oscillation (PDO) temporarily influence how and where heat is stored or released. The oceans' interaction with the atmosphere further amplifies this process, as winds, evaporation, and precipitation regulate the exchange of heat between the two systems. Figure 24.2 shows the Change in Sea Surface Temperature over the year 1901–2022.

This map shows how average sea surface temperature around the world changed between 1901 and 2022. It is based on a combination of direct measurements and satellite measurements. A black "+" symbol in the middle of a square on the map means the trend shown is statistically significant. White areas did not have enough data to calculate reliable long-term trends.

24.2.3 Ocean Temperature Trends Over the Past Century

Ocean temperature records show a clear upward trend in recent decades, coinciding with the Industrial Revolution and increased greenhouse gas emissions. Since the late nineteenth century, average global ocean temperatures have risen by approximately 0.8 °C (Fig. 24.3), with much of this warming occurring since the 1970s (IPCC 2019). Data from deep-sea buoys, satellites, and historical ship measurements indicate that this warming is not limited to the surface; it extends to depths of over 700 m.

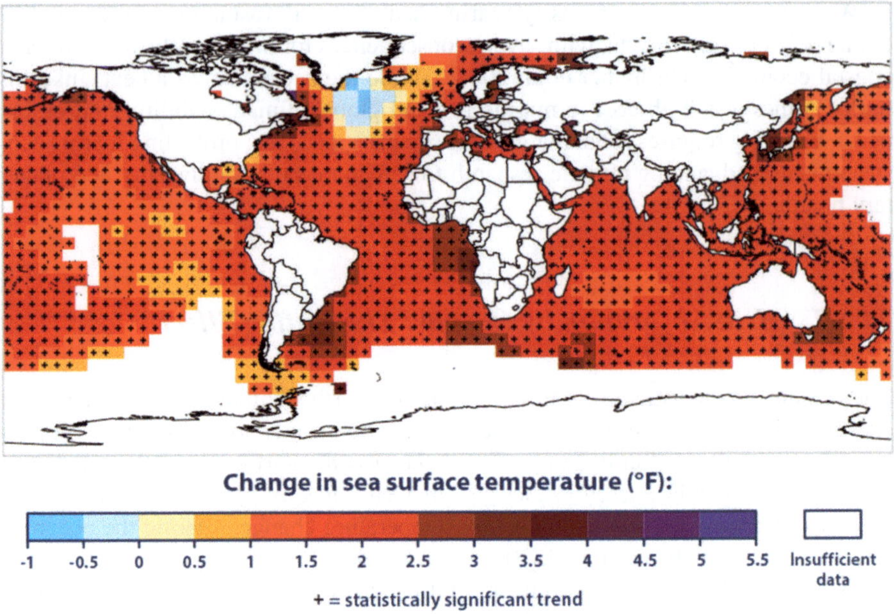

Fig. 24.2 Change in sea surface temperature, 1901–2022 (www.epa.gov)

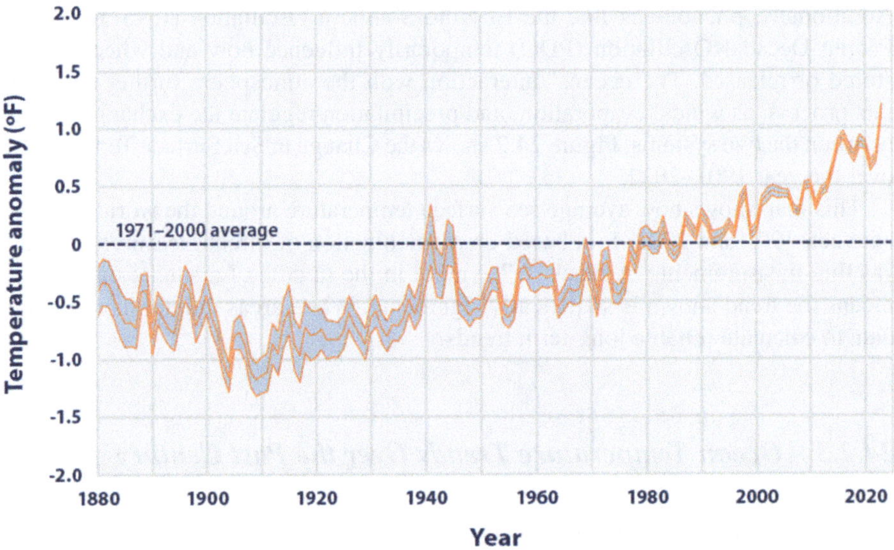

Fig. 24.3 Average global sea surface temperature, 1880–2023 (www.epa.gov)

The period between 1993 and 2019 witnessed an unprecedented rise in ocean heat content, with studies estimating that the oceans absorbed more than 90% of the excess heat generated by human activities (Roemmich and Gilson 2015). This trend underscores the critical role of oceans in buffering the climate system. However, the heat absorption has also caused thermal expansion, contributing to rising sea levels and intensifying extreme weather events.

This graph illustrates the changes in the average ocean surface temperature since 1880. The average from 1971 to 2000 is used as a baseline in this graph to illustrate change. The shape of the data across time would remain unchanged if a new baseline era was selected. Based on the quantity of measurements taken and the accuracy of the techniques employed, the shaded band displays the range of data uncertainty.

24.2.4 Regional Differences in Warming Patterns

Climate warming does not uniformly elevate temperatures throughout all regions of the ocean. Certain regions experience accelerated warming, resulting in varied impacts on marine species and ecosystems. A combination of natural variability, geographical factors, and human-induced climate change drives regional differences in temperature changes.

The Indian Ocean is among the most rapidly warming oceans globally. In recent decades, its surface temperature has increased rapidly. This warming influences the monsoon precipitation upon which numerous countries in South Asia rely for agriculture and water supply. It adversely affects coral reefs and diminishes fish populations, which are crucial for sustenance and employment in the area. Elevated sea temperatures result in increased frequency and intensity of cyclones, which can devastate coasts and populations. The Pacific Ocean is vast and exhibits uneven warming patterns. In certain regions, such as the western Pacific, water temperatures are rising more rapidly, influencing climatic phenomena such as El Niño and La Niña. These alterations can precipitate floods, droughts, and severe meteorological phenomena in numerous regions globally. Elevated Pacific temperatures jeopardize coral reefs and compel fish to migrate to colder regions, so hurting fishing operations. The Atlantic Ocean is undergoing significant transformations. Certain regions of the North Atlantic are experiencing warming, but other parts, particularly adjacent to Greenland, are undergoing cooling. This irregular pattern may impede the ocean currents that regulate the climate in Europe and North America. Elevated temperatures in the Atlantic correlate with intensified hurricanes and the reduction of sea ice in the Arctic. The Southern Ocean encircling Antarctica is seeing a slower rate of surface warming, although its deeper waters are undergoing increased warmth. This can cause the melting of ice sheets from beneath and elevate sea levels. It also influences the microscopic organisms that constitute the foundation of the food chain, potentially affecting larger animals such as whales and seals.

The Arctic Ocean, for instance, is warming at a rate nearly four times faster than the global average due to a process known as Arctic amplification. As sea ice melts, it exposes darker ocean surfaces, which absorb more solar energy compared to reflective ice. This feedback loop accelerates warming in the region, with significant implications for marine ecosystems and global weather patterns (Polyakov and Timmermann 2017). In contrast, parts of the Southern Ocean have exhibited slower warming rates. This phenomenon is attributed to the unique interactions between deep water currents and atmospheric circulation, which bring cold water to the surface and temporarily offset warming. However, even these regions are not immune to long-term temperature increases, as evidenced by shifts in sea ice extent and ecosystem changes.

Tropical and subtropical regions are also experiencing significant warming, leading to coral bleaching and other ecosystem disruptions. In some cases, warming patterns are intensified by localized factors such as pollution and overfishing, which weaken ecosystems' ability to adapt. Regional variations in warming highlight the complexity of ocean systems and their sensitivity to both natural and anthropogenic influences.

24.3 Impacts on Oceanic Physical and Chemical Properties

Human-driven climate change is reshaping ocean properties, impacting marine ecosystems and global climate systems. Warming has led to ice sheet loss, declining Arctic Sea ice and rising permafrost temperatures. As oceans absorb excess heat from greenhouse gas emissions, changes in temperature, salinity, and stratification alter circulation, nutrient availability, carbon storage, and marine life health.

The Earth's oceans are undergoing significant changes due to human activities, particularly climate change. These changes are reshaping the physical and chemical properties of the oceans, affecting marine ecosystems and global climate systems (IPCC 2019).

With mass loss from ice sheets and glaciers, decreases in snow cover, and Arctic sea ice extent and thickness, as well as an increase in permafrost temperature, global warming has caused the cryosphere to shrink significantly over the past few decades (IPCC 2021a, b).

Ocean warming is not only a significant threat to marine ecosystems but also has profound effects on the physical and chemical properties of the oceans. As global temperatures rise due to increased greenhouse gas emissions, the oceans absorb much of the excess heat, which leads to changes in temperature, salinity, stratification, and other important characteristics (Levitus et al. 2012; Rhein et al. 2013). These changes affect ocean circulation, the availability of nutrients, the ability of oceans to store carbon, and the overall health of marine life (Bindoff et al. 2019a, b; Doney et al. 2012a, b).

24.3.1 Changes in Sea Surface Temperature and Stratification

Ocean stratification refers to the layering of ocean water based on differences in temperature, salinity, and density. Typically, the ocean is divided into layers: the surface layer (mixed layer), which is warmer and less dense, and deeper layers that are cooler and denser. The process of stratification plays a critical role in regulating ocean circulation, the movement of nutrients, and the distribution of marine life. The sea surface temperature (SST) is one of the most visible indicators of climate change in the oceans, with significant changes observed over recent decades. As global temperatures rise, the ocean absorbs a large portion of the excess heat, which directly influences the SST. This increase in temperature, along with other factors like changes in precipitation and ice melt, has also led to notable changes in ocean stratification, the process by which the ocean is layered according to temperature and density. These changes have important consequences for ocean circulation, marine ecosystems, and global climate patterns.

24.3.2 Ocean Acidification and Its Links to Warming

The oceans absorb approximately 30% of anthropogenic carbon dioxide (CO_2) emissions, which has led to a decrease in seawater pH, a phenomenon known as ocean acidification (OA) (Fig. 24.4). Since the industrial revolution, the average pH of surface seawater has dropped from 8.2 to 8.1, representing a 30% increase in acidity (NOAA 2023).

Ocean warming and acidification are two interconnected processes driven by the same underlying factor: the increase in atmospheric CO_2 due to human activities. As the planet warms, a greater amount of heat is absorbed by the oceans, raising sea surface temperatures (SST). This means that as SST rises, the rate of acidification could accelerate, further threatening marine ecosystems (Gattuso et al. 2015).

Warmer waters absorb CO_2 more efficiently, and the solubility of gases like CO_2decreases as temperatures rise. While this may seem like a mitigation mechanism, it actually accelerates ocean acidification in specific regions. For example, in the Polar Regions, where surface water temperatures are rising the fastest, more CO_2 is absorbed, exacerbating acidification (Kawamiya et al. 2017). This creates a cycle where warming accelerates acidification, which in turn has negative effects on marine life.

One of the critical impacts of ocean acidification is the reduction in the availability of carbonate ions (CO_2^{2-}), which are essential for marine organisms that build calcium carbonate ($CaCO_2$) shells and skeletons, such as corals, molluscs, and certain plankton species. As the concentration of hydrogen ions increases due to acidification, carbonate ions react with them to form bicarbonate ions, reducing the availability of carbonate ions. Ocean warming further complicates this process by

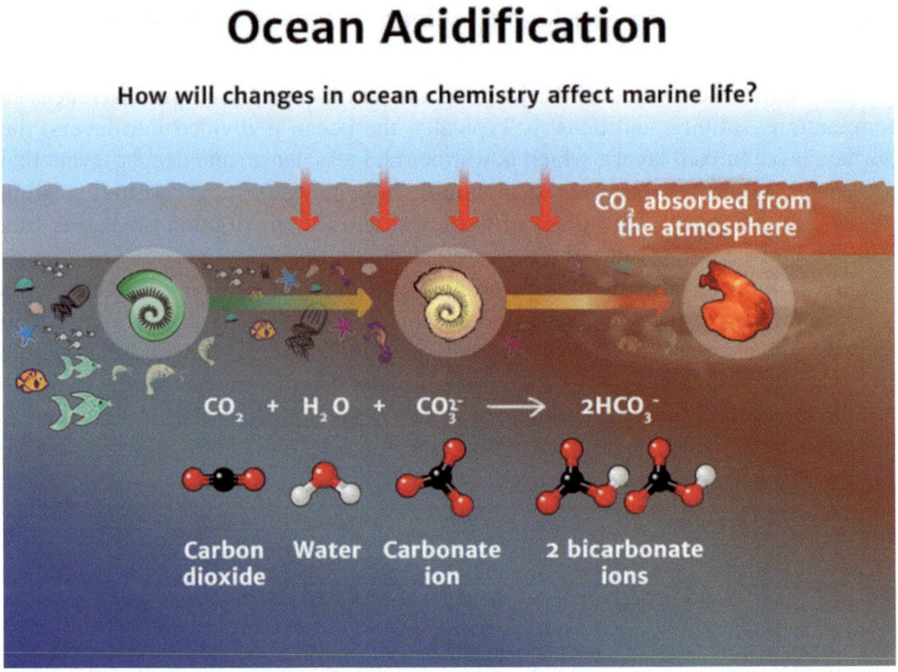

Fig. 24.4 Ocean acidification, impact of dissolved CO_2. (Source: https://solarschools.net/knowledge-bank/climate-change/ocean-acidification)

reducing the capacity of seawater to hold carbonate ions, which can be detrimental to the organisms that rely on them for shell and skeleton formation (Orr et al. 2005).

Deoxygenation refers to the process of oxygen loss in ocean waters, and it has become an increasingly critical issue as human-induced climate change intensifies. The oceans are experiencing a gradual reduction in dissolved oxygen levels, particularly in deeper layers, and this trend is expected to continue. The primary drivers of deoxygenation include global warming, nutrient pollution, and changes in ocean circulation. Deoxygenation poses significant threats to marine habitats, particularly in areas known as "dead zones," where oxygen levels are too low to support most forms of marine life. As oxygen levels decrease, the entire marine ecosystem is disrupted, with severe consequences for biodiversity, fisheries, and ecosystem services.

Ocean acidification poses severe risks to calcifying organisms such as corals, molluscs, and certain plankton species. These organisms rely on carbonate ions to build their shells and skeletons. However, acidification reduces the availability of these ions, weakening their structures and impacting marine food webs (Kroeker et al. 2013).

24.3.3 Shifts in Ocean Currents and Thermohaline Circulation

Thermohaline circulation, often referred to as the "global conveyor belt," is a large-scale ocean circulation pattern driven by differences in temperature (thermo) and salinity (haline) across the world's oceans. These differences in water temperature and salinity affect water density, and the circulation is powered by the sinking of cold, dense water in the polar regions, and the upwelling of warmer water in other parts of the ocean. This process helps transport heat, nutrients, and gases across vast distances in the ocean, influencing climate and weather patterns across the globe (Fig. 24.5).

In the North Atlantic, for example, warm water from the tropics is carried northward by currents like the Gulf Stream. As this water cools in the polar regions, it becomes denser and sinks, drawing in more surface water and continuing the circulation pattern. This process helps moderate temperatures in northern latitudes and is a key component in regulating Earth's climate (Broecker 1991).

As the global climate warms, ocean currents are experiencing significant changes. These shifts are largely driven by the increase in atmospheric temperatures, melting ice, and changes in precipitation patterns. The main drivers of these shifts include warming temperature and ocean stratification, melting polar ice, and changes in precipitation patterns.

Global warming is leading to an increase in ocean surface temperatures, which in turn causes the ocean's upper layers to become warmer and less dense. This reduces the vertical mixing of ocean waters, making it harder for surface waters to sink and contribute to thermohaline circulation. The stratification of the ocean can weaken the strength of currents like the Gulf Stream and other important currents that regulate climate and weather patterns. For example, a study showed that the

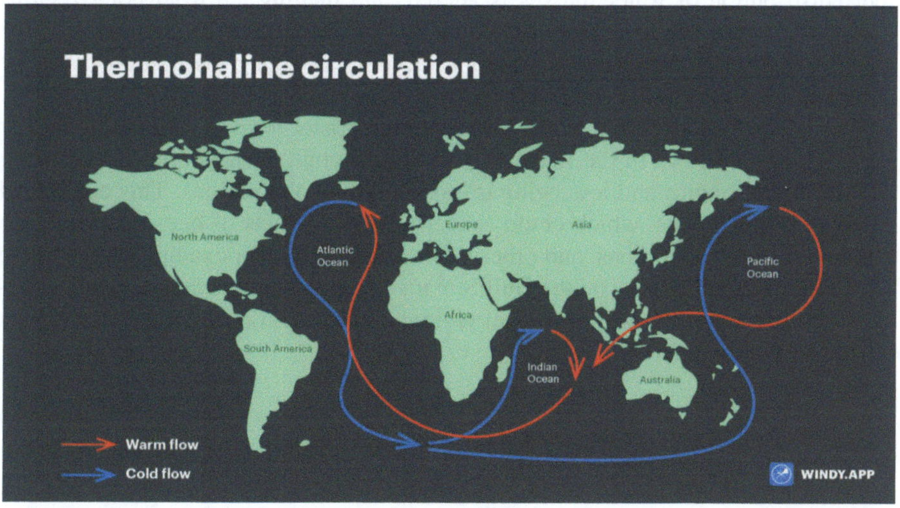

Fig. 24.5 Thermohaline circulation. (Source: Windy.app)

Gulf Stream has slowed by about 15%–20% over the past century, and this slowdown could intensify as temperatures rise (Caesar et al. 2018).

Melting ice from glaciers and ice sheets, particularly in the Arctic and Antarctic regions, is adding large volumes of freshwater to the oceans. Freshwater is less dense than saltwater, so the influx of freshwater can disrupt the delicate balance that drives thermohaline circulation. In the Arctic, the melting of sea ice is reducing the salinity of surface waters, making it more difficult for these waters to sink and continue the circulation. This can weaken the "global conveyor belt," potentially affecting ocean circulation patterns and regional climates (Rahmstorf 2002).

Climate change is also altering global precipitation patterns, with some regions experiencing more intense rainfall while others becoming drier. Increased rainfall can add more freshwater to the oceans, further diluting seawater and weakening the salinity-driven components of thermohaline circulation. In regions like the North Atlantic, where the sinking of cold, salty water plays a critical role in driving ocean currents, this could lead to a slowdown in the circulation pattern and disruptions to climate systems that rely on it (Lenton et al. 2008).

The changes in ocean currents and thermohaline circulation due to climate change are not isolated but have broad implications for global climate systems, marine ecosystems, and human activities. Some of the most notable impacts include:

Ocean currents play a significant role in shaping weather patterns by redistributing heat and moisture around the planet. A slowdown in the thermohaline circulation could lead to shifts in weather patterns, including cooler temperatures in Europe, as the Gulf Stream weakens. This could result in harsher winters, changes in precipitation, and altered growing seasons, which would have significant consequences for agriculture and local economies (Rahmstorf et al. 2005).

The slowdown of thermohaline circulation could also contribute to rising sea levels. As warm surface waters are no longer transported as efficiently to the poles, the redistribution of water masses is altered, and melting ice from polar regions could exacerbate sea-level rise. Additionally, the weakening of currents like the Gulf Stream could cause ocean waters to accumulate along the eastern coasts of continents, further raising sea levels in these areas (Griffies et al. 2009).

Ocean currents also play a role in redistributing heat from the equator to higher latitudes. A slowdown in thermohaline circulation could result in more heat being trapped in the tropics and less being transported to cooler regions. This could lead to hotter tropical seas, which could exacerbate the impacts of ocean warming on coral reefs, fish populations, and other marine organisms. For example, areas in the North Atlantic, which rely on the Gulf Stream to regulate temperatures, could experience a sharp cooling, while tropical regions might experience more intense heat (Brix and Wallman 2013).

24.4 Effects on Marine Life

The warming of oceans due to climate change is causing widespread changes in marine ecosystems, affecting a diverse range of species and habitats (Fig. 24.6). As sea temperatures rise, these impacts extend to various levels of marine life, from microscopic plankton to large marine mammals.

24.4.1 Plankton and Primary Producers

24.4.1.1 Shifts in Phytoplankton Distribution and Productivity

Phytoplankton are the foundational primary producers in ocean ecosystems, playing a vital role in global carbon cycling and marine food webs. These microscopic organisms form the basis of the marine food chain by converting solar energy into biomass through photosynthesis. However, climate change is driving shifts in their distribution and productivity, with significant consequences for ocean ecosystems and the global carbon cycle.

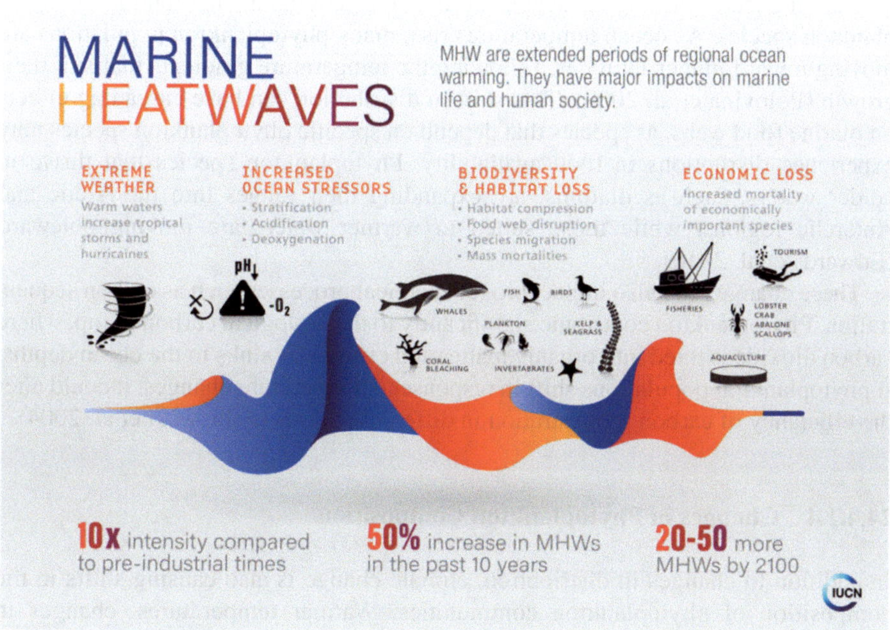

Fig. 24.6 Marine heatwaves: A serious threat to marine biodiversity. (Source: www.iucn.org)

24.4.1.2 Temperature and Phytoplankton Growth

Phytoplankton productivity is highly influenced by water temperature, as temperature affects metabolic rates, nutrient availability, and the physical environment of the oceans. Rising ocean temperatures, due to global warming, can alter the growth and distribution patterns of phytoplankton (Boyd and Ellwood 2014). Warmer waters can accelerate metabolic processes, increasing phytoplankton growth in some areas. However, excessive warming may reduce productivity in other regions by creating unfavorable conditions, such as nutrient depletion or stratification of the water column (Barton et al. 2016).

In tropical regions, phytoplankton may face nutrient limitations due to increased surface water stratification caused by warmer temperatures. This stratification prevents nutrients from deeper waters from reaching the surface, reducing the productivity of phytoplankton in these areas (Behrenfeld et al. 2006). In contrast, cooler polar regions may see an increase in phytoplankton productivity due to increased nutrient availability and favorable light conditions (Arrigo et al. 2008).

24.4.1.3 Shifts in Geographic Distribution

Climate-induced warming is causing shifts in the geographic distribution of phytoplankton species. As ocean temperatures rise, many phytoplankton populations are moving toward higher latitudes, following the temperature gradients that suit their growth (Polovina et al. 2008). This shift in distribution can have cascading effects on marine food webs, as species that depend on specific phytoplankton species may experience disruptions in food availability. Phytoplankton species that thrive in colder waters, such as diatoms, are expanding their ranges into the Arctic and Antarctic regions, while those suited to warmer waters are moving poleward (Edwards et al. 2010).

These changes can also impact biogeochemical processes such as carbon sequestration. Phytoplankton contribute significantly to the biological carbon pump, where carbon dioxide is fixed into organic matter and eventually sinks to the ocean depths. If phytoplankton populations shift in response to temperature changes, it could alter the efficiency of carbon sequestration in different regions (Falkowski et al. 2004).

24.4.1.4 Changes in Phytoplankton Composition

In addition to changes in distribution, climate change is also causing shifts in the composition of phytoplankton communities. Warmer temperatures, changes in nutrient availability, and ocean acidification are driving species turnover, with some groups of phytoplankton, such as diatoms and coccolithophores, experiencing a decline in abundance, while others, such as flagellates, may become more dominant (Rivkin and Anderson 2016). This change in community structure can have far-reaching implications for the marine food web, as different phytoplankton groups provide different types of nutritional resources for higher trophic levels (Litchman and Klausmeier 2015).

For example, diatoms, which are rich in omega-3 fatty acids, are important food sources for zooplankton and fish. If diatom populations decline and are replaced by other phytoplankton species, this could alter the nutrient composition of marine food webs, potentially affecting fish populations and the marine animals that depend on them (Doney et al. 2012a, b).

24.4.1.5 Impact on Marine Food Webs

Shifts in phytoplankton productivity and composition directly affect marine food webs. As primary producers, phytoplankton are the base of the food web, supporting zooplankton, fish, and higher trophic levels. A reduction in phytoplankton productivity or changes in their composition can have cascading effects on the entire marine ecosystem. For instance, if phytoplankton populations decline due to warming-induced nutrient limitations or acidification, this could lead to a reduction in the available food supply for herbivores, which would then impact the entire food chain, including commercially important fish stocks (Burkholder et al. 2007).

In addition to changes in productivity, altered phytoplankton species composition can influence the efficiency of the biological carbon pump, the process by which carbon is drawn down from the atmosphere and stored in the deep ocean. Some phytoplankton species, like diatoms, contribute more to carbon sequestration due to their larger size and ability to sink to the ocean depths after they die (Boyd et al. 2007). Changes in the relative abundance of these species could thus affect the ocean's role in mitigating climate change by reducing the efficiency of carbon storage.

Climate change is causing significant shifts in the distribution and productivity of phytoplankton, with broad implications for marine ecosystems and global biogeochemical cycles. As ocean temperatures rise, phytoplankton are moving to new areas, altering species composition and productivity patterns. These changes disrupt marine food webs and the efficiency of carbon sequestration, affecting both biodiversity and the ability of oceans to mitigate climate change. Continued monitoring and research into these shifts are essential to understanding the full extent of their impacts and developing strategies to protect marine ecosystems.

24.4.2 Fish and Invertebrate Species

Fish and invertebrates are vital components of marine ecosystems, serving as key links in the food chain and as critical resources for human consumption and biodiversity. Climate change, particularly ocean warming, is significantly impacting these species, driving changes in their migration patterns, reproduction, growth rates, and overall survival. These alterations in marine life not only affect ecological balance but also have socioeconomic repercussions for fisheries and food security.

24.4.2.1 Altered Migration Patterns and Habitat Shifts

Fish species, particularly those that are migratory, are strongly influenced by changes in water temperature and other oceanographic conditions. Many species, such as commercially important fish like cod, haddock, and tuna, follow predictable migration routes tied to temperature, feeding, and breeding conditions. However, as ocean temperatures rise, these fish are shifting their ranges toward cooler, higher-latitude waters (Pörtner and Poloczanska 2014). For example, studies have shown that North Atlantic fish populations, such as cod and haddock, have been moving northward in response to rising sea temperatures (Drinkwater 2005). This shift can cause disruptions in fishing industries that depend on stable fish stocks, as well as changes in local ecosystems, as species composition in certain regions changes.

In addition to shifts in latitude, fish are also moving to deeper waters in response to warming. This vertical migration can lead to changes in the structure of marine food webs, as deeper waters tend to be less productive due to lower light levels and reduced nutrient availability (Olsen and Reimer 2019). As fish move to deeper and cooler waters, the ecological interactions between species may also change, which can lead to cascading effects throughout the food chain.

24.4.2.2 Changes in Reproduction and Growth Rates

Ocean warming has a significant impact on the reproductive cycles of many fish species. Temperature changes can affect the timing of spawning, the survival rates of larvae, and the development of eggs. For instance, many fish species rely on specific temperature ranges to time their reproductive events, with warmer waters potentially triggering earlier spawning, which may not always coincide with the availability of food sources (Munday et al. 2011). This mismatch between spawning time and food availability can result in lower survival rates for larvae and juvenile fish.

Increased ocean temperatures can also affect the growth rates of fish. As water temperatures rise, metabolic rates in fish generally increase, leading to faster growth in some species. However, in cases of extreme temperature rise, the opposite effect may occur, with fish growth rates slowing or even stalling due to stress or changes in food availability (Bönisch et al. 2015). The overall health and size of fish populations can be influenced by these changes in growth patterns, which can impact the stability of marine ecosystems and fisheries that rely on these species.

24.4.2.3 Impacts on Fish Behavior and Physiology

Ocean warming disrupts fish behavior and physiology by altering migration, navigation, and predator avoidance. Temperature, salinity, and oxygen changes can impair feeding and increase predation risk (Doney et al. 2012a, b). Thermal stress weakens immune systems, making fish more disease-prone (Pörtner 2002). Reduced oxygen levels, especially in deep waters, force fish to flee hypoxic zones, shifting species distributions (Vaquer-Sunyer and Duarte 2008).

24.4.2.4 Impacts on Invertebrates

Marine invertebrates, such as crustaceans, molluscs, and echinoderms, are also highly sensitive to changes in ocean temperature. Many invertebrates, particularly shell-forming species like corals, oysters, and clams, are directly impacted by ocean acidification, a consequence of increased atmospheric CO_2 levels. As CO_2 dissolves in seawater, it forms carbonic acid, which lowers the pH of ocean water and makes it more difficult for organisms to form calcium carbonate shells (Orr et al. 2005). Ocean acidification, combined with warming waters, can lead to slower growth, thinner shells, and increased mortality rates in these species (Gazeau et al. 2013).

The combination of ocean warming and acidification can also impact the behavior and development of invertebrates. For example, warmer waters can affect the development of invertebrate larvae, leading to changes in their morphology, behavior, and survival rates (Bownes et al. 2016). This can disrupt marine food webs, as many invertebrates serve as a primary food source for fish and other marine animals. Additionally, warmer waters may alter the reproductive timing and success of invertebrates, further affecting population dynamics (Hughes et al. 2011).

The impacts of ocean warming on fish and invertebrate species are complex and far-reaching. These species are experiencing shifts in migration patterns, reproductive cycles, growth rates, and overall behavior, all of which can disrupt marine ecosystems and pose challenges for fisheries and food security. Addressing these challenges requires comprehensive management strategies to protect vulnerable species and their habitats. This includes reducing greenhouse gas emissions, conserving marine biodiversity, and enhancing our understanding of how climate change affects marine life at different levels of the food web.

24.4.3 Coral Reefs

24.4.3.1 Coral Bleaching and Mortality

Coral reefs are diverse ecosystems that support marine life, local economies, and coastal protection. However, climate change threatens them, primarily through coral bleaching. Bleaching occurs when rising sea temperatures stress corals, causing them to expel symbiotic algae (zooxanthellae), which provide food and color. Though bleached corals remain alive, prolonged stress weakens them, increasing disease risk and mortality. Even a 1–2 °C temperature rise can trigger bleaching, a phenomenon worsening due to global warming and greenhouse gas emissions (Hoegh-Guldberg 1999; Hughes et al. 2017).

24.4.3.2 Impact of Coral Bleaching

When corals bleach, they lose their primary food source, leading to significant stress. While they can survive for short periods without their symbiotic algae, prolonged bleaching can lead to mortality. The following are the key impacts of coral bleaching:

Loss of Biodiversity Coral reefs support about 25% of all marine species. When corals die due to prolonged bleaching, the habitat they provide disappears, causing a decline in marine biodiversity. Fish, invertebrates, and other marine organisms that depend on corals for shelter and food face a loss of their environment (Spalding et al. 2001).

Increased Vulnerability to Disease Bleached corals are less able to defend themselves against diseases. Pathogens that are otherwise controlled in healthy corals can take hold and cause damage, leading to coral death (Harvell et al. 2002).

Decreased Growth and Reproduction Corals that experience bleaching may suffer from stunted growth and reduced reproduction rates. Bleaching affects the coral's ability to build and maintain its skeleton, making them more fragile and less likely to survive long-term (Berkelmans and van Oppen 2006).

24.4.3.3 Coral Mortality

Coral mortality occurs when corals cannot recover from bleaching. If temperatures normalize quickly, some may recover, but prolonged stress leads to death. Dead reefs, or "coral graveyards," lose biodiversity and resilience. Widespread coral die-offs impact marine ecosystems and coastal communities, threatening fisheries, tourism, and coastal protection (Hughes et al. 2017).

24.4.3.4 Impacts on Reef-Associated Species

Coral reefs support diverse marine life, providing habitat and food for numerous species. Climate stressors, such as mass bleaching, disrupt these ecosystems, leading to biodiversity loss and potential fishery collapse (Williams et al. 2018). As reefs degrade, algae and sponges may dominate, altering reef structure and reducing resilience. Algae overgrowth can outcompete corals, hindering recovery and shifting the ecosystem from coral- to algae-dominated, with long-term ecological consequences (Hughes et al. 2007). Table 24.1 summarizes climate-related threats to marine ecosystems.

Table 24.1 Climate-related threats to major marine ecosystems and some consequences

Ecosystem	Threat	Consequence
Coastal		
Salt marshes/ Mangroves	Rising sea levels	Habitat migration inland or loss; changes in productivity
	Rising temperatures and CO_2	Altered growth rates; shifts in photosynthesis; dominance of C3 over C4 species
	Increased storm frequency	Physical damage; disruption or flooding of channels; changes in salinity
Estuaries	Rising sea levels	Habitat shift inland; changes in basin structure
	Increased storm frequency	Fluctuating nutrient and sediment input due to variable freshwater inflows
	Rising temperatures	Increased stress from desiccation at low tide
	Mitigation (flooded barrages)	Loss of ecosystem integrity
Rocky Substrates		
Rocky intertidal	Rising sea levels	Altered zonation patterns; increased pressure on vertical infrastructure
	Rising temperatures	Increased thermal stress and desiccation; shifts in species distribution
	Increased storm frequency	Higher exposure; shift from grazers to filter feeders; changes in trophic control
Kelp forests	Rising temperatures	Physiological stress on growth and photosynthesis; shifts in distribution
	Increased storm frequency	Negative impact on upper and near-shore kelp limits
Coral reefs	Rising temperatures	Coral bleaching; contraction of distribution ranges; algae overgrowth
	Ocean acidification	Reduced calcification; weakening of reef structure
	Rising sea levels	Reefs submerged; habitat loss; abandonment of atolls reducing fishing pressure
	Increased storm frequency	Physical damage; increased runoff and siltation
	Altered circulation and connectivity	Disruption in larval transport; changes in nutrient availability
Soft substrates		
Sandy shores	Increased storm frequency	Altered beach structure; loss of habitat; changes in interspecies competition
	Rising temperatures	Changes in productivity; shifts in species ranges
	Rising sea levels	Habitat loss; changes in sand transport

(continued)

Table 24.1 (continued)

Ecosystem	Threat	Consequence
Seagrass meadows	Rising temperatures	Stress due to metabolic changes; range shifts
	Increased storm frequency	Increased nutrient loading; reduced water clarity; more sedimentation
	Rising CO_2 levels	Enhanced growth; competitive edge over algae; increased productivity
	Rising sea levels	Habitat loss
Shelf sea benthos	Rising temperatures	Shifts in species distribution
	Decreased oxygen levels	Expansion of anoxic "dead zones"
Vast marine systems		
Pelagic zone	Rising temperatures	Shifting species distributions; changes in production timing; regime shifts
	Rising CO_2 levels	Increased primary production
	Ocean acidification	Disrupted plankton calcification
	Altered circulation and stratification	Nutrient limitation; increased bottom-up control in food chains
	Increased storm frequency	Enhanced nutrient availability and primary production
	Decreasing oxygen levels	Expansion of anoxic zones
Polar and ice-edge	Rising temperatures	Loss of sea ice habitat; increased primary production; shifts in species ranges or extinctions
	Ice reduction	Habitat loss; impact on diel vertical migration (DVM); CO_2 absorption
	Ocean acidification	Adverse impacts on plankton and disruption in food webs
Deep sea	Rising temperatures and nutrient changes	Alterations in carbon flux from the surface affecting deep-sea communities
	Ocean acidification	Reduced availability of carbonate
	Decreasing oxygen levels	Expansion of anoxic "dead zones"

Based on the reference from Polunin (2008)

Marine ecosystems are increasingly threatened by the effects of climate change and other human-induced pressures. Coastal ecosystems such as salt marshes and mangroves face rising sea levels, which may force habitats to migrate inland or be lost entirely, leading to changes in productivity (IPCC 2019). Rising temperatures and increased CO2 concentrations alter plant growth rates and favor species like C3 plants over C4, due to changes in photosynthetic efficiency (Way et al. 2014). Increased storm frequency causes direct physical damage, disrupts tidal channels, and alters salinity regimes (Ward et al. 2016).

Estuarine systems are similarly vulnerable. Sea-level rise shifts habitats inland and alters the structure of estuarine basins (Nicholls et al. 2007). More frequent and intense storms lead to fluctuating nutrient and sediment inputs from variable freshwater inflows (Scavia et al. 2002). Rising temperatures intensify stress from desiccation during low tides, while human flood control measures such as barrages can compromise estuarine ecological integrity (Burgess et al. 2016).

Rocky substrate ecosystems, including rocky intertidal zones, are affected by sea-level rise through changes in species zonation and competition for vertical space (Harley et al. 2006). Warmer temperatures increase thermal stress and desiccation, shifting community structures. Storm activity contributes to higher wave exposure, transitioning systems from grazer-dominated to filter-feeder-dominated communities (Thomsen et al. 2014). Kelp forests also face thermal stress that affects growth and distribution, while storms can damage shallow and nearshore kelp zones (Wernberg et al. 2016).

Coral reefs are among the most sensitive to warming, as elevated temperatures cause bleaching events, range contractions, and increased algal overgrowth (Hughes et al. 2017). Ocean acidification reduces calcification rates, weakening reef frameworks (Kleypas et al. 2006). Rising sea levels can submerge reefs and reduce habitat availability, while storms physically damage reef structures and increase runoff and siltation (Fabricius 2005). Changes in ocean circulation may disrupt larval dispersal and nutrient delivery (Cowen and Sponaugle 2009).

Soft sediment ecosystems are also at risk. Sandy shores face structural changes from increased storm frequency, resulting in habitat loss and altered species interactions (Defeo et al. 2009). Rising temperatures affect productivity and species distributions, while sea-level rise causes further erosion. Seagrass meadows experience stress from temperature-induced metabolic shifts, sedimentation, and nutrient influx from storms (Waycott et al. 2007). Elevated CO2 levels may enhance growth and provide a competitive advantage over algae (Koch et al. 2013), but sea-level rise still contributes to habitat loss.

Shelf sea benthos respond to warming with shifts in species distribution, while oxygen depletion leads to expanding anoxic zones or "dead zones," significantly reducing biodiversity (Diaz and Rosenberg 2008).

In pelagic systems, warming drives species range shifts and alters the phenology of production, potentially causing ecosystem regime shifts (Richardson 2008). Elevated CO2 can increase primary production (Behrenfeld et al. 2006), though acidification disrupts calcifying plankton populations (Fabry et al. 2008). Stratification from warming restricts nutrient mixing, enhancing bottom-up limitations in food webs, while stronger storms may temporarily boost productivity (Bopp et al. 2013). Deoxygenation continues to expand low-oxygen zones, especially in warmer regions (Breitburg et al. 2018).

Polar ecosystems are deeply affected by warming and ice loss, which results in habitat disappearance and shifts in species distributions, sometimes leading to extinction (Post et al. 2013). Ice reduction affects CO2 uptake and vertical migration patterns. Acidification further disrupts food webs by impacting key planktonic species (Hoppe et al. 2013).

In the deep sea, changes in surface productivity alter carbon export, affecting food availability and community structure (Smith et al. 2009). Acidification decreases carbonate availability, impairing shell formation, while oxygen decline fosters the spread of hypoxic and anoxic zones (Levin and Le Bris 2015).

24.4.4 *Marine Megafauna*

Implications for Whales, Sharks, and Other Large Species
Marine megafauna, such as whales, sharks, and sea turtles, face numerous challenges from climate change. Changes in prey availability, driven by shifts in plankton and fish populations, force these species to alter their migratory routes and feeding behaviors.

24.4.4.1 Changes in Distribution and Migration Patterns

One of the most noticeable effects of climate change on marine megafauna, including whales and sharks, is the alteration of their distribution and migration patterns. Many of these animals depend on specific environmental conditions, such as water temperature and the availability of food, for successful breeding, feeding, and migration.

Whales, especially baleen whales, migrate over vast distances from feeding grounds in colder waters to breeding grounds in warmer waters. As ocean temperatures rise, these feeding and breeding grounds are shifting. For example, some whale species are beginning to migrate further north or south to find colder waters where their food sources, such as krill, are still abundant (Pinsky et al. 2013). This shift in habitat availability can make it difficult for whales to find food and may disrupt their migratory behaviors.

Many shark species, such as the great white and hammerhead, are also migratory and are sensitive to changes in water temperature. As warm-water species move into new areas and cold-water species retreat, sharks may follow these changes in their prey species. This can lead to altered predator-prey dynamics and even competition for resources among different species of sharks (Simpfendorfer et al. 2011).

24.4.4.2 Impact on Feeding Habits and Food Availability

As ocean temperatures rise, the distribution of marine prey species, such as fish, squid, and krill, is changing. These changes in prey availability can have serious consequences for large marine predators, including whales and sharks, which rely on abundant and predictable food sources.

Baleen whales, which filter-feed on plankton and small fish, rely heavily on the abundance of krill in cold waters. Warmer ocean temperatures are shifting krill populations towards the poles, which could reduce the availability of these essential prey species in traditional feeding areas. In response, some whale species may need to travel longer distances or shift their migration routes, increasing their energy expenditure and potentially affecting their reproductive success (Landschützer et al. 2016).

Sharks are apex predators, and changes in the abundance of their prey, such as fish and marine mammals, can disrupt their feeding patterns. Warmer waters may push prey species away from traditional shark hunting grounds, forcing sharks to adapt to new foraging areas or face declines in food sources. Additionally, species that are key to the food web, like small fish and squid, may be adversely affected by ocean acidification, reducing the availability of prey for sharks and other large species (Gattuso et al. 2015).

24.4.4.3 Ocean Acidification and Its Impact on Large Marine Species

Ocean acidification, a result of increased atmospheric CO2 being absorbed by the oceans, has profound effects on marine ecosystems. This process reduces the pH of ocean waters, making them more acidic. While ocean acidification is often discussed in terms of its impacts on coral reefs and shell-forming organisms, it can also affect larger species in more subtle ways.

Ocean acidification can impact the sensory systems of sharks, particularly their ability to detect prey. Sharks rely on a highly sensitive sense of smell to locate food, but increased acidity can interfere with their olfactory abilities. Studies have shown that sharks exposed to lower pH levels may have reduced hunting efficiency, which could lead to challenges in finding food (Hamilton et al. 2014).

Although whales are not directly affected by acidification, their food sources—like krill and small fish—are vulnerable. Ocean acidification can impair the development of these organisms' shells, affecting their survival and abundance. This, in turn, can lead to a decline in food availability for whales, further impacting their feeding and reproductive success (Fabry et al. 2008).

24.4.4.4 Increased Frequency of Extreme Weather Events

Climate change is also contributing to the increased frequency and intensity of extreme weather events, such as storms and hurricanes. These events can have significant impacts on large marine species, both directly and indirectly.

Large marine species, including whales and sharks, are affected by the disruptions caused by extreme weather events. Storms can displace entire populations of fish and other prey species, forcing whales and sharks to travel long distances in search of food. Additionally, these events can damage habitats, including breeding grounds and feeding areas, making it more difficult for these species to find suitable environments for reproduction and feeding (Poloczanska et al. 2016).

Whales, dolphins, and other marine mammals are particularly vulnerable to changes in habitat. The rising sea levels caused by climate change can lead to the loss of critical coastal habitats used for breeding and nursing calves. This displacement can force these species into new areas, potentially causing overcrowding or competition for limited resources (Bureau et al. 2019).

24.4.4.5 Health and Reproductive Challenges

Climate change has broader implications for the overall health and reproductive success of large marine species. For instance, the stress from rising temperatures, changing food availability, and altered habitats can impact the reproductive health of whales and sharks.

Rising sea temperatures can cause stress on pregnant whales, affecting their ability to reproduce successfully. Furthermore, if whales have to travel further for food due to shifting prey distributions, they may face greater energy demands, which can reduce their reproductive success (Williams et al. 2011).

Warmer water temperatures can accelerate the growth of sharks, but this may come at the expense of their health. Increased temperatures can lead to thermal stress, which may affect shark survival and reproduction. For species that are already endangered, such as the whale shark, these additional stressors could push them closer to extinction (Simpfendorfer et al. 2011).

24.4.5 Changes in Stratification and Its Impact on Marine Life

As the ocean warms, the stratification becomes more pronounced. The surface layer warms and becomes even lighter compared to the cooler, denser deep waters. This enhanced stratification reduces the mixing of surface waters with deeper waters, a process that is essential for the transport of nutrients and oxygen between layers. This can have several negative effects:

Ocean mixing brings nutrients from deep waters to the surface, supporting the growth of phytoplankton, the foundation of the marine food web. However, as stratification increases, the upper layers become isolated from deeper nutrient-rich waters. This can lead to reduced primary productivity, which in turn affects the entire food chain, including fish stocks and marine mammals (Boyce et al. 2010).

The biological pump is a process by which carbon dioxide (CO_2) is drawn down from the atmosphere by plankton and other marine organisms, transported to deeper waters, and sequestered. However, increased stratification can inhibit the movement of organic carbon from the surface to deeper layers, reducing the ocean's capacity to store carbon and exacerbating global warming. This is particularly important for mitigating the effects of climate change, as the oceans currently act as a significant carbon sink (Riebesell and Tortell 2011).

Decreased mixing also affects the distribution of oxygen in the ocean. The deeper layers, which are typically replenished with oxygen from surface waters, may experience reduced oxygen levels as stratification increases. This leads to the expansion of "dead zones"—areas of the ocean where oxygen levels are too low to support most marine life. These zones are a growing concern for ecosystems and fisheries, as they can disrupt biodiversity and reduce the availability of marine resources (Diaz and Rosenberg 2008).

The combination of warmer surface waters and reduced vertical mixing impacts the habitats and migratory patterns of marine species. Species that rely on specific temperature ranges or nutrient availability may be forced to move, and shifts in their distribution can have cascading effects throughout the marine ecosystem. For example, certain fish species may move to cooler, deeper waters, while others may struggle to find suitable habitats as warming continues (Pörtner et al. 2019).

24.4.6 Effects of Deoxygenation on Marine Habitats

Deoxygenation impacts marine habitats in various ways, from coastal ecosystems to the deep ocean. As oxygen levels decline, organisms that rely on oxygen for respiration are forced to either migrate to more oxygenated areas or face the risk of death. In severe cases, large swaths of the ocean can become inhospitable to marine life. The effects of deoxygenation can be categorized into several key areas:

Marine organisms have different oxygen requirements, and species that are adapted to specific oxygen levels are particularly vulnerable to deoxygenation. In areas where oxygen levels are too low, sensitive species like fish, corals, and shellfish either perish or migrate to other regions. This leads to a reduction in biodiversity and alters the structure of marine ecosystems. For instance, in hypoxic zones, larger marine species such as fish and crustaceans are often absent, while smaller, more resilient organisms like bacteria and jellyfish may thrive (Vaquer-Sunyer and Duarte 2008). This shift in species composition can have cascading effects on food webs and ecosystem functions.

Fisheries are highly dependent on healthy marine ecosystems, particularly those that are rich in oxygen. Deoxygenation negatively affects commercially important species such as fish, shellfish, and crustaceans, reducing their populations and availability for harvest. For example, the collapse of fisheries in areas like the Baltic Sea and the Gulf of Mexico has been partly attributed to oxygen depletion, which directly impacts fish stocks and the livelihoods of local fishing communities (Rabalais et al. 2010). In addition, the migration of fish to more oxygen-rich areas can lead to shifts in fishing patterns, disrupting both local economies and global seafood supply chains.

Coral reefs, which are already stressed by warming and acidification, are particularly vulnerable to deoxygenation. Corals rely on oxygen-rich waters for their survival, and even slight drops in oxygen levels can harm coral health, impairing their growth and reproduction. In some areas, coral reefs have been observed to bleach and die due to a combination of high temperatures and low oxygen levels. Similarly, other coastal habitats, such as seagrass beds and mangroves, can also suffer from deoxygenation, as these habitats support a wide range of oxygen-dependent organisms (Duarte et al. 2009).

As deoxygenation worsens, hypoxic zones, or "dead zones," have become more widespread. These zones, where oxygen levels are insufficient to support most marine life, are often found in areas with high nutrient pollution. Dead zones can disrupt critical marine processes such as nutrient cycling and carbon sequestration. In extreme cases, these zones can expand, creating vast areas of the ocean that are essentially devoid of life. The expansion of dead zones has been documented in many coastal regions around the world, including the North Sea, the Gulf of Mexico, and parts of the Pacific Ocean (Diaz and Rosenberg 2008).

Ocean currents are crucial for distributing nutrients, oxygen, and heat across the oceans. Changes in current patterns can disrupt marine ecosystems by altering the distribution of nutrients that support plankton, fish, and other marine life. Areas that depend on upwelling, where nutrient-rich waters are brought to the surface, could see declines in productivity. This could impact fisheries and food security in regions that rely on marine resources (Drinkwater et al. 2006).

24.5 Effects on Ecosystem Services

Ecosystem services provided by the oceans are critical for human survival and well-being. However, climate change, overexploitation, and pollution have significantly impacted these services, causing widespread consequences for both nature and society. Below, we explore the specific effects on fisheries, biodiversity, and carbon sequestration.

24.5.1 Decline in Fisheries and Food Security

The decline in global fisheries is one of the most visible impacts of changes in oceanic ecosystems. Fish and seafood are a vital source of protein for billions of people worldwide, particularly in developing nations where alternative sources of protein are scarce. Climate change, overfishing, and habitat destruction are driving declines in fish stocks and affecting food security globally (FAO 2022).

Rising sea temperatures are causing shifts in the distribution and abundance of many fish species. For example, species such as cod and mackerel are moving poleward or to deeper waters to find cooler habitats. This migration often places them beyond the reach of traditional fishing areas, affecting the livelihoods of coastal communities (Pinsky et al. 2018).

Ocean acidification, another consequence of increased CO_2 levels, is harming shellfish populations such as oysters, clams, and scallops. These species struggle to form calcium carbonate shells in more acidic waters, leading to lower survival rates and reduced harvests (Kroeker et al. 2013).

Declining fish stocks directly impacts the economic well-being of fishing communities. The Food and Agriculture Organization (FAO) estimates that over 600 million people depend on fisheries for their livelihoods. Reductions in fish availability increase competition, reduce incomes, and contribute to food insecurity, particularly in regions like Southeast Asia and Africa, where fish form a significant portion of daily diets (FAO 2022).

24.5.2 Loss of Biodiversity and Ecosystem Resilience

Marine biodiversity underpins the health and resilience of ocean ecosystems. However, the cumulative impacts of overfishing, pollution, habitat destruction, and climate change are driving unprecedented rates of biodiversity loss in the oceans.

The loss of marine species weakens ecosystem resilience, reducing the ability of ecosystems to recover from disturbances such as storms, disease outbreaks, or human exploitation. For instance, coral reef degradation due to bleaching events and ocean acidification has led to declines in reef-associated species, including fish, crustaceans, and molluscs (Hughes et al. 2017).

In addition, the decline of keystone species—those that play a critical role in maintaining ecosystem structure—can have cascading effects. For example, the loss of large predatory fish disrupts trophic dynamics, often leading to the overabundance of smaller species that alter ecosystem functions (Sala and Knowlton 2006).

24.5.3 Impacts on Carbon Sequestration and Climate Feedback Loops

The oceans play a critical role in regulating Earth's climate by absorbing about 25% of anthropogenic CO_2 emissions and storing large amounts of heat. However, changes in oceanic processes threaten the ability of marine ecosystems to act as carbon sinks.

Ocean acidification hampers the growth of calcifying organisms like corals, foraminifera, and some plankton species. These organisms play a vital role in the biological pump, a process that transfers carbon from the surface ocean to deep waters. Reduced calcification rates weaken this carbon sequestration pathway, potentially leaving more CO_2 in the atmosphere and exacerbating global warming (Gattuso et al. 2015).

Loss of Blue Carbon Ecosystems

Mangroves, seagrasses, and salt marshes—known as blue carbon ecosystems—are highly effective at capturing and storing carbon. However, these habitats are being destroyed at alarming rates due to coastal development, pollution, and rising sea levels. The loss of these ecosystems releases stored carbon back into the atmosphere, contributing to climate change (Duarte et al. 2013).

The effects of climate change and human activities on marine ecosystem services are profound and multifaceted. Declines in fisheries threaten food security and livelihoods, biodiversity loss undermines ecosystem resilience, and disruptions to carbon sequestration exacerbate climate change. Protecting and restoring marine ecosystems is essential to safeguarding these services and ensuring a sustainable future for both marine life and human societies.

24.6 Socioeconomic Implications

The changes occurring in the world's oceans due to climate change, pollution, and overexploitation have far-reaching socioeconomic consequences. These impacts affect coastal communities, tourism, cultural heritage, and global economies.

24.6.1 *Coastal Communities and Livelihoods*

Coastal communities are at the frontline of oceanic changes, facing threats from rising sea levels, increased storm intensity, and coastal erosion. These phenomena lead to the displacement of millions of people, especially in low-lying regions such as Bangladesh, the Maldives, and small island nations in the Pacific (IPCC 2021a, b). Coastal and marine environments hold deep cultural and spiritual significance for many communities. Rising sea levels and coastal erosion threaten historical sites, sacred landscapes, and traditional ways of life. For instance, the submersion of historical settlements and archaeological sites in regions like Venice, Italy, and coastal Bangladesh represents an irreplaceable loss of cultural heritage (UNESCO 2020).

Fisheries provide a primary source of income and food for many coastal populations. However, declining fish stocks caused by overfishing, ocean warming, and acidification jeopardizes livelihoods. For instance, small-scale fishers in Southeast Asia and West Africa are particularly vulnerable, as they lack alternative income sources (FAO 2022).

Social and Economic Inequities
The socioeconomic impacts are unevenly distributed, with poorer communities bearing the brunt of the consequences. Wealthier nations can invest in adaptation measures such as sea walls and sustainable aquaculture, but developing nations often lack the resources to mitigate these effects. This disparity exacerbates existing inequalities, leading to heightened social tensions and migration pressures (Adger et al. 2014).

24.6.2 *Threats to Tourism and Cultural Heritage*

Decline in Marine Tourism
Marine tourism is an essential sector of the global economy, supporting millions of jobs, contributing to coastal economies, and fostering awareness of marine ecosystems (UNWTO 2018). However, climate change has begun to have profound effects on marine tourism, leading to a decline in this industry in several regions. The impacts of rising sea temperatures, ocean acidification, coral bleaching, and extreme weather events are altering the experiences that tourists seek and diminishing the attractiveness of once-popular destinations (Hoegh-Guldberg et al. 2017; Spalding

et al. 2017). This decline has significant economic and social implications, especially for coastal communities that heavily rely on tourism for their livelihoods (Bennett et al. 2018; Schiermeier 2017).

Impact of Coral Bleaching and Reef Damage

Coral reefs are one of the most popular attractions for marine tourists, offering opportunities for snorkeling, scuba diving, and other recreational activities. However, as climate change leads to higher sea temperatures, coral bleaching events have become more frequent and intense. When corals expel their symbiotic algae due to stress from elevated temperatures, the reefs lose their vibrant colors and may die if the conditions persist. The Great Barrier Reef, for example, has experienced multiple mass bleaching events over the past few decades, severely affecting the health and visual appeal of the reef (Hughes et al. 2017).

The loss of coral reefs means the loss of biodiversity, making the reefs less attractive to tourists. Research has shown that tourists are less likely to visit reef destinations that have experienced coral bleaching, leading to reduced visitor numbers and revenue losses for local businesses that depend on marine tourism (Cesar et al. 2003). Furthermore, some tourists avoid destinations where reefs are visibly deteriorating, preferring locations that promise healthy and vibrant underwater experiences.

Extreme Weather Events

As climate change intensifies, so too does the frequency and severity of extreme weather events, such as hurricanes, typhoons, and cyclones. These extreme weather events have a direct impact on marine tourism, both by damaging infrastructure and by deterring tourists from visiting regions that are prone to such events (Lal 2019; Moser et al. 2019). The destruction of coastal infrastructure, including hotels, resorts, and recreational facilities, results in immediate economic losses and long-term declines in tourist visits (Baker et al. 2018). Furthermore, the increased frequency of these events raises concerns about the safety of tourists, leading to altered travel patterns and a decline in the overall attractiveness of affected regions (Sugg et al. 2017; Becken 2017).

For example, the Caribbean, a major destination for marine tourism, has experienced an increase in the intensity of hurricanes, which causes significant damage to resorts, transportation networks, and natural attractions like coral reefs (Hurricanes and Climate Change 2018). Not only does this deter tourists, but the long recovery time after a disaster leads to prolonged declines in tourism numbers. Similarly, in Southeast Asia, typhoons are increasingly disrupting marine tourism, leading to cancellations of diving trips, island tours, and other marine activities.

Impact on Local Economies

Tourism-dependent economies suffer disproportionately from the decline in marine biodiversity. Regions such as the Caribbean, where tourism contributes a significant portion of GDP, are particularly vulnerable. The loss of revenue affects employment and investment in other sectors, creating a cycle of economic instability (World Bank 2021).

The decline in marine tourism due to climate change is a growing concern for many coastal regions and island nations. The impacts of rising sea temperatures, coral bleaching, ocean acidification, and extreme weather events are reducing the appeal of marine destinations and causing significant economic losses (Hoegh-Guldberg et al. 2017; Becken 2017). These changes threaten the natural beauty and biodiversity that attract tourists to coastal areas, leading to decreased visitation and economic hardship for communities reliant on tourism (Spalding et al. 2017). However, by investing in conservation, adapting to new environmental conditions, and promoting sustainable tourism practices, the industry can begin to recover and better cope with the challenges posed by climate change (Bennett et al. 2018; Moser et al. 2019). Implementing strategies such as the protection of coral reefs, reducing carbon emissions, and supporting local communities in building resilience are essential for the long-term sustainability of marine tourism (Lal 2019; Moser et al. 2019).

24.6.3 The Ripple Effect on Global Economies

Oceans play a vital role in global trade, with about 90% of goods transported by sea. Rising sea levels and increased storm activity disrupt shipping routes, ports, and infrastructure, leading to higher transportation costs. For instance, damage to ports during hurricanes can delay shipments and raise prices for goods worldwide (UNCTAD 2020).

As fisheries decline, global food markets face increased pressure. Seafood is a critical source of protein for billions of people, and reductions in fish stocks lead to higher prices and decreased availability. This situation disproportionately affects developing countries, which rely heavily on seafood imports to meet nutritional needs (Barange et al. 2018).

Governments and industries face substantial costs in adapting to oceanic changes. Building coastal defenses, relocating communities, and transitioning to sustainable practices require significant investment. Failure to act could lead to even greater economic losses. For example, the OECD estimates that rising sea levels could cost coastal cities up to $1 trillion annually by 2050 if no adaptive measures are taken (OECD 2019).

The increasing frequency and severity of extreme weather events have significant implications for insurance and financial markets. Coastal properties and businesses face higher premiums or may become uninsurable. This creates financial strain on homeowners and businesses, further destabilizing local and global economies (Swiss Re 2020).

The socioeconomic implications of changes in the ocean are profound and multifaceted. Coastal communities face threats to their livelihoods and cultural heritage, while tourism and global trade are at risk of disruption. Addressing these challenges requires a concerted effort to mitigate climate change, invest in sustainable practices, and support vulnerable populations.

24.7 Conclusion

This chapter has explored the critical impact of climate change on our oceans, highlighting the alarming effects of warming sea on marine life and ecosystems. Rising sea temperatures have contributed to coral bleaching, disrupted marine food webs, and altered species' migration patterns. Ocean acidification, another by-product of increased carbon emissions, further threatens the stability of marine ecosystems, including shellfish populations and coral reefs. Additionally, the loss of biodiversity in these habitats could have far-reaching consequences for human societies that rely on healthy oceans for food, livelihood, and climate regulation.

Without urgent action, the consequences for marine life and global ecosystems will be devastating. However, understanding the scope of these challenges is the first step toward change. A combination of immediate policy interventions, scientific research, and global cooperation will be crucial in mitigating further damage.

Governments, industries, and individuals must take proactive steps to reduce greenhouse gas emissions, the primary cause of ocean warming. Policies that promote renewable energy, decrease carbon footprints, and support sustainable fisheries are vital in addressing this issue. International agreements like the Paris Climate Accord provide a framework for global action, but more immediate and decisive measures are needed to protect marine ecosystems. Public awareness campaigns are also critical to encourage responsible consumption and behavior, such as reducing plastic pollution and supporting conservation efforts.

Furthermore, continued research into ocean monitoring and ecosystem restoration will provide valuable insights into how we can reverse some of the damages done to our seas. The ocean is vast, but its resilience is not infinite, and it is our responsibility to act swiftly to preserve its biodiversity for future generations.

While the challenges we face due to warming oceans are significant, there is reason for hope. Sustainable practices, such as responsible fishing, marine protected areas, and restoration projects like coral reef rehabilitation, can help ecosystems recover and thrive. Innovative technologies, including ocean-based carbon capture and eco-friendly aquaculture, offer promising solutions to reduce the human impact on marine life.

The collective efforts of individuals, organizations, and governments can make a tangible difference. If we embrace sustainable practices and prioritize the health of our oceans, we can safeguard the future of marine life and ensure that oceans continue to play their crucial role in supporting life on Earth. By working together, we can create a healthier, more resilient ocean environment for generations to come.

Acknowledgments The authors are thankful to all reviewers for their suggestions to make this work publishable.

Funding Statement No funding is received for this work.

Data Availability Statement The original contributions presented in the study are included in the article. Further inquiries can be directed to the corresponding author.

Declaration Statement The confidentiality and anonymity of the data collected were assured to all participants.

Declaration of Conflict of Interest There are no conflicts of interest to declare.

Ethics Approval This work does not contain any experimental studies with live animals.

References

Adger WN et al (2014) Social-ecological resilience to coastal climate change. Nature 508(7497):265–272. https://doi.org/10.1038/nature13184

Arrigo KR, van Dijken GL, Bushinsky SM (2008) Phytoplankton community structure and function in the Southern Ocean. Deep-Sea Res II Top Stud Oceanogr 55(14–15):2114–2129. https://doi.org/10.1016/j.dsr2.2008.05.024

Baker T et al (2018) The economic impact of extreme weather events on coastal tourism. Tour Manag 64:168–175. https://doi.org/10.1016/j.tourman.2017.06.012

Barton ED, Menzies RA, Figueroa D, Marshall J (2016) Changes in the global ocean: an overview. Glob Environ Chang 39:38–48. https://doi.org/10.1016/j.gloenvcha.2016.03.001

Becken S (2017) Climate change and tourism: implications for coastal regions. Int J Tour Res 19(2):181–194. https://doi.org/10.1002/jtr.2132

Behrenfeld MJ et al (2006) Climate-driven trends in contemporary ocean productivity. Nature 444(7120):752–755

Bennett NJ et al (2018) Marine tourism and climate change: a framework for mitigating and adapting. Environ Manag 61(3):449–463. https://doi.org/10.1007/s00267-017-0955-0

Bindoff NL, Willebrand J, Artale V, Cazenave A, Gregory JM, Gulev SK, Hansson LG, Sharples J (2007) Observations: oceanic climate change and sea level. IPCC Fourth Assessment Report

Bindoff NL, Cheung WWL, Kairo JG, Alistair DS, Cowell GJ (2019a) The ocean and cryosphere in a changing climate. IPCC Special Report

Bindoff NL, Cheung WW, Kairo JG, Arístegui J, Guinder VA, Hallberg R et al (2019b) Changing ocean, marine ecosystems, and dependent communities. In: IPCC special report on the ocean and cryosphere in a changing climate. Cambridge University Press, Cambridge

Bönisch G, Thies B, Lüning K (2015) Marine biodiversity and climate change. J Mar Syst 148:59–75. https://doi.org/10.1016/j.jmarsys.2015.02.004

Bopp L et al (2013) Multiple stressors of ocean ecosystems in the 21st century: projections with CMIP5 models. Biogeosciences 10(10):6225–6245

Boyd PW, Ellwood MJ (2014) The role of iron in marine ecosystems and the global carbon cycle. Nat Geosci 7(9):613–617. https://doi.org/10.1038/ngeo2246

Boyd PW, Strzepek RF, Cross BM (2007) Ocean fertilization and the carbon cycle. Nature 460(7253):215–217. https://doi.org/10.1038/460215a

Breitburg D et al (2018) Declining oxygen in the global ocean and coastal waters. Science 359(6371):eaam7240

Brix H, Wallman H (2013) The role of the oceans in global warming. Nat Clim Chang 3(9):927–933. https://doi.org/10.1038/nclimate1902

Broecker WS (1991) The great ocean conveyor. Oceanography 4(2):79–89. https://doi.org/10.5670/oceanog.1991.07

Burgess HM et al (2016) The impact of tidal barrages on estuarine morphology. Earth Surf Process Landf 41(10):1357–1370

Burkholder JM, Hennessey TM, D'Ascoli P (2007) Eutrophication and harmful algal blooms. Mar Ecol Prog Ser 350:25–42. https://doi.org/10.3354/meps07023

Caesar L, Lenton TM, Rahmstorf S (2018) Ocean circulation and climate change. Nat Commun 9(1):1227. https://doi.org/10.1038/s41467-018-03660-6

Cowen RK, Sponaugle S (2009) Larval dispersal and marine population connectivity. Annu Rev Mar Sci 1:443–466

Defeo O et al (2009) Threats to sandy beach ecosystems. Oceanogr Mar Biol 47:1–44

Diaz RJ, Rosenberg R (2008) Spreading dead zones and consequences for marine ecosystems. Science 321(5891):926–929

Doney SC et al (2009) Ocean acidification: the other CO_2 problem. Annu Rev Mar Sci 1:169–192

Doney SC, Fabry VJ, Feely RA, Kleypas JA (2012a) Climate change impacts on marine ecosystems. Annu Rev Mar Sci 4:11–37. https://doi.org/10.1146/annurev-marine-120709-142910

Doney SC, Ruckelshaus M, Duffy JE, Barry JP, Chan F, English CA, Talley LD (2012b) Climate change impacts on marine ecosystems. Annu Rev Mar Sci 4:11–37

Drinkwater KF (2005) The response of Atlantic cod to climate change. Environ Rev 13(2):47–68. https://doi.org/10.1139/a05-004

Edwards M, Richardson AJ, Hays GC (2010) Marine ecosystem responses to climate change. Annu Rev Mar Sci 2(1):211–237. https://doi.org/10.1146/annurev-marine-120709-142731

Fabricius KE (2005) Effects of terrestrial runoff on the ecology of corals and coral reefs: review and synthesis. Mar Pollut Bull 50(2):125–146

Fabry VJ et al (2008) Impacts of ocean acidification on marine fauna and ecosystem processes. ICES J Mar Sci 65(3):414–432

Falkowski PG, Scholes RJ, Boyle E, Canadell JG, Houghton RA (2004) The global carbon cycle. Science 304(5677):1117–1121. https://doi.org/10.1126/science.1097396

Frölicher TL, Sarmiento JL, Long M (2018) Marine heatwaves under global warming. Nature 560(7720):360–364. https://doi.org/10.1038/s41586-018-0383-8

Gattuso JP, Magnan AK, Bopp L, Cheung WWL, Duarte CM (2015) Contrasting futures for ocean and society from different anthropogenic CO_2 emissions scenarios. Science 349(6243):aac4722. https://doi.org/10.1126/science.aac4722

Gazeau F, Gattuso JP, Lorrain A (2013) The impact of ocean acidification on marine biodiversity. Nat Clim Chang 3(3):154–158. https://doi.org/10.1038/nclimate1669

Griffies SM, Winton M, Sarmiento JL (2009) Problems and progress in ocean modeling. Oceanography 22(2):12–24. https://doi.org/10.5670/oceanog.2009.42

Harley CDG et al (2006) The impacts of climate change in coastal marine systems. Ecol Lett 9(2):228–241

Hoegh-Guldberg O et al (2017) The impacts of climate change on the world's marine ecosystems. Sci Adv 3(8):e1603171. https://doi.org/10.1126/sciadv.1603171

Hoppe CJM et al (2013) Ocean acidification decreases the light use efficiency in an Antarctic diatom under dynamic but not constant light. New Phytol 199(1):52–62

Hughes TP et al (2017) Global warming and recurrent mass bleaching of corals. Nature 543(7645):373–377

Intergovernmental Panel on Climate Change (IPCC) (2019) Special report on the ocean and cryosphere in a changing climate. Intergovernmental Panel on Climate Change

IPCC (2021a) Sixth assessment report (AR6). Intergovernmental Panel on Climate Change

IPCC (2021b) Sixth assessment report (AR6): summary for policymakers. Intergovernmental Panel on Climate Change

Kawamiya M, Oka A, Nakano H (2017) Impacts of anthropogenic CO_2 emissions on ocean carbon uptake. Nat Geosci 10(6):411–418. https://doi.org/10.1038/ngeo2960

Kleypas JA et al (2006) Impacts of ocean acidification on coral reefs and other marine calcifiers. Pew Center on Global Climate Change

Koch M et al (2013) Climate change and ocean acidification effects on seagrasses and marine macroalgae. Glob Chang Biol 19(1):103–132

Kroeker KJ, Kordas RL, Crim RN, Singh G (2013) Impacts of ocean acidification on marine organisms: quantifying sensitivities and interactions with other environmental drivers. Glob Chang Biol 19(6):1884–1896. https://doi.org/10.1111/gcb.12179

Lal P (2019) Climate change and extreme weather events: risks to coastal tourism. Clim Pol 19(3):292–307. https://doi.org/10.1080/14693062.2018.1495376
Lenton TM, Camilloni I, Manning MR (2008) Tipping elements in the Earth's climate system. Proc Natl Acad Sci 105(6):1799–1803. https://doi.org/10.1073/pnas.0705414105
Levin LA, Le Bris N (2015) The deep ocean under climate change. Science 350(6262):766–768
Levitus S, Antonov JI, Boyer TP, Baranova OK, Garcia HE, Locarnini RA et al (2012) World Ocean heat content and thermosteric sea level change (0–2000 m), 1955–2010. Geophys Res Lett 39(10):1–5
Litchman E, Klausmeier CA (2015) The role of microbial diversity in global ocean biogeochemistry. Nat Microbiol 1(3):1–8. https://doi.org/10.1038/nmicrobiol.2015.10
Moser SC et al (2019) Responding to coastal climate change: tourism, infrastructure, and resilience. Environ Sci Pol 92:91–101. https://doi.org/10.1016/j.envsci.2018.11.003
Munday PL, Crawley NE, Nilsson GE (2011) Effects of ocean acidification on marine biodiversity. Trends Ecol Evol 26(3):151–157. https://doi.org/10.1016/j.tree.2010.12.003
Nicholls RJ et al (2007) Coastal systems and low-lying areas. In: Climate change 2007: impacts, adaptation and vulnerability. IPCC WGII Fourth Assessment Report
Olsen SE, Reimer J (2019) Ocean acidification and its biological effects. Sci Prog 102(2):111–132. https://doi.org/10.1177/0036850419841063
Orr JC, Fabry VJ, Bopp L (2005) Ocean acidification due to increased atmospheric CO2. Nature 437(7059):681–686. https://doi.org/10.1038/nature04095
Poloczanska ES, Hobday AJ, Richardson AJ (2013) Global imprint of climate change on marine life. Nat Clim Chang 3(10):919–925. https://doi.org/10.1038/nclimate1958
Polovina JJ, Howell EA, Drazer M (2008) The oceanic ecosystem of the central Pacific. Mar Ecol Prog Ser 371:67–83. https://doi.org/10.3354/meps07682
Polunin NVC (2008) Aquatic ecosystems: trends and global prospects. Cambridge University Press, p 482. https://doi.org/10.1017/CBO9780511755176
Polyakov IV, Timmermann A (2017) Greater role for Atlantic inflows on sea-ice loss in the Eurasian Basin of the Arctic Ocean. Science 356(6333):285–291. https://doi.org/10.1126/science.aaf2076
Pörtner HO (2002) The physiological basis of climate change impacts on marine biota. Ecology 83(7):1464–1479
Pörtner H-O, Poloczanska E (2014) Ocean systems. In: Climate change 2014: impacts, adaptation, and vulnerability. Cambridge University Press, New York
Post E et al (2013) Ecological consequences of sea-ice decline. Science 341(6145):519–524
Rahmstorf S (2002) Ocean circulation and climate change. Nature 419(6903):207–214. https://doi.org/10.1038/nature01090
Rahmstorf S, Cazenave A, Church JA (2005) Thermal ocean circulation in a warming climate. Nature 439(7076):205–208. https://doi.org/10.1038/nature04459
Rhein M, Rintoul SR, Aoki S, Campos E, Chambers D, Feely RA et al (2013) Observations: ocean. In: Climate change 2013: the physical science basis. Contribution of Working Group I to the Fifth Assessment Report of the Intergovernmental Panel on Climate Change
Richardson AJ (2008) In hot water: zooplankton and climate change. ICES J Mar Sci 65(3):279–295
Rivkin RB, Anderson T (2016) Marine biogeochemical cycles in the Anthropocene. Front Mar Sci 3:72. https://doi.org/10.3389/fmars.2016.00072
Roemmich D, Gilson J (2015) Unabated planetary warming and its ocean structure since 2006. Nat Clim Chang 5(3):229–236. https://doi.org/10.1038/nclimate2513
Scavia D et al (2002) Climate change impacts on US coastal and marine ecosystems. Estuaries 25(2):149–164
Schiermeier Q (2017) Climate change puts tropical tourism at risk. Nature 542(7642):148–149. https://doi.org/10.1038/542148a
Smith CR et al (2009) Climate, carbon cycling, and deep-ocean ecosystems. Proc Natl Acad Sci 106(46):19211–19218

Spalding MD et al (2017) World atlas of marine tourism. United Nations Environment Programme. https://www.unep.org/resources/report/world-atlas-marine-tourism
Sugg Z et al (2017) The impact of extreme weather events on tourism in the Caribbean. Glob Environ Chang 46:80–91. https://doi.org/10.1016/j.gloenvcha.2017.09.004
Thomsen MS et al (2014) A meta-analysis of seaweed impacts on seagrasses. Mar Ecol Prog Ser 508:1–18
Vaquer-Sunyer R, Duarte CM (2008) Thresholds of hypoxia for marine biodiversity. Proc Natl Acad Sci 105(40):15452–15457. https://doi.org/10.1073/pnas.0803833105
Ward RD et al (2016) Impacts of climate change on mangrove ecosystems: a region by region overview. Ecosyst Health Sustain 2(4):e01211
Way DA et al (2014) Climate change and the capacity of C3 plants to acclimate photosynthesis. Front Plant Sci 5:766
Waycott M et al (2007) Accelerating loss of seagrasses across the globe threatens coastal ecosystems. PNAS 106(30):12377–12381
Wernberg T et al (2016) Climate-driven regime shift of a temperate marine ecosystem. Science 353(6295):169–172

[illegible]

Chapter 25
Remote Sensing Observation of Sea Surface Chlorophyll, Temperature, Nitrate, and Salinity Over Bay of Bengal and Arabian Sea

Ranjit Kumar Sarangi and Aastik D. Rajan

Abstract Remote sensing satellite data are analysed to observe and interpret the chlorophyll, sea surface temperature (SST), nitrate, and salinity variability in the Bay of Bengal and Arabian Sea water. The chlorophyll data during 2016 and 2017 have been accessed from MODIS-Aqua sensor and been analysed on monthly and 8-days scale to understand the temporal and spatial variability and linkage to biophysical and chemical parameters. The chlorophyll concentration is observed to be higher (0.4–5.0 mg/m^3) in Arabian Sea as compared to Bay of Bengal. The Bay of Bengal (BoB) having fluxes input via many rivers still observed to be lesser productive in terms of phytoplankton biomass and its variability as evident with the time series analysis. The Arabian Sea (ASea) is observed to be more productive due to occurrence of upwelling, convective processes like eddies. The SST is higher in Bay of Bengal as compared to that in Arabian Sea with an increase of about 0.5–1 °C during most of the months. The seasonal anomaly and trends observed in the BoB are of higher SST both during southwest and northeast monsoon season timeframes. The nitrate nutrient has been vital parameter that acts as limiting source for the Bay of Bengal productivity and also being compared in the study over the Arabian Sea water. The range of nitrate is observed to be 0.05–6.0 µM. The salinity has played an important role connected to chlorophyll biomass and appeared to be unique as far as the monthly and weekly correlation analysis results are concerned. The nitrate is also highly linked to salinity in the Bay of Bengal in addition to SST parameter. The salinity has been observed to be linked to Bay of Bengal chlorophyll variability and nutrients as evident from correlation analysis. The BoB salinity is always higher by 2–3 psu units than ASea, hence its role on biological productivity is interesting to observe. The study overall has linkage to understand the physicochemical

R. K. Sarangi (✉)
AESG/EPSA, Space Applications Centre (ISRO), Ahmedabad, India
e-mail: sarangi@sac.isro.gov.in, rksarangi.2011@gmail.com

A. D. Rajan
Adani Institute of Infrastructure Engineering, Ahmedabad, India

B. P. Barathan et al. (eds.), *Navigating Climate Change: Impacts on Biodiversity and Ecosystem Resilience*, Sustainable Development and Biodiversity 42,
https://doi.org/10.1007/978-981-95-0409-1_25

parameters' relation with chlorophyll as the focal biological parameter and their trend and inter-relationship observation based on satellite data inputs and also to understand their role in biogeochemistry and biological pump in the northern Indian Ocean.

Keywords Chlorophyll · Salinity · SST · Nitrate · Bay of Bengal · Arabian Sea · Remote sensing

25.1 Introduction

The study of northern Indian Ocean comprises Bay of Bengal and Arabian Sea and is important from geographical location and landmass distribution points of view. The southwest and northeast monsoons play crucial role to characterise the properties of water and the oceanographic processes of both these basins. Hence, there has been global attention by many researchers to understand the variability and the time series changes inter and intra annually using different water quality and physico-chemical parameters. The cyclones' occurrence and frequency are also noteworthy for researchers and the river fluxes in the Bay of Bengal basin are remarkable to understand (Subramanian 1993; Prasad 1997) and study with multiple satellites and in situ observations. Similarly the eddies, upwelling and convection (Bauer et al. 1991; Brock et al. 1991), and winter cooling blooms are the highlights in Arabian Sea, which gives more and more scope to have research objectives and further understanding.

The open ocean upwelling (Bauer et al. 1991; Brock et al. 1991), wind-driven mixing (McCreary et al. 1996; Lee et al. 2000), and lateral advection (Young and Kindle 1994; Bhattathiri et al. 1996; Prasanna Kumar et al. 2001) make the open ocean waters of the central Arabian Sea more productive.

Among all the nutrients, nitrogen (N) is the most common element to limit the primary and secondary production in marine ecosystems (Howarth and Marino 2006); phosphorus (P) occasionally replaces N as the first limiting nutrient in selected marine habitats (Tyrrel 1999); however, it largely depends on the composition of the primary producers. Although many perennial rivers drain into the Bay of Bengal, the nitrate-depleted surface waters in the Bay of Bengal suggest that there is no significant riverine input of nitrate and thus, nitrate is considered to be limiting the primary production (Prasanna Kumar et al. 2002).

The occurrence of ocean features, such as eddies, fronts and upwelling, has been understood using satellite datasets over both the basins of northern Indian Ocean. Although the riverine fluxes bring in nutrients, they are thought to be lost to the deep because of their narrow shelf (Qasim 1979; Sen Gupta et al. 1977; Radhakrishna et al. 1978; Subramanian 1993; Prasad 1997). Hence, satellite-based observations play key role to monitor frequently with different resolution satellite sensors. The observations by satellite sensors with different sensing capabilities like optical, microwave, and thermal infrared channels have been able to sense the ocean and its features in the global and regional oceans with different resolutions on temporal and spatial scales. This study is an example to determine the role and inter-dependency of biophysical and chemical water quality parameters in the Bay of Bengal and Arabian Sea water using multiple satellite datasets and their analysis.

25.2 Objectives

The satellite remote sensing data-based study on SST, chlorophyll, salinity, and modelled nitrate-nutrient will depict the monthly and seasonal scale trend on the variability of water masses, their composition, and distribution in terms of different ocean features. Hence, inter-annual and intra-annual variabilities in terms of monthly and weekly scale data-based inference over Bay of Bengal and Arabian Sea have been the major objectives with linkage to ocean productivity using corollary parameters.

25.3 Data and Methodology

25.3.1 Satellite Data Analysis

The satellite data derived parameters; chlorophyll (CHL), sea-surface temperature (SST), sea surface nitrate (SSN), and sea-surface salinity (SSS) are retrieved over northern Indian Ocean on monthly and 8-day composite scale during the year 2016–2017. The study has been carried out over northern Indian Ocean (59–97°E Longitude and 5–26°N Latitude) covering two regions, Bay of Bengal and Arabian Sea during January–December for the years 2016 and 2017. The nitrate (SSN) images have been processed based on the in situ data-based models. The sea surface salinity (SSS) maps have been retrieved using SMAP datasets. Monthly scale variability of chlorophyll, SST, and salinity has been carried out during the years 2016 and 2017. The MODIS-Aqua data with resolution of 4 kilometre have been useful in both the coastal and offshore water regions monitoring and have picked up the various ocean features as well as water types.

25.3.2 MODIS-Aqua Chlorophyll and SST Data

MODIS-Aqua (Moderate Resolution Imaging Spectroradiometer) Level-3 chlorophyll and SST datasets were processed for 2 years monthly 4 km resolutions. The CHL and SST images were retrieved using ENVI- 4.4, SEADAS, and python softwares.

Morel and Maritorena (2001) proposed and identified the OC-3 empirical algorithm (also known as Ocean Chlorophyll 3), which has been operated on MODIS-Aqua ocean colour data. The algorithm operates with five coefficients and has following mathematical form:

$$C = 10 \wedge \left(0.283 - 2.753 \times R + 1.457 \times R^2 - 0.659 \times R^3 - 1.403 \times R^4\right),$$

Where $R = \log_{10}$ [(R_{rs}443 > R_{rs} 488)/R_{rs}551] and R_{rs} is remote sensing reflectance.

The MODIS-Aqua sensor data of resolution 4-kilometer have been used and provide SST with accuracy of 0.2 °C [http://modis.gsfc.nasa.gov] (Sarangi et al. 2011).

25.3.3 Modelling of Nitrate Using In Situ and Satellite Data

The SST relation-based nitrate algorithm was initially developed in the Arabian Sea (Sarangi 2011). Subsequently, the SST and chlorophyll datasets have been regressed with the nitrate three dimensionally (Sarangi et al. 2011, 2015). The nitrate algorithm is developed using in situ datasets, sampling duration with all total 834 points ($R^2 = 0.647$). The nitrate algorithm implies the following equation which has been obtained from parabolid function fit (Sarangi et al. 2011, 2015; Poornima et al. 2015):

$$\text{Nitrate} = 1.6854 - 0.0250 \times \text{SST} + 1.0473 \times \text{Chl} + 0.0005 \times \left(\text{SST}\right)^2 - 0.0009 \times \left(\text{Chl}\right)^2$$

The modelled nitrate image has been processed by using the MODIS-Aqua satellite-derived SST and chlorophyll data and nitrate algorithm developed using in situ data of SST, nitrate, and chlorophyll for paraboloid function during post-monsoon season with the help of ERDAS IMAGINE Ver. 9.2 image analysing software.

25.3.4 SMAP Salinity Data

The ocean salinity data from the NASA satellite Soil Moisture Active Passive (SMAP) mission L-band (1.41 Ghz) microwave radiometer-derived Level-3 monthly averaged product (resolution 0.25 × 0.25 degree) data have been accessed during 2016 and 2017. The data have been processed using ENVI- 4.4 software to generate sea surface salinity (SSS) images for BoB and AS. For the basin scale measurement studies, the salinity data resolution of 25 kilometer are optimal, which is not the perfect resolution as far as the near shore coastal water regions are concerned. Still the data are useful in interpreting the salinity gradient ranges and the water-mass boundaries and to distinguish between coastal and offshore waters.

25.4 Results and Discussion

25.4.1 Phytoplankton Variability in Relation to SST, Nitrate, and Salinity

Monthly scale variability of sea surface chlorophyll, temperature (SST), and nitrate using Modis-Aqua and ocean salinity using SMAP satellite datasets has been studied during 2016 and 2017 (Figs. 25.1 and 25.2). The SST images show the seasonality of temperature variability. The high SST and warmer water (29–31 °C) have been observed in the southern part of the Bay of Bengal and Arabian Sea. During the months January–May, the SST observed to be warmer. But the coastal water observed to be cooler with SST range 24–27 °C. During the winter cooling months, the chlorophyll dense patches are observed in the northwest Arabian Sea and extended till March month. The Arabian Sea salinity has been observed to be higher and uniform (32–36 psu) during all months of the year. But Bay of Bengal salinity is drastically different than Arabian Sea and ranges observed to be 24–31 psu. The estuarine and near coastal water salinity is lesser too, 22–24 psu. There is difference in distribution of phytoplankton types in both the basins. Hence, salinity difference is also playing the role in biogeography and phytoplankton habitat and hence the productivity and chlorophyll biomass quantification. The high salinity and warmer water of Arabian Sea is enhancing the Arabian Sea productivity even if monsoon impact is high in Bay of Bengal and river fluxes are high too. The nitrate nutrient flux has been observed higher in the Arabian Sea, probably the surface nutrient enhancement is taking place due to upwelling, convection, and other processes like entrainment which enhances the nutrient in Arabian Sea as compared to Bay of Bengal nutrient distribution and coastal fluxes mainly due to rivers.

The monsoon impacts have been observed in the monthly scale images during 2016 and 2017. The southwest monsoon has shown increase in chlorophyll biomass (0.5–2.0 mg/m^3) in the Arabian Sea and Bay of Bengal with different patches and features in coastal and offshore regions. Similarly, the northeast monsoon impact

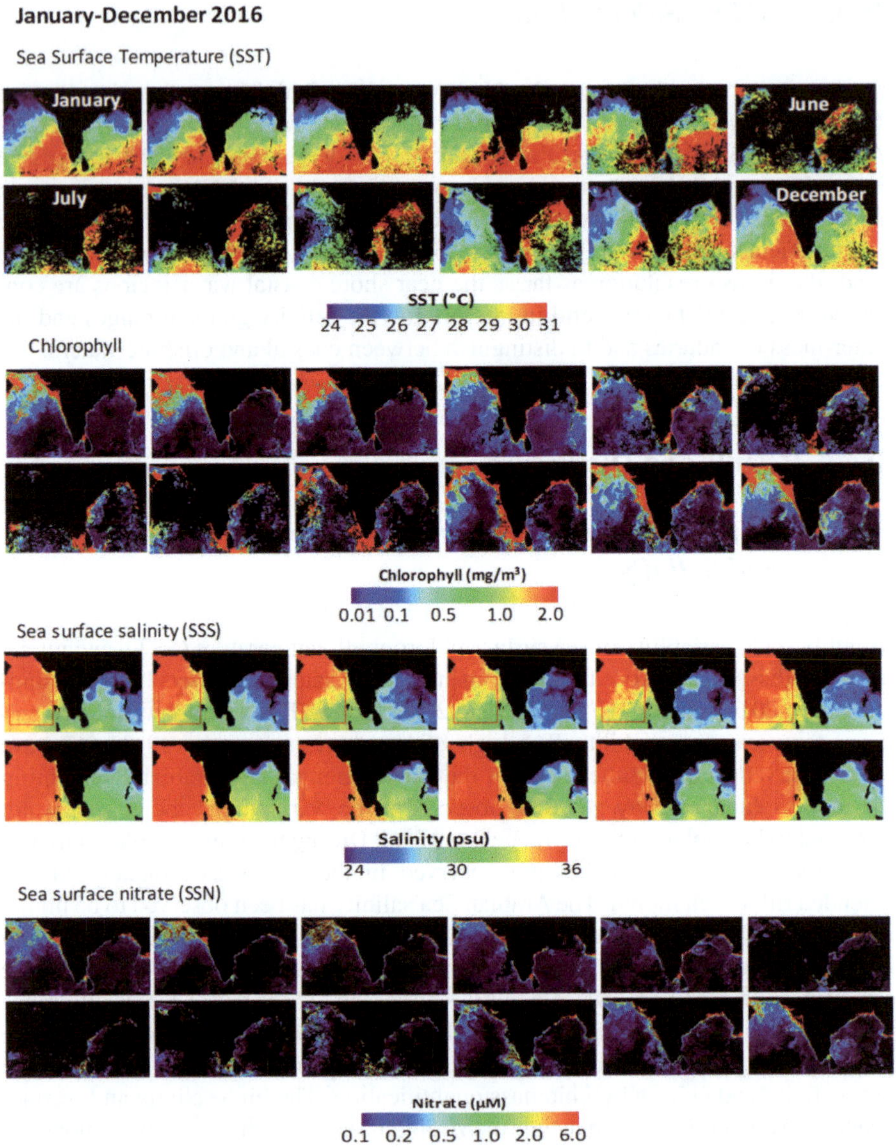

Fig. 25.1 Monthly scale variability of satellite derived sea surface temperature, chlorophyll, salinity and nitrate in the Bay of Bengal and Arabian Sea water during 2016

has been observed in the southwest Bay of Bengal off Tamilnadu coast and offshore water. With the impact of southwest monsoon and the current system, the high saline Arabian Sea water has observed to be intruded in the southern Bay of Bengal, as observed during August–October 2016 and 2017 (Figs. 25.1 and 25.2). The nitrate features with concentration (0.1–2.0 μM) have been observed in the coastal

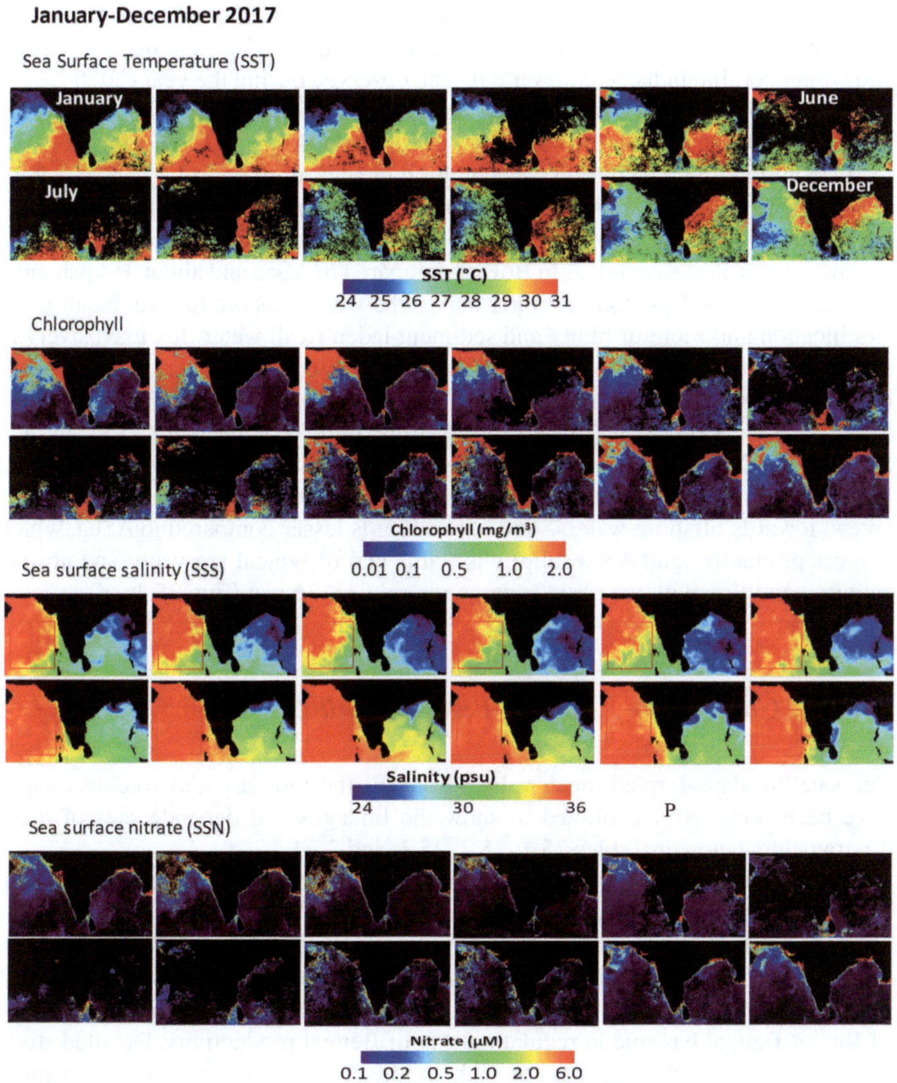

Fig. 25.2 Monthly scale variability of satellite derived sea surface temperature, chlorophyll, salinity and nitrate in the Bay of Bengal and Arabian Sea water during 2017

water, maybe there is need for basin scale nutrient models and its linkage to productivity instead of local scale algorithms. Detailed analysis would provide the linkages of ocean productivity with physicochemical parameters. More robust algorithms are needed to monitor the accuracy of products and different time scale variability in Indian water using satellite datasets.

To understand the intra-annual trend of the variability of ocean productivity in terms of ocean chlorophyll concentration, the monthly satellite image-derived mean values are plotted and chlorophyll seen higher trend (around 1.5 mg m^{-3}) in the Arabian Sea as compared to BoB peaked upto 0.8 mg m^{-3}. The southeast monsoon and winter cooling induced chlorophyll peak observed during the year (2016–2017) monthly chlorophyll observation. The SST images show the seasonal trends over the BoB and ASea and the BoB SST is higher than the ASea SST mean by about 0.5 to 1C in all months. The nitrate concentration images also showed seasonal pattern in its distribution; peak observed post-southwest monsoon, September December, may be due to the impact of upwelling, cyclones, convection, etc., in the ocean basins.

The salinity is always lesser in BoB as compared to ASea and about 3–4 psu units in all months. The lesser saline water makes the water less productive due to more stratification and more turbidity and sediment laden fresh water flux exclusively to northern and eastern BoB. Rather, the ASea productivity is higher due to convection, upwelling the deepening of mixed layer depth (MLD), etc. (Fig. 25.3a–d). Similarly, the weekly scale images observed the trend in chlorophyll, SST, nitrate, and salinity. The BoB having influenced by many rivers and their fluxes still showed less nutrients; it means more sediments are being brought to coastal water and get spread towards offshore waters. And, upwelling is lesser compared to ASea, which is more productive and ASea salinity is stable and of typical sea water and normal salinity condition with the basin average ranging 35–36 psu (Fig. 25.4a–d).

25.4.2 Correlation Analysis-Based Result

The satellite date-derived mean values from all the monthly and weekly images have been retrieved and plotted to show the linkages and dependencies of these 4-parameters below in Tables 25.1, 25.2, 25.3, and 25.4.

The regression analysis-based correlation coefficient result showed that Arabian Sea chlorophyll and productivity is depending on SST ($R = -0.60$); other parameters did not show good correlation as observed from monthly data (Table 25.1). In case of Bay of Bengal monthly data correlation, the parameter salinity has shown better correlation ($R = -0.58$) with the chlorophyll (Table 25.2). Hence, the salinity of Bay of Bengal has role in regulating Bay of Bengal productivity. Detailed study with seasonal and decadal pattern with in situ trend studies would reveal more information.

Similarly, the weekly data-based correlation result revealed that nitrate has better relation ($R = 0.652$) with Bay of Bengal SST as with earlier studies (Sarangi 2011; Sarangi et al. 2011; Poornima et al. 2015). Also, the other parameters like salinity showed far better correlation with nitrate ($R = 0.849$). The salinity has also shown good relation with chlorophyll and SST with R values 0.486 and 0.483, respectively. Hence, Bay of Bengal salinity has a crucial relation in basin scale oceanography in terms of the biological productivity. The weekly scale Arabian Sea has not produced so noticeable correlation. But the chlorophyll has shown good inverse relation with SST ($R = -0.544$) and the salinity and SST relation observed with good correlation R, −0.43. The bottom saline water by process of upwelling, convection, and

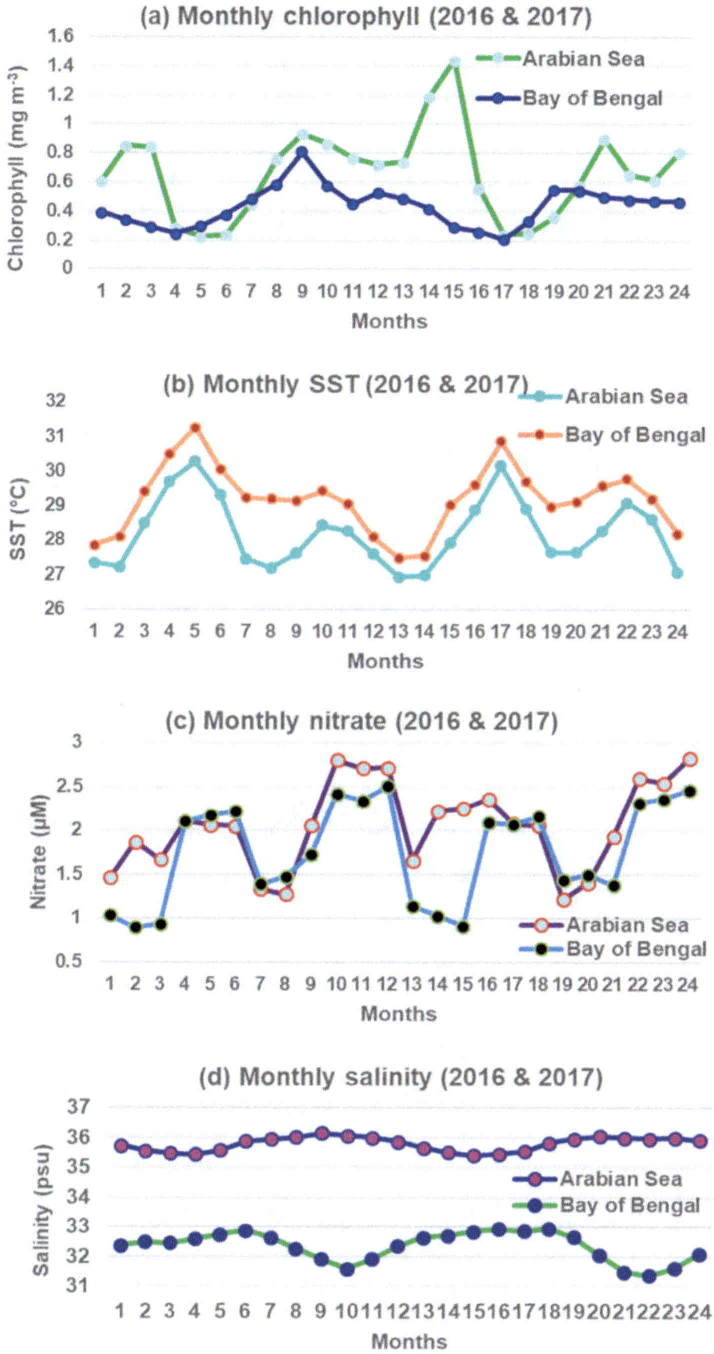

Fig. 25.3 (**a–d**) Monthly scale variability of chlorophyll, SST, nitrate and salinity in the Arabian Sea and Bay of Bengal during 2016–2017

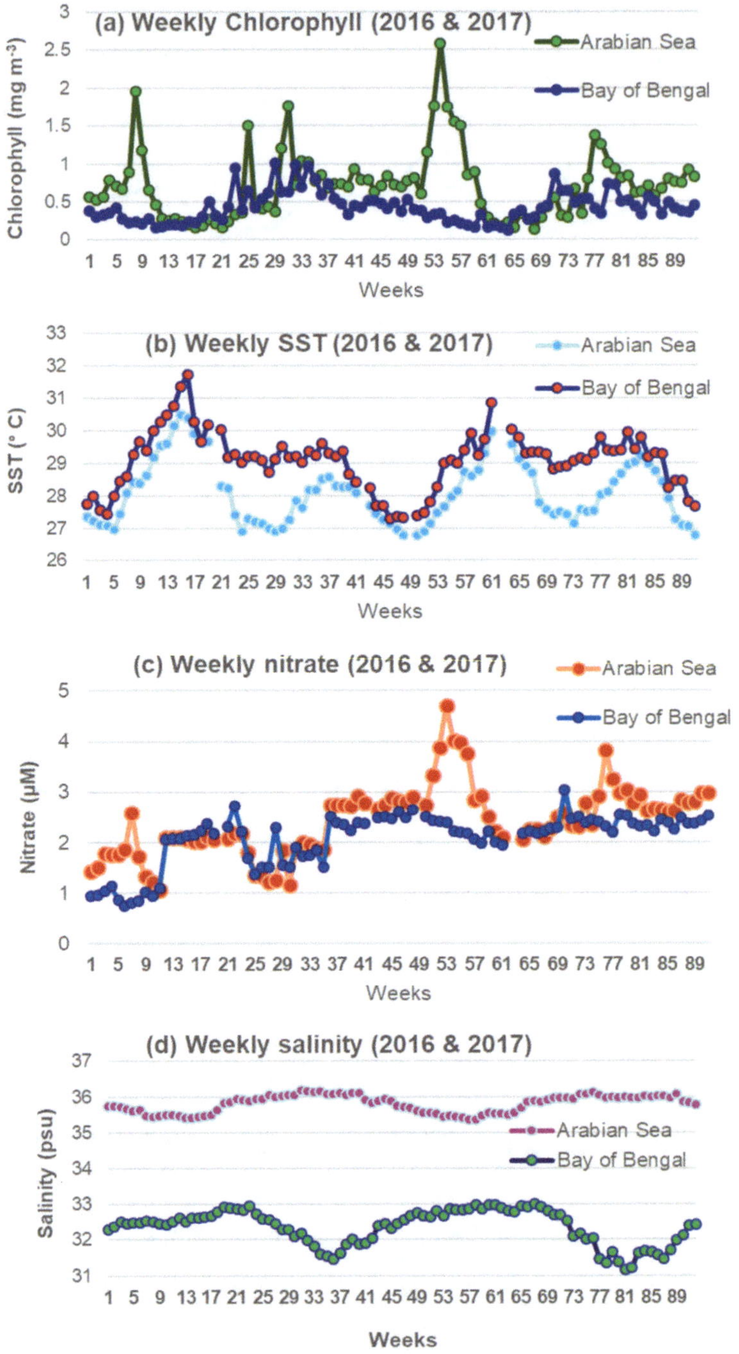

Fig. 25.4 (**a–d**) Weekly (8-day) scale variability of chlorophyll, SST, nitrate and salinity in the Arabian Sea and Bay of Bengal during 2016–2017

Table 25.1 Monthly mean data correlation result for Arabian Sea

	ASea Chl	*ASea SST*	*ASea NO3*	*ASea SSS*
ASea Chl	1			
ASea SST	**−0.60037**	1		
ASea NO3	0.219298	0.258649	1	
ASea SSS	−0.0517	−0.25924	0.026721	1

Table 25.2 Monthly mean data correlation result for Bay of Bengal

	BoB Chl	*BoB SST*	*BoB NO3*	*BoB SSS*
BoB Chl	1			
BoB SST	−0.35849	1		
BoB NO3	0.053945	**0.434955**	1	
BoB SSS	**−0.58887**	0.078987	−0.26362	1

Table 25.3 Weekly mean data correlation result for Arabian Sea

	ASea chl	*ASea SST*	*ASea NO3*	*ASea SSS*
ASea chl	1			
ASea SST	**−0.54432**	1		
ASea NO3	0.103813	0.298804	1	
ASea SSS	−0.06517	**−0.43098**	−0.00845	1

Table 25.4 Weekly mean data correlation result for Bay of Bengal

	BoB Chl	*BoB SST*	*BoB NO3*	*BoB SSS*
BoB Chl	1			
BoB SST	−0.24621	1		
BoB NO3	0.409201	**0.652166**	1	
BoB SSS	**0.486114**	**0.483536**	**0.849047**	1

entrainment may be maintaining the SST and salinity relation within the water mass; this may be a topic of sound research aspect.

25.4.3 Productivity and Environmental Parameters

Oceanographers often refer to this process as the biological pump, as it pumps carbon dioxide (CO_2) out of the Surface Ocean and atmosphere and into the voluminous deep ocean (Volk and Hoffert 1985). Of the organic matter produced by phytoplankton (NPP), most is respired back to dissolved inorganic forms within the surface ocean and thus recycled for use by phytoplankton (Eppley and Peterson 1979). In the early 1900s, oceanographer Alfred Redfield found that plankton build their biomass with C: N: P stoichiometric ratios of ~106:16:1, to which we now refer as the Redfield ratios (Redfield 1958). Phytoplankton growth limitation has traditionally been interpreted in the context of Liebig's Law of the Minimum, which states that plant growth will be as great as allowed by the least available resource, the "limiting nutrient" that sets the productivity of the system (de Baar 1994). Change in salinity has an important role in the ecosystem function. It has been reported that the optimum salinity conditions for phytoplankton growth are phytoplankton group-specific (Brand 1984). Many previous studies suggested that salinity gradient in the estuary systems in temperate to tropical regions potentially modifies phytoplankton community composition due to the difference in optimum salinity (Yamaguchi et al. 1997; Földer et al. 2010; Balzano et al. 2011).

The relationship between chlorophyll and biomass is changeable, depending on the physiology of phytoplankton; for example, phytoplankton adapted to lower light and/or higher nutrients tend to have a higher cellular concentration of chlorophyll (Geider et al. 1997). For a given NPP, small variations in grazing can lead to large proportional changes in phytoplankton biomass (Landry and Hassett 1982). Compared to nutrient-bearing regions, nutrient-deplete regions (e.g. the subtropical gyres) have a larger fraction of chlorophyll below the depth that can be sensed by the satellite (Smith 2007). Hence, productivity measurement and linkage to ecological parameters are vital and need to be established using models and satellite datasets for synoptic and synergistic observations.

25.5 Future Research and Direction

Multi-parameters-based in situ and satellite remote sensing will boost our knowledge on ocean processes and associated features in the different water types and regimes in regional and global scale oceanography. Hence, the parameters relevant to ocean productivity; chlorophyll, nutrients, SST, wind, currents, salinity, etc., will be vital to study in upcoming satellite missions with better repetivity and high resolution datasets and also synoptically and synergistically.

Future work is likely to focus more on geographic information system (GIS) in connection with remote sensing datasets and techniques. This will be able to have automation of feature tracking algorithms like ocean eddies, currents, and fronts from chlorophyll, SST, and salinity images. Hence, satellite data-based synoptic scale imaging has more demand in current scenario. The data comprising different resolutions will enable the study in understanding local scale to mesoscale features and its dynamics. The machine learning techniques are capable to simulate ocean datasets and to process model-derived outputs and do the predictions as well. The different resolution satellites and derived parameters' fusion based on their features will be helpful in the time series analysis and process studies.

25.6 Summary and Conclusion

The study revealed the capability of multiple sensor and multi-platform-based datasets for retrieving ocean biology and physicochemical parameters on synoptic scale. The monthly scale surface chlorophyll, SST, salinity, and nitrate have been retrieved over the Bay of Bengal and Arabian Sea. The seasonality in chlorophyll biomass in monthly scale has been observed and interpreted. The winter cooling has been observed both over the Bay of Bengal and Arabian Sea basins, with decrease in SST by about 3–4 °C has been observed in the coastal waters. The southwest and northeast monsoon-induced productivity has been observed in terms of chlorophyll distribution. The Bay of Bengal observed lesser salinity during summer and pre-monsoon months (22–26 psu), February–June as compared to other months (26–30 psu). The fresh water flux region has been observed in the northern Bay of Bengal with the help of SMAP data and also being linked to less productive water using the MODIS-Aqua-derived chlorophyll, temperature gradient zones, and modelled nitrate images. Hence, the synoptic and synergistic measurement of ocean productivity using satellite data from optical, thermal infrared, and microwave sensor has the potential to interpret the ocean productive zones in the northern Indian Ocean. The nitrate is observed to be having good role in enhancing productivity in coastal water. The offshore nitrate variation has not been observed with its features. Linkages of winter cooling and productivity enhancement in Bay of Bengal and Arabian Sea have been observed. The correlation analysis proved the role and linkage of salinity in Bay of Bengal phytoplankton productivity and also to the nitrate concentration. This is due to the riverine fluxes with high nutrients inducing to have more productivity in the coastal water and fresh water flux relations as well. The phytoplankton productivity, time series study, and understanding the biogeochemistry in a broad aspect need to be studied with multiple satellite missions with technological advancements.

Acknowledgements The authors are thankful to Dr. Bimal Bhattacharya, Group Director, AESG, Deputy Director, EPSA and Director, Space Applications Center, Ahmedabad, for necessary guidance and facilities for carrying out the work. Authors acknowledge NASA/MODIS server for Aqua datasets and PODAAC website for providing SMAP salinity datasets.

Declarations Informed Consent Statement Not applicable.

Data Availability Statement Data are available from the authors on request.

Ethics Approval Not applicable.

Conflict of Interest The author declares no conflict of interest.

References

Balzano S, Sarno D, Kooistra W (2011) Effects of salinity on the growth rate and morphology of ten Skeletonema strains. J Plankton Res 33:937–945

Bauer S, Hitchcock G, Olson D (1991) Influence of monsoonally-forced Ekman dynamics upon the surface layer depth and plankton biomass distribution in the Arabian Sea. Deep-Sea Res 38:531–553

Bhattathiri PMA, Pant A, Sawant S, Gauns M, Matondkar SGP, Mohanraju R (1996) Phytoplankton production and chlorophyll distribution in the eastern and central Arabian Sea in 1994–1995. Curr Sci 71:857–862

Brand LE (1984) The salinity tolerance of forty-six marine phytoplankton isolates. Estuar Coast Shelf Sci 18(5):543–556

Brock J, McClain C, Luther M, Hay W (1991) The phytoplankton bloom in the northwestern Arabian Sea during the southwest monsoon of 1979. J Geophys Res 96:20623–20642

de Baar HJW (1994) von Liebig's law of the minimum and plankton ecology (1899–1991). Prog Oceanogr 33(4):347–386. https://doi.org/10.1016/0079-6611(94)90022-1

Eppley R, Peterson B (1979) Particulate organic matter flux and planktonic new production in the deep ocean. Nature 282:677–680. https://doi.org/10.1038/282677a0

Földer S, Jaschinski S, Wells G, Burns CW (2010) Dominance and compensatory growth in phytoplankton communities under salinity stress. J Exp Mar Biol Ecol 395:223–231. https://doi.org/10.1016/j.jembe.2010.09.006

Geider RJ, Machintyre HL, Kana TM (1997) A dynamic model of phytoplankton growth and acclimation: responses of the balanced growth rate and chlorophyll a: carbon ratio to light, nutrientlimitation and temperature. Mar Eco Prog Ser 148:187–200

Howarth RW, Marino R (2006) Nitrogen as the limiting nutrient for eutrophication in coastal marine ecosystems: evolving views over three decades. Limnol Oceanogr 51(1, part 2):364–376

Landry MR, Hassett RP (1982) Estimating the grazing impact of marine micro-zooplankton. Mar Biol 67:283–288

Lee C, Jones B, Brink K, Fischer A (2000) The upper-ocean response to monsoonal forcing in the Arabian Sea: seasonal and spatial variability. Deep-Sea Res II 47:1177–1226

McCreary JP, Kohler KE, Hood RR, Olson DB (1996) A fourcomponent ecosystem model of biological activity in the Arabian Sea. Prog Oceanogr 37:193–240

Morel A, Maritorena S (2001) Bio-optical properties of oceanic waters: a reappraisal. J Geophys Res 106:7163–7180

Poornima D, Sarangi RK, Shanthi R, Thangaradjou T, Chauhan P (2015) Seasonal nitrate algorithms for nitrate retrieval using OCEANSAT-2 and MODIS-AQUA satellite data. Environ Monit Assess 187:176

Prasad TG (1997) Annual and seasonal mean buoyancy fluxes for the tropical Indian Ocean. Curr Sci 73:667–674

Prasanna Kumar S, Madhupratap M, Dileep Kumar M, Muraleedharan PM, de Souza SN, Gauns M, Sarma VVSS (2001) High biological productivity in the central Arabian Sea during summer monsoon driven by Ekman pumping and lateral advection. Curr Sci 81:1633–1638

Prasanna Kumar S, Muraleedharan PM, Prasad TG, Gauns M, Ramaiah N, de Souza SN, Sardesai S, Madhupratap M (2002) Why is the Bay of Bengal less productive during summer monsoon compared to theArabian Sea? Geophys Res Lett 29(24):2235.https://doi.org/10.1029/2002GL016013

Qasim SZ (1979) Primary production in some tropical environments. In: Dunbar MJ (ed) Marine production mechanisms, international biological Programme 20. Cambridge University Press, pp 31–69

Radhakrishna K, Devassay VP, Bhargava RMS, Bhattathiri PMA (1978) Primary production in the northern Arabian Sea. Indian J Mar Sci 7:271–275

Redfield AC (1958) The biological control of chemical factors in the environment. Am Sci 46:205–221

Sarangi RK (2011) Remote-sensing-based estimation of surface nitrate and its variability in the southern peninsular Indian waters. Int J Oceanogr 2011:16

Sarangi RK, Thangaradjou T, Kumar AS, Balasubramanian T (2011) Development of nitrate algorithm for the southwest Bay of Bengal water and its implication using remote sensing satellite datasets. IEEE J Sel Top Appl Earth Obs Remote Sens 4(4):983–991

Sarangi RK, Thangaradjou T, Poornima D, Shanthi R, Kumar AS, Balasubramanian T (2015) Seasonal nitrate algorithms for the southwest Bay of Bengal water using in situ measurements for satellite remote-sensing applications. J Coast Res 31(2):398–406

Sen Gupta R, De Sousa SN, Joseph T (1977) On nitrogen and phosphorous in the western Bay of Bengal. Indian J Mar Sci 6:107–110

Smith VH (2007) Using primary productivity as an index of coastal eutrophication: the units of measurement matter. J Plankton Res 29(1):1–6

Subramanian V (1993) Sediment load of Indian Rivers. Curr Sci 64:928–930

Tyrrel T (1999) The relative influences of nitrogen and phosphorus on oceanic primary production. Nature 400:525–531

Volk T, Hoffert MI (1985) Ocean carbon pumps: analysis of relative strengths and efficiencies in ocean-driven atmospheric CO_2 changes. J Geophys Res. Washington DC 32:9. https://doi.org/10.1029/GM032p0099

Yamaguchi M, Shigeru I, Nagasaki K, Matsuyama Y, Uchida T, Imai I (1997) Effects of temperature and salinity on the growth of the red tide flagellates Heterocapsa circularisquama (Dinophyceae) and Chattonella verruculosa (Raphidophyceae). J Plankton Res 19(8):1167–1174. https://doi.org/10.1093/plankt/19.8.1167

Young DK, Kindle JC (1994) Physical processes affecting availability of dissolved silicate for diatom production in the Arabian Sea. J Geophys Res Oceans 99(C11):22619–22632

Chapter 26
Climate Change and Invasive Species: Suitability of *Acacia auriculiformis* in Kasaragod District, Kerala

Kavya Jeevan, Sara Grace, Kannankodantavida Manjusha, Shalu George, Nadirsha P. S. Nawab, Anbazhagi Muthukumar, and Muthukumar Muthuchamy

Abstract Invasive alien species (IAS) are any non-native species that extend out of their natural habitat and threaten the native biodiversity of that new ecosystem. The increase in human population, improved facilities for trade and transport, and degradation of natural habitats have facilitated invasion. Climate change, another cause for creating an imbalance in ecosystems, has indirectly or directly accelerated the invasion. *Acacia auriculiformis* is a plantation crop introduced to the tropics in the twentieth century, which continues to be sustained in our country as an invasive species. The study used the Maximum Entropy model to identify the current and future distribution of *Acacia auriculiformis* in the Kasaragod district and the major factors influencing its habitat suitability. The research identified that a decrease in temperature seasonality, an increase in mean temperature in the wettest quarter, and precipitation in the coldest quarter will increase the probability of the species' presence and pose a greater threat to native biodiversity. The Shared Socio-economic pathways (SSPs) 245 and 585 represent future global warming under current and higher greenhouse gas emission conditions, respectively. The projection of habitat suitability for both the future climate change scenarios inferred that the suitability of Kasaragod district for the species will decrease. However, the adaptability of invasives to harsh climatic conditions draws our attention to Ranipuram National Park, which will provide moderately suitable conditions for the species in future. Thus, this study calls for immediate restoration and conservation efforts in this biodiversity-rich area; as previous research has proved, an area with healthy natural vegetation is less prone to invasion.

K. Jeevan · S. Grace · K. Manjusha · S. George · N. P. S. Nawab · A. Muthukumar · M. Muthuchamy (✉)
Climate Change and Remediation Research Lab (CCRRL), Department of Environmental Science, Central University of Kerala, Kasaragod, Kerala, India
e-mail: mmuthukumar@cukerala.ac.in

B. P. Barathan et al. (eds.), *Navigating Climate Change: Impacts on Biodiversity and Ecosystem Resilience*, Sustainable Development and Biodiversity 42,
https://doi.org/10.1007/978-981-95-0409-1_26

Keywords *Acacia auriculiformis* · Invasion ecology · Biodiversity decline · Climate change · Future climate scenarios · Invasive alien species · Management strategies · Maximum Entropy model

26.1 Introduction

26.1.1 Invasive Species

The species whose introduction to an ecosystem outside its natural habitat can threaten the native biodiversity of that ecosystem is called Invasive alien species (IAS) (Convention on Biological Diversity 2009). Intentional and accidental human activity alters the distribution of species, introducing them to new habitats, such as altered and degraded habitats that are vulnerable to biological reorganisation (Rahel and Olden 2008). The expansion of the human population and increased mobility of people and their goods have increased the chances of species' movement across the globe. IAS is characterised by rapid multiplication and growth, high dispersal ability, high adaptability to new environmental and nutrition conditions, and is favoured by the absence of natural predators or competitors that can control its population (Convention on Biological Diversity 2009). Their rapid spread can disrupt ecological balances, harm native species, and lead to substantial economic burdens (IPBES 2023).

26.1.2 Pathways of Invasion

The pathways of invasion by these species, as reported by Convention on Biological Diversity (2009), include ship ballast water, pets released into the wild and aquarium trade, hitchhikers, introduction of species for agricultural, forestry and aquacultural purposes, ornamental plants, and introduction of biological agents for pest control and tourism. Regarding an increase in trade, travel, and business, the spread and establishment of alien species has increased over time (Khuroo et al. 2012). During the fifteenth and nineteenth centuries, many Europeans crossed the African and Asian continents for the establishment of alien species in the places where they colonised for food and energy (Pyšek et al. 2008; Khuroo et al. 2012). India has a wide trade relationship with Europeans back then. Nomadic tribes are also considered to be a factor for the establishment of invasive species, such as food grain contaminants, as packing materials, and also intentionally introduce many exotic plant species for the establishment of botanical gardens, for ornamental purposes, etc. (Suresh 2012). During colonisation, the British East India Company introduced almost 3200 plant species to India for the establishment of the royal botanical garden (Royle 1840). World's largest wildlife survey reveals that 66% of India's natural system is threatened by invasive plant species. Historical propagation of invasive plants, change in the soil characteristics, and climatic conditions are the major drivers for the facilitation of the invasion. India has one of the world's highest population density, so there will be a shift in the lifestyle, food, and infrastructure, which will intensify the already existing unchecked invasion even more. Using the data collected from national scale invasive plant assessment, the researchers introduced a new model that is to identify the sites where the invasion is in its initial stage with priority for the protected areas. Protecting these sites requires less investment and intervention. (British ecological society 2023). According to the study of Rai and Singh (2024), Western Ghats, Eastern Ghats, and Central Himalayan regions in India appear to be more susceptible to invasion and invasive species thrive easily in landscapes outside of protected areas (Heller and Zavaleta 2008). Moreover, the readiness of vacant niches also promotes the introduction and establishment of exotic ones created due to ecosystem disturbances (Kühn and Klotz 2008). As a result, species invasion has emerged as a cutting-edge ecological concern (Didham et al. 2005).

26.1.3 Major Threats of Invasion

Globally, there are enormous environmental dangers and financial losses due to ecological issues and growing biological invasion threats (Wendlandt et al. 2022). Studies have found that invasive species are among the top drivers of global biodiversity loss through predation, competition, and disease transmission (Defenders of Wildlife 2023). More than 37,000 alien species have been introduced by human activities and presently 200 alien species are unprecedently increasing annually. In a global scale, the economy, water security, and food security are profoundly and negatively affected by the alien species. Invasive species will cause a dramatic and sometimes irreversible negative impact to the biodiversity and the ecosystem including locals. Invasive species are capable of altering the characters of soil and water (Helen et al. 2014). In addition to posing a serious threat to biodiversity, biological invasions have a big impact on socioeconomic factors like culture, religion, human health, safety, livelihoods, and other aspects of human well-being. From an ecological perspective, invasion of exotic species forms a reason for native species' extinction and habitat loss (Pejchar and Mooney 2009). Invasive plants, however, have the ability to change soils in ways that harm native plants' mycorrhizal relationships (Grove et al. 2017). Conversely, researches showed that the invasive species either promotes economic benefit or strengthens the ecosystem functioning (Wallach et al. 2015).

The study of invasives started very recently, although the first study was conducted by Elton many years ago (Elton 1958). For the better understanding of invasive species, it is very important to collect the data about the landscape vulnerability for getting a better insight (Vicente et al. 2013). Environmental parameters such as sunlight, water, soil nutrients, and pH decide where and how an invasive species will propagate and establish (Daehler 2003). Thus, identifying the factors that determine the invasion of the alien species is important. Climate change, pollution, habitat loss, and anthropogenic disturbances have increased the impacts and threats caused by these alien species (Convention on Biological Diversity 2009).

26.1.4 Climate Change and Invasive Species

Invasion became a global environmental agenda when the invasion was linked with climate change; all these led to focus mainly on the spread, impact, and management of invasive (Davis 2011). Climate change can facilitate invasion by introducing new invasive species, altering species' hierarchies in ecosystems that may result in new dominants, and expanding invasion pathways through climate-induced stress (Masters and Norgrove 2010). The direct consequences of climate change along with species and population responses are making ecosystems more susceptible to biological invasions (Weiskopf et al. 2020). Bellard et al. (2018) underlined that climate change is expected to have a major role in the distribution of invasive alien species in future (Bellard et al. 2018). Many studies predicted the range shifts (Elith et al. 2010; Wang et al. 2022) and continental accumulation (Seebens et al. 2021) of invasive alien species in associated with climate change utilising modelling techniques. The extinction of native species is a widely explored area of research, but less or little attention has been given to the idea of which species will replace them. Climate change can accelerate the invasion by degrading the existing ecosystem and providing favourable conditions for IAS (IPCC, 2022). As climate change provides new, suitable habitats for invasive species establishment, it amplifies the synergistic effects of biological invasions (Bellard et al. 2016; Bradley et al. 2019). Therefore, the elements related to climate change that may have an impact on the dispersal of invasive species must be taken into account when establishing management plans for them. To aid in decision-making, bioclimatic modelling tools offer quantitative scenarios of how climate change may affect the distribution of species (Pereira et al. 2010). Potential distributions of invasive species were predicted using species distribution modelling, which is based on the geographic association between species presence places and climate conditions (Bradley et al. 2019; Roura-Pascual 2020; Villemant et al. 2011; Bellard et al. 2016).

A comprehensive list of invasive species is necessary for the prevention and management of invasives (Reddy 2008). Current status of invasion, factors supporting it, and the biogeographic information are necessary to base the invasion biology (Sankaran and Sajeev (2012)).

26.1.5 Acacia auriculiformis: *An Afforestation Intervention to Tropics*

The *Acacia auriculiformis* (known as Earleaf acacia) is native to Australia, Indonesia, and Papua New Guinea and most of its introduction to other regions across the globe occurred in the twentieth century (Schmerbeck and Naudiyal 2014). During the 1980s, several researches were conducted for assessing the nitrogen fixing potential of *Acacia auriculiformis* and their adaptability in Asian countries (Brewbaker 1987; Brewbaker et al. 1983, 1984). In India, the study conducted by Chakraborty and Chakraborty (1989) reported that *Acacia auriculiformis* have great potential to restore degraded land by enhancing the nitrogen, organic carbon, and potassium levels in the soil. Also, they could increase the acidic pH of the soil to slightly alkaline and improve the water holding capacity of the soil. Furthermore, this plant was observed to sustain health even in poor nutrient conditions and poorly managed land (Pinyopusarerk 1990).

Its adaptability to harsh pH conditions and saline areas made it more suitable in tropics. It also tolerates polluted air and waterlogging conditions, which later dominate the landscape after any degradation, suppressing the revival of the native species which are intolerant to these conditions (Heringer et al. 2019; Pinyopusarerk 1990). According to Duponnois et al. (2013), *Acacia auriculiformis* can alter the microbial dynamics of the soil (mycorrhizal fungi and rhizobia), inhibit the growth of native species, and be capable of changing both the soil and leaf nitrogen (Meiro-neto et al. 2018). If *Acacia auriculiformis* are introduced to dry or water-limited area, they can alter the seasonal pattern, because they are capable of absorbing a huge amount of water from underground (Siddiq and Cao 2016).

Although the *Acacia auriculiformis* is considered a major threat, previous research has mentioned a few positive effects offered by this species on the soil and recommends them for restoration of highly degraded land (e.g., mining sites). *A. auriculiformis* significantly increases the soil organic carbon and cation exchange capacity of the soil and the symbiotic relationship of this plant with rhizobia increases the total nitrogen of the soil (Kasongo et al. 2009; Rodríguez-Echeverría et al. 2011).

However, this invasive alien species has a negative impact on endangered species. The forest department already started cutting down the acacia trees in the Trivandrum district due to excess use of water, and they are taking control of the natural habitat there by causing a drastic reduction in the growth of grass. Reduction in grass negatively affects the herbivore community (Times of India 2022).

26.1.6 Objectives of the Study

Biological invasions are one of the top priority areas for more study and assessment, according to the Governmental Platform on Biodiversity and Ecosystem Services (IPBES) (2015). The detrimental effects of invasion have been widely explored in previous studies. The investigations on factors affecting the invasion have poured some light on the need for understanding the influence of climate change on invasion. The current study tries to explore and define the influence of climatic variables on the current and future distribution of *Acacia auriculiformis.*

26.2 Materials and Methods

26.2.1 Study Area

Kasaragod district, which is known as the crown of Kerala, is located between 11°18'N to 12°48'N latitude and 74°52'E to 75°26' E longitude. The location map and physiography of Kasaragod are shown in Fig. 26.1. The district extends up to 1,99,166 ha, with a vast midland area, coastal line, and 5625 ha forest. This region has a rich biodiversity and natural heritage. The soil can be classified as sandy, red ferruginous loam of lateritic origin and laterite (District Urbanisation Report 2011). The district has a tropical monsoon climate with a mean annual rainfall of 3350 mm. The average maximum and minimum temperatures recorded are 37 °C and 17 °C, respectively (https://kasargod.nic.in/en/climate/). Kasaragod district poses similar temperature and rainfall conditions to the native range of *Acacia auriculiformis* which was reported by Pinyopusarerk (1990). Moreover, this district is facing a rapid land degradation (Karunakaran 2014) that might support invasion.

26.2.2 Data Collection and Variable Selection

Species occurrence data were collected during field visits, where Acacia auriculiformis trees were identified using plant identification keys and their coordinates recorded. A total of 64 locations were obtained, thinned to 56 points with a minimum distance of 1 km between individuals to align with the resolution of climate data. The thinning was done using the 'spin' package in RStudio. The data points were converted into a single CSV file for introduction into the MaxEnt software.

For analysing the current habitat suitability of *Acacia auriculiformis,* current climate data were downloaded from dataset; Climatologies at high resolution for earth's land surface areas (CHELSA). This dataset, which spans the years 1979–2013, includes high geographic resolution monthly climatologies of mean, maximum, and minimum temperatures as well as mean precipitation. These

Fig. 26.1 Location and physiography map of the study area

high-resolution (~1 km) datasets can be applied in ecology, agriculture, and meteorology (https://chelsa-climate.org/).

To reduce the error caused by multicollinearity, the correlation coefficient and Variance Inflation Factor (VIF) were considered, where the selected variables had a correlation less than 0.7 and VIF less than 5.

To observe the response of the species to future climate, two climate change scenarios, Shared Socio-economic Pathways (SSP) 245 and SSP 585, were considered. In SSP 245, the greenhouse gas emission is about the same as the current condition, and the global average temperature tends to reduce with human intervention. In the SSP 585 scenario, there will be a high greenhouse gas emission where temperatures are projected to rise in the near future (Riahi et al. 2017). The climatic scenarios were considered during the time periods of 2021–2040, 2041–2060, 2061–2080, and 2081–2100 (Kishoreet al. 2024) and the data were obtained from WorldClim v2.1 (Fick and Hijmans 2017).

26.2.3 Habitat Suitability Modelling

The best model for the prediction of habitat suitability was selected using Akaike information criteria (AIC). AIC was used to evaluate model performance by penalising overfitting and selecting the simplest model with optimal predictive accuracy (Muscarella et al. 2014). The selected variables were introduced into MaxEnt as ASCII format. The model was run for five repetitions using a random subset of the distribution points, with 30% designated as testing data and the remaining 70% as training data. The Jackknife test was employed to determine the relative effect of the selected predictor factors (Padalia et al. 2014).

The receiver operating characteristic curve (ROC) method was used to test the accuracy of the model. Area under curve (AUC) value is the area where the *A. auriculiformis* and ROC curve is enclosed. The value of AUC ranges from 0 to 1, where an AUC value between 0.5 and 0.6 indicates model failure, whereas 0.6–0.7 indicates the simulation effect is poor, 0.7–0.8 indicates the simulation effect is average, 0.8–0.9 indicates the simulation effect is good, and finally 0.9–1 indicates the simulation effect is excellent (Swets, 1988).

The area was classified into three categories based on the Maximum Test Sensitivity Plus Specificity (MTSPS); 0-0.0348 as not suitable, 0.03481–0.3153 as moderately suitable, and 0.3153–1 as suitable (Li et al. 2019). The area under each category was obtained using DIVA-GIS 7.5.0.

26.3 Results and Discussion

Species that can cross from their natural habitat and are capable of establishing themselves in a new environment are known as alien species (Saxena 1991). Invasion can happen either by accident or purposefully to exploit the ecological niche (Kaphle and Pokheral 2014). The spread of the invasive species can be intentional or accidental trade, tourism especially importing goods such as grains, seeds, fruits, flowers, etc. can aid in invasion (Usher 1988). For the establishment of invasive species, the major drivers for the adaptation and establishment are sunlight, water requirement, soil nutrients, and climate and in addition to that propagule pressure (Lockwood 2005). The current study was conducted to identify the area under potential threat of invasion by *Acacia auriculiformis***.** The regions of Kasaragod with suitable climate conditions for the species were identified.

26.3.1 Current Suitability and Influencing Factors

Identifying the current distribution of invasive species helps to identify areas that are at risk of ecosystem degradation and biodiversity decline (Convention on Biological Diversity 2009). The current suitability of *Acacia auriculiformis* is depicted in Fig. 26.2, where the suitability of the area is categorised as Suitable, Moderately suitable, and Unsuitable. Figure 26.3 shows that approximately 1, 18,173 ha of land in Kasaragod district is suitable for *Acacia auriculiformis*, while approx. 73,879 ha is moderately suitable, and 3296.67 ha is unsuitable.

The variables considered for the model are provided in Table 26.1; these include Temperature seasonality (Bio 4), Mean temperature of the wettest quarter (Bio 8), Annual precipitation (Bio 12), Precipitation of the driest month (Bio 14), Precipitation of the warmest quarter (Bio 18), and Precipitation of the coldest quarter (Bio 19). The obtained MaxEnt model had an AUC of 0.96, which indicates that the model is excellent at predicting species distribution.

The response curves of the habitat suitability to environmental variables given in Fig. 26.4 show how the variable influences the suitability. As observed in the current study, the temperature seasonality and precipitation of the driest month have a negative impact on the species. While the region's mean temperature during the wettest quarter, annual precipitation and precipitation during the coldest quarter have a positive effect on the species. The habitat suitability of the region was highest in the regions with 150 mm of precipitation during the warmest month. Temperature seasonality is a major determinant of the northern extent of plants in the northern hemisphere (Qian et al. 2022). Additionally, water availability is another limiting factor pertaining to the distribution of terrestrial plants (Lambers and Oliveira 2019). In the current study, the decrease in temperature seasonality, increase in mean temperature of the wettest quarter, and precipitation of the coldest quarter will elevate the probability of the species presence to 0.85, 0.95, and 0.9, respectively. However,

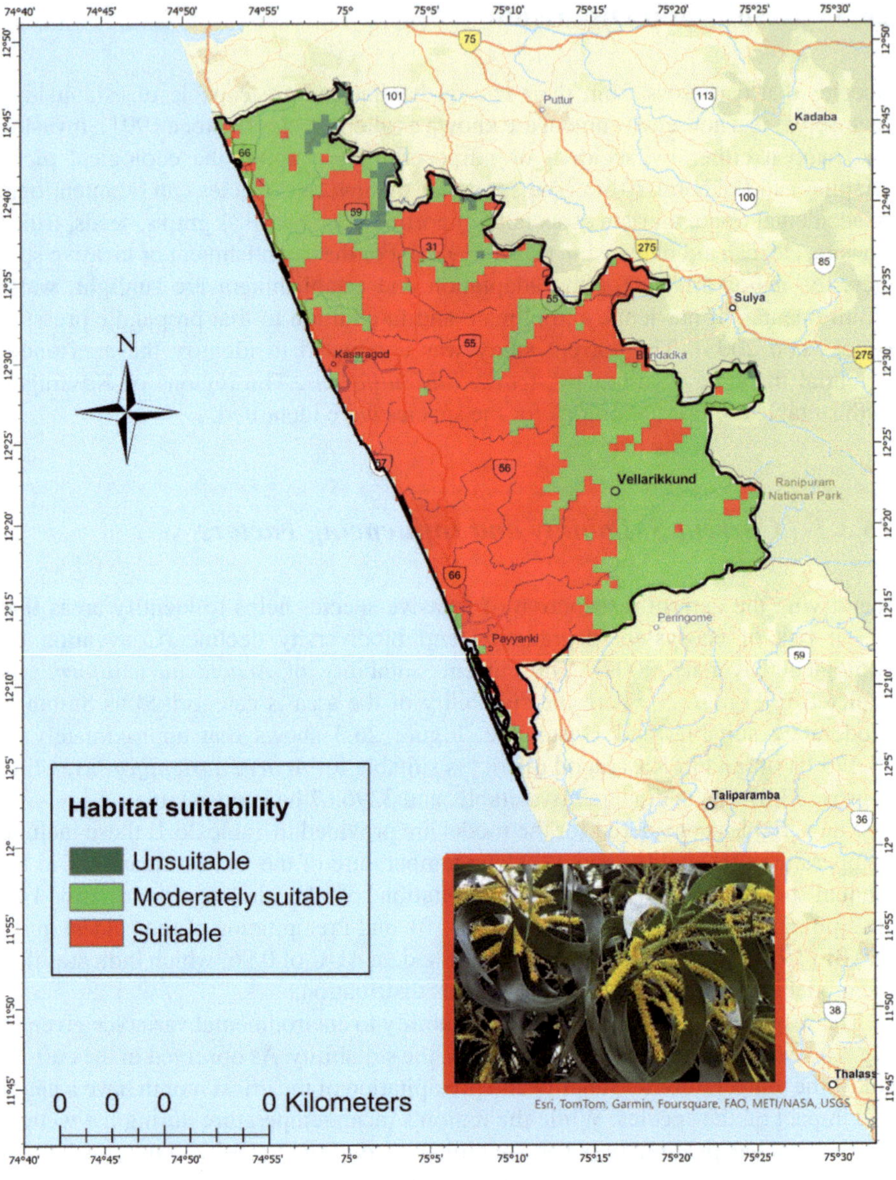

Fig. 26.2 Current habitat suitability of *Acacia auriculiformis* in Kasaragod, Kerala

annual precipitation, precipitation of the driest month, and precipitation of the warmest month have a minimal influence on the probability of the species' occurrence.

In the present study, the probability of species presence was found to be higher at the regions where the mean temperature of the wettest quarter of the study area is

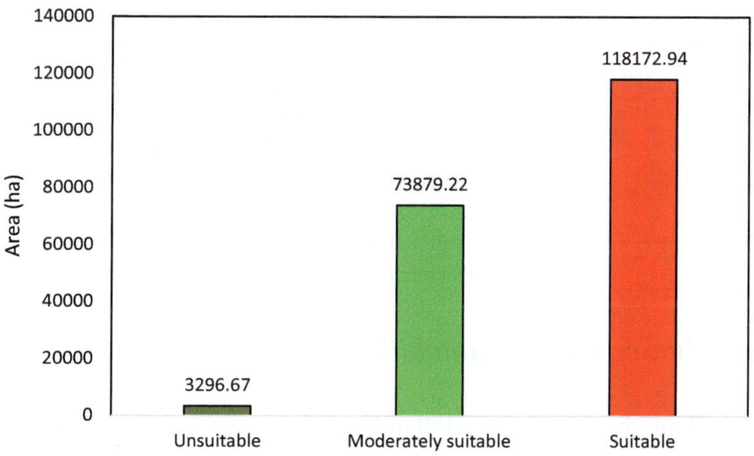

Fig. 26.3 Area (ha) under each suitability classification in Kasaragod district, Kerala

similar to that of the species' native range, which was also reported by Booth and Jovanovic (1988). The study conducted by Uehira and Shimono (2024) has also reported that the decrease in temperature seasonality of Asian countries will positively influence the suitability of the area for invasive species. Booth (1987) identified annual precipitation as a factor for habitat suitability for *Acacia auriculiformis*, but the study also reported the species were prominent near water bodies even if not enough precipitation was available. The study by Brewbaker et al. (1983) concluded that *Acacia auriculiformis* is widely adaptable to areas with minimum mean annual precipitation of 750 mm.

Table 26.2 shows the percentage contribution of each environmental variable in predicting habitat suitability. The results reveal that precipitation of the driest month (Bio 14) and precipitation of the warmest quarter (Bio 18) have the major percent contribution with 52.6% and 22.5%, respectively. The variable importance suggested by Jackknife test results in Fig. 26.5 also shows that Precipitation of the driest month (Bio 14) and Precipitation of the coldest quarter (Bio 19) have the most significant importance, followed by Precipitation of the warmest quarter (Bio 18). The study by Uehira and Shimono (2024) also reported that precipitation of the coldest quarter is a major factor for the habitat suitability of invasive species. Other major factors influencing the habitat suitability of invasive species in the study were Temperature seasonality, Annual Mean Temperature, and Temperature Annual Range.

Table 26.1 Selected environmental factors for predicting the habitat suitability of *Acacia auriculiformis* in Kasaragod district

Environmental variables	Description
Bio 4	Temperature seasonality (° C)
Bio 8	Mean temperature of the wettest quarter (° C)
Bio 12	Annual precipitation (mm)
Bio 14	Precipitation of the driest month(mm)
Bio 18	Precipitation of the warmest quarter (mm)
Bio 19	Precipitation of the coldest quarter (mm)

26.3.2 Future Habitat Suitability

The predictions given by the MaxEnt model indicate that the future climatic conditions of the Kasaragode district are not suitable for *Acacia auriculiformis*. Booth (2021) has categorised the countries in the world based on the climatic similarity they show with the native regions of Acacias, where India has been categorised as not similar. Figure 26.6 depicts the future habitat suitability for *Acacia auriculiformis* in Kasaragod under climate change scenarios SSP 245 and SSP 585 during different time periods. The results indicate that the areas suitable for the species will decrease to 0% for both scenarios in the future. The habitat suitability modelling of *Acacia auriculiformis* by Heringer et al. (2019) also suggest that the areas which are suitable for the species will shrink in future. Under SSP 245, the area which is moderately suitable for *Acacia auriculiformis* will be 0.74% and 0.78% for the period of 2021–2040 and 2041–2060, respectively, while it was 0% for 2061–2080 and 2081–2100. The moderately suitable area for the species during 2021–2040 will be 0.78% under SSP 585, which will decrease to 0% after 2040. Even though the projected decline in suitable areas for *Acacia auriculiformis* shows a lesser probability of the spread of this species, the species still exists as a threat due to its invasive behaviour, which will alter the ecosystem dynamics.

26.3.3 Possible Threat and Major Area for Concern

Invasive species are a major threat due to their high adaptive capacity to adverse environmental conditions (Schmerbeck and Naudiyal 2014) and dominance in the ecosystem (Heringer et al. 2019). The world's largest wildlife survey reveals that 66% of India's natural system is threatened by invasive plant species. Historical propagation of invasive plants, changes in soil characteristics, and climatic conditions are the major drivers for the facilitation of the invasion. The invasion has caused extensive loss in agriculture and forest ecosystems in India (Sankaran and Sajeev (2012).

In the current research, the area which shows moderate suitability for the species is located near the Ranipuram National Park, which is rich in biodiversity. The

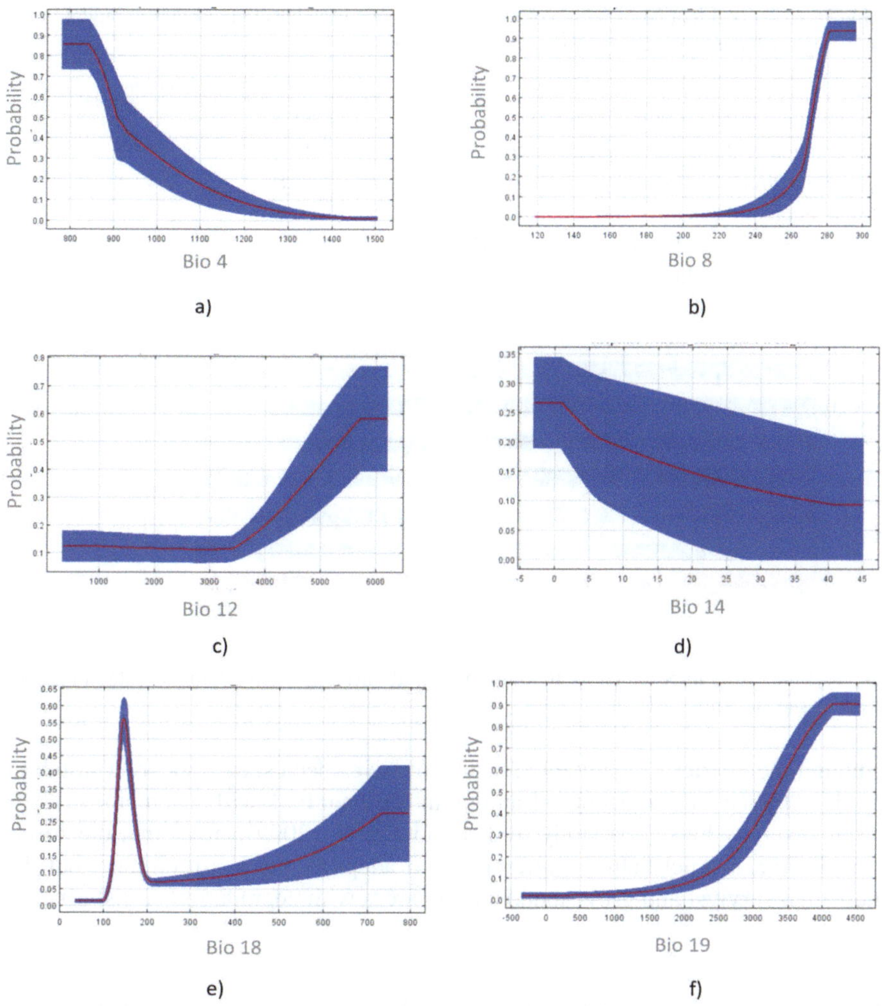

Fig. 26.4 Response curve showing the relationship between environmental variables and the probability occurrence of *Acacia auriculiformis:* (**a**) temperature seasonality (Bio 4), (**b**) mean temperature of the wettest quarter (Bio 8), (**c**) annual precipitation (Bio 12), (**d**) precipitation of the driest month (Bio 14), (**e**) precipitation of the warmest month (Bio 18), (**f**) precipitation of the coldest quarter (Bio 19)

favourable environmental conditions provided by the area will accelerate the spread of this invasive species and degrade the ecosystem (Kotiluoto et al. 2009). The annual average rainfall required for this species ranges from 940 to 3430 mm in its native range (Doran and Turenbull 1997). The research by Schmerbeck and Naudiyal (2014) reported that the species will not suit areas with an average annual rainfall below 1000 mm. The results of the current study also show the increase in suitability with an increase in average annual rainfall, especially beyond 3000 mm. Thus,

Table 26.2 The percent contribution of the environmental variables to the MaxEnt model for predicting the habitat suitability of *Acacia auriculiformis*

Variable	Per cent contribution
Precipitation of the driest month (mm)	52.6
Precipitation of the warmest quarter (mm)	22.5
Annual precipitation (mm)	7.8
Precipitation of the coldest quarter (mm)	6.9
Mean temperature of the wettest quarter (° C)	5.7
Temperature seasonality (° C)	4.5

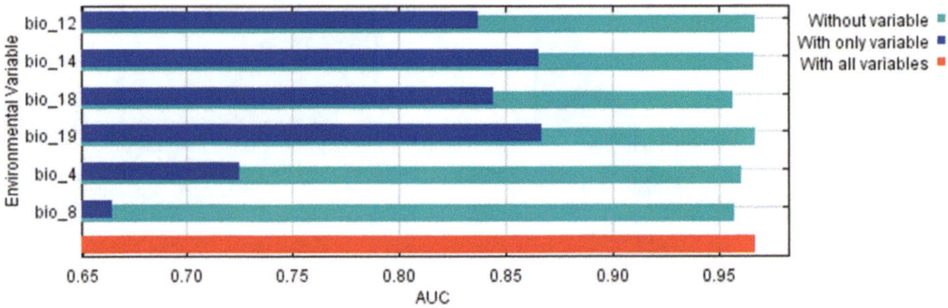

Fig. 26.5 Relative importance of the environmental variables in the habitat suitability model based on the jackknife test

although the district is unsuitable for this invasive species, the moderate suitability provided by the area adjacent to Ranipuram National Park will be a major stress for the biodiversity here. The study by Kotiluoto et al. (2009) shows that the species don't have a distinct habitat preference and adapt to any harsh climate, but they show high survival and grow abundantly in areas with degradation or disturbances.

26.3.4 Management Strategies and Future Recommendations

India has one of the world's highest population densities, so there will be a shift in the lifestyle, food, and infrastructure, which will intensify the already existing unchecked invasion even more. Using the data collected from National-scale invasive plant assessment, the researchers introduced a new model that identifies the sites where the invasion is in its initial stage with priority for the protected areas. Protecting these sites requires less investment and intervention (British Ecological Society 2023).

The convention on Biological Diversity (CBD) mandates steps to stop new invasive species from being introduced and to increase efforts to control or destroy existing invasive species. One of the current strategic plans of the CBD is the 20 Aichi Biodiversity Targets, Target 9 of which includes the identification and

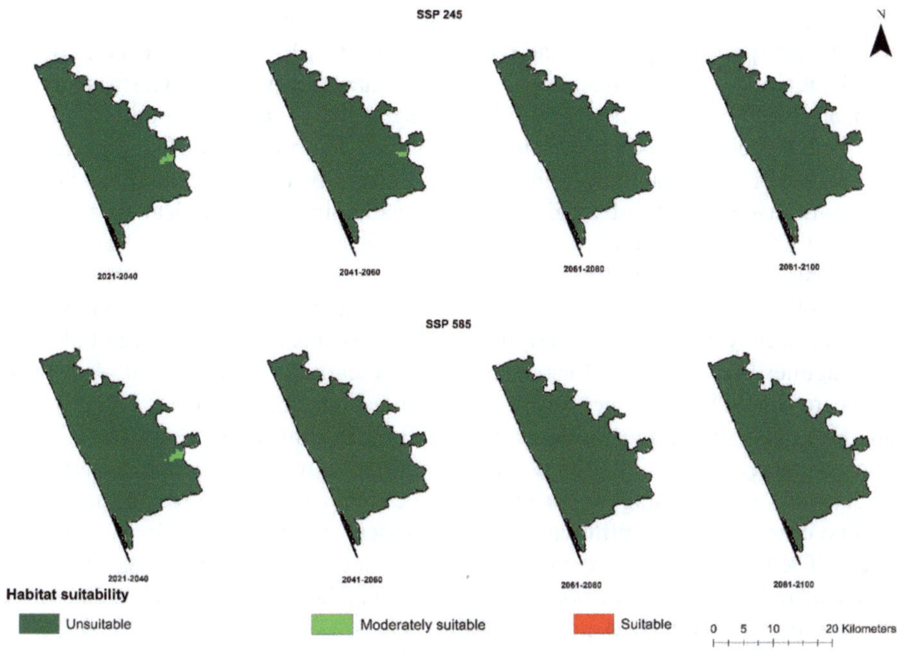

Fig. 26.6 Future habitat suitability of *Acacia auriculiformis* in Kasaragod, Kerala

prioritisation of invasive species and their dispersal pathways as the priority actions to manage biological invasions by 2020 (CBD 2010). The pace of species imports across geographic regions and within taxonomic groups, however, does not appear to be declining, suggesting that previous efforts were insufficient (Seebens et al. 2021).

Early detection, risk assessment, pathway control, vector management, and innovative mitigation and restoration measures appear to be the most promising areas for addressing invasion challenges (Pyšek and Richardson 2010). Conservation or restoration of ecosystems, native biodiversity, and ecosystem functions should be the main objective of such initiatives. Available technical options, projected impacts, risk management of targeted species, success rate, and the level of stakeholder engagement and public concern are all important considerations in a comprehensive invasion management strategy (Hulme 2006). It was believed that the invasion of alien species would be less in natural habitats as compared to the disturbed habitats. However, recent studies have shown that the survival strategy of invasive species has increased and helped them adapt to the natural ecosystem (Sankaran and Sajeev 2012).

The parties to the Convention on Biological Diversity 2010 held at Nagoya, Japan, adopted a special strategy for prioritising, eradicating, and controlling measures to prevent the pathway and establishment of invasive species (Goyal et al. 2018). The lack of baseline data on species introduction across geographic regions and within taxonomic groups is making it difficult for several developing countries, to formulate pertinent policies and programs (MFSC 2014; McGeoch et al. 2016).

The intentional use of resilience-based governance approaches (Garmestani and Benson 2013) that can coordinate the management of invasive species at scales pertinent to ecosystems, ecosystem function, and the provision of ecosystem services—rather than at anthropocentric scales like political and jurisdictional boundaries—is made possible by gaining an understanding of biological invasions in terms of ecological resilience. The ability of ecosystems to rearrange and regain their basic structure, functions, and processes—that is, to recover—when subjected to stress factors like prolonged and severe drought and disturbances like wildfires is known as ecosystem resilience (Holling 1973). It points to the existence of alternative regimes. The potential for recovery and the capacity to withstand plant invasions are shown by resilience and resistance indicators (Chambers et al. 2023). The adaptive cycle of complex systems illustrated by Holling 1986 showed how a system can experience a shift after passing a disturbance threshold, which is marked by a relatively abrupt change in structuring processes and reorganisation into a new regime dominated by a different set of processes, structures, functions, and feedbacks. The biotic and abiotic characteristics of an ecosystem, as well as ecological mechanisms that restrict the population expansion of an invading species, determine resistance to invasion (D'Antonio and Thomsen 2004).

The *Acacia auriculiformis* is a shade-intolerant species (Kotiluoto et al. 2009), thus the regeneration and establishment of this invasive species can be inhibited by sustaining more shade in the areas. In a study conducted by Elton (1958), he stated that greater the diversity, the risk of invasion will be less. Thus, promoting the restoration of the degraded areas prone to invasion will also reduce the risk of invasion and ensure successful restoration (Heringer et al. 2019).

26.4 Conclusion

Invasive species is recognised as a major cause for biodiversity decline and ecosystem degradation. The movement of plants through sea, land, and air further increased the possibilities of the spread of invasive species. Even though proper quarantine systems are present, the spread of alien invasive species is still happening and posing threats. Climate change, which is another major concern in the current era, is positively influencing the invasion by providing favourable factors directly or indirectly. The current study noticed that the major variables contributing to the habitat suitability model of *Acacia auriculiformis* are Precipitation of the driest month (Bio 14) and Precipitation of the coldest quarter (Bio 19). However, the decrease in temperature seasonality, increase in mean temperature of the wettest quarter, and precipitation of the coldest quarter will increase the probability of the species presence. The AUC value (0.96) given by the MaxEnt model indicates that the chosen model is excellent in predicting the habitat suitability of the species. The model suggests that 60.5% land of Kasaragod district is suitable for *Acacia auriculiformis* in the current period, which will decline to 0% in the future. But the moderate suitability provided by the Ranipuram national park, biodiversity rich area of Kasaragod district, in the near future under climate change scenarios SSP 245 and 585 is alarming. Several attempts have been made worldwide to protect the natural ecosystems from invasion. Still, invasion continues to pose a major threat due to its higher adaptability to any harsh environment. Considering healthy native vegetation inhibits spread of invasive species, restoration of the natural ecosystem and prevention of its degradation should be given first priority to eradicate invasion.

Acknowledgement The facilities provided by the Central University of Kerala for the execution of the study and the financial assistance in the form of a junior research fellowship provided by the Kerala State Council for Science, Technology and Environment to Mrs. Kavya Jeevan are duly acknowledged.

Disclosures and Declarations All authors of the study consent to publish the book chapter.

Conflict of Interests The authors declare that they have no conflict of interest

Ethics Approval Not applicable

Data Availability Statement The data used for the study will be available on request.

Funding Statement No funders to disclose

References

Bellard C, Leroy B, Thuiller W, Rysman JF, Courchamp F (2016) Major drivers of invasion risks throughout the world. Ecosphere 7(3):01241

Bellard C, Jeschke JM, Leroy B, Mace GM (2018) Insights from modeling studies on how climate change affects invasive alien species geography. Ecol Evol 8(11):5688–5700

Booth TH (1987) Selecting Acacia species for testing outside Australia. In: Turenbull JW (ed) Australian acacias in developing countries: proceedings of an international workshop. Australian Centre for International Agricultural Research, Canberra, pp 74–76

Booth TH (2021) A problem with variable selection in a comparison of correlative and process-based species distribution models: comments on Higgins et al., 2020. Ecol Evol 11(19):13609–13612

Booth TH, Jovanovic T (1988) Assaying natural climatic variability in some Australian species with fuelwood and agroforestry potential. Commonw For Rev 67(210):27–34

Bradley BA, Laginhas BB, Whitlock R, Allen JM, Bates AE, Bernatchez G et al (2019) Disentangling the abundance–impact relationship for invasive species. Proc Natl Acad Sci 116(20):9919–9924

Brewbaker JL (1987) Performance of Australian acacias in Hawaiian nitrogen-fixing tree trials. In: Turenbull JW (ed) . Australian acacias in developing countries, Forestry Training Centre, pp 16.180–16.184

Brewbaker JL, Halliday J, Lyman J (1983) Economically important nitrogen fixing trees. Winrock International Institute for Agricultural Development, p 35

Brewbaker JL, Van Den Beldt R, MacDicken K (1984) Fuelwood uses and properties of nitrogen-fixing trees. Pesq Agropec Bras 19:193–204

British Ecological Society (2023) World's largest wildlife survey highlights threats from invasive plant species in India. https://wwwbritishecologicalsociety.org. Assessed 1 Dec 2024

Chakraborty RN, Chakraborty D (1989) Changes in soil properties under Acacia auriculiformis plantations in Tripura. Indian For 115(4):272–273

Chambers JC, Brown JL, Bradford JB, Board DI, Doherty KE, Crist MR, Schlaepfer DR, Urza AK, Short KC (2023) Combining resilience and resistance with threat-based approaches for prioritizing management actions in sagebrush ecosystems. Conservation Science and Practice, p e13021

Convention on Biological Diversity (2009) Invasive Alian species: a threat to biodiversity. Secretariat of the Convention on Biological Diversity wwwcbdint. Assessed 25 Nov 2024

Convention on Biological Diversity (2010) Strategic plan for biodiversity 2011–2020 and the Aichi biodiversity targets. https://wwwcbdint/sp. Assessed 25 Nov 2024

D'Antonio CM, Thomsen M (2004) Ecological resistance in theory and practice. Weed Technol 18(sp1):1572–1577. https://doi.org/10.1614/0890-037X(2004)018

Daehler CC (2003) Performance comparisons of co-occurring native and alien invasive plants: implications for conservation and restoration. Annu Rev Ecol Evol Syst 34(1):183–211

Davis MA (2011) Invasive species 50 years after Elton: a cautionary tale. In: Fifty years of invasion ecology, p 269

Defenders of Wildlife (2023) How do invasive species affect biodiversity and how can they be controlled? Retrieved from https://defenders.org. Assessed 9 Dec 2024

Didham RK, Tylianakis JM, Hutchison MA, Ewers RM, Gemmell NJ (2005) Are invasive species the drivers of ecological change? Trends Ecol Evol 20(9):470–474

District Urbanisation Report (2011) Department of town and country planning. Government of Kerala

Doran JC, Turenbull JW (1997) Australian trees and shrubs: species for land rehabilitation and farm planting in the tropics. Australian Centre for International Agricultural Research, Canberra, p 394

Duponnois R, Baudoin E, Sanguin H, Thioulouse J, Le Roux C, Tournier E, Galiana A, Prin Y, Dreyfus B (2013) Is the introduction of Australian acacias to rehabilitate degraded ecosystems free from environmental risks? Bois For Trop 318(4):59–65

Elith J, Kearney M, Phillips S (2010) The art of modelling range-shifting species. Methods Ecol Evol 1(4):330–342

Elton CS (1958) The Ecology of Invasions by Animals and Plants. T. Methuen and Co, London. http://doi.org/10.1007/978-1-4899-7214-9

Fick SE, Hijmans RJ (2017) WorldClim 2: new 1km spatial resolution climate surfaces for global land areas. Int J Climatol 37(12):4302–4315

Garmestani AS, Benson MH (2013) A framework for resilience-based governance of social-ecological systems. Ecol Soc 18:9

Goyal N, Esler KJ, Sharma GP (2018) What drives performance potential of Lantana camara L. (sensu lato) in the invaded range? Trop Ecol 59:57–68

Grove SE, Haubensak KA, Gehring CA, Parker IM (2017) Mycorrhizae, invasions, and the temporal dynamics of mutualism disruption. J Ecol 105(6):1496–1508

Helen R, Peyton J, Aldridge DC, Bantock T, Blackburn TM, Britton R, Walker KJ (2014) Horizon scanning for invasive alien species with the potential to threaten biodiversity in Great Britain. Glob Chang Biol 20(12):3859–3871

Heller NE, Zavaleta ES (2008) Biodiversity management in the face of climate change: a review of 22 years of recommendations. Biol Conserv 142(1):14–32

Heringer G, Bueno ML, Meira-Neto JAA, Matos FAR, Neri AV (2019) Can Acacia mangium and Acacia auriculiformis hinder restoration efforts in the Brazilian Atlantic Forest under current and future climate conditions? Biol Invasions 21(9):2949–2962

Holling CS (1973) Resilience and stability of ecological systems. Annu Rev Ecol Evol Syst 4(1):1–23

Holling CS (1986) The resilience of terrestrial ecosystems: local surprise and global change. In: Clark WC, Munn RE (eds) Sustainable development of the biosphere. Cambridge University Press, Cambridge, UK, pp 292–317

Hulme PE (2006) Beyond control: wider implications for the management of biological invasions. J Appl Ecol 43(5):835–847

Intergovernmental Panel on Climate Change (2022) Climate change 2022: impacts, adaptation and vulnerability. Working group II contribution to the sixth assessment report of the intergovernmental panel on climate change. Retrieved from https://www.ipcc.ch/report/ar6/wg2/. Assessed 10 Dec 2024

Intergovernmental Science-Policy Platform on Biodiversity and Ecosystem Services (2023) Summary for policymakers of the thematic assessment report on invasive alien species and their control. Retrieved from https://www.ipbes.net. Assessed 12 Dec 2024

Kaphle K, Pokheral CP (2014) Invasive alien species management through utilization for application in domesticated livestock species raising. In: Proceedings: invasive alien species management. National Trust for Nature Conservation Biodiversity Conservation Centre, Chitwan, p 232

Karunakaran N (2014) Crop diversification and environmental conflicts in Kasaragod District of Kerala. Agric Econ Res Rev 27(2):299

Kasongo RK, Van Ranst E, Verdoodt A, Kanyankagote P, Baert G (2009) Impact of Acacia auriculiformis on the chemical fertility of sandy soils on the Batéké plateau, D.R Congo. Soil Use Manage 25(1):21–27

Khuroo AA, Reshi ZA, Malik AH, Weber E, Rashid I, Dar GH (2012) Alien flora of India: taxonomic composition, invasion status and biogeographic affiliations. Biol Invasions 14:99–113

Kishore BSPC, Kumar A, Saikia P (2024) Understanding the invasion potential of Chromolaena odorata and Lantana camara in the Western Ghats, India: An ecological niche modelling approach under current and futureclimatic scenarios. Ecol Info, 79:102425.https://doi.org/10.1016/j.ecoinf.2023.102425

Kotiluoto R, Ruokolainen K, Kettunen M (2009) Invasive Acacia auriculiformis Benth. In different habitats in Unguja, Zanzibar. Afr J Ecol 47(1):77–86

Kühn I, Klotz S (2008) From ecosystem invasibility to local, regional and global patterns of invasive species. In: Biological invasions. Springer, Berlin/Heidelberg, pp 181–196

Lambers H, Oliveira RS (2019) Plant water relations. In: Lambers H, Oliveira RS (eds) Plant physiological ecology. Springer, pp 187–263

Li W, Yu Y, Liu P, Tang R, Dai Y, Li L, Zhang L (2019) Identifying climate refugia and its potential impact on small population of Asian elephant (Elephas maximus) in China. Global Ecol Conserv 19:e00664

Lockwood JULIE (2005) Predicting which species will become invasive: what's taxonomy got to do with it. In: Phylogeny and conservation, pp 365–386

Masters G, Norgrove L (2010) Climate change and invasive alien species. CABI working paper 1, p 30

McGeoch MA, Genovesi P, Bellingham PJ, Costello MJ, McGrannachan C, Sheppard A (2016) Prioritizing species, pathways, and sites to achieve conservation targets for biological invasion. Biol Invasions 18:299–314

Meira-Neto JAA, da Silva MCNA, Tolentino GS et al (2018) Early Acacia invasion in a sandy ecosystem enables shading mediated by soil, leaf nitrogen and facilitation. Biol Invasions 20:1567–1575.https://doi.org/10.1007/s10530-017-1647-2

Ministry of Forests and Soil Conservation (MFSC) (2014) National biodiversity strategy and action plan: 2014–2020. Government of Nepal

Muscarella R, Galante PJ, Soley-Guardia M, Boria RA, Kass JM, Uriarte M, Anderson RP (2014) ENM eval: an R package for conducting spatially independent evaluations and estimating optimal model complexity for MAXENT ecological niche models. Methods Ecol Evol 5(11):1198–1205

Padalia H, Srivastava V, Kushwaha SPS (2014) Modeling potential invasion range of alien invasive species, Hyptis suaveolens (L.) Poit. In India: comparison of MaxEnt and GARP. Eco Inform 22:36–43

Pejchar L, Mooney HA (2009) Invasive species, ecosystem services and human Well-being. Trends Ecol Evol 24(9):497–504

Pereira J, Pinto A, Almeida JP, Pinto S, Oliveira AG, De Carvalho M (2010) Home telemonitoring of non-invasive ventilation decreases healthcare utilisation in a prospective controlled trial of patients with amyotrophic lateral sclerosis. J Neurol Neurosurg Psychiatry 81(11):1238–1242

Pinyopusarerk K (1990) Acacia auriculiformis: an annotated bibliography. Winrock International Institute of Agricultural Development, Australian Centre for International Agricultural Research, p 154

Pyšek P, Richardson DM, Pergl J, Jarošík V, Sixtová Z, Weber E (2008) Geographical and taxonomic biases in invasion ecology. Trends Ecol Evol, 23(5):237–244.https://doi.org/10.1016/j.tree.2008.02.002

Pyšek P, Richardson DM (2010) Invasive species, environmental change and management, and health. Annu Rev Environ Resour 35(1):25–55

Qian H, Zhang Y, Ricklefs RE, Wang X (2022) Relationship of minimum winter temperature and temperature seasonality to the northern range limit and species richness of trees in North America. J Geogr Sci 32(2):280–290

Rahel FJ, Olden JD (2008) Assessing the effects of climate change on aquatic invasive species. Conserv Biol 22:521–533

Rai PK, Singh JS (2024) Ecological insights and environmental threats of invasive alien plant *Chromolaena odorata*: prospects for sustainable management. Weed Biol Manag 24(1):15–37. https://doi.org/10.1111/wbm.12286

Reddy CS (2008) Catalogue of invasive alien flora of India. Life Sci J 5(2):84–89

Riahi K, Van Vuuren DP, Kriegler E, Edmonds J, O'Neill BC, Fujimori S, Bauer N, Calvin K, Dellink R, Fricko O, Lutz W, Popp A, Cuaresma JC, Kc S, Leimbach M, Jiang L, Kram T, Rao S, Emmerling J et al (2017) The shared socioeconomic pathways and their energy, land use, and greenhouse gas emissions implications: an overview. Glob Environ Chang 42:153–168

Richardson DM, Pyšek P (2007) Elton, CS 1958: the ecology of invasions by animals and plants. Global Environ Change 31(6):659–666

Rodríguez-Echeverría S, Le Roux JJ, Crisóstomo JA, Ndlovu J (2011) Jack-of-all-trades and master of many? How does associated rhizobial diversity influence the colonization success of Australian Acacia species? Divers Distrib 17(5):946–957

Roura-Pascual N (2020) Drivers of future alien species impacts: an expert-based assessment. Glob Chang Biol 26(9):4880–4893

Royle JA, Cam E (2011) Hierarchical modeling of an invasive spread: the Eurasian Collared-Dove *Streptopelia decaocto* in the United States. Ecol Appl 21(1):290–302

Sankaran KV, Sajeev TV (2012) Needs assessment and formulation of technical co-operation programme (TCP) project for forest invasive species (fis) in south India, Sri Lanka and Maldives

Saxena KG (1991) Biological invasions in the Indian subcontinent: review of invasion by plants. In: Ramakrishnan PS (ed) Ecology of biological invasion in the tropics. International Scientific Publications, New Delhi, pp 53–73

Schmerbeck J, Naudiyal N (2014) In: Roloff A, Weisgerber H, Lang UM, Stimm B (eds) Enzyklopädie der Holzgewächse: Handbuch und Atlas der Dendrologie, Wiley-VCH VerlagGmbH and Co. KGaA. https://doi.org/10.1002/9783527678518

Seebens H, Bacher S, Blackburn TM, Capinha C, Dawson W, Dullinger S et al (2021) Projecting the continental accumulation of alien species through to 2050. Glob Chang Biol 27(5):970–982

Siddiq Z, Cao KF (2016) Increased water use in dry season in eight dipterocarp species in a common plantation in the northern boundary of Asian tropics. Ecohydrology 9:871–881.https://doi.org/10.1002/eco.1689

Suresh TA, Sajeev TV, Sankaran KV (2012) Are alien invasive plants a threat to forests of Kerala. *KFRI* occasional papers, Forest Health Programme Division. Kerala Forest Research Institute, Peechi

Swets JA (1988) Measuring the accuracy of diagnostic systems. Science (New York, N.Y.), 240(4857): 1285–1293.https://doi.org/10.1126/science.3287615

Times of India (2022, December 19) Forest dept starts felling acacia trees. The Times of India. (Accessed on 8 December 2024)

Uehira K, Shimono Y (2024) Evaluation of climate conditions and ecological traits that limit the distribution expansion of alien Lolium rigidum in Japan. NeoBiota 96:89–104

Usher MB (1988) Biological invasions of nature reserves: a search for generalizations. Biol Conserv 44(1–2):119–135

Vicente JR, Fernandes RF, Randin CF, Broennimann O, Gonçalves J, Marcos B et al (2013) Will climate change drive alien invasive plants into areas of high protection value? An improved model-based regional assessment to prioritise the management of invasions. J Environ Manag 131:185–195

Villemant C, Barbet-Massin M, Perrard A, Muller F, Gargominy O, Jiguet F, Rome Q (2011) Predicting the invasion risk by the alien bee-hawking yellow-legged hornet Vespa velutina nigrithorax across Europe and other continents with niche models. Biol Conserv 144(9):2142–2150

Wallach AD, Ripple WJ, Carroll SP (2015) Novel trophic cascades: apex predators enable coexistence. Trends Ecol Evol 30:146–153

Wang A, Melton AE, Soltis DE, Soltis PS (2022) Potential distributional shifts in North America of allelopathic invasive plant species under climate change models. Plant Divers 44(1):11–19

Weiskopf SR, Rubenstein MA, Crozier LG, Gaichas S, Griffis R, Halofsky JE, Hyde KJ, Morelli TL, Morisette JT, Muñoz RC, Pershing AJ, Peterson DL, Poudel R, Staudinger MD, Sutton-Grier AE, Thompson L, Vose J, Weltzin JF, Whyte KP (2020) Climate change effects on biodiversity, ecosystems, ecosystem services, and natural resource management in the United States. Sci Total Environ 733:137782

Wendlandt CE, Gano-Cohen KA, Stokes PJN, Jonnala BNR, Zomorrodian AJ, Al-Moussawi K, Sachs JL (2022) Wild legumes maintain beneficial soil rhizobia populations despite decades of nitrogen deposition. Oecologia 198(2):419–430. https://doi.org/10.1007/s00442-022-05116-9

Chapter 27
The Role of Climate Change in Shaping Copepod Communities and Invasive Species Dynamics: An Ecological Perspective

R. Nandakumar, M. Jayaprakashvel, and M. Sindhuja

Abstract This chapter aims to highlight the effects of climate change on the invasion and displacement of copepod populations. Copepods are integral components of marine food web that serve as predators and prey. Their dietary preferences range from herbivorous to carnivorous and can even lead parasitic lives. Diel vertical migration, an important aspect of copepods, significantly contributes to carbon sequestration and nutrient cycling. Copepods can exhibit varied response to community structure, abundance, and distribution changes, making them excellent bioindicators. With the stress of increasing global warming and ocean acidification, their physiology, phenology, and distribution will change along with the rest of the ecosystems and be commercially affected by copepods as primary prey. These impacts enable invasion of new species, further worsened by increased temperatures and maritime trade. Emerging genomic studies and eDNA metabarcoding offer better strategies for managing and efficiently monitoring of these invasions. Effective strategies and collaborative efforts to consider combined effects of climate change and invasive species are necessary to mitigate these issues.

Keywords Copepods · Climate change · Invasive species · Marine ecosystem

R. Nandakumar (✉) · M. Jayaprakashvel
Department of Marine Biotechnology, Academy of Maritime Education and Training (AMET), Chennai, Tamil Nadu, India

M. Sindhuja
Department of Computing Technologies, School of Computing, College of Engineering and Technology, SRM Institute of Science and Technology, Kattankulathur, Tamil Nadu, India

B. P. Barathan et al. (eds.), *Navigating Climate Change: Impacts on Biodiversity and Ecosystem Resilience*, Sustainable Development and Biodiversity 42,
https://doi.org/10.1007/978-981-95-0409-1_27

27.1 Introduction

27.1.1 The Critical Role of Copepods in Marine Ecosystems

Primary producers such as phytoplankton typically form the base of aquatic food webs. Zooplanktons are secondary producers that link phytoplankton and higher trophic levels through energy transfer. Copepods are diverse type of zooplankton found in almost every environment, from freshwater lakes to the open ocean (Mauchline 1998). Copepods play an important role in both food webs and biogeochemical cycles. Their vertical migrations, whereby they move to deep water, play an important role in the vertical movement of organic matter through the water column (Turner 2015). This activity, known as the "biological pump," is vital to the global carbon budget and to dilute the effects of climate change by removing carbon dioxide from the atmosphere (Karl et al. 1996).

Copepods possess significant ecological adaptability, with dietary habits ranging from herbivory to carnivory and most species leading parasitic lives (Boxshall and Halsey 2004). Copepods are excellent bioindicators of environmental change due to their flexibility and diversity. They respond to climate change, ocean acidification, and pollution in a variety of ways including change in community structure, abundance, and distribution (Hays et al. 2005). Hence, implementing successful conservation programs requires research on copepods' response to environmental changes. Further, studying copepod populations is critical to understand changes and mitigation required in an ecosystem (Fig. 27.1).

27.1.2 Climate Change and Invasive Species: A Dual Threat

27.1.2.1 Synergistic Impacts of Warming, Acidification, and Species Invasions

Since 1900, global mean surface temperatures have risen by about 0.7 °C each century, with a faster rate of 0.16 °C per decade since 1970. The consequences of global warming and ocean acidification are accelerated by anthropogenic activity, leading to disruption of marine ecosystems. On the other hand, rising temperatures favors the distribution of invasive species by making conditions suitable for their survival (Barton et al. 2016). Warmer waters, for example, have allowed the invasive copepod *Oithona davisae* to spread northward in the North Sea, potentially outcompeting native species for resources (Barton et al. 2016). Ocean acidification can impact skeletogenesis and metabolic processes, increasing mortality and making them more vulnerable to invading species (Kroeker et al. 2013). While copepods' shells are not calcified, acidification can still affect their egg production, growth, and food source.

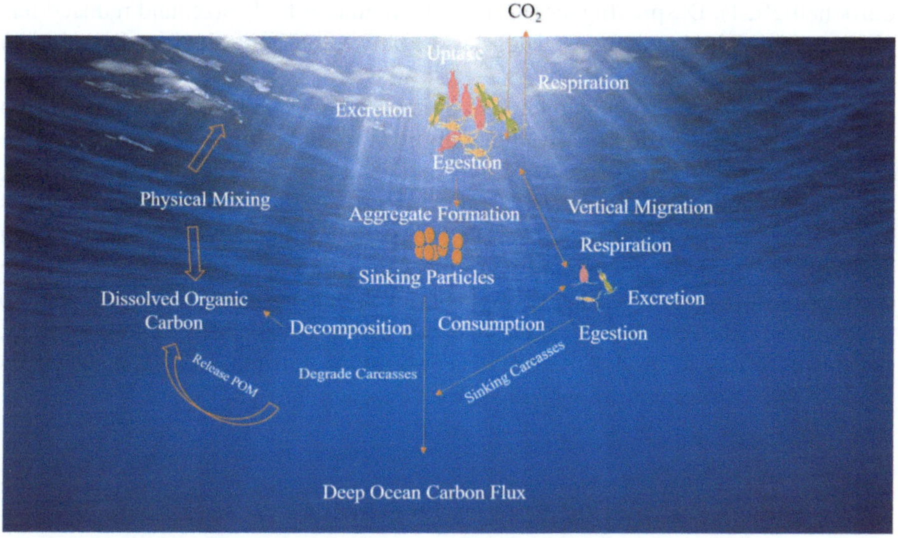

Fig. 27.1 The role of copepods in the marine environment due to climate change

27.1.2.2 Why Copepods Are Ideal Model Organisms

Changes in copepod populations can have varying consequence at higher trophic levels, thus, making them ideal models for studying the consequences of climate change and invasive species. Furthermore, their short generation times make them excellent for adaptation and evolution studies (Hirst and Bunker 2003). Their sensitivity to temperature, pH, and other environmental changes makes them useful markers of ecosystem health (Richardson 2008).

Their diversity, small size, and transparent bodies allow precise studies on their behavior, reproduction, and physiological responses to stimuli. They are also extremely useful in experimental investigations since they can be cultured and harvested in laboratory conditions. These reasons highlight their importance in studies related to climate-driven changes.

27.2 Climate Change Impacts on Copepod Populations

27.2.1 Physiological Responses to Warming Oceans

According to Q10 principle, warmer temperatures generally boost metabolic rate. As a result, metabolic processes use the majority of the energy, negatively affecting other growth parameters. However, research indicates that their heat tolerance might vary over generations. For example, species *Paracartia grani* raised for over 18 generations demonstrated a 1–1.3 °C increase in temperature tolerance (Juan

Carbonell 2024). Despite higher survival, their smaller body size, and reduced fertility at optimal temperatures, suggests negative consequence in adaptation to warmer waters.

Rising temperatures can deplete energy reserves if food supply does not match metabolic demand. Additionally, fluctuation in salinity along with temperature changes furthers metabolic process Lagerspetz and Kivivuori (1970). For instance, the estuarine copepod, *Acartia tonsa*, previously exposed to low salinity conditions showed lower thermal tolerance in temperature-related experiments (Von Weissenberg et al. 2022). Global warming can shift important events in the life cycle of living organisms, like mating and development. These shifts may lead to inconsistency between life cycle of copepods and the availability of food sources. Organisms failing to survive these effects generally get displaced from the environment, in turn disrupting ecological balance Schoener et al. (2019).

27.3 Invasive Copepod Species Dynamics

27.3.1 Pathways of Invasion

Maritime trade can effectively aid the spread of non-native organisms, mainly through ballast water. This water filled with copepods can bypass natural barriers and easily occupy new environments. Certain situations where high water temperature and salinity fluctuation can provide niche conditions to migratory species to outcompete indigenous species. In addition to the human impact, this kind of climate-driven conditions will facilitate the migration of the invasive species.

27.3.2 Traits of Successful Invasive Copepods

Rapid reproduction is a key feature of invasive species, allowing them to establish population size large enough to dominate native species. In addition, their high fecundity and thus short time required for new generations make them adapt to the new location, where they can further exploit the resources. Invasive copepods through their diet flexibility that is, the ability to feed on a variety of food sources, can establish their dominance in food web. Wonham et al. (2001) observed that the majority of calanoid and cyclopoid copepods initially present in ballast water did not survive the ship voyage, whereas harpacticoids did. Harpacticoids being grazers can survive on flexible diets, while cyclopoids and calanoids rely on phytoplankton, whose productivity was hindered by dark conditions during the voyage. Further, their adaptability to a variety of temperatures has the additional potential to invade new areas. So, the best way forward is a multilevel approach following a strict system of water ballast, the development of new mechanisms, and the involvement of researchers in climate change issue analysis. The authors admit this and they are indicating that protection of aquatic ecosystems from invasive copepods demands collective scientific, policy, and industry efforts.

27.4 Ecosystem Disruption and Cascading Effects

27.4.1 Food Web Restructuring

Zooplankton are a vital part of the food chains in the seas. Consequently, changes in their variety, quantity, and distribution are considerably important. Results from various surveys keep stressing the point that the influence of climate is the leading cause of the problem. Many species migrate in search of favorable place with optimal temperature. Such migrations can result in substitution of cold-water species by warm-water species, thereby changing the entire community. One more aspect of global warming that is the change in zooplankton seasonal events, such as time of reproduction or growth. The latest studies have shown that the quantity of zooplankton may not correspond with the number of phytoplankton, a primary source of their food (Hallegraeff 2010). Such shifts in food availability will lead to less successful zooplankton reproduction rate and negatively influence higher trophic levels. Despite this, there are reports saying that communities are increasingly composed of small-sized zooplankton which might be connected with higher metabolic rates and food quality changes. Thus, the efficiency of energy transfer through the food web network will decrease (Heneghan et al. 2016).

These alterations can vary based on hydrography, existing community structure, and other stress in synergy. However, endpoint seems to result in inefficient energy transfer to higher trophic levels and ineffective carbon sequestering. Understanding these changes is crucial for predicting the health of marine ecosystems, managing fisheries sustainably, and assessing the ocean's role in climate change.

27.4.2 Impacts on Higher Trophic Levels (Fish, Seabirds, Whales)

The effects of climate change on marine organisms can be both direct and indirect. Organisms at a higher trophic level are heavily dependent on the productivity and community structure of the food web. So, changes at the base of the food web can create significant challenges for top predators. Warmer waters directly impact fish physiology, affecting its reproductive cycle and success. Sea-level rise and storms can destroy nesting habitats of seabirds that rely on specific locations. Most seabirds depend on the availability and distribution of small fish as food source (Grémillet and Boulinier 2009). Migration of fish due to climate change forces seabirds to expend more energy in search of food. Similarly, certain species of whales rely on a steady supply of prey at designated feeding grounds. Movement of those prey forces whales to alter their migration routes and potentially impact their energy budgets and reproductive health. For ice-dependent creatures, the rapid loss of sea ice means a considerable loss of critical habitat for resting, reproducing, and obtaining food sources.

These combined effects on these higher trophic levels can potentially disrupt entire ecosystem. Hence, understanding these interactions and their impacts on various organisms becomes vital. Recent research highlights the interconnected effects and their link to the overall of the marine environment under climate change (Fig. 27.2).

27.5 Emerging Tools and Collaborations

27.5.1 Genomic Insights into Species' Climate Responses

Genomic studies are helping us understand how organisms are genetically changing to survive rapid climate change. By observing genes of different groups of a species in various environments over time, scientists can find the specific genes that help them adjust. These genes often help them handle environmental stress. Hence, finding the genes that help them adapt to these environmental variation is crucial. Similarly, environment with constant fluctuations can serve as hubs for genetic research. Studies on corals have identified genes involved in the heat stress response, oxidative stress management, and interactions with symbiotic algae. In fish, genomic analyses have revealed adaptations related to thermal tolerance limits, cardiovascular performance at higher temperatures, and salinity tolerance. Identifying the specific genes and pathways helps in predicting resilient populations.

Despite these advances, significant challenges still remain. Many adaptive traits are influenced by numerous genes with small individual effects, making them harder to detect than single-gene adaptations. Distinguishing true genetic adaptation from phenotypic plasticity (nongenetic flexibility) requires careful experimental design and often long-term monitoring. Understanding the genomic architecture of adaptation is key to forecasting biological responses to ongoing climate change.

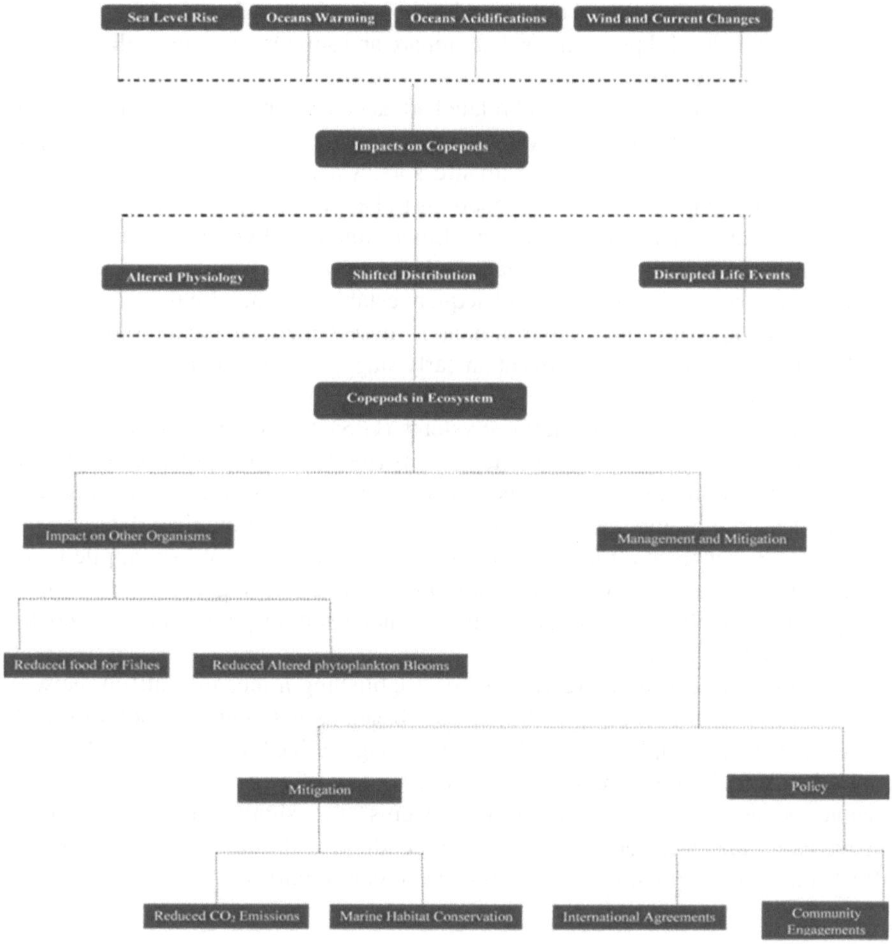

Fig. 27.2 Climate change mitigations by copepod populations

27.5.2 *The Role of Computing Technologies in Analyzing Climate Change Effects on Copepod Populations and Invasive Species Dynamics*

The impact of computing technologies on environmental science is particularly pronounced in the assessment of the needs and impacts of climate change on copepod communities, as well as on the processes of species invasions. High-performance computing has improved the ability of scientists to quickly and accurately assess and act on large datasets, such as from ocean sensors, satellite imagery, and genetic sequencing. Computational procedures allow researchers to model temporal ecosystems, predict patterns in the distribution of species, and consider effects from warming sea temperatures and ocean acidification on copepods. Computational procedures yield insightful projections that prepare and support climate resilience planning for marine biodiversity.

Genetics and bioinformatics and related advanced computational methodologies can give researchers greater clarity to the adaptation strategies copepods can employ when they live in an environment with stressors. Machine learning methods can be used to identify genetic markers in copepods linked to thermal toleration, reproductive viability, and tolerance to invasives. Environmental DNA (DNA) metabarcoding, a computational pipeline that analyzes large numbers of sequences, can be used for the monitoring of invasion and subsequent establishment, with the emergence of invasive copepods into new environments. These beneficial technologies can be used to detect ecological problems at an early stage, so rapid intervention and management can occur.

Notably, Geographic Information Systems (GIS) and remote sensing technologies are increasingly being used to track the distribution of copepod populations and the establishment of invasive species. These systems may overlay physical oceanographic data (e.g., temperature, salinity, and currents) on biological data and produce holistic representations of marine ecosystem status. When applied with forecasting tools, these platforms are important for identifying at-risk areas, monitoring ecological change, and developing a management response in marine conservation and fishery management.

Systems described here are critical for establishing a meaningful link between intricate ecological processes and their user-based assessments and solutions. The tools enable interdisciplinary collaboration among marine biology, climatology, and data science, facilitating comprehensive assessments of ecosystem risks. As climate change persistently alters marine ecosystems, the significance of computing resources escalates in monitoring, comprehending, and alleviating the ecological effects on copepod populations and associated communities.

27.6 Mitigation and Management Strategies

From the collection of Environmental DNA samples, it has become relatively easier to identify different species through DNA metabarcoding as a new species. This helps to manage detection of invasive geographies during the primary stages of invasion, even when their population is small. Below the surface of the water, on land, or even in the air, fragments of eDNA can be used to track invasive species. Migrating through different climates efficiently monitors the movement of invasive life forms. New research highlights how effective DNA metabarcoding is in tracking the spread of invasive species in a changing climate. For instance, Borrell et al. (2017) by examining the eDNA archives from blue mussel eDNA capsules have allowed scientists to retrace the invasion history of the barnacle *Austrominius modestus* over 40 years. This study has revealed important population dynamic patterns such as growth spurts, declines due to cold winter periods, and later reinvasions associated with warming periods.

A study by Stern and Lee (2020) tested a hypothesis that environmental fluctuations in the native range could promote successful invasions by imposing balancing selection on key traits and maintaining the genetic variation that enables rapid adaptation in novel habitats. This study tested the hypothesis by performing whole-genome sequencing of multiple independent invasive freshwater and native saline populations of the copepod *Eurytemora affinis* complex. The study found that invasive populations have repeatedly responded to selection through the parallel use of the same single-nucleotide polymorphisms and genomic loci, to a much greater degree than expected. The results clearly prove that fluctuating habitats can promote invasive success and that balancing selection might serve as an important mechanism for adaptation in nature.

However, using DNA metabarcoding for monitoring invasive species comes with its own difficulties. Deficiencies in species identification can arise from a lack of primary databases, reference DNA databases, or the quality of the available database. Also, numerous abiotic factors tend to influence the circulation and decay rates of eDNA transport in various environments which makes measuring the species' volume and distribution complicated. Some contemporary models pay more attention to estimating potential risk and exposure alongside vulnerabilities, integrating the change in climate features—such as rise in temperature, precipitation changes, sea-level rise, occurrence of extreme weather events, and their intensity and frequency—into the model. Such models are useful in dynamically adapting to new changes, hence enabling effective risk management and proactive mitigation plans. The use of global and regional climate models to project possible future climatic conditions and their impacts has been recently enhanced by integrating climate scenarios. Knowledge related to climate change, combined with ongoing research and development, can further strengthen the importance of DNA metabarcoding in monitoring and managing invasive species.

27.7 Conclusion

To sum up, copepods are efficient markers of change in marine ecosystems due to their interconnectedness with the impacts of climate change and invasive species. The changes in marine food webs reveal the impacts of physiological stresses, change in distribution patterns, and the trophic mismatch synergistically to these impacts. The management of these complex interactions can benefit from the use of genomic tools and eDNA metabarcoding; however, their success relies on proper monitoring and collaboration. Climate-driven changes have put marine ecosystem under great risk, hence, mitigation and control strategies must incorporate predictions for the climate combined with effects of elevated ocean temperatures and invasive species.

Acknowledgments The authors are thankful to the AMET University for providing facilities.

Disclosures and Declarations All authors of the study consent to publish the book chapter

Conflict of Interests The authors declare that they have no conflict of interest

Ethics Approval Not applicable

References

Barton AD, Irwin AJ, Finkel ZV, Stock CA (2016) Climate change and the biogeography of oceanic diatoms. Proc Natl Acad Sci 113(11):2963–2968

Borrell YJ, Miralles L, Do Huu H, Mohammed-Geba K, Garcia-Vazquez E (2017) DNA in a bottle—rapid metabarcoding survey for early alerts of invasive species in ports. PLoS One 12(9):e0183347

Boxshall GA, Halsey SH (2004) An introduction to copepod diversity, vol 1 & 2. The Ray Society, Andover

Grémillet D, Boulinier T (2009) Spatial ecology and conservation of seabirds facing global climate change: a review. Mar Ecol Prog Ser 391:121–137

Hallegraeff GM (2010) Ocean climate change, phytoplankton community responses, and harmful algal blooms: a formidable predictive challenge 1. J Phycol 46(2):220–235

Hays GC, Richardson AJ, Sims DW (2005) Climate change and marine plankton. Trends Ecol Evol 20(6):337–344

Heneghan RF, Everett JD, Blanchard JL, Richardson AJ (2016) Zooplankton are not fish: improving zooplankton realism in size-spectrum models mediates energy transfer in food webs. Front Mar Sci 3:201

Hirst AG, Bunker AJ (2003) Growth of marine planktonic copepods: Global rates and patterns in relation to chlorophyll a, temperature, and body weight 1. Limnol Oceanogr 48(5):1988–2010

Juan Carbonell CD (2024) Thermal compensation in marine copepods: long-term ecophysiological implications in Paracartia grani. Ph.D. Thesis, University of Barcelona

Karl DM, Michaels AF, Knauer GA (1996) Seasonal variability of particulate organic carbon flux in the deep-water region of the Hawaiian Islands. Deep-Sea Res II Top Stud Oceanogr 43(2–3):539–568

Kroeker KJ, Kordas RL, Crim RN, Singh GG (2013) Meta-analysis reveals negative yet variable effects of ocean acidification on marine organisms. Ecol Lett 16(12):1488–1496

Lagerspetz KYH, Kivivuori L (1970) Salinity-induced changes in the thermal resistance of *Acartia clausi* Giesbr. (Copepoda, Crustacea). Ann Zool Fenn 7(4):231–234

Mauchline J (1998) The biology of calanoid copepods. Academic Press, San Diego

Richardson AJ (2008) Climate change and copepods. J Mar Syst 74(1–2):1–5

Schoener DS, Batten SD, крайняя OB, Skjoldal HR, Beaugrand G (2019) Rapid warming drives species-level divergence in the phenology of a North Sea copepod. Glob Chang Biol 25(1):244–254

Stern DB, Lee CE (2020) Evolutionary origins of genomic adaptations in an invasive copepod. Nat Ecol Evol 4(8):1084–1094

Turner JT (2015) Zooplankton fecal pellets, marine snow, phytodetritus and their role in the biological pump. Prog Oceanogr 130:205–248

Von Weissenberg E, Mottola G, Uurasmaa TM, Anttila K, Engström-Öst J (2022) Combined effect of salinity and temperature on copepod reproduction and oxidative stress in brackish-water environment. Front Mar Sci 9:952863

Wonham MJ, Walton WC, Ruiz GM, Frese AM, Galil BS (2001) Going to the source: role of the invasion pathway in determining potential invaders. Mar Ecol Prog Ser 215:1–12

Chapter 28
Climate Change and Environmental Instability: The Rise of Invasive Species and Desertification

Muhammad Shaheryar, Dilshad Safiullah, Hafiz Saqib Hayat, Muhammad Nadeem Shah, Muhammad Talha Akram, Muhammad Umer Iqbal, Azhar Hayat, and Muhammad Adnan Naseem

Abstract Climate change is a serious global problem affecting nature, agriculture, and human life. This study looks at how climate change, invasive species, increasing soil salinity, and desertification are connected. Human activities like deforestation, industrial pollution, and burning fossil fuels have caused global temperatures to rise. This has changed rainfall patterns, sea levels, and weather conditions, making it easier for invasive species to spread. These species grow rapidly, compete with native plants and animals, and disrupt ecosystems. Rising sea levels and changes in rainfall have increased soil salinity, making farmland less fertile. More than 1060 million hectares of land worldwide suffer from high salt levels, reducing crop yields by up to 58% in some cases. In dry regions, desertification is also getting worse, making the land less productive and harming local communities. The spread of invasive plants like *Prosopis juliflora* and *Verbesina encelioides* in arid lands further damages soil and reduces biodiversity. To address these problems, we need better land and water management, the planting of drought-resistant trees, and early detection of invasive species. Modern technology, such as satellite monitoring and species tracking, can help predict and prevent environmental damage. Governments and international organizations must work together to reduce the impact of climate change and protect natural resources. This study highlights the need for a simple but effective approach to saving ecosystems, protecting biodiversity, and ensuring food security for future generations.

M. Shaheryar · M. T. Akram
Institute of Agronomy, Faculty of Agricultural Sciences and Technology, Bahauddin Zakariya University, Multan, Punjab, Pakistan

D. Safiullah
New Mexico Highlands University, Las Vegas, NM, USA

H. S. Hayat (✉)
Department of Agronomy, Faculty of Agriculture and Environmental Sciences, Muhammad Nawaz Shareef University of Agriculture, Multan, Punjab, Pakistan
e-mail: saqib.hayat@mnsuam.edu.pk

B. P. Barathan et al. (eds.), *Navigating Climate Change: Impacts on Biodiversity and Ecosystem Resilience*, Sustainable Development and Biodiversity 42,
https://doi.org/10.1007/978-981-95-0409-1_28

Keywords Climate change · Invasive species · Soil salinity · Desertification · Biodiversity loss

28.1 Background/Overview of Climate Change

Climate change is defined as the gradual variation of temperature or erratic changes in it. Anthropogenic activities such as the burning of fossil fuels, the unsustainable agricultural practices, the industrial development, and allied activities have contributed to the enhancement of Greenhouse gases (GHGs) especially; CO_2, CH_4, and N_2O in the atmosphere in general (IPCC 2021). These gases embrace carbon dioxide which traps heat thereby increasing the natural green barrel effect, hence leading to global warming. This anthropogenic input to climatic condition has created upset in Earth's climatic circuits and precipitated a string of environmental as well as socioeconomic difficulties.

Over the millennia since the late nineteenth century, the global temperature has been raised by about 1.1 °C with the last 40 years having a faster rate of rising temperatures (NASA 2023). This warming trend is linked with several effects which include the melting of glaciers and of more proportionate polar ice caps which leads to an appreciable rise in sea levels. The IPCC (2019) says that the global sea levels have risen by 0.2 m in the period 1901–2018 and will rise by 0.3–1.1 m by 2100 based on differing Special Report on Emissions Scenarios (SRESs). These changes expose severe threats on coastal ecosystems and human settlements, small island nations, and densely populated coastal cities.

Apart from the increasing temperature, the conditions of climate change have extended the frequency and magnitude of extreme weather conditions. Extended periods of high temperatures, intense precipitation, storms, as well as lengthy periods of no rainfall have also increased substantially as heatwaves, heavy hoar frosts, hurricanes, and heavy showers that in turn inflict severe effects on people and nature (UNFCCC 2020). In July 2020, a record was set as it became the fifth month in history to have a named storm. The 2020 Atlantic hurricane season was also one of the

M. N. Shah
Institute of Agronomy, Faculty of Agricultural Sciences and Technology, Bahauddin Zakariya University, Multan, Punjab, Pakistan

Department of Agriculture, Government College University, Lahore, Punjab, Pakistan

M. U. Iqbal
Department of Plant Pathology, Faculty of Agriculture & Environmental Sciences, Muhammad Nawaz Shareef University of Agriculture, Multan, Punjab, Pakistan

A. Hayat
Department of Statistics, Bahauddin Zakariya University, Multan, Punjab, Pakistan

M. A. Naseem
Department of Agriculture Business and Marketing, Faculty of Agricultural Sciences and Technology, Bahauddin Zakariya University, Multan, Punjab, Pakistan

most active ever, mainly because warmer ocean temperatures provided more energy for storms (WMO 2021). In the same way, extended dry seasons, for instance in the Sub-Saharan area, have deepened water and food deficit, which pose a risk to millions of people (FAO 2020).

Consequently, the regional effects of climate change are complex and depend on the specific context. For example, South Asia is currently facing environmental degradation, unpredictable monsoon rains, and accelerated glacial melting in the Himalayas, all of which threaten water access and agriculture (Rasul et al. 2019). Some other regions are experiencing heat waves, severe dust storms, and prolonged drought conditions. Higher latitudes are warming at a much faster rate due to both the melting ice cap and the thawing of permafrost which generates changes in ecosystems, as well as releasing methane, a potent GHG. Seawater intrusion into freshwater is a global issue in coastal regions which impacts agriculture, aquaculture, and water for human consumption (Nicholls et al. 2021).

The future path of climate change is dependent on international responses to reduction of greenhouse gas emissions or adaptation policies. Failure to achieve deep cuts in global GHG emissions means that the average global temperature could increase by 2.7 °C or more this century, which will cause extensive loss of species, further rise in temperature could aggravate it and make extreme heat, and disruption of food and water systems is the new normal (IPCC 2021). In response to these challenges, Paris Agreement launched in 2015 in Paris, France, intends to ensure that global mean temperature does not rise beyond 2 °C above preindustrial levels, let alone 1.5 °C. To achieve this goal, radical shifts in the energy sectors, along with related urban and land-use systems and consumption patterns, are needed (Rogelj et al. 2018).

That again underlines an important fact that there is no better time than now to address the issue of climate change. Financial commitments in clean technologies, afforestation, and climate-smart agriculture are important towards achieving the emission goals and depth. Furthermore, how to promote the most effective international cooperation and how to strengthen policy support to minimize the adverse effects of climate change on vulnerable people? If people of the world come together and respond positively and with purpose in coordination, then the harsher aspects of this shared problem can be restrained and a better world can be encouraged.

28.2 Introduction

Global warming is a major agent of ecological and agricultural change affecting all aspects of ecosystem, soil, and land use management. Some of its impacts include: providing means for invasive species and enabling and promoting other changes in soil salinity and the rates of processes of desertification. All these interrelated threats constitute significant risks to food production, food security, and sustainability of ecosystems around the globe.

Invasive species are most problematic in climate change-affected areas as indicated in this paper. Fluctuations in temperature, water, and rainfall, and even strange weather patterns help invasive species find suitable environment to migrate, germinate, and reproduce. These species invade the habitat of indigenous plants and food crops, negatively alter ecosystem functions, and raise biological susceptibility to pathogens and parasitoids. Scientific evidence suggests climate change will further escalate the problem of Biological Invasion (BI), with plants, diseases, and pests continuing to extend into new or larger habitats, adding more pressure to agricultural production (Turbelin and Catford 2021).

Climate change makes agricultural sustainability worse as saline conditions appear from climatic uncertainties. Studies also pointed out that a number of deposition factors such as sea levels, the changes in precipitation regimes as well as the increase in rates of evaporation are creating salinities in many territories. This phenomenon not only has a negative impact on the productivity of plants but also forms habitats where given halophytic species can develop and spread causing additional deterioration of the already affected agricultural soils. They aggravate resource pull and are life risking in subsistence or marginal farming-based regions.

Erosion is another problem that is caused by changes in climate, mainly affecting the areas with desert-like climate. Fertile soil degradation is a result of high temperature, longer periods of dry seasons, and poor land practices. Some invasive plant species, for instance, contribute to further desertification due to competition with native vegetation, while at the same time there are more complicated shifting dynamics at regional level. For instance, it is postulated that while warming may lessen desertification in certain regions, for instance, the drylands of North China, different climatic states might even reverse land degradation trends (Yang et al. 2022).

28.2.1 Understanding Climate Change: Causes and Global Effects

Climate change is driven by increased greenhouse gas emissions from fossil fuel use, deforestation, industrialization, and unsustainable farming. These gases, including CO_2, CH_4, and N_2O, trap heat in the atmosphere, altering the Earth's energy balance. By the end of the century, global temperatures could rise by 2.7 °C, with faster warming in the polar regions due to the albedo effect (Heshmati 2021). Climate change leads to shifting climate zones, altered rainfall patterns, and more frequent extreme weather events like droughts, floods, and hurricanes. Rising sea levels, caused by ocean expansion and glacier melt, threaten coastal communities through habitat loss, saltwater intrusion, and stronger storms. Water shortages from reduced glacial melt are affecting millions, while changes in weather patterns disrupt species migration and biodiversity, endangering coral reefs and rainforests (Walker 2018). For humans, climate change worsens food and water insecurity, particularly in vulnerable regions. Droughts reduce crop yields, leading to economic losses and food shortages.

Health risks are rising due to vector-borne diseases, heat-related illnesses, and poor air quality. Climate-induced displacement is also increasing, forcing millions from their homes. Vulnerable groups, such as children, indigenous populations, and economically disadvantaged communities, face the greatest risks. While some regions may benefit from longer growing seasons and milder winters, these gains do not outweigh the widespread negative impacts, particularly where adaptive capacity is low. Global solutions require reducing emissions, protecting natural resources, and implementing policies to safeguard at-risk populations. Addressing both the causes and consequences of climate change demands international cooperation.

28.2.2 The Concept of Invasive Species in Changing Climates

Global warming and the introduction of invasive species threaten biodiversity by disrupting ecosystems and increasing pressure on global species stocks. Invasive aquatic plants interfere with ecosystem structures by competing for resources, leading to the displacement of native species and a decline in essential ecosystem services like pollination and water regulation. As climate change expands the range of invasive species, previously unaffected areas become vulnerable, reducing the effectiveness of conventional management strategies. For example, warmer winters hinder the control of invasive European rabbits in New Zealand, while habitat suitability models show fluctuating availability for *Rhododendron ponticum* in Wales, depending on climate change intensity (Latham et al. 2015; Manzoor et al. 2020).

Agricultural systems face compounding threats, as climate change enhances the spread of invasive weeds, pests, and pathogens, directly reducing crop yields and increasing management costs. Drought-weakened crops become more susceptible to invasive pests, while species like *Parthenium hysterophorus* thrive in warmer, CO_2-rich environments, outcompeting crops and worsening food insecurity (Mokany et al. 2014). Controlling invasive species under a changing climate requires integrated, adaptive, and region-specific strategies. Effective measures include biological control, environmentally sustainable pest management, precision agriculture, and smart farming technologies. The conservation of seed-dispersing animals is also critical, as they support natural regeneration by spreading native and agriculturally valuable plant species (Hellmann et al. 2010).

To address these challenges, collaboration among researchers, policymakers, farmers, and communities is essential. Data-sharing, integrated research networks, and citizen science can enhance early detection and improve management of invasive species. Public education and awareness campaigns are crucial for engaging stakeholders from local to global levels. The interconnectedness of climate change and invasive species raises concerns about the effectiveness of current management strategies in ensuring future food security and ecological balance. Developing robust, adaptable systems is key to mitigating these threats and sustaining agricultural production in the face of climate change.

28.2.3 Interlinkages between Climate Change, Salinity, and Desertification

Global warming, water logging, and soil erosion are the linked phenomena that augment the pace of environmental deterioration mainly in the arid zones. Global warming causes evapotranspiration through heat and changes in precipitation and intensity of weather that reduce soil moisture and accumulate salts where the water table is shallow or groundwater is poor. As sea levels rise, coastal areas face increased threats due to the intrusion of saline water into freshwater aquifers, which decreases soil fertility and pollutes water used for irrigation (Corwin 2020). The pie chart (Fig. 28.1) also draws the attention towards the climate change contributing 40% and salinity and desertification contributing 30%, respectively, towards land degradation. Climate change influences excessive rainfall and drought, while salinity and desertification affect the health of soil and diminish the quality of fertilized soil and food production for growth in arid and coast regions. Integrated strategies are required urgently in dealing with these challenges that are closely related with one another.

28.2.3.1 Climate Change and Desertification Feedback Loops

Desertification does not only occur due to climate change, but also intensifies it due to alterations in the surface-atmosphere exchanges. Vegetation loss resulting from desertification raises surface reflectivity thus causing low precipitation rates within

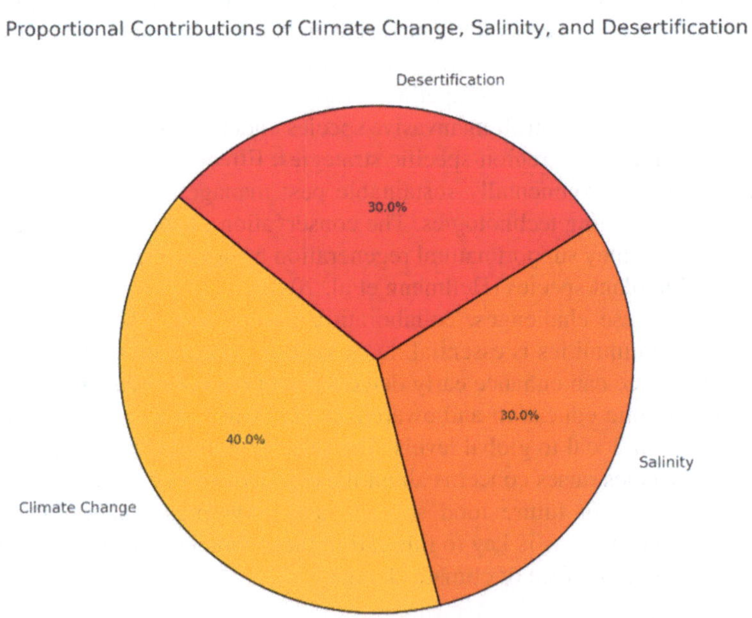

Fig. 28.1 Proportional contributions of climate change, salinity, and desertification

the region, hence aridification. The emission of dust from marginal land creates an input into the existing atmospheric radiative forcing and cloud formation which in turn enhance global warming impact, also threatens the carbon cycle: By decreasing the amount of organic matter in the soils, reducing the land's carbon sink capacity, and emitting carbon stored in the ecosystem (D'Odorico et al. 2012).

Such processes entwine to develop complex feedback relationships at once. For example, while reclaiming saline land reduces vegetation cover, exposes soil to wind and water erosion—a factor which fosters desertification. The process of desertification leads to the decrease of the land productivity and, therefore, aridification and desertification continue. Here, climate change through aridification aggravates these impacts by favoring land desertification.

28.2.3.2 Socioeconomic Impacts of Climate-Salinity-Desertification Interactions

Due to climate change, soil salinization, and land desertification, there are numerous socioeconomic impacts, particularly affecting rural communities and farmers Salinization of soil decreases arable land and crop production and thus is a major threat to food security in many countries. New costs from a salinity control program include soil adjustments for salinity, better irrigation systems and, at the same time, low income from reduced production. Coastal areas depending on freshwater aquifers are most affected because, as salinization progresses, water must be extracted from more costly desalination processes or other sources.

Those within the developing regions are most affected. Whenever land with potential for productive agriculture is removed from circulation in these high poverty counties, poverty, forced migration, and social disruption follow. Land desertification will force families to abandon unproductive land in search of more fertile areas, increasing population density in other regions and putting pressure on the limited resources of already populous urban areas.

28.2.3.3 Strategies for Mitigation and Adaptation

Therapeutic management of the triad of climate change, salinity, and desertification needs a comprehensive, continuous, and transdisciplinary intervention. Although conservation agriculture practices involve minimum soil turning through tilling, use of practices like crop rotation has an added advantage of correcting the problem of compacted soil and increased salinity. There are practices, like drip, which help to conserve water and do not allow salts to accumulate within the root zone (Corwin 2020). It is essential to protect and rehabilitate freshwater aquifers from the harmful effects of saltwater intrusion, particularly in coastal regions. Reforestation and afforestation eradicate soil erosion, enhance carbon storage, and reduce the hazards of desertification (D'Odorico et al. 2012). There should be sale of salt-tolerant crops and Halophytes in areas of high salinity to support irrigated agriculture. Strengthening of local knowledge and information, skills development and promotion of learning technology, and providing appropriate access to finance (Ouhamdouch et al. 2016). Establishment of indicators to detect variation in salinity and levels of desertification and have the means of taking corrective measures. Developing international treaties for better codes of conduct, like United Nations Convention to combat desertification (UNCCD) among others. Mainstreaming LDN targets in National Sustainable Development Plans and Climate Change action plans. Approaches between climate change, salinity, and desertification requires an integrated, adaptive, and multidisciplinary approach. Effective strategies include:

(a) *Sustainable Land and Water Management*:

- Conservation agriculture practices, such as reduced tillage and crop rotation, can improve soil structure and minimize salinity risks.
- Efficient irrigation techniques, such as drip irrigation, reduce water use and limit salt buildup in the soil profile (Corwin 2020).
- Protecting and restoring freshwater aquifers to prevent saltwater intrusion, particularly in coastal areas.

(b) *Restoration of Degraded Lands*:

- Reforestation and afforestation initiatives stabilize soil, improve carbon sequestration, and reduce the risk of desertification (D'Odorico et al. 2012).
- Cultivating salt-tolerant crops and halophytes in salinity-prone areas to maintain agricultural productivity.

(c) *Building Climate Resilience*:

- Enhancing the adaptive capacity of local communities through education, technology transfer, and access to financial resources (Ouhamdouch et al. 2016).
- Developing monitoring and early warning systems to track changes in salinity and desertification and enable proactive responses.

(d) *Policy and International Collaboration*:

- Strengthening global commitments, such as the United Nations Convention to Combat Desertification (UNCCD), to coordinate efforts across countries.
- Integrating land degradation neutrality (LDN) targets into national development plans and climate adaptation policies (Patil et al. 2023).

28.2.3.4 The Need for Holistic Solutions

Thus, it shows the interconnections between climate change, salinity, and desertification and the necessity of an integrated approach to solve them one after the other. It is important to understand the feedback between these processes to come up with Southern sustainable solutions that look for the solution to land degradation and building of resilience. It will be important for scientists, policymakers, and local communities to work together in order to implement management adaptions that will help maintain and conserve the world's biodiversity and the livelihoods of peoples in vulnerable parts of the globe.

28.3 Climate Change and Invasive Species

Climate change intensifies global food security threats, while invasive agricultural species further strain ecosystems, farming systems, and policies. Together, these factors worsen the impact of invasive species, making control efforts more challenging for farmers, conservationists, and policymakers. Climate change expands the range of invasive pests and weeds, allowing them to colonize new habitats while also weakening conventional control methods. The effects of climate change on invasive agricultural species vary by species and climate scenario. Some invasive terrestrial plants, particularly those from temperate regions, may decline in certain environments, while others expand into newly favorable areas (Nikkel et al. 2023). In contrast, aquatic invasive plants are less affected due to the stability of their habitats. Additionally, climate change alters the timing of biological events, disrupting interactions between invasive species and their natural enemies, which may weaken biological control strategies and cause unintended ecological effects (Gong et al. 2020).

Addressing these challenges requires integrating research with practical management, fostering cooperation among governments, organizations, and stakeholders to improve awareness and policy responses. Effective solutions include early identification of new invasive species and the development of adaptive strategies to protect food security, biodiversity, and sustainable agriculture. Table 28.1 highlights key invasive species benefiting from climate change. Signal crayfish and red swamp crayfish spread across Europe and North America, disrupting freshwater ecosystems. *Ageratum houstonianum* in Nepal and cogongrass in the southeastern USA displace native vegetation and disrupt fire regimes. Cheatgrass in North America increases wildfire frequency and outcompetes native grasses. These cases illustrate how climate change creates favorable conditions for invasive species, amplifying

Table 28.1 Climate change and invasive species

Invasive species	Region of expansion	Key impacts
Signal crayfish	Europe and North America	Disrupts freshwater ecosystems, outcompetes native species
Red swamp crayfish	Europe and Mediterranean	Alters aquatic food webs, damages vegetation
Ageratum houstonianum	Nepal	Reduces native vegetation, alters nutrient cycles
Cheatgrass	North America	Increases wildfire frequency, displaces native grasses
Cogongrass	Southeastern USA	Alters fire regimes, reduces native biodiversity

their ecological and socioeconomic impacts. Robust management solutions are needed to control their spread and protect native biodiversity.

28.3.1 Mechanisms Driving the Spread of Invasive Species

Invasiveness of biological entities, particularly species, is regulated by a cocktail of fixed attributes, ecological factors, and consequently their invasiveness, establishment, and dominance. Knowledge of these mechanisms is important for the formation of successful measures to prevent damages to the environment and the economy. Spread pathways of invasive species can be broadly categorized by the means that those species use to disperse as well as the way those species interact with their habitats.

28.3.1.1 Biological Traits, Dispersal Mechanisms, and Environmental Influences on Invasive Species

Invasive species success depends on key biological traits such as dispersal distance, fecundity, survival, and reproductive age. Among these, dispersal distance is crucial as it dictates colonization speed and range expansion (Coutts et al. 2010). Rapid seed production enables many invasive plants to dominate disturbed habitats, while species with limited dispersal and longer lifespans rely on stable conditions for persistence. Environmental factors like disturbance and nutrient availability further influence invasive species behavior, classifying them as "drivers" that alter ecosystems or "passengers" that thrive in already disturbed areas (Wilson and Pinno 2012). Some species, such as *Bromus inermis* and *Cirsium arvense*, dominate succession patterns, while others monopolize resources without directly modifying ecosystems.

The interaction between biological traits and environmental conditions complicates predictions of invasive species spread. Climate change, through land-use changes and extreme weather, facilitates invasions by increasing habitat suitability and resource availability (Gong et al. 2020; Nikkel et al. 2023). Higher nutrient levels in altered environments often give invasive species a competitive edge over natives (Harper et al. 2020). Human and natural interferences, particularly climate-driven extreme weather events, create new opportunities for invasive species to expand (Stephens et al. 2019). Addressing these challenges requires adaptive management strategies that consider biological and environmental factors, alongside increased awareness and policy interventions to mitigate the effects of invasive species (Heshmati et al. 2019).

28.3.2 *Impact of Changing Climates on Ecosystem Balance*

Global warming and the invasion of species both apply heavy pressure on ecosystems; they influence both structure and processes in all possible ways. This is because both these driving forces often occur simultaneously and in combination, thereby increasing their impacts on native biotic diversity and ecosystem function. Climate change is likely to act to the advantage of many invaders over natives, therefore, offering such species an additional opportunity to deal a "double whammy" to native species and further upset the ecosystems (Meek et al. 2012). These two threats are all the more worrisome as invasive species are already good at benefiting from disturbed ecosystems; climate change can worsen the conditions that allow invasive species to thrive.

28.3.2.1 Climate Change, Species Expansion, and Ecosystem Impacts

One of the most immediate and substantial effects of climate change on invasive species is the increase in global temperatures. Warming climates are expected to expand the geographical distribution of many invasive species as previously unsuitable conditions become favorable. For instance, *Pacifastacus leniusculus* and *Procambarus clarkii*, invasive crayfish species, are likely to benefit from increasing temperatures, allowing them to outcompete native species and alter freshwater ecosystems (Zhang et al. 2019). Similarly, tropical and subtropical waterweeds such as *Salvinia molesta* and *Eichhornia crassipes* will expand into temperate zones where rising temperatures facilitate their growth and reproduction (Hoveka et al. 2015).

Climate change impacts on invasive species vary by ecosystem type. Terrestrial systems often experience increased growing seasons, higher resource availability, and reduced competition from stressed native species, which enhances invasive species' competitive success and decreases overall species richness. Conversely, aquatic ecosystems may experience a decline in the effectiveness of some invasive species due to factors like lower oxygen availability in warmer waters and altered hydrological patterns (Stephens et al. 2019). Despite most invasive plants being expected to thrive under changing climates, their success is not uniform across species or spatial scales. Some temperate invasive species may struggle to survive in rising temperature regimes, highlighting the need for species and ecosystem-specific studies to predict outcomes accurately (Hoveka et al. 2015; Lamsal et al. 2018).

As climate change alters ecosystem dynamics and competitive balances, predicting the spread of invasive species remains challenging. Some species may find new opportunities, while others may struggle with shifting climate factors. This variability underscores the importance of targeted research to develop accurate models for predicting future invasive species distributions.

28.3.2.2 Implications for Ecosystem Functioning and Management Strategies

The interaction between climate change and invasive species threatens essential ecosystem services, including provisioning, regulation, and biodiversity support. Invasive species disrupt trophic structures and ecosystem resilience, reducing key services such as pollination, water purification, and carbon storage. For example, replacing native vegetation with invasive weeds diminishes habitats for pollinators and reduces forage availability. In aquatic ecosystems, invasive species can disrupt nutrient cycling and food chains, further threatening biodiversity. These disruptions affect not only biodiversity but also human communities that depend on ecosystem services.

Addressing these challenges requires proactive management strategies that integrate climate adaptation with invasive species control. Monitoring changes in invasive species distributions and predicting potential future invasions under climate change scenarios are crucial. Early intervention strategies, including habitat restoration and species-specific control measures, can prevent long-term ecological damage. Policymakers, researchers, and local communities must collaborate to implement targeted management approaches that consider shifting ecological conditions. Additionally, further research should focus on how changes in phenology, resource availability, and disturbances affect interactions between invasive and native species (Macinnis-Ng et al. 2021; Rahel and Olden 2008). Closing these knowledge gaps and applying integrated management approaches may help mitigate the combined impacts of climate change and invasive species on ecosystems and biodiversity.

28.3.3 Examples of Invasive Species Thriving Under Climate Change

Climate change is providing more accommodating conditions for many invasives to spread, to have larger populations, and thus generate more harms ecologically and economically. The increase in average temperatures and evaporation rates, and changes in precipitation regime and incidence of more frequent extreme weather conditions give the desired environmental conditions to many invasive species. These changes are remodelling ecosystems globally, increasing the potential of transport of neobiota including plants, aquatic organisms, and vertebrates in regions which were once unsuitable. Some of the following are invasive species that are benefiting from climate change and the troubles they bring to ecosystems and human systems.

28.3.3.1 Aquatic Species Expanding Their Ranges

Climate change effects pose special risk on aquatic ecosystems since global warming and changes in hydrologic cycles offer favorable conditions for invasive species.

- *Signal Crayfish* (*Pacifastacus leniusculus*): This North American neophyte is increasing in range both in North America and Europe as a result of warmer temperatures. Climate change has led to changes in the temperatures of water; which makes original inhospitable habitats to be socially acceptable for the signal crayfish displacing the native crayfish species and disturbing the freshwater communities (Zhang et al. 2019). Through direct predation of native aquatic fauna and changes to sediment and water quality, the species poses a considerable threat to the ecosystem of river systems.
- *Red Swamp Crayfish* (*Procambarus clarkii*): The red swamp crayfish, native to the southern United States, could spread across Europe under climate change, according to the study. It is native from North Africa to Western Asia and it is an invasive pest in Mediterranean and subtropical climates interfering with food chains, predating native fish, and damaging water plants. Due to climate warming, its range increases the negative impact on biotic and abiotic diversity of water objects in Europe (Verlinden et al. 2013).

28.3.3.2 Invasive Plants Benefiting from Warming Temperatures

Climate change is therefore able to bring about an increase in the rate of invasion by plant species through the expansion of their growing period and increase in CO_2 levels and conditions that can top up their collapsibility.

- *Ageratum* (*Ageratum houstonianum*): In Nepal, Ageratum houstonianum was selected as one of six invasive alien plants that are likely to see some formidable climatically suitable habitat in the future. Thus, at the end of the period 2050, the area of suitable living conditions may increase by 30–70%, depending on the climate scenarios, and become a serious threat to native fauna and agricultural production (Shrestha et al. 2018). It is worst in disturbed habitats, lowering the canopied native vegetation and changing the local nutrient balances.
- *Cheatgrass* (*Bromus tectorum*): Although cheatgrass is an invasive plant species in North American, it is adapting well to the warmer and drier climate favored by climate change. Because it can colonize easily in disturbed habitats and is highly flammable, conebur is considered the primary causal agent of rising fire rate and increased severity in the western United States. Grazing by sheep and cattle has enabled climate change to lengthen its growing season and thus gain competitive advantage over the native grasses.
- *Cogongrass* (*Imperata cylindrica*): Cogongrass is native to Southeast Asia and is a highly effective weed in the southeastern United States which is gradually sprawling. Its growth and reproduction have been favored by increasing temperatures and raised levels of CO_2, and the species has invaded the regions with

temperatures below its preferred optimum. Cenchrus ciliaris (formerly known as *Pennisetum ciliare*), commonly called buffelgrass, is highly competitive with agriculture, forestry, and native vegetation and also affects fire regimes.

28.3.3.3 Invasive Vertebrates Benefiting from Climate Change

This also applies to the species of the vertebrates whereby these species have also responded to the new climatic changes in regard to the new warming as well as new ecosystems to occupy and exercise dominance.

- *European Rabbit* (*Oryctolagus cuniculus*): Over the past ten years, the European rabbit has become a major invasive pest in New Zealand, with climate change greatly undermining the success of control measures. Control measures have been further reduced owing to shorter duration of winter, through which the rabbit population can freely expand during the relatively warmer seasons (Latham et al. 2015). This species was a menace to farmers through crop and native plant grazing which led to crop destruction, soil erosion, and thus destruction of their habitats.
- *Cane Toad* (*Rhinella marina*): First released into Australia for the purpose of controlling pests, the cane toad has been able to extend its geographical distribution at higher thermal regimes. The general population of the knobby frog has increased in northern Australia from the effects of high temperature and large amount of rainfall as these create favorable conditions in what used to be unsuitable habitat. Juvenile cane toads release poison when threatened which is dangerous to native predators; they directly alter dynamics of the ecosystem.

28.3.3.4 Species with Mixed Responses to Climate Change

Consequently, other invasive species remain to be sensitive to climate change, depending on regional climatic conditions and competition.

- *Japanese Knotweed* (*Fallopia japonica*): This plant species is found to be highly adaptive and can grow in various environmental changes, but its vulnerability to climate changes differs. In certain areas, the species has expanded into warmer zones and higher elevations than previously observed, while in other regions, rising heat stress has hindered its growth along the altitudinal gradient. Such variability underlines the necessity of the specificity of the invasive species' behavior prognosis under climate change circumstances.
- *Tropical Aquatic Plants*: Self-sowing aquatic plants, for instance, Salvinia molesta and Eichhornia crassipes, are anticipated to spread in the temperate climate since increasing temperatures mean more favorable environments. But, in some instances, inadequate water supply constraining their growth in vegetation due to shifts in rainfall distribution proves that climate's effects are relative (Hoveka et al. 2015).

28.3.3.5 Implications for Ecosystem Management

The examples above show how invasive species are likely to be on the increase due to the changing climatic conditions. An increase in the numbers and distribution of invasive species affects ecosystem services and native biological diversity and has flow-on effects on agriculture, forestry, and water resources. At the same time, the influence of climate change progresses significantly and that is why it is supposed that the distribution of invader species will increase significantly resulting in increased difficulties in management.

To resolve such questions, long-term active and sensitive management measures are needed. Through the use of the parameters involved in the modeling process, the endangered areas can easily be identified and as such the invasion by the invader species can easily be prevented. Moreover, international cooperation, monitoring programs, and public awareness campaigns can play a vital role in combating the spread of invasive species and minimizing their impacts on both ecosystems and human systems.

The information about the invasive species, on the one hand, and the impacts of climate change, on the other, is very important so as to be able to conserve ecosystems and promote the sustainable use of the available agricultural and natural resources as well as the diversity of species.

28.4 Impacts of Saline Conditions Due to Climate Change

Global warming is increasing soil salinity worldwide, affecting farming, nature, and people's lives. Around 1.06 billion hectares of land are impacted, with the problem worsening due to rising sea levels, higher temperatures, changing rainfall, and poor irrigation (Eswar et al. 2021). This threatens food security, especially in coastal areas where both salinity and flooding are growing. Farmers are adapting by practicing both farming and aquaculture or moving to other areas (Chen and Mueller 2018). Interestingly, higher salinity increases migration, but flooding does not have a big impact on farming or movement. Rising sea levels also affect disease spread by expanding saltwater areas, increasing salt-tolerant mosquitoes, and shifting freshwater mosquitoes to salty waters. This could spread diseases like malaria and dengue in coastal areas (Ramasamy and Surendran 2012). Salinity also changes soil microbes, reducing species diversity and altering ecosystem functions (Li et al. 2024). This affects nutrient cycles and the environment. To manage these issues, better farming practices, disease control, and ecosystem management will be necessary.

28.4.1 Causes of Increasing Soil Salinity: Climate and Human Activities

Climate change and the human impact on it also affect the increase in salinity: climate change consequences, including high temperature, fluctuating precipitation, and rising sea level influence natural-deep salinization processes (Eswar et al. 2021). The following human-induced factors aggravate the development of soil salinization: inappropriate methods of irrigation, poor drainage facilities, and the utilization of water with salinity levels for agricultural purposes (Hopmans et al. 2021). Excessive or insufficient application increases the problem through imbalance of nutrients, nitrogen (N) and phosphorus (P); their interactions depend on soil characteristics, climate, and management practices. These are worsened by climate change by changes in nutrient ratios coupled with changes in soil conditions that lead to a higher impact on crop yields (Murtaza et al. 2022).

28.4.1.1 Climate Change and Human Activities as Contributors to Soil Salinity

Soil salinity is increasingly driven by both climate change and human activities, leading to severe land degradation. Rising temperatures, fluctuating precipitation, and rising sea levels accelerate soil salinization. Higher temperatures increase evapotranspiration, worsening salt accumulation in the soil (Eswar et al. 2021). In coastal regions, rising sea levels infiltrate shallow water aquifers, degrading soil quality. Additionally, erratic and reduced rainfall limits the natural leaching of salts, further intensifying salinity. Climate-induced dryness expands arid lands, accelerating desertification, particularly in semiarid regions (Stavi et al. 2021).

Human activities also play a significant role in soil salinity. Poor irrigation practices, low-quality water use, inefficient drainage, and unsustainable water abstraction contribute to secondary salinization (Hopmans et al. 2021). Agricultural land clearing, overgrazing, and farming on fragile slopes disturb soil properties, making them more prone to salt accumulation (Munns et al. 1999). Additionally, weak supervision and inadequate management further degrade soil conditions (Teshaev et al. 2020).

Climate change and human activities interact, creating a feedback loop that exacerbates soil salinity. Prolonged droughts caused by climate change often lead to over-irrigation, which, instead of alleviating the issue, contributes to salt buildup in the soil. This combined effect highlights the urgent need for sustainable agricultural practices and efficient water management to prevent the ongoing degradation of arable land and ensure food security (Buono 2020; Stavi et al. 2021).

28.4.2 *Effects of Salinity on Agricultural Productivity*

Salinity is a globally experienced problem affecting soil health and farm land productivity and in many ways threatening food security all over the world. Glacial makes up 20% of the cultivated and a shocking 33% of irrigated land, with irrigation in the rise as the support of intensive farming practices. Salinity has enormous economic impacts as all crops are affected worldwide reducing farm income and in many cases the costs of remedial actions through soil correction or changes in irrigation practices can be expensive.

28.4.2.1 Impact of Soil Salinity on Agriculture and Food Security

Soil salinity has significant consequences for agricultural productivity, affecting crop yields, soil health, and global food security. High salinity levels reduce crop yields, with some sensitive crops experiencing a decline of up to 58% (Haj-Amor et al. 2022). This is due to osmotic stress, which prevents plants from effectively absorbing water, as well as the toxic effects of excessive sodium and chloride ions. Many vegetables, which have a low tolerance to salinity, experience growth reductions at even mild salinity levels of 1–2.5 dS m^{-1}.

Salinity impacts different crops and growth stages in various ways. Vegetables are particularly sensitive to salt stress at multiple growth stages, while cereals and oilseeds, though more tolerant, still suffer yield reductions as salinity increases. Seedlings and early vegetative plants are more vulnerable than mature crops, making it essential to develop crop-specific strategies to combat salinity and maintain agricultural productivity.

Beyond plant growth, salinity disrupts soil health and biogeochemical processes. Saline soils host fewer microorganisms and reduced microbial diversity, impairing nutrient cycling and lowering soil productivity (Singh et al. 2024). High salinity also affects enzyme activity, slowing down organic matter decomposition and nutrient availability. Additionally, salinity influences greenhouse gas emissions by altering microbial metabolism, potentially increasing nitrous oxide emissions, which contribute to climate change. These factors make soil salinity a broader ecological issue, not just a challenge for farming.

Global food security is increasingly threatened by rising soil salinity, particularly in regions that rely on irrigation. With a growing global population, maximizing agricultural output per unit area is crucial. However, salinity-induced production declines hinder these efforts, making it more difficult to meet future food demands.

28.4.2.2 Mitigation Strategies and Technological Innovations

Addressing the complex effects of salinity on agriculture requires sustainable solutions. Remote sensing and GIS technologies enable large-scale soil salinity assessments, facilitating timely management. Nanotechnology, through improved nutrient use efficiency and soil remediation, can also help mitigate salinity stress (Mishra et al. 2023). Additionally, agronomic practices such as cultivating salt-tolerant crops, adopting efficient irrigation methods like drip irrigation, and applying soil amendments such as gypsum can reduce salinity impacts.

Integrating these strategies with global sustainability frameworks, such as the United Nations Sustainable Development Goals, can strengthen resilience against soil salinity. This paper emphasizes that the impact of salinity extends beyond yield losses to include soil degradation, biodiversity decline, and sustainability challenges. Addressing these issues requires a combination of advanced technology, sustainable agricultural practices, and global policy interventions. Long-term food security depends on effectively managing both the causes and consequences of soil salinity.

28.4.3 Role of Invasive Species in Saline Environments

Harmful strains have trailed the effects of invasive species on ecosystems for quite some time and even have been lately influenced by climate change. Coastal areas, estuaries, and saline agriculture lands, which are immensely sensitive to both biotic and abiotic factors, are majorly under threat from impacts of climate change and bio invasion. Since the temperatures are continuously increasing across the globe, this poses a trend that many invasive species will thrive and are likely to cause enhanced threat to the native saline species through resource competition. This paper provides insight into the invasion of these habitats by NIS and understanding them remains vital in formulating the best management strategies.

28.4.3.1 Climate Change, Salinity, and Invasive Species

Climate change is shifting invasive species' ranges, especially in saline habitats like estuaries and coastal marshes. Rising temperatures, sea level rise, and altered rainfall patterns create conditions where invasive species outcompete natives. For example, the fish *Menidia beryllina* thrives in warm, salty waters, displacing native *Delta Smelt* (*Hypomesus transpacificus*) (Komoroske et al. 2020). The impact varies by species and habitat. Salt-tolerant plants and invasive halophytes can dominate, altering ecosystems and reducing biodiversity (Gong et al. 2020). Invasive species may also increase sedimentation, disrupt water flow, and create hypoxic conditions, further stressing native species.

Climate change and salinity interact, worsening each other's effects. Higher temperatures increase evaporation, raising soil and water salinity, while erratic rainfall reduces natural salt leaching (Shrestha et al. 2018). In regions like Southeast Asia, saltwater intrusion expands saline areas, making it easier for invasive species to spread (Yin et al. 2022). Managing saline ecosystems is challenging as invasive species with high adaptability replace natives (Beaury et al. 2019). Site-specific and species-specific conservation strategies, along with mapping and modeling, are needed to protect ecosystems. Future research should explore the long-term effects of invasive species on nutrient cycles, greenhouse gas emissions, and ecosystem health to develop effective management solutions.

28.5 Climate Change and Desertification

Global warming and desertification make each other worse. As the climate gets hotter and drier, dry areas lose even more water, leading to more desertification. Lack of rain, heat, and droughts, combined with human activities like deforestation and overgrazing, turn fertile land into deserts. Desertification also changes local climate conditions, creating a cycle that keeps worsening the problem. However, some places have found ways to stop or even reverse desertification. China, for example, has successfully controlled desert expansion through better land management (Lyu et al. 2020). In northern China, areas below 40° N may even benefit from climate change by preventing further desert spread (Yang et al. 2022). Overall, climate change increases desertification, harming both nature and people (Huang et al. 2020). But local efforts and smart policies can help restore land. Sustainable land management and ecosystem rehabilitation are key solutions. Controlling nutrients like nitrogen and phosphorus also improves soil quality, helping maintain agriculture even in changing climates (Murtaza et al. 2022).

28.5.1 Understanding Desertification: A Climate-Induced Phenomenon

Desertification is a slow process of destruction of the vegetated land characterized by climate factors and human endeavors. It is for this reason that climate change can drive desertification due to increased temperatures, poor rainfall and increased wind activity especially in the arid and semiarid regions (Shoba and Ramakrishnan 2016). It emerges that climate change represents 20.6–46.6% of changes in vegetation cover and desertification, although its impact varies considerably depending on the location (Feng et al. 2015).

28.5.1.1 Climate Change and Desertification Reversal

Sometimes, it has been proven that climate change enables reforestation of the desert. A number of authors outlined that climate change factors like enhanced hydrothermal conditions including better access to the soil moisture and decreased wind speeds are the key prerequisites for the desertification recovery in certain regions (Xu et al. 2018). For example, between 2001 and 2010, climate change contributed positively to reversing 48.4% of desertification in northwest China. However, due to DLLU, it also had a negative impact, leading to a 21.7% expansion of desertified areas during the same period. This reveals the fact that climate interacting with desertification is operational and contextually dependent here.

28.5.1.2 Interplay Between Climate Change and Human Activities

Desertification and climate change have strong relations with human activities like deforestation, land use change, and unsustainable agriculture. Some of these factors that contribute to climate change impacts are human induced, which tend to enhance the rates of soil degradation as well as reduction of vegetation. Hence, it implies that climatic and anthropogenic factors of desertification differ in terms of their impacts in space and time, and thus, any successful attempts in combating the problem must be localized (Zhou et al. 2014).

28.5.1.3 Strategies for Addressing Desertification

In fighting desertification, insight into climate and anthropogenic factors is crucial. Accurate quantification of these drivers over the long-term monitoring process is essential if an evaluation of the relative contribution of each factor is to be made. Both the climatic and sociodemographic data can be used for specific and effective interventions in the management of land in a bid to fight desertification and restore degraded environments (Feng et al. 2015).

28.5.2 Role of Invasive Species in Accelerating Desertification

IAS have been identified as major players in the process due to the disturbance of native ecosystems and the enhancement of processes leading to desertification. These species are usually brought in by oblivious human activity or accidentally and they upset the delicate natural order of the ecosystems causing further desertification. Its capacity for its ability to exclude native plants and alter the physical characteristics of soil and water and hydrological cycle makes invasive species important agents of land degradation in sensitive areas.

28.5.2.1 Invasive Species and Their Role in Desertification

Invasive species significantly contribute to desertification by competing with native vegetation and disrupting ecosystem functions. They outcompete native plants for water, nutrients, and light, leading to a decline in biodiversity (Thiele et al. 2009). The loss of native vegetation removes ground cover, increasing soil erosion from wind and water. This weakens the land's ability to retain moisture, making arid and semiarid regions more vulnerable to desertification. Invasive species also alter ecosystem processes, further degrading land quality. Some invaders, such as rhizome plants, reduce soil fertility and organic matter, while others increase soil salinity, making the land unsuitable for native plant growth or agriculture. Additionally, invasive species that consume large amounts of water disrupt natural hydrological cycles, increasing water shortages are a major contributing factor to the spread of desertification. Their rapid spread and adaptability make invasive species even more dangerous. Many invasive plants and animals reproduce quickly, tolerate a wide range of environmental conditions, and colonize degraded areas efficiently. For example, the harlequin ladybird (*Harmonia axyridis*) spreads rapidly and displaces native species (Brown et al. 2018). These characteristics allow invasive species to dominate landscapes, accelerating land degradation and desertification.

28.5.2.2 Complexity of Invasive Species' Impacts and Research Needs

While invasive species are generally seen as harmful, their effects on desertification are complex and context-dependent. Some studies suggest that their impact varies based on climate, soil type, and land use. In certain cases, invasive species may not always worsen desertification, and their influence depends on regional conditions (Bates et al. 2020). This complexity highlights the need for region-specific studies to fully understand their role in land degradation. Current research indicates that invasive species disrupt ecosystems and contribute to desertification, but further studies are necessary to determine their exact impact across different climates and geographic regions. Targeted management strategies are essential to prevent and control invasions, protect native biodiversity, and mitigate desertification risks. Future research should focus on tailored solutions to minimize the damage caused by invasive species while preserving ecological balance.

28.5.3 Case Studies: Regions Affected by Desertification and Species Spread

Vectors such as desertification and invasion of undesirable flora and fauna are therefore closely related depending on the level of regional climate and human interferences. Such processes are illustrated by the case-studies of different regions, thus showing that ecosystem degradation is a rather complicated process.

Table 28.2 Case Study Regions Affected by Desertification and the Spread of Invasive Species

Region	Key drivers	Impact
Balochistan, Pakistan	Climatic factors and unsustainable practices	Land degradation, reduced vegetation cover
Minqin Oasis, China	Long-term arid climate and water scarcity	Desertification due to water component changes
Punjab, India	Invasive plant *Verbesina encelioides* and climatic stress	Biodiversity loss, ecosystem degradation

Table 28.2 includes descriptions of certain regions influenced by desertification and species spread, major driving forces, and their effects. Climatic conditions and uncontrolled exploitation of land resource in Balochistan, Pakistan, have resulted in land degradation and less vegetation cover. Sustainable impacts are found in the Minqin Oasis, China, where long-term and severe arid conditions and water shortage have been the main causes, which lead to desertification due to the modification of the water item in the ecosystem. On the other hand, in Punjab-India, *V. encelioides* and climate change affect the losing biological diversity and degrading environment. These case studies show that climatic changes, together with other factors, lead to land degradation and decrease in ecosystems, necessitating humanitarian regional approaches to control desertification.

28.5.3.1 Balochistan Province, Pakistan

Perhaps, Balochistan's contribution is unique where a 21-year research investigated climate and human contributions to both land degradation and restoration. In the case of restoration, climatic parameters like precipitation and hydrothermal conditions influenced to some extent the works carried out in the area including the synergy with more proper management of the lands. Land degradation is mainly caused by climate factors, but human activities also play a role. Irregular farming practices and excessive use of natural resources make the problem worse. Temperature, actual evapotranspiration, potential evapotranspiration, and mean annual precipitation were identified as the leading climate variables that influenced vegetation distribution (Chen et al. 2023). This particular case shows that the attempts to combat desertification are closely followed by climate-related problems, and the efforts made by people are defeated by climate.

28.5.3.2 Minqin Oasis, Northern China

Specifically, the synoptic observations from the Minqin Oasis in northern China suggest that over the past 2000 years, the principal cause factors in desertification of the land have been changes in the water element of the phase and the arid conditions. These trends have been further amplified by other local geomorphological processes. While people increased their activities since the 1940s, it can be stated that the desertification process in the considered area reflects long-term climatic fluctuations features rather than the results of recent anthropogenic impacts (Zhang et al. 2007). This case shows that even as climatic trends continue to drive desertification for centuries, humankind has imposed its agency to increase the extent of land deterioration.

28.5.3.3 Punjab, Northwestern India

In the Punjab drylands, *Verbesina encelioides* competes successfully for resources and has fundamentally transformed the plant communities, and most importantly, decreased plant species richness. The most devastating impacts of this invasive plant are always realized in arid and semiarid regions because climate stress factor such as aridification is likely to increase. *V. encelioides* not only outcompetes native species but also diminishes overall habitat resilience and disrupts ecosystem functions. These impacts raise important questions about the potential amplification of such effects under future climate change scenarios. (Mehal et al. 2023). This case also shows how invasive species are agents of biotic homogenization and agents of desertification particularly in sensitive areas.

28.6 Interconnections Between Salinity, Desertification, and Invasive Species

One of the factors that contribute most to desertification is soil salinity because of the large concentrations in arid and semiarid areas. Annually, salinization accounts for 0.3–1.5 million hectares of arable land lost in various parts of the world. This process leads to soil deterioration, crop yields' reduction, reduces the numbers and activities of soil organisms, and encourages desertification (Harper et al. 2020). For instance, in the Horqin Sandy Land of northern China, the degree of desertification is directly linked to the losses of SOC and TN, respectively. In some cases, salinity has negative impacts on plants; however, plants that have a high tolerance to salt or Halophytes provide potential for redeveloping high salinity land and using water sources such as wastewater. But the release of such species has to be regulated in a way in order to avoid probable invasive impacts on the existing population. In conclusion, salinity encourages desertification; however, certain strategies such as farming halophytes can be used for developing suitable land use methods appropriately. Solving these problems needs the use of comprehensive measures based on proper management of soils, introduction of efficient water use, as well as the correct choice of the plantings, facilitating the restoration of ecosystems (Plaut et al. 2013).

28.6.1 Climate-Driven Synergistic Effects

Interactions between the climate changes and other environmental stressors are very synergistic, meaning that the impacts of the climate change and the stressors on ecosystems are greater than the sum of the individual effects. These interactions increase the impacts of ecosystem degradation and complicate natural change management and understanding. Research conducted across these ecosystems brings out these synergistic effects and varying results, thus underlining the need for context-dependent research.

28.6.1.1 Synergistic Impacts in Coastal Ecosystems

Climate factors as well as fishing activities and nutrient input in the coastal areas are deemed threats that affect the coral reef and related resources. It has been ascertained that these stressors are cumulative in their effects and the combined effects of all these candidates are far worse than the total of individual impact of each factor. For instance, heat flux, temperature rise, increases in wind speed, and fishing pressure influences the hard corals and coralline algae, which are critical elements in reef structures and resistance (Ellis et al. 2019). These results demonstrate that coral reefs are threatened both by multiple anthropogenic impacts and climate change, and that their stewardship requires a complex solution.

28.6.1.2 Synergistic Effects in Alpine Lake Ecosystems

For example, in the alpine lakes on the Tibetan Plateau, more extreme climate change and nutrient deposition increase the ecological effects of global warming. All these stress factors collectively facilitate algal growth and dynamics of community structure, thus tending to select the fast-growing algal species under high nutrient and warming conditions as mentioned by Zhang et al. (2023). The warming effects are most significant in those environments characterized by a high rate of nutrient deposition, underscoring the importance of nutrient control in minimizing the impacts of climate change on these marginal ecosystems. They also explain the synergy between Clim/CHG and nutrients with implications on water quality, abundance of fauna and flora, and carbon stocks in aquatic ecosystem.

28.6.1.3 Contradictory Findings in Lake Ecosystems

Perversely, some stressor-stressor interactions contradict fundamental ideas of climate change, about other ecosystems. For instance, research on shallow lakes shows that while warming and reduced water transparency due to eutrophication or browning reduce warming to the water column, it cools bottom waters instead. This cooling effect improves the rate of carbon sequestration in these lakes, unlike the impacts of global warming, which are often assumed to increase rates of organic matter mineralization and greenhouse gas emissions (Bartosiewicz et al. 2019). Examples like the above show that ecosystem response to multiple stressors is counterintuitive and hence necessitates elaborate investigations.

28.6.2 Ecosystem Degradation and Resource Competition

Resource competition and ecosystem degradation are both closely related, and climate change could be classified as the deciding factor that enforces these activities. There is a strong and significant positive relationship between resource competition and ecosystem resilience and degradation. Competition over the natural resource leads to selective pressures that help improve the functionality of ecosystems through evolution-based positive frequencies; however, invasions that have been upset by giants and anthropogenic influences supported by climate change alter the dynamics leading to degradation.

28.6.2.1 Climate Change, Resource Competition, and Human Impact

Competition for natural resources is a key factor shaping ecosystems, influencing species interactions and evolutionary processes. Species compete for water, nutrients, and light, with intraspecific competition improving resource use efficiency and interspecific competition boosting biodiversity and productivity (Loreau 1998). Over time, this enhances nutrient cycling and increases ecosystem's carrying capacity. However, climate change disrupts these processes by making resources scarcer, forcing species to relocate or decline. Competition varies across scales; while it limits growth at microscales, diversity at macroscales promotes stability (Zhang et al. 2024).

Climate change alters resource availability, affecting competition. Rising temperatures and shifting rainfall patterns disrupt species interactions, often making one species dominant over another, leading to ecosystem imbalances. Extended dry seasons intensify competition for water, weakening ecosystem resilience to environmental stress. Invasive species further disrupt resource competition, outcompeting native species for light, water, and nutrients. They alter soil properties and water cycles, reducing biodiversity and degrading ecosystem services like soil fertility and water retention (Carboni et al. 2021). Climate change accelerates invasions by creating favorable conditions for invasive plants, worsening resource scarcity, and ecosystem instability.

Human activities such as deforestation, overgrazing, and unsustainable agriculture amplify resource depletion, weakening ecosystems. CO_2 emissions and poor resource management degrade environments, making ecosystems less competitive and more vulnerable to climate stress (Hemendra et al. 2023). Governance, technological advancements, and policy interventions play crucial roles in mitigating these impacts. The combined effects of climate change and human activities create a feedback loop that accelerates ecosystem degradation and resource scarcity. Weak ecosystems struggle to withstand climate shocks like heat waves and droughts. Addressing these challenges requires integrated resource management, sustainable development policies, and technological innovations to enhance ecosystem resilience and long-term sustainability.

28.6.3 Impact on Biodiversity and Agricultural Sustainability

Returning again to the central research question, it is evident that agricultural practices pose both opportunities for and threats to biodiversity and sustainability. Hyped by massive production, extensive agriculture frequently results in bio diversity reduction and environmental deterioration; however, extensive practices provide effective solutions for sustainable production. These complexities are then amplified by climate change which increases the pressure on biodiversity and decreases the stability of agriculture.

28.6.3.1 Biodiversity Loss and the Role of Sustainable Agriculture

Intensive agriculture is a major cause of biodiversity loss due to practices such as mono-cropping, chemical fertilizer use, deforestation, and habitat conversion (Turley and Brudvig 2016). As global populations grow, the demand for agricultural land increases, leading to habitat destruction and displacement of species. Climate change worsens this by altering habitats and reducing species' ability to adapt, making biodiversity decline more severe (Kok et al. 2018). Sustainable agricultural practices offer a solution by reducing the negative impact of farming on biodiversity. Methods such as organic farming, agroforestry, crop diversification, and minimal chemical use support diverse species and improve soil health (Underwood et al. 2011). These approaches enhance soil fertility, water retention, and nutrient availability, benefiting both ecosystems and food production. Additionally, genetic diversity within and between crops increases resilience to climate extremes, pests, and diseases, improving food security (Frison et al. 2011). Sustainable farming also has economic and social benefits. Practices like conservation agriculture and integrated pest management lower input costs, increasing farmers' profits, especially in developing countries (Makate et al. 2017). A multifunctional landscape approach integrates ecological, social, and economic sustainability, balancing agricultural production with environmental conservation (Dudley and Alexander 2017).

28.7 Climate Change, Salinity, and Desertification

Climate change, salinity, desertification, and invasive species are interconnected environmental challenges that significantly impact ecosystems, agriculture, and biodiversity. Rising global temperatures accelerate soil degradation, while increased salinity and desertification reduce land productivity and threaten water security. Simultaneously, invasive species exploit these changing conditions, further destabilizing ecosystems and reducing the resilience of native flora and fauna (Nikkel et al. 2023; Gong et al. 2020).

The interactions between these factors create complex feedback loops that worsen environmental degradation. Climate change alters rainfall patterns, intensifies droughts, and raises sea levels, leading to increased soil salinity and desertification. As landscapes become more arid, invasive species outcompete native plants, disrupting ecological balance and agricultural stability (D'Odorico et al. 2012; Corwin 2020). Salinity weakens plant resilience, making ecosystems more vulnerable to biological invasions and further contributing to biodiversity loss.

Addressing these challenges requires an integrated approach combining sustainable land management, climate adaptation strategies, and proactive invasive species control. Conservation agriculture, improved irrigation techniques, and reforestation efforts can mitigate salinity and desertification impacts (Musyinye 2018; Ouhamdouch et al. 2016). Additionally, global cooperation and policies supporting adaptive ecosystem management, such as the United Nations Convention to Combat Desertification (UNCCD), play a crucial role in fostering resilience against these environmental threats (Patil et al. 2023). By understanding these interconnections, policymakers and conservationists can develop targeted solutions to mitigate long-term environmental and socioeconomic impacts.

28.7.1 Interlinkages Between Climate Change, Salinity, and Desertification

Global warming, water logging, and soil erosion are interconnected phenomena that accelerate environmental degradation, particularly in arid regions. Rising temperatures and altered precipitation patterns contribute to excessive evaporation, reducing soil moisture and increasing salt accumulation where groundwater is shallow or of poor quality. Coastal areas are especially vulnerable, as sea-level rise causes saline water intrusion into freshwater aquifers, reducing soil fertility and contaminating irrigation sources (Corwin 2020). Climate change influences excessive rainfall and drought, while salinity and desertification degrade soil health, diminishing the quality of arable land and food production. Integrated strategies are essential to address these interrelated challenges (Fig. 28.2).

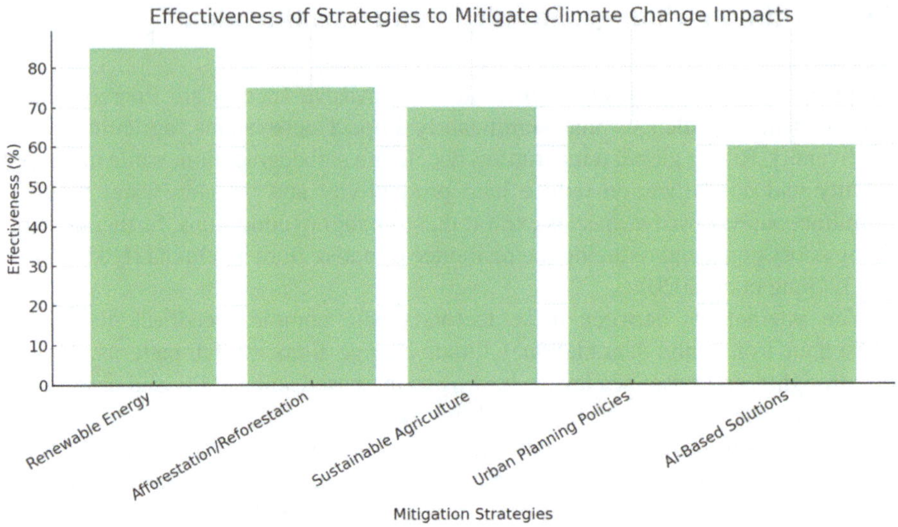

Fig. 28.2 Effectiveness of strategies to mitigate climate change impacts

28.7.1.1 Climate Change and Desertification Feedback Loops

Desertification is both a consequence of and a contributor to climate change. Vegetation loss due to desertification increases surface reflectivity, reducing precipitation rates and intensifying aridification. Dust emissions from degraded land contribute to atmospheric radiative forcing, exacerbating global warming. Additionally, desertification alters the carbon cycle by reducing soil organic matter, decreasing the land's carbon sink capacity, and releasing stored carbon into the atmosphere (D'Odorico et al. 2012). Climate change further aggravates these effects by promoting conditions favorable for land degradation.

28.7.1.2 Socioeconomic Impacts of Climate-Salinity-Desertification Interactions

Salinization reduces arable land and crop yields, posing a severe threat to food security. Controlling soil salinity requires costly interventions, including improved irrigation and soil amendments. Coastal regions are particularly affected, as salinization depletes freshwater resources, forcing reliance on expensive desalination processes. In developing regions, land degradation exacerbates poverty, forces migration, and leads to social instability. As fertile land becomes unproductive, communities are displaced, increasing pressure on urban areas with limited resources (Musyinye 2018).

28.7.1.3 Strategies for Mitigation and Adaptation

Addressing the interconnected challenges of climate change, salinity, and desertification requires multidisciplinary approaches. Conservation agriculture practices, such as reduced tillage and crop rotation, improve soil structure and minimize salinity risks. Efficient irrigation techniques like drip irrigation reduce water usage and limit salt accumulation (Corwin 2020). Reforestation stabilizes soil, improves carbon sequestration, and reduces desertification risks (D'Odorico et al. 2012). The use of salt-tolerant crops in high-salinity regions can sustain agricultural productivity (Fig. 28.3). Strengthening adaptive capacity through education, technology transfer, and financial support is crucial (Ouhamdouch et al. 2016). Global cooperation, including commitments like the UNCCD, can facilitate large-scale mitigation efforts (Patil et al. 2023).

28.7.2 Climate Change and Invasive Species

Climate change intensifies global food security threats, while invasive agricultural species further strain ecosystems and policies. Climate change expands the range of invasive pests and weeds, allowing them to colonize new habitats while also weakening conventional control methods. The effects of climate change on invasive species vary by species and climate scenario. Some invasive terrestrial plants, particularly those from temperate regions, may decline in certain environments, while others expand into newly favorable areas (Nikkel et al. 2023). Aquatic

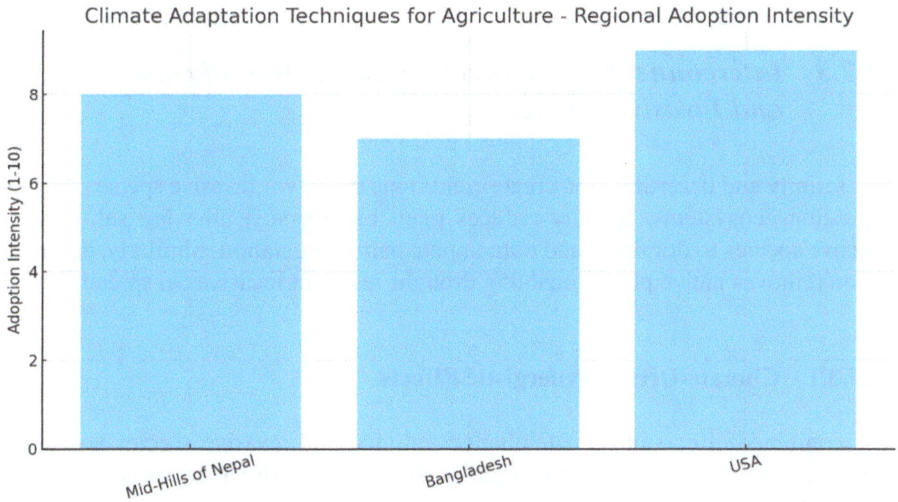

Fig. 28.3 Climate adaptation techniques for agriculture—regional adoption intensity

invasive plants, however, are less affected due to their stable habitat conditions. Climate change also disrupts biological interactions, weakening natural control systems and causing unforeseen ecological consequences (Gong et al. 2020).

28.7.2.1 Mechanisms Driving the Spread of Invasive Species

Biological invasions are influenced by species traits, environmental conditions, and climate interactions. Dispersal ability, reproductive rates, and environmental adaptability determine an invasive species' success. Some species thrive in disturbed habitats, altering ecosystem dynamics. For example, invasive plants like *Bromus tectorum* alter fire regimes, increasing wildfire frequency, while aquatic invasive plants like *Eichhornia crassipes* disrupt water flow and oxygen levels (Coutts et al. 2010; Wilson and Pinno 2012).

28.7.2.2 Impact of Changing Climates on Ecosystem Balance

Climate change provides favorable conditions for many invasive species. Rising temperatures expand habitable zones, allowing species such as *Pacifastacus leniusculus* and *Procambarus clarkii* to outcompete native species in freshwater systems (Zhang et al. 2019). However, climate change impacts vary across ecosystems. While terrestrial invasive species benefit from longer growing seasons, aquatic invasives may face oxygen depletion due to warming waters (Stephens et al. 2019). The complex interplay between climate change and invasive species underscores the need for tailored management strategies.

28.7.3 Interconnections Between Salinity, Desertification, and Invasive Species

Soil salinity and desertification create conditions that favor invasive species, further degrading ecosystems. Salinity reduces plant biodiversity, allowing salt-tolerant invasive species to dominate and outcompete native vegetation. Similarly, desertification removes native plants, enabling drought-resistant invasives to spread.

28.7.3.1 Climate-Driven Synergistic Effects

The combined effects of climate change, salinity, and invasive species accelerate land degradation. Warming temperatures and sea-level rise increase soil salinity, weakening native plant resilience. Invasive species exploit these changes, further destabilizing ecosystems (Harper et al. 2020).

28.7.3.2 Resource Competition and Ecosystem Degradation

Invasive species outcompete native plants for water and nutrients, accelerating soil depletion. *Prosopis juliflora* invades arid regions, reducing water availability for native species. Similarly, invasive grasses alter fire regimes, worsening desertification (Carboni et al. 2021).

28.7.3.3 Management of Invasive Species in Saline and Desertified Areas

Addressing the combined threats of salinity, desertification, and invasive species requires early detection and adaptive management. Sustainable land use practices, such as agroforestry and controlled irrigation, can mitigate these effects. Biological control and habitat restoration strategies must be region-specific to ensure effectiveness (Heshmati et al. 2019).

By integrating climate change mitigation, land management, and invasive species control, it is possible to enhance ecosystem resilience and sustain agricultural productivity in affected regions.

28.8 Conclusion

Climate change, invasive species, salinity, and desertification are major threats to ecosystems, farming, and biodiversity. Rising temperatures, changing rainfall, and higher sea levels make it easier for invasive species to spread, increase soil salinity, and speed up desertification. These changes harm soil quality, reduce farm productivity, and threaten food security, especially in dry areas. Invasive species take advantage of changing conditions, outcompeting local species and damaging ecosystems. This is worse in areas already struggling with salinity and desertification, leading to soil degradation, water issues, and loss of biodiversity. Salinity increases due to higher temperatures, rising sea levels, and less rainfall. Poor land use and irrigation methods make it harder to manage. Solutions include subsurface drainage, agroforestry, and growing salt-tolerant crops. Desertification reduces vegetation, lowers soil fertility, and increases erosion. Climate change and human activities make it worse. Solutions like planting trees, restricting grazing, and conserving water help, but their social impacts must be considered. Addressing these problems requires a mix of climate action, ecosystem protection, and economic development. Policies should promote sustainable land use, drought-resistant plants, and early detection of invasive species. Solutions must be adapted to local conditions, blending traditional knowledge with modern science. In conclusion, tackling climate change, salinity, invasive species, and desertification requires cooperation between scientists, policymakers, and communities. With the right strategies, we can restore land, protect ecosystems, and balance development with conservation.

Acknowledgments The authors would like to express their gratitude to all individuals and institutions who contributed to the completion of this research. Special thanks to the research teams, environmental scientists, and data analysts whose insights and expertise were invaluable. We also appreciate the contributions of peer reviewers and mentors who provided critical feedback that enhanced the quality of this work.

Funding Statement The author(s) received no financial support for the research, authorship, or publication of this chapter.

Data Availability Statement The data used in this study are available upon request from the corresponding author. Publicly available datasets were also used and are cited accordingly within the manuscript.

Declarations The author(s) confirm that this chapter is their original work and has not been published elsewhere.

Conflict of Interest The authors declare no conflict of interest in relation to this study.

Ethics Approval This research did not involve human or animal subjects; hence, no ethical approval was required.

Informed Consent Statement Not applicable as this research did not involve human participants.

References

Bartosiewicz M, Lehmann MF, Przytulska A, Lapierre J, Maranger R, Laurion I (2019) Hot tops, cold bottoms: synergistic climate warming and shielding effects increase carbon burial in lakes. Limnol Oceanogr Lett 4(5):132–144. https://doi.org/10.1002/lol2.10117

Bates OK, Bertelsmeier C, Ollier S (2020) Smaller climatic niche shifts in invasive than non-invasive alien ant species. Nat Commun 11(1):5213. https://doi.org/10.1038/s41467-020-19031-1

Beaury EM, Allen JM, Laginhas BB, Fusco EJ, Morelli TL, Pasquarella VJ, Bradley BA, Jackson MR (2019) Incorporating climate change into invasive species management: insights from managers. Biol Invasions 22(2):233–252. https://doi.org/10.1007/s10530-019-02087-6

Brown PMJ, Roy HE, Roy DB, Harrower C, Rorke SL, Dean HJ (2018) Spread of a model invasive alien species, the harlequin ladybird *Harmonia axyridis* in Britain and Ireland. Sci Data 5(1):180239. https://doi.org/10.1038/sdata.2018.239

Carboni M, Isaac ME, Cadotte MW, Livingstone SW (2021) Invasion drives plant diversity loss through competition and ecosystem modification. J Ecol 109(10):3587–3601. https://doi.org/10.1111/1365-2745.13739

Chen J, Mueller V (2018) Coastal climate change, soil salinity and human migration in Bangladesh. Nat Clim Chang 8(11):981–985. https://doi.org/10.1038/s41558-018-0313-8

Chen X, Zhou N, Wang Y, Chen Y, Fu S (2023) NDVI-based assessment of land degradation trends in Balochistan, Pakistan, and analysis of the drivers. Remote Sens 15(9):2388. https://doi.org/10.3390/rs15092388

Corwin DL (2020) Climate change impacts on soil salinity in agricultural areas. Eur J Soil Sci 72(2):842–862. https://doi.org/10.1111/ejss.13010

Coutts SR, Van Klinken RD, Buckley YM, Yokomizo H (2010) What are the key drivers of spread in invasive plants: dispersal, demography or landscape? And how can we use this knowledge to aid management? Biol Invasions 13(7):1649–1661. https://doi.org/10.1007/s10530-010-9922-5

D'Odorico P, Bhattachan A, Davis KF, Ravi S, Runyan CW (2012) Global desertification: drivers and feedbacks. Adv Water Resour 51(51):326–344. https://doi.org/10.1016/j.advwatres.2012.01.013

Buono D (2020) Can biostimulants be used to mitigate the effect of anthropogenic climate change on agriculture? It is time to respond. Sci Total Environ 751:141763. https://doi.org/10.1016/j.scitotenv.2020.141763

Dudley N, Alexander S (2017) Agriculture and biodiversity: a review. Biodiversity 18(2–3):45–49. https://doi.org/10.1080/14888386.2017.1351892

Ellis JI, Coker DJ, Carvalho S, Hewitt J, Hariprasad D, Curdia J, Krokos G, Roth F, Jamil T, Kürten B, Jones BH, Hoteit I, Anlauf H (2019) Multiple stressor effects on coral reef ecosystems. Glob Chang Biol 25(12):4131–4146. https://doi.org/10.1111/gcb.14819

Eswar D, Karuppusamy R, Chellamuthu S (2021) Drivers of soil salinity and their correlation with climate change. Curr Opin Environ Sustain 50:310–318. https://doi.org/10.1016/j.cosust.2020.10.015

Feng Q, Ma H, Wang X, Jiang X, Cao S (2015) What has caused desertification in China? Sci Rep 5(1):15998. https://doi.org/10.1038/srep15998

Food and Agriculture Organization (FAO) (2020) The state of food security and nutrition in the world 2020: transforming food systems for affordable healthy diets. Retrieved from http://www.fao.org/publications

Frison EA, Cherfas J, Hodgkin T (2011) Agricultural biodiversity is essential for a sustainable improvement in food and nutrition security. Sustainability 3(1):238–253. https://doi.org/10.3390/su3010238

Gong X, Chen Y, Wang T, Jiang X, Hu X, Feng J (2020) Double-edged effects of climate change on plant invasions: ecological niche modeling global distributions of two invasive alien plants. Sci Total Environ 740:139933. https://doi.org/10.1016/j.scitotenv.2020.139933

Haj-Amor Z, Araya T, Kim D-G, Bouri S, Lee J, Ghiloufi W, Yang Y, Kang H, Jhariya MK, Banerjee A, Lal R (2022) Soil salinity and its associated effects on soil microorganisms, greenhouse gas emissions, crop yield, biodiversity, and desertification: a review. Sci Total Environ 843:156946. https://doi.org/10.1016/j.scitotenv.2022.156946

Harper RJ, Dell B, Ruprecht JK, Sochacki SJ, Smettem KRJ (2020) Chapter 7 – Salinity and the reclamation of salinized lands. In: Soils and landscape restoration. Elsevier, pp 193–208. https://doi.org/10.1016/b978-0-12-813193-0.00007-2

Hellmann JJ, Nadelhoffer KJ, Iverson LR, Ziska LH, Matthews SN, Myers P, Prasad AM, Peters MP (2010) Climate change impacts on terrestrial ecosystems in metropolitan Chicago and its surrounding, multi-state region. J Great Lakes Res 36:74–85. https://doi.org/10.1016/j.jglr.2009.12.001

Heshmati HM (2021) Impact of climate change on life. IntechOpen. https://doi.org/10.5772/intechopen.94538

Heshmati I, Khorasani N, Riazi B, Shams-Esfandabad B (2019) Forthcoming risk of *Prosopis juliflora* global invasion triggered by climate change: implications for environmental monitoring and risk assessment. Environ Monit Assess 191(2):72. https://doi.org/10.1007/s10661-018-7154-9

Hopmans JW, Jiao H, Kisekka I, Lavado RS, Grattan SR, Taleisnik E, De Jong Van Lier Q, Qureshi AS, Li B, Wang G, Skaggs TH, Javaux M, Rengasamy P, Raats PAC, Lazarovitch N, Minhas PS, Ben-Gal A, Assouline S, Munns R (2021) Critical knowledge gaps and research priorities in global soil salinity. In: Advances in agronomy. Elsevier, pp 1–191. https://doi.org/10.1016/bs.agron.2021.03.001

Hoveka LN, Bezeng BS, Yessoufou K, Boatwright JS, Van Der Bank M (2015) Effects of climate change on the future distributions of the top five freshwater invasive plants in South Africa. S Afr J Bot 102:33–38. https://doi.org/10.1016/j.sajb.2015.07.017

Huang J, Zhang G, Guan X, Guo R, Zhang Y, Wei Y (2020) Global desertification vulnerability to climate change and human activities. Land Degrad Dev 31(11):1380–1391. https://doi.org/10.1002/ldr.3556

Intergovernmental Panel on Climate Change (IPCC) (2019) Special report on the ocean and cryosphere in a changing climate. Cambridge University Press. https://doi.org/10.1017/9781108866093

Intergovernmental Panel on Climate Change (IPCC) (2021) Climate change 2021: the physical science basis. Cambridge University Press. https://doi.org/10.1017/9781009157896

Kok MTJ, Alkemade R, Bakkenes M, Van Eerdt M, Janse J, Mandryk M, Kram T, Lazarova T, Meijer J, Van Oorschot M, Westhoek H, Van Der Zagt R, Van Der Berg M, Van Der Esch S, Prins A-G, Van Vuuren DP (2018) Pathways for agriculture and forestry to contribute to terrestrial biodiversity conservation: a global scenario-study. Biol Conserv 221:137–150. https://doi.org/10.1016/j.biocon.2018.03.003

Komoroske LM, Brander SM, Britton M, Whitehead A, Fangue NA, Roach JL, Connon RE, Verhille C, Jeffries KM (2020) Transcriptional flexibility during thermal challenge corresponds with expanded thermal tolerance in an invasive compared to native fish. Evol Appl 14(4):931–949. https://doi.org/10.1111/eva.13172

Lamsal P, Aryal A, Atreya K, Kumar L (2018) Invasive alien plant species dynamics in the Himalayan region under climate change. Ambio 47(6):697–710. https://doi.org/10.1007/s13280-018-1017-z

Latham ADM, Latham MC, Cieraad E, Tompkins DM, Warburton B (2015) Climate change turns up the heat on vertebrate pest control. Biol Invasions 17(10):2821–2829. https://doi.org/10.1007/s10530-015-0931-2

Li M, Zhou W, Sun M, Shi W, Lun J, Zhou B, Hou L, Gao Z (2024) Decoupling soil community structure, functional composition, and nitrogen metabolic activity driven by salinity in coastal wetlands. Soil Biol Biochem 198:109547. https://doi.org/10.1016/j.soilbio.2024.109547

Loreau M (1998) Ecosystem development explained by competition within and between material cycles. Proc R Soc Lond B Biol Sci 265(1390):33–38. https://doi.org/10.1098/rspb.1998.0260

Lyu Y, Shi P, Zhang G, Han G, Guo L, Hu X, Liu L (2020) Desertification control practices in China. Sustainability 12(8):3258. https://doi.org/10.3390/su12083258

Macinnis-Ng C, Curran TJ, Waipara N, Dickinson KJ, Mcintosh AR, White RS, Clark CD, Boudjelas S, Stanley MC, Richardson SJ, Peltzer DA, Perry GL, Monks JM, Clearwater MJ, Nelson N (2021) Climate-change impacts exacerbate conservation threats in island systems: New Zealand as a case study. Front Ecol Environ 19(4):216–224. https://doi.org/10.1002/fee.2285

Makate C, Makate M, Mango N (2017) Sustainable agriculture practices and livelihoods in pro-poor smallholder farming systems in southern Africa. Afr J Sci Technol Innov Dev 9(3):269–279. https://doi.org/10.1080/20421338.2017.1322350

Manzoor SA, Griffiths G, Lukac M (2020) Land use and climate change interaction triggers contrasting trajectories of biological invasion. Ecol Indic 120:106936. https://doi.org/10.1016/j.ecolind.2020.106936

Meek MH, May B, Wintzer AP, Wetzel WC (2012) Climate change likely to facilitate the invasion of the non-native hydroid, *Cordylophora caspia*, in the San Francisco Estuary. PLoS One 7(10):e46373. https://doi.org/10.1371/journal.pone.0046373

Mehal KK, Kalia N, Singh HP, Kaur A, Sharma A, Kohli RK, Batish DR (2023) Modelling the ecological impact of invasive weed *Verbesina encelioides* on vegetation composition across dryland ecosystems of Punjab, northwestern India. Environ Monit Assess 195(6):725. https://doi.org/10.1007/s10661-023-11299-2

Mishra AK, Sinha T, Kumar M, Das R, Sharma S, Arora P, Biswal B, George Kerry R (2023) Promising management strategies to improve crop sustainability and to amend soil salinity. Front Environ Sci 10:962581. https://doi.org/10.3389/fenvs.2022.962581

Mokany K, Prasad S, Westcott DA (2014) Loss of frugivore seed dispersal services under climate change. Nat Commun 5(1):3971. https://doi.org/10.1038/ncomms4971

Munns R, Cramer GR, Ball MC (1999) Interactions between rising CO_2, soil salinity, and plant growth. In: Carbon dioxide and environmental stress. Elsevier, pp 139–167. https://doi.org/10.1016/b978-012460370-7/50006-1

Murtaza G, Sheryar A, Hayat HS, Bukhari A, Shaheryar M, Ghaffor K, Shah MN, Mehran M, Akram HMB (2022) Effect of nitrogen and phosphorus application rate on peanut (*Arachis hypogaea* L.) phenology, yield, and soil nutrient status. J Arable Crop Mark 4(2):143–153. https://doi.org/10.33687/jacm.004.02.4569

NASA (2023) Global temperature time series. Retrieved from https://climate.nasa.gov

Nicholls RJ, Hinkel J, Lincke D, van der Pol T (2021) Global investments for coastal adaptation to sea level rise. Nat Commun 12:1841. https://doi.org/10.1038/s41467-021-21868-8

Nikkel E, Clements DR, Anderson D, Williams JL (2023) Regional habitat suitability for aquatic and terrestrial invasive plant species may expand or contract with climate change. Biol Invasions 25(12):3805–3822. https://doi.org/10.1007/s10530-023-03139-8

Ouhamdouch S, Souhel A, Paula C, Bahir M (2016) Vulnerability and impact of climate change processes on water resources in semi-arid areas. In: Essaouira basin (Morocco). Springer, pp 719–736. https://doi.org/10.1007/978-3-319-30127-3_53

Patil M, Dalal PH, Salifu E, Iyer KKR, Dave TN (2023) Biostabilization of soils as a sustainable pathway for anti-desertification: present and future perspectives. Mater Today Proc. https://doi.org/10.1016/j.matpr.2023.04.216

Plaut Z, Edelstein M, Ben-Hur M (2013) Overcoming salinity barriers to crop production using traditional methods. Crit Rev Plant Sci 32(4):250–291. https://doi.org/10.1080/07352689.2012.752236

Ramasamy R, Surendran SN (2012) Global climate change and its potential impact on disease transmission by salinity-tolerant mosquito vectors in coastal zones. Front Physiol 3:198. https://doi.org/10.3389/fphys.2012.00198

Rasul G, Sharma B, Singh SP (2019) Climate change, water resources, and food security in South Asia: a nexus perspective from the Hindu Kush Himalaya. Environ Dev 31:63–77. https://doi.org/10.1016/j.envdev.2019.01.003

Rogelj J, Shindell D, Jiang K, Fifita S, Forster P et al (2018) Mitigation pathways compatible with 1.5°C in the context of sustainable development. In: Global warming of 1.5°C. IPCC. Retrieved from https://www.ipcc.ch

Shoba P, Ramakrishnan SS (2016) Modeling the contributing factors of desertification and evaluating their relationships to the soil degradation process through geomatic techniques. Solid Earth 7(2):341–354. https://doi.org/10.5194/se-7-341-2016

Shrestha UB, Sharma KP, Devkota A, Siwakoti M, Shrestha BB (2018) Potential impact of climate change on the distribution of six invasive alien plants in Nepal. Ecol Indic 95:99–107. https://doi.org/10.1016/j.ecolind.2018.07.009

Singh A, Minkina T, Kriemadis A, Papadakis M, Yesayan A, Agrawal S, Petropoulos D, Ghazaryan K, Rajput VD, Alexiou A, Zhao Y (2024) Nanoparticles in revolutionizing crop production and agriculture to address salinity stress challenges for a sustainable future. Discov Appl Sci 6(6):317. https://doi.org/10.1007/s42452-024-06009-7

Stavi I, Priori S, Thevs N (2021) Soil salinity and sodicity in drylands: a review of causes, effects, monitoring, and restoration measures. Front Environ Sci 9:712831. https://doi.org/10.3389/fenvs.2021.712831

Stephens KL, Patten MA, Dantzler-Kyer ME, Souza L (2019) Differential responses to global change of aquatic and terrestrial invasive species: evidence from a meta-analysis. Ecosphere 10(4):e02680. https://doi.org/10.1002/ecs2.2680

Teshaev N, Mamadaliyev B, Khasanov S, Ibragimov A (2020) The soil-adjusted vegetation index for soil salinity assessment in Uzbekistan. InterCarto InterGIS 26(3):324–333. https://doi.org/10.35595/2414-9179-2020-3-26-324-333

Thiele J, Kollmann J, Otte A, Markussen B (2009) Impact assessment revisited: improving the theoretical basis for management of invasive alien species. Biol Invasions 12(7):2025–2035. https://doi.org/10.1007/s10530-009-9605-2

Turbelin A, Catford JA (2021) Invasive plants and climate change. In: Climate change. Elsevier, pp 515–539. https://doi.org/10.1016/b978-0-12-821575-3.00025-6

Turley NE, Brudvig LA (2016) Agricultural land-use history causes persistent loss of plant phylogenetic diversity. Ecology 97(9):2240–2247. https://doi.org/10.1002/ecy.1443

Underwood T, McCullum-Gomez C, Harmon A, Roberts S (2011) Organic agriculture supports biodiversity and sustainable food production. J Hunger Environ Nutr 6(4):398–423. https://doi.org/10.1080/19320248.2011.627301

United Nations Framework Convention on Climate Change (UNFCCC) (2020) Climate change and extreme weather events. Retrieved from https://unfccc.int

Verlinden M, Van Kerkhove A, Nijs I (2013) Effects of experimental climate warming and associated soil drought on the competition between three highly invasive West European alien plant species and native counterparts. Plant Ecol 214(2):243–254. https://doi.org/10.1007/s11258-012-0163-9

Walker J (2018) The influence of climate change on waterborne disease and *Legionella*: a review. Perspect Public Health 138(5):282–286. https://doi.org/10.1177/1757913918791198

Wilson SD, Pinno BD (2012) Environmentally-contingent behaviour of invasive plants as drivers or passengers. Oikos 122(1):129–135. https://doi.org/10.1111/j.1600-0706.2012.20673.x

World Meteorological Organization (WMO) (2021) State of the global climate 2020. Retrieved from https://public.wmo.int

Xu D, Wang Z, Song A, Li D, Ding X (2018) Assessing the relative role of climate change and human activities in desertification of North China from 1981 to 2010. Front Earth Sci 13(1):43–54. https://doi.org/10.1007/s11707-018-0706-z

Yang J, Wang C, Yang K (2022) How desertification in northern China will change under a rapidly warming climate in the near future (2021–2050). Theor Appl Climatol 151(1–2):935–948. https://doi.org/10.1007/s00704-022-04315-x

Yin Y, Liu Q, Wu Y, Li X, Pan X, He Q (2022) Predicting current potential distribution and the range dynamics of *Pomacea canaliculata* in China under global climate change. Biology 11(1):110. https://doi.org/10.3390/biology11010110

Zhang X, Wang X, Yan P (2007) Re-evaluating the impacts of human activity and environmental change on desertification in the Minqin Oasis, China. Environ Geol 55(4):705–715. https://doi.org/10.1007/s00254-007-1022-1

Zhang Z, Liu X, Capinha C, Usio N, Weterings R, Yokota M, Li Y, Landeria JM, Zhou Q (2019) Impacts of climate change on the global potential distribution of two notorious invasive crayfishes. Freshw Biol 65(3):353–365. https://doi.org/10.1111/fwb.13429

Zhang C, Shen J, Yang X, Lin Q, Xue B, Kong X, Cheng L, Zhang K, Zhao C (2023) Synergistic effects of climate warming and atmospheric nutrient deposition on the alpine lake ecosystem in the south-eastern Tibetan Plateau during the Anthropocene. Front Ecol Evol 11:1119840. https://doi.org/10.3389/fevo.2023.1119840

Zhang H, Gao Y, Zheng X, Hu Y, Zhou X, Fang Y, Li Y, Xie L, Ding H (2024) Neighborhood diversity promotes tree growth in a secondary forest: the interplay of intraspecific competition, interspecific competition, and spatial scale. Plants 13(14):1994. https://doi.org/10.3390/plants13141994

Zhou W, Gang C, Zhou F, Li J, Dong X, Zhao C (2014) Quantitative assessment of the individual contribution of climate and human factors to desertification in northwest China using net primary productivity as an indicator. Ecol Indic 48:560–569. https://doi.org/10.1016/j.ecolind.2014.08.043

Chapter 29
Changing Coastal Landscapes: The Role of Climate Change in Biodiversity Loss and Desertification

Ashok Kumar Behera and Rachna Chandra

Abstract India's vast coastline measuring 11,098 km during 2023–2024 supports an array of coastal and marine biodiversity. Climate change has led to significant transformations in the Indian coastal landscapes, increased desertification, and biodiversity losses. Desertification, biodiversity loss, and soil degradation are often interlinked. This chapter explores the intricate relationship between climate-induced changes like sea level rise, precipitation patterns, temperature variations, increased cyclonic activities, and their impact on coastal ecosystems. The results indicate that habitat fragmentation, saltwater intrusion, and sediment dynamics aggravate species extinction rates and ecosystem degradation. The present study employs a secondary data analysis approach, amalgamating quantitative and qualitative research methods to evaluate the impact of climate change on coastal biodiversity and desertification. By integrating secondary data sources and policy reviews, the findings provide a holistic knowledge of coastal vulnerabilities and potential mitigation measures. Multidisciplinary approach involving ecological restoration, coastal zone management, and climate resilient infrastructure is quintessential in conserving coastal ecosystems. This chapter synthesizes existing research on impact of climate change on Indian coastal biodiversity and examines how desertification is transforming these fragile landscapes.

Keywords Coastal biodiversity · Climate change · Desertification · Indian coastline · Sea level rises

A. K. Behera
Department of Soil Science, Odisha University of Agriculture & Technology (OUAT), Bhubaneswar, Odisha, India

R. Chandra (✉)
Terrestrial Ecology Division, Gujarat Institute of Desert Ecology (GUIDE), Bhuj-Kachchh, Gujarat, India

B. P. Barathan et al. (eds.), *Navigating Climate Change: Impacts on Biodiversity and Ecosystem Resilience*, Sustainable Development and Biodiversity 42,
https://doi.org/10.1007/978-981-95-0409-1_29

29.1 Introduction

India's vast coastline measuring 11,098 km during 2023–2024 spans nine coastal states and four union territories (SoI Report 2024). This vast expanse is home to diverse ecosystems like coral reefs, estuaries, mangrove forests, sandy beaches, lagoons, and coastal wetlands (Rahman 2013). The coastal environment also plays a crucial role in the socioeconomic framework of India, supporting millions of people engaged in fisheries, agriculture, tourism, and industrial activities (Kumar et al. 2017; Upadhyay and Mishra 2020). However, in past few decades, it has emerged as a dominant factor influencing coastal dynamics, causing biodiversity loss, and accelerating desertification along India's shores (Subramanian et al. 2023; Roy et al. 2024).

One of the foremost persistent impacts of climate change on India's shore regions is the increasing frequency and intensity of extreme weather events, such as cyclones, storm surges, and heavy rainfall (Roy et al. 2023). Bay of Bengal and the Arabian Sea have witnessed a significant rise in the occurrence of high-intensity tropical cyclones, which cause extensive damage to coastal habitats and lead to mass human populations' relocation (Mohapatra et al. 2021; Rajeev and Mishra 2023). These climatic disturbances exacerbate coastal erosion, destroy critical habitats like mangroves and coral reefs, and alter sediment transport mechanisms, resulting in long-term environmental degradation. Mangrove ecosystems, acting as natural barriers against storm surges and coastal erosion, are being lost and, leading to habitat fragmentation and biodiversity loss (Subramanian et al. 2023; Panja et al. 2025).

Additionally, rising sea levels pose another major threat to India's coastal biodiversity. Global warming has accelerated the melting of glaciers, leading to sea level rise, which has severe consequences for low-lying coastal areas (Das and Swain 2024). Coastal inundation leads to habitat loss and contributes to salinization of freshwater resources, negatively impacting agriculture (Mahajan et al. 2021; Mandal 2019).

Coral reefs, particularly those found in the Lakshadweep, Andaman and Nicobar Islands, and Gulf of Mannar, are among the most threatened marine ecosystems in India (Saroj et al. 2016). Rising sea temperatures have triggered widespread coral bleaching, wherein corals release their symbiotic algae due to temperature disturbances, ultimately leading to coral mortality (Jayakumar and Uttam 2024). Ocean acidification, as a consequence of high carbon dioxide, deteriorates coral structures, adversely impacting their capability to recover from environmental stressors. The decline of coral reef ecosystems has severe implications for marine biodiversity, affecting numerous fish and invertebrate species crucial for local fisheries and livelihoods (Hill and Hoogenboom 2022; Obura et al. 2022).

Coastal wetlands and estuaries, feeding and breeding grounds for marine organisms and migratory birds, are also facing degradation due to anthropogenic activities and climate change (Singh et al. 2021; Burger 2023). Urban expansion, industrialization, and pollution have resulted in habitat loss, eutrophication, and decline in water quality (Sarker et al. 2021). Additionally, major riverine estuaries such as the Ganges, Mahanadi, and Godavari are experiencing altered hydrological regimes caused by upstream damming, reduced freshwater inflow, and increased siltation, all of which affect the delicate balance of coastal biodiversity (Mishra et al. 2022). Pollution from industrial discharge and untreated sewage in estuaries like the Hooghly, Godavari, and Mahanadi further exacerbates threats to aquatic biodiversity.

Following the degradation of critical marine ecosystems, desertification is emerging as a significant concern along India's coasts (Jain et al. 2024). Coastal desertification is driven by multiple factors, including soil degradation, loss of vegetation cover, and increased salinity intrusion. Moreover, unchecked coastal development, sand mining, and deforestation further accelerate desertification processes, impacting productive land and agricultural output (Nisar 2021; Naorem et al. 2023). The transformation of fertile coastal lands into barren landscapes poses severe implications for local stakeholders reliant on these resources (Mir 2025).

Scientific research and technological advancements have also played a crucial role in assessing the impacts of climate change on coastal ecosystems (Obura et al. 2019; Danovaro et al. 2020). Remote sensing and Geographic Information System (GIS) tools have enabled the mapping of coastal changes, identifying areas most vulnerable to erosion, habitat loss, and desertification (Rahman 2023). Furthermore, climate modelling and predictive analyses have provided valuable insights into future scenarios, assisting policymakers and conservationists in creating adaptive strategies to mitigate climate change impacts (Wang et al. 2024).

India's coastal landscapes are undergoing rapid and alarming transformations due to rising sea levels, habitat destruction, coastal erosion, and increased cyclonic activity posing severe threats to terrestrial and aquatic ecosystems (Jain et al. 2024; Roy et al. 2024). The degradation of crucial ecosystems such as coral reefs, mangroves, estuaries, and wetlands threatens livelihood as well as biodiversity. However, climate change-induced alterations are accelerating coastal degradation, biodiversity loss, and desertification. Rising sea levels, increased cyclonic activity, coastal erosion, and habitat destruction threaten terrestrial and marine ecosystems (Nichols et al. 2019; Roy et al. 2023). This review synthesizes existing literature on the impact of climate change on Indian coastal biodiversity and examines how desertification is transforming these fragile landscapes. Strengthening conservation efforts, promoting ecosystem-based adaptation strategies, and enhancing climate resilience are essential for preserving India's rich coastal biodiversity. Through collaborative efforts and proactive conservation measures, mitigating the impacts of climate change and ensuring the long-term sustainability of India's coastal ecosystems are conceivable.

29.2 Study Area

29.2.1 Indian Coastline

India's coastline, along the Bay of Bengal and the Arabian Sea, encompasses a diverse range of ecosystems, including mangrove forests, estuaries, coral reefs, sandy beaches, lagoons, and delta. The coastal states and union territories, including West Bengal, Andhra Pradesh, Odisha, Maharashtra, Gujarat, Karnataka, Tamil Nadu, Goa, Kerala, Lakshadweep, and Andaman & Nicobar Islands, host unique ecological habitats that support extensive biodiversity (Fig. 29.1). Sundarbans, world's largest mangrove ecosystem and a UNESCO World Heritage Site in West Bengal, is a habitat to several species, including estuary crocodiles, and Royal Bengal Tigers. Situated between Sri Lanka and Tamil Nadu, the Gulf of Mannar is another important area that is well known for its marine biodiversity and abundant coral reef systems. Odisha's Chilika Lake, Asia's largest coastal lagoon, supports a dynamic ecosystem and is a crucial breeding ground for fish and a major stopover for migratory birds. The estuarine regions of the Mahanadi, Godavari, and Krishna Rivers play a crucial role in nutrient cycling and sustaining fisheries. Gahirmatha Beach, Rushikulya River mouth, and Devi River mouth in Odisha being the world's

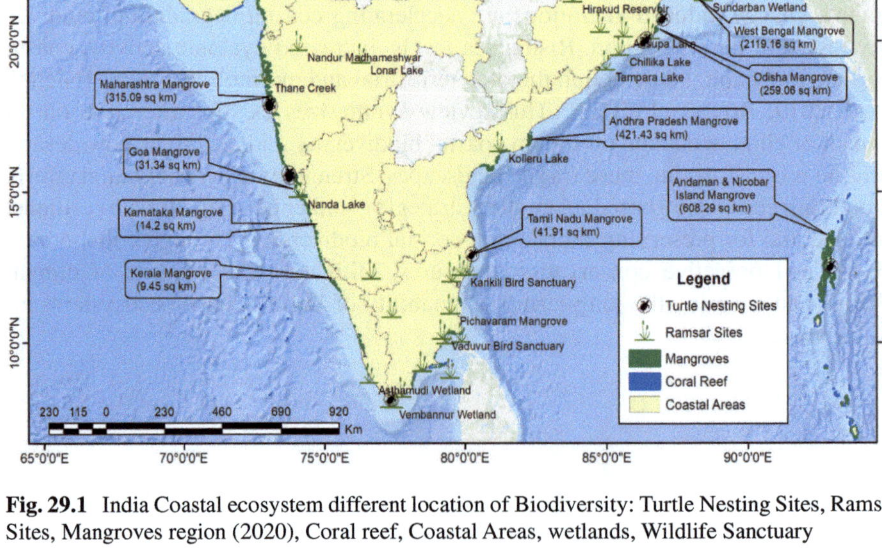

Fig. 29.1 India Coastal ecosystem different location of Biodiversity: Turtle Nesting Sites, Ramsar Sites, Mangroves region (2020), Coral reef, Coastal Areas, wetlands, Wildlife Sanctuary

largest turtle nesting site serve as crucial breeding sites for endangered Olive Ridley Turtle. The western coastline, stretching from Gujarat to Kerala, features rocky shores, tidal mudflats, and extensive estuarine systems that support mangroves and seagrass beds. The coral atolls of Lakshadweep and the fringing reefs of the Andaman and Nicobar Islands harbor some of India's most pristine marine ecosystems, although they are vulnerable to climate change. The Indian coastal area is not only ecologically important, but also supports major economic activities, including fisheries, tourism, shipping, and agriculture. However, climate change, industrialization, and rapid urbanization are threat to coastal ecosystems. Sustainable coastal management and conservation approaches are important to mitigate biodiversity loss and safeguard their resilience.

29.3 Methodology

This study employed a secondary data analysis approach (Andrews et al. 2012; Whiteside et al. 2012; Johnston 2014), focusing on analyzing climate data and synthesizing existing work on impacts of climate change on ecosystems and community adaptation strategies. The methodology integrates secondary data analysis and a synthesis of peer-reviewed research, reports, and satellite-derived datasets. By employing these approaches, this study provides a comprehensive assessment of changing coastal landscapes in India and their environmental implications.

29.3.1 Secondary Data Collection and Analysis

The methodology involved gathering data from multiple secondary sources, including the World Bank Climate Change knowledge Portal, National Aeronautics and Space Administration (NASA), India Meteorological Department e-Atlas (IMD), Indian-Water Resource Information System (WRIS), and the Intergovernmental Panel on Climate Change (IPCC). These platforms provided extensive climate datasets on parameters such as extreme weather events, precipitation, temperature, and sea-level rise, which were essential for analyzing trends and regional climate variability. Quantitative data were carefully examined to identify significant trends, fluctuations, and patterns across various geographical regions. The data were cross-checked between multiple platforms to ensure consistency and reliability. Comparison was made to assess the vulnerability of various regions to climate change, particularly focusing on biodiversity loss and ecosystem degradation. This approach allowed the documentation of areas most at risk from climate change impacts and helped assess the level of adaptation required.

29.3.2 Literature Review and Writing Process

In addition to the quantitative data analysis, a comprehensive desk-based literature review was undertaken. This review focused on synthesizing information from peer-reviewed journal articles, government reports, and technical documents published between 2001 and 2024. The goal was to extract both qualitative and quantitative insights on the impacts of climate change on biodiversity and to explore adaptation strategies used by communities. Thematic analysis was used to categorize adaptation responses, ranging from livelihood diversification to the construction of climate-resilient infrastructure. The methodology ensured comprehensive information of the climate change impacts, and adaptation strategies without primary data collection.

29.3.3 Climate Change Projections and Scenario Analysis

To evaluate future risks, climate projection models such as CMIP6 (Coupled Model Intercomparison Project Phase 6), ACCESS-CM2 (Australian Community Climate and Earth System Simulator-CM2), Multi-Model Ensemble models, World Bank Climate Change knowledge Portal, and IPCC were reviewed. These models provided insights into expected temperature increases, precipitation shifts, and sea level rise scenarios across Indian coastal states. This helped in contextualizing how climate-induced changes might influence future land-use patterns, biodiversity conservation efforts, and human settlements along the coast.

29.3.4 Assessment of Biodiversity Decline and Desertification Trends

The biodiversity assessment was conducted through an extensive review of case studies focusing on mangrove loss, coral reef degradation, and species migration due to habitat shifts. Studies on mangrove resilience and restoration from the Sundarbans and Pichavaram mangrove forests were compared to identify key conservation challenges. Research on coral bleaching in Lakshadweep and the Gulf of Mannar was examined to understand long-term ecosystem changes due to rising sea surface temperatures.

29.4 Result and Discussion

29.4.1 Climate Change Factors'Impact on Indian Coastal Region

Climate change including temperature, rainfall, cyclones, and sea level changes significantly impact coastal biodiversity in India (Senapati and Gupta 2014; Roy et al. 2023). Rising temperatures and rainfall contribute to coral bleaching and altered breeding cycles of marine species affect estuarine ecosystems, leading to fluctuations in water quality and salinity, which influence fish populations and mangrove health (Mama et al. 2021; Woesik et al. 2022). Cyclones, which have increased in intensity and frequency due to climate change, cause severe coastal erosion, habitat destruction, and salinity ingression, threatening marine and coastal biodiversity (Swain 2022). Sea level rise leads to submergence of coastal habitats, loss of nesting sites for turtles and seabirds, and salinization of freshwater ecosystems, affecting both marine and human communities (Von Holle et al. 2019). The combined impact of these climate factors poses a serious threat to India's rich coastal biodiversity, emphasizing the urgent need for conservation strategies, sustainable coastal management, and climate adaptation measures.

29.4.1.1 Temperature and Precipitation in India

Figure 29.2 presents the monthly climatology of precipitation and surface air temperature across India for the period 1991–2023, highlighting the seasonal variations influenced by the Indian monsoon system. The average maximum, mean, and minimum surface air temperatures exhibit a distinct annual cycle, with a steady increase from January to June, peaking during the summer months. The maximum surface temperature rises from around 24 °C in January to approximately 36 °C in May–June, while the mean surface temperature follows a similar trend, maximum at around 28 °C during the same period. The minimum surface temperature, which starts at around 10 °C in January, also reaches its highest point in June (~24 °C) before gradually decreasing towards the end of the year. These temperature variations indicate a strong seasonal influence.

The precipitation pattern, represented by blue bars, reflects a typical monsoonal trend, with the dry season extending from November to April and a rapid increase in rainfall from May onwards. The monsoon peaks in July with approximately 280 mm of rainfall, and high precipitation levels continue through August and September. The winter months, particularly December and January, experience minimal rainfall, making them the driest period. The correlation between high temperatures and the onset of monsoonal precipitation is evident, as the extreme summer heat during May and June leads to increased evaporation and atmospheric instability. This cyclic climatic pattern is crucial for India's agriculture, water

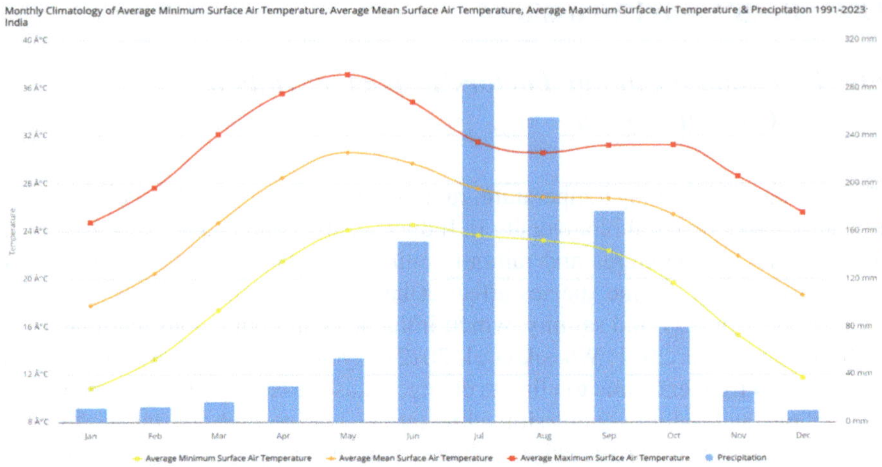

Fig. 29.2 Monthly climatology of surface air temperature and precipitation in India for the period 1991–2023

resources, and overall environmental balance, emphasizing the role of monsoonal variations in shaping the country's climate.

29.4.1.2 Temperature Change in Coastal Region

Indian coastal region has witnessed increased average temperature in past decades. Reports from IPCC and the IMD indicate a rise of 0.2 °C per decade in coastal temperatures (IPCC 2021; Rajeevan et al. 2023). The temperature trend in Indian coastal states, as depicted in the graph, shows a noticeable increase in surface temperature during 1995 to 2025 (Fig. 29.3). The data, derived from NASA POWER, indicates a consistent rise in temperature across all states, with fluctuations due to seasonal and interannual variations. The dataset presents year-wise temperature variations across coastal states of India from 2001 to 2022, showing fluctuating trends with some regions experiencing gradual increase while others remaining stable. Tamil Nadu (Chennai) and Gujarat (Bhuj) consistently recorded higher temperatures, often exceeding 29 °C, whereas Odisha and West Bengal had the lowest temperatures, generally below 27 °C. Peaks were observed in several regions between 2016 and 2019, followed by slight declines in 2020–2022, possibly due to climate variability or localized environmental changes. Kerala, Goa, and Maharashtra displayed minor fluctuations, while Andhra Pradesh showed relatively stable temperatures. The highest temperatures were recorded in Kerala, Tamil Nadu, and Maharashtra, while regions such as Odisha and Gujarat (Bhuj) showed significant variability. The data suggest a warming trend in several coastal states, which could be attributed to urbanization, industrialization, and climate change impacts.

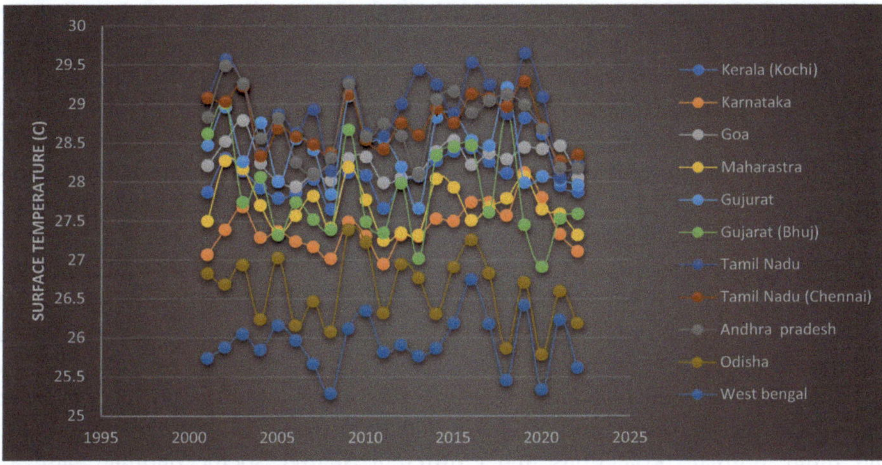

Fig. 29.3 Surface temp. of Indian Coastal region

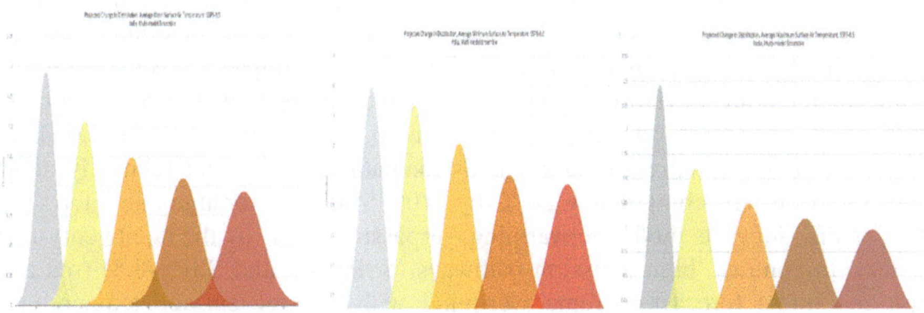

Fig. 29.4 (**a**) Project change distribution Avg. Mean surface Air temperature in India, (**b**) Project change distribution Avg. Maximum surface Air temperature in India, (**c**) Project change distribution Avg. Minimum surface Air temperature in India: CCKP Projected Data Set

Figure 29.4a illustrates the projected changes in the distribution of average mean surface air temperature over India under the SSP5–8.5 scenario, based on a multi-model ensemble. The y-axis shows the probability distribution and x-axis represents temperature in °C. Different colors indicate different time periods. The historical reference period (1995–2014) had an average temperature centered around 24.5 °C. However, by 2020–2039, the distribution shifts to a higher mean, reaching approximately 25.5 °C. This warming continues into 2040–2059, with the mean temperature surpassing 26 °C. By 2060–2079, it reaches around 27 °C, and by 2080–2099, the temperature distribution peaks near 28.5 °C. The projected change in India's average maximum surface air temperature under the SSP5–8.5 scenario shows (Fig. 29.4b) a consistent rise throughout the twenty-first century. The historical reference period (1995–2014) had peak temperatures around 29.5 °C to 30 °C, but by 2020–2039, this shifts to 30.5 °C to 31 °C, increasing by about 1 °C. By 2040–2059, the peak moves to 31.5 °C to 32 °C, and by 2060–2079, it reaches

32.5 °C, marking a 3 °C rise. By 2080–2099, temperatures are anticipated to be 33.5 °C to 34.5 °C, an alarming increase of 4–5 °C. This progressive temperature increase will have severe implications for India's coastal biodiversity and desertification processes. Figure 29.4c shows that the historical reference period (1995–2014) had minimum temperatures centered around 19 °C, but by 2020–2039, this shifts to approximately 20 °C, indicating a 1 °C rise. By 2040–2059, the minimum temperature further increases to 21 °C, followed by 22 °C in 2060–2079. By the end of the century (2080–2099), the minimum temperature is projected to reach 24 °C, reflecting an overall 5 °C increase from historical levels. This rise in night time temperatures can lead to increased heat stress, reduced agricultural productivity, altered monsoon patterns, and higher energy demands for cooling, accentuating the need for adaptation approaches.

Figure 29.5 illustrates the projected departure from natural variability of average mean surface air temperature in India during 1950–2100, based on historical data and future climate projections under different Shared Socioeconomic Pathway (SSP) scenarios. The y-axis represents average mean surface air temperature (°C) and x-axis denotes time. Historical data, shown in black, reveal a gradual increase in temperature until 2015, after which various SSP scenarios project different warming trends. The shaded regions mark the period when projected temperatures exceed natural variability, indicating significant climate change impacts. According to the figure, temperature deviations from historical norms become noticeable around 2015–2018, depending on the emission scenario. SSP1–1.9 and SSP1–2.6 (light blue and blue lines), representing low-emission pathways, project a relatively stable temperature rise, reaching below 28 °C by 2100. SSP2–4.5 (orange), an intermediate scenario, suggests moderate warming, surpassing 29 °C by the late twenty-first century. In contrast, high-emission pathways (SSP3–7.0 and SSP5–8.5) (red and dark red lines) forecast extreme warming, with average surface temperatures

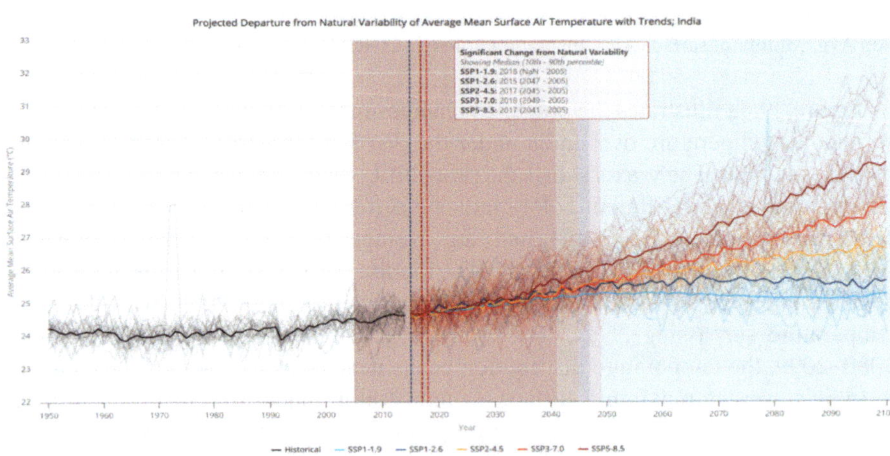

Fig. 29.5 Projected Departure from Natural Variability of Average Mean Surface Air Temperature in India from 1950 to 2100. https://climateknowledgeportal.worldbank.org

exceeding 30 °C by 2080 and potentially reaching 32 °C or more by 2100. The wide range of temperature projections highlights increasing uncertainty and climate variability, particularly under high-emission scenarios (IPCC 2021). The data underscore the critical role of global climate policies, as lower-emission pathways could limit warming and mitigate severe consequences such as heatwaves, water stress, and ecosystem disruptions.

29.4.1.3 Temperature Projections and Their Uncertainties

The temperature projections discussed in the present study are based on outputs from a Multi-Model Ensemble approach under the SSP5–8.5 high-emission scenario. Multi-Model Ensembles average the outputs from multiple global climate models (GCMs) to produce a more robust estimate of future temperature trends across India. While this approach reduces the biases associated with any single model, important uncertainties still remain. Different models exhibit varying climate sensitivities, particularly in how they simulate feedbacks like cloud cover, aerosols, and land-atmosphere interactions (IPCC 2021). Additionally, the SSP5–8.5 scenario assumes continued high fossil fuel use and minimal mitigation, representing a worst-case pathway. Some researchers argue that actual future emissions could be lower, meaning SSP5–8.5 may overstate warming projections if strong global mitigation occurs. Although the ensemble mean provides a stable projection of overall warming, individual model spread leads to uncertainty regarding the precise magnitude and rate of temperature rise, especially at regional scales like coastal India. Therefore, while consistent warming trends are expected, caution must be exercised in interpreting the absolute temperature values.

29.4.1.4 Precipitation Variability in India Coastal Region

Coastal India experiences erratic rainfall patterns, with an increased frequency of heavy precipitation events. Coastal states are highly influenced by monsoon variability, cyclonic activity, and climatic changes. Figure 29.6 presents yearly precipitation trends (mm/day) during 2001–2022 in Indian coastal regions, including West Bengal, Tamil Nadu (Chennai), Goa, Karnataka, Kerala (Kochi), Maharashtra, Gujarat (Bhuj), Odisha, and Andhra Pradesh. Data from NASA Power Data Viewer show that Maharashtra and Karnataka exhibit the highest precipitation, peaking around 2013 and 2020, with rainfall exceeding 4000 mm/day. Kerala also shows significant fluctuations, with notable peaks in 2005, 2010, and 2018. Odisha and West Bengal display relatively moderate trends but have increasing rainfall patterns post-2015. Gujarat (Bhuj) and Tamil Nadu (Chennai) demonstrate lower precipitation levels, with occasional peaks. Maharashtra's high rainfall aligns with the occurrence of major monsoons, while Odisha and West Bengal likely experience fluctuations due to Bay of Bengal storm systems. Tamil Nadu's variation reflects northeast monsoon impacts. The overall increasing trend in coastal precipitation

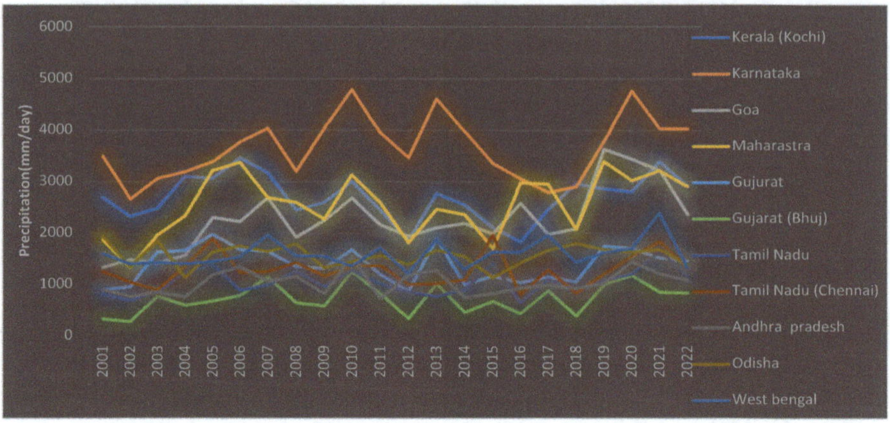

Fig. 29.6 Yearly precipitation Trends in Indian Coastal States (2001–2022). https://power.larc.nasa.gov/

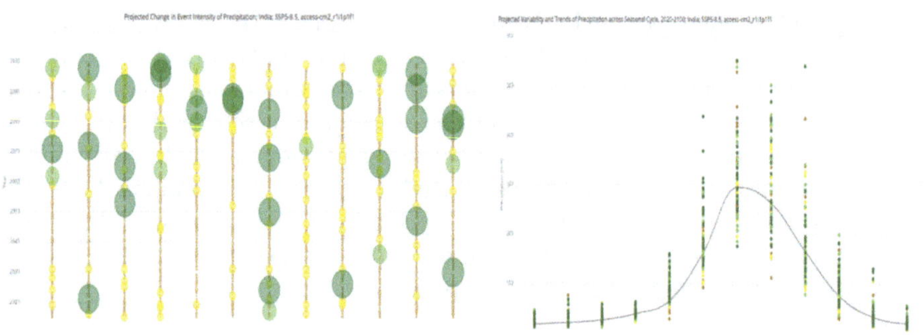

Fig. 29.7 (**a**) The projected changes in the intensity of precipitation events across India under the SSP5–8.5 scenario, based on the ACCESS-CM2 model, (**b**) the project variability and trends of precipitation across the seasonal cycle (2020–2100) under SSP5–8.5 for India, Based on the ACCess-CM2 model. https://climateknowledgeportal.worldbank.org

suggests intensified monsoonal activity and shifting climate patterns, warranting further research into long-term coastal weather changes. The peaks observed in 2013 and 2020 correspond to heavy monsoon periods and extreme weather events like floods and cyclones.

Figure 29.7a illustrates the probable variations in the intensity of precipitation in India under the SSP5–8.5 scenario, based on the ACCESS-CM2 model. The y-axis represents the years from 2020 to 2100, while the x-axis represents the months of the year. Different colors and bubble sizes indicate the standard deviation (SD) of precipitation intensity changes. Brown and yellow circles represent lower intensity changes (less than 1 SD and 2 SD, respectively), while light green indicates moderate increases (less than 2.5 SD), and dark green represents extreme increases (greater than 2.5 SD). The trend shows a clear upsurge in extreme precipitation, particularly in the monsoon months (June to September) as the twenty-first century progresses.

Figure 29.7b projected variability and trends of precipitation across the seasonal cycle (2020–2100) under SSP5–8.5 for India**,** based on the ACCESS-CM2 climate model, and suggested significant shifts in rainfall distribution, intensification of monsoonal patterns, and increased variability. Comparing the historical reference period (1995–2014) with the future timeframes (2020–2039, 2040–2059, 2060–2079, 2080–2099), the data reveal a notable increase in monsoon precipitation (June–September), with heavier and more concentrated rainfall events due to enhanced atmospheric moisture content and warming-induced convective activity. This could heighten risks of flash floods, soil erosion, and crop damage in states like Odisha, West Bengal, and Kerala, which experience high monsoon dependency. However, greater interannual variability indicates the possibility of extreme wet and dry years, making long-term water resource planning more challenging. Additionally, the post-monsoon (October–December) and pre-monsoon (March–May) are likely to become drier, leading to longer drought periods, reduced groundwater recharge, and water scarcity issues, particularly affecting western India (Gujarat, Maharashtra).

Figure 29.8 illustrates the projected changes in precipitation levels in India, comparing historical trends with future projections under different SSP scenarios. The x-axis represents the years from 1950 to 2100, while the y-axis indicates precipitation levels in mm. The black line and shaded gray region represent historical precipitation data until approximately 2015, showing relatively stable trends with natural variability. Beyond 2015, projections for multiple SSP scenarios (SSP1–1.9, SSP1–2.6, SSP2–4.5, SSP3–7.0, and SSP5–8.5) are represented by colored lines and shaded regions, indicating possible future precipitation trends based on different levels of greenhouse gas emissions.

The red-shaded background highlights the period where projected changes depart significantly from natural variability, emphasizing the increasing uncertainty

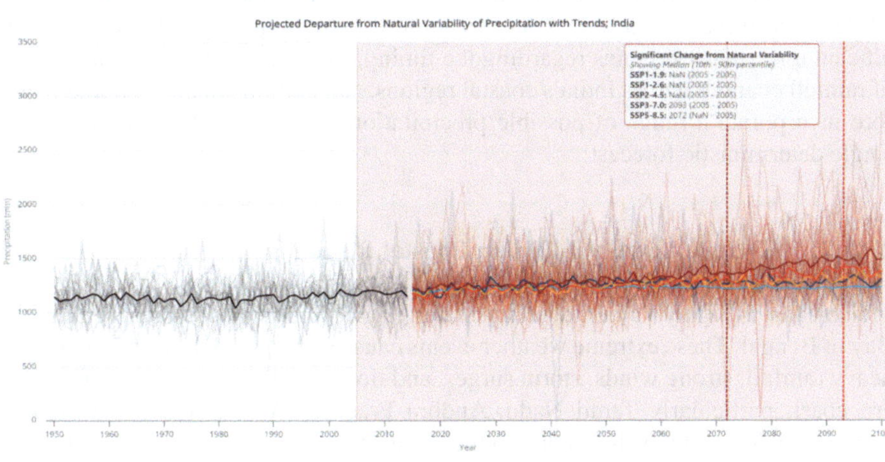

Fig. 29.8 Illustrates the projected changes in precipitation levels in India. https://climateknowledgeportal.worldbank.org

and variability in future precipitation patterns. The legend provides details on when significant changes from natural variability are expected to occur under different SSP scenarios. Notably, for SSP3–7.0, this significant change is projected to occur around 2093, while for SSP5–8.5, it is expected around 2072. Other SSP pathways do not indicate a significant shift from natural variability within the timeframe of the projection. The spread of the projections widens considerably over time, particularly for high-emission scenarios (SSP3–7.0 and SSP5–8.5), suggesting greater variability and uncertainty in future precipitation patterns. The darkest red lines and shading in SSP5–8.5 indicate the most pronounced changes, with increased precipitation variability and higher mean values. This suggests that under high-emission scenarios, India could experience more extreme fluctuations in rainfall, potentially leading to more frequent and intense floods or droughts.

29.4.2 Precipitation Projections and Their Uncertainties

Precipitation projections in this study are primarily derived from the ACCESS-CM2 model under the SSP5–8.5 scenario. Although ACCESS-CM2 has improved resolution over earlier models, precipitation remains one of the most challenging climate variables to simulate accurately. Models often struggle with capturing localized convective processes, monsoon variability, and extreme rainfall events, particularly in complex topographies like the Indian subcontinent (IPCC 2021). Precipitation responses to warming are also highly sensitive to regional factors such as sea surface temperature anomalies (e.g., ENSO, IOD) and land-atmosphere feedbacks. ACCESS-CM2, like other GCMs, may exhibit biases in monsoon onset timing, spatial distribution of rainfall, and extreme event frequency. Downscaling from global models to regional impacts introduces additional uncertainties. As a result, while broad trends suggest an intensification of extreme precipitation events, significant uncertainty remains regarding the timing, intensity, and geographic pattern of rainfall changes across India's coastal regions. Adaptation planning should therefore incorporate a range of possible precipitation outcomes rather than rely on a single deterministic forecast.

29.4.2.1 Cyclone Activity in Indian Coastal Region

Indian coast is highly vulnerable to cyclones originating in the Arabian Sea and the Bay of Bengal. These extreme weather events cause widespread devastation through heavy rainfall, strong winds, storm surges, and flooding (Ray et al. 2019). The eastern coast, particularly Tamil Nadu, Andhra Pradesh, West Bengal, and Odisha, experiences frequent cyclones due to the warmer waters of Bay of Bengal. In contrast, the western coast, including Gujarat, Maharashtra, and Kerala, is less frequently affected but still witnesses severe cyclones (Rao et al. 2020a). Figure 29.9a represents historical cyclone tracks and depressions over the Indian subcontinent

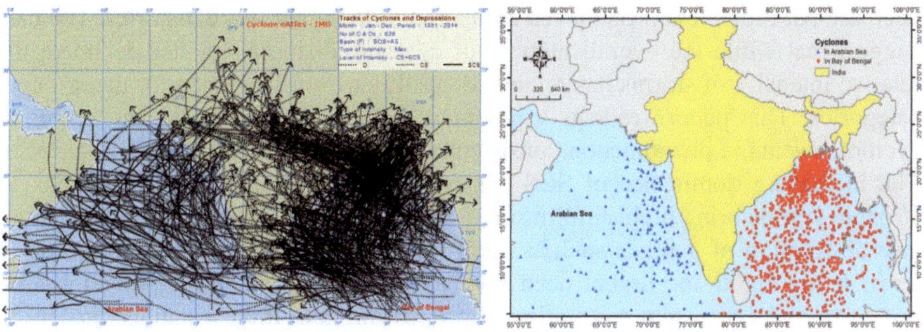

Fig. 29.9 (**a**) Cyclone tracks over the Indian sea region from 1891 to 2014 (Source: India Meteorological Department e-Atlas). (**b**) Cyclone paths over the Indian sea region from 1900 to 2020

from 1891 to 2014, as recorded by the IMD and Fig. 29.9b. The majority of cyclone paths originate over the Bay of Bengal, especially impacting the eastern coastal states while comparatively fewer cyclones form over the Arabian Sea. These cyclone tracks highlight the region's most vulnerable to repeated cyclone strikes, particularly low-lying delta areas, estuaries, and mangrove forests. The clustering of cyclone paths along these zones indicates repeated disturbance to coastal ecosystems, soil erosion, and biodiversity loss over time. The western coast, although historically less impacted, is witnessing an emerging trend of increased cyclonic activity, which could further destabilize fragile ecosystems in Gujarat, Maharashtra, and Kerala. Understanding the spatial concentration of historical cyclone tracks is critical for designing targeted climate adaptation plans, restoring coastal vegetation like mangroves, and strengthening early warning systems in cyclone-prone areas. Almost 1295 cyclonic disturbances were reported during 1900–2020; of which, 273 were severe cyclonic storms (SCS = 21.08%), 292 were cyclonic storms (CS = 22.54%), and 731 were depressions (D = 56.44%). Maximum cyclonic disturbances, i.e., 20, happened in the year 1925 and 1975, whereas 19 cyclonic disturbances happened in the year 1937, 1939, 1941, 1944, and 1982 and 18 number of cyclonic disturbances happened in the year 1947, 1948, 1961, 1966, and 1972, respectively.

Particularly, a prominent decrease in D and CS cyclones can be noticed, for example, from 1980 to 1989 to from 2010 to 2020 in Fig. 29.10. It also shows the decadal distribution of cyclonic disturbances over the North Indian Ocean from 1900 to 2020, categorized into Depressions (D), Cyclonic Storms (CS), and Severe Cyclonic Storms (SCS). The graph indicates that the total number of Depressions (blue bars) was significantly higher throughout the early to mid-twentieth century, peaking between 1940 and 1949, but has shown a marked declining trend over subsequent decades, as represented by the downward blue dotted line. Cyclonic Storms (orange bars) also exhibit a gradual decrease over time, with a corresponding downward orange dotted trend line. In contrast, Severe Cyclonic Storms (grey bars) have maintained a relatively stable trend, with the grey dotted line indicating a slight decrease but much less pronounced compared to the other two categories. The data suggest that while the overall number of cyclonic disturbances has declined, the relative intensity of storms has remained steady or even increased for the stronger categories. This pattern reflects a potential shift toward fewer but more intense cyclonic events, a phenomenon consistent with global climate change projections. The increasing dominance of SCS has critical implications for coastal India, as stronger storms bring higher storm surges, greater coastal erosion, and more extensive destruction of habitats such as coral reefs, mangroves, and estuarine ecosystems. It highlights the urgent need for early warning systems, improved coastal defenses, and ecosystem restoration efforts to mitigate the escalating impacts of intense tropical cyclones on coastal biodiversity and human settlements.

Decadal percentage frequency distribution of the cyclonic disturbances exhibits that during initial 40 years (1900–1939), the share of CS remained at an average of nearly 30%, which has reduced to 12% during 1940–2020 (Fig. 29.9b). The most perceptible decline occurred over the most recent sixty years (1960–2020), which registered on normal just 12% of CS in the decades 1990–1999 and 2010–2020, but the years between 2000 and 2009 had seen an upsurge in CS to 24%. While long-term average of SCS was at 8%, decadal % frequency of SCS changed over the years. Initially during 1900–1919, SCS frequency averaged nearly 21%; in the following forty years (1920–1959), it reduced to an average of 13%, while the most recent sixty years (1960–2020) experience an increase of normal SCS (to 26%). The latest ten years, 2010–2020, had the most noteworthy commitment (29%) from SCS along the coast of Bay of Bengal and Arabian Sea for the whole review time frame (1900–2020). The ten year period of 2000–2010 was different as it registered the

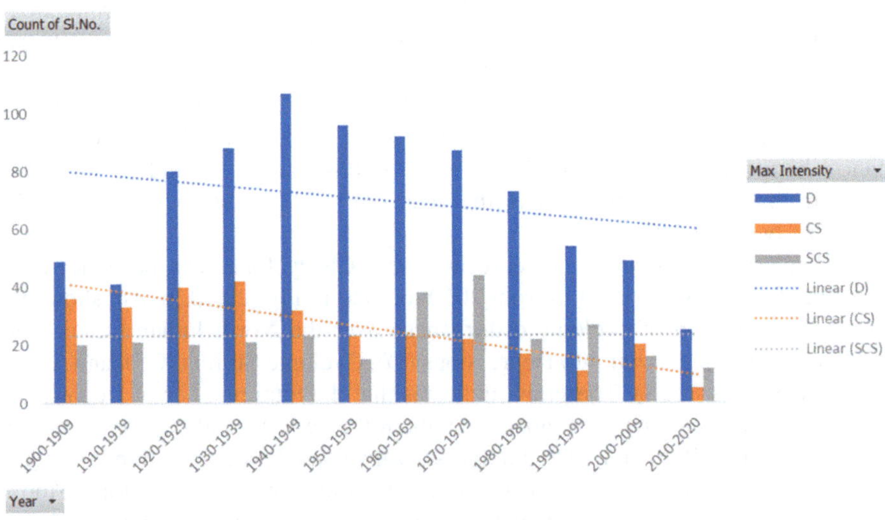

Fig. 29.10 Decadal frequencies of cyclonic disturbances [SCS = Severe Cyclonic Storms, CS = Cyclonic Storms, and D = Depression] along the BOB & AS coast for 1900–2020; the dotted shaded lines portray direct pattern of cyclonic disturbances throughout a similar time

highest percentage of depression which is D = 72% (Fig. 29.11). Historically, Depressions (D) dominated, contributing between 43% and 72% of the total disturbances. Over time, however, the share of Cyclonic Storms (CS) has steadily declined, from around 34–35% during the early decades (1900–1920) to just 12% in the most recent decades (2000–2020). Meanwhile, the proportion of SCS has generally increased, rising from 19 to 22% in the early twentieth century to 29% in the last two decades. This shift indicates a changing trend where, although the total number of disturbances might be declining, the relative intensity of cyclonic events is growing. A larger proportion of SCS points toward stronger, more destructive systems affecting the Indian coastline. These high-intensity storms can lead to devastating consequences, including the destruction of mangrove forests, submergence of low-lying areas, saltwater intrusion into freshwater ecosystems, and permanent loss of critical biodiversity habitats.

Figure 29.12 presents the simulated relative frequency distribution (in percentage) of cyclone categories for the Indian Ocean, Indian Exclusive Economic Zone (EEZ), and India (landfalls) under historical (1951–2014) and projected SSF 4.5 (2035–2064) conditions. In the Indian Ocean, the percentage of tropical storms slightly increases from 44.89% (historical) to 46.27% (projection), while Category 1 and 2 cyclones show minimal variations. However, Category 3 storms decrease from 11.77% to 11.15%, and Category 4 storms drop from 11.33% to 10.78%, whereas Category 5 storms increase from 4.31% to 4.58%. In the Indian EEZ, the frequency of tropical storms remains relatively stable (54.66% to 54.44%), but a slight decline is observed in Category 1 storms (20.12% to 19.46%). Category 3 and 4 storms experience minor increase, while Category 5 cyclones rise from 1.64% to 1.78%. For India landfalls, the frequency of tropical storms increases from 60.83% to 61.89%, and Category 1 storms slightly decline. The relative frequency of intense

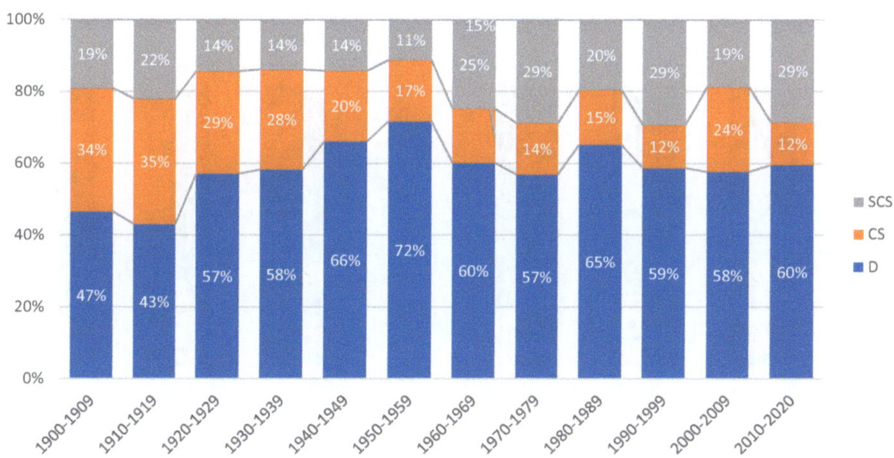

Fig. 29.11 Decadal percentage (%) contribution of cyclonic disturbances [D = Depression, CS = Cyclonic Storms and SCS = Severe Cyclonic Storm] along the BOB & AS coast (1900–2020) 120 years

cyclones (Category 3 and 4) remains largely unchanged, while Category 5 storms show a marginal increase from 3.35% to 3.55%. These results suggest a stable trend for tropical storms, with minor shifts in higher-intensity cyclone categories in the projected climate scenario.

Figure 29.13 illustrates the percentage change in the number of cyclones across different intensity categories between the historical period (1951–2014) and the projected SSP2–4.5 period (2035–2064) for the Indian Ocean, Indian Exclusive Economic Zone (IEEZ), and India (landfalls). In the Indian Ocean, the changes are relatively minor, with a slight increase observed in tropical storms and Category 3 cyclones, while Categories 1, 2, and 4 show minimal shifts and Category 5 remains nearly constant. The error bars across categories indicate a considerable range of uncertainty, especially for higher-intensity cyclones. For the Indian EEZ, the change in cyclone frequency is more variable. Tropical storms and Category 1 cyclones remain mostly stable, while Category 2 and 3 storms show a modest increase. Notably, Category 4 and 5 cyclones exhibit higher percentage changes, with Category 5 showing a particularly large potential increase, though accompanied by wide uncertainty bounds, highlighting variability in projections for the most intense storms. In the case of India landfalls, there is a more pronounced increase across all cyclone categories. Tropical storms, Category 1, and especially Categories 2 through 5 all show positive growth, with the percentage change becoming significant as storm intensity increases. Category 5 storms show the largest projected increase in frequency, accompanied by substantial uncertainty, suggesting that while the most intense cyclones may become more frequent, model confidence varies. These patterns underscore a growing concern for the Indian coastline, where increases in landfalling high-intensity cyclones could have serious implications for disaster preparedness and coastal resilience planning in a warming climate.

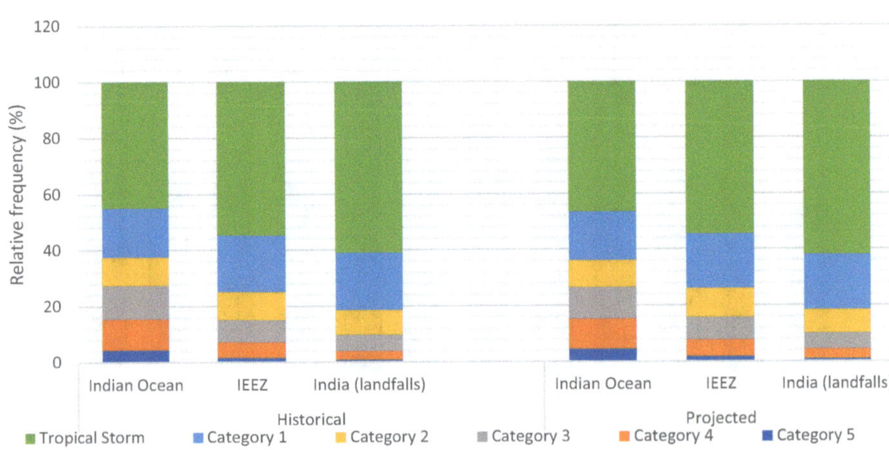

Fig. 29.12 Relative frequency distribution (in percentage) of cyclone categories for the Indian Ocean, Indian Exclusive Economic Zone (IEEZ) and India (landfalls) under historical (1951–2014) and projected SSF 4.5 (2035–2064)

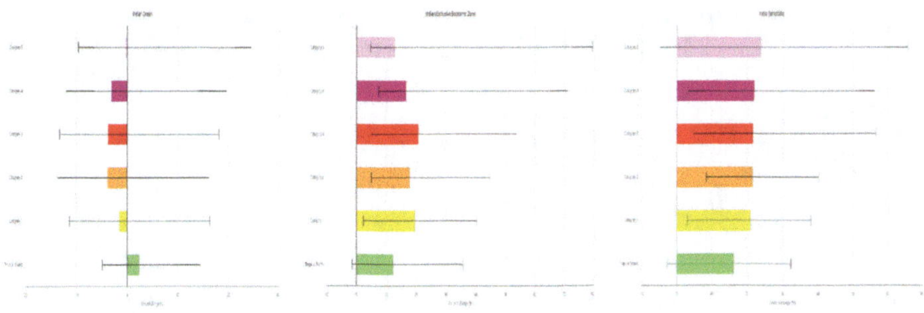

Fig. 29.13 Percentage change in the number of cyclones across different intensity categories between the historical period (1951–2014) and the projected SSP2-4.5 period (2035–2064)

29.4.2.2 Sea Level Changes Across the Coastal Regions in India

Figure 29.14 illustrates the total sea level change in India from 2001 to 2023, analyzed on a monthly basis, with different colors representing seasonal variations. The data exhibit an overall increasing trend, indicating a steady rise in sea levels over the years due to factors such as seasonal influences, ocean currents, and climatic events like monsoons and storms. From 2001 to 2010, sea level fluctuations remained within ±0.05 meters, showing moderate variation, but after 2010, a more pronounced rise was observed, peaking around 2020 due to global warming, glacial melt, and thermal expansion of seawater. The 2010–2015 period saw a clear upward trend, with values exceeding 0.05 m, particularly during monsoon months (June–September), likely influenced by increased precipitation and storm surges. From 2015 to 2020, the rise in sea levels accelerated sharply, with peaks exceeding 0.1 m, especially around 2019–2020, highlighting the intensifying impact of climate change. Between 2020 and 2023, sea levels remained high but showed greater variability, with both peaks and sudden drops, reflecting the complex interplay of climate change, oceanic patterns, and regional monsoon effects. Figure 29.15 presents a donut chart illustrating factors contributing to sea level rise. Sterodynamics is a largest contributor, accounting for 55% of the total rise, indicating the impact of ocean expansion due to warming. Glaciers and the Greenland Ice Sheet each contribute 15%, highlighting the significance of melting ice masses. The Antarctic Ice Sheet accounts for 10%, while terrestrial (land) water storage and vertical land motion contribute the least, at 5% each. These proportions emphasize the dominant role of thermal expansion and melting of ice in global sea level rise, posing challenges for coastal regions worldwide.

The IPCC projects a sea level rise of 3–8 mm per year in the Indian Ocean. Figure 29.16 shows that the sea level rise at Diamond Harbour (West Bengal) shows a clear upward trend, with an accelerating increase over time. The observed data indicate a gradual rise in the early twentieth century, which has steepened post-1950, aligning with global warming trends driven by thermal expansion and ice melt. Future projections under the SSP3–7.0 scenario estimate a total sea level rise of approximately 0.75 m by 2100, with increasing uncertainty over time. Similarly, the

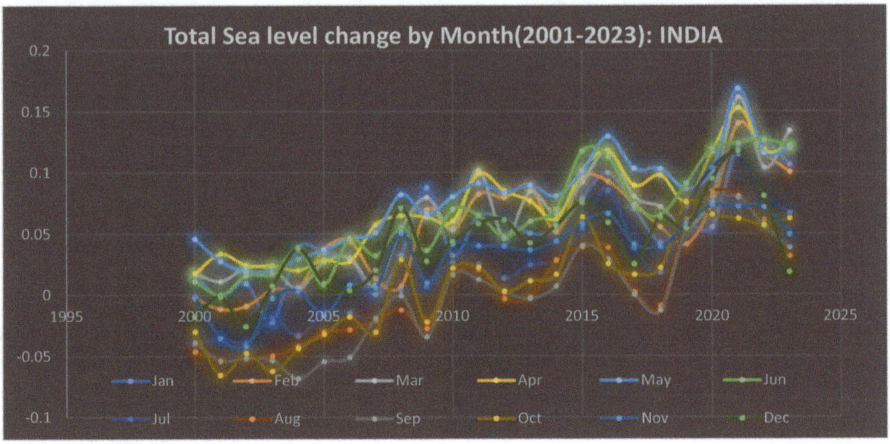

Fig. 29.14 Sea level change in India from 2001 to 2023

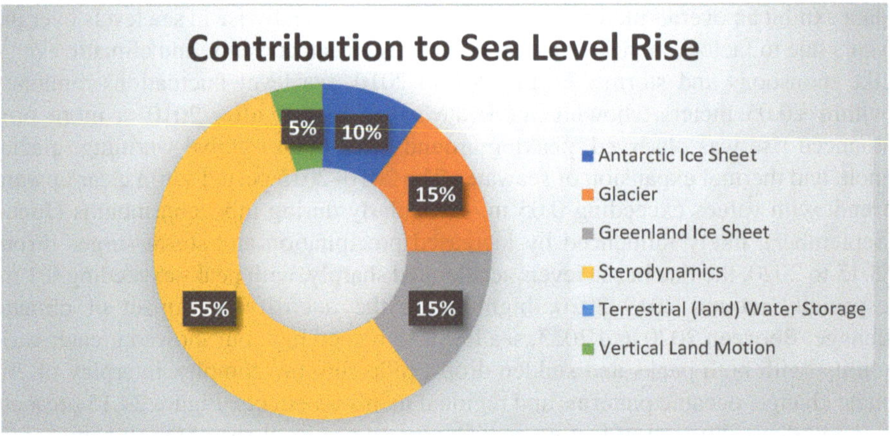

Fig. 29.15 Contribution of various factors to sea level rise

sea level rise at Paradip follows a steady upward trajectory, becoming more pronounced after the 1950s, with an expected rise of 0.59 m by 2100. At Visakhapatnam, the trend is continuous, with acceleration over time, leading to a projected rise of 0.54 m by 2100. Other key coastal locations also face significant risks. In Chennai (Tamil Nadu), the sea level is projected to rise by 0.57 m, while Cochin (Kerala) is expected to see a more substantial increase of 0.71 m, posing a severe threat to low-lying areas**.** Mormugao (Goa) and Mumbai (Maharashtra) are projected to experience sea level rises of 0.63 m and 0.58 m, respectively, which could exacerbate coastal erosion and flooding in these highly populated regions. Bhavnagar (Gujarat) is projected to see one of the highest increases, reaching 0.82 m by 2100, highlighting significant vulnerability, while Okha (Gujarat) is expected to experience a 0.60-m rise. These projections underscore the urgent need for coastal resilience

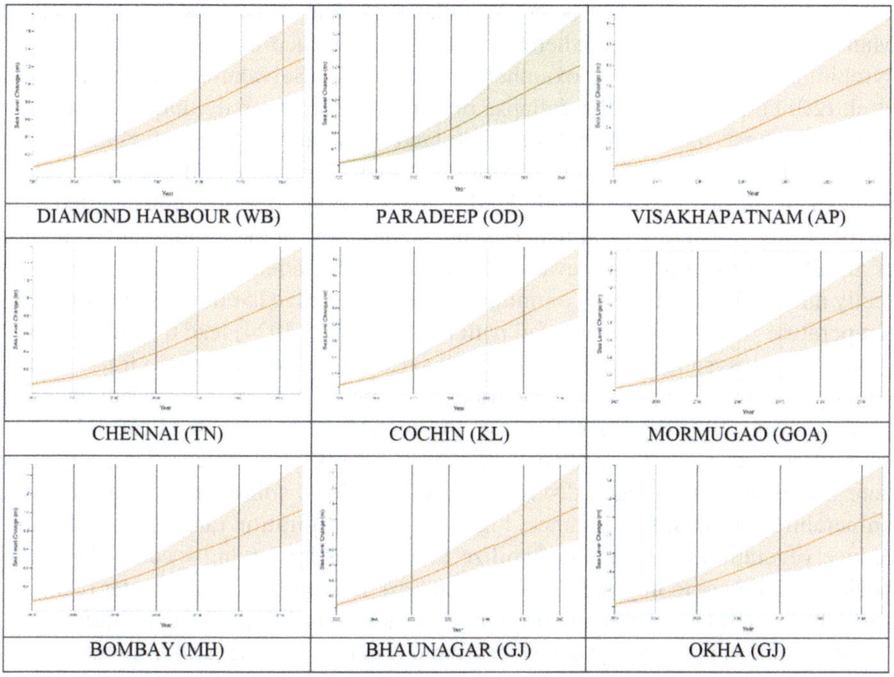

Fig. 29.16 IPCC (Intergovernmental Panel on Climate Change) projects a sea level rise at Indian Coastal region

measures, adaptation strategies, and climate mitigation efforts to protect infrastructure, ecosystems, and communities from long-term sea-level rise impacts across India's coastline.

29.4.2.3 Impact of Climate Changing Patterns on Indian Coastal Region

India's coastal regions are facing unprecedented changes due to climate-induced factors. Rising temperatures are altering the ecosystem dynamics, leading to increased heat stress on marine and terrestrial biodiversity. For example, higher temperatures accelerate evapotranspiration, reducing soil moisture and promoting arid conditions (Singh et al. 2023). This contributes to desertification, typically as a result of drought, deforestation, or inappropriate agriculture especially in semiarid and coastal regions. Warming has also led to the mortality of mangrove forests (Friess et al. 2022), crucial ecosystems that protect shorelines and serve as nurseries for marine life.

Direct impacts of climate change include higher temperatures and sea surface warming. Hoegh-Guldberg et al. (2018) highlighted that global coral reef ecosystems are undergoing rapid changes due to increased sea surface temperatures, with more than 50% of coral cover lost in certain regions due to prolonged bleaching events. Coral bleaching has been particularly evident in regions like the Gulf of Mannar and Lakshadweep, where increasing sea surface temperatures disturb the symbiotic relationship of zooxanthellae and corals (Selvakumar 2014). Pandolfi et al. (2011) documented the collapse of coral reef biodiversity under temperature rise. In contrast, indirect impacts include land use changes driven by climate pressures, such as altered agricultural patterns, urban migration, and coastal development. Das et al. (2020) highlighted that warming oceans have shifted fish distributions, impacting local fisheries and economies. In the Sundarbans, Samanta et al. (2021) observed that elevated temperatures degraded mangrove health, indirectly influencing land use as communities seek alternative livelihoods. Additionally, temperature-driven shifts in rainfall patterns (Roy et al. 2024) lead to water scarcity, promoting ecosystem degradation through agricultural expansion into fragile lands. Higher land temperatures also accelerate desertification. In India's coastal belts and semiarid zones like Rajasthan, Gujarat, and central India, increased heat and irregular rainfall are intensifying land degradation (Maity et al. 2024). Desertification here is driven by prolonged droughts, vegetation loss, soil erosion, and increased vulnerability to fires. For instance, higher evapotranspiration rates dry out the soil, reduce vegetation cover, and destabilize food chains by forcing species like blackbuck and chinkara to migrate (Rao et al. 2020b). Diminished groundwater recharge, caused by irregular rainfall and droughts, further fuels desert-like conditions.

Precipitation extremes also have profound impacts. Intense rainfall events lead to flooding, coastal erosion, and salinity ingression, damaging wetlands and estuaries (Lorrain-Soligon et al. 2023). Forests and terrestrial biodiversity are also vulnerable to shifting rainfall patterns. Deb et al. (2018) found that increased monsoon precipitation may support the growth of tropical rainforests in wetter regions; prolonged dry spells in central and southern India threaten dry deciduous forests with degradation. Unpredictable rainfall patterns could fragment habitats, endangering species reliant on stable environments. Fern et al. (2023) found that wetlands and freshwater biodiversity are at risk of extreme flood-drought cycles, disrupting aquatic ecosystems. The loss of wetlands and reduction in groundwater recharge further deplete freshwater sources, exacerbating water scarcity and promoting desert-like conditions. Changes in hydrological regimes could hinder breeding success in amphibians and migratory birds that depend on seasonal water availability. Moreover, excessive rainfall can worsen water pollution by washing sediments, nutrients, and pollutants into rivers and lakes (Coffey et al. 2019). Heavy rainfall leads to soil erosion and the washing away of nutrient-rich topsoil, while prolonged dry spells cause land desiccation (Ahmed et al. 2014). In states like Odisha and Gujarat, inconsistent rainfall is accelerating land degradation.

Cyclone activity has intensified, bringing frequent and powerful storms that directly devastate coastal ecosystems. High winds uproot vegetation, strip fertile soil, and promote land degradation (Swain 2022). The destruction of mangroves and wetlands (Temmerman et al. 2023) removes natural storm buffers, increasing vulnerability to storm surges, erosion, and saltwater intrusion. It is harder for ecosystems to recover, hastening the process of desertification (Wei et al. 2022). The 2020 Cyclone Amphan and 2021 Cyclone Yaas highlighted how weakened ecosystems increase human vulnerability (Barik and Ratha 2024). Without robust coastal vegetation, the process of desertification is hastened, as protective ecosystems fail to regenerate.

Rising sea levels, driven by climate change, are another major direct impact. The IPCC projects a sea level rise of 3–8 mm per year in the Indian Ocean, which poses a severe threat to coastal biodiversity. India's coastal regions are experiencing rising sea levels at an alarming rate, exacerbating the risks of habitat submergence, increased salinity intrusion, and coastal erosion (Roy et al. 2023; Subramanian et al. 2023). In the Sundarbans, sea levels are rising at 4–6 mm per year (Mondal et al. 2021; Paul and Paul 2022), causing habitat submergence and the decline of critical species like the Royal Bengal Tiger and estuarine crocodile (Barik 2024). Cities like Chennai (3–5 mm/year) and Mumbai (2–4 mm/year) are also experiencing frequent coastal flooding (Sudipta et al. 2021; Thamizoli and Rengalakshmi 2022), while in Kerala, saltwater intrusion has degraded paddy fields and freshwater systems (Praveen and Kunnampalli 2024). Saltwater encroachment is rendering large tracts of land unproductive in regions such as the Sundarbans and Kachchh (Id et al. 2023). Mishra et al. (2023) reported that sea level rise resulted in the loss of turtle nesting sites along the coasts of Odisha and Tamil Nadu, endangering species like the Olive Ridley Turtle. Additionally, coastal erosion and habitat loss have displaced both wildlife and human communities (Nicholls 2010). Rising sea levels continue to erode coastal biodiversity, necessitating urgent mitigation and adaptation strategies (Roy et al. 2023).

The cumulative effects of temperature rise, altered precipitation patterns, intensified cyclonic activity, and sea-level rise are accelerating desertification in India's coastal zones. Loss of vegetation cover, increased soil salinity due to saltwater intrusion, and reduced groundwater availability are transforming once-productive ecosystems into barren landscapes (Chandra et al. 2023). Without the protective functions of mangrove forests and wetland ecosystems, the vulnerability of coastal areas to extreme climatic events has increased, leading to further land degradation (Alongi 2018; Adame et al. 2021; Abdullahi et al. 2023). If current trends persist, the coastal regions of Tamil Nadu, Andhra Pradesh, and Gujarat could experience widespread desertification, jeopardizing the livelihoods of millions who depend on these ecosystems.

Integrated Coastal Zone Management (ICZM) is often discussed as a solution but faces significant implementation challenges. Practical challenges include inadequate coordination between multiple agencies, conflicting land use policies, insufficient funding, lack of community participation, and poor enforcement of environmental regulations (Fang 2018; Gaonkar 2020). For example, efforts to conserve mangroves can conflict with industrial development or port construction projects. The decentralized governance structure in India may lead to jurisdictional overlaps and gaps in coastal management efforts. Moreover, building local capacity and ensuring the long-term monitoring of coastal health are still emerging priorities that require greater focus for successful ICZM outcomes.

29.4.3 Impact of Climate Change on Biodiversity

Climate change is exerting profound impacts on global ecosystems, leading to alarming rates of biodiversity loss across terrestrial, freshwater, and marine environments. Quantitative assessments reveal the severity of these impacts. According to the WWF Living Planet Report 2024 (WWF 2024), monitored populations of mammals, birds, amphibians, reptiles, and fish have declined by an average of 73% between 1970 and 2020. This decline highlights the widespread disruption of ecological systems driven by habitat loss, climate extremes, pollution, and other anthropogenic factors. Similarly, the International Union for Conservation of Nature (IUCN) Red List has evaluated more than 166,000 species globally, with over 46,000 species categorized as threatened, including species classified as Vulnerable, Endangered, or Critically Endangered (IUCN 2024). The spatial data from the IUCN Red List further demonstrate that biodiversity hotspots regions with exceptionally high species richness and endemism are under severe threat, particularly in tropical forests, coral reefs, and freshwater wetlands.

The IPCC Sixth Assessment Report emphasizes that among over 6116 projections of biodiversity responses to climate change, approximately 44% of species in biodiversity hotspots face high extinction risks, with 24% of species facing very high extinction risks by 2100 if high-emission scenarios continue (IPCC 2022). Climate-related factors such as rising temperatures, shifting rainfall patterns, increased frequency of droughts, and ocean acidification are already altering species' distributions, population dynamics, and community structures. For instance, coral reef ecosystems, vital for supporting marine biodiversity, have experienced significant declines, with mass coral bleaching events linked to ocean warming. The World Health Organization (WHO) notes that current species extinction rates are estimated to be 10 to 100 times higher than the natural background rate, primarily driven by human-induced environmental changes, including climate change (WHO 2020).

Furthermore, these biodiversity losses have cascading effects on ecosystem services, such as food security, water purification, carbon sequestration, and disease regulation, all of which are crucial for human survival. The degradation of ecosystems reduces their resilience, making them less capable of adapting to ongoing climatic shifts. Consequently, urgent conservation actions, informed by robust scientific data, are necessary to mitigate biodiversity loss and build ecosystem resilience. Incorporating strategies like habitat restoration, climate-adaptive conservation planning, and sustainable land-use practices are critical. By integrating global datasets such as the WWF Living Planet Index, the IUCN Red List assessments, and IPCC projections, a clearer and more alarming picture emerges of how climate change is reshaping life on Earth. Quantitative evidence thus underlines the pressing need to address climate impacts not only for preserving biodiversity, but also for ensuring long-term ecosystem stability and human well-being.

29.4.4 Mitigation and Adaptation Strategies for Enhancing Coastal Resilience

Mitigation and adaptation strategies need to focus on restoring coastal vegetation, improving water management, and implementing sustainable land-use practices (Chandra et al. 2023). Large-scale mangrove restoration can help stabilize coastal soil and prevent erosion, while climate-resilient agriculture can enable local communities to adapt to changing conditions (Hernández-Delgado 2024). Improved water management, including rainwater harvesting and desalination projects, can mitigate freshwater scarcity (Ssekyanzi 2024).

In addition to addressing physical environmental changes, social and economic adaptation strategies are crucial for strengthening community resilience. Coastal communities in India, for example, are increasingly adopting livelihood diversification, shifting from traditional fishing and farming to alternative sources of income such as ecotourism, handicrafts, and aquaculture (FAO 2018; IPCC 2022). Community-based coastal management initiatives, including participatory mangrove restoration projects in Odisha and Gujarat, empower local populations to conserve ecosystems while securing economic benefits (World Bank 2020).

Infrastructure development has also been a critical focus, with government initiatives like the National Adaptation Fund for Climate Change (NAFCC) supporting the construction of climate-resilient infrastructure such as elevated housing, cyclone shelters, and saline-resistant agriculture systems (MoEFCC 2021). Financial tools, including microinsurance schemes and climate risk financing, are emerging as important measures to protect vulnerable households against the economic impacts of extreme weather events (UNDP 2020). Social strategies further strengthen resilience, notably through the incorporation of traditional ecological knowledge and the development of community-based early warning systems for cyclones and flooding (UNDRR 2019). Importantly, gender-inclusive adaptation approaches ensure that women and marginalized groups have equitable access to resources, decision-making processes, and resilience-building initiatives (UN Women 2021). Strengthening early warning systems, scaling up climate-resilient infrastructure, and fostering community participation through supportive policy frameworks are essential for building long-term resilience. Together, these integrated strategies are vital for reducing vulnerability, protecting livelihoods, and preventing further coastal desertification and biodiversity loss.

29.5 Conclusion

The findings of this study highlight the alarming rate at which climate change is driving land degradation in coastal zones, with dryland areas experiencing accelerated desertification and significant biodiversity loss. Rising sea levels, changing precipitation patterns, and human-induced environmental pressures continue to threaten coastal ecosystems. The study emphasizes that unless immediate and effective adaptation and mitigation strategies are implemented, the ecological integrity of these landscapes will continue to deteriorate. A key takeaway from this research is the necessity for an integrated approach involving ecological restoration, sustainable land-use planning, and policy interventions. Conservation initiatives, such as habitat restoration and coastal buffer zone establishment, must be prioritized. Additionally, the involvement of local communities in climate resilience efforts is crucial to ensuring sustainable coastal management. Moving forward, interdisciplinary collaboration between scientists, policymakers, local stakeholders, and the integration of traditional ecological knowledge will be essential in developing innovative and adaptive strategies to combat coastal degradation. Future research should focus on long-term monitoring of coastal changes, the effectiveness of adaptation and mitigation policies, and the socioeconomic implications of climate-induced environmental shifts. By fostering proactive climate action, societies can work towards preserving coastal biodiversity, preventing land degradation, and mitigating desertification where dry land conditions prevail, thereby ensuring the stability and health of coastal ecosystems for future generations.

Acknowledgments The authors are thankful to reviewers and editors for their time, expertise, and valuable feedback, which helped to improve the quality and clarity of the manuscript.

Declarations The author(s) confirmed that this chapter is their original work and has not been published elsewhere.

Informed Consent Statement Not applicable as this research did not involve human participants.

Data Availability Statement The data used in this study are available upon request from the corresponding author. Publicly available datasets are also used and are cited accordingly within the manuscript.

Ethics Approval This research did not involve human or animal subjects; hence, no ethical approval was required.

Conflict of Interest The authors declare no conflict of interest in relation to this study.

Funding The author(s) received no financial support for the research, authorship, or publication of this chapter.

References

Abdullahi MB, ElNaggar A, Omar M, Murtala AI, Lawal M, Mosa A (2023) Land degradation, causes, implications and sustainable management in arid and semi-arid regions: a case study of Egypt. Egypt J Soil Sci 63(4):659–676. https://doi.org/10.21608/ejss.2023.230986.1647

Adame MF, Reef R, Santini NS, Najera E, Turschwell MP, Hayes MA, Masque P, Lovelock CE (2021) Mangroves in arid regions: ecology, threats and opportunities. Estuar Coast Shelf Sci 248:106796. https://doi.org/10.1016/j.ecss.2020.106796

Ahmed Z, Dongwei G, Sameh KAE, Ghulam M, Sikandar A (2014) An overview of global desertification control efforts: key challenges and overarching solutions. Soil Use Manag 40. https://doi.org/10.1111/sum.13154

Alongi DM (2018) Impact of global change on nutrient dynamics in Mangrove forests. Forests 9(10):1–13. https://doi.org/10.3390/f9100596

Andrews L, Higgins A, Andrews M (2012) Classic grounded theroy to analyse secondary data: reality and reflections. Ground Theor Rev 11(1):12–26

Barik P (2024) Climate change effects on the sundarban mangrove ecosystem and tiger population dynamics in the Indian Region

Barik A, Ratha DKC (2024) Building resilience: safeguarding Odisha from natural calamities. Int J Polit Sci Govern 6(1):177–183. https://doi.org/10.33545/26646021.2024.v6.i1c.320

Burger J (2023) Estuarine and coastal birds, climate change, and sea-level rise. Climate change and estuaries. CRC Press, pp 507–534. eBook: ISBN9781003126096

Chandra S, Chatterjee U, Chakrabortty R, Roy P (2023) Progress in disaster science anthropogenic drivers induced desertification under changing climate: issues, policy interventions, and the way forward. Progr Disast Sci 20(November):100303. https://doi.org/10.1016/j.pdisas.2023.100303

Coffey R, Paul MJ, Stamp J, Hamilton A, Johnson T (2019) A review of water quality responses to air temperature and precipitation changes 2: nutrients, algal blooms, sediment, pathogens. J Am Water Resour Assoc 55(4):844–868. https://doi.org/10.1111/1752-1688.12711

Danovaro R, Fanelli E, Aguzzi J, Billett D, Carugati L, Corinaldesi C, Anno A, Gjerde K, Jamieson AJ, Kark S, McClain C, Levin L, Levin N, Ramirez E, Thomsen L, Van Dover CL, Yasuhara M (2020) Ecological variables for developing a global deep-ocean monitoring and conservation strategy. Nat Ecol Evolut 4(2):181–192. https://doi.org/10.1038/s41559-019-1091-z

Das A, Swain PK (2024) Navigating the sea level rise: exploring the interplay of climate change, sea level rise and coastal communities in India. Environ Monit Assess 196(11):1010. https://doi.org/10.1007/s10661-024-13191-z

Das I, Lauria V, Kay S, Cazcarro I, Arto I, Fernandes JA, Hazra S (2020) Effects of climate change and management policies on marine fisheries productivity in the North-East Coast of India. Sci Total Environ 724:138082. https://doi.org/10.1016/j.scitotenv.2020.138082

Deb JC, Phinn S, Butt N, McAlpine CA (2018) Climate change impacts on tropical forests: identifying risks for tropical Asia. J Trop For Sci 30(2):182–194. https://doi.org/10.26525/jtfs2018.30.2.182194

Fang Q (2018) Marine functional zoning: a practical approach for integrated coastal management (ICM) in Xiamen. Ocean Coast Manag

FAO (2018) Climate-smart agriculture: building resilience to climate change. Food and Agriculture Organization of the United Nations. Retrieved from https://www.fao.org/3/i5175e/i5175e.pdf

Fern RR, Baron MD, England AE, Giese JC, Kraai KJ, Lancaster JD, Oldenburger SL, Shipes JC, Wilson BC, Wyckoff SR (2023) The state of Texas wetlands: a review of current and future challenges. Texas Water J 14(1):136–174. https://doi.org/10.21423/TWJ.V14I1.7163

Friess DA, Adame MF, Adams JB, Lovelock CE (2022) Mangrove forests under climate change in a 2°C world. Wiley Interdiscip Rev Clim Chang 13(4):1–15. https://doi.org/10.1002/wcc.792

Gaonkar S (2020) An assessment of coastal zone regulations and judicial approach for conservation, protection and sustainable development of the coast. July

Hernández-Delgado EA (2024) Coastal restoration challenges and strategies for small island developing states in the face of sea level rise and climate change. Coasts 4(2):35–286. https://doi.org/10.3390/coasts4020014

Hill TS, Hoogenboom MO (2022) The indirect effects of ocean acidification on corals and coral communities. Coral Reefs 41(6):1557–1583. https://doi.org/10.1007/s00338-022-02286-z

Hoegh-Guldberg O, Kennedy EV, Beyer HL, McClennen C, Possingham HP (2018) Securing a long-term future for coral reefs. Trend Ecol Evol 33(12):936–944. https://doi.org/10.1016/j.tree.2018.09.006

Id PT, Id JL, Id ES, Liou Y, Nguyen K (2023) Saltwater intrusion and climate change impact on coastal agriculture:1–5. https://doi.org/10.1371/journal.pwat.0000121

IPCC (2021) Technical summary. Contribution of Working Group I to the sixth assessment report of the Intergovernmental Panel on Climate Change. In Climate change 2021: the physical science basis

IPCC (2022) Climate change 2022: impacts, adaptation, and vulnerability. contribution of Working Group II to the sixth assessment report. Cambridge University Press. Retrieved from https://www.ipcc.ch/report/ar6/wg2/

IUCN (2024) The IUCN red list of threatened species. International Union for Conservation of Nature

Jain S, Srivastava A, Khadke L (2024) Global-scale water security and desertification management amidst climate change. Environ Sci Pollut Res 31:58720–58744. https://doi.org/10.1007/s11356-024-34916-0

Jayakumar T, Uttam KS (2024) Habitat degradation in coral reef ecosystems and mangroves: current status and management measures. Sustainable management of fish genetic resources. Springer, Singapore, pp 111–149. https://doi.org/10.1007/978-981-97-5250-8_7

Johnston MP (2014) Secondary data analysis: a method of which the time has come. Qual Q Methods Libr (QQML) 3:619–626

Kumar B, Bhagyalekshmi V, Riyas A (2017) Climate change, fisheries and coastal ecosystems in India. J Aquat Biol Fish 5:7–17. www.cdc.noaa.gov

Lorrain-Soligon L, Robin F, Bertin X, Jankovic M, Rousseau P, Lelong V, Brischoux F (2023) Long-term trends of salinity in coastal wetlands: effects of climate, extreme weather events, and sea water level. Environ Res 237:116937. https://doi.org/10.1016/j.envres.2023.116937

Mahajan GR, Das B, Morajkar S, Desai A, Murgaokar D, Patel KP, Kulkarni RM (2021) Comparison of soil quality indexing methods for salt-affected soils of Indian coastal region. Environ Earth Sci 80(21):1–19. https://doi.org/10.1007/s12665-021-09922-x

Maity SS, Shaw RP, Maity R (2024) Climate change may cause oasification or desertification both: an analysis based on the spatio-temporal change in aridity across India. Theor Appl Climatol 155(2):1167–1184. https://doi.org/10.1007/s00704-023-04686-9

Mama AC, Bodo WKA, Ghepdeu GFY, Ajonina GN, Ndam JRN (2021) Understanding seasonal and spatial variation of water quality parameters in Mangrove estuary of the Nyong river using multivariate analysis (Cameroon Southern Atlantic Coast). Open J Mar Sci 11(03):103–128. https://doi.org/10.4236/ojms.2021.113008

Mandal U (2019) Global climate change and human interferences as risk factors, and their impacts on geomorphological features as well as on farming practices. (Issue Jan 2024). https://doi.org/10.1007/978-3-030-00680-8

Mir YH (2025) Overview of land use and land cover change and its impacts on natural resources. In: Ecologically mediated development. Sustainable development and biodiversity, vol 41. Springer, Singapore. https://doi.org/10.1007/978-981-96-2413-3_5

Mishra M, Acharya T, Chand P, Santos CAG, da Silva RM, Beja SK, Kar D (2022) Response of long-to short-term tidal inlet morphodynamics on the ecological ramification of Chilika Lake, the tropical Ramsar wetland in India. Sci Total Environ 807:150769. https://doi.org/10.1016/j.scitotenv.2021.150769

Mishra M, Acharya T, Santos CAG, da Silva RM, Kar PK, Mohanty PK, Mahapatra S (2023) Impact assessment of severe cyclonic storm Asani on the nesting grounds of Olive Ridley turtle, Rushikulya Estuary and spit in Odisha state, India. Ocean Coast Manag 238:106572. https://doi.org/10.1016/j.ocecoaman.2023.106572

MoEFCC (2021) National Adaptation Fund for Climate Change (NAFCC) progress report. Ministry of Environment, Forest and Climate Change, Government of India. Retrieved from http://moef.gov.in

Mohapatra S, Harish VSKV, Dwivedi G (2021) Climate change, cyclone and rural communities: understanding people's perceptions and adaptations in rural eastern India. Mater Today Proc 49(March):412–417. https://doi.org/10.1016/j.matpr.2021.02.384

Mondal I, Thakur S, Ghosh P, De TK (2021) Assessing the impacts of global sea level rise (SLR) on the mangrove forests of Indian Sundarbans using geospatial technology. Geogr Inf Sci Land Resour Manag:209–227. https://doi.org/10.1002/9781119786375.ch11

Naorem A, Jayaraman S, Dang YP, Dalal RC, Sinha NK, Rao CS, Patra AK (2023) Soil constraints in an arid environment-challenges, prospects and implications. Agronomy 13(1). https://doi.org/10.3390/agronomy13010220

Nicholls RJ (2010) Sea-level rise and its impact on coastal zones:1517. https://doi.org/10.1126/science.1185782

Nichols CR, Zinnert J, Young DR (2019) Degradation of coastal ecosystems: causes, impacts and mitigation efforts. Coast Res Libr 27(June):119–136. https://doi.org/10.1007/978-3-319-75453-6_8

Nisar M (2021) Impact of agriculture on soil health. In: Agricultural waste (Issue August 2023). https://doi.org/10.1201/9781003105046-3

Obura DO, Aeby G, Amornthammarong N, Appeltans W, Bax N, Bishop J, Brainard RE, Chan S, Fletcher P, Gordon TAC, Gramer L, Gudka M, Halas J, Hendee J, Hodgson G, Huang D, Jankulak M, Jones A, Kimura T, Wongbusarakum S (2019) Coral reef monitoring, reef assessment technologies, and ecosystem-based management. Front Mar Sci 6(SEP):1–21. https://doi.org/10.3389/fmars.2019.00580

Obura D, Gudka M, Samoilys M, Osuka K, Mbugua J, Keith DA, Porter S, Roche R, Hooidonk R, Ahamada S, Araman A, Karisa J, Komakoma J, Madi M, Ravinia I, Razafindrainibe H, Yahya S, Zivane F (2022) Vulnerability to collapse of coral reef ecosystems in the Western Indian Ocean. Nat Sustain 5(2):104–113. https://doi.org/10.1038/s41893-021-00817-0

Pandolfi JM, Connolly SR, Marshall DJ, Cohen AL (2011) Projecting coral reef futures under global warming and ocean acidification. Science 333(6041):418–422. https://doi.org/10.1126/science.1204794

Panja A, Bhattacharjee P, Bhattacharjee S, Kumar D, Makarana G, Kushwaha M, Yadav MR, Kumar R, Vishnu DR (2025) The nexus between environmental damage, poverty, and climate change in hard-to-reach areas: a somber tale of the 21st century. In: Environmental nexus for resource management. CRC Press, pp 188–222. eBook ISBN9781003358169

Paul AK, Paul A (2022) Adjustment of the coastal communities in response to climate variability and sea level rise in the Sundarban, West Bengal, India. https://doi.org/10.1007/978-3-030-91010-5

Praveen D, Kunnampalli J (2024) Evaluating the impacts of anticipated sea level rise, climate change and land use land cover scenarios on the rice crop in Alappuzha, Kerala and strategies to build climate responsive agriculture. Int J Disast Resilience Built Environ 15(4):755–775. https://doi.org/10.1108/IJDRBE-05-2023-0066

Rahman MJ (2013) Coastal and marine fisheries management. In: Coastal and marine fisheries management (Issue December). Coastal and Marine Fisheries Management in SAARC Countries, p 136

Rahman MAE (2023) An overview of land degradation, desertification and sustainable land management using GIS and remote sensing applications. In: Rendiconti Lincei, vol 34, Issue 3. Springer, p 767. https://doi.org/10.1007/s12210-023-01155-3

Rajeev A, Mishra V (2023) Increasing risk of compound wind and precipitation extremes due to tropical cyclones in India. Environ Res Clim 2(2):021004. https://doi.org/10.1088/2752-5295/accf2e

Rajeevan M, Anil Nair S, Tirkey S, Goswami T, Kumar N (2023) Heat and cold waves in India processes and predictability, 1 April

Rao AD, Upadhaya P, Ali H, Pandey S, Warrier V (2020a) Coastal inundation due to tropical cyclones along the east coast of India: an influence of climate change impact. Nat Hazards 101(1):39–57. https://doi.org/10.1007/s11069-020-03861-9

Rao S, Siripurapu KK, Kairaly MM (2020b) Pastoralist-wildlife corridors for unrestricted movement of pastoralists – livestock and wildlife and unhindered ecological funcitons and ecosystem services of grasslands in India. https://doi.org/10.13140/RG.2.2.19471.61601

Ray K, Pandey P, Pandey C, Dimri AP, Kishore K (2019) On the recent floods in India. Curr Sci 117(2):204–218. https://doi.org/10.18520/cs/v117/i2/204-218

Roy P, Pal SC, Chakrabortty R, Chowdhuri I, Saha A, Shit M (2023) Effects of climate change and sea-level rise on coastal habitat: vulnerability assessment, adaptation strategies and policy recommendations. J Environ Manag 330(January):117187. https://doi.org/10.1016/j.jenvman.2022.117187

Roy P, Pal SC, Chakrabortty R (2024) Climate change and geo-environmental factors influencing desertification: a critical review. Environ Sci Pollut Res 2024. https://doi.org/10.1007/s11356-024-32432-9

Samanta S, Hazra S, Mondal PP, Chanda A, Giri S, French JR, Nicholls RJ (2021) Assessment and attribution of mangrove forest changes in the Indian Sundarbans from 2000 to 2020. Remote Sens 13(24):1–23. https://doi.org/10.3390/rs13244957

Sarker B, Keya NK, Mahir IF, Nahiun MK, Shahida S, Khan AR (2021) Surface and ground water pollution: causes and effects of urbanization and industrialization in South Asia. Sci Rev 73:32–41. https://doi.org/10.32861/sr.73.32.41

Saroj J, Gautam RK, Joshi A, Tehseen P (2016) Review of coral reefs of India: distribution, status, research and management. Int J Sci Environ Technol 5(5):3088–3098

Selvakumar PUSM (2014) Review on influence of climate alterations on corals and associated fishes for Indian scenario. J Climatol Weather Forecast 02(02). https://doi.org/10.4172/2332-2594.1000114

Senapati S, Gupta V (2014) IPA-under creative commons license 3.0 Climate change and coastal ecosystem in India: issues in perspectives. Int J Environ Sci 5(3):530–543. https://doi.org/10.6088/ijes.2014050100047

Singh P, Yadav D, Pandian ES (2021) Link between air pollution and global climate change. Glob Clim Change (Issue January). https://doi.org/10.1016/B978-0-12-822928-6.00009-5

Singh P, Ao B, Yadav A (2023) Global climate change and environmental refugees: nature, framework and legality. In: Global climate change and environmental refugees: nature, framework and legality (Issue February 2023). https://doi.org/10.1007/978-3-031-24833-7

SoI Report (2024) Length of Coastline of India Background, 4–5

Ssekyanzi G (2024) Sustainable solutions for mitigating water scarcity in developing countries: a comprehensive review of innovative rainwater storage systems

Subramanian A, Nagarajan AM, Vinod S, Chakraborty S, Sivagami K, Theodore T, Sathyanarayanan SS, Tamizhdurai P, Mangesh VL (2023) Long-term impacts of climate change on coastal and transitional eco-systems in India: an overview of its current status, future projections, solutions and policies. RSC Adv 13(18):12204–12228. https://doi.org/10.1039/d2ra07448f

Sudipta C, Kambekar AR, Arnab S (2021) Impact of climate change on sea level rise along the coastline of Mumbai City, India. July. ISNI:0000000091950263

Swain D (2022) Tropical cyclones and coastal vulnerability: assessment and mitigation. In: Geospatial technologies for land and water resources management. Water science and technology library, vol 103. Springer, Cham. https://doi.org/10.1007/978-3-030-90479-1_30

Temmerman S, Horstman EM, Krauss KW, Mullarney JC, Pelckmans I, Schoutens K (2023) Marshes and mangroves as nature-based coastal storm buffers. Annu Rev Mar Sci 15:95–118. https://doi.org/10.1146/annurev-marine-040422-092951

Thamizoli P, Rengalakshmi R (2022) Sea level change and the livelihood security of coastal communities in Tamil Nadu, Peninsular India. In: Handbook on climate change and disasters. Edward Elgar Publishing, pp 432–452. https://doi.org/10.4337/9781800371613.00042

UN Women (2021) Gender and climate change. United Nations entity for gender equality and the empowerment of women. Retrieved from https://www.unwomen.org/en/what-we-do/climate-change

UNDP (2020) Climate risk insurance and risk financing mechanisms. United Nations Development Programme. Retrieved from https://www.undp.org/publications/climate-risk-insurance-and-risk-financing

UNDRR (2019) Global assessment report on disaster risk reduction 2019. United Nations Office for Disaster Risk Reduction. Retrieved from https://gar.undrr.org/

Upadhyay DK, Mishra M (2020) Blue economy: emerging global trends and India's multilateral cooperation. Marit Aff:30–45. https://doi.org/10.1080/09733159.2020.1785087

Von Holle B, Irish JL, Spivy A, Weishampel JF, Meylan A, Godfrey MH, Dodd M, Schweitzer SH, Keyes T, Sanders F, Chaplin MK, Taylor NR (2019) Effects of future sea level rise on coastal habitat. J Wildl Manag 83(3):694–704. https://doi.org/10.1002/jwmg.21633

Wang Z, Wang T, Zhang X, Wang J, Yang Y, Sun Y, Xiaohua (2024) Biodiversity conservation in the context of climate change: facing challenges and management strategies. Sci Total Environ:173377. https://doi.org/10.1016/j.scitotenv.2024.173377

Wei Y, Zhen L, Du B (2022) The evolution of desertification control and restoration technology in typical ecologically vulnerable regions. J Resour Ecol 13(5):775–785. https://doi.org/10.5814/j.issn.1674-764x.2022.05.003

Whiteside M, Mills J, McCalman J (2012) Using secondary data for grounded theory analysis. Aust Soc Work 65(4):504–516. https://doi.org/10.1080/0312407X.2011.645165

WHO (2020) Biodiversity and health. World Health Organization

Woesik R, Shlesinger T, Grottoli AG, Toonen RJ, Vega Thurber R, Warner ME, Marie Hulver A, Chapron L, McLachlan RH, Albright R, Heron SF, Huang D, Humanes A, Krueger T, Zaneveld J (2022) Coral-bleaching responses to climate change across biological scales. Glob Chang Biol 28(14):4229–4250. https://doi.org/10.1111/gcb.16192

World Bank (2020) Building resilience in coastal communities: lessons from India. The World Bank. Retrieved from https://www.worldbank.org/en/news/feature/2020/01/30/building-resilience-in-coastal-communities-lessons-from-india

WWF (2024) Living planet report 2024: breaking boundaries. World Wide Fund for Nature

Conclusion

Summary of Key Findings

Climate changes are drastically altering both marine and terrestrial ecosystems contributing towards the loss of biodiversity and diminishing the resilience within ecosystems. Some of the major consequences are the degradation of habitats owing to increased temperatures, changes in precipitation, and more frequent and intense weather events—with coral reefs and tropical forests being the most impacted. Many species are moving to cooler regions, but most do not have the means to adapt to the change and relocate, which increases their chances of extinction. Climate change alters the timing of ecological interactions, which in turn changes the timing of phenological events, allowing invasive species and native biodiversity threatening pathogens to spread unchecked. Furthermore, the systematic decline in populations alongside fragmented habitats further diminishes the eroding genetic diversity within a species. These conserved species become trapped and unable to modify themselves to changes in the environment. The change in biodiversity affects many ecosystems including important services like water purification and carbon storage which are crucial to humans and stabilizing the climate.

Future Directions for Research and Conservation

For effective policies, future research and conservation undertakings should take mitigation and monitoring provisions on climate changes' impacts on biodiversity in a more integrated multifaceted approach as model frameworks for understanding biodiversity processes and projecting scenario shifts are defined. Projected climate adaptive management, such as the construction of wildlife corridors and the restoration of altered ecosystems, will assist in the proficiency of ecosystem connectivity

B. P. Barathan et al. (eds.), *Navigating Climate Change: Impacts on Biodiversity and Ecosystem Resilience*, Sustainable Development and Biodiversity 42,
https://doi.org/10.1007/978-981-95-0409-1

enhancement to pattern species movement and adaptation. Furthermore, nature-based strategies for biodiversity that utilize ecosystem restoration and conservation for climate change mitigation and adaptation will also be added alongside integrated policies for biodiversity carbons conservation. Specialized actions concerning conserving the genetic diversity of some vulnerable taxa through ex-situ management and assisted movement will also be useful. Collaboration is important to solve issues dealing with climate change and biodiversity loss and thus encourages cross-curricular and international interactions among the researchers, policy developers, and the locals.

Implications for Policy and Management

The Invasive Aliens policy requires equal attention to climate change as it requires financing if it is to succeed. Nature-based solutions as well require attention. Maintaining Ecosystem Services and Managed Retreat require these areas defined and also dominantly situated to the climate-polar refuge-areas containing migration species. Expansion and preservation also need to be enlisted to address other problems. Furthermore, considering the ecosystem's antagonistic factors and improvement factors like over-usage of the resource or use of the land needs it requires multi-faced attention to build and sustain resilience in the ecosystem. Anthropocentric issues like representation and accessibility play vital roles in fostering participatory advocacy, which rallies support for protecting biodiversity against climate change. This involves engaging people who are knowledgeable about local resources and aligns with a broader, decentralization approach to sustainability. Societies need to be complementary and responsive, or in other words they must be politically supported at the global level and act relentlessly at all scales until biological balance, ecosystem health, and sustainable use of resources to secure a humane future.

MIX
Papier aus verantwortungsvollen Quellen
Paper from responsible sources
FSC® C105338

If you have any concerns about our products,
you can contact us on
ProductSafety@springernature.com

In case Publisher is established outside the EU,
the EU authorized representative is:
Springer Nature Customer Service Center GmbH
Europaplatz 3, 69115 Heidelberg, Germany

Printed by Libri Plureos GmbH
in Hamburg, Germany